Ecology

Theories and Applications

FOURTH EDITION

Peter Stiling

University of South Florida

Prentice Hall
Upper Saddle River, NJ 07458

Library of Congress Cataloging-in-Publication Data

Stiling, Peter D.
Ecology/Peter Stiling.--4th ed.
p. cm.
Includes bibliographical references (p.).
ISBN 0-13-091102-X
1. Ecology. I. Title.

QH541.S674 2002
577--dc21 2001034364

Executive Editor: *Teresa Ryu*
Editor in Chief: *Sheri L. Snavely*
Development Editor: *Ellen Smith*
Project Manager: *Travis Moses-Westphal*
Vice President of Production and Manufacturing: *David W. Riccardi*
Executive Managing Editor: *Kathleen Schiaparelli*
Assistant Managing Editor: *Beth Sturla*
Assistant Managing Editor, Media: *Alison Lorber*
Media Production Support: *Elizabeth Wright*
Director of Creative Services: *Paul Belfanti*
Director of Design: *Carole Anson*
Art Director: *Jonathan Boylan*
Assistant to Art Director: *John Christiana*
Interior Design: *Joe Sengotta*
Cover Designer: *Stacey Abraham*
Cover Illustration: *Dream Walker (Arctic Fox) by Thomas D. Mangelsen*
Managing Editor, Audio/Video Assets: *Grace Hazeldine*
Manufacturing Manager: *Trudy Pisciotti*
Assistant Manufacturing Manager: *Michael Bell*
Marketing Manager: *Jennifer Welchans*
Assistant Marketing Manager: *Anke Braun*
Photo Editor: *Beth Boyd*
Photo Researcher: *Christina Pullo*
Photo Coordinator: *Michelina Viscusi*
Illustrator: *RM Blue Studios (www.RMBlueStudios.com)*
Editorial Assistants: *Colleen Lee; Nancy Bauer*
Proofreader: *Michael Rossa*
Copy Editor: *Brian Baker*

Printed in the United States of America
10 9 8 7 6 5 4 3 2 1

ISBN 0-13-091102-X

Pearson Education LTD., *London*
Pearson Education Australia PTY, Limited, *Sydney*
Pearson Education *Singapore*, Pte. Ltd
Pearson Education North Asia Ltd, *Hong Kong*
Pearson Education Canada, Ltd., *Toronto*
Pearson EducaciÛn de Mexico, S.A. de C.V.
Pearson Education—Japan, *Tokyo*
Pearson Education Malaysia, Pte. Ltd

Brief Contents

About the Author

The author on Honeymoon Island, Florida, at one of his field sites.

Peter Stiling is a professor of biology at the University of South Florida at Tampa. He has taught classes in ecology, environmental science, and community ecology, and in 1995 he received a teaching award in recognition of classroom excellence in these areas. Dr. Stiling obtained his Ph.D. from University College, Cardiff, Wales, and completed postdoctoral research at Florida State University. It was while he was teaching ecology at the University of the West Indies in Trinidad that the idea for this book was conceived. Dr. Stiling's research interests include plant–insect relationships, parasite–host relationships, biological control, restoration ecology, and the effects of elevated CO_2 levels on native communities. He has published many scientific papers in journals such as *Ecology*, *Oikos*, and *Oecologia* and was an editor for *Oecologia*. His field research has been supported by the National Science Foundation, the U.S. Fish and Wildlife Service, the U.S. Department of Agriculture, and the Nature Conservancy.

Contents

CHAPTER 21

CHAPTER 22

Preface

Ecology is a fascinating science. It is the most valuable discipline for learning what causes the distribution of plants and animals on Earth. Knowledge of ecology is vital in taking conservation measures and in attempting to restore the planet after the ravages of pollution. Ecology provides the conceptual framework upon which environmental science is built. Indeed, ecology is to environmental science as physics is to engineering. That is, an engineer cannot build a bridge without knowing the physical principles underlying its construction, and similarly, environmental scientists cannot understand the environment without a sound knowledge of ecological principles.

Ecology is also a broad discipline. It borrows from many areas: from mathematics to build models of population growth, from physiology to understand how organisms live in their environments, from geology to understand soils, from chemistry to understand the chemical defenses of plants, and from genetics to understand the extinction of species. From all these fields, ecology has emerged as a science vital to the very preservation of our planet. Ecology is now a household word; this book will help you understand its every facet.

The changes to the fourth edition are substantial. The introduction has been completely rewritten to introduce students to the disciplines of evolutionary and behavioral ecology, population ecology, community ecology, and ecosystems ecology, the four pillars upon which this book is built. The introduction also discusses the methods used in ecological studies, giving examples of how investigators proceed from observation through experimentation to analysis. The section on evolutionary and behavioral ecology has been completely stripped down and refocused on ecology.

The population ecology section maintains its focus on factors affecting population growth, but more space has been allotted to mutualism and commensalism and to parasitism, in order to give balance to the chapters on competition and predation. The community ecology section has been simplified: Complex subjects such as ordination have been omitted, and the focus is now more on worked examples, using actual data sets. In the ecosystems section, I have again reworked all chapters. Here, Chapter 22, on nutrients, addresses the role of chemicals in the distribution of organisms and does not center on the nutrient cycles themselves.

There have been many pedagogical changes. Each chapter starts with its own "Road Map"—a set of brief statements that give a one-sentence outline of what each section in the chapter is about. Many of the tables and figures have been replaced and redrawn using voice balloons, so that the student is instantly alerted to the main point each graphic is making. Nearly 50% of the diagrams in this edition are new, in addition to the scores of new color photographs, each carefully chosen to illustrate a particular point raised in the text.

The format of the chapters in the fourth edition remains similar to that of the last edition. Each chapter begins with an explanation of a concept, followed by examples well illustrated with data, figures, and tables. For example, in Chapter 13, a discussion of indispensable mortality is followed by how this concept relates to sea turtle conservation and how the protection of a few adult sea turtles may actually be much more profitable than protecting dozens of eggs on the beach. Following the examples is a synthesis of the preceding material in the section, with details of a review or a mathematical model, or both, that tell us where the preponderance of the evidence lies and which concept or theory is best supported. A summary at the end of the chapter reiterates the main points and should be a valuable study aid.

Instructor's Resource CD-ROM (0-13-092639-6) and Transparency Acetates (0-13-061637-0)

Prentice Hall's commitment to a four-color format for this edition of *Ecology: Theories and Applications* has enabled us to make the diagrams, data graphics, and photographs easier to interpret, and the overall presentation brighter and more accessible. These images are available to the instructor for presentation purposes on an easy-to-use

Instructor's Resource CD-ROM or as Transparency Acetates. The Instructor's Resource CD-ROM contains every piece of line art from the text formatted for a clear on-screen lecture hall presentation as well as PowerPoint lecture presentations for each chapter. Also available are 150 transparencies. They are labeled with large, boldfaced type for easy reading in the classroom. Professors can receive the Instructor's Resource CD-ROM or the Transparency Acetates by contacting their local Prentice Hall representative or Prentice Hall faculty services at (800) 526-0485.

Student Companion Website (www.prenhall.com/stiling)

The companion website for *Ecology: Theories and Applications* has been revised and expanded. Each unit now features Case Studies that challenge students to pose questions, formulate hypotheses, design experiments, analyze data, and draw conclusions. Many tutorials include real data sets from current ecological research projects that students are asked to interpret and analyze. The website also features extensive links to other ecology sites as well as chapter self quizzes that can be submitted to the instructor.

Acknowledgments

I have had many new people help me in this fourth edition, but my first vote of thanks should go to my editor, Teresa Ryu. She raised the production values enabling the book to enjoy full color throughout. My Developmental Editor, Ellen Smith, worked tirelessly with me to make every word and every illustration count for the student. Her attention to detail and quality shows on every page, as she read the text and asked questions like a student. Travis Moses-Westphal, Project Manager, was instrumental in engineering the media and supplements package with much enthusiasm. Brian Baker was meticulous in his copyediting and went well beyond the normal duties of a copy editor, even calling for new illustrations if he felt the need! Robin Manasse labored long and hard to turn my chicken-scratch figures into graceful works of art. She often improved the clarity of my original figures and never complained when I sent her work back with new modifications. Shari Toron, Production Editor, seamlessly wove together the art, photographs, figures, and copyedited manuscript into a bright new ecology text. Finally, I am grateful to the following reviewers of the third edition, who suggested so many of the improvements herein:

Gregory H. Adler, *University of Wisconsin, Oshkosh*
Clifford Amundsen, *University of Tennessee*
Gerardo B. Camilo, *St. Louis University*
Mitchell B. Cruzan, *University of Tennessee*
Richard Deslippe, *Texas Tech University*
James Drake, *University of Tennessee*
Leo S. Luckinbill, *Wayne State University*
Nancy McCreary Waters, *Lafayette College*
Peter Meserve, *Northern Illinois University*
L. Maynard Moe, *California State University, Bakersfield*
John Mull, *Weber State University*
Katharine Nash Suding, *University of Colorado, Boulder*
Christopher Paradise, *Davidson College*
Craig Plante, *College of Charleston*
Frank Romano, III, *Jacksonville State University*
Anthony M. Rossi, *University of North Florida*
Jan Savitz, *Loyola University of Chicago*
Benjamin Steele, *Colby-Sawyer College*
William R. Teska, *Furman University.*

Once again, Jacqui Stiling, out of all these people, deserves special credit for her untiring help in preparing the manuscript and supervising the wonderful task of acquiring permissions!

Peter D. Stiling
University of South Florida

SECTION I

Introduction

Life is not distributed evenly on Earth, but rather in patches. Ecology seeks to explain this phenomenon. Here, in the Okavango Delta, Botswana, papyrus reedbeds and other vegetation are distributed throughout the region. Islands may begin as termite mounds. Physical factors, such as water levels, may affect the distribution of plants as may herbivores. *(Gregory G. Dimijian, Photo Researchers, Inc.)*

CHAPTER 1

Why and How to Study Ecology

Road Map

1. Ecology is the study of that which limits the abundance of plants and animals. We can study ecology at the level of individual behavior, populations, communities, or ecosystems.
2. Observing natural systems affords an insight into their workings, but field and laboratory experiments provide the most rigorous tests of ecological ideas.
3. Ecological measurement must be made at the spatial and temporal scale appropriate to the question being asked.

1.1 What Is Ecology?

Ecology involves the study of individuals, populations, communities, and ecosystems.

Ecology is the study of interactions among organisms and between organisms and their environment. Ecology concerns that which governs the population densities of plants and animals and that which determines the number of species in an area. Ecological knowledge is of vital importance to human endeavors. For example, in 1970, after 11 years and an expenditure of $1 billion, the construction of the Aswan High Dam, shown in Fig. 1.1, was completed. Located in southern Egypt on the world's longest river, the Nile (Fig. 1.2), the dam is the largest of its kind in the world. It contains more than four times the capacity of Lake Mead, the reservoir behind Hoover Dam, which is the largest dam in the United States. The Aswan High Dam was projected to provide several years of irrigation reserve, to add 526,000 hectares (ha:1.3 million acres) to the arable lands of Egypt, to produce 10 billion kilowatts of electric power annually, and to protect the country from catastrophic floods. Today, the dam does produce more than 50 percent of Egypt's electrical power. The reservoir of water saved the rice and cotton crops during the droughts of 1972 and 1973. It has facilitated the cultivation of two or three crops annually rather than one, thus increasing productivity and national annual income from agriculture by 200 percent. Three hundred eighty thousand ha of desert are being irrigated for the first time.

However, it has been argued that in many ways the dam stands as a monument to ecological ignorance. First, the incidence of schistosomiasis (a debilitating parasitic disease caused by a tropical flatworm in the area) increased from 47 to 80% because the parasite's secondary hosts, snails, reproduce year-round in the reservoir and thus are no longer reduced in numbers by drought. Second, diminished flow of the Nile into the Mediterranean decreased phytoplankton blooms and fish harvest in the discharge area. The catch of sardines alone dropped from 15,000 tons annually to 500 tons, and yields from the new fishing areas behind the dam are low. Third, reduced silt deposition along the floodplain has increased the need for commercial fertilizers to the tune of $100 million annually. The new fertilizer plants use much of the hydroelectric power produced by the dam. Fourth, because water is readily available, many farmers overwater their land, and as the water evaporates, salt is left behind. In 1986, almost half the irrigated area in

Figure 1.1 The Aswan High Dam, Egypt, permits continued irrigation of important agricultural areas.

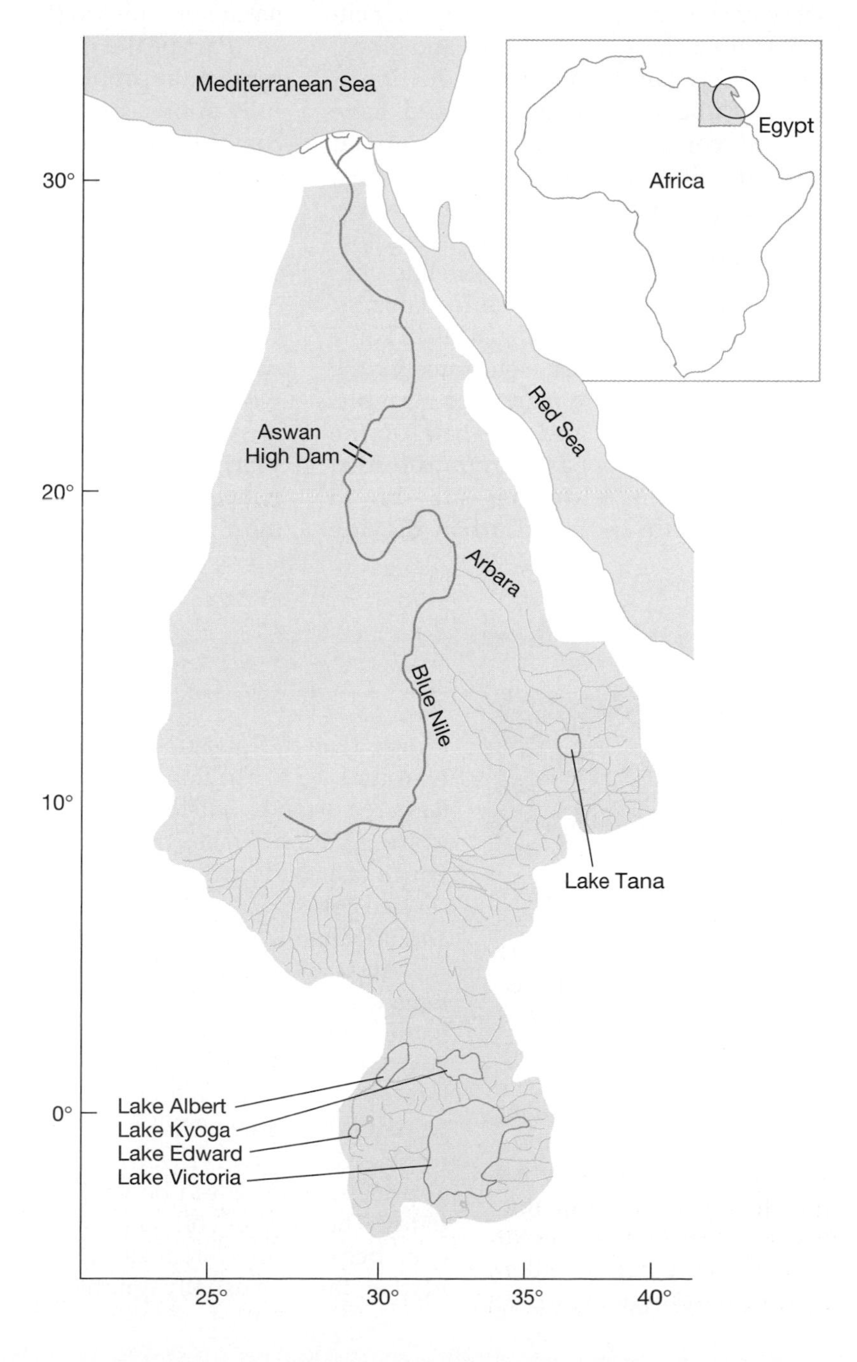

Figure 1.2 The Nile River and its tributaries and the location of the Aswan High Dam.

Egypt was affected by salt. Ecological studies, had they been done, could have predicted such effects of the dam. Ecology embraces such fields as parasite–host interactions and expected yields from fish harvests; thus, ecological studies would have been very timely in the case of the Aswan High Dam.

Ecologists are among the best-equipped scientists to study natural systems. This fact is especially important now that the effects of humans have begun to change the entire globe. Before 1960, ecologists were few in number, and their activities were dominated by taxonomy, natural history, and speculation about observed patterns. Their equipment included sweep nets, quadrats, and specimen jars. Since that time, ecologists have become active in investigating environmental change on regional and global scales. They have embraced reductionist analyses and experimentation and have adapted concepts and methods derived from agriculture, physiology, biochemistry, genetics, physics, chemistry, and mathematics (Grime, 1993). Their equipment now includes portable computers, satellite-generated images, and chemical autoanalyzers.

The challenge for ecologists is to come together and agree on solutions to the world's ills, which are many. For example, between one-third and one-half of the Earth's land surface has been transformed by human action. Acid rain is carried from one country to another. Carbon dioxide pumped out by the industrial centers of developed nations has increased atmospheric levels of the gas worldwide, from the poles to the equator. More nitrogen is fixed by humanity than by all natural sources combined. Pesticides, powerful human-made poisons, have been detected in human breast milk and in the tissues of penguins—both of which were completely unintended targets. About one-quarter of the bird species on earth have been driven to extinction. Now, more than ever, there is a strong impetus to understand how natural systems work, how humans change those systems, and how in the future we can reverse the changes. Ultimately, ecologists will be instrumental in determining every phase of the services that nature provides to the world.

Part of the cause of the world's environmental problems has been a failure to fully appreciate those services. This shortcoming leads to undervaluation and abuse of the land and its flora and fauna. A recent paper in the journal *Nature* by an economist, Robert Constanza, and his colleagues (Constanza *et al.*, 1997) made the first attempt to correctly value natural systems. The researchers came to the conclusion that the world's ecosystems are worth more than the global gross national product, $19 trillion. (See "Applied Ecology: The Value of the World's Natural Services.") If we were to pay for the services provided by ecosystems, most goods would jump hugely in price.

Applied Ecology

The Value of the World's Natural Services

Ecologists and economists have joined forces to estimate the value of the services that the world's ecosystems provide (Constanza *et al.*, 1997). There are many direct goods (such as food) and indirect services (such as the assimilation of waste) that ecological systems provide (Table 1) on a renewable basis. (Nonrenewable fuels and minerals are excluded.) Many ecosystems provide more than one service. For example, swamps are important in flood control, water supply and waste treatment and as habitat (Fig. 1). In turn, most services are provided by more than one ecosystem; for instance, many systems are involved in nutrient recycling.

Robert Constanza and his colleagues estimated that the world's ecosystems provide at least U.S.$33 trillion worth of services annually (Table 2). This staggering figure is more than the total global gross national product, which is around U.S.$19 trillion per year. The majority of the services are currently outside the market system and include gas regulation, waste treatment, and nutrient cycling, which at U.S.$17 trillion, was by far the most expensive service performed, but even if we eliminate this, the total annual value would still be a whopping U.S.$16 trillion.

About 63 percent of the estimated value of natural systems is contributed by marine systems, with most of the value coming from coastal systems (U.S.$10.6 trillion per year). About 37 percent of the estimated value came from terrestrial systems, mainly forests (U.S.$4.7 trillion) and wetlands (U.S.$4.9 trillion).

If ecosystem services were actually paid for, the global price system would be very different from what it is today. The price of commodities would skyrocket. However, because ecosystem services are largely outside the market, they are usually ignored or grossly undervalued.

TABLE 1 Partial list of the world's ecosystem services.

Service	Example
Indirect	
Atmospheric gas regulation	Maintenance of carbon dioxide, ozone, and sulfur dioxide levels
Climate regulation	Maintenance of carbon dioxide, nitrogen dioxide, methane, and CFC levels
Disturbance regulation	Storm protection, flood control
Waste treatment	Sewage purification
Soil erosion control	Retention of topsoil, reduction in siltation of lakes
Nutrient recycling	Maintenance of nitrogen, phosphorus, carbon, and other elemental cycles
Direct	
Water supply	Irrigation, provision of water for industry
Pollination	Pollination of crops
Biological control	Regulation of pest populations
Refuges	Maintenance of habitats for wildlife
Food production	Production of crops, maintenance of livestock
Raw materials	Provision of renewable fuels and timber
Genetic resources	Maintenance of plants and animals for medicines and provision of genes for plant resistance
Recreation	Ecotourism
Cultural	Aesthetic value

Figure 1 What is the value of this swamp at Skuppernog Creek, Kettle Moraine State Forest, Wisconsin? Is the swamp useless land or a valuable asset? A recent assessment by Robert Constanza and his colleagues (1997) valued swamps such as these at $19,580 per hectare because of their function in regulating water supply, treating waste, and providing habitat. (*Terry Donnelly, Tom Stack*)

TABLE 2 Valuation of the world's ecosystem services (U.S. billion $ per year) and value per hectare (U.S. $). *(After Constanza et al., 1997).*

Biome	Total Global Value ($U.S. Billion)	Total Value (Per Ha) ($U.S.)	Main Services Provided
Open ocean	8,381	252	Nutrient cycling
Estuaries	4,100	22,832	Nutrient cycling
Sea grass, algal beds	3,801	19,004	Nutrient cycling
Coral reefs	375	6,075	Recreational, disturbance regulation
Coastal shelf	4,283	1,610	Nutrient cycling
Tropical forest	3,813	2,007	Nutrient cycling, raw materials
Temperate forest	894	302	Climate regulation, waste treatment
Grasslands	906	232	Waste treatment, food production
Tidal marsh	1,648	9,990	Waste treatment, disturbance regulation
Swamps	3,231	19,580	Water supply, disturbance regulation
Lakes, rivers	1,700	8,498	Water regulation
Desert	0	0	None
Tundra	0	0	None
Ice, rock	0	0	None
Cropland	128	92	Food production
Urban	0	0	None
Total	33,268		

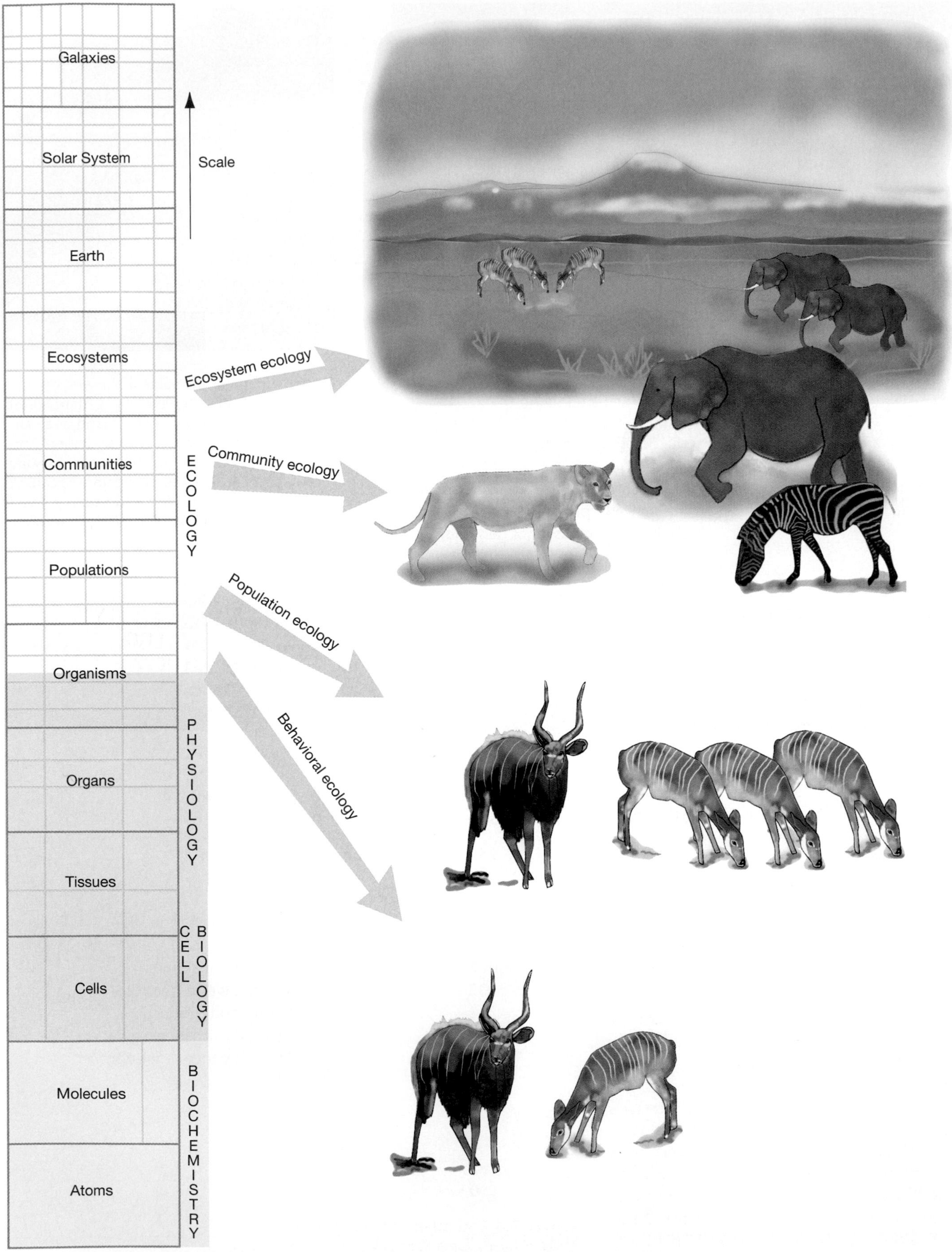

Figure 1.3 The organizational levels of ecology—behavioral, population, community, and ecosystem ecology—are on a larger scale than the biological disciplines of biochemistry, cell biology, and physiology.

In general, ecologists work in one of four broad areas: behavioral ecology, population ecology, community ecology, and ecosystems ecology. The relative organizational levels of these areas are illustrated in Fig. 1.3. The four areas correspond to the four main sections used in this textbook. Behavioral ecology focuses on the ecology of the individual organism and how its behavior, including its interaction with other individuals, affects its reproductive success and the population density. Here, we strive to explain why many organisms act selfishly rather than for the "good of the herd" or the "good of the species." In population ecology, groups of interbreeding individuals, called populations, are studied, with the emphasis on factors that affect population growth and determine population size. Such things as competition, predation, and herbivory are important in population ecology. In community ecology, the focus is on biodiversity and what influences the numbers of species in an area. This information is particularly important to conservation biologists, who are trying to maintain as many species as possible on Earth. Finally, in ecosystem ecology, we view the community as a user of nutrients and energy, and we examine nutrient availability and energy flow. Such studies are vital to a full understanding of phenomena like the carbon cycle and the effects of pesticides on food webs.

Behavioral, population, community, and ecosystem ecology all have important implications in the real world. However, ecology is not environmental science. If we can make an analogy, ecology is to environmental science as physics is to engineering. Both physics and ecology provide the theoretical framework on which to pursue more applied studies. Engineers rely on the principles of physics to build bridges; environmental scientists rely on ecological principles to save the planet. First, ecology describes the framework necessary to understand how populations are affected by features of the physical environment, such as temperature and moisture, and, later, by other organisms. Through ecology, we learn how plants compete with one another, how herbivores reduce plant populations, and how natural enemies affect prey populations. Only then can we examine the effects of humans on the environment, including pollution, global warming, and the introduction of exotic species.

Behavioral ecology.

Behavioral ecology concerns how behavior contributes to the survivorship, reproduction, and population growth of species. The behavior of animals is at the core of our philosophy of nature. If we see species cooperating with one another and acting in selfless, even altruistic ways, we tend to accept nature as neat and harmonious. If we see species constantly competing, we might argue that nature involves a fierce struggle for survival. Frequently, what appear to be altruistic acts are, in fact, examples of selfish behavior.

Many species of caterpillars can essentially be regarded as social, traveling together *en masse* like herds of ungulates. Prominent among such species are forest tent caterpillars in the genus *Malacosoma* (Fig. 1.4). These caterpillars are infamous for residing in silken tents and defoliating trees in deciduous forests. Initially, the group behavior of tent caterpillars appears counterproductive. Caterpillars need food and must grow as quickly as possible, because they are especially vulnerable to a wide array of predators, especially birds. Wouldn't group living promote intense competition for food and attract even more predators than solitary individuals? How can we explain this behavior? In the case of the tent caterpillar, the group constructs a silken tent to which layers are progressively added. During the daytime, this tent provides a safe refuge from predators. Group living enables the caterpillars to fabricate a large tent, which single larvae could not easily construct. At night, the caterpillars leave the tent to forage. They

Figure 1.4 Tent caterpillars, *Malacosoma*, in their tents. Why do some species live in dense congregations that promote competition? Behavioral ecology addresses these and other questions.

must locate a tree whose branches have not been defoliated and remember where it is so that they can return to forage on subsequent days. Here again, group living becomes an advantage, because the caterpillars lay down silk trails between the tent and the leaves so that they can relocate the leaves. In addition, the silk trails are marked with pheromones to draw in colony mates. The denser the concentration of caterpillars, the stronger is the trail. Group living is therefore advantageous in this species. Furthermore, nest mates of tent caterpillars are often brothers and sisters, having hatched from the same batch of eggs. In this case, constructing a nest provides a safe haven for siblings. Protecting siblings in turn helps propagate a caterpillar's own genes because its siblings carry copies of many of those genes. Thus, seemingly altruistic acts are actually selfish in nature, since they promote the survival of one's own genes in a population. Knowing that such cooperation is actually selfish colors the way we think about nature. We shall discuss other such behaviors in an evolutionary context in the next few chapters.

Population ecology.

Populations represent groups of individuals of a single species, all of the members of which can interbreed. To determine what controls the abundance of species in nature, we need to examine how populations grow and how they are limited by food, competitors, and natural enemies. Much of the theory of ecology is built upon the ecology of populations. Knowing what affects populations can help us combat species extinctions, lessen species endangerment, and maximize sustainable yields in fisheries and forests.

The introduction of exotic species is one of the most pressing ecological problems in the world. Indeed, in the United States alone, 142 species of exotic vertebrates have self-sustaining populations in the wild, including many species, such as ring-necked pheasants, brought over by hunters and others, such as house sparrows, some pigeons, and parrots, brought over by pet owners. Of the 300 most invasive exotic weeds in the United States, over half, including purple loosestrife and Japanese honeysuckle in the northeast, kudzu in the southeast, Chinese tallow in the south, and leafy spurge in the Great Plains, were brought in for gardening, horticulture, or landscape purposes.

Invading exotic plants were thought to succeed often because they escaped their natural enemies, mainly insect herbivores that remained in the country of origin and were not transported to the new locale. However, new research on the population ecology of *Centaurea diffusa*, an invasive Eurasian plant in Montana, by Ragan Callaway and Erik Aschehoug (2000), suggests a totally different reason for their success: The exotic species secrete powerful root exudates called *allelochemicals* that kill the roots of North American species, allowing *Centaurea* to proliferate. In their native Eurasia, the local plants have evolved together with *Centaurea* and are not so susceptible to the allelochemicals, so that *Centaurea* does not overrun entire plant populations, as it does in North America. If this scenario is correct, the whole strategy of controlling exotic weeds by importing their natural enemies is called into question.

To support their claims, Callaway and Aschehoug performed some experiments. They collected seeds of three native Montana grasses—*Festuca idahoensis*, *Koeleria cristata*, and *Agropyron spicata*—and grew them on their own and together with the exotic *Centaurea* competitor from the Caucasus Mountains of the Republic of Georgia. Sure enough, *Centaurea* depressed the biomass of the native grass (Fig. 1.5). When the experiments were repeated with similar grasses from the Republic of Georgia (*F. ovina, K. laerssenii*, and *A. cristatum*), the Georgia species were affected, but not nearly so much as the U.S. species. Furthermore, when activated carbon was added to the soil, it absorbed the chemical excreted by the *Centaurea* roots, permitting the U.S. grass species to increase in biomass. There was still some competitive depression of biomass, probably because not all of the exudate was absorbed, but the competitive mechanism of a toxic root exudate was confirmed. Interestingly, the Georgia species probably secrete their own allelochemicals to compete with *Centaurea*. When the activated charcoal was added, it probably absorbed these chemicals, too, allowing *Centaurea* to compete even more strongly and further depressing the biomass of the other species. This study on the population biology of an exotic plant could change the way we think about controlling introduced species in the future.

Community ecology.

Community ecology concerns the biodiversity of life on earth. It focuses on why certain areas have high numbers of species (are

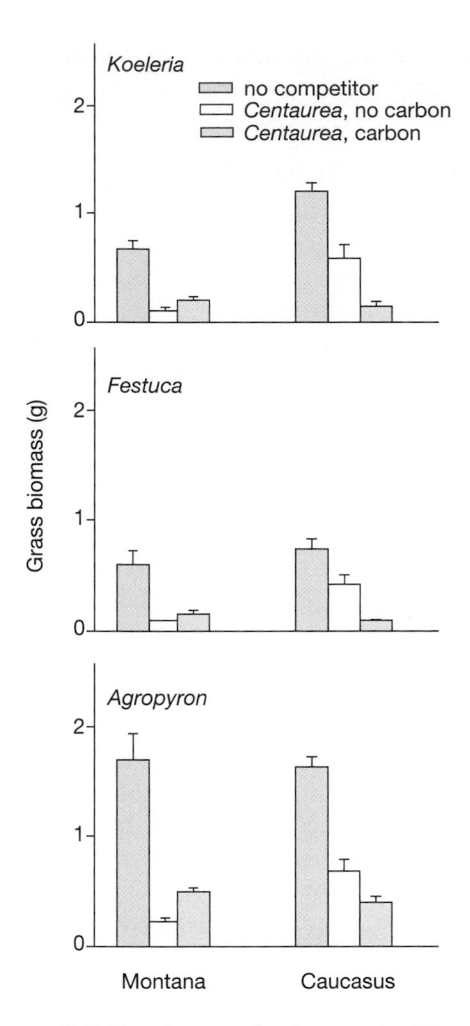

Figure 1.5 Plant biomass for three native Montana grass species—*Koeleria, Festuca,* and *Agropyron*—when planted alone (with no competitors) and with the exotic *Centaurea* from the Republic of Georgia. *Centaurea* depressed the native grass biomass, but less so when activated carbon was added to the soil. Activated carbon absorbs the toxic root exudates from *Centaurea*, lessening its effects on the other species. When the experiment was performed with *Koeleria, Festuca,* and *Agropyron* native to the Republic of Georgia, the effects of *Centaurea* were not as strong, indicating that native Georgia grasses have evolved to compete better with *Centaurea*. They probably do this by secreting their own toxic root exudates, which were absorbed by the activated charcoal, thus allowing *Centaurea* to depress their biomass more. (*Redrawn from Callaway and Aschehoug, 2000.*)

species rich), compared with other areas that have low numbers of species (are species poor). We are interested in species richness for two reasons: in its own right, for we wish to preserve species-rich areas; and because there might be a link between species richness and community function. It is generally thought that species-rich communities perform better, perhaps in yield of plant material or in resistance to disturbance, than species-poor communities.

During the latter half of the 1980s, the reduction of the Earth's biological diversity emerged as a critical issue and was perceived as a matter of public policy (U.S. Congress, 1987). Humans benefit not just from individual species, but also from whole communities. Forests soak up carbon dioxide, maintain soil fertility, and retain water, preventing floods. The loss of biodiversity could disrupt a community's ability to carry out such functions. This consideration bears on the debate about the level of diversity necessary to carry out the world's ecosystem functions. In the 1950s, the British ecologist Charles Elton (1958) proposed a linear relation between diversity (here meant as numbers of species) and stability (here meant as community persistence in an unchanged state) (Fig. 1.6). His *diversity–stability hypothesis* proposed that the more species there are in a community, the more stable the community is. In 1981, Stanford ecologists Paul and Anne Ehrlich proposed the *rivet hypothesis*. The diversity of life, said the Ehrlichs, is like the rivets on an airplane, with each species playing a small, but critical, role in keeping the plane airborne. The loss of a rivet weakens the plane and causes it to lose a little airworthiness. The loss of a few rivets could probably be tolerated, but the loss of more rivets would prove critical. In an extension of this idea, the so-called

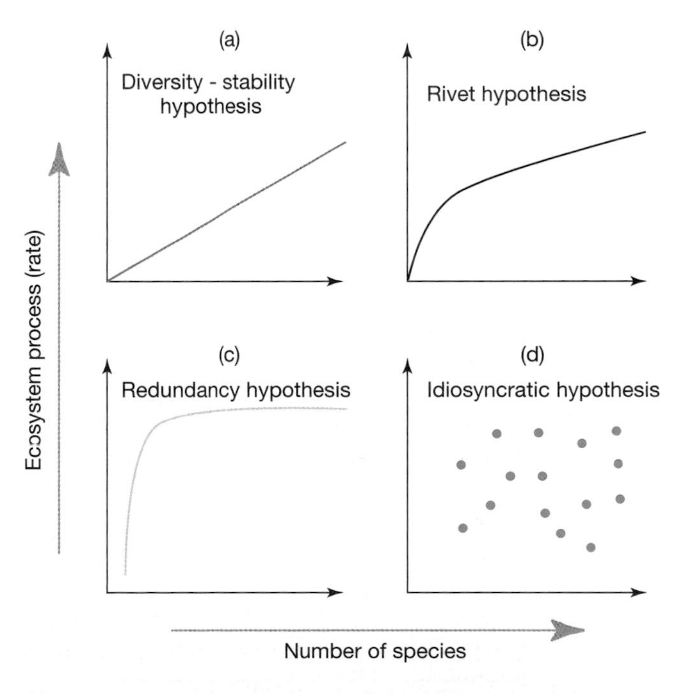

Figure 1.6 Four hypotheses (a–d) for the functional role of species diversity in ecosystems. (*After Johnson et al., 1996.*)

redundancy hypothesis (or *passenger hypothesis*) was proposed a decade later by Australian ecologist Brian Walker (1992). According to this hypothesis, most species are more like passengers on a plane: They take up space, but do not add to the airplane's airworthiness, which is affected by the activity of just a few species—perhaps the "pilot," "copilot," "flight attendants," and so on. In this scenario, the loss of a noncritical species doesn't affect community function, but the loss of a species of critical importance does. To round out the alternatives, British ecologist John Lawton (1994) included the possibility that there is no relationship between community richness and ecosystem function—a possibility he called the *idiosyncratic model*. Determining which model is correct is very important, as our understanding of the effect of species loss can greatly affect the way we manage our environment.

Although the idiosyncratic model and the linear diversity–stability model are not found to obtain widely, the rivets and redundancy hypotheses are still hotly debated, and the evidence is insufficient to decide between the two. However, the redundancy hypothesis has much empirical support. For example, American forest communities are still functional despite the loss of the passenger pigeon and the Carolina parakeet. On a larger scale, it is noteworthy that the productivities (amounts of plant biomass) of temperate forests of the Northern Hemisphere are virtually identical in spite of huge differences in tree species richness: East Asia has 876 species, North America 158, Europe 106. One rule of thumb that has been proposed is that average annual forest productivity tops out in the range of 10 to 40 species, but the presence of more tree species may ensure a supply of "backups" should some of the most productive species fail through insect attack or disease. This scenario has actually happened—consider, for example, the demise of the American chestnut or the elm tree, both of whose numbers in American forests were much lessened by disease in the mid-20th century, whereas the forests continued to function without them. In the section on community ecology, we shall examine how and at what level changes in species richness affect community functions such as biomass production and nutrient uptake efficiency. (The diversity–ecosystem-efficiency connection is very important and will be discussed in more detail in Chapter 15.)

Like all scientific disciplines, ecology employs specific, formalized methods to achieve advances in understanding.

Ecosystems ecology.

Ecosystems ecology is concerned with the passage of energy and nutrients through communities and what effects energy and nutrients have on those communities. Humans have greatly altered various nutrient cycles on Earth, including the carbon and sulfur cycles. They have also greatly modified the nitrogen cycle: Because so much fertilizer is made and applied to agricultural systems, more nitrogen-rich material is artificially added to ecosystems than through natural sources (Vitousek *et al.,* 1997). Among the many consequences of this modification is increased plant biomass. However, the increased amount of nitrogen in the soil greatly reduces species richness in a community, because it favors a few species of dominant plants that do well in nitrogen-rich soils. Concomitantly, the herbivore community becomes less rich, because only the relatively few species that feed on these dominant plants are favored. This concatenation of events was elegantly shown to occur by Nick Haddad, John Haarstad, and David Tilman in nitrogen-loading experiments in a Minnesota prairie (Haddad et al., 2000). These scientists sampled 54 4m by 4m plots that had had nitrogen added to them at various rates for 14 years. Both plant species richness and insect species richness dropped as the amount of nitrogen added increased (Fig. 1.7). This finding is of concern because it is unlikely that fertilizer use in agricultural areas will decrease in the future. Hence, we should expect to see a dramatic decline in species richness in agricultural habitats in the future.

1.2 Ecological Methods

How do we go about performing an ecological study? Suppose you are employed by the Food and Agricultural Organization (FAO) of the United Nations, which operates internationally out of Rome, Italy. You are charged with finding out what causes outbreaks of locusts, a type of grasshopper that periodically increases its numbers, attaining huge densities in Africa and other parts of the world, where it destroys crops and other vegetation. To investigate the problem, first of all, you might draw up an interaction web

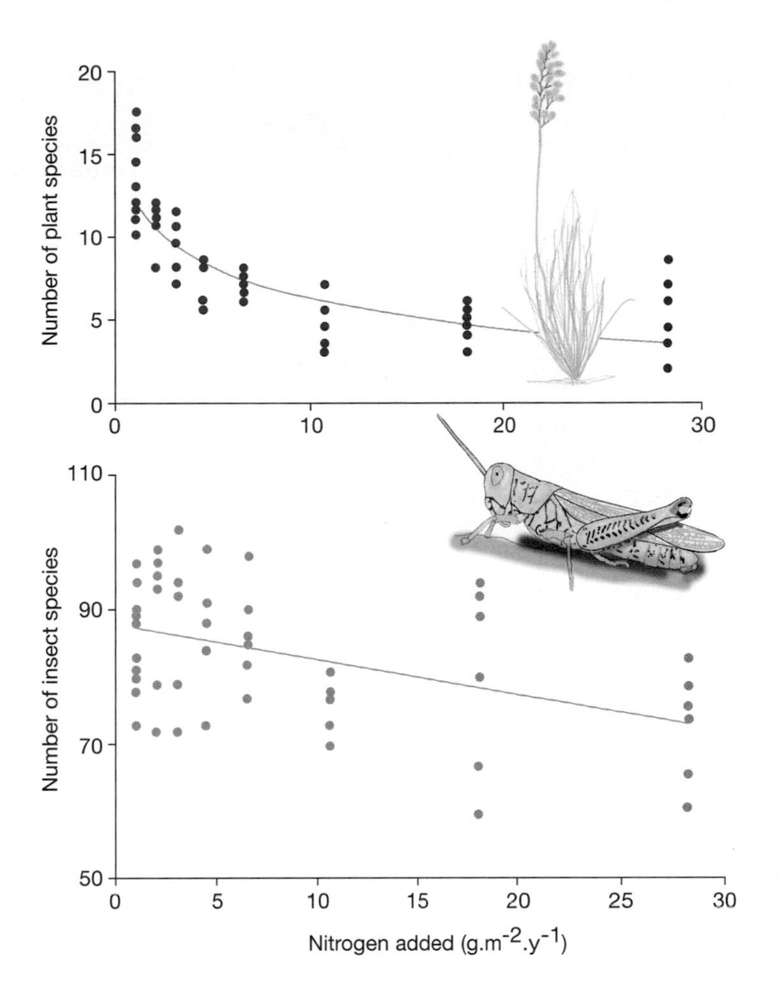

Figure 1.7 Drop in plant and insect species richness with added nitrogen (grams per square meter per year) in Minnesota prairie plots. (*Redrawn from Haddad et al., 2000.*)

of all the factors that could influence locust population size (Fig. 1.8). The interactions that influence the population are many and varied and include the following:

Natural enemies

1. Bird or other vertebrate predators.
2. Insects that parasitize locusts, using them as hosts for their developing larvae.
3. Bacterial parasites.

An increase or a decrease in predation or parasitism could cause fluctuations in locust numbers.

Competitors

1. Other insects that eat the same types of food.
2. Vertebrate grazers.

An increase or decrease in competitors for the same food would affect the availability of that food and so, indirectly, the numbers of locusts.

Host plants

1. Quality of host plants.
2. Quantity of host plants.

An increase or decrease in the number of host plants, which could result from changes in rainfall patterns in the area, might greatly affect the density of locusts there. More plant biomass often means more available food and therefore higher locust numbers. Alternatively, the quality of the host plants may be improved, as it is when fertilizer is added to a crop, and increased food quality could trigger an outbreak of locusts.

Physical factors

Locust numbers may be directly controlled by physical variables such as temperature. Warmer temperatures might allow more locust eggs to hatch. Heavy rainfall may drown more locust eggs, which are laid in the ground.

Already you have a vast array of possibilities to investigate. This is partly why ecology is such a difficult science: We have to follow so many leads. We have to be plant chemists to assay plant quality, parasitologists to examine the effect of parasites, climatologists to examine the effects of environmental extremes such as cold or hot temperatures, and community biologists to

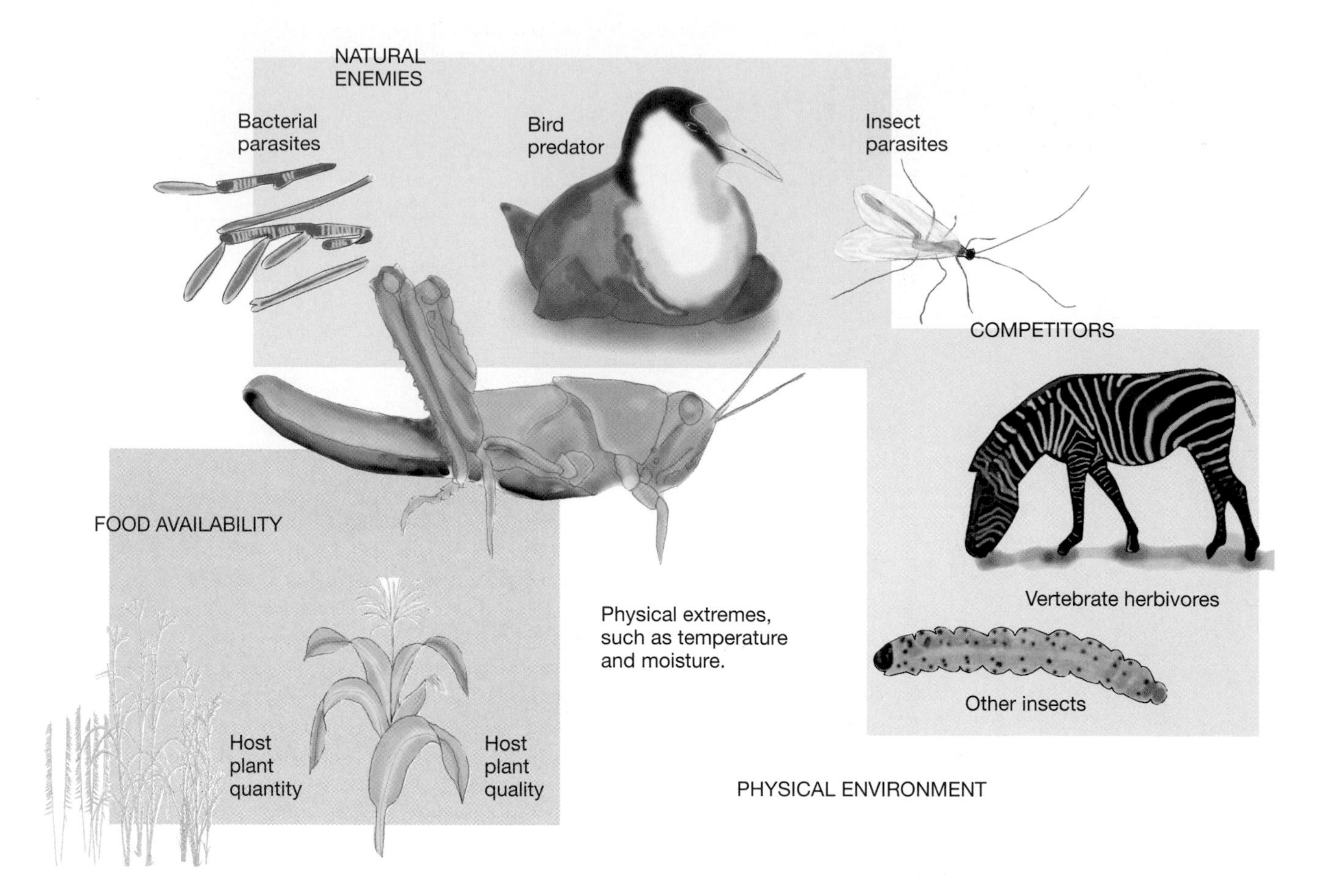

Figure 1.8 Interaction web of the factors that could influence locust population size: parasites and predators, competitors for food, quality and quantity of host plants, and physical variables such as temperature and moisture.

be aware of biotic interactions such as competition and predation from other taxa.

Where is the best place to start? Most ecological studies start with careful observation. We can observe the fluctuations of locusts and see whether their populations vary with fluctuations in other phenomena, such as levels of parasitism, numbers of predators, or food supplies. Suppose we found that locust numbers were affected by bird predation levels: As predation levels decreased, locust numbers increased. If we graphed the data we obtained, we might get a plot like that depicted in Fig. 1.9(a). This would give us some confidence that locust numbers were determined by predation levels. In fact, we would have so much confidence, that we could draw a line of best fit to represent the relationship between the two parameters of locust density and predation, as in Fig. 1.9(b). To draw this line, we would enter the data into a computer, and a special statistical software package would determine the line of best fit. In this case, the points are highly clustered around the line. However, if the points were not highly clustered, as in Fig. 1.9(c) we would have little confidence that predation affects locust density, and we would not draw a line of best fit. In actuality, there are many statistical tests we can use to determine whether the two variables are related, as in Fig. 1.9(b), or not, as in Fig. 1.9(c). In this text, unless otherwise stated, most graphs like Fig. 1.9(b) imply a meaningful relationship between the two variables. We call this type of relationship *significant*, and the analysis we perform is called *correlation*. If locust density is linearly related to predation, we say that locust density is *correlated* with predation. We shall not cover the mathematics behind statistical tests such as correlation or the fitting of best lines to the data, but you will see many such data sets graphed throughout the book.

We have to be careful when we use correlations. For example, it could be that there is also a correlation between locust abundance and food abundance: High numbers of locusts could be associated with large,

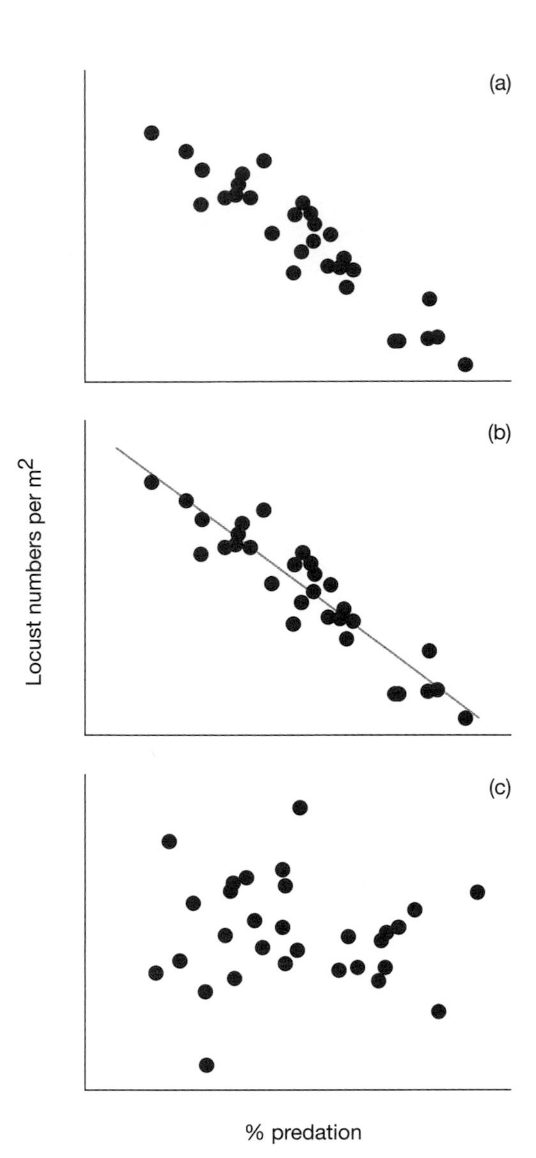

Figure 1.9 How locust numbers might be correlated with predation. In (a), higher locust numbers are found in nature where predator levels are lowest. We might draw a line of best fit (b) to represent this relationship. In (c), the relationship might be so weak that we would not have much confidence that it was linear.

dense plants (and hence more food). We might then conclude that locust density is controlled by food availability. However, this correlation could be spurious. It would look like plant size affects locust density because large plants provide abundant food. However, in reality, predation would still be the most important factor affecting locust density: Locusts would be abundant because large food plants would provide them with refuge from bird predators, which would be unable to attack them in the interior of dense bushes. Thus, correlation does not always imply causation. This is why, after observations, we often turn to experiments.

In our locust example, if we want to see whether predation is as important as our observational data suggest, we should do an experiment. Ideally, it would be good to remove predators from locust populations, because, theoretically, if predators are having a huge effect, then removing them should cause an increase in locust numbers. So we would have two groups: a group of locusts with predators removed—called the *treatment group*—and a group of locusts with nothing done to them, so that the predators are still there—called the *control group*. Any differences we observe in locust population density would then be due solely to differences in predation. Such a treatment might be achieved by putting a cage made of chicken wire over and around some bushes containing locusts, so that birds are denied access to the locusts. We would have to be careful to similarly disturb the control, say, by putting chicken wire around the edges of some bushes, leaving just the tops open for access to predatory birds. We could look at locust survivorship over the course of one generation of locust and predator.

We would perform these experiments many times simultaneously, a strategy called *replication*. The number of replicates might be 5, 10, or even more. At the end of the experiments, we would add up the total number of living locusts and take an average. In the treatment group, the resulting number of locusts in each of 10 replicates might be 5, 4, 7, 8, 12, 15, 13, 6, 8, and 10, so the average number surviving would be 8.8. In the control groups, which still allow predators access, the numbers surviving might be 2, 4, 7, 5, 3, 6, 11, 4, 1, and 3—an average of 4.6. We usually refer to the average as the *arithmetic mean*, or simply the *mean*. Without predators, the mean number of locusts surviving would therefore be almost double the average number with predators. This finding would give us confidence that predators were indeed the cause of the changes in locust numbers that we observed. The results of such experiments can be illustrated graphically by a bar graph (Fig. 1.10). Again, there are statistical techniques for determining whether these differences are statistically significant and whether the treatment is likely to have caused our observed effects. Among such tests are *t*-tests and analysis of variance, or ANOVA. We shall not examine the mechanics of these tests, but in this book, when treatment and control groups are presented

Experiments that involve the deliberate manipulation of a single factor provide the most rigorous tests of hypotheses.

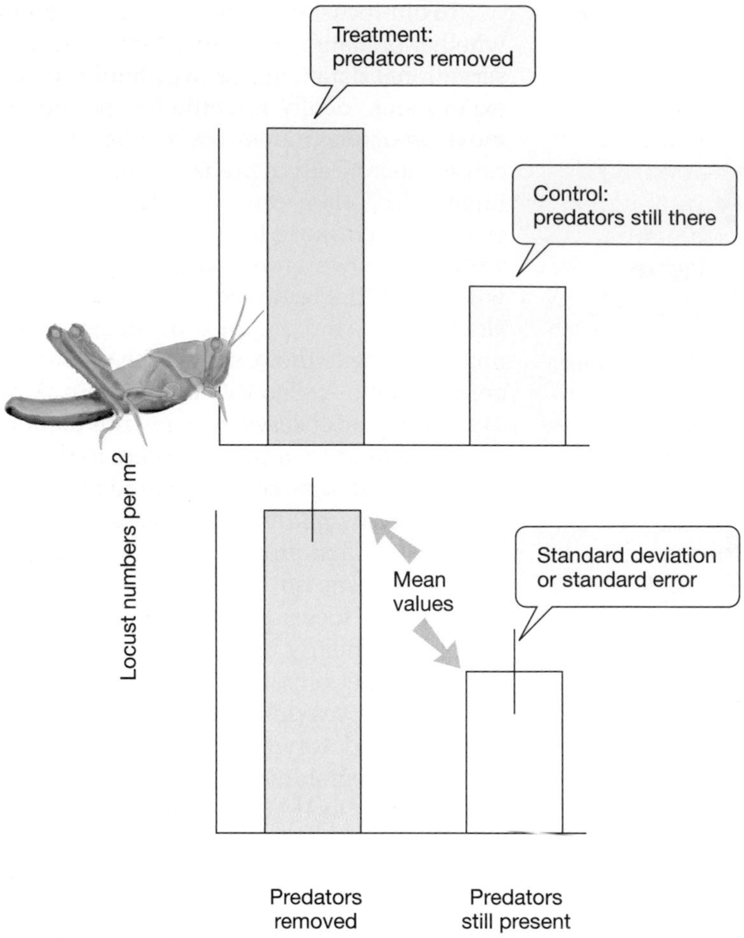

Figure 1.10 A graphical display of how a predator-removal experiment might be reported. In the upper panel, the two bars represent the average number of locusts where predators are removed (the treatment group) and the average number where predators are not removed (the control group). In the lower panel, the vertical bars (standard deviations) give an indication of how tightly the individual replicate results are clustered around the mean. The smaller the bars, the tighter is the cluster, and the more confidence we have in the result.

as differing, the difference is presumed to be statistically significant unless stated otherwise. (Incidentally, the real story concerning locusts is that weather—in particular, rain—is the most important factor affecting their numbers. Moist soil allows eggs to hatch and provides water for germinating plants, affording a ready source of food for young hatching locusts.)

You might notice that in some bar graphs there are little vertical lines around the mean. These lines represent a measure of the spread of the points around the mean and are called the *standard error of the mean* or the *standard deviation*. The smaller the bars, the tighter the replicate values are around the mean and, usually, the more significant the differences are between the treatment and control groups.

Experiments can be classified into three main types: laboratory experiments, field experiments, and natural experiments. Laboratory experiments provide the most exact regulation of factors such as light, temperature, and moisture, while varying only the factor of interest, such as increasing nitrogen availability to plants by adding fertilizer. However, the biotic community represented in a laboratory experiment is simplified, so conclusions based on laboratory results are limited. Laboratory experiments are best used to study the physiological responses of individual plants or animals rather than the dynamics of reproducing populations.

Field experiments are conducted outdoors and have the advantage of operating on natural rather than artificially contrived, communities. The most commonly used manipulations include the local elimination of competitors, predators, or herbivores and the local introduction of such species. The density of the species of interest can then be monitored to see whether it increased or decreased with or without competitors or predators, relative to controls. Darwin used a field experiment to demonstrate that the introduction of grazing animals increases number of plant species on a lawn. Diversity is increased because grazers often eat the most common species, preventing these species from outcompeting others. Field experiments commonly manipulate species through the use of tools like cages or fences to keep predators or herbivores in or out. Such manipulations are unlikely to be generated by nature itself.

Sometimes, natural extremes like severe droughts, freezes, or floods provide the opportunity to study the effects of water deprivation, cold temperatures, or abundant rainfall in a field setting. Such natural experiments are usually the sole technique for following the trajectory of a perturbation beyond a few decades. The weather is frequently shown to be of vital importance in influencing the population densities of many species, but we cannot easily manipulate the weather. Natural experiments involving volcanic explosions or hurricanes commonly provide the only data on these subjects. Furthermore, natural experiments often have general implications for other communities, because they sample from a wider range of natural variation among sites than do field experiments.

There is no best type of experiment; the choice depends on what one is investigating. The strengths and weaknesses of the different types of experiments are outlined in Table 1.1. For example, the spatial scale of laboratory experiments is likely to be limited to the size of a constant-temperature laboratory room, around 0.01 ha, and that of field experiments to usually less than 1 ha. Natural experiments, however, may be virtually unlimited in scale and often take place on large islands or even a continent.

One problem with experiments is that they take a lot of effort. Time and money must be taken into consideration, and the experiment must be carefully planned; and even then, often only one factor is tested at a time. These constraints frequently lead to low levels of replication and then to what is known as a *type I error*: the declaration of a hypothesis to be false when it is actually true. For example, suppose that fertilization increases plant growth, but we can detect such an increase only by performing 10 replicates of an experiment in which we add fertilizer to plants. If we perform only 5 replicates, our standard errors are bigger than if we use 10 replicates. Bigger standard errors reduce confidence in the results, and we cannot conclude that they are meaningful or significant. Now imagine that we had 100 of these fertilization experiments reported in the literature, 90 with an insufficient number of replicates (perhaps 5) and 10 with sufficient numbers of replicates (10 or more). If we summarized the literature, we would say that 90 studies failed to find a significant effect of fertilizer on plants and 10 found a significant effect, *even though these results were purely a reflection of the sample size.*

If most of the studies that we investigate have low statistical power, the failure to demonstrate the phenomenon will be perpetuated. The few studies with good statistical power that demonstrate the phenomenon will be outweighed by the hordes of badly designed experiments that fail to show it. One technique for detecting the "true" strength of poorly replicated experiments is *metaanalysis*, a method of combining the results from different experiments. Metaanalysis weights different studies primarily on the basis of their sample size. The method was popularized in ecology by Jessica Gurrevitch from the State University of New York at Stony Brook on

Metaanalysis is a technique used to combine results from different experiments to allow us to make inferences from bigger data sets.

TABLE 1.1 Strengths and weaknesses of different types of experiments in ecology. *(Abbreviated from Diamond, 1986.)*

	Laboratory Experiment	Field Experiment	Natural Experiment
1. Regulation of independent variables	High	Medium	None
2. Maximum temporal scale	Low	Low	High
3. Maximum spatial scale	Low	Low	High
4. Scope (range of manipulations)	Low	Medium	High
5. Realism	Low	High	High
6. Generality	Low	Medium	High

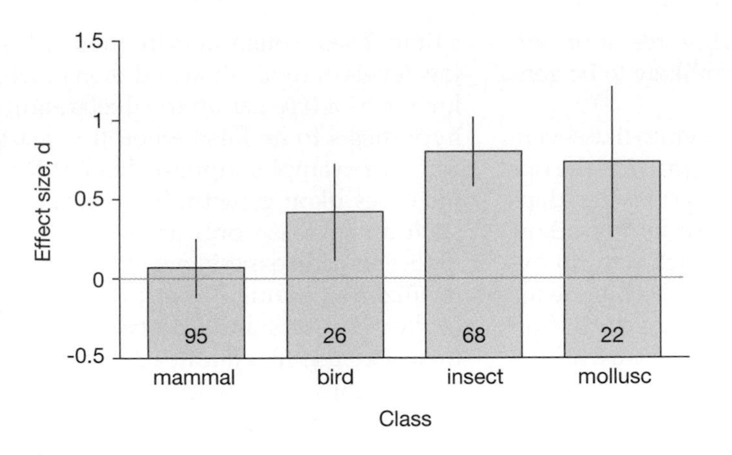

Figure 1.11 A metaanalysis of the effect of herbivore identity on its impact on plant biomass. The size of the effect size is measured as the variable *d*. Each point is the mean of a given number of studies shown in the bar. The effect of insects is the strongest, and because the 95% confidence interval (the vertical bar) does not overlap that of mammals, the effect of insects is significantly different from that of mammals. (*Modified from Bigger and Marvier, 1998.*)

Long Island (Gurrevitch et al., 1992). In metaanalysis, the data are not reanalyzed; rather, the results from a number of different studies are examined to see whether, together, they demonstrate an effect that is significant. Metaanalysis starts by estimating the size of the effect of a treatment from every experiment and then pooling all the effects together to get one overall effect size, usually called *d* (Fig. 1.11). The individual effect sizes are weighted by the number of replicates performed for each experiment. Thus, an experiment with only a few replicates of a treatment would not be weighted as heavily as an experiment with 10 or 20 replicates. As well as the effect sizes, some measure of the variation in experimental results is noted by drawing bars around the mean. This technique is analogous to drawing error bars around the means of standard experimental results. Such bars represent something called *95% confidence intervals.* If the bars do not overlap each other on different treatments, then the differences found are probably statistically significant. If the bars overlap each other, the treatments generally are not statistically significant. If the bars overlap the zero value on the *y*-axis, then the effect probably is not significant. Metaanalysis is being incorporated more frequently into ecology, and several different metaanalyses are mentioned in this book, so a good grasp of the concept is valuable.

Ecological phenomena occur on a variety of spatial and temporal scales.

Sometimes it is very difficult or impossible to do an experiment. In that case, we might turn to the use of mathematical models. Suppose we thought that exotic predators caused the demise of an endangered species, such as a panda bear. We couldn't in all good conscience experimentally expose panda bears to predators to see whether the predators decreased the population of panda bears, because we don't really want to kill any pandas. Instead, we might try to mathematically model what would happen. We could construct a model that incorporated the density of pandas, the density of the predators, and the predators' feeding rate. We could also see whether our model predicted the results from our locust experiments, in which we were able to empirically examine the effects of predators. If it did, then it might be able to predict what would happen to panda bears. We would then have an explanation for the effects of predators in two systems and perhaps a general explanation for the effects of predators in *all* systems. Models can often give us valuable signposts as to how natural systems might work, what further data we need to collect to verify our ideas, or what further observations we have to make.

1.3 Spatial Scale

Because laboratory experiments, field experiments, and natural experiments are performed on different scales, it has become obvious that ecologists must address the issue of using the correct spatial scale in planning their experiments. The size of the study area and the duration of an investigation can limit what one observes in an ecological system. John Wiens and colleagues (Wiens *et al.*, 1986) stated that studying ecology is analogous to what it would be like to study chemistry if the chemist were only a few angstroms long and lived for only a few microseconds. If the chemist were no larger or longer lived than the molecules and processes he or she was studying, the overall course of chemical reactions would be difficult to distinguish from the random collisions of molecules. Ecological phenomena, especially on larger scales, often occur on time spans of hundreds of years, and we must be careful to detect them. Although scale is, of course, a continuum, four major points on that continuum can be recognized:

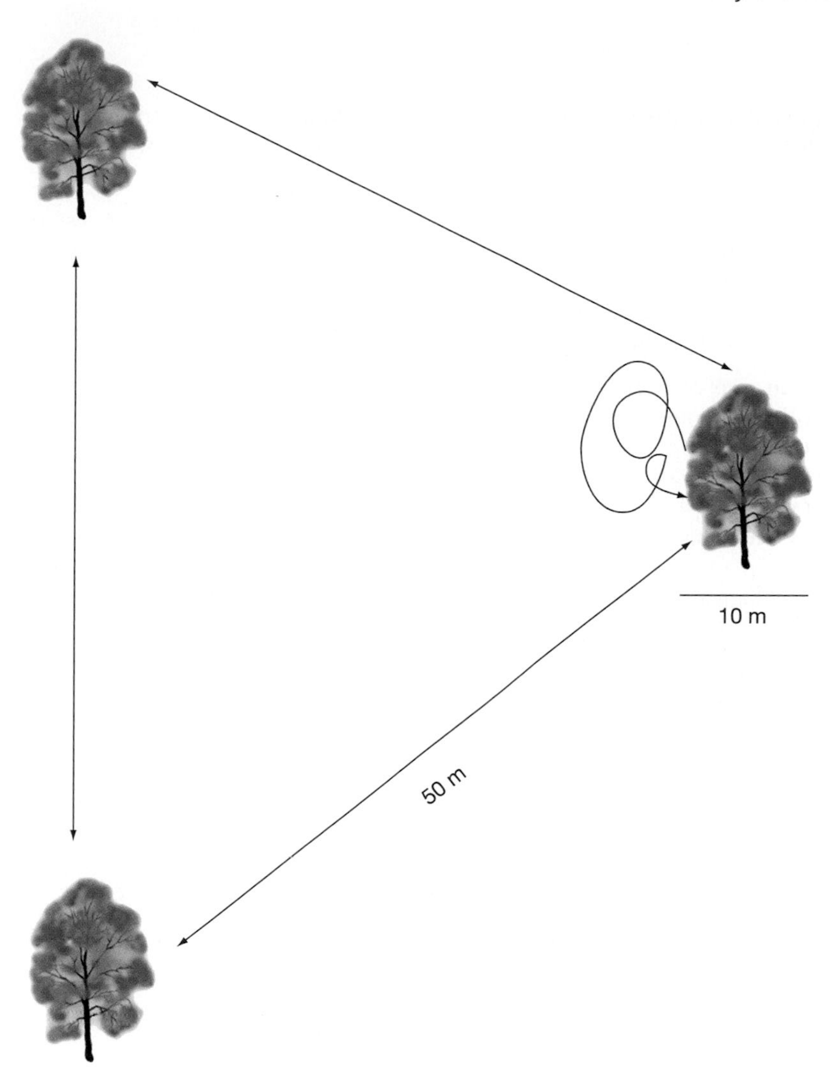

Figure 1.12 The effect of spatial scale on ecological investigations. If the ambit of a parasitic wasp is only 10 m, then it will not be able to respond to differences in densities of caterpillars, its prey, between trees spaced 50 m apart. The correct spatial scale to examine wasp responses to caterpillar density is then within trees, not between them.

1. A space occupied by a single individual sessile organism or a space in which a mobile organism spends its entire life (the realm of behavioral ecology).
2. A local patch occupied by many individuals of one species (a population).
3. A space large enough to contain many populations of different species (a community).
4. A biogeographic scale large enough to encompass a community and its nutrients and energy cycles (an ecosystem).

Investigations on these different scales yield answers to different questions, and it is important to realize this point.

John Addicott and his colleagues at the University of Alberta (Addicott *et al.*, 1987) suggested that the correct scale of investigation depends on the question being asked. There will never be a single ecological scale for a given organism, but rather a number of scales, each appropriate to a different process. The correct scale might depend on the ambit, or range of movement, of the organisms involved. For example, imagine 3 trees, each 50 meters (m) apart (Fig. 1.12). Each tree contains caterpillars and, in addition, wasps that parasitize the caterpillars. In a study of how the wasps attack the caterpillars, one might examine all 10 trees to see whether the wasps aggregate on the tree with the most caterpillars. However, if the wasps can fly only a few meters, they would not be able to fly from tree to tree; they would be able to respond only to differences in density of caterpillars *within* a tree. The correct scale of ecological investigation would then be within trees, not between them.

Just as with a spatial scale, the appropriate choice of temporal scale depends on the phenomenon and the species to be studied. One would choose a relatively short time over which to study behavioral responses, a longer one for population dynamics, and a still longer scale for studies of ecosystem processes such as nutrient cycling (Wiens *et al.*, 1986).

Summary

Ecology is concerned with understanding why plants and animals occur in certain areas, what controls their numbers, and what controls the numbers of species present in a given area. The natural world is integral to our survival, so to understand it is of paramount importance.

1. A knowledge of ecology is important for understanding such endeavors as conserving species, preserving biodiversity, maintaining ecosystem functions, preserving soil fertility, studying the effects of pollutants, preventing global warming, and calculating sustainable yields in fisheries, forests, and hunting grounds.

2. In general, ecologists work in one of four broad areas: behavioral ecology, population ecology, community ecology, and ecosystems ecology. In behavioral ecology, we study how behavior and evolution can affect species, particularly in terms of their numbers and extinction rates. In population ecology, we emphasize the factors that limit population densities. Community ecology concerns species richness in an area, and ecosystems ecology examines the flow of nutrients and energy through communities.

3. Progress in understanding ecological systems can be gained through different ecological methods, including observations of natural systems, field or laboratory experiments, mathematical modeling, and reviews and syntheses of existing studies.

4. Great care must be taken to ensure that investigations are conducted at the right spatial and temporal scales. The scale of an investigation can have a large influence on its conclusions.

In performing studies to answer ecological questions, great care must be taken in making pertinent observations that establish a pattern and in performing appropriate experiments. In each phase, the spatial and temporal scales relevant to the question at hand must be used.

Discussion Questions

1. What is the difference between ecology and environmental science?

2. What are ecological methods? How do we apply them to ecological questions?

3. Think about five ecological questions you could ask about a local park, a forest, or even your backyard and the information you would need to answer those questions. Do your questions relate to behavioral, population, community, or ecosystems ecology, or do they cross categories?

S E C T I O N II

Evolution and Behavioral Ecology

Conservation of many endangered species, such as this siberian tiger, may depend on a knowledge of genetics and how to overcome loss of genetic diversity, as well as how to preserve suitable habitats. A knowledge of evolutionary and behavioral ecology is therefore vital to modern ecologists. *(Tom & Pat Leeson, Photo Researchers, Inc.)*

CHAPTER 2

Genetics and Ecology

Road Map

1. Many species inhabit certain areas because of their evolutionary past, not because of current conditions.
2. Mutations and chromosomal rearrangements create genetic variation sufficient to result in a wide variety of species on earth.
3. Genetic variation can be measured by allozyme analysis or by DNA sequencing.
4. Genetic variation in populations can be reduced through inbreeding, genetic drift, neighborhood effects, or a harem mating system.

2.1 Evolutionary History

Given that ecology is concerned with explaining the distribution and abundance of plants and animals around the world and with the control of their numbers, it follows that evolutionary ecology is an important part of the discipline. For example, one may argue about what controls penguin numbers in the Southern Hemisphere, but a nagging question remains: Why are there no penguins in the Northern Hemisphere? The answer is not insufficient food or too many predators; penguins simply evolved in the Southern Hemisphere and have never been able to cross the tropics to colonize northern waters.

South America, Africa, and Australia all have similar climates, ranging from tropical to temperate, yet each of these continents is characterized by its distinct inhabitants. South America is inhabited by sloths, anteaters, armadillos, and monkeys with prehensile tails. Africa possesses a wide variety of antelopes, zebras, giraffes, lions, baboons, the okapi, and the aardvark. Australia, which has no native placental mammals except bats, is home to a variety of marsupials, such as kangaroos, koala bears, Tasmanian devils, and wombats, as well as the duck-billed platypus and the echidna, two egg-laying monotremes. A plausible explanation of this state of affairs is that each region supports the fauna best adapted to it, but introductions have proved this explanation incorrect: European rabbits introduced into Australia proliferated rapidly, and *Eucalyptus* from Australia grows well in California. The best explanation is provided by evolution: Some species simply evolved in different places. A complementary explanation is geological phenomena, which can account for the distributions of some plants and animals. For example, during the glacial periods of the Pleistocene, lakes covered much of what is now desert in the U.S. southwest. With the retreat of the ice, the lakes disappeared. Desert pupfishes *(Cyprinodon)* and other aquatic organisms were once widely dispersed throughout the Death Valley region, but now occur only in isolated springs. Similarly, tapir fossils are found widely around the globe, and their present spotty distribution in South America and Malaysia merely represents the relict populations of once-widespread groups. A knowledge of evolutionary ecology is clearly of paramount importance to understanding contemporary distributions of species.

2.2 Genetic Mutation

Increases in the number of species present on Earth arise chiefly from mutations that occur during the copying of DNA, the genetic material making up every living thing on the planet. Two kinds of mutations arise during DNA replication: *gene*, or *point, mutations* (the most important means of enriching the **gene** pool) and *chromosomal* mutations (the most important means of rearranging genes).

A point mutation results from a "misprint" in DNA copying called a *transition* (Fig. 2.1). Most point mutations are thought to involve changes in the nucleotide bases that make up the DNA base pairs (adenine, thymine, guanine, and cytosine) at single locations—for example, from adenine to guanine or from adenine to thymine. Two such changes that transform the sequence GAA to CUA combine to substitute the amino acid valine for glutamic acid, causing the abnormal beta chains we see in sickle-cell hemoglobin. Because of the redundancy of the genetic code, about 24 percent of the substitutions within codons (three bases) do not change amino-acid sequences and thus do not alter the organism's phenotype.

More drastic changes in amino-acid sequences are caused by *frameshift mutations*, in which an addition to (or deletion from) the amino-acid triplet sequence alters the whole reading frame, leading to severe and often fatal mutations.

When the order of base pairs within the gene is unaffected, but the order of genes on the chromosome is altered (Fig. 2.2), the chromosomes can undergo any of four types of changes: deletions, duplications, inversions, and translocations. A *deletion* is the simple loss of part of a chromosome and is the most common source of new genes. A deletion is often lethal, unless, as in some higher organisms, many genes are duplicated.

Chromosomal mutations do not actually add to or subtract from the variability of the gene pool; they merely rearrange it and make certain adaptive gene combinations easier for the population to maintain. When two chromosomes are not perfectly aligned during crossing-over, the result is one chromosome with a deficiency of genes and one with a *duplication* of genes. The duplication may be advantageous in that greater amounts of enzymes may be coded for. In yeasts, for example, an increase in acid monophosphatase enables cells to more efficiently exploit low concentrations of phosphate in the medium in which the cells are growing.

An *inversion* occurs when a chromosome breaks in two places and then re-fuses with the segment between the breaks turned around so that the order of its genes is reversed with respect to that on the unbroken chromosome. Such breaks probably occur during the prophase of cell division, when the chromosomes are long and slender and often bent into loops. In a *translocation*, two nonhomologous chromosomes break simultaneously and exchange segments.

Chromosomal mutations introduce genotypic variability by altering the sequence of genes on a chromosome.

Point mutations involve changes in the sequence of nucleotide bases.

New species originate from point mutations and chromosomal rearrangements.

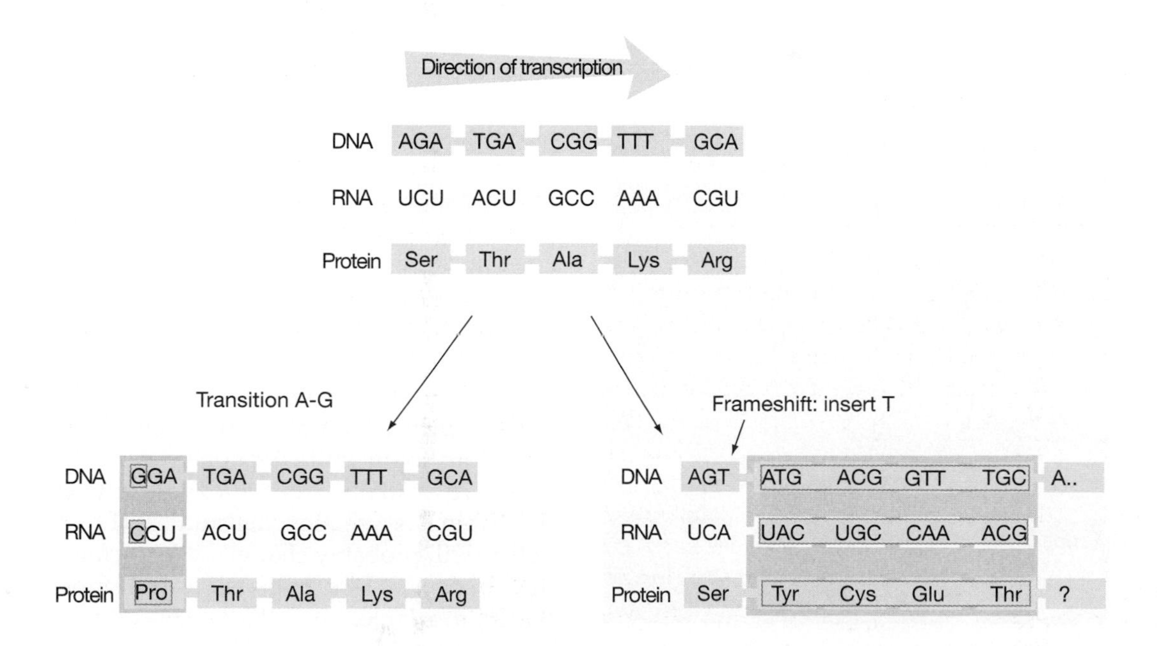

Figure 2.1 Types of point mutation occurring during the copying of DNA. Point mutations, or transitions, involve changes in nucleotide bases at a single location. Frameshift mutations alter whole sequences and are often fatal. Boxes indicate a change in the base, codon, or amino acid.

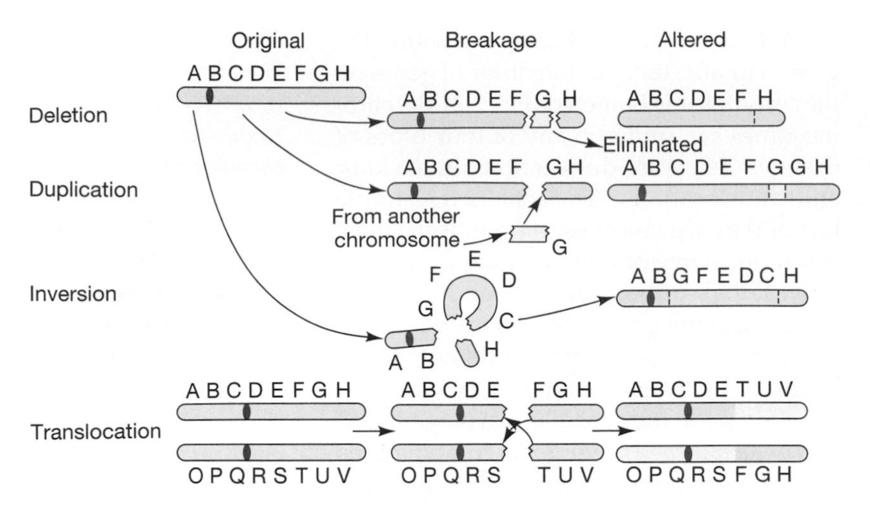

Figure 2.2 Chromosome breakage and reunion can give rise to four principal structural changes: deletion, duplication, inversion, and translocation. Each letter represents a gene.

2.3 Measuring Genetic Variation

Recent technological advances have improved the ability of scientists to estimate how much genetic variation exists in nature.

Genetic diversity is central to the breeding success of most populations. Therefore it is of great value to know how much genetic variation exists in nature. For a long time, it was impossible to tell how many loci exhibited genetic variation—no clues are given in the phenotype. In the 1960s it was realized that most loci actually code for proteins, especially enzymes. Different forms of, say, alcohol dehydrogenase are coded for by different alleles. Two individuals with the same form of enzyme are presumably genetically identical at that **locus**. Consequently, variation in gene loci can be found through a search for variation in enzymes. The most common technique for distinguishing different genetic forms of enzymes (allozymes) is gel electrophoresis. In this technique, specially prepared samples of tissue are placed in a porous gel, and an electrical potential is applied, causing the electrically charged enzymes to migrate through the gel along the lines of electrical force. Because slightly different forms of the same amino acid differ in charge, they migrate at different rates and separate into bands at different distances from the original specimen. These bands become visible when the gel is first flooded with a substrate on which the enzyme acts and then stained with a substance that colors the reaction products (Fig. 2.3). The differences in enzymes made visible by electrophoresis have proved to be of genetic origin, and experienced workers have identified much genetic variation among individuals in populations whose phenotypes looked identical.

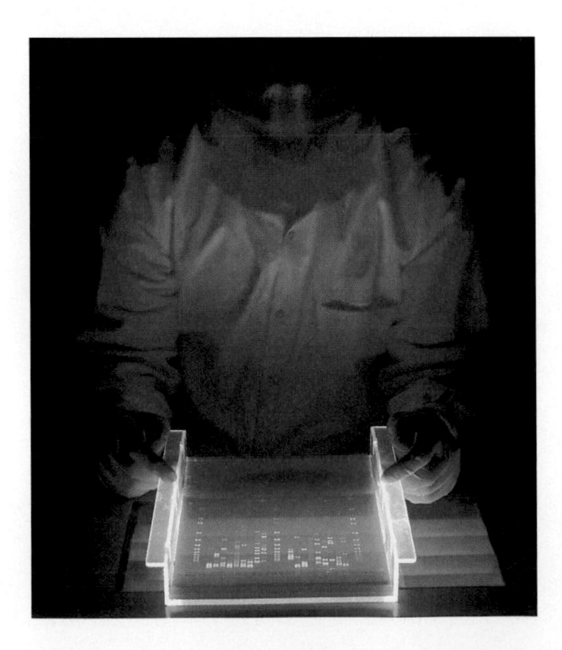

Figure 2.3 Researcher examines an agarose gel on which samples are separated according to migration rates during the application of an electric current.

A yet more powerful (and more expensive) method for assessing genetic variation is to sequence a portion of DNA itself. DNA sequencing has been made possible by the advent of the polymerase chain reaction (PCR) technique, which can be used to make millions of copies of a particular region of DNA. It is now possible to obtain DNA sequence data from a wide variety of organisms much more quickly than could be done previously. Because PCR techniques can amplify even minute amounts of DNA, we can obtain data from very small organisms that contain little tissue. We can also use tiny samples of tissue from larger organisms without having to kill or otherwise injure them. For example, a drop of blood, a hair root, or a feather are now adequate material for DNA sequence-based work. The technique has obvious importance in dealing with rare and endangered species, and it is therefore useful in conservation ecology. In fact, some conservation biologists are curious as to whether the new advances in DNA technology will permit endangered species to be cloned. (See "Applied Ecology: Can Cloning Help Save Endangered Species?") At present, however, DNA sequencing is more expensive and generally more difficult than allozyme techniques. Therefore, many more data on species variation are available from allozyme studies than from DNA sequencing studies. The situation will probably change as the technology becomes less expensive and more widely available.

Applied Ecology

Can Cloning Help Save Endangered Species?

In 1997, Ian Wilmut and his colleagues at Scotland's Roslin Institute published their famous paper on the cloning of the now-famous sheep Dolly (Fig. 1), from mammary cells of an adult ewe. Conservation biologists are curious as to whether the same technology could be used to save endangered species. However, there are a number of technical hurdles to overcome:

1. We would have to develop an intimate knowledge of every species' reproductive cycle. For sheep, this was routine, but eggs of different species, even if they could be harvested, often require different nutritive media in laboratory cultures.
2. We would have to identify surrogate females that would carry babies to term, because it is desirable to leave natural mothers available for subsequent breeding. Closely related species would therefore have to be tried.
3. Cloning is currently so expensive that conservation resources might be better spent elsewhere—for example, buying up habitat.
4. Cloning might not be able to do much to increase the genetic variation of a population. However, if it were possible to use cells in collections from long-deceased animals, then these clones would theoretically reintroduce lost genes into the population.

Figure 1 Dolly, a sheep cloned by researchers in Scotland in 1997. Can conservation biologists use the same technology to save endangered species? (*Najlah Feanny, Stock Boston.*)

Although no one knows if cloning would work to save endangered species all the time, it is encouraging that in November 2000 an Iowa farm cow called Bessie gave birth to an endangered species called an Asian guar. The guar, an oxlike animal native to the jungles of India and Burma, was cloned from a single skin cell taken from a dead guar.

The first assays of genetic variation by protein electrophoresis were published in 1966 by Harris, who reported on variation in humans, and by Lewontin and Hubby, working on *Drosophila*. The results were something of a surprise: In both *Drosophila* and humans, the average population was variable at about 30 percent of its loci. Moreover, it has since been shown that the enzyme products of some genes have similar electrophoretic mobility, so that variation would not be distinguished by the gel technique. Therefore, existing estimates of variation in nature could err on the low side.

Although mutations may be accelerated by human-made radiation, ultraviolet light, or substances such as colchicine, in nature such mutagens (usually in the form of weak cosmic rays) are too rare to produce many mutations. Nevertheless, it has been conservatively estimated that mutations occur in nature at the rate of one per gene locus in every hundred thousand sex cells. Since higher organisms contain at least 10 thousand gene loci, about 1 out of 10 individuals carries a newly mutated gene at one of its loci. Of course, most mutations are deleterious—they arise by chance, and the genes cannot know how and when it is good for them to mutate. Only one out of a thousand mutations may be beneficial, and thus, 1 in 10 thousand individuals carries a useful mutation per generation. If we estimate that there are a hundred million individuals per generation and 50 thousand generations during the evolutionary life of a species, then 500 million useful mutations would be expected to occur during that span. It has been estimated that only 500 mutations may be necessary to transform one species into

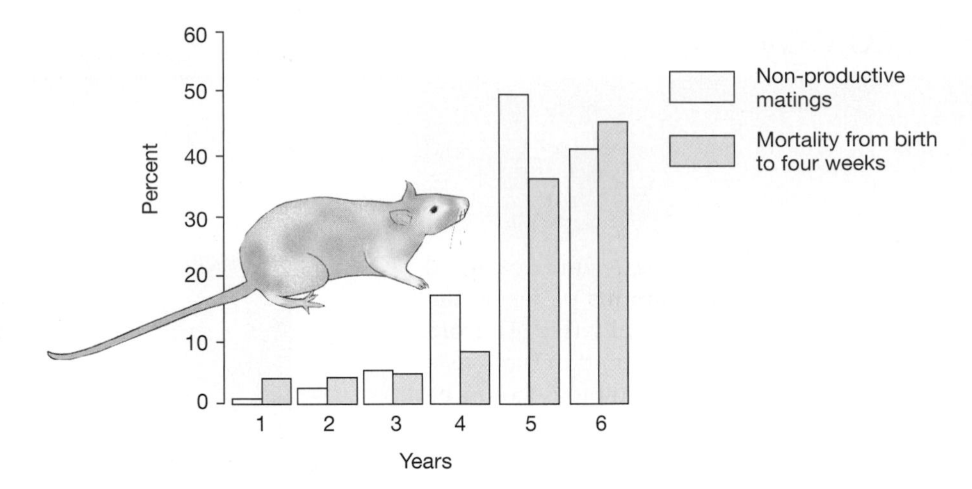

Figure 2.4 Inbreeding in rats. Both the percentage of matings that resulted in no births and the mortality rate of young rats increased the longer inbreeding continued. *(From data in Lerner, 1954, after work by the German Scientist J. Ritzema Bos in 1894.)* [*Note*: The six years 1887–1892 spanned about 30 generations of parent–offspring matings and sibling matings.]

Only one in a million of all of its useful mutations needs to become established in a population in order for the observed numbers of species to have evolved.

A number of processes, most of which are associated with small population sizes, act to reduce genetic variation.

Matings between closely related individuals can decrease the genetic variation of a population.

another, so only 1 in a million of the useful mutations needs to be established in a population in order to provide the genetic basis of observed rates of evolution.

The chief factor limiting genetic variation is therefore *not* the rate of new mutations; instead, variation is limited mainly by gene recombination and the structural patterns of chromosomes. It is not surprising, then, that some scientists have argued that genetic drift alone is sufficient to cause genetic change in species with effective population sizes of under 5 million (Charlesworth, 1984), a figure that is probably greater than the population of many mammals. Russell Lande (1976) showed mathematically how even very weak selection on horses' teeth—two selective deaths out of 1 million individuals per generation—would be sufficient to explain the dramatic evolution of the animal's teeth through the ages.

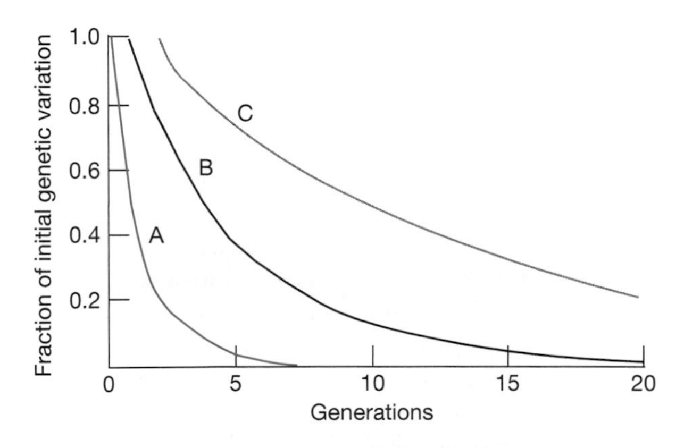

Figure 2.5 The decrease in genetic variation is faster the greater the inbreeding. Systems of mating shown are exclusive self-fertilization (curve *A*), sibling mating (curve *B*), and double-first-cousin mating (curve *C*). (*Redrawn from Crow and Kimura, 1970.*)

2.4 Genetic Variation and Population Size

An understanding of genetic variation is important in conservation biology in order to help manage populations. Reduced genetic variation can arise from three factors that are all a function of population size: inbreeding, genetic drift, and neighborhoods. Reduced genetic variation can greatly impair a population's growth and can jeopardize the recovery of endangered species.

Inbreeding.

Inbreeding, which is mating among close relatives, occurs commonly in social animals. In the laboratory, the survivorship of offspring declines as populations become more inbred, as was seen in a six-year study of rats carried out in the 19th century (Fig. 2.4). The litter size of the inbred rats declined from 7.5 in 1887 to 3.2 in 1892. Generally, the more inbred the population, the quicker the genetic variation in the population drops. This was shown by James Crow and Motoo Kimura (1970), who mathematically examined the loss in genetic variation with time for various types of inbreeding: sibling matings, first-cousin matings, and self-fertilization (in plants) (Fig. 2.5). The most severe form of inbreeding is self-fertilization, or selfing, in plants.

Later data on the effects of inbreeding were provided by Ralls and Ballou (1983), who showed the effects of inbreeding on juvenile mortality in captive populations of mammals (Fig. 2.6). Ungulates, primates, and small mammals exhibited a higher mortality from inbred matings than from non-inbred matings.

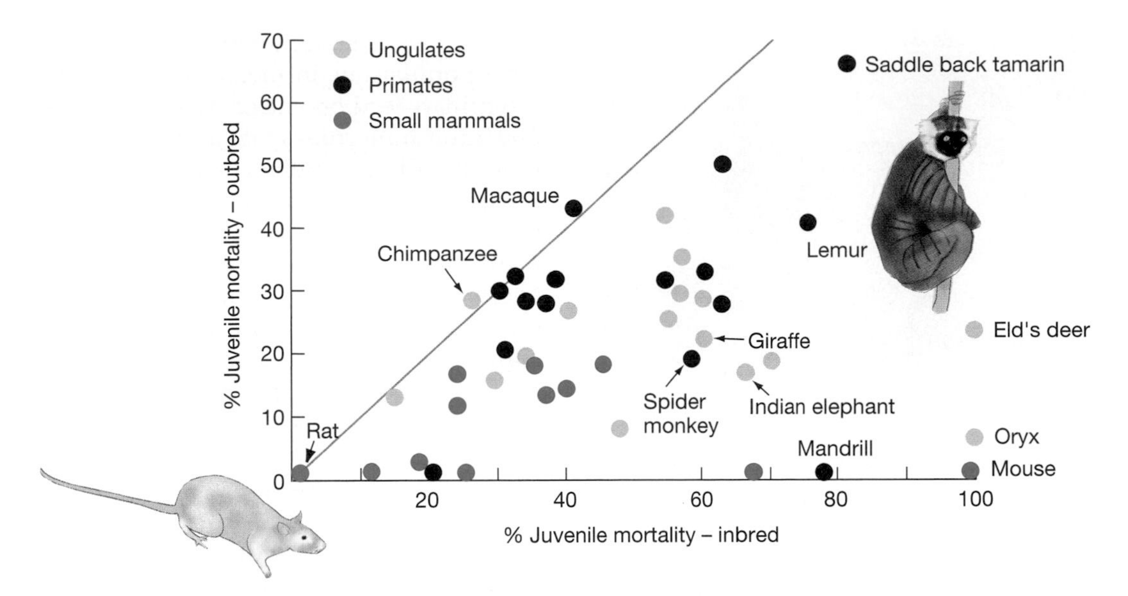

Figure 2.6 The effects of inbreeding on juvenile mortality in captive populations of mammals. Each point compares the percentage of juvenile mortality for offspring of inbred and noninbred matings. The line indicates equal levels of mortality under the two breeding strategies. Points above the line represent higher mortality from noninbred matings; nearly all points fall below the line, indicating higher mortality from inbred matings. The distance of a point below the line indicates the strength of the effect of inbreeding. (*From data in Ralls and Ballou, 1983.*)

The effects of inbreeding are more extreme in small populations. Calculations by Monroe Strickberger of the University of Missouri at St. Louis (1986) showed that the smaller the population size *N*, the faster genetic variation declines (Fig. 2.7). This result has important consequences in the real world, where plant and animal populations are constantly declining because of shrinking habitats. As a result, conservation biology has become particularly concerned with the genetics of small populations. A rule of thumb is that a population of at least 50 individuals is necessary to prevent the deleterious effects of inbreeding for the immediate future.

One of the best examples of the effects of inbreeding in the field concerns the greater prairie chicken, *Tympanuchus cupido*. The birds have a spectacular mating display, inflating bright orange air sacs on their throats, stomping their feet, and spreading their tail feathers (Fig. 2.8). Unfortunately, prairie chickens now live in only a fraction of their original habitat in the U.S. Midwest. For example, in Illinois, the prairies were home to millions of prairie chickens, but as the prairies were converted to farmland, the range of the bird shrank to only two counties: Marion and Jasper. In Illinois, the population of prairie chickens decreased from 25,000 in 1933 to 2,000 in 1962, 76 in 1990, and less than 50 in 1994. Even though prairie chicken habitat was restored in the 1970s and hunting had been banned since the 1940s, the population continued to decline. Studies by Ronald Westemeier and colleagues (1998) showed that the populations had become so small that only five or six males existed by 1994. The genetic variation in the population had dwindled, a fact that was reflected in the steady reduction in

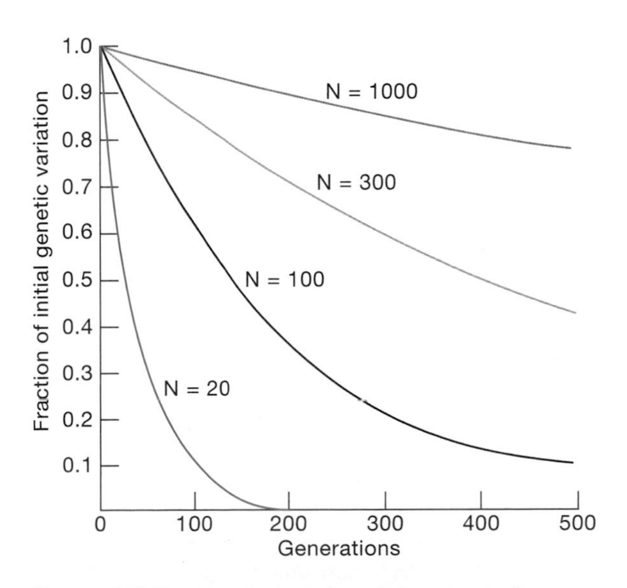

Figure 2.7 Decrease in genetic variation due to finite population size. The smaller the population size *N*, the faster is the decline in genetic variation in the population. (*Modified from Strickberger, 1986.*)

Figure 2.8 The male prairie chicken inflates air sacs, stomps on the ground, and erects special feathers when on its mating ground, in an effort to attract females.

the hatching of eggs over the years (Fig. 2.9). In 1990, fewer than 40% of the eggs hatched. The prairie chicken population had entered a downward spiral from which it could not naturally recover. Luckily, salvation was at hand: Conservation biologists trapped prairie chickens in Kansas and Nebraska and moved them to Illinois, bringing an infusion of new genetic material into the population. This translocation resulted in an increase in the hatching rate of eggs, back to 90%.

Finally, in 1998, a group of Finnish scientists proved what conservation biologists had suspected for some time: Inbreeding brought about by a small population size increases the risk of extinction in nature (Saccheri *et al.*, 1998). In Finland, the Glanville fritillary butterfly (*Melitaea cinxia*, Fig. 2.10) exists in numerous small, isolated local populations in meadows where the caterpillars feed on one or two host plants. The adult butterflies mate and lay eggs in June, and the caterpillars hatch and feed in conspicuous family groups of 50–250, making large-scale counting of their numbers in many meadows relatively easy. Caterpillars overwinter from August until March, with the survivors continuing to feed in the spring and pupating in May. Yearly censuses revealed larvae present in about 400 meadows in an area of 3,500 km². Many populations were small, consisting of one group of caterpillars—the offspring of just one pair of butterflies. In 42 of the populations, the genetic variation was determined by a molecular technique called miscrosatellite analysis. Seven of the 42 populations studied became extinct between 1995 and 1996; all seven had a lower population size and genetic variation than the survivors. Laboratory studies showed how just one generation of brother–sister mating, which might take place in one small group of caterpillars, increased inbreeding and reduced the hatching of eggs by up to 46%. Furthermore, inbred females laid fewer eggs than noninbred females, and the survival of the larvae was reduced. If we wish to examine how populations of the Glanville fritillary butterfly grow in nature, we have to examine the evolutionary ecology of the population. A small population size can have deleterious effects on population growth because of genetic effects.

In small populations, genetic variation is lost just by chance.

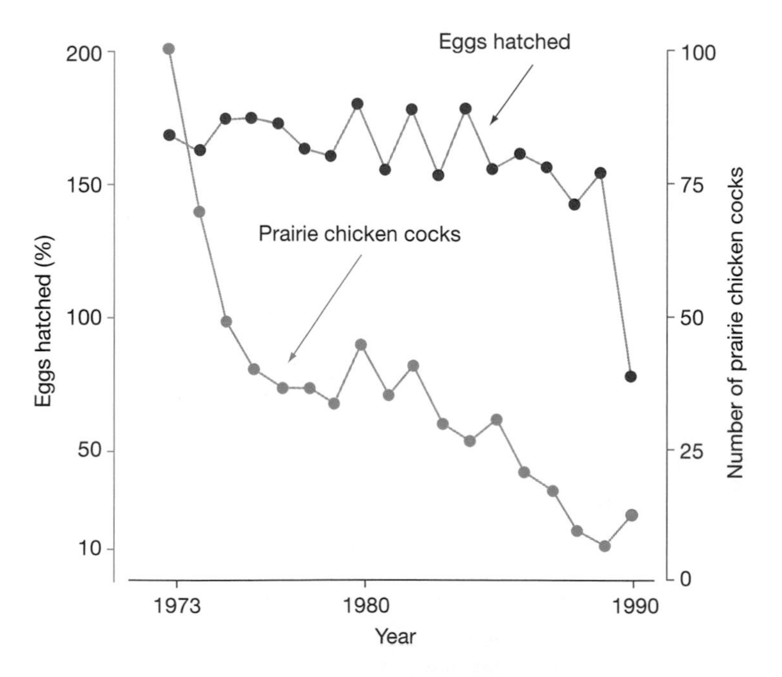

Figure 2.9 Decrease in hatching of prairie chicken eggs on booming grounds in Jasper County, Illinois, 1963–1997. As the number of cocks dropped, so inbreeding increased, resulting in fewer eggs hatching. (*Modified from Westemeier et al., 1998.*)

Genetic drift.

In small populations, some individuals, purely by chance, will fail to mate, often because they fail to find a mate. If an individual that fails to mate possesses a rare gene, the genetic information encoded in that gene may not be passed on to the next generation, resulting in a loss of genetic variation from the population. The likelihood of an allele being represented in just one or a few individuals is higher in small than in large populations, so small, isolated populations are particularly vulnerable to this type of reduction in genetic variation, called *genetic drift*. Such isolated populations will lose a percentage of their original variation over time, approximately at the rate of $1/2N$ per generation, where N is the number of individuals in the population (the population size). Thus, if

$$N = 500, \text{ then } \frac{1}{2N} = \frac{1}{1{,}000} = 0.001, \text{ or}$$

0.1% genetic variation lost per generation,

and if

$$N = 50, \text{ then } \frac{1}{2N} = \frac{1}{100} = 0.01, \text{ or}$$

1% genetic variation lost per generation.

Accordingly, a population of 500 will retain 99.9% of its genetic variation in a generation, whereas a population of 50 will retain 99.0%. Such losses seem insignificant, but they are magnified over many generations. Thus, after 20 generations, a population of 500 will still retain over 98% of its original variation, but a population of 50 will retain only 81.79%. For organisms that breed annually, this could mean a substantial loss in variation over a very few years. Once again, the effect becomes more severe as the population size decreases. A rule of thumb is that a population of at least 500 is necessary to decrease the effects of genetic drift. Thus, the "50/500" rule has entered the literature as a "magic" number in conservation theory. (Fifty is the critical size to prevent excess inbreeding and 500 is the critical size to prevent genetic drift; see Simberloff, 1988).

Robert Lacey (1987) showed that the effects of genetic drift could be countered by the immigration of individuals into a population. Even the relatively low rate of one immigrant every generation would be sufficient to counter genetic drift in a population of 120 individuals (Fig. 2.11). One of the best tests of the idea that 50 or 500 individuals may be a minimum number for conservation purposes was provided by Joel Berger's (1990) study of 120 bighorn sheep *(Ovis canadensis)* populations in the U.S. southwest. The striking observation was that 100 percent of the populations with fewer than 50 individuals became extinct within 50 years, while virtually all of the populations with more than 100 individuals persisted for that length of time (Fig. 2.12).

Figure 2.10 The Glanville fritillary butterfly, *Melitaea cinxia:* Inbreeding increases the risk of extinction in this species.

Neighborhoods and effective population size.

Even in large populations, the effective population size may actually be smaller than the number of individuals in the population if individuals mate only within a neighborhood. For example, by marking individual deer mice (*Peromyscus maniculatus*), Howard (1949) showed that at least 70 percent of males and 85 percent of females breed within 150 m of their birthplaces. Even migratory species, such as birds or salmon, often return to the vicinity of their birthplaces to breed. Levin (1981) argued that even for plants, the effective size of local populations is often in the dozens or hundreds because of limited pollen and seed dispersal. If only half the individuals in a population breed, half the variation is lost per generation.

Genetic variation tends to decline in populations in which individuals mate primarily with close neighbors.

Furthermore, even within a neighborhood, some individuals do not reproduce. This is especially true in animals with a

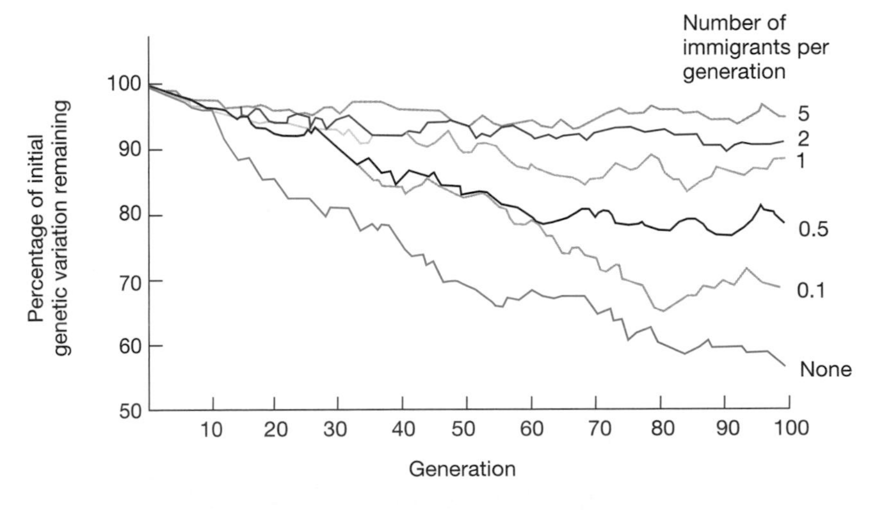

Figure 2.11 The effect of immigration on genetic variation in 25 simulated populations of 120 individuals each. Even the low rate of one immigrant per generation can prevent the loss of heterozygosity through genetic drift. (*After Lacey, 1987.*)

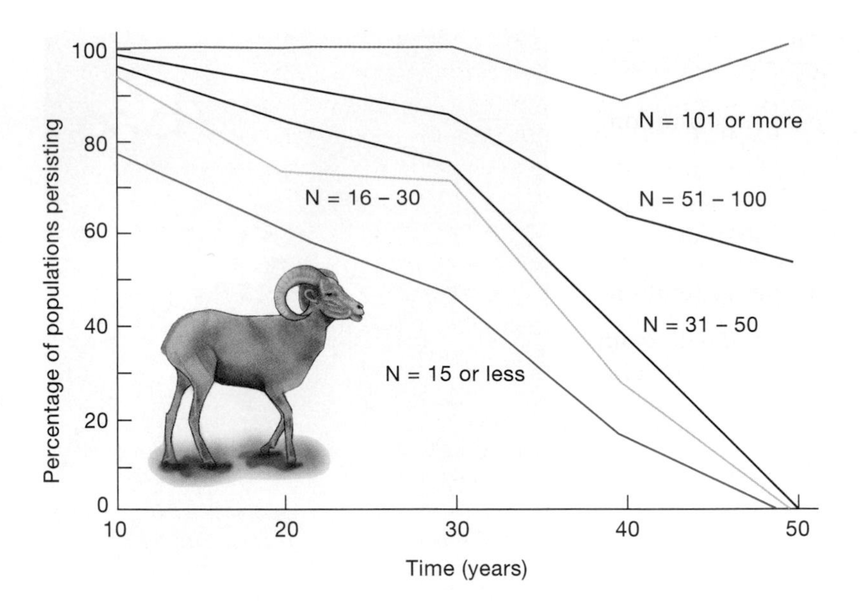

Figure 2.12 The relationship between the size of a population of bighorn sheep and the percentage of populations that persist over time. The numbers on the graph indicate the population size (*N*); almost all populations with more than 100 sheep persisted beyond 50 years, while populations with fewer than 50 individuals died out within 50 years. (*After Berger, 1990.*)

harem mating structure in which only a few dominant males breed. For example, in elephant seals, dominant bulls control harems of females, and a few males command all the matings. If a population consists of breeding males and breeding females, the effective population size is given by

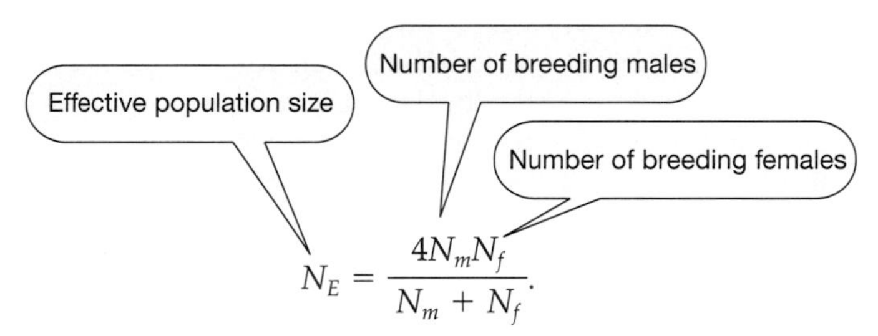

$$N_E = \frac{4N_mN_f}{N_m + N_f}.$$

Where N_m is the actual number of breeding males and N_f is the actual number of breeding females. Thus, in a population of 500 with a 50:50 sex ratio and all individuals breeding,

$$N_E = \frac{4 \times 250 \times 250}{250 + 250} = 500.$$

However, if 250 females bred with 10 males,

$$N_E = \frac{4 \times 10 \times 250}{10 + 250}$$
$$= \frac{10{,}000}{260}$$
$$= 38.5$$

or 7 percent of the actual population size.

A knowledge of the effective population size is vital to ensure the success of many conservation projects, notable among which are reserves designed to protect grizzly bear populations in the contiguous 48 states of the United States. The grizzly bear (*Ursus arctos horribilis*) has declined in numbers from an estimated hundred thousand in 1800 to fewer than a thousand at present. The range of the species is now less than 1 percent of its historical range, and the grizzly is restricted to six separate populations in four states (Fig. 2.13). Research by Fred Allendorf of the University of Montana (1994) indicated that populations as large as a hundred would likely lose genetic variation because effective population sizes of grizzlies are only about 25 percent of actual population sizes. Thus, even fairly large isolated populations, such as the 200 bears in Yellowstone National Park, are vulnerable to the harmful effects of loss of genetic variation. Allendorf and his colleagues proposed that

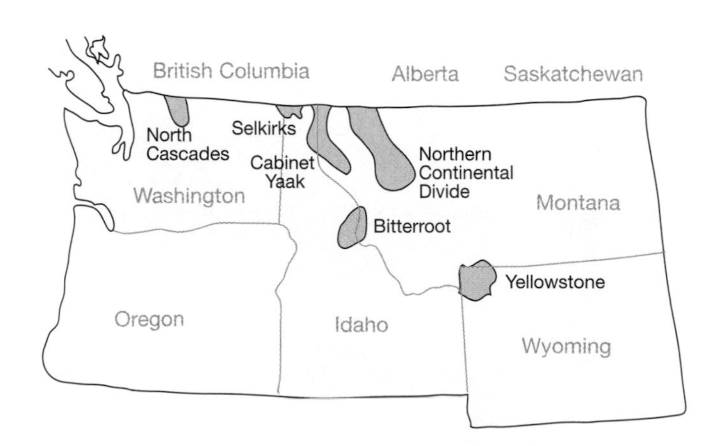

Figure 2.13 The current range of the grizzly bear in the continental United States has contracted to just six populations in four states as the grizzly population has shrunk from 100,000 before the west was settled to about 1,000 alive today.

an exchange of bears between populations in the wild or even between those populations and zoo collections would help tremendously in promoting genetic variation. An exchange of just two bears between populations would greatly reduce the loss of genetic variation. Allendorf has concluded that viable populations of grizzly bears can be preserved only by maintaining large reserves, by promoting artificial gene flow among reserves, and by promoting gene flow between zoos and reserves. A combination of various techniques, then, appears necessary for the preservation of the grizzly bear (and probably for the preservation of many large animals in the United States). A knowledge of the genetics and breeding of the grizzly is vital for safeguarding its population in the continental United States.

Summary

To conserve rare species, a knowledge of the genetic structure of populations is vital. Small populations can lose genetic variation in a number of ways, and genetic variation not passed on to the next generation is lost forever.

1. New species result from an accumulating number of gene and chromosome mutations. There are easily enough mutations in nature to account for the vast numbers of species we see today.

2. Genetic variation in populations is reduced by inbreeding, genetic drift, and neighborhoods. The effects of all these factors increase in severity with decreasing population size. The 50/500 rule suggests that 50 individuals is the critical size to prevent inbreeding in a population and 500 individuals is the critical size to prevent genetic drift.

3. Humans can move individuals from one wild population to another to simulate natural dispersal. Some models suggest that moving even one or two individuals per generation would counteract genetic drift.

4. The effective population size of a species can be reduced by a harem mating structure or by territoriality. Populations of elephant seals and grizzly bears may suffer in just this regard. Population sizes in reserves would therefore have to be even bigger than otherwise expected in order to counteract the effect.

Conservation biologists can use their knowledge of how genetic variation is decreased to help them maintain viable populations. Such genetic management of populations must be used in conjunction with preservation of the organisms' habitat.

Discussion Questions

1. A small population size is obviously detrimental to a population's genetic variation. With this in mind, why is habitat fragmentation especially detrimental to populations, and can linking conservation areas by corridors or siting them close together help alleviate the problem?

2. Just as inbreeding diminishes a population's variability, so can outbreeding reduce a population's fitness, especially if the population is highly adapted to its local environment. (Outbreeding is breeding between individuals that are relatively unrelated.) Consider, for example, insects that feed on trees on which local populations do much better on a given tree than other populations do. By what mechanism do you think this reduction in fitness works, and what implications does it have for conservation biology?

3. In 1986, the California condor had declined to only 27 individuals (14 females and 13 males), all contained in the San Diego Wild Animal Park. Since then, over 150 condors have been bred, with 88 having been released back into the wild. What genetic problems do you think might be encountered in trying to reestablish this population in nature? Meretsky et al. (2000) discusses some of the successes and pitfalls conservation biologists have encountered in this project.

C H A P T E R 3

Extinction

Road Map

1. The current rate at which species are becoming extinct is faster than the historical rate of extinction as revealed by the fossil record.
2. The main causes of species extinction are, in order, the introduction of exotic species into a habitat, destruction of habitats, and hunting.
3. Currently, most endangered species are at risk of extinction mainly because of the loss of their habitat and, to a lesser extent, hunting and the introduction of exotic species.
4. Certain characteristics of species, including how much they reproduce, how long they live, what they eat, and how common they are, can influence their likelihood of extinction.

To make the best decisions that will protect species from extinction, we first need to understand those species' rates of extinction. In general, we can compare current rates of extinction with those from the fossil record to see which are higher. We can learn what factors threaten species with extinction. We can examine why species have become extinct in the past, and we can find out where extinctions are most common. Armed with this type of information, ecologists can help avert the extinction crisis.

3.1 The Extinction Crisis

Species become extinct when all of their individuals die without producing progeny. Species disappear in a different sense when their lineage is transformed over evolutionary time or when they divide into two or more separate lineages (called pseudoextinction). The relative frequency of true extinction and pseudoextinction in evolutionary history is not yet known. The extinction of species is a natural process. Fossil records show that the vast majority of species that have ever existed are now extinct, with extinct species outnumbering living species by a factor of perhaps 1,000:1.

How long do species last on Earth? The average life span of a species in the fossil record is around 4 million years, which would give, at a very rough estimate, a background extinction rate of 2.5 species each year out of a total number of species of around 10 million. However, it can be argued that the fossil record is heavily biased toward successful, often geographically wide-ranging, species, which undoubtedly persist far longer than average. The fossil record is also biased toward vertebrates and marine mollusks, both of which fossilize well because of their hard body parts. If background extinction rates were 10 times higher than the rates perceived from the fossil record, then extinctions would be expected to occur at a rate of one every hundred years among the 4 thousand or so living mammals today and one every 50 years among birds. It is indisputable that recent extinction rates have been far higher than this. Therefore, we should try to determine how human activity has increased the number of species that are becoming extinct.

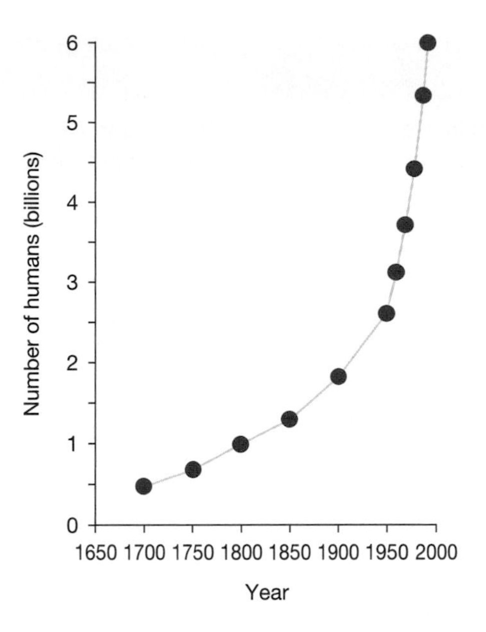

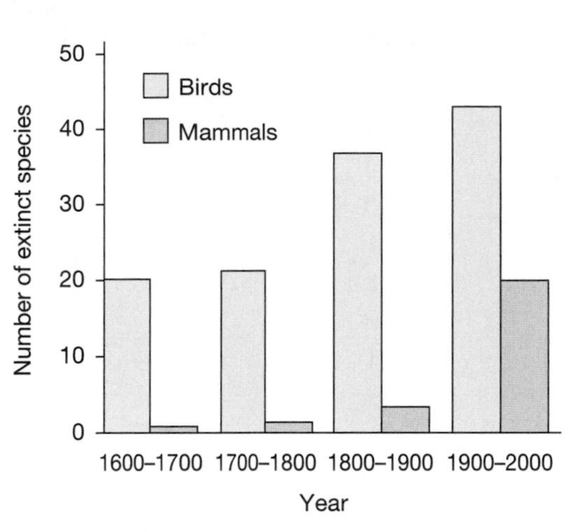

Figure 3.1 Population growth and animal extinctions. (left) Geometric increase in the human population. (right) Increasing numbers of extinctions in birds and mammals. These figures suggest that, as human numbers increase, more and more species are exterminated.

Humans have caused extinctions by introducing exotic species, by altering habitats, and by hunting.

It is easy to suggest that humans have been the overwhelming cause of recent extinctions. The arrival of humans on previously isolated continents, around 13 thousand years ago in the case of Australia and 11 thousand years ago or possibly earlier on North and South America, seems to have coincided with large-scale extinctions among certain taxa. Australia lost nearly all its species of very large mammals, giant snakes, and reptiles and nearly half its large flightless birds around that time. Similarly, North America lost 73 percent and South America 80 percent of their genera of large mammals around the time of the arrival of the first humans. The probable cause was hunting, but the fact that the climate changed at around this same time leaves the door open for natural changes as a contributing cause of these extinctions. However, the rates of extinctions on islands in the more recent past seem to confirm the devastating effects of humans. Polynesians who colonized Hawaii in the fourth and fifth centuries A.D. appear to have been responsible for exterminating around 50 of the 100 or so species of endemic land birds in the period between their arrival and that of the Europeans in the late 18th century. A similar impact probably was felt in New Zealand, which was colonized some 500 years later than Hawaii. There, an entire avian megafauna consisting of huge land birds was exterminated by the end of the 18th century. This extinction was probably accomplished through a combination of direct hunting and large-scale destruction of habitats through burning.

On Madagascar, the giant elephant bird, the largest bird ever recorded; 20 species of lemur, most of them larger than any surviving species; and two giant land tortoises have become extinct within the last 15 hundred years. In the Caribbean, at least two ground sloths became extinct before Europeans arrived at the end of the 15th century. Once again, climate change, particularly progressive desiccation, may have played a part. However, it is indisputable that recorded extinctions have increased dramatically in recent years, just as the population of humans has skyrocketed (Fig. 3.1).

3.2 Patterns of Extinction

Extinction rates are higher on islands than on mainland.

Certain generalized patterns of extinction emerge on examination of the data. Perhaps one of the most important is the preponderance of extinctions on islands versus continental areas. Although both mainland and islands have similar overall numbers of recorded extinctions, there are fewer species on islands than on continents, making the percentage of taxa extinct on islands greater than that on continents (Table 3.1). Perhaps the extinction rates on islands is so high because many species effectively consist of single populations on isolated islands. Adverse factors are thus likely to affect the entire species and bring about its extinction. Also, species on islands may have evolved in the absence of terrestrial predators and may often be flightless. Tameness, flightlessness, and reduced reproductive rates appear to be major contributory factors to

TABLE 3.1 Recorded extinctions, 1600 to present, on mainland areas (larger than 10^6 km², the size of Greenland), islands, and the ocean.

Taxa	Mainland	Island	Total	Approximate Number Of Species	Percentage Of Taxa Extinct Since 1600
Mammals	30	51	81	4,000	2.0
Birds	21	92	113	9,000	1.3
Reptiles	1	20	21	6,300	0.3
Amphibians	2	0	2	4,200	0.05
Fish	22	1	23	19,100	0.12
Invertebrates	49	48	97	more than 1,000,000	0.02
Vascular plants	380	219	599	250,000	0.24
Total	505	431	936		

the extinction of such species, especially when novel predators are introduced. This trend becomes particularly apparent when one compares the threats to Hawaiian birds and plants with those to U.S. continental birds and plants. A much higher proportion of Hawaiian birds and plants is threatened by exotic species and diseases—especially the latter in the case of birds—than on the mainland (Fig. 3.2).

When we look at the possible causes of extinction, we find that most often no cause has been assigned (Fig. 3.3). We simply do not know why most species have become extinct, especially those that did so decades or centuries ago. Of those causes that have been assigned, introduced animals and direct destruction of habitat by humans have been major factors, being implicated in 17 percent and 16 percent of cases, respectively. These same factors are equivalent to 39 percent and 36 percent, respectively, of the extinctions whose causes are known. Hunting and deliberate extermination also con-

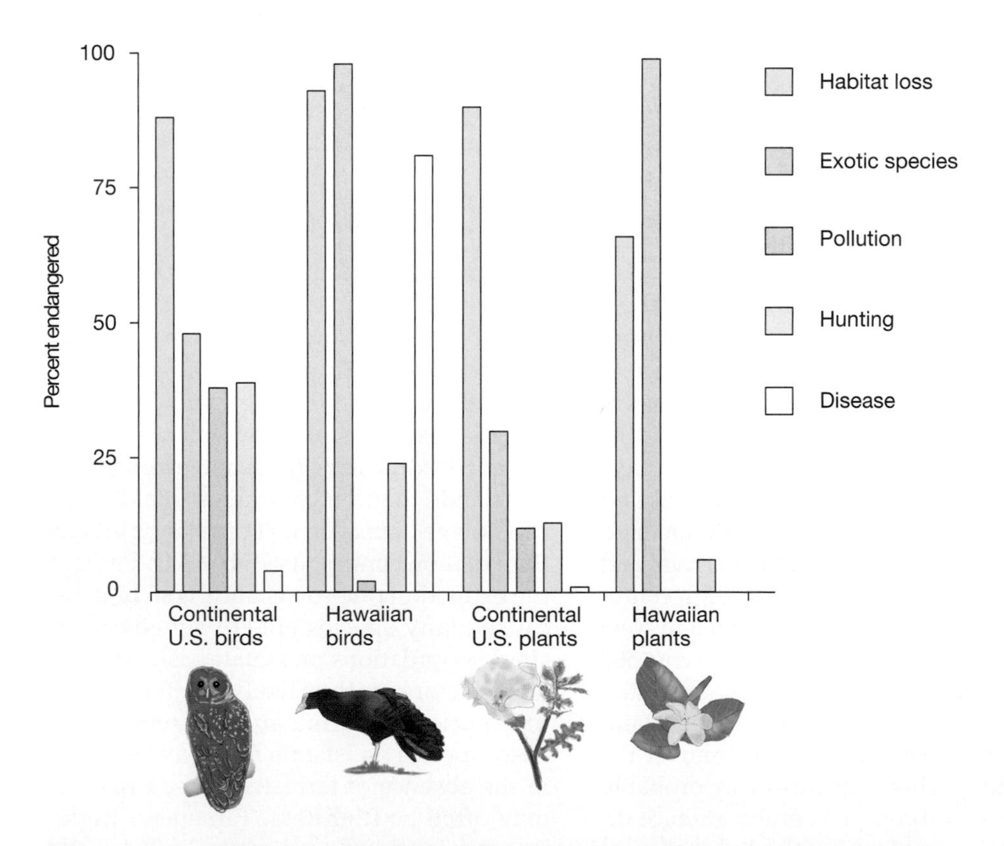

Figure 3.2 Percentages of endangered birds and plants in Hawaii and in the continental United States. Island species are particularly threatened by habitat loss and, in the case of birds, disease. *(From data in Wilcove et al., 1998.)*

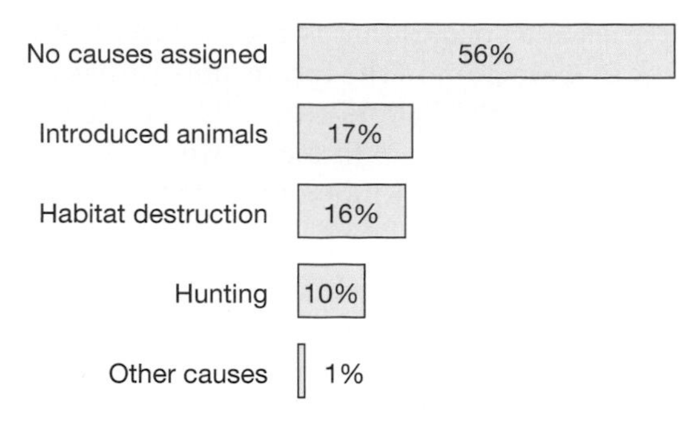

Figure 3.3 The causes of extinction of animals, based on a knowledge of 484 extinct species. (*From data in World Conservation Monitoring Centre, 1992.*)

tribute significantly (a combined 23% of extinctions with known causes).

We can break up the "introduced species" category a little further. The effects of introduced species can be attributed to competition from introduced species, predation from introduced species, or disease and parasitism from introduced species. Competition may exterminate local populations, but it has not yet been clearly shown to extirpate entire populations of rare species. As regards predation, there have been many recorded cases of extinction. Introduced predators such as rats, cats, and mongooses have accounted for at least 112 of 258 recorded extinctions of birds on islands, or 43.4% of those extinctions (Brown, 1989). Parasitism and diseases brought by introduced organisms are also important in causing extinctions. Avian malaria in Hawaii, facilitated by the introduction of mosquitoes, is thought to have contributed to the demise of 50 percent of local Hawaiian birds. Similarly, the American chestnut tree and European and American elm trees have had their numbers diminished by introduced plant diseases, although neither species has yet become extinct.

Habitat destruction, like deforestation, is a prime cause of the extinction of species. More subtle alterations of habitat by events such as modification of climate due to pollution have not yet been shown to have caused any extinctions. By contrast, direct exploitation, particularly the hunting of animals, has caused many extinctions. Steller's sea cow (Fig. 3.4a), a huge species of manatee-like

(a)

(b)

(c)

(d)

Figure 3.4 Extinct species: (a) Steller's sea cow, (b) the dodo, (c) the passenger pigeon, and (d) the Carolina parakeet. The first two species were essentially hunted to death, whereas the last two succumbed to a combination of hunting and habitat destruction.

mammal, was hunted to extinction in the Bering Straits only 27 years after its discovery in 1740. The dodo (Fig. 3.4b) was hunted to death on Mauritius soon after the island became a Dutch colony in 1644. In the United States, the passenger pigeon was the most common bird in the entire bird population of North America. Incredibly, hunting (Fig. 3.4c) was the primary reason for its eventual extinction by 1900, although habitat destruction also played a large part. The Carolina parakeet suffered the same fate (Fig. 3.4d). The list of species hunted to extinction goes on and on.

3.3 Endangered Species

Ecological data on species' vulnerability to extinction can be used to rank species for conservation purposes.

Knowing why species have become extinct helps us to recognize the threats that are likely to endanger species that are alive today. This knowledge may make it easier to protect vulnerable species. An endangered species is a species that is thought to be at risk of extinction in the foreseeable future.

Most of the factors currently threatening species with extinction are anthropogenic in nature. Among these factors are the following:

1. Habitat loss or modification. For example, forests are cut down to develop grazing land for cattle (pastoral development) or to establish commercial plantation forests, and grasslands are converted to agricultural fields (cultivation and settlement).
2. Hunting for meat and fur, or the taking of animals for the pet trade.
3. Accidental or deliberate introduction of exotic species, which may compete with, prey on, or hybridize with native species.
4. Deliberate eradication, persecution, or disturbance of species considered to be pests. This factor is perhaps most serious in the case of predatory species such as wolves and tigers.
5. Incidental takes, particularly the drowning of aquatic reptiles and mammals in fishing nets.
6. Disease, both exotic and endemic, which is exacerbated by the presence of large numbers of domestic livestock or introduced species.

In many cases, individual species are faced by several of these threats simultaneously. Some understanding of the relative importance of the different types of threat, as measured by the frequency of their occurrence, has been achieved from an examination of threats facing the terrestrial mammals of Australia and the Americas. Threats to terrestrial mammals have been fully investigated over a large area. Of the 119 species of Australian and North and South American mammals that are considered endangered, 75 percent are threatened by more than one factor and, of these, 27 face four or more threats. The major threat, which affects 76 percent of species, is habitat loss and modification (Fig. 3.5). Habitats are modified for many reasons, of which the most frequent is the cultivation of virgin habitats and settlement by people. Half the species are affected by overexploitation, the most significant form of which is hunting for meat. Introduced predators and competitors affect 18 percent of the endangered species. The most serious other factor is lim-

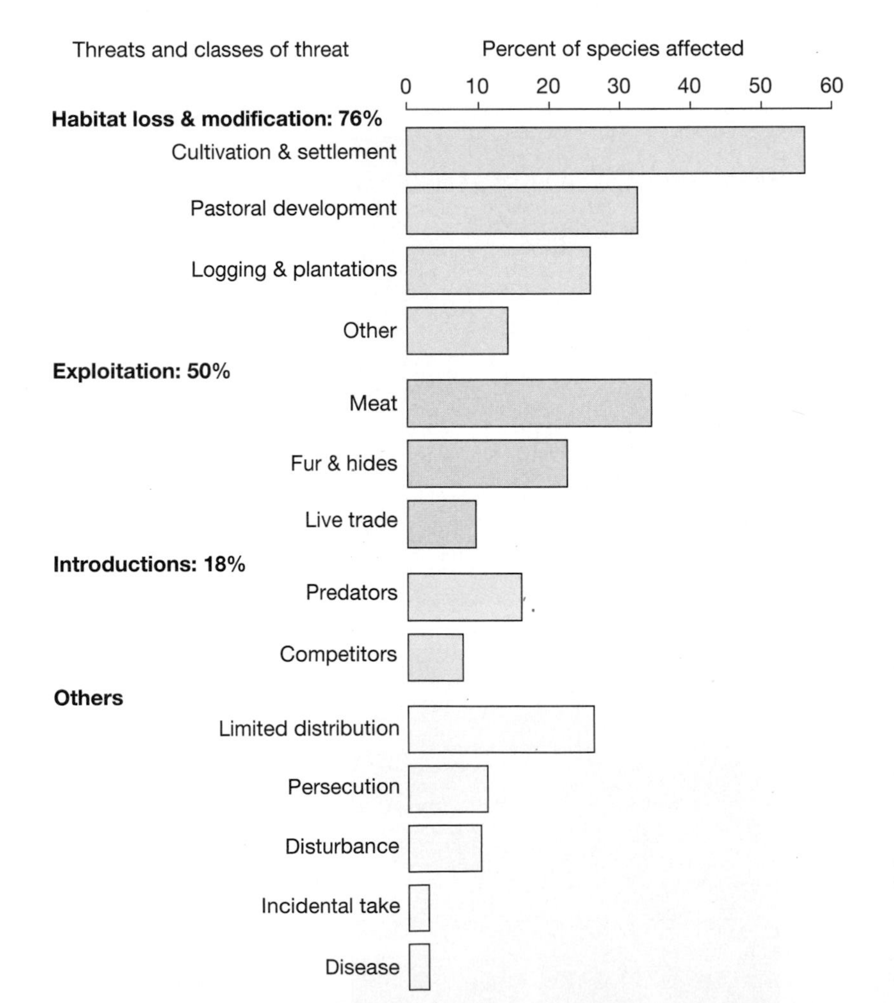

Figure 3.5 The factors that threaten mammals in Australia and the Americas. The most important threat is habitat loss and modification. Species can suffer from multiple threats. (*Reproduced from World Conservation Monitoring Centre, 1992.*)

ited distribution, which affects one-quarter of the species. A species that is limited in its distribution is naturally rare and has a low population size, making it vulnerable to extinction. For example, species may exist only on small islands. Interestingly, the factors that threaten endangered species today are somewhat different from those which are known to have caused extinctions in the past (e.g., introduced species).

Hunting affects some endangered animals more than others. Many fur-bearing animals, including the chinchilla, the giant otter, many species of cats, and some species of monkeys, have declined to very low population sizes because their pelts are prized (Figs. 3.6a–b). Populations of valuable timber species, including the West Indian mahogany in the Bahamas and the caoba (mahogany) of Ecuador, have been severely depleted, and the Lebanese cedar has been reduced to a few scattered remnants of forest (Figs. 3.6c–d). Overexploitation is a more selective threat to the survival of species than is habitat loss and threatens primarily vertebrates and certain taxa of plants and insects. More specifically, carnivores, ungulates, primates, sea turtles, showy tropical birds, and timber species have been overharvested. Many species of butterflies and orchids have been overharvested for commercial interests, too, and rare plants have been threatened by collectors.

Recently, David Wilcove, a senior ecologist at the Environmental Defense Fund in Washington, DC, and his colleagues (1998) categorized threats to imperiled plants and animals in the United States (Fig. 3.7). They recognized five categories of threats: habitat destruction, alien species, overharvesting, disease (caused by both alien and native pathogens) and pollution (including siltation for freshwater species). The sample size was an impressive 1,880 species, with many plants, mammals, birds, reptiles, amphibians,

Figure 3.6 Endangered species. The chinchilla (a) and giant otter (b) are prized for their pelts, while the West Indies mahogany (c) and Lebanese cedar (d) have been reduced to scattered remnants in forests because of their value to the timber trade.

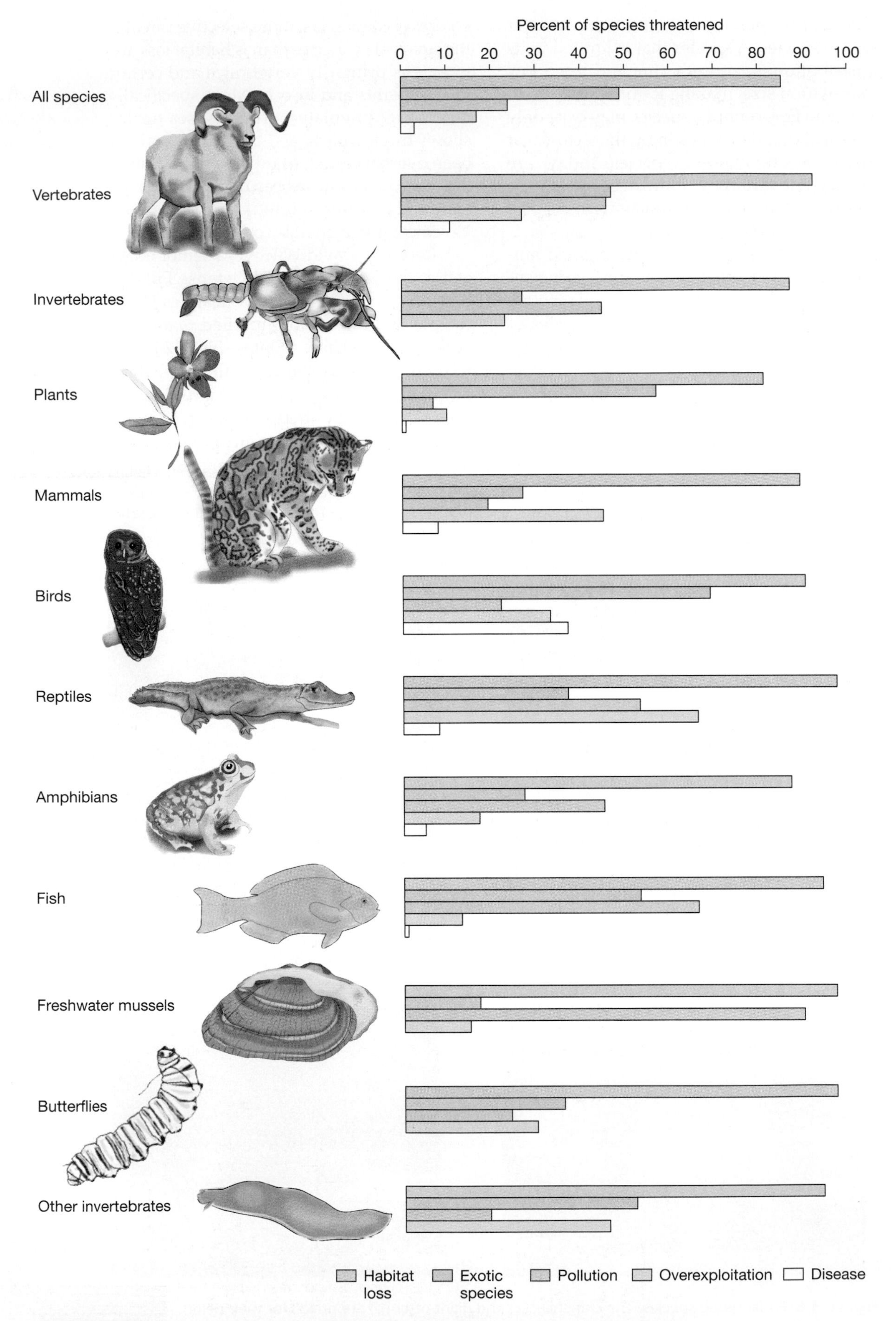

Figure 3.7 Percentages of species in different taxa threatened by various causes. Habitat degradation is of paramount importance to all groups, but pollution is important in freshwater habitats, and overexploitation is of concern to mammals, birds, and reptiles. Species can suffer from multiple threats, so categories do not sum to 100%. *(From data in Wilcove et al., 1998.)*

fish, and invertebrates included. The results were similar to those of the World Conservation Union regarding birds of the world and mammals of Australia and the Americas, in that habitat degradation was by far the most important threat (threatening 85% of species). However, exotic species threatened about half of the endangered species because exotic fish are a major problem in aquatic habitats. Interestingly, pollution was also a major problem for aquatic species such as freshwater fish, mussels, and amphibians. Overall, pollution threatens 46% of vertebrates and 45% of invertebrates, and is of minor importance only for plants (7%). Overexploitation of mammals, birds, and reptiles is considerable.

We can get some idea of the relative threat to different taxa by using a smaller database of some 835 animal species listed as endangered in 1990 by the World Conservation Union. The class with the greatest number of endangered species is fish, but, expressed as a percentage of total species, mammals have the highest percentage of threatened species (Fig. 3.8). Within the mammals, some orders seem to have a disproportionately high number of endangered species. For example, two out of two elephant species are threatened; four out of four species of manatees and dugongs are threatened; and primates, carnivorous cats, and antelopes are also highly threatened with 53 percent, 32 percent, and 31 percent, respectively, of their constituent species listed. Although the last three orders combined contain only some 20.6 percent of the world's mammalian species, they account for just over half of the endangered species. Vertebrates are probably more vulnerable to extinction than invertebrates, because vertebrates are much larger and require more resources and larger ranges. For example, the Florida panther population is now extremely low—about 70 animals—but there may have been only about 16 hundred of the animals at their maximum, before humans arrived.

Endangered species are not evenly distributed among geographical areas.

We can also break down the data to show which geopolitical areas contain the most endangered species (Table 3.2). The majority of threatened mammals occur mainly in tropical countries, with the highest numbers recorded from Madagascar (53), Indonesia (49), Brazil (40), and China (40). Such countries may have large numbers of endangered animals simply because they have more animals in general; therefore, if the same percentage were threatened with extinction in each country, a country having a larger number of animals to begin with would have a higher number endangered. We can obtain some idea of what a country might be expected to have in terms of numbers of endangered species by drawing a graph of that country's area against the number of endangered animals (Fig. 3.9). Bigger countries should have more animals than smaller ones and so should have more endangered species. The line describes the relationships between area and number of endangered species. Madagascar and Indonesia, in particular, have more endangered animals in relation to their area than would be predicted statistically (they are represented by points above the line), whereas the United States, for example, has fewer. It may come as a surprise to find that the United States heads the list of endangered reptiles, amphibians, and fishes (Table 3.2), but this is probably because other countries monitor those taxa less well and so have not fully documented the number of endangered species in them.

Endangered species are not evenly distributed among taxonomic groups.

In the United States, the greatest numbers of endangered species occur in Hawaii, southern California, southern Appalachia, and the southeastern coastal states. Many of these endangered species are restricted to very small areas. For example, 48% of

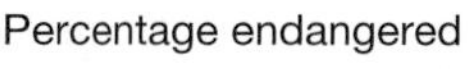

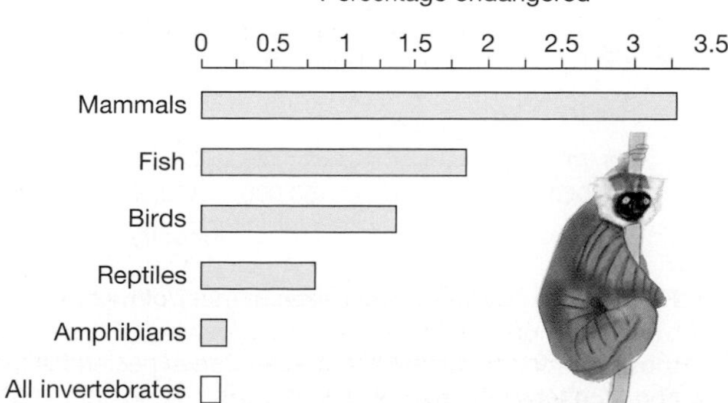

Figure 3.8 Percentage of known species classed as endangered. Mammals are clearly the most endangered taxonomic group.

TABLE 3.2 Countries with the most endangered species, by biological class. *(From World Conservation Monitoring Centre, 1992.)*

Mammals		Birds		Reptiles		Amphibians		Fishes	
Country	No. of Species	Country	No. of Species	Country	No. of Species	Country	No. of Species	Country	No. of Species
Madagascar	53	Indonesia	135	USA	25	USA	22	USA	164
Indonesia	49	Brazil	123	India	17	Italy	7	Mexico	98
Brazil	40	China	83	Mexico	16	Mexico	4	Indonesia	29
China	40	India	72	Bangladesh	14	Australia	3	South Africa	28
India	39	Colombia	69	Indonesia	13	India	3	Philippines	21
Australia	38	Peru	65	Malaysia	12	New Zealand	3	Australia	16
Zaire	31	Ecuador	64	Brazil	11	Seychelles	3	Canada	15
Tanzania	30	Argentina	53	Colombia	10	Spain	3	Thailand	13
Peru	29	USA	43	Madagascar	10	Yugoslavia	2	Sri Lanka	12
Vietnam	28							Cameroon	11

endangered plants and 40% of endangered arthropods in the United States are restricted to a single county. Princeton Professor Andy Dobson and his colleagues (1997) examined the relation between the density of endangered species in each state and the intensity of human economic and agricultural activities, the climate, the topology, and vegetative cover in the state. They found that two variables—agricultural output and rainfall—explained 80 percent of the variation in density of endangered species. Agricultural output was the key variable for plants, mammals, birds, and reptiles. The higher the agricultural output, the more agricultural land there was, and the less land available for wild species. Lower rainfall meant arid conditions, lower numbers of plant species, and hence lower numbers of animal species. Of course, rainfall is a natural threat to endangered species, whereas agricultural production is due to humans. Accordingly, we cannot do much about rainfall, but we can address the agricultural issue if we have the political will to do so.

University of Ottawa professors Jerome Kerr and David Currie (1995) did a similar analysis comparing 90 countries and using six indices of human activities (Table 3.3). Human population density explained the

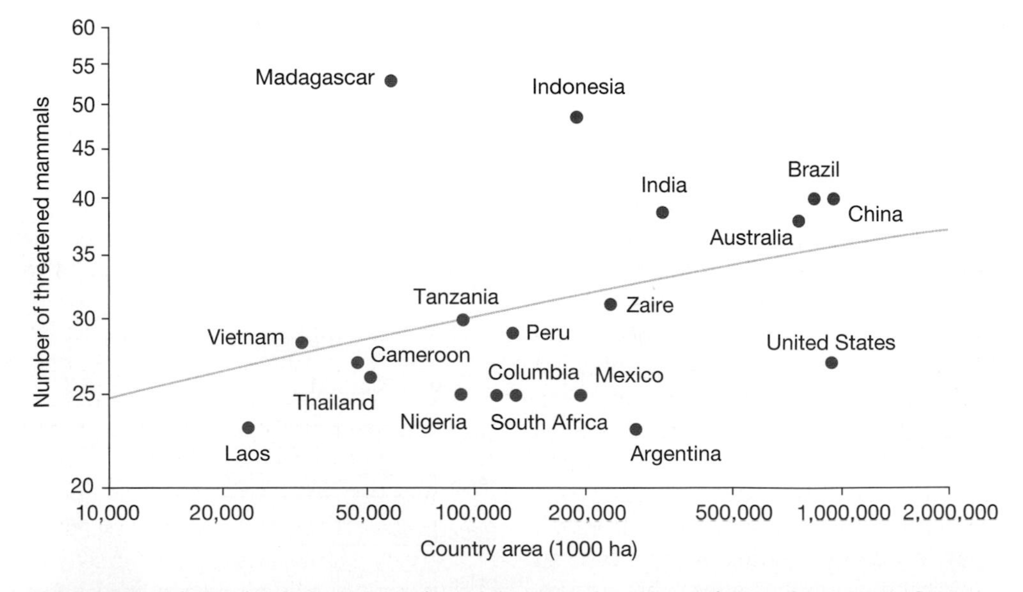

Figure 3.9 Relationship between number of threatened species and area of a country. Countries with more threatened species than would be expected from their area fall above the line; those with fewer threatened species than expected fall below. *(Reproduced from World Conservation Monitoring Centre, 1992.)*

TABLE 3.3 Factors shown to influence susceptibility to extinction among birds and mammals. *(After Kerr and Currie, 1995.)*

Factor	Rationale
Human population	High populations lead to habitat loss, which increases numbers of threatened species.
Per capita GNP	Subsistence hunting, uncontrolled sport hunting, and poaching are more extensive in poorer countries, and damaging land-use practices are less likely to be regulated.
Extent of protected area	Areas established to safeguard endangered species may permit higher population densities of those species.
Total cropland	Loss of habitat attributable to agricultural activity increases numbers of endangered species.
Birthrates	High birthrates are related to high overall human population.
Per capita industrial CO_2 emissions	Greater industrialization may result in higher pollution levels, threatening species' well-being.

most variation in the proportion of endangered species of birds, and per capita GNP explained the most variation for mammals. In addition, the mammalian population density was found to correlate strongly with the extent of protected area per country.

3.4 Species Characteristics and Extinction

In order to further predict what type of human influences are critical in the extinction of wild species, some knowledge of those species' traits that may be correlated with high levels of extinction is desirable. At least seven characteristics have been proposed as factors affecting a species' sensitivity to extinction (Fig. 3.10):

1. *Rarity*. Generally, rare species are more prone to extinction than common ones. This generalization is not as intuitive as might first be thought. For example, a very common species might be highly susceptible to even the slightest change in climate, whereas a rare species, although possessing very low numbers of individuals, may be more resistant to climatic change and thus more persistent in evolutionary time. But what is rarity? Again, this is not as obvious as one might think. Deborah Rabinowitz (1981) showed that the concept may depend on three factors: geographic range, breadth of habitat, and local population size. A species is often termed rare if it is found in only one geographic area or one type of habitat, regardless of its density there. A species that is widespread, but at very low density, can also be regarded as rare. Conservation by habitat management is much easier, and more likely to succeed, for species restricted to one area or type of habitat than for widely distributed, but rare, species.
2. *Ability to disperse*. Species that are capable of migrating between fragments of habitat, such as between mainland areas and islands, may be more resistant to extinction. Even if one small population becomes extinct in one area, it may be "rescued" by immigrating individuals from another population. (We shall discuss this population structure, called a *metapopulation*, further in Chapter 13.)
3. *Degree of specialization*. It is often thought that organisms which are specialists are more likely to become extinct. For example, pandas, which feed on only a single species of bamboo, are at high risk of extinction. By contrast, animals that have a broader diet may be able to switch from one type of food to another if their habitat is altered or even lost and are thus less prone to extinction. Plants that can live in only one type of soil may be more prone to extinction than plants that prosper in many types of soil.
4. *Population variability*. Species with relatively stable populations may be less prone to extinction than others. For example, some species, especially those in northern forests, show pronounced cycles. Lemmings reach very high numbers

The probability of a species becoming extinct can be linked to the species' characteristics.

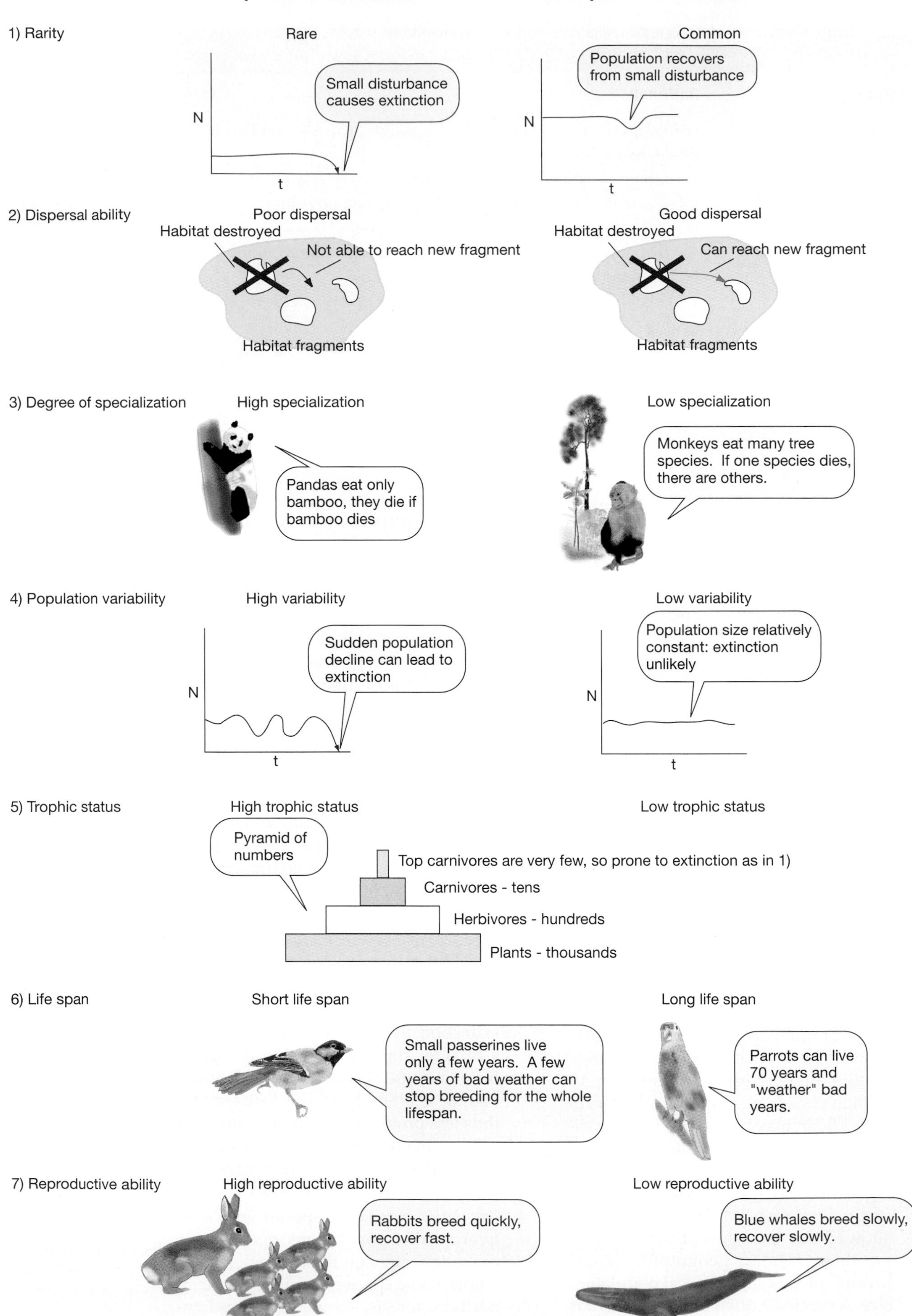

Figure 3.10 Characteristics that make species particularly vulnerable to extinction.

in some years, and their population crashes in others. It is thought that these species might be more likely to become extinct than others.

5. *Trophic status* (animals only). Animals occupying higher trophic levels usually have small populations. For example, birds of prey and Florida panthers are far fewer in number than their prey, and, as noted earlier, rare species may be more vulnerable to extinction.
6. *Life span*. Species with a naturally short life span may be more likely to become extinct. Imagine two species of birds, one of which lives for 70 or 80 years and begins breeding at age 10, such as a parrot, and the other, which is about the same size, but begins breeding at age 2 or 3 and lives only to the age of 10. The parrot, with its 70-year life span, may be able to "weather the storm" of a fragmented habitat for 10 years without breeding. The parrot can begin breeding again when the habitat becomes favorable. By contrast, species with a naturally short life span are not able to wait out unfavorable periods.
7. *Reproductive ability*. Species that can reproduce and breed quickly may be more likely to recover after severe population declines than those which cannot. Thus, it is thought that organisms with a high rate of increase, especially small organisms like bacteria, insects, and small mammals, are less likely to become extinct than larger species, such as elephants, whales, and redwood trees. The passenger pigeon laid only one egg per year, a reproductive rate that probably contributed to its demise.

In order to minimize the threats to endangered species and increase their chance for survival, we must stop destroying habitats, cut back on hunting, and prevent the release of exotic species in habitats to which they are not native. To identify species that are at risk of extinction before they reach endangered status, we must understand many aspects of their biology, including their abundance, how widely distributed they are, how specialized they are, and their reproductive rate and life span. The behavior of a species can also affect its reproduction and abundance and, therefore, its susceptibility to extinction. In the next two chapters, we shall explore the behavior and life history strategies of organisms.

Summary

In the conservation crisis of the 21st century, we are concerned with preserving as many species as possible. To this end, knowledge of what causes species to become extinct and what currently threatens other species with extinction is vital.

1. Historically, the three most important known causes of extinction, in order of importance, are introduced species (39% of cases in which the cause is known), habitat destruction (36%), and direct exploitation (23%). Introduced predators such as rats and cats account for the extinction of many ground-nesting birds and lizards on islands. Hunting has also caused the extinction of many species, with hunting for meat or fur contributing to the demise of many vertebrates.

2. Today, habitat destruction, such as deforestation and the conversion of grassland and forest to cropland or pasture, is the prime threat to most endangered species. Habitat loss is the chief factor threatening mammalian species with extinction.

3. At least seven characteristics of species may affect their sensitivity to extinction: rarity, ability to disperse, degree of specialization, population variability, trophic status, longevity, and intrinsic rate of population increase.

Slowing the destruction of habitats, easing hunting pressure, and reducing the release of exotic species will help preserve endangered species. A thorough knowledge of the biology of species is also likely to tell us which of them are particularly vulnerable to extinction. Armed with that knowledge, we can act to reduce future extinction rates and secure populations of endangered species.

Discussion Questions

1. Which types of organisms do you think deserve priority in conservation efforts and why?

2. What ecological information would you need in order to list a species as endangered?

3. If we are concerned with protecting rare species, should we also be concerned with protecting subspecies (or races) or even individual populations? Discuss the pros and cons of this issue.

CHAPTER 4

Group Selection and Individual Selection

Road Map

1. Group selection is not a common or potent force in nature. Individual selfishness is much more common.
2. Altruism in nature is common, and can usually be explained by kin selection.
3. Living in groups conveys benefits, and a group's size is often the result of a trade-off between the advantages and disadvantages of group living.

4.1 Group and Individual Selection

Some ecologists have argued that populations regulate themselves at a level that can be sustained without the ravages of starvation—that is, at levels below which competition becomes important. In this way, wasteful squabbling and fighting for food is avoided. Nature is neat, tidy, and harmonious. Several pieces of evidence led early ecologists to believe that many populations do regulate their numbers. First, many birds are territorial, and a population in a given area is often below a size that could be supported by a given amount of prey. Second, any increase in the number of organisms within a given area results in emigration, which again operates to maintain populations about a mean level. Experimental manipulations show that some of these mechanisms operate at low densities in the absence of limitations on food or other resources. Third, many species, in particular birds, show high variation in their reproductive rates, and some birds fledge more young than others. In the past, this variation was regarded as evidence of the adjustment of fecundity to balance mortality in the population. In the late 1940s, however, two ornithologists, David Lack and Alexander Skutch, raised the argument of self-regulation versus external regulation. The argument was precipitated by data which showed that songbirds typically lay a clutch of four to six eggs in temperate regions of North America and Europe and only two or three in the tropics. Lack argued that birds in the tropics couldn't gather enough resources to fledge more than two to three young, so the availability of resources provided a limitation on reproduction. Skutch suggested that tropical populations were self-regulated to ensure that no resources were wasted. The trend of clutch size increasing away from the tropics is general, affecting virtually all groups of birds in all regions of the world.

In 1962, the self-regulation viewpoint was championed by British ecologist V. C. Wynne-Edwards, who articulated the full concept of self-regulation in a book called *Animal Dispersion in Relation to Social Behaviour*. The premise was that most groups of individuals purposely controlled their rate of consumption of resources and their rate of breeding to ensure that the group would not become extinct—a theory known as **group selection**. Individuals in successful groups would not tend to act selfishly. Groups that consisted of selfish individuals would overexploit their resources and die out. The idea of self-regulation, or group selection, is straightforward, is intellectually satisfying, and would seem to represent

Group selection implies that individuals should not overutilize their resources for the good of the group.

Individual selection entails an "every one for themselves" philosophy and seems more likely than group selection in nature.

what nature ought to do: avoid the grisly clashes and potentially damaging confrontations of competition. Why fight to the death if a contest can be settled amicably or, indeed, if species never overstepped their limits so that confrontations never occurred? This is perhaps how humans would like to be. In the late 1960s, however, the idea of self-regulation came under severe attack, and some of the data that supported the group-selection camp was equally well explained by **individual selection**. Leading the charge was the biologist G. C. Williams, in his book *Adaptation and Natural Selection* (1966). Williams' arguments against group selection were as follows:

1. *Mutation.* Consider a species of bird in which a pair lays only two eggs and there is no overexploitation of resources. Two eggs would then ensure a replacement of the parent birds, but would prevent a population explosion, which would be safe for the species. Suppose the tendency to lay two eggs is inherited as a group-selection trait. Now consider a **mutant** that lays three eggs, the equivalent of a "cheater" in this scenario. If the population is not overexploiting its resources, there should be sufficient food for all three young to survive. In that case, the three-egg genotype will eventually become more common than the two-egg genotype. The principle would hold for even larger broods, like four eggs or five eggs, and broods would tend to increase until they became so large that the parents could not look after all their young, causing an increase in infant mortality. Egg production is energetically costly to female birds, and the energy is wasted if the young fail to survive. Thus, the clutch size in nature evolves so as to maximize the number of surviving offspring. Field studies of great tits in Wytham Woods, England, for example, show a median clutch size of eight to nine eggs because adult birds cannot reliably supply sufficient food for more than that number chicks to survive (Fig. 4.1).
2. *Immigration.* Even in a population in which all pairs laid two eggs and no mutations occurred to increase the clutch size, "selfish" individuals that laid more could still migrate in from other areas. In nature, populations are rarely sufficiently isolated to prevent immigration.

Figure 4.1 Great tits, *Parus major,* limit their clutch size to the amount of young they can reliably feed. Here there are only four surviving nestlings.

3. *Individual selection.* For group selection to work, some groups must die out faster than others. In nature, groups do not become extinct very frequently; individuals nearly always die more frequently than groups, so individual selection will be the more powerful evolutionary force.
4. *Resource prediction.* For group selection to work, individuals must be able to assess and predict future availability of food and the population density within their own habitat. To date, however, there is little evidence that they can. For example, it is difficult to imagine that songbirds are able to predict the future supply of caterpillars they feed to their young and adjust their clutch size accordingly.

Individual selfishness seems a more plausible effect of natural selection. Any reduction in population sizes by means of **self-regulation** is likely to come from intraspecific competition, in which individuals selfishly strive to command as much of a resource as they can. Indeed, we often see animals in nature acting in their own selfish interest. For example, male lions kill existing cubs when they take over a pride. The advantage of infanticide for the male lion is that, without their cubs, females come into

Figure 4.2 Male Hanuman langur monkeys in India, can act agressively especially toward young which they may kill when taking over groups of females from other males.

Altruistic behavior does occur in nature and is often associated with kin selection.

estrus much faster—in 9 months as opposed to 25—than they would if the cubs were spared—hastening the day when males can father their own offspring. A male's reproductive life in the pride is only two to three years before he in turn is supplanted by a younger, stronger male. Infanticide ensures that the male will father more offspring and that the genes governing this tendency will spread by natural selection (Bertram, 1975). The mothers of the deceased cubs derive no advantage from the infanticide, but being smaller than the males, they are powerless to stop them. This episode of infanticide is not an isolated example: Male langur monkeys in India (Fig. 4.2) also kill infants when they take over groups of females from other males. Sarah Hrdy (1977) showed how the elimination of nursing offspring allowed the mothers to mate with the new male and have his babies. In those rare instances in nature when males of a species take care of the offspring, females are known to commit infanticide, too. Male giant water bugs take care of egg masses, protecting them against predators (Fig. 4.3). Females sometimes attack the egg masses and stab them with her mouthparts. If enough eggs are destroyed, the male guarding the eggs will sire a new clutch with the female egg stabber and guard the new clutch (Ichikawa, 1995).

Figure 4.3 Male giant water bugs, *Belostomatidae*, protect egg masses laid by the female from predators by carrying them on their backs.

In human society, individuals also rarely act for the good of the group and instead tend to act selfishly. This causes many environmental problems, such as overgrazing and overfishing, a phenomenon known as the tragedy of the commons (See "Applied Ecology: The Tragedy of the Commons.")

If selfishness is more common than group selection, then we must look for explanations of phenomena like **altruism** and the existence of nonbreeding castes.

4.2 Altruism

Although natural selection favors individual rather than group selection, it is still common to see apparent cooperation. Animals of the same species groom one another, hunt communally, and give warning signals to each other in the presence of danger. How can this altruistic behavior be explained by natural selection?

All offspring have copies of their parents' genes, so parents taking care of their young are, in the process, caring for copies of their own genes. Genes for altruism toward one's young will therefore become more numerous, because offspring have copies of those same genes. In meiosis, any given gene has a 50-percent chance of entering an egg or sperm. Thus, each parent contributes 50 percent of its genes to its offspring. The probability that a parent and offspring will share a copy of a particular gene is a quantity r called the **coefficient of relatedness**. By similar reasoning, brothers or sisters are related by an amount $r = 0.5$, grandchildren to grandparents by 0.25, and cousins to each other by 0.125 (Fig. 4.4). The famed British evolutionary ecologist W. D. Hamilton (1964) was the first to realize the implication of relatedness for the evolution of altruism. If an organism can pass on its genes through parental care, it can also pass them on by caring for siblings, nieces, nephews, and cousins. So even if you have no offspring of your own, you can pass on your genes via nieces, nephews, or cousins. This means you have a vested interest in protecting the offspring of your brothers and sisters.

The term **inclusive fitness** is used to designate the total copies of genes passed on through all relatives, nieces, nephews, and cousins, as well as sons and daughters. Selection for behavior that lowers an indi-

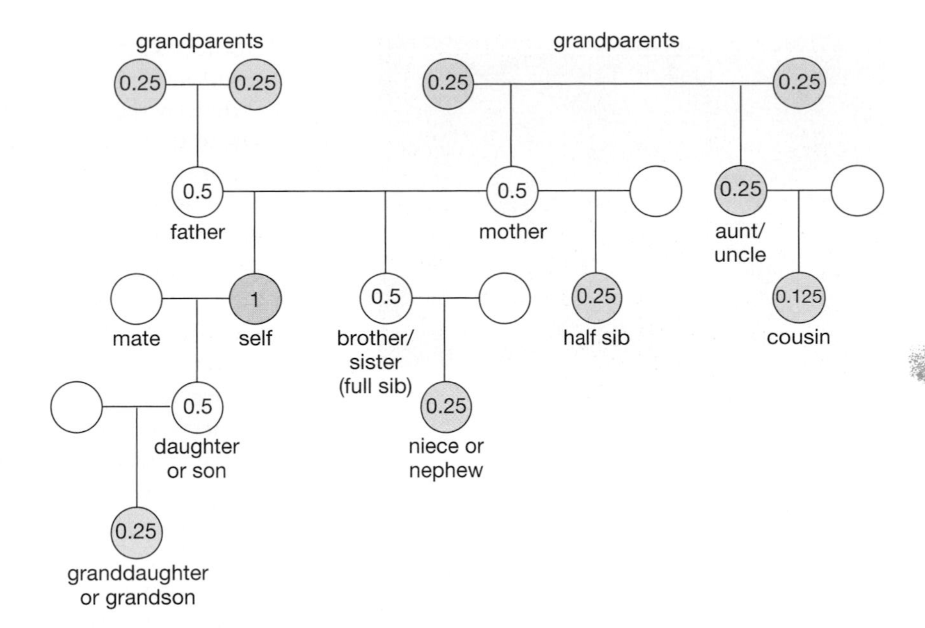

Figure 4.4 Degree of genetic relatedness to oneself in a diploid organism. Open circles represent completely unrelated individuals.

vidual's own chance of reproduction, but raises that of a relative, is known as **kin selection**. Hamilton proposed that the conditions under which a gene for altruism will spread by kin selection can be quantified as follows: If the donor sacrifices C offspring, for which the recipient gains B offspring, then the gene causing the donor to act in this way will spread if

$$rB - C > 0,$$

where r is the coefficient of relatedness of donor to recipient. Therefore, your own life is equivalent in genetic value to two sisters, two brothers, or eight cousins.

Individuals may display altruism toward relatives because doing so maximizes their inclusive fitness.

Kin selection.

Many insect larvae–especially caterpillars–are soft-bodied creatures. They rely on having a bad taste or poison to deter predators and advertise this condition with bright warning colors. Caterpillars with such colors are known as **aposematic** and are in stark contrast to other caterpillars that match the color of the leaves they feed on. (We say these camouflaged caterpillars are **cryptic**.)

Applied Ecology

The Tragedy of the Commons

The University of Santa Barbara ecologist Garrett Hardin wrote many famous ecological and environmental pieces, but perhaps none as famous as a 1968 essay in *Science* called the "Tragedy of the Commons." In his article, Hardin invited us to imagine a "commons": a piece of pastoral land owned by no one, but used by all to graze their cattle. Each person seeks to maximize the profit from his or her herd of cattle. Say there are 10 ranchers, each with 100 cattle, on a piece of land that can support 1,000 cattle. We can call 1,000 cattle the *carrying capacity* of the land. Once there are more than 1,000 cattle on the land, it deteriorates: The grass is overgrazed, and the pasture becomes of little value for grazing. To maximize profit, a rancher increases her herd by one animal. The profit to that individual is clear: a positive value of +1 for the one extra cow. But the herd is now over the limit that the land can sustain: The number of animals is now 1,001, not 1,000. However, the negative value of −1 cow is shared by all the ranchers. Each one of their cows suffers a little, an amount probably not noticeable. The logic is clear: Adding one cow is a good strategy for one individual rancher, with a positive value of +1 and a negative value of −0.1, since the negative is shared by all 10 ranchers. The problem is that soon all the ranchers realize the value of this strategy, so that all add cows to the commons, which then becomes overgrazed and useless. This is the tragedy of the commons.

The upshot of this tale is that the benefits of the environment often accrue to single individuals, but the cost of using the environment is borne by the whole population. This disparity holds true for commons in agriculture, grazing land, and other habitats such as forests and the open sea. Exploitation of the land for the good of a few at the expense of the majority must stop if we are to prevent degradation of the environment.

Figure 4.5 Altruistic behavior. *Datana* caterpillars exhibit a bright, striped warning pattern to advertise their bad taste to predators. All the larvae in the group are likely to be the progeny of one egg mass from one adult female moth, so the death of the one caterpillar it takes to teach a predator to avoid the pattern benefits that caterpillar's close kin. *(Photo by P. Stiling.)*

For example, noxious *Datana* caterpillars, which feed on oak and other trees, have bright red and yellow stripes (Fig. 4.5). Of course, unless it is born with an innate avoidance of a given type of prey, a predator has to kill and attempt to eat one of the caterpillars to learn to avoid similar individuals in the future. It is of no personal use to the unlucky caterpillar to be killed. However, animals with warning colors often aggregate in kin groups (Fig. 4.6), so the death of one individual is most likely to benefit its relatives, such as siblings, and its genes will be preserved. Even some solitary species have warning colors, suggesting a direct benefit as well: If the unlucky larva isn't killed by an attack, it will probably not be chosen again.

Another example of altruism is seen in lions. Lionesses tend to remain within the pride, whereas the males leave. As a result, lionesses within a pride are related, on average, by $r = 0.15$. Females all come into heat at the same time, one individual probably influencing the others' estrous cycles by means of pheromones. The result is the simultaneous birth of cubs, and females exhibit the apparently altruistic behavior of suckling other females' cubs. Because the females are related, the kin-selection hypothesis accounts for this behavior.

Another common example of altruism is the raising of an alarm call by "sentries" when a group of social animals is exposed to a predator. By this behavior, the individual raising the alarm is drawing attention to itself and increasing its risk of being attacked. In many groups, those animals living near a "sentry" are most likely to be its sisters or sisters' offspring and will bolt down their burrows. Thus, the altruistic act of alarm calling could be held to be favored by kin selection. However, as is often the case in behavioral ecology, there are alternative explanations for behaviors. In this case, "sentries" can be subordinate individuals who are driven to the edge of the group where the risk is higher and they are forced to be alert for their own safety. Nonetheless, at least for these individuals, some selfish motive is involved, because if a predator fails to catch any prey in a particular area, it is less likely to return, and the number of attacks over the "sentry's" lifetime is likely to be diminished, reducing the threat to its own life.

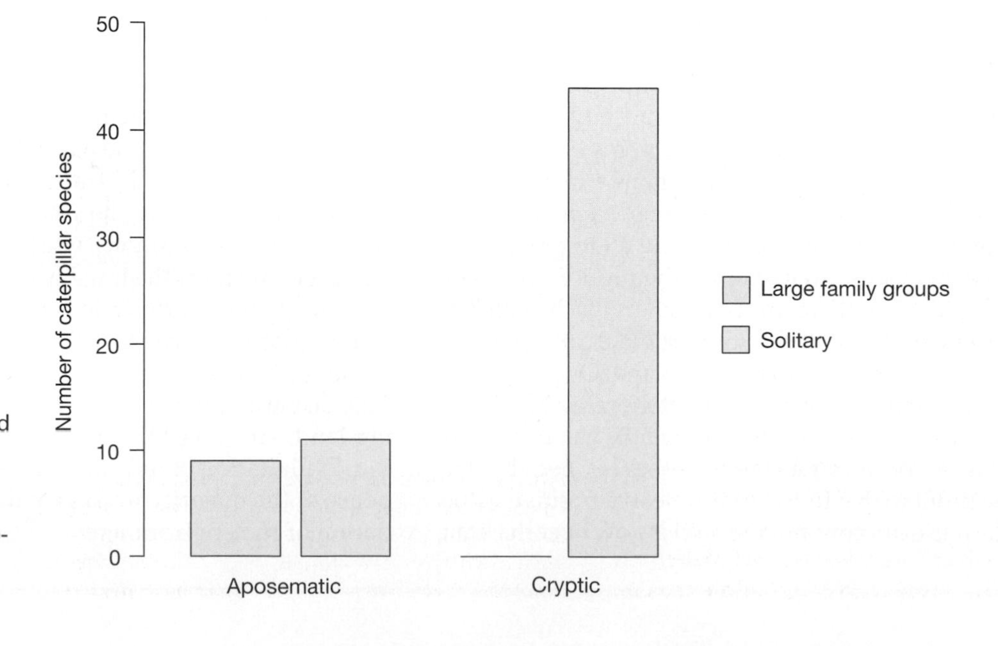

Figure 4.6 Brightly colored species of caterpillars of British butterflies are more likely to be aggregated in family groups than are cryptic species. *(From data in Harvey et al., 1983.)*

Determining which hypothesis is right and working out the exact costs and benefits in a system of kin selection can be a nightmare and has often been a stumbling block in behavioral ecology. To get a precise measure of natural selection, one must be able to calculate an individual's contribution to the gene pool, called its **inclusive fitness**. An individual's inclusive fitness is the contribution of that individual, plus 0.5 times its number of brothers and sisters, 0.125 times its number of cousins, and so on.

Altruism between unrelated individuals.

Altruistic acts between unrelated individuals have a "you scratch my back, I'll scratch yours" aspect to them, often called *reciprocal altruism*. Sometimes, unrelated individuals will occupy the same territories as breeding individuals and help the parents raise offspring by foraging for additional food for the young. This happens in mongooses (Rood, 1990). By comparing the nesting or brooding success of breeding pairs with helpers with those whose helpers were removed, researchers were able to show that helpers do significantly increase parents' fitness. The motive in most situations seems to be a sort of reciprocal altruism. In these situations, the habitat is usually saturated with breeders, and helpers probably could not obtain a territory for themselves. What they do is help increase the size of the territory of a breeding pair; later, they are able to carve off a fragment of that territory for themselves. Alternatively, they take over from one member of the breeding pair after that member dies.

Unrelated individuals may engage in altruistic acts if there is a high probability that the altruism will be reciprocated.

For many predators, reciprocal altruism in the form of social hunting allows bigger game to be caught: The benefits of a large kill outweigh the cost of having to share the meat. Most acts of altruism between unrelated individuals, whether in defense, attack, or mating, are exhibited by individuals living in social groups.

Altruism in social insects.

Perhaps the most extreme sort of altruism is the evolution of sterile castes in social insects. In some species of these insects, some females, known as workers, rarely reproduce themselves, but instead help the queen to raise offspring, a phenomenon called **eusociality**. The differentiation of one species of social insect into different-sized castes can be very impressive, with huge soldier castes and tiny worker castes (Fig. 4.7). Of the 263 living genera of ants known worldwide, 44 possess species with caste systems. The explanation of the peculiar system of castes again lies primarily in the particular genetics of most social-insect reproduction (Table 4.1). Females develop from fertilized eggs and are diploid, which is the normal condition of most organisms. Males develop from unfertilized eggs and are **haploid**. Male gametes are formed without meiosis, so every sperm is identical. Thus, each daughter receives an identical set of genes from her father. Half of a female's genes come from her **diploid** mother, so the total relatedness of sisters is 0.5 (from the father) + 0.25 (from the mother) = 0.75. Such a genetic system is called **haplodiploidy**. Thus, females are more related to their sisters than they would be to their own offspring. It is therefore advantageous to stay in the nest or hive and to try to produce new reproductive sisters.

The story becomes a little bit complicated when the interests of the queen are incorporated into the theory. Queens are equally related to their sons and their daughters; $r = 0.5$ in each case. To maximize her reproductive potential, a queen should produce as many sons as daughters, a 50:50 sex ratio. If she did, then sterile worker females would spend as much time rearing brothers (to which they are related only by 0.25) as sisters. The average relatedness of a female to her siblings would then be 0.5, and the female would do equally well to breed on her own. From the workers' viewpoint, it is far better to have more sisters, and in this conflict with the queen, they appear to have won, because in any colony there are many more females than males.

Figure 4.7 A soldier Amazonian termite. Many types of social insects, including ants, wasps, bees, and termites, have sterile castes that do not reproduce themselves, but are strictly confined to other functions, in this case defense. *(Photo by P. Stiling.)*

Altruism in social insects may arise from the unique genetics of their reproduction.

TABLE 4.1 Degree of genetic relatedness among relatives in haplo–diploid organisms such as ants, wasps, and bees.

	Daughter	Son	Mother (queen)	Father
Daughter	0.75			
Son	0.25	0.5		
Mother (queen)	0.5	0.5	1	
Father	0.5	0	0	1

Elegant though these types of explanations are, they do not provide the whole picture: Large social colonies exist in termites, too, but theirs is not a haplodiploid genetic system, but a purely diploid one. Furthermore, mole rats—mammals living underground in South Africa—have a division of labor based on castes, but the animals are, of course, diploid (Fig. 4.8; see also Jarvis *et al.*, 1994). There is only one breeding female, the queen; she suppresses reproduction in other females by producing a chemical in her urine that is passed around the colony when its inhabitants groom after visits to a communal toilet. The other castes perform different types of work. One caste, the "frequent workers," appears to do most of the burrowing work; another, the "infrequent workers," consists of heavier individuals that do some of the work; and even larger individuals, the nonworkers, rarely work at all. These are often male and may be a reproductive caste. There is even a special dispersing morph: large, fat males that are disinclined to mate with their own queen, but with a strong tendency to leave home and mate with members of other colonies (O'Riain, Jarvis, and Faulkes, 1996). These males are equivalent to the dispersing queens of ant, bee, and wasp colonies. How do we explain these social systems?

Richard Alexander, curator of the Museum of Zoology at the University of Michigan, suggested that it is the particular *lifestyle* of animals, not genetics, that promotes eusociality (Alexander, 1974). He argued that in a normal diploid organism, females are related to their daughters by 0.5 and to their sisters by 0.5, so it matters little to them whether they rear sibs or daughters of their own. He predicted that mammals could exhibit a castelike society under the following conditions:

1. When the individuals of the species are confined in nests or burrows.
2. When food is abundant enough to support a high concentration of individuals in one place.
3. When adults exhibit parental care.
4. When there are mechanisms by which mothers can manipulate other individuals.
5. When "heroism" is possible, whereby individuals give up their lives and, by so doing, can save the queen.

These factors can immediately account for eusociality in the termites as well. In mole rat colonies, whose burrows become as hard as cement, a heroic effort by a mole rat blocking the burrow effectively stops a predator (commonly a rufous-beaked snake) because predators cannot rip open the surrounding substrate. Self-sacrifice by a "worker" does translate into a genetic gain. The worker dies, but its genes are passed on by fellow nest mates, many of whom share genes because they have the same mother, the queen. In mole rats, the queen does indeed manipulate the colony members; all workers and nonworkers, whether male or female, develop teats during their pregnancies—a testament to the power of their pheromonal cues. The superabundant food comes in the form of tubers of the plant *Pyrenacantha kaurabassana*, which weigh up to 50 kg and can provide food for a whole colony. The mole rats harvest the tubers from below, and often the tubers remain only half eaten and can regenerate in time. Why eusociality has developed in this, but not in other, systems of burrowing rodents, such as prairie dogs, is open to speculation. At this stage, all we can say is that eusociality is promoted by haplodiploid mating systems and species, like termites and mole rats, that defend fortresses. However, not all species that could become eusocial have adopted such a life system. Furthermore, even diploid termites and naked mole rats contain very closely related individuals because of inbreeding

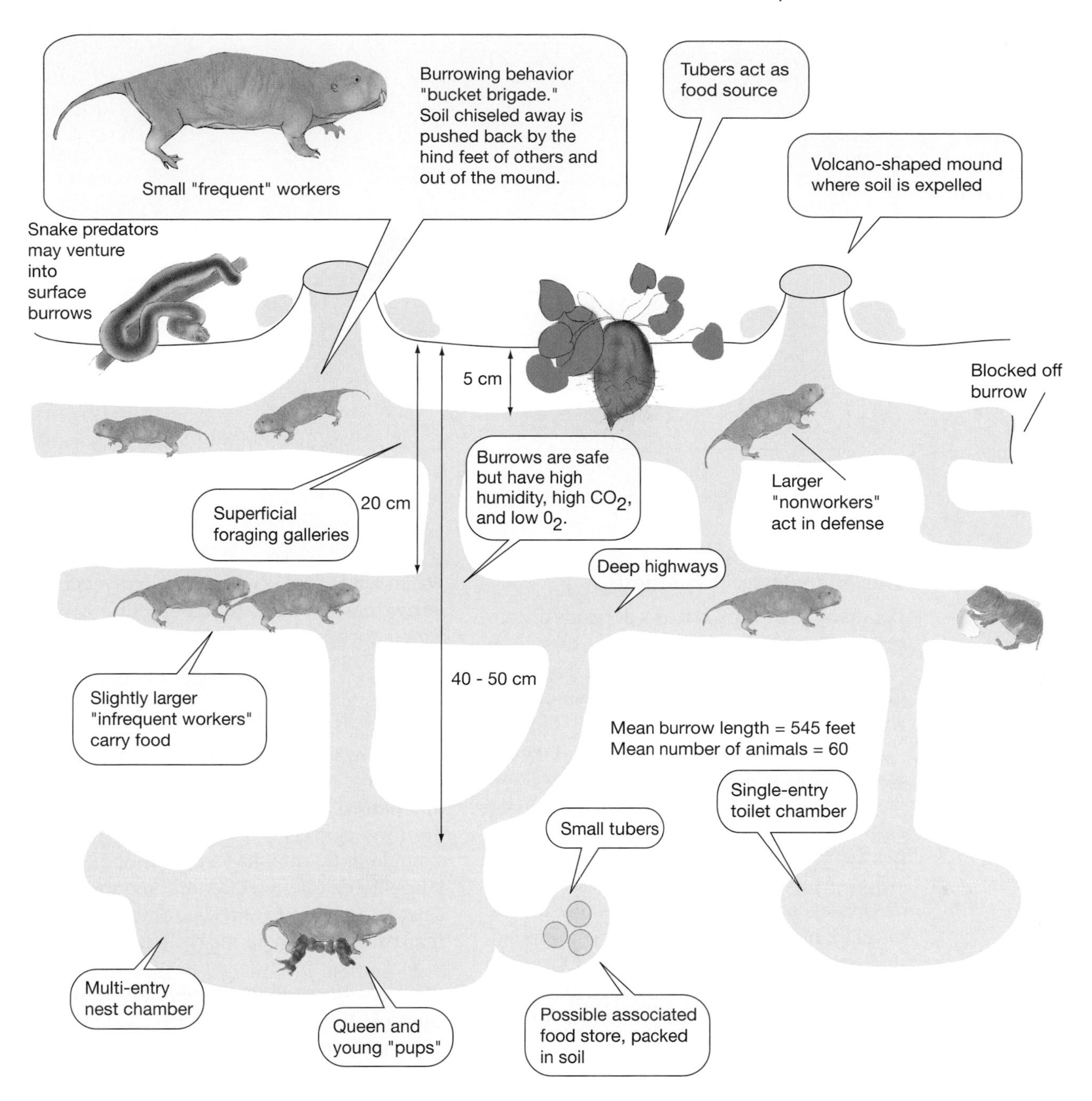

Figure 4.8 Cross section of naked mole rat colony.

(Reeve *et al.*, 1990), and this provides an additional reason that females might help queens within colonies.

4.3 Group Living

Individual selection can explain the occurrence of altruism in nature, but many species live together in dense shoals, flocks, or herds, which would surely only promote intense competition between individuals. If the central concern of ecology is to explain the distribution patterns of plants and animals, we must be able to provide a framework from which to understand the social behavior of animals. If dense congregations promote intense competition, there must be some significant advantages of group living to compensate. Many of these advantages relate to group defense against predators.

Guppies (*Poecilia reticulata*), were first discovered in Trinidad in the 19th century by the Reverend P. L. Guppy. In their native habitat, guppies live in tighter groups when they are in streams in which predators are more common (Fig. 4.9), suggesting that being in a group helps an individual to avoid becoming a meal. Group living could reduce predator success in several different ways.

Organisms that live in dense groups may incur a cost in competing for resources, but group living can also provide a number of benefits.

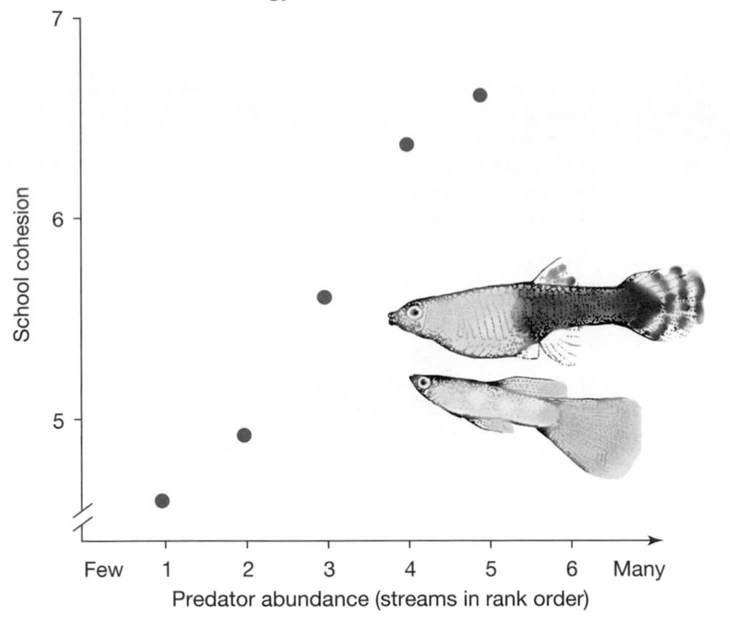

Figure 4.9 Variation in group size may be related to defense against predators. Guppies (*Poecilia reticulata*) from streams with many predators live in tighter schools than those from streams with few predators. Each dot is a different stream, and "cohesion" was measured by a count of the number of fish in grid squares on the bottom of the rank. (*Modified from Seghers, 1974.*)

Large groups may be more successful at detecting predators.

An individual in a large group has a lower probability of being eaten if a predator attacks than does a solitary individual or one in a small group.

The "many-eyes" hypothesis.

For many predators, success depends on surprise. If prey is alerted to an attack, the predator's chance of success is lowered (Fig. 4.10). Goshawks (*Accipiter gentilis*) are less successful in attacks on large flocks of pigeons (*Columba palumbus*) mainly because the birds in a large flock take to the air when the hawk is still some distance away. If each pigeon occasionally looks up to scan for a hawk, the bigger the flock, the more likely it is that one bird will spot the hawk early. (This is known as the "many-eyes theory.") Once this one pigeon takes flight, the others in the flock are alerted, and they fly away, too (Fig. 4.11). Of course, "cheating" is a possibility because some birds might never look up, relying on others to keep watch while they keep feeding. However, the individual that happens to be scanning when a predator approaches is the one most likely to escape. This tends to discourage "cheating."

The selfish-herd theory.

Normally, predators take only one of their prey per attack. Thus, an individual antelope in a herd of a hundred has only a one-in-a-hundred chance of being attacked, whereas a single individual has a one-in-one chance. Large herds may well be attacked more, but a herd is unlikely to attract one hundred times more attacks than an individual. Associated with herds is a tendency to prefer the middle, because predators are likely to attack prey on the edge of the group. This was

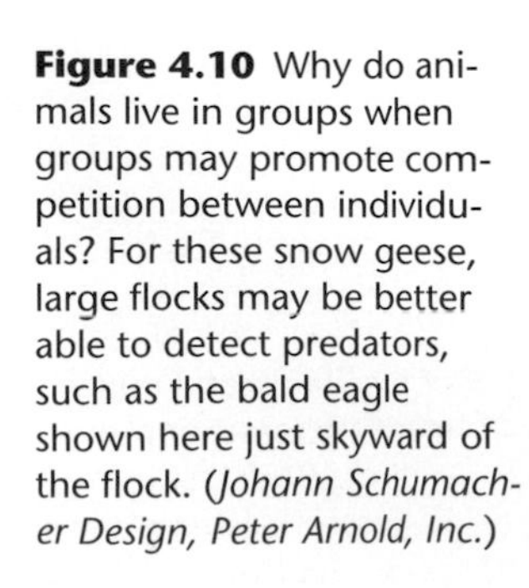

Figure 4.10 Why do animals live in groups when groups may promote competition between individuals? For these snow geese, large flocks may be better able to detect predators, such as the bald eagle shown here just skyward of the flock. (*Johann Schumacher Design, Peter Arnold, Inc.*)

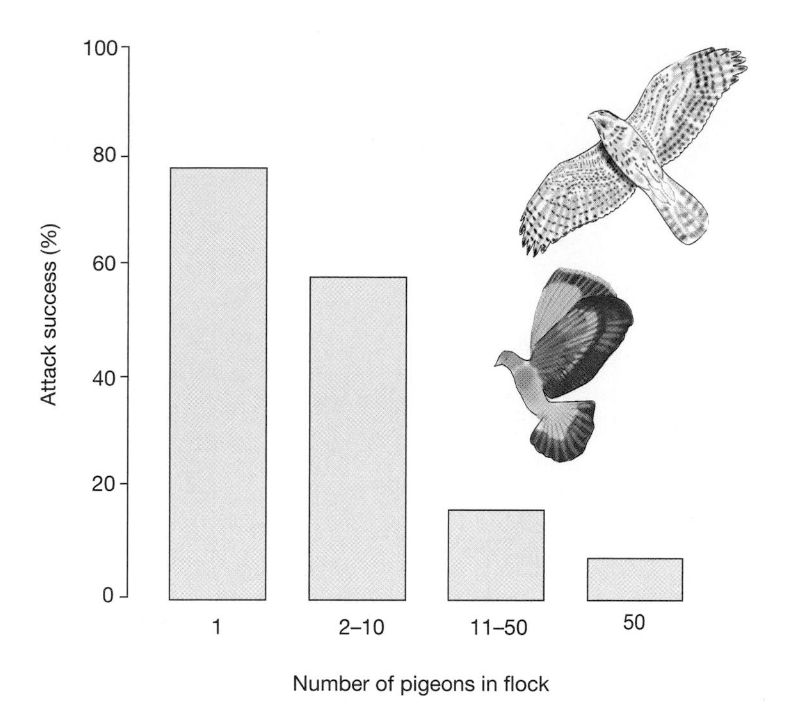

Figure 4.11 The larger the flock of pigeons, the more likely it is that they will see an approaching predatory goshawk, and the more likely it is that they will flee, decreasing the goshawk's rate of success in attacking. (*Redrawn from Kenward, 1978.*)

termed the "geometry of the selfish herd" by W. D. Hamilton in 1971. Part of the explanation of this type of defense may be the difficulty of visually tracking large numbers of prey. Throw two balls to a friend, and the chances are that he or she will drop them both, whereas one ball can be tracked much more easily. A similar phenomenon may operate in predator–prey interactions, with peripheral individuals being easier to isolate visually. It may also be physically difficult to get to the center of a group, and many herds tend to bunch close together when they are under attack. Furthermore, large numbers of prey are able to defend themselves better than single individuals, which usually flee. For example, nesting black-headed gulls mob crows remorselessly and reduce the crows' success at stealing the gulls' eggs (Kruuk, 1964).

In the end, group size may be the result of a trade-off between the advantages and disadvantages of group living. The operation of just two conflicting variables—competition for food and the presence of predators—on the size of bird flocks shows how this works (Fig. 4.12). Increased predation rates increase flock size because more vigilance is needed. Increased food levels decrease flock size because birds can afford to spend more time fighting. Tom Caraco provided a field test of this model in Arizona (Caraco, 1979b). He scored the percentage of time spent feeding, scanning, and fighting by different flock sizes of yellow-eyed junco birds (*Junco phaeonotus*) feeding

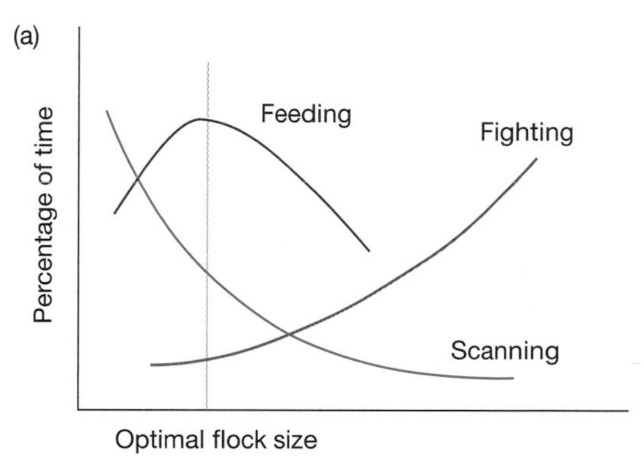

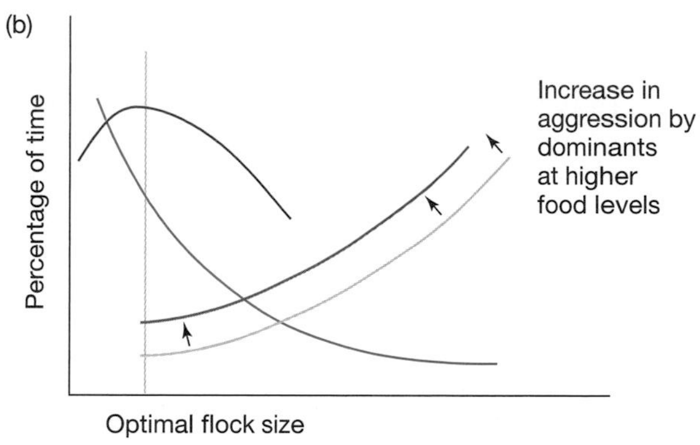

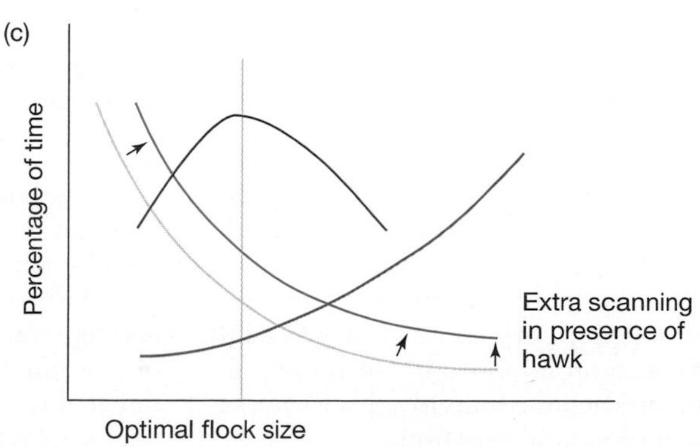

Figure 4.12 A model of optimal flock size. This model assumes that birds do only one of three things: feed, fight, or scan. The time spent feeding is that left over after the birds have finished scanning for predators or fighting over territory or food. (a) The trade-off between fighting and scanning for predators. As flock size increases, birds spend more time fighting and less time scanning. An intermediate flock size gives the maximum proportion of time feeding. (b) The effect of an increase in resources on flock size. When food is more plentiful, birds can afford to spend more time fighting. The optimal flock size therefore decreases. (c) The effect of an increase in predation on flock size. When the risk of predation is increased by the presence of a hawk, scanning levels increase, and the optimal flock size is increased. (*Modified from Krebs and Davies, 1981.*)

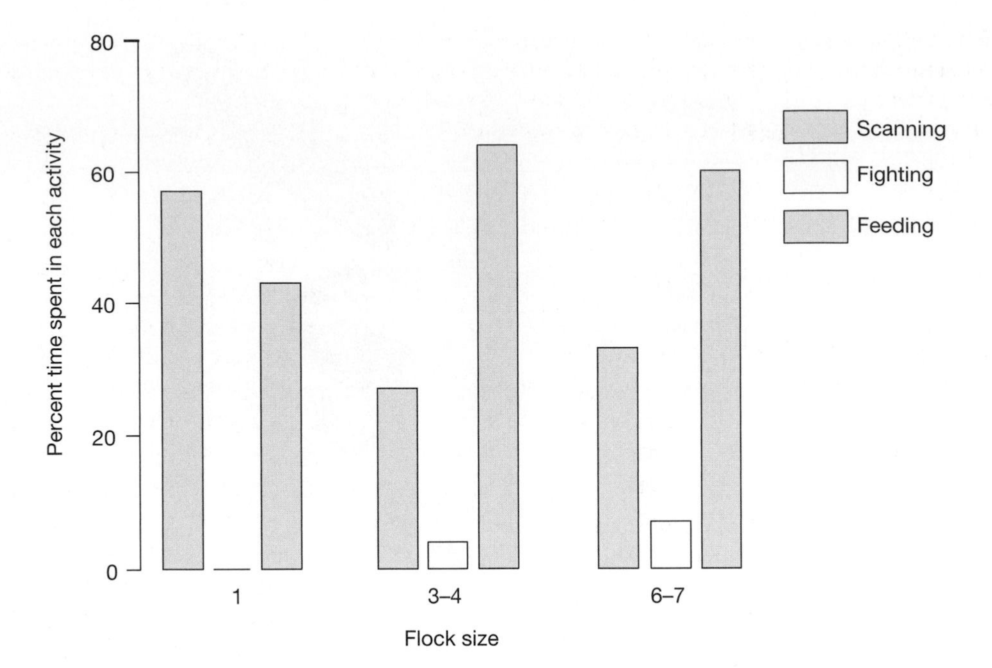

Figure 4.13 The increase in fighting and decrease in scanning of yellow-eyed juncos with increasing flock size yields the highest rate of feeding at intermediate flock size. *(From data in Caraco et al., 1980.)*

in a canyon. A tame hawk was flown across the canyon periodically to simulate the risk of predation. As predicted, as the flock size increased, the time spent scanning decreased, and the time spent fighting increased. The time spent feeding was highest at an intermediate flock size (Fig. 4.13). Then the hawk was removed. Caraco found that the average flock size decreased from 7.3 birds per flock when the hawk was present to 3.9 when it was absent. This again supported the idea that group size would increase in the presence of predators. Many other variables are likely to influence flock size at the same time. Such logic may help us explain why we see particular flock or herd sizes in nature: They are a result of a trade-off between selective pressures.

An understanding of how individual selection works helps us understand many different types of behavior in nature, such as apparent altruism, caste systems, and flock sizes. Individual selection helps explain why some organisms occur in groups while others remain solitary. Individual selection convinces us that nature is generally selfish and is probably red in tooth and claw. As well as different behaviors, plants and animals exhibit different reproductive and life history strategies, and in the next section we shall see how these affect a species' population growth.

Summary

Do populations deliberately underutilize their resources to ensure a food supply for future generations? If they did, then behaving for the good of the species would be more likely than selfish behavior designed to maximize gains to individuals.

1. In the past, some ecologists believed that populations of species in nature were maintained at constant or equilibrium levels by group selection—self-regulation by individuals in a population, to prevent the overexploitation of resources. Attractive though the idea of group selection is, it has several flaws, including failing to account adequately for mutation, immigration, and resource prediction. Individual selfishness seems a more plausible result of natural selection.

2. If individual selection is more likely than group selection, how can we explain acts of altruism between individuals? One popular explanation that is invoked is kin selection—selection for behavior that lowers an individual's own chances of survival and reproduction, but raises that of a relative (possessed of some of the same genes). Kin selection is invoked to explain the unusual caste systems of social insects, especially those with a haplodiploid mating system.

3. Eusociality and cooperation are found not only in haplodiploid organisms: Termites and naked mole rats exhibit a caste system, too. In 1974, Richard Alexander predicted that certain ecological factors, namely, confinement to burrows, high food concentrations, parental care of offspring, mechanisms for mothers to manipulate other individuals, and the opportunity for "heroism," could be sufficient to produce a castelike society.

4. While ecological or evolutionary factors can explain the existence of altruism and social behavior, group size can often be seen to be a compromise between large groups that lessen the impact of predation and small groups that lessen the impact of competition.

Selfishness is more likely than group selection to occur in the wild. Therefore, we should see much evidence of competition and fighting over resources, rather than a harmonious nature.

Discussion Questions

1. If kin selection occurs in nature, how do you think animals recognize their kin?

2. If it is equally valuable for a female to raise her own young or help raise sisters (both having an $r = 0.5$), why do we see most females preferentially raising their young?

3. Unrelated vampire bats that roost communally often rest next to the same neighbors every night and sometimes regurgitate meals of blood to their hungry neighbors. How can you explain this apparent act of altruism?

4. How, if at all, can animal behavior be important to how we design nature reserves?

CHAPTER 5

Life History Strategies

Road Map

1. Some species breed repeatedly throughout their lifetimes and are called iteroparous, while others breed just once in their lifetimes and are called semelparous.
2. In growing populations, there are many young individuals; in declining populations, there are few young.
3. Mating systems are classified according to the number of partners with which each male and each female mates.
4. There is a continuum of life history strategies, from *r*-selected species, which have a high reproductive output, to *K*-selected species, which are good competitors and exist at *K*, the carrying capacity of the environment.

An organism's life history largely concerns its lifetime pattern of growth and reproduction. For example, why is it that some plants—especially those we regard as weeds—produce vast numbers of tiny seeds, while some trees produce fewer, bigger seeds? Why do some species breed once in their lifetime, in one big burst, while others breed repeatedly and at more moderate levels? Many ecologists believe that an understanding of such differences in life history strategies is vital to understanding how populations grow.

5.1 Reproductive Strategies

Some organisms produce all of their offspring in a single reproductive event. This pattern is called **semelparity** and occurs in organisms as diverse as salmon, bamboo, and yucca plants (Fig. 5.1), each individual of which reproduces once only and then dies. The contrasting pattern is repeated reproduction, or **iteroparity**, wherein organisms reproduce in successive years or breeding seasons. Semelparous organisms may live for many years before reproducing, as do the yuccas, or they may be **annual** plants that develop from seeds, flower, and drop their own seeds within a year.

Among iteroparous organisms, there is much variation in the number of clutches and in the number of offspring per clutch. Many species, such as temperate birds and temperate forest trees, have distinct breeding seasons that lead to distinct generations. In a few species, individuals reproduce repeatedly and at any time of the year. This behavior is known as continuous iteroparity and is exhibited by some tropical species, many parasites, and, of course, humans.

Why do species have either semelparous or iteroparous reproductive strategies? Why do species bother to have repeated reproduction? The answer may lie in environmental uncertainty: If the survival of juveniles

(a)

(b)

Figure 5.1 (a) This soaptree yucca plant, *Yucca elata,* grows for many years before it flowers and produces seed. After that one reproductive event, the parent plant dies. (b) Such a strategy is called semelparity and contrasts with iteroparity, whereby organisms breed repeatedly.

is very poor and unpredictable, selection favors repeated reproduction and a long reproductive life to spread the risk of reproducing over a longer period and to increase the chance that juveniles will survive in at least some years. This reproductive strategy is often referred to as "bet hedging" (Stearns, 1976). In the plant world, if the environment is stable, then annuals prevail, because the organism, such as an herb, can devote more of its energy to making seeds, and not expend all of its energy on maintaining its own body. Under favorable circumstances, annuals produce more seeds than do trees, which have to invest a lot of energy in maintenance. However, when the environment becomes stressful, annuals run the risk of not being able to maintain their population. They must rely on some seeds successfully lying dormant and germinating after the stress has ended.

Reproductive strategies differ among organisms, with some breeding continuously, others in discrete intervals, and still others just once in their lifetimes.

5.2 Age Structure

Growing populations have a large number of young.

The reproductive strategy employed by an organism has a strong effect on the subsequent age structure of a population. Semelparous organisms often produce groups of same-aged young, called **cohorts**, that grow at similar rates. Iteroparous organisms have a great many young of different ages because the parents reproduce every year. The age structure of populations can be characterized by specific age categories, such as years in mammals, or other categories, like eggs, larvae, or pupae in insects or size classes in plants.

An increasing population should have a large number of young, whereas a decreasing population should have few young. The loss of age classes can have a profound influence on a population's future. In an overexploited fish population, for instance, the older reproductive age classes are often removed. If the population experiences reproductive failure for one or two years, there will be no young fish to move into the reproductive age class to replace the fish that were removed, and the population can collapse. In other populations, the younger age classes may be removed. For example, in regions where populations of white-tailed deer are high, they overgraze the vegetation and eat many young trees, leaving only the older trees, the foliage of which is too tall for the deer to reach (Fig. 5.2). This can have disastrous effects on future recruitment, for while the forest might consist of healthy, mature trees, when those trees die, they will not be replaced. Overgrazing by Key deer in the Florida Keys has had just such an effect (Stiling and Barrett, unpublished data). The removal of deer predators, such as panthers and bobcats, often allows deer numbers to skyrocket and survivorship of young trees in forests to plummet.

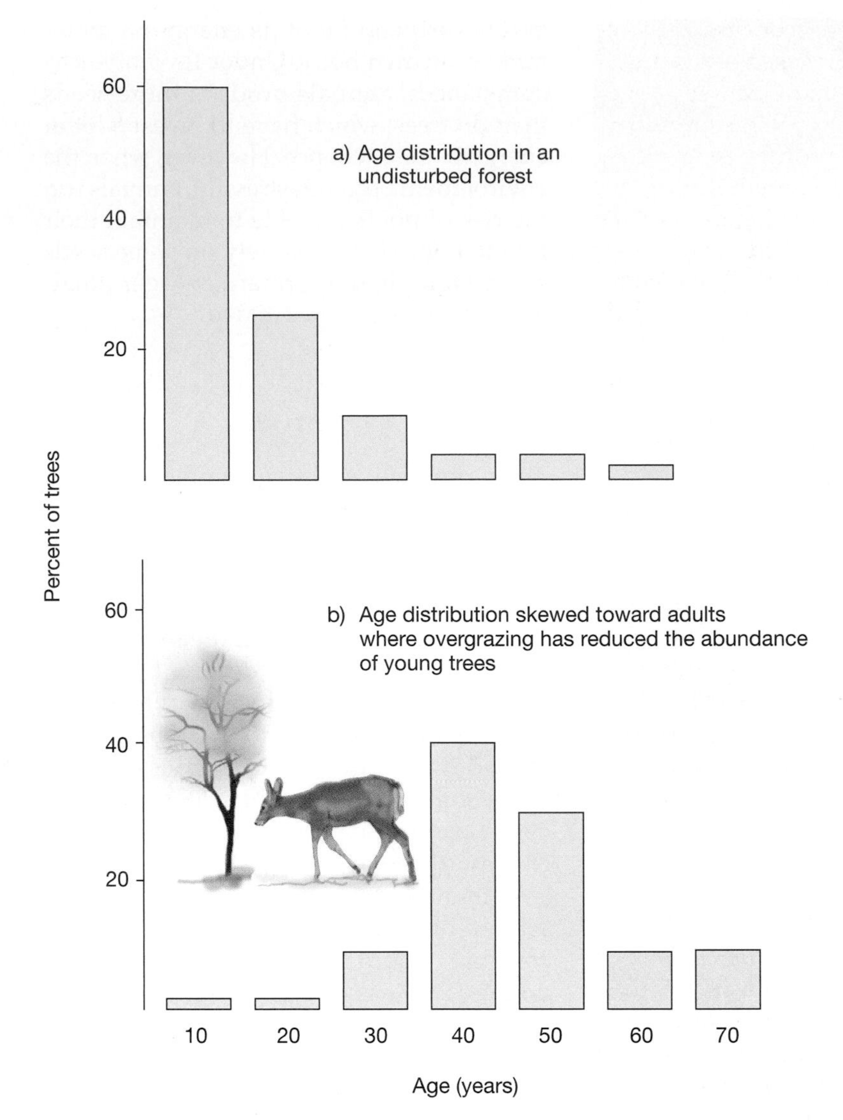

Figure 5.2 Theoretical age distribution of two populations of tree. (a) Age distribution of undisturbed forest with numerous young trees, many of which die as the trees age and compete with one another for resources, leaving relatively few big, older trees. (b) A stand in which most young trees have been killed, leaving mainly trees in the older age classes. This commonly happens in overgrazed areas and is indicative of a stressed population.

5.3 Mating Systems

Mating systems may be monogamous or polygamous.

A population's *sex ratio* refers to the proportions of the two sexes, male and female, in the population. Sex ratios are of interest to pure and applied ecologists alike. For example, hunters prefer to take male deer (bucks), so they would be pleased to have a deer population skewed toward males. However, too many males would severely limit a herd's growth, making the job of game managers very difficult.

In nature, most males seem superfluous, because one male could easily fertilize all the females in a local area. For example, five milliliters (ml) of human semen contain enough sperm to fertilize twice the female population of the United States. If one male can fertilize so many females, why are there few species with a sex ratio of, say, 1 male to 20 females? Instead, the sex ratio is more often about 1:1. The answer lies with selfish genes. If a population contained 20 females for every male, then a parent whose children were exclusively sons could expect to have 20 times the number of grandchildren produced by a parent with daughters. Selection would favor sons, and males would become prevalent in the population. In that case, females would be at a premium, and parents having female offspring would get more grandchildren. Such constraints operate on the numbers of both male and female offspring, keeping the sex ratio to about 1:1.

The most common exception to the 1:1 sex ratio is a situation in which one male dominates in breeding, and other males therefore become superfluous because they are never likely to leave any offspring. This phenomenon is most likely to occur in species with low powers of dispersal, in which brothers and sisters stay in the same place and inbreeding is frequent. In this kind of situation, such as in parasitic Hymenoptera whose broods develop and mate in or around patches of host insect larvae, biased sex ratios of about 20 females to 1 male can be found. Here, the females mate once, but store sperm, and the female can control the sex ratio by using sperm to produce females or by not using it and thereby producing males. This state of affairs is called a haplodiploid mating system, since males are haploid, but females are diploid. In the mite *Acarophenox*, which gives birth to live young, the brood size is about 20, and each brood contains only one son. The male mates with his sisters inside the mother and dies before he is born (Hamilton, 1967) (Fig. 5.3).

There are several different types of mating systems among animals. In **monogamy**, each individual mates exclusively with one partner over at least a single breeding cycle and sometimes for longer. Monogamy is most common among birds, about 90% of which are monogamous. Monogamy is thought to be prevalent in birds because eggs and chicks take a considerable amount of parental care. Most eggs need to be incubated by one parent if they are to hatch. The chicks require almost continual feeding, and without both partners, most would starve. It is therefore in the male's best interest to help fledge his young, because he would have few surviving offspring if he did not. Monogamy is very rare among other vertebrates.

In contrast to monogamy is **polygamy** ("many marriages," from the Greek *poly* = many and *gamos* = marriage), in which individuals mate with more than one partner. In **polygyny** (many females), one male mates with more than one female, but females mate with only one male. Many males, therefore, do not mate at all. In **polyandry** (many males), an individual female mates with several males, but males mate with only one female.

Figure 5.3 A vivparous mite of the family *Acarophenacidae*. Here brothers mate with sisters while both are still inside the body of the mother.

Polygyny.

Physiological constraints often dictate that female organisms must care for their young, because they are the ones most often left "holding the baby." In mammals, males are able to desert, and most mammals have polygynous mating systems in which males mate with several females, but each female mates with only one male (Fig. 5.4). In cases in which females are able to shed their young at an early stage—for example, in some fish and amphibians—different strategies are adopted. In most fish, the female deposits her eggs, and then the male fertilizes them. *Females* are thus able to desert, leaving the male "holding the babies," sometimes literally, as in mouthbreeders. In those fish in which fertilization takes place internally, however, it usually falls to the female to provide parental care.

Figure 5.4 This red deer stag mates with more than one female, so its mating system is called polygynous.

In polygynous mating systems, one male mates with many females.

Polygyny can be influenced by the spatial or temporal distribution of breeding females. In cases where all females are sexually receptive at the same time, there is little opportunity for a male to garner all the females for himself. Monogamous relationships are more common in these situations, and Knowlton (1979) suggested that female synchrony has evolved specifically to enforce monogamy on males. When female reproductive receptivity is spread out over weeks or months, there is much more opportunity for males to mate with more than one female. In two contrasting examples, females of the common toad, *Bufo bufo*, all lay their eggs within a week, and males generally have time to mate with only one female, whereas female bullfrogs (*Rana catesbeiana*) have a breeding season of several weeks, and males may mate with as many as six females in a season.

Resource-based Polygyny. When some critical resource is patchily distributed and in short supply, certain males have a great opportunity to dominate the resource and to breed with more than one visiting female. For example, males of a species of bird called the orange-rumped honeyguide (*Indicator xanthonotus*) defend bees' nests because the species feeds on wax. When a female comes to feed, the male mates with

Resource-based polygyny occurs when males hold valuable territories that contain resources the females need.

her, exchanging food for sex. The more bees' nests he can defend, the more females he will attract (Cronin and Sherman, 1977).

In the lark bunting (*Calamospiza melanocorys*), which mates in North American grasslands, males arrive at the grasslands first, compete for territories, and then display with special courtship flight patterns and songs to attract females. The major source of nestling death in this species is overheating from too much exposure to the sun. Prime territories are those with abundant shade, and some males with shaded territories attract two females, even though the second female can expect no help from the male in the process of rearing young. Males in some exposed territories remain bachelors for the season. Supplementing open areas with plastic strips to provide shade turns them from bad to good territories (Pleszczynka, 1978). Predation is another strong selective pressure that acts in a similar manner to force polygynous relationships on birds in which females choose males with safe territories, away from predators.

From the male's point of view, resourced-based polygyny is advantageous; from the female's point of view, there are drawbacks. Although, by choosing dominant males, a female may be gaining access to good resources, she may also have to share those resources with other females. In the yellow-bellied marmot (*Marmota flaviventris*), males attract more females if they defend the best burrow sites. For a female, sharing a burrow means less success per female (Fig. 5.5). Although it is best for a female to be with a monogamous male, it is best for the male to mate with as many females as possible. A compromise is often evident in which about two females are usually observed per territory.

Sometimes males simply defend a group of females as a harem without bothering to command a conventional resource-based territory. This pattern is more common when females naturally occur in groups or herds, perhaps to avoid predation. Usually, the largest and strongest males command all the matings, but being a harem master is often so exhausting, that males manage to remain at the top only for a year or two. In the elephant seal (*Mirounga* sp.) named for the enlarged proboscis of the male, the females haul up onto the beach to give birth and gain safe haven for the pups away from marine predators. In this situation, males are able to command substantial harems. To defend a harem, a male must constantly lumber across the beach to fight other males, and pups are frequently squashed in the process. Because the offspring are likely to have been fathered by the previous year's dominant bulls, the havoc matters little to the current males.

In one form of polygynous mating, neither resources nor harems are defended. In some instances, particularly in birds and

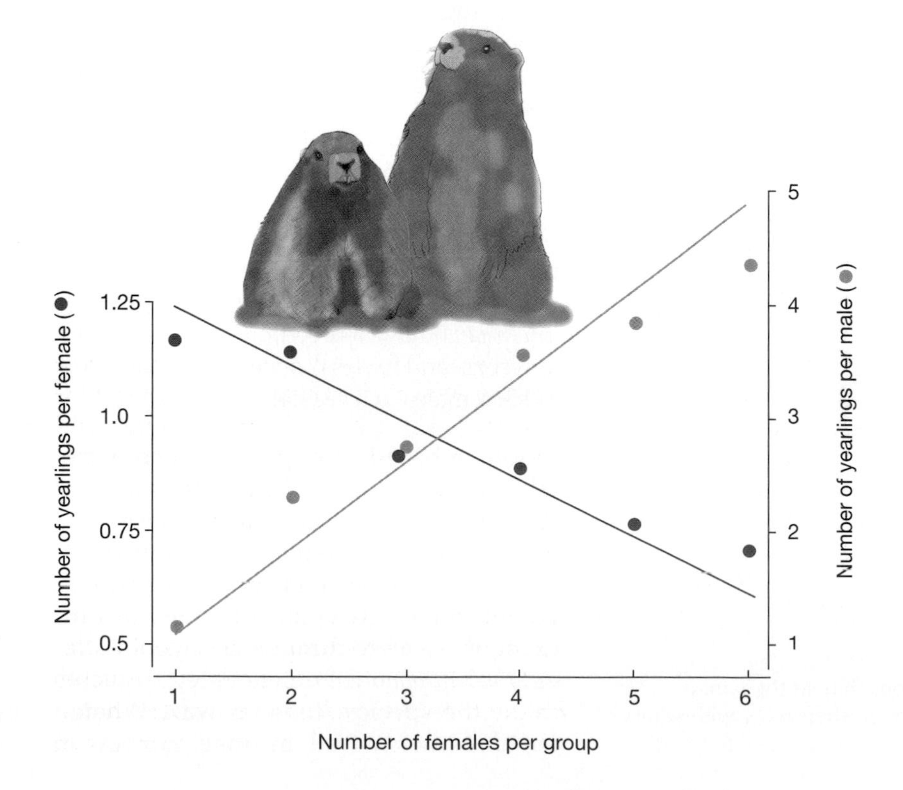

Figure 5.5 Reproductive success in yellow-bellied marmots in relation to the number of females on a male's territory. The success rate per female declines with an increasing number of females. The success of the male, on the other hand, increases with an increasing number of females. In nature, most burrows contain two to three females. (*Modified from Armitage, 1986.*)

mammals, males display in specific communal courting areas called **leks**. The females come to these areas specifically to find a mate. Females choose their prospective mates after the males have performed elaborate dancing displays. Most females choose the same male, so a few successful males perform the vast majority of the matings. In the white-bearded manakin of South America, one male accounted for 75% of the 438 copulations at a 150-m² lek containing as many as 10 males. A second male mated 56 times (13% of matings), while six others mated only a total of 10 times (Lill, 1974) (Fig. 5.6).

Figure 5.7 The American jacana (*Jacana spinosa*) has elongated toes to permit it to walk on surface vegetation.

Polyandry.

In most systems in which one individual mates with more than one individual of the opposite sex, the polygamous sex is the male. The opposite condition, polyandry, in which the female is polygamous, is much rarer. Nevertheless, it is practiced by a few species of birds. In the Arctic tundra, the summer season is short, but very productive, providing a bonanza of insect food for two months. For the spotted sandpiper, *Actitis macularia*, the productivity of the breeding grounds is so high that the female becomes rather like an egg factory, laying five 20-egg clutches in 40 days. Her reproductive success is limited not by food, but by the number of males she can find to incubate the eggs, and females compete for males, defending territories where the males sit (Oring, 1985). Polyandry is also exhibited in species where egg predation is high, because males need to guard the nests and only females can replace the young. For example, polyandry occurs in the American jacana, or lily trotter (*Jacana spinosa*), a bird of the wetlands that uses its elongated toes to walk on floating lily leaves. The female jacana is large, to enhance her egg-laying capacity (Fig. 5.7). Females have large territories with several nests, each attended by a male. The males incubate the eggs while the females move about their territory mating with the different males, laying eggs, and defending their territories against other females. Females also break up fights between squabbling males. If a male tries to desert his nest to mate with another female, he runs the risk of losing his eggs to predators.

In polyandrous mating systems, one female mates with many males.

Figure 5.6 Male long-tailed manakins at a lek in Central America. Males congregate in areas low down in the forest canopy, pulling the leaves off branches to create a clear display area in which they jump about. Females, shown at lower right, also visit the leks and choose their prospective mates.

5.4 Life History Strategies

r and K selection represent the two extremes of a continuum of life history strategies.

The reproductive strategies discussed so far—iteroparity vs. semelparity and polygyny vs. polyandry—have important implications for the reproductive success of individuals and, indeed, of populations and species. However, these reproductive strategies can be viewed in the context of a much bigger picture in which life history strategies incorporate not only reproductive strategies, but strategies of survival, habitat usage, and competition with other organisms.

Life history strategies make up a continuum that encompasses what are called *r*-selected species at one end and *K*-selected species at the other (Fig. 5.8). The parameters *r* and *K* refer to population growth equation parameters and will be discussed in the next chapter. Species that are *r* selected have a high rate of per capita population growth, *r*, but poor competitive abilities. An example is a weed that quickly colonizes vacant habitats (such as barren land), passes through several generations, and then disappears. Weeds produce huge numbers of

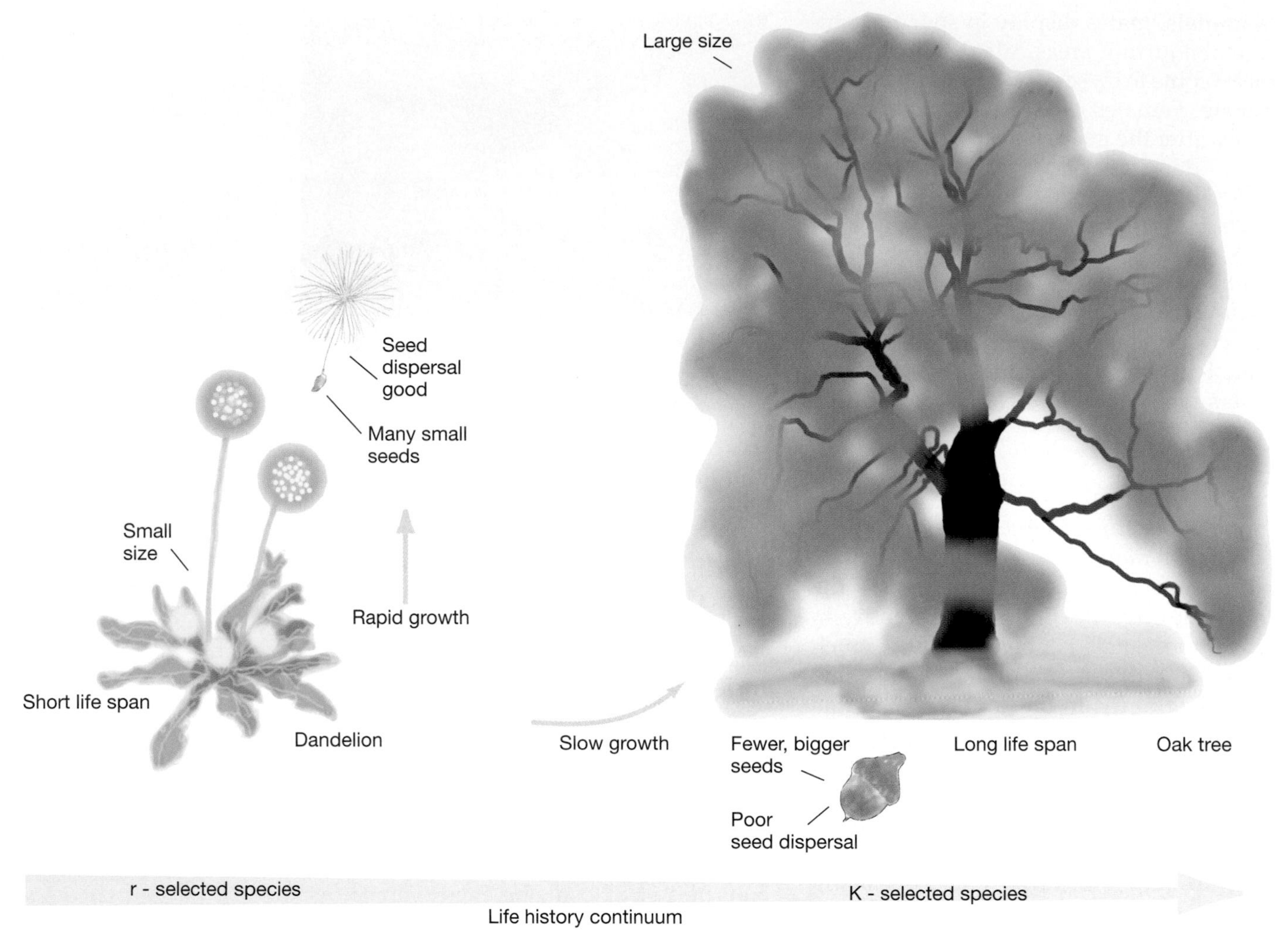

Figure 5.8 Life history traits of a dandelion and an oak tree illustrate some of the differences between *r*- and *K*-selection strategies.

tiny seeds and therefore have high values of *r*. In contrast, *K*-selected species tend to increase more slowly to the carrying capacity *K* of the environment. An example is a tree that exists in a mature forest. While *K*-selected species have a relatively low value of *r*, they tend to compete well and eventually displace the weedy species by overtopping them.

The *r–K* classification was originally proposed by the biologists Robert McArthur and E. O. Wilson (1967), who were interested in the different abilities of species to colonize islands. *r*-selected species tend to disperse well, even to the islands McArthur and Wilson were interested in. Broadly speaking, *K*-selected species have relatively low reproductive allocations and are iteroparous, while *r*-selected species have a high investment in seeds and are semelparous. The *r*- and *K*-selection continuum brings together various life history attributes, including reproductive strategy, habitat selection, ability to disperse, and population size (Table 5.1).

A good way of thinking about *r*-selected species is, again, to think of a weed. Weeds exist in disturbed habitats, such as gaps in a forest canopy where trees have blown down, allowing light to penetrate to the forest floor. Weeds grow quickly and reach reproductive age early, devoting much energy to a large number of seeds that disperse widely, often to more light gaps in the forest. Weedy species remain small and have a brief life span, perhaps passing a few generations in the light gap before it closes.

A good way of thinking about *K*-selected species is to think of a mature forest tree. It exists in non-disturbed native forest habi-

TABLE 5.1 Some of the correlates of *r* and *K* selection.

Life History Feature	*r* Selection	*K* Selection
Development	Rapid	Slow
Reproductive rate	High	Low
Reproductive age	Early	Late
Body size	Small	Large
Reproductive type	Single reproduction (semelparous)	Repeated reproduction (iteroparous)
Length of life	Short	Long
Competitive ability	Weak	Strong
Survivorship	High mortality of young	Low mortality of young
Population size	Variable, usually well below carrying capacity	Fairly constant in time, at or near carrying capacity of the environment
Dispersal ability	Good	Poor
Type of habitat	Disturbed	Not disturbed

tats. Forest trees grow slowly and reach reproductive age late, having to devote much energy to growth and maintenance. *K*-selected species grow large and shade out *r*-selected species, eventually outcompeting them. Such trees live a long time and produce seeds repeatedly every year when mature. These seeds are bigger than those of *r*-selected species: Consider the acorns of oaks versus the seeds of dandelions. The large size of acorns contains a big food reserve that helps them get started, whereas dandelion seeds must rely on whatever nutrients they can gather from the soil.

There is a trade-off between the size and number of seeds: A plant can produce a few big seeds or lots of small ones. While the weed–tree examples are a useful way to think about the *r*–*K* continuum, we must remember that animals can be *r* or *K* selected, too. We can think of insects as small, *r*-selected species that produce many young and have short life cycles. On the other hand, big mammals like elephants grow slowly, have few young, and reach large sizes, typical of *K*-selected species. *K*-selected species are therefore particularly at risk of extinction. (See "Applied Ecology: Life Histories and Risk of Extinction.")

Applied Ecology

Life Histories and Risk of Extinction

Life history theory may be useful in helping us determine which types of species are more at risk of extinction and therefore should become a priority in conservation. Almost without fail, every life history attribute of a *K*-selected species sets it at risk of extinction. First, *K*-selected species tend to be bigger, so they need more habitat to live in. Florida panthers, for example, need huge tracts of land to establish their territories and hunt for deer. There is room for only about 22 panthers on publicly owned land in South Florida. Privately owned land currently supports another 50 panthers. Second, *K*-selected species tend to have fewer offspring, so their populations cannot recover as fast from a disturbance such as fire or overhunting. Condors, for instance, produce only a single chick every other year. Third, *K*-selected species breed at a later age, so their generation time and time to grow from a small population to a larger population is long. As an example, gestation in elephants takes 22 months, and elephants require at least seven years to become sexually mature. Finally, in *K*-selected species, population sizes are often small, and therefore, individuals run a high risk of inbreeding. Florida panthers are again illustrative: They are known to be inbred and possess distinctive features, such as a kinked tail and a cowlick of hair on their back, that other panther subspecies lack. Large *K*-selected trees such as the giant sequoia, large terrestrial mammals like elephants, rhinoceros, and grizzlies, and large marine mammals such as blue whales and sperm whales are all competitively dominant species that run a substantial risk of extinction.

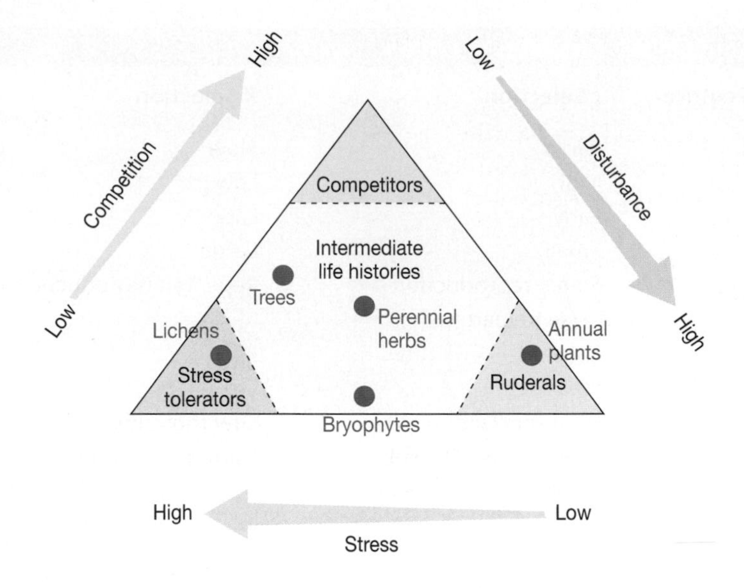

CHARACTERISTIC	COMPETITORS	RUDERALS	STRESS TOLERATORS
Life form	• Large herbs, shrubs, or trees	• Small herbs	• Lichens, herbs, shrubs
Leaf size	• Large	• Large	• Small
Life span	• Long	• Short	• Long
Seed production	• Small	• Large	• Small
Growth rate	• Rapid	• Rapid	• Slow
Palatability	• Various	• High	• Low
Vegetative spread	• Yes	• No	• Yes
Leaf litter production	• High	• Low	• Low

Figure 5.9 Expected plant life histories, expressed in the form of a model in which stress, disturbance, and competition are the important factors. (*Based on data in Grime, 1979.*)

Alternatives to the r and K continuum have been proposed. For plants, Phillip Grime, from the University of Sheffield in England (1977, 1979), proposed a trio of strategies: ruderals, competitors, and stress tolerators. *Ruderals* (a botanical term for weeds) are adapted to cope with habitat disturbance, including that by herbivores. *Competitors* are adapted to live in highly competitive, but benign environments like the tropics. *Stress tolerators* are adapted to cope with severe environmental conditions, such as salt marshes and deserts. Not only plants, but animals, can be classified according to this scheme: Desert animals are stress tolerators, having long lives and burrows to escape extreme conditions; lions are competitors, displacing other predators; and insects are ruderals. The scheme may be represented in a triangular fashion (Fig. 5.9), implying a trade-off among the three strategies. Thus, one cannot be a good stress tolerator and a good competitor at the same time. Craig Loehle (1987) did find evidence of some trade-offs between North American trees. For example, slow-growing hardwoods like hickory and oak lived a long time, just under 200 years, whereas fast-growing species such as box elder and cottonwood died young—often less than 100 years old. This existence of a trade-off supported Grime's ideas.

More recently, Johnathan Silvertown and his colleagues (Silvertown, Franco, and McConway, 1992; Silvertown *et al.*, 1993) provided a demographic interpretation of Grime's ideas in which the contributions of fecundity, growth, and survival to the overall reproductive rate r were emphasized. Species could either survive a long time as adults, grow large, or produce a lot of seeds, but not all three. Again, a demographic trade-off was implied. Individuals could then be placed on triangular diagrams with fecundity, growth, and survival as axes, from which it became apparent that species of similar life-forms (e.g., woody plants) and from particular environments (e.g., forests) tend to be found together on the diagrams (Fig. 5.10). Semelparous herbs would emphasize high fecundity, whereas woody trees would emphasize longevity or survival, with low, but repeated, levels of reproduction each year. In this scheme, the contributions from growth, survival, and fecundity could not sum to more than unity. Classification systems based on r selection vs. K selection, competitors vs. stress tolerators vs. ruderals, and growth vs. survival vs. fecundity provide a good theoretical framework that should guide research on life history strategies in the future.

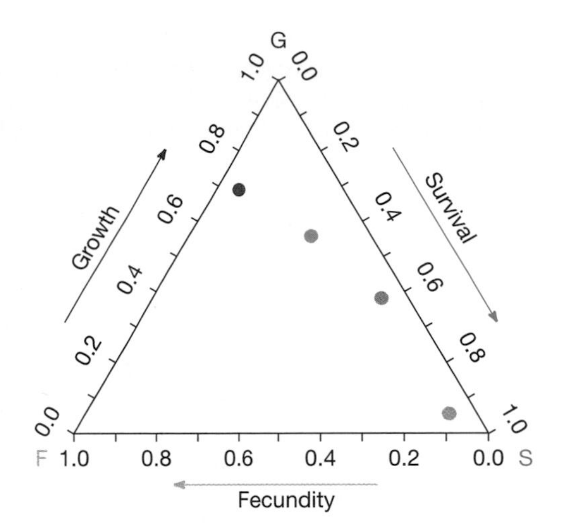

Figure 5.10 The distribution of species of perennial plants in the growth–survival–fecundity triangle. Note that growth, survival, and fecundity values are defined such that they always sum to unity. (a) Semelparous herbs. (b) Iteroparous herbs of open habitats. (c) Iteroparous forest herbs. (d) Woody trees.

Summary

There are many different types of life history strategies that organisms can employ to try to maximize their population growth.

1. An organism's life history largely concerns its lifetime pattern of growth and reproduction. Organisms that produce all their offspring in a single reproductive event, like salmon or yucca plants, are called semelparous. This contrasts with iteroparous organisms, such as trees and mammals, which live a relatively longer time and reproduce repeatedly.

2. Reproductive strategy has a strong effect on the age structure of a population. A low ratio of young to adults indicates a population in decline. A high ratio of young to adults indicates a growing population.

3. A ratio of one male to one female is the expected sex ratio in most populations. In polygynous mating systems, males mate with more than one female; in polyandrous systems, females mate with more than one male. In monogamous systems, each individual (male or female) has only one mate. Polygamy is often based on the availability of resources: Females (or males) have higher reproductive success on a territory with abundant resources, even though they may have to share it with others of the same sex.

4. The effects of mating systems, habitat selection, and dispersal can be synthesized under the banner of the r–K continuum. Organisms that are r selected are poor competitors, but have a high per capita population growth rate r. Such species often disperse well and colonize new habitats before being outcompeted by K-selected species, which are good competitors and often exist in mature habitats, close to the carrying capacity K, where they can outcompete most species. Other life history strategies can be classified in terms of competitors, ruderals, and stress tolerators or by the growth–longevity–fecundity triangle.

Mating systems, reproductive age, size at reproduction, relative investment in seeds or young, ability to disperse, speed of development, and maximum body size are all life history attributes that can differ among species according to the type of habitat they live in. These differences can have profound effects on the ability of species to survive and compete in different ecological situations.

Discussion Questions

1. What particular life history strategies are possessed by successful exotic invaders like kudzu in the southeast or zebra mussels in the Great Lakes? Can knowing their life histories help us in the war against exotics?

2. How could you test the idea that there is a trade-off between life history strategies? What would happen if you plant salt marsh plants like *Spartina* grass or mangroves in a terrestrial habitat vs. a freshwater habitat?

3. In some species males are much bigger than females, a property called sexual dimorphism. Speculate about the types of animals in which sexual dimorphism would and would not occur.

SECTION III

Population Ecology

Wild lupines bloom in Acadia National Park, Maine. What is the most important source of mortality for these plants—competition with other plants, herbivory by insects or vertebrates, or cold temperatures? Such questions abound in nature, and understanding what limits populations is the province of population ecology.

CHAPTER 6

Population Growth

Road Map

1. Time-specific and age-specific life tables are two ways of tabulating how a population's age structure varies over time. In time-specific life tables, a one-time census of the population is taken—a snapshot in time. In age-specific life tables, an entire cohort of a population is followed from birth to death.
2. Fecundity schedules show how female fecundity changes with age. If we know a population's female fecundity and female mortality at each age, we can estimate the population's future growth.
3. In deterministic models, individuals are treated as precise, unvarying numbers in population growth equations. Geometric and logistic models of population growth describe the shapes of population growth curves. In geometric growth, resources are assumed to be unlimited, and *J*-shaped growth curves result. In logistic models, resources are assumed to be limited, and *S*-shaped curves result. The effects of time lags on population growth allow oscillations around the mean. Chaos appears to be random population growth, but is actually generated by precise deterministic growth models and time lags.
4. In stochastic models, individual births and deaths are allowed to vary around means, and the models produce more variable results.

Population growth is at the core of modern applied ecology. Growth affects how common or rare a species is, which in turn affects the rate of population harvest: How many fish can we trawl from the sea or how many trees can we cut from a forest and still ensure a healthy population for future use? Should we expend more effort on saving rarer species with slower growth rates? To understand how a population grows and what limits growth, we have to consider the precise age structure of the population. We need to know especially how many females are in the population, what their age-specific fertility rate is, and how long females can be expected to live. This information allows us to calculate how many offspring are produced by females during their lives and, therefore, how fast a population is growing. We can use the rate of growth of a population, often called the *net reproductive rate,* in precise deterministic models that let us see how fast populations grow or decline. We can also use stochastic models, which add biological realism regarding prospects for growth in different situations.

6.1 Life Tables

With data on the numbers of individuals alive at each age in a population it is possible to construct life tables that show precisely how the population is structured in terms of age. The construction of life tables is termed *demography,* a subject that shows us how populations grow. The basic data used in a life table are the number of individuals in each age class. These data can be obtained either by following a group of individuals (a *cohort*) from birth until all members of the cohort have died (producing an age-specific life table) or from a snapshot of the age structure of the population at a single point in time (producing a time-specific, or static, life table).

Life tables are a tool for characterizing a species' life history strategies by estimating how the likelihood of mortality and reproduction change with age for individuals of the species.

Time-specific life tables provide a snapshot of a population's age structure from a sample at a given time.

Time-specific life tables are useful in examining populations of long-lived animals, say, herds of elephants, because following a cohort of such individuals from birth to death would be impractical. An example of a time-specific life table, prepared from a collection of skulls of known ages, for Dall mountain sheep (see Fig. 6.1) living in Mount McKinley (now Denali) National Park, Alaska, is shown in Table 6.1. The data were originally collected by Adolf Murie in 1937. Murie was interested in the effects of wolf predation on sheep, and he believed that wolves were the main cause of sheep dying. Murie collected 608 sheep skulls by picking them up off the ground. He could determine their age by counting annual growth rings on the horns. Thus, the collection of skulls gave a snapshot of how old the animals were when they died. Edward Deevey (1947) put Murie's data in the form of a life table that listed each age class and the number of skulls in it. From the data, Deevey could calculate other useful parameters, such as rates of mortality in each age class and expectations of life. In the life table, letters symbolize the values given in the columns:

x = age class or interval (years).

n_x = the number of survivors at beginning of age interval x. By

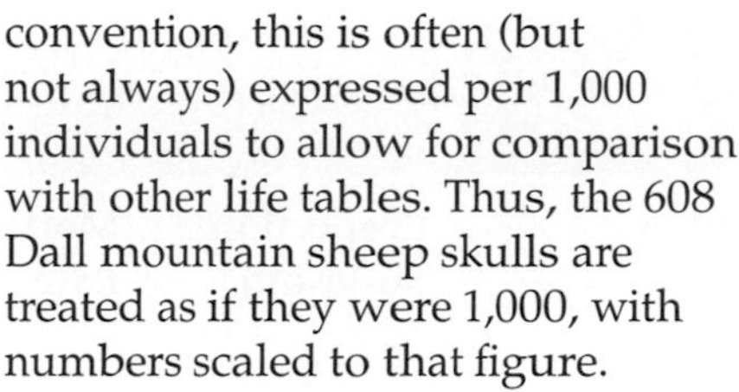

convention, this is often (but not always) expressed per 1,000 individuals to allow for comparison with other life tables. Thus, the 608 Dall mountain sheep skulls are treated as if they were 1,000, with numbers scaled to that figure.

d_x = number of organisms dying between the beginning of age interval x and the beginning of age interval $x + 1$.

l_x = proportion of organisms surviving to the beginning of age interval x

q_x = rate of mortality between the beginning of age interval x and the beginning of age interval $x + 1$.

e_x = the mean expectation of life for organisms alive at age x. Determining this figure is the goal of actuaries, who deal with human life tables. Ecologists borrowed the use of e_x from the insurance industry.

Beginning with the age class (x) and the number of individuals at the beginning of that age class (n_x), one can calculate all the other variables in the life table, including the number of deaths occurring in the age class x, given by

deaths = number alive in this year − number alive in next year, or, symbolically,

$$d_x = n_x - n_{x+1}.$$

For example, in the sixth age class, $n_{5-6} = 734$ indicating that there were that many individuals alive at the beginning of

Figure 6.1 Dall mountain sheep, *Ovis dalli*, in McKinley Park, Alaska. Collections of skulls, together with accurate age estimates of the skulls, permitted the construction of an accurate life table for this species. *(Bean, DRK Photo.)*

TABLE 6.1 Time-specific life table for the Dall mountain sheep (*Ovis dalli*), based on the known age at death of 608 sheep dying before 1937 (both sexes combined). Data are expressed per 1,000 individuals. The column *Tx* has no real biological meaning.

Age Class	Number Alive	Number Dying	Proportion Surviving	Mortality Rate	Average Number Alive in Age Class	*Tx*	Life Expectancy
x	n_x	d_x $= n_x - n_{x+1}$	l_x $(= n_x/n_0)$	q_x $(= d_x/n_x)$	L_x $(= n_x + n_{x-1})/2$	$\sum_x^{\infty} L_x$	e_x $(= T_x/n_x)$
0–1	1,000	199	1.000	.199	900.5	7,053	7.0
1–2	801	12	0.801	.015	795	6,152.5	7.7
2–3	789	13	0.789	.016	776.5	5,357.5	6.8
3–4	776	12	0.776	.015	770	4,581	5.9
4–5	764	30	0.764	.039	749	3,811	5.0
5–6	734	46	0.734	.063	711	3,062	4.2
6–7	688	48	0.688	.070	664	2,351	3.4
7–8	640	69	0.640	.108	605.5	1,687	2.6
8–9	571	132	0.571	.231	505	1,081.5	1.9
9–10	439	187	0.439	.426	345.5	576.5	1.3
10–11	252	156	0.252	.619	174	231	0.9
11–12	96	90	0.096	.937	51	57	0.6
12–13	6	3	0.006	.500	4.5	6	1.0
13–14	3	3	0.003	1.00	1.5	1.5	0.5

the age class. Since only 688 individuals were alive at the beginning of the next age class ($n_{6-7} = 688$), we know that 46 individuals died during the unfolding of the sixth age class.

Next, values of n_x are used to calculate survivorship, which is the proportion of individuals in a cohort that survive to each age class. Survivorship, or l_x, is calculated by dividing the number of individuals in an age class by the number in the original cohort; that is,

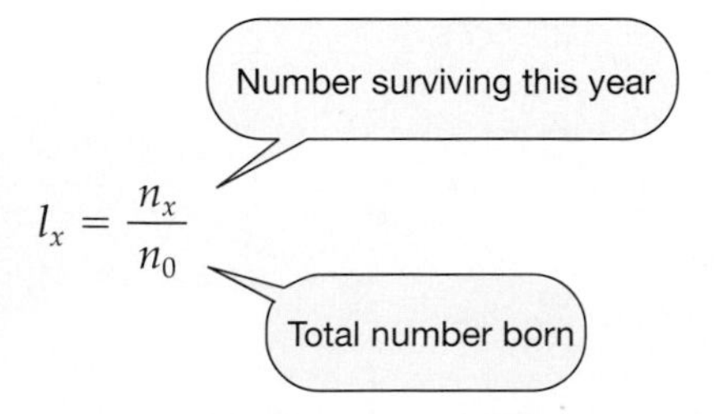

Survivorship to the sixth age class (five to six years) is therefore 734/1,000, or 0.734. This value tells us that over three-quarters of those individuals born will survive to at least five years old. The mortality rate within each age class is estimated by the variable q_x, which is calculated as

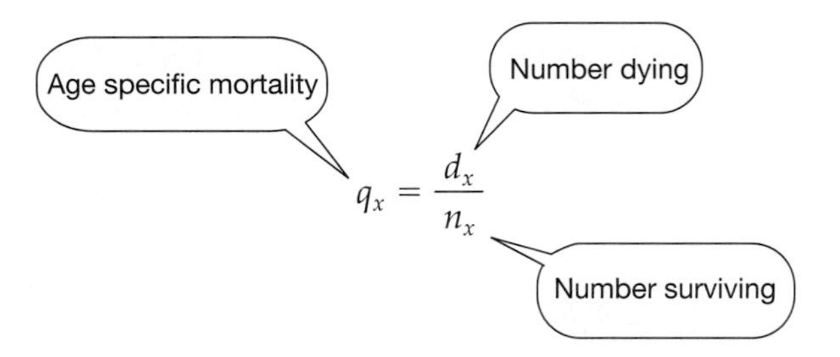

This equation simply expresses the number of deaths during the unfolding of an age class (d_x) as a proportion of the number of individuals that were alive at the beginning of that age class (n_x) and so indicates the fraction of individuals that died as the age class unfolded. For example, in the sixth age class (five to six years), there were 46 deaths out of 734 individuals, so $q_x = 46/734$, or 0.063. This mortality rate allows us to determine whether mortality increases with age, decreases with age, or is independent of age. In the Dall mountain sheep, q_x is high in the first age class, meaning that a lot of young sheep die. Then q_x drops dramatically and remains low until about age 8, when it increases quickly.

Another parameter that is commonly estimated is life expectancy, which is the average number of additional age classes an individual can expect to live at each age. This calculation requires two intermediate steps. First, the average number of individuals alive in each age class must be calculated. Since n_x is the average number of individuals alive at the beginning of each age class and n_{x+1} is the number alive at the beginning of the next age class, the number alive in an age class can be estimated by the relationship

$$L_x = \frac{n_x + n_{x+1}}{2}$$

A numerical example, from the sixth and seventh rows of Table 6.1, is

$$L_5 + \frac{n_5 + n_6}{2} = \frac{734 + 688}{2} = 711.$$

The second step in determining life expectancy is to calculate a quantity called T_x. This is a purely intermediate step and, unlike the other columns in a life table, is without real biological meaning. T_x is calculated by summing the values of L_x in age class x and all subsequent (older) age classes. Thus,

$$T_x = \sum_x^\infty L_x.$$

The value of T_x for the five-year-old sheep (in the sixth age class) is therefore

$$T_6 = L_6 + L_7 + L_8 + L_9 + L_{10} + L_{11} + L_{12} + L_{13},$$

or

$$711 + 644 + 605.5 + 505 + 345.5 + 174 + 51 + 4.5 + 1.5 = 3{,}062.$$

The sheep's life expectancy can now be calculated as

$$e_x = \frac{T_x}{n_x}.$$

The value of e_x indicates the average number of additional age classes an individual can expect to live at each age, so

$$e_6 = \frac{3{,}062}{734} = 4.2.$$

Therefore, five-year-old sheep can be expected to live, on average, for another 4.2 years.

Taken as a whole, these calculations allow important features of a population to be quantified: What is the life expectancy of individuals in this population? How does survivorship change with age? When in an individual's life is the mortality rate highest?

The number alive in each age class are commonly plotted against age to give a survivorship curve, as shown in Fig. 6.2. Here, n_x is expressed on a logarithmic scale, because we are interested in rates of change with time, not change in absolute numbers. For example, if we start with 1,000 individuals, and 500 are lost in one year, the logarithm of the decrease is

$$\log_{1,000} - \log_{500} = 3.0 - 2.7 = 0.3.$$

If we start with 100 individuals, and 50 are lost, the logarithm of the decrease is

$$\log_{100} - \log_{50} = 2.0 - 1.7 = 0.3.$$

In both cases, the rates of change are identical, even though the absolute numbers are different. Plotting the n_x data on a logarithmic scale ensures that, no matter what the starting size of the population, the rate of change of one survivorship curve can easily be compared with that of another species. For the Dall mountain sheep, there is an initial decline in survivorship as young lambs are

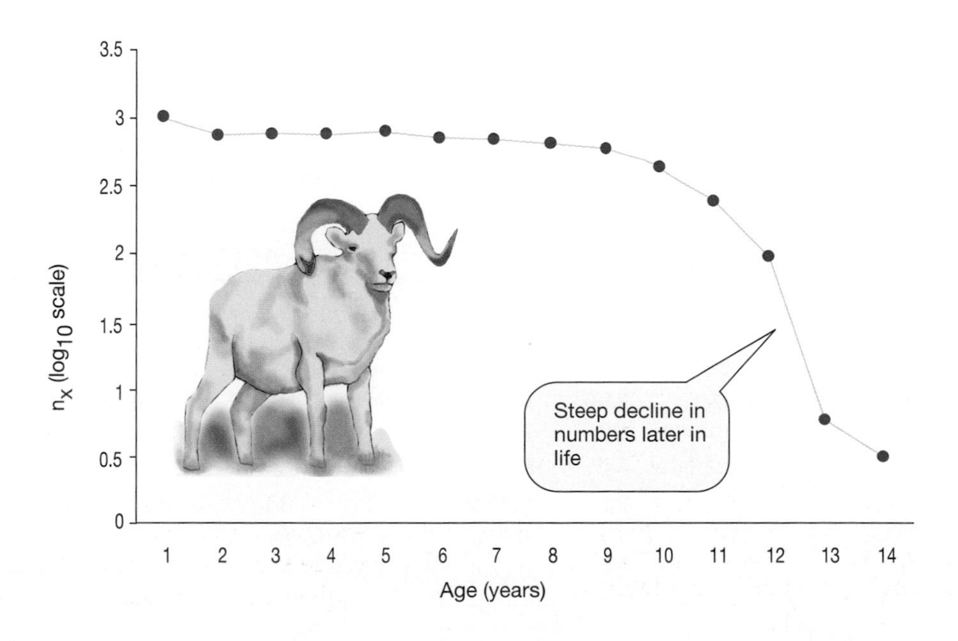

Figure 6.2 Time-specific survivorship curve for the Dall mountain sheep (*Ovis dalli*), showing the number of sheep alive in each age class plotted against age in years. The number (n_x) is expressed as a logarithm to report the rate of survival independently of the actual population size. *(Based on data from Deevey, 1947.)*

lost, and then the curve flattens out, indicating that the sheep survive well through about age 7 or 8. Then old age takes its toll, and sheep numbers decline rapidly. The main cause of deaths of both young lambs and older sheep is wolf predation. Survivorship curves for many other mammals, such as the African buffalo and the zebra, were also obtained by skull collection. These data yield curves that show similar patterns of early deaths, a period of several years of low mortality, and an increase in mortality in old age.

Despite the value of time-specific life tables, some of their underlying assumptions limit their accuracy. Paramount among these is the assumption that equal numbers of offspring are born each year. For example, if the rate of mortality of two-year-old Dall mountain sheep were identical to the rate of four-year-old sheep, but there were more two-year-old sheep born because of a favorable climate in a particular year, then more skulls of two-year-old sheep would be found later on, which would imply a higher rate of mortality of two-year-olds. Often, there is no independent method for estimating the birthrate of each age class. Perhaps for this reason, another type of life table, the age-specific life table, is more commonly reported. Of 31 life tables examined during the research for this book, 26 were age specific and only 5 were time specific.

Age-specific life tables follow an entire cohort of individuals from birth to death.

For organisms with short life spans—for example, an annual plant or an insect that completes its life span within a year—a snapshot, time-specific life table may not give the correct picture, because it will be severely biased toward the stage which is common at that moment. For example, one could inadvertently perform sampling when all the caterpillar stages of a butterfly are feeding and there are no eggs, pupae, or adults. In these cases, *age-specific life tables*, which follow one cohort or generation, are used. Population censuses must be conducted frequently, but only for a limited time (usually less than a year for insects or annual plants). In such cases, the age classes may be weeks or months, not years. For other organisms, like many birds or small mammals, an annual or bi-annual census for a few years will suffice. An age-specific life table for the American robin is shown in Table 6.2 and represented graphically in Fig. 6.3.

As a result of both time-specific and age-specific demographic techniques, three general types of survivorship curves can be recognized (Fig. 6.4): Type I, in which most individuals are lost when they are older; Type II, in which there is an almost linear rate of loss; and Type III, in which a large fraction of the population is lost in the juvenile stages. Type I curves are often observed in vertebrates and other organisms that exhibit parental care and protect their young. In actuality, there is usually a small dip in survivorship at young ages in Type I curves because some of the very young are lost to predators. Type II curves are exhibited by many birds and some invertebrates, such as hydra. Type III curves are often exhibited by invertebrates, such as insects, many plants (especially weeds), and marine invertebrates that do not exhibit parental care. For example, barnacles release millions of young into the sea, but most drift off and are eaten by predators. Only a few survive and settle in the rocky intertidal basins (although, once there, they show excellent survivorship).

What is the difference in accuracy between a time-specific and a cohort life table? It is very difficult to get data for both types of table for most populations in order to compare the two techniques, but it is possible for humans, because, on the one hand, we can use birth and death records for people born in, say, 1930 and follow them for their entire

TABLE 6.2 Life table for the American robin, *Turdus migratorius*, from data in Farner (1945). Based on returns of 568 birds banded as nestlings.

Age (yr)	n_x	d_x	l_x	q_x	L_x	T_x	e_x	Log n_x
0–1	1,000	503	1.00	0.503	748.5	1377	1.37	3
1–2	497	268	0.497	0.54	363	628.5	1.26	2.70
2–3	229	130	0.229	0.57	164	265.5	1.16	2.36
3–4	99	63	0.099	0.64	67.5	101.5	1.03	1.99
4–5	36	26	0.036	0.72	23	34	0.94	1.56
5–6	10	4	0.010	0.4	8	11	1.10	1
6–7	6	6	0.006	1.0	3	3	0.50	0.78

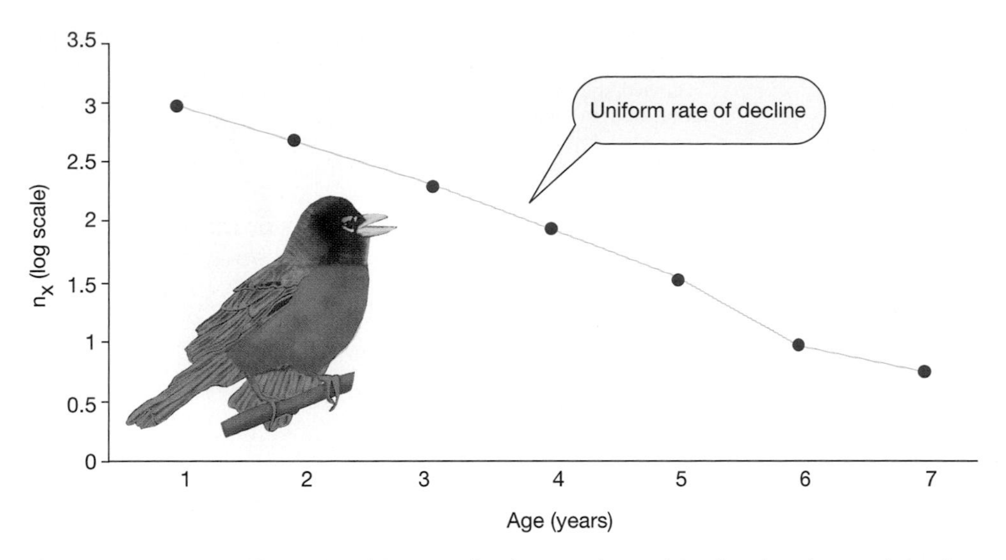

Figure 6.3 Age-specific survivorship curve for the American robin, showing the number of robins alive in each age class plotted against their age in years (based on data in Table 6.2).

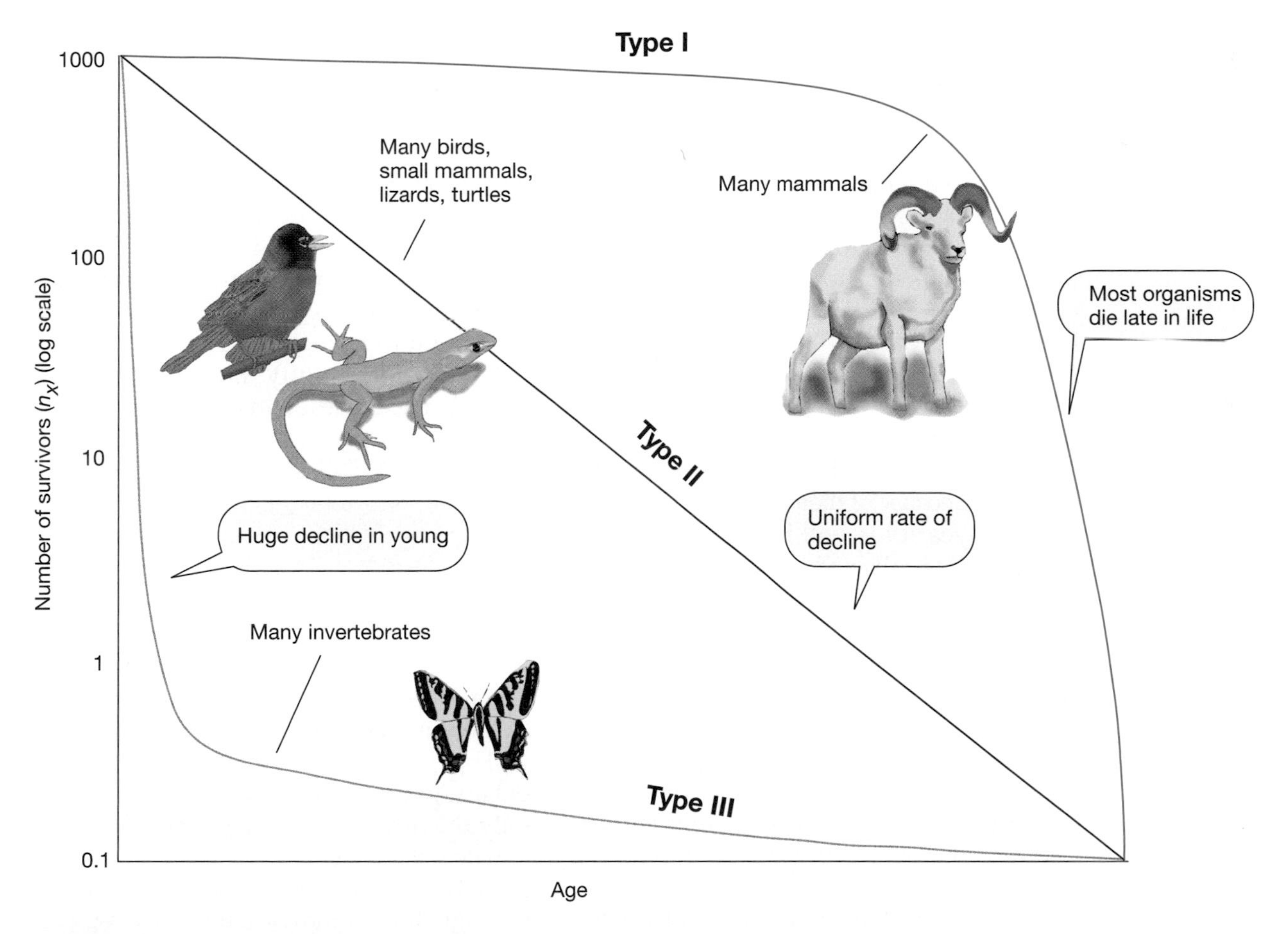

Figure 6.4 Idealized survivorship curves. Type I includes many large mammals that care for their young: buffalo, hippopotamuses, zebras, Dall mountain sheep, elk, red deer, and humans. In reality, there is often an initial dip in survivorship at very young ages, as observed in the Dall mountain sheep. Type II includes many birds (robins, starlings, blackbirds, herring gulls, and pheasant), many small mammals (snowshoe rabbits, squirrels, warthogs and some deer), and salamanders, turtles, and some lizards. Type III includes many insects (grasshoppers, gypsy moths, and scale insects), organisms with pelagic juvenile stages (barnacles, oysters, and mollusks), and many fish and annual plants, including grasses.

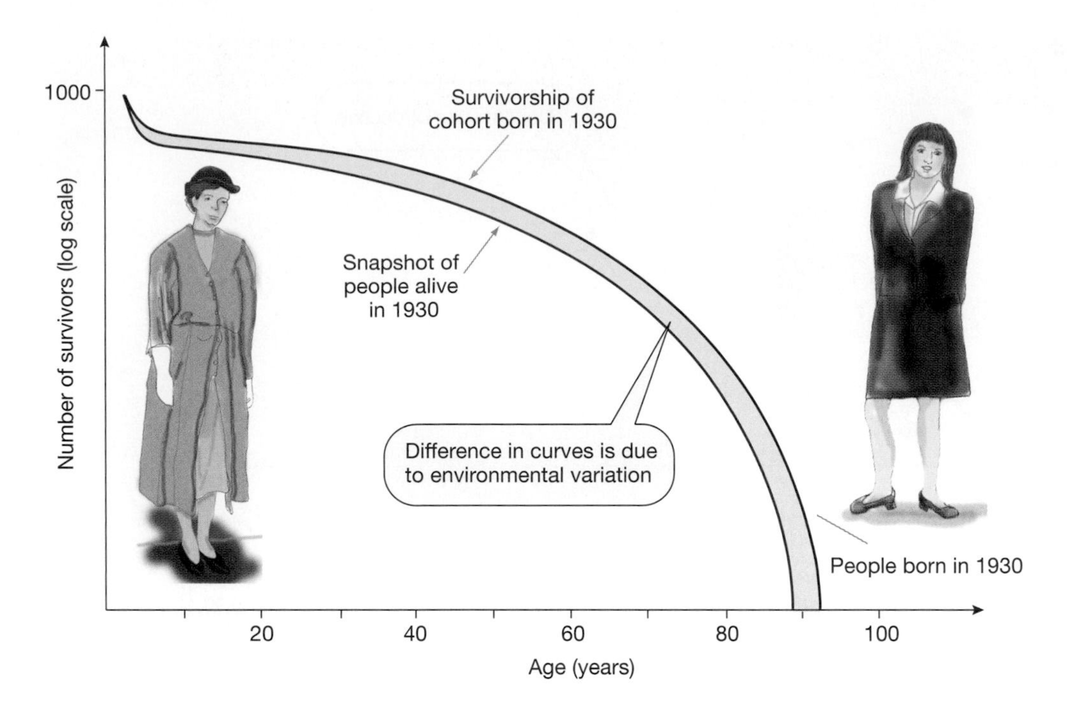

Figure 6.5 Hypothetical comparison of cohort survivorship of humans born in 1930 with static or time-specific survivorship curves for 1930.

lives, and on the other hand, we can also take a snapshot or cross section of a hypothetical population at 1930 to get a time-specific survivorship curve (Fig. 6.5). The 1930 time-specific curve actually relates to people born before 1930, while the cohort table gives us information on death rates after 1930. In a 1930 time-specific survivorship curve, 70-year-old people would have been born in 1860, and fewer would have survived to 1930, compared with 70-year-old people born in 1930 and still living in 2000, because modern medicine and diet have improved. Thus, static curves ignore environmental variation, because, if their predictions of how many 70-year-olds will be alive in 2000 is based on data collected from 70-year-olds in 1930, then it ignores health and dietary improvements. In the natural world, environmental variation includes years of good and bad climate or food supply.

6.2 Reproductive Rate

Life tables and fecundity schedules allow calculation of the population's net reproductive rate.

In order to use life table data to calculate population growth, one other component is needed: age-specific birthrates, or fecundity. **Fecundity** is defined as the number of female offspring produced by each breeding female. The resultant life table, complete with this information, is often termed a *fecundity schedule.* Fecundity schedules describe reproductive output and survivorship of breeding individuals only. Table 6.3 shows a fecundity schedule for a pride of female lions. Survivorship, l_x, is calculated as before from the number of females surviving to each age class. Age-specific fecundity, m_x, is the average number of female offspring produced per female at each age class. Fecundity schedules not only allow age-specific trends in reproduction to be examined, but also allow the population's net reproductive rate, R_0, to be calculated. R_0 is calculated by multiplying survivorship (l_x) by age-specific fecundity (m_x) and summing over all age classes. Thus, we have

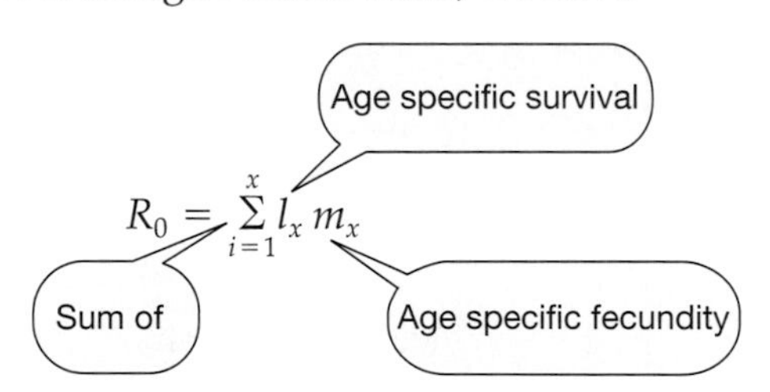

$$R_0 = \sum_{i=1}^{x} l_x m_x$$

The net reproductive rate is defined, then, as the number of breeding individuals that will be produced by each breeding individual in a population. It takes into account not only the number of female offspring that a female will produce at each age, but also the probability that each female will survive to that age. If each individual exactly replaces herself, then $R_0 = 1$, and the population is stationary. If each individual more than replaces herself, $R_0 > 1$, and the population is increasing. If each individual does not replace herself, then $R_0 < 1$, and the population is declining. In the example illustrated in Table 6.3, R_0 is 1.33, indicating that, on average, each fe-

TABLE 6.3 Fecundity schedule for female lions (x represents survivorship of females and m_x the number of female offspring produced in each age class).

X (Years)	Survivorship l_x	Age-Specific Fertility m_x	$l_x m_x$
0–1	1.00	0.00	0.00
1–2	0.75	0.00	0.00
2–3	0.58	0.14	0.08
3–4	0.58	0.57	0.33
4–5	0.50	0.50	0.25
5–6	0.42	0.00	0.00
6–7	0.33	1.25	0.42
7–8	0.25	0.67	0.17
8–9	0.25	0.33	0.08
9–10	0.17	0.00	0.00
10–11	0.08	0.00	0.00
			$R_0 = 1.33$

male will produce 1.33 female offspring, and hence, the population is increasing.

For some species, such as plants, it is difficult to determine which female give rise to which offspring. Here, the calculation of age-specific fecundity (m_x) is facilitated by a knowledge of the total number of eggs, seeds, or young deposited (F_x), divided by the total number of reproducing individuals (n_x) such that

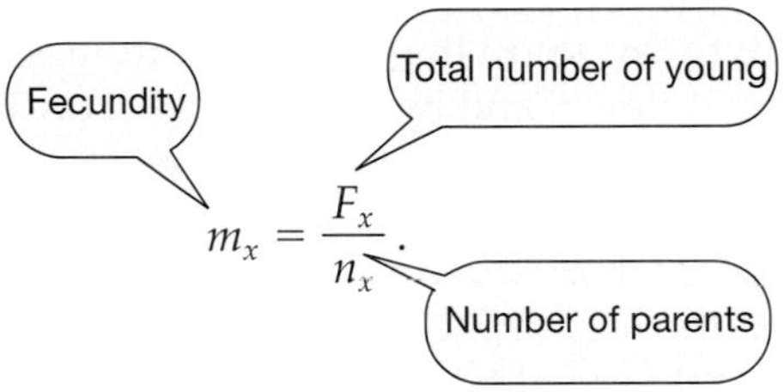

$$m_x = \frac{F_x}{n_x}.$$

A worked-out example is shown in Table 6.4 for the plant *Phlox drummondi*

TABLE 6.4 Fecundity schedule for *Phlox drummondi* (from data in Leverich and Levin, 1979).

Age Interval (Days) x	Number n_x	Mortality Rate q_x	Proportion Surviving Today l_x	F_x	m_x	$l_x m_x$
0–63	996	0.329	1.000	—*	—	—
63–124	668	0.375	0.671	—	—	—
124–184	295	0.105	0.296	—	—	—
184–215	190	0.014	0.191	—	—	—
215–264	176	0.004	0.177	—	—	—
264–278	172	0.005	0.173	—	—	—
278–292	167	0.008	0.168	—	—	—
292–306	159	0.005	0.160	53.0	0.33	0.05
306–320	154	0.007	0.155	485.0	3.13	0.49
320–334	147	0.043	0.148	802.7	5.42	0.80
334–348	105	0.083	0.105	972.7	9.26	0.97
348–362	22	0.022	0.022	94.8	4.31	0.10
362–	0	—	0.000	—	—	—
						$= \Sigma\, 2.41$

*means no offspring at this stage

Figure 6.6 *Phlox drummondi*, a colorful roadside plant, common in many areas of the United States.

(Fig. 6.6). In the case of the phlox population, R_0 was 2.4, meaning that there was a 2.4-fold increase in the size of the population over one generation. If this were true every year, then we would be up to our eyes in phlox. The year shown was evidently a good one. In bad years, R_0 would be much less than unity, but many years' data would probably reveal a more balanced picture. The reasons for good and bad years may include bad weather, herbivory, or disease. (We will address each of these factors later.)

6.3 Deterministic Models: Geometric Growth

Deterministic models predict population growth on the basis of intrinsic properties of the population, such as its growth rate, current size, and carrying capacity. Geometric or exponential growth occurs in populations that face no competition for resources.

Once we know a population's net reproductive rate R_0, we can begin to predict how the population will grow. First, we consider the simplest type of growth: that of an annually breeding species with no limits to growth at the present time. We determine growth by knowing the population size now, at time t, and by knowing R_0. Consider an insect, like a butterfly, that breeds once a year and has a life span of one year. Because no females survive more than one year, the size of the population of females the next year, at time $t + 1$, is given by

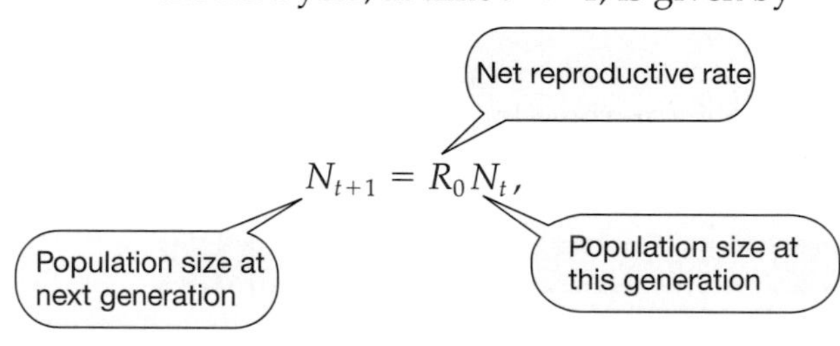

$$N_{t+1} = R_0 N_t,$$

where N_t is the population of females at generation t, N_{t+1} is the population of females at generation $t + 1$, and R_0 is the net reproductive rate, or the number of females produced per female per generation. If $R_0 = 1.5$ and $N_t = 500$, then $N_{t+1} = 1.5 \times 500 = 750$. At the second generation, N_{t+2}, the size of the population would be $750 \times 1.5 = 1{,}125$. Similarly, the size of the population at $N_{t+3} = 1{,}687.5$, at $N_{t+4} = 2{,}531.25$, and, after five generations, at $N_{t+5} = 3{,}796.875$.

Clearly, much depends on the value of R_0: If $R_0 < 1$, the population becomes extinct, if $R_0 = 1$, the population remains constant, and if $R_0 > 1$, the population increases. When $R_0 = 1$, the population is often referred to as being at *equilibrium*, with no changes in population density. If $R_0 = 2.0$, the population doubles every generation. This staggering rate of growth is illustrated by the following riddle: Suppose you had a pond that was being overgrown by an aggressive weed which doubled its numbers every day ($R_0 = 2.0$). If the weed went unchecked, it would cover the pond in 30 days. The weed's growth is very slow—almost negligible—to begin with, so you decide not to remove the weed until the pond is half covered. When will that be? On the 29th day, of course. You have one day to save your pond! Even if R_0 is only fractionally above 1, population increase is rapid (Fig. 6.7), and a characteristic *J*-shaped curve results. We refer to this characteristic curve as *geometric growth*.

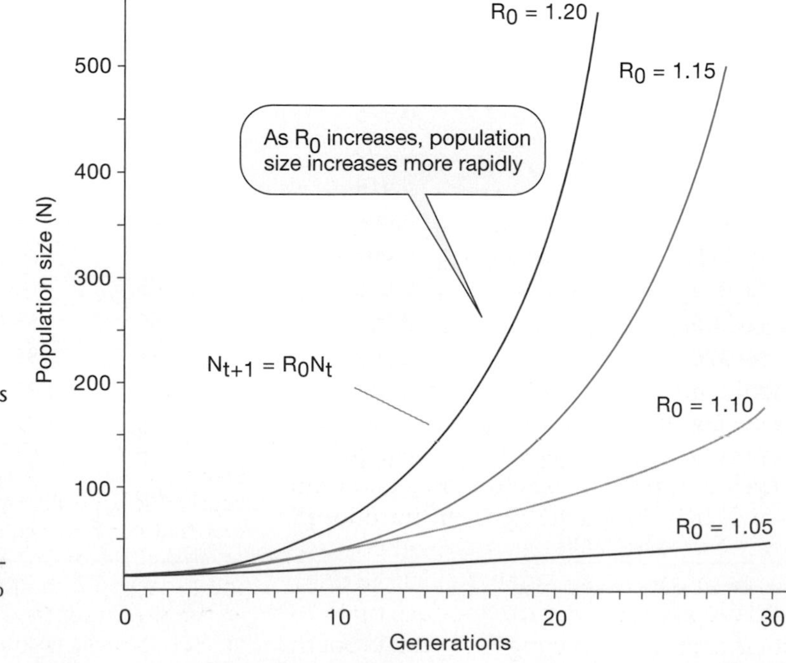

Figure 6.7 Four examples of geometric population growth for populations with discrete generations. $N_0 = 10$. Even with low net reproductive rates, such as $R_0 = 1.05$ the population increases over 50% in 10 generations and 100% by 30 generations.

How do field data fit this simple model of geometric growth? Clearly, population growth cannot go on forever, as envisioned by the geometric growth formula: As anyone can see, we are not knee deep in Dall mountain sheep, lions, or phlox. Something limits population growth—maybe resources. But initially at least, in a new, expanding population, geometric growth is possible. For example, when exotic pest animals are introduced into new environments or when species are protected under new conservation measures, geometric growth may occur. When reindeer were introduced onto the Pribilof Islands, off Alaska, in 1911, they showed the classic *J*-shaped curve characteristic of geometric growth, at least initially (Fig. 6.8a). After that, the population crashed, probably because it had used up all its resources, particularly lichens. The reindeer had no predators on the island.

Another example of geometric growth comes from elephant seals (Fig. 6.8b). Northern elephant seals were nearly hunted to extinction in the late 19th century, because their blubber was in demand. About 20 surviving animals were found off Mexico on Isla Guadalupe in 1890, and the population was protected. Recolonization of old habitats followed, and the actual growth of the elephant-seal population followed a geometric curve in many areas. Breeding began on Año Nuevo Island off the coast of Santa Cruz, California, in 1961 and the number of pups increased geometrically (Fig. 6.8c).

The growth of protected annually breeding populations recovering from overexploitation in the past often provides some of the best fits of actual population growth curves to the geometric model. Whooping cranes in Texas showed a similar increase in recovering from near extinction in 1940 after they were protected from hunters and after their overwintering habitat on the Texas coast was protected. African buffalo in the Serengeti showed geometric growth after recovering from an outbreak of disease called rinderpest in 1890.

The growth of some exotic species introduced into novel habitats also seems to fit the geometric pattern. The rapid expansion of rabbits after their introduction into South Australia in 1859 is a case in point. On Christmas 1859, Thomas Austin received two dozen European rabbits from England (into which they had actually been introduced from France by William the Conqueror centuries earlier). Rabbit gestation lasts a mere 31 days, and each doe can produce up to 10 litters of at least six young each year. In South Australia, the rabbits had essentially no enemies. Even when two-thirds of the population was shot for sport, which was the purpose of the introduction, the population grew into the millions in a few short years. Sixteen years later, rabbits were reported on the west coast of Australia, having moved

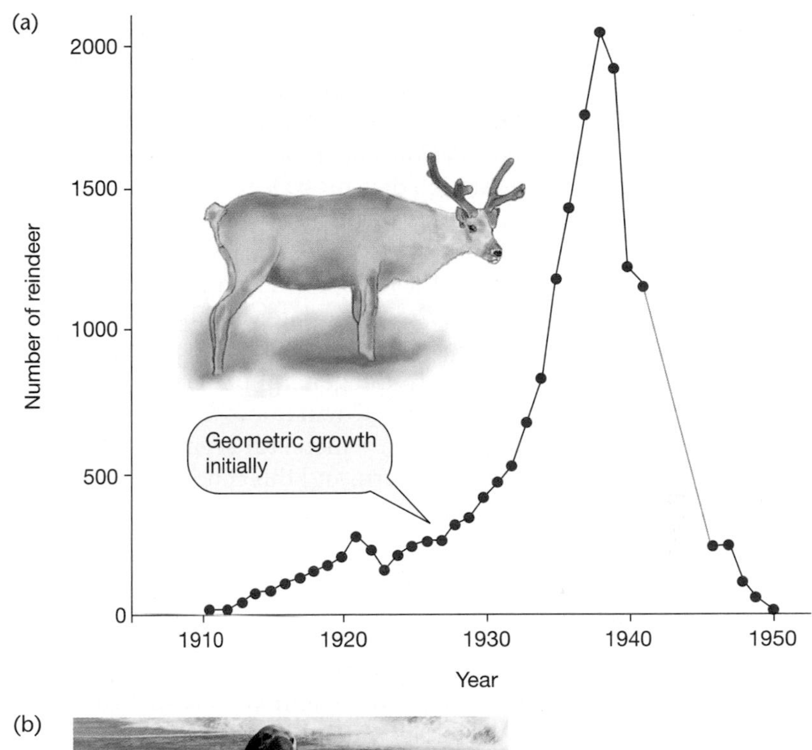

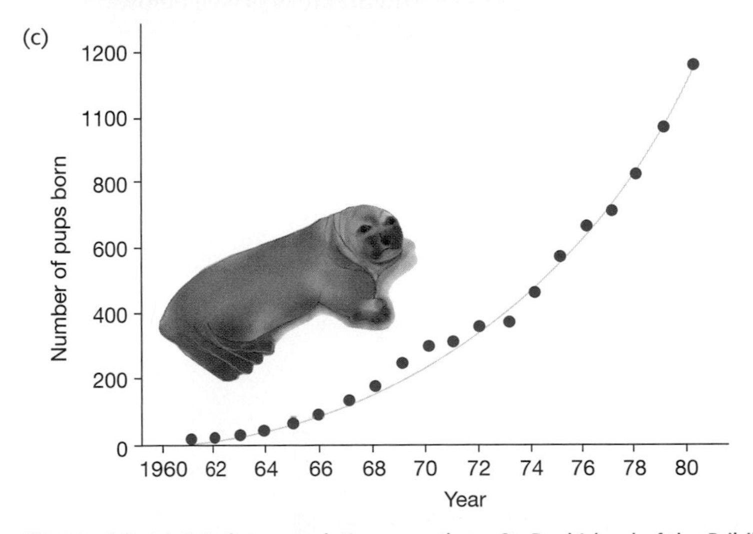

Figure 6.8 (a) Reindeer population growth on St. Paul Island of the Pribilof Islands, Bering Sea, from 1911, when about 20 of the deer were introduced, until 1950. Initially, the population follows the classic *J*-shaped curve characteristic of geometric growth. Later, however, the population crashed after its food supply was overexploited. *(Redrawn from Scheffer, 1951.)* (b) *Mirounga angustirostris,* large bull, small cow, and darker pup in the foreground. (c) The growth of northern elephant seals on Año Nuevo Island, near Santa Cruz, California, following recolonization and first breeding in 1961, follows a geometric pattern. *(Reproduced from LeBoeuf and Kaza, 1981.)*

over 1,100 miles across the continent, despite efforts to stop them by means of huge, thousand-mile-long fences (Fenner and Ratcliffe, 1965).

In general, however, in over 200 cases of mammalian introductions, most have failed to grow in an explosive or geometric pattern beyond the initial increase. After the growth period, populations of most introduced species level off or, more likely, decline (De Vos, Manville, and Van Gelder, 1956). This leveling off or decline occurs because R_0, the population's net reproductive rate, decreases at high densities: Death rates increase and birthrates decrease due to food shortages, attacks by predators, or epidemics. The effect of these changes in reproductive rates will be examined later in this section when we look at logistic growth. A prominent exception to the general pattern of declining population growth rate is the global human population, which still seems to fit a geometric pattern, and this continued growth is a source of concern to many.

Human population growth.

It has been very roughly estimated that up until the beginning of agriculture and the domestication of animals, about 10,000 B.C., the average rate of population growth was about 0.0001 percent per year. After the establishment of agriculture, the world's population grew to about 300 million by 1 A.D. and 800 million by the year 1750, but still the average growth rate was well below 0.1 percent per year. The modern period of rapid population growth may be regarded as having started at about 1750. Average annual growth rates climbed to about 0.5 percent between 1750 and 1900, 0.8 percent in the first half of the 20th century, and 1.7 percent in the second half. Advances in medicine and nutrition are certainly responsible for a large part of the growth in reproductive rates and the longevity of humans. In this relatively tiny period of human history (equivalent to the last one minute on a 24-hour clock), the world's human population increased from 0.8 billion in 1750 to 6 billion in 1998. Thus, in less than 0.1 percent of human history, more than 80 percent of the increase in human numbers has occurred. This rapid rise in human growth rates in recent times is illustrated in Fig. 6.9.

It was estimated that in 1995 the world's population was increasing at the rate of three people every second. In 1984, United Nations projections estimated that the most likely course of world population was growth toward an eventual stabilization at around 10 billion people by the year 2100. However, human growth rates have not fallen as quickly as was anticipated. United Nations 1995 projections point to a world population stabilizing at around 11.6 billion toward the year 2150, although nobody knows for sure whether that will happen (Cohen 1995, 1997). Interestingly, the population sizes of domestic animals associated with humans have also increased geometrically, at a rate even faster than that of humans. For example, the num-

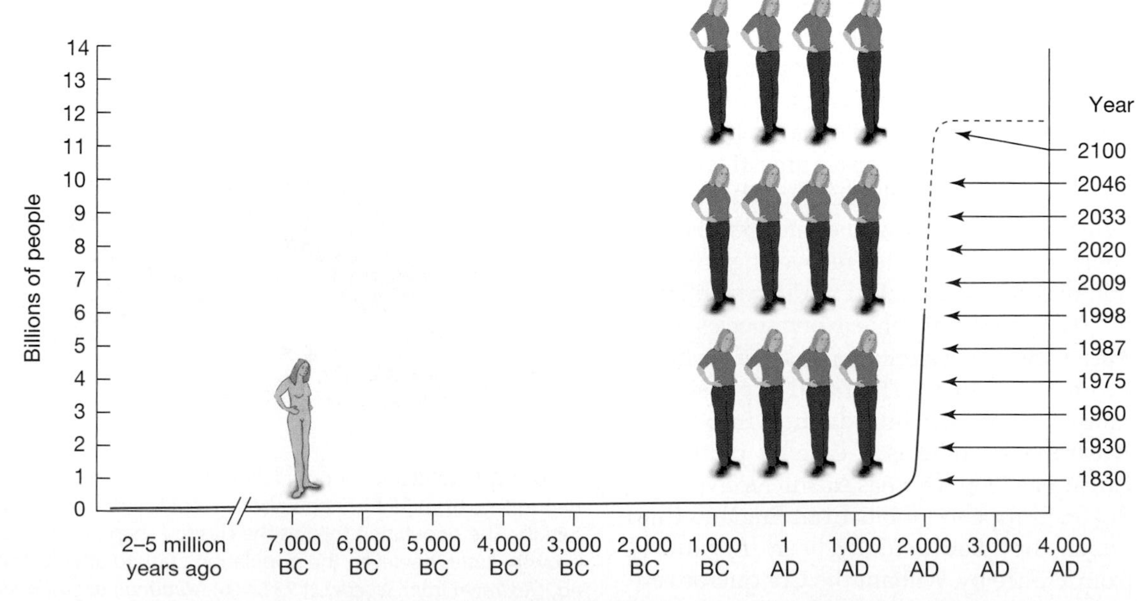

Figure 6.9 The world population explosion. For most of human history, the population grew very slowly, but in modern times it has suddenly "exploded." Where and when it will level off is the subject of much debate. At the moment, this is still one of the best examples of geometric growth.

TABLE 6.5 Trends in world population and its rate of increase in recent decades.

	Population (millions)				Rate of Increase (%)			
Region	1960	1970	1980	1990	1960–1965	1970–1975	1980–1985	1990–1995
World	3,019	3,698	4,450	5,292	1.99	1.96	1.74	1.71
Developed regions	945	1,049	1,136	1,205	1.19	0.86	0.65	0.48
Developing regions	2,075	2,649	3,314	4,087	2.35	2.38	2.10	2.06
Africa	281	363	481	648	2.48	2.69	2.95	3.01
Latin America	218	285	362	448	2.80	2.48	2.19	1.94
Northern America	199	226	252	276	1.49	1.06	1.00	0.71
Asia	1,667	2,101	2,583	3,108	2.19	2.27	1.86	1.82
Europe	425	460	484	498	0.91	0.58	0.32	0.22
U.S.S.R.	214	243	266	288	1.49	0.96	0.84	0.68

bers of chickens more than doubled in the decade prior to 1996, when it reached 17 million, and the combined weight of the 4.3 million large domestic animals maintained by humans is greater than the weight of the human population.

In the developed countries, the average annual rate of human population increase has slowed considerably, from 1.19 percent in 1960–1965 to 0.48 percent in 1990–1995. In the developing world, however, average annual growth rates rose from 2.35 percent in 1960–1965 to 2.38 percent in 1970–1975, before dropping to 2.1 percent in 1980–1985. The results of these growth rates are that, in developed nations, the population appears to have stabilized at around 1 billion, but the number of people in developing countries is still increasing (Table 6.5). Overall, world population is rising; human numbers reached 6 billion in 1998. If human population growth fuels pollution and the use of resources, then developing countries, particularly those in Africa, are clearly most at risk. However, even though African countries have a high rate of population increase, Africa's total population, at about 650 million people, is far below the 3 billion that inhabit Asia. Thus, in terms of the global population level, even a small decrease in the rate of population increase in Asia would reduce the absolute numbers of humans more than would a substantial decrease in the rate of population increase in Africa.

Global population growth can also be examined by looking at fertility rates—specifically, the average number of live children that would typically be borne by a woman during her lifetime (Table 6.6). Theoretically, the replacement rate is, of course, 2.0—two children born to replace their mother and father, who later die; but because of natural mortality, the actual replacement rate is 2.1. Again, the fertility rate differs considerably between geographic areas. In Africa, the total fertility rate of 6.0 in 1990 has scarcely declined since the 1950s, when it was around 6.6 children per woman. In Latin America and Southeast Asia, fertility rates have declined considerably from the 1970s and are now at around 3.3 and 3.2, respectively. In 1960–1965, the global average

TABLE 6.6 The average number of children per woman in her reproductive lifetime; 1990 data from women just out of reproductive age.

Region	Fertility Rate	Region	Fertility Rate
Africa	6.0	Former U.S.S.R.	2.3
South America	4.0	Europe	2.0
Asia	4.0	United States	1.9
Oceania	3.0	World	3.0
North America	3.0		

Applied Ecology

Human Population Growth and the Use of Contraceptives

A 1992 Johns Hopkins University study indicated that in developed countries over 70 percent of couples in which the woman was of reproductive age were using some form of contraception. In developing countries, the percentage was estimated to be around 45 percent. In Africa, only an estimated 14 percent of couples used contraceptives, whereas in Asia around 50 percent and in Latin America about 57 percent practiced contraception. Some of the most startling reductions in fertility have taken place in China. In the 1950s and 1960s, the total fertility rate in that country was six children per woman. During the 1970s, the government offered family planning services, and incentives were also offered to couples who agreed to limit themselves to one child. In 1990, an estimated 75 percent of the Chinese population used birth control, and the fertility rate had dropped to 2.2. Government-supported family-planning programs have been implemented in an increasing number of countries during the past two decades. In 1976, only 97 governments provided direct support for family planning, compared with 125 in 1988. Concomitantly, the number of governments limiting access to family planning fell from 15 to 7. However, as of 1989, there were still 31 countries in the developing world in which couples had virtually no access to modern family-planning methods. One could also argue that reductions in birthrates have not occurred in some developing countries because women wish to have large families, but numerous surveys show that women in developing countries would like to have fewer children than they already have. For example, young women 15 to 19 years in virtually every country outside sub-Saharan Africa, desire fewer than three children; some figures are Thailand, 1.8; Colombia, 2.1; and Brazil, 2.2. However, in sub-Saharan Africa, the desired number in most countries surveyed is five or more.

It is interesting that, despite the concerns in many parts of the world about human population, in some Western European countries the concern is that growth rates are not high enough! Since about 1965, the "baby boom" in these countries, together with other developed countries, such as Australia, Canada, New Zealand, and the United States, has turned into a "baby bust," and total fertility rates have fallen below the national replacement level of 2.1 children per woman. A rate below the replacement rate could eventually produce smaller populations that would severely affect the political strength and economic structure of some nations, and that prospect alarms a number of politicians.

---

number of children born per woman was almost 5. In 1990, the average was 3.3, but this is still greater than the 2.1 needed for zero population growth. In more developed regions, the fertility rate has declined steadily in recent times and is now expected to remain fairly stable, at around 1.9, for the next few decades. Part of the reason for the differences among developed and developing countries has been linked to the use of contraceptives. (See "Applied Ecology" box.)

Overlapping generations.

For many species, including bacteria, cockroaches, and many plants and animals in warm climates, reproduction occurs not once a year, but continually, and generations overlap. For such species, the rate of increase is essentially geometric, like those seen in Fig. 6.7, but, because of its persistent nature, it must be described by a differential equation. We can best understand this difference by an analogy to interest rates. If we have a population of one thousand organisms increasing at a finite rate of 10 percent per year, then the size of the population at the end of one year will be 1,000 + (10% × 1000) = 1,100. The following table give the corresponding values for subsequent years:

Year	*N*
0	1,000
1	1,100
2	1,210
3	1,331
4	1,464
5	1,610

If we repeated these calculations with a finite rate of increase of 5 percent every six months, our numbers would be a little different. After six months, the population size would be 1,050, after one year 1,102, two years 1,215, three years 1,340, and so on. If we repeated the calculations every day or every hour, our values would be slightly different again. Because many populations are continuous breeders, this algorithm is quite appropriate. Dividing a year into one thousand short periods, we have a population size of 1,000.1 in the first thousandth of the year, 1,000.2 in the second thousandth, and so on. If we apply the algorithm for one thousand time intervals, we will end up with 1,105 organisms at the end of the first year.

The differential equation that describes the population growth of continuously reproducing populations with overlapping generations is

$$\frac{dN}{dt} = rN = (b - d)N,$$

where

- N = population size
- t = time
- r = per capita rate of population growth,
- b = instantaneous birthrate
- d = instantaneous death rate
- dN = rate of change in numbers
- dt = rate of change in time
- dN/dt = rate of population increase

So if $r = 0.1$ and $N = 1{,}000$ then $dN/dt = 0.1 \times 1{,}000 = 100$. This means that 100 individuals are added to the population in a unit of time, such as a year or a month.

Once again, plots of population growth against time would yield *J*-shaped curves, with the steepness of the curve depending on the value of r. Expressed in a different way, a plot of the natural logarithm of N (the logarithm of N to the base e), written ln N, against time would give a straight line, with the slope equal to r (Fig. 6.10).

In the foregoing equation, r (the per capita rate of population growth) is analogous to R_0 (the net reproductive rate). In populations with a stable age distribution (i.e., populations in which the proportions of different age classes remain the same from year to year),

$$r = \frac{(\ln R_0)}{T_c}$$

where T_c is the generation time—the time taken for new offspring to grow and have their own offspring. This equation essentially means that, for a continuously breeding population, the net reproductive rate divided by the generation time gives an approximation of the instantaneous rate of population growth.

We can also use the geometric growth model to estimate the doubling time for a population growing at a certain rate. According to the rules of calculus,

$$\int_{dt}^{dN} = \int^{rN},$$

or

$$N_t = N_0 e^{rt},$$

where e is the mathematical constant 2.71828, the base of natural logarithms.

Thus,

$$\frac{N_t}{N_0} = e^{rt}.$$

If the population doubles, then $N_t/N_0 = 2$, so that

$$2.0 = e^{rt},$$

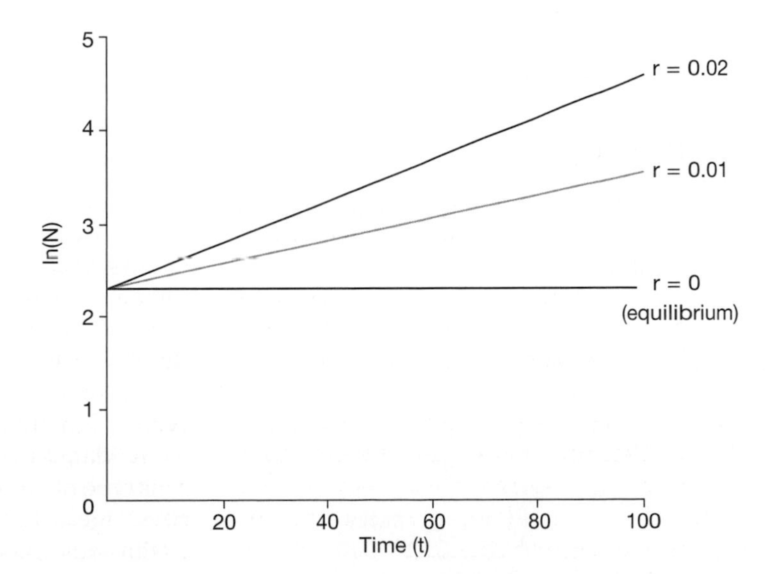

Figure 6.10 Exponential growth for populations with overlapping generations. The starting population is $N = 10$. Here, population growth is plotted on a semilogarithmic set of axes, with the natural logarithm (with base e) being used. The result is a straight line with slope equivalent to r. If a logarithmic scale is not used, then the same *J*-shaped curve as in Fig. 6.7 results. A semilogarithmic plot is a convenient method used to measure r.

or

$$\ln(2.0) = rt,$$
$$0.69315 = rt,$$

and

$$t = \frac{0.69315}{r},$$

where

t = time for population to double its size

and

r = rate of population growth per capita.

A few values instantiating this relationship are given for illustration:

r	t
0.01	69.3
0.02	34.7
0.03	23.1
0.04	17.3
0.05	13.9
0.06	11.6

Thus, if a human population is increasing at an instantaneous rate of only 0.03 per year, its doubling time would be about 23 years if a geometric increase prevails. Even very low rates of population increase give very short doubling times. In other organisms, doubling times can vary from as little as 17 minutes for *E. coli* bacteria to 47 days for rats, 2 years for cows, and as much as 25.3 years for a southern beech tree, *Nothofagus*.

Logistic growth occurs in populations in which resources are or can be limiting.

For many species, resources become limiting as populations grow. The intrinsic rate of natural increase decreases as these resources are used up. Thus, a more appropriate equation to explain population growth under these conditions is

Per capita rate of population growth

Population size

Rate of population change

$$\frac{dN}{dt} = rN\frac{(K - N)}{K} \quad \text{or} \quad \frac{dN}{dt} = rN\left(1 - \frac{N}{K}\right)$$

Carrying capacity

K is the upper asymptote or maximal value of N, commonly referred to as the *carrying capacity* of the environment at the equilibrium level of the population, and $(K - N)/K$ represents the proportion of unused resources remaining. In essence, this equation means that the larger the population size N, the closer it is to the carrying capacity K, and the fewer the available resources for population growth. At large values of N, $(K - N)/K$ becomes smaller and smaller, so population growth is very small. At small values of N, $(K - N)/K$ is close to unity, and population growth is large. The equation, called the *logistic equation*, was discovered by the Belgian mathematician P. F. Verhulst in 1838. In the logistic equation, one would know or be given K, the carrying capacity, which would come from intense field and laboratory work in which one would determine the amount of resources—say, food—needed by each individual and then determine the amount of available food in the wild. N would be determined from field censuses, and r would also come from field censuses of births and deaths per unit time. Let's plug some numbers into the equation to see how it works. If K = 1,000, N = 900, and r = 0.1, then

$$\frac{dN}{dt} = rN\frac{(1{,}000 - 900)}{1{,}000}$$
$$= 0.1 \times 900\frac{(100)}{1{,}000}$$
$$= 90 \times 0.1$$
$$= 9.$$

Population growth is thus 9 individuals per unit time. However, if K = 1,000, N = 500, and r = 0.1, then

$$\frac{dN}{dt} = rN\frac{(1{,}000 - 500)}{1{,}000}$$
$$= 0.1 \times 500(0.5)$$
$$= 25.$$

So population growth is greater when populations are lower than maximal and more resources remain. Of course, if the population (N) is smaller—say, only 100 individuals—population growth will also be smaller, in this case 9 individuals again, simply because the original population is small. When this type of growth is represented graphically, a sigmoidal, S-shaped curve results (Fig. 6.11). This type of population growth was later applied independently by Pearl and Reed (1920) to describe human population growth

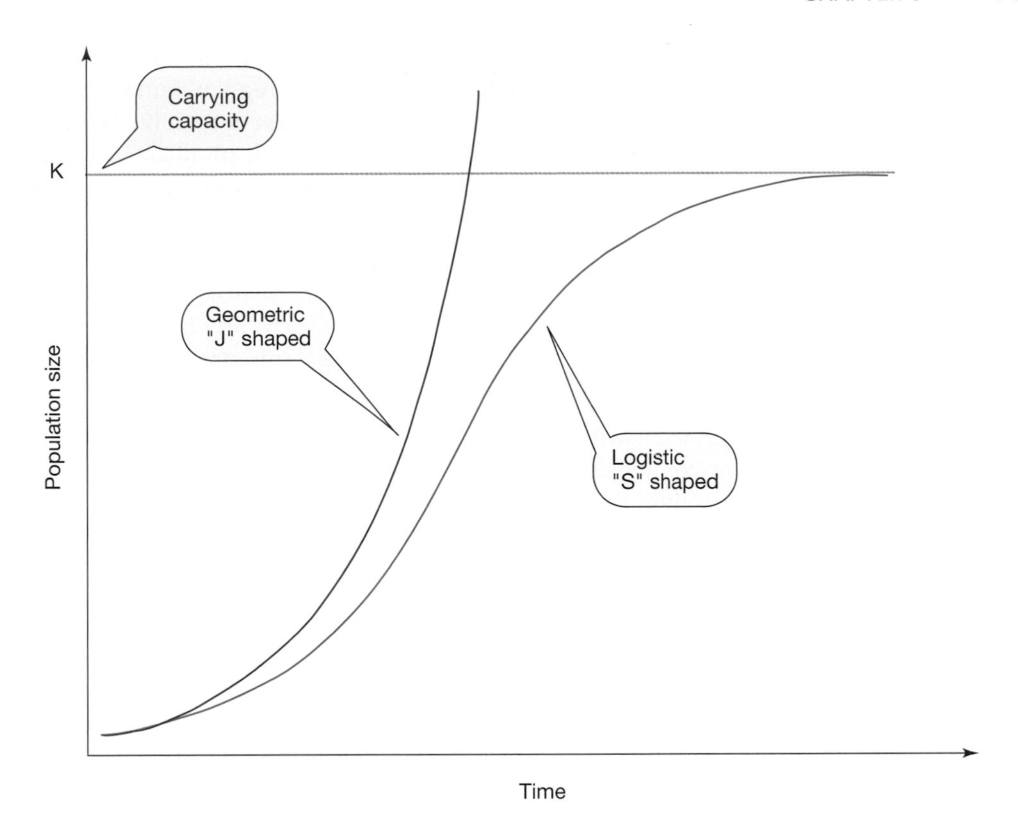

Figure 6.11 Geometric or exponential growth (*J*-shaped curve) in an unlimited environment and logistic growth (*S*-shaped curve) in a limited environment.

in the United States. Thus, it is also sometimes known as the *Verhulst–Pearl logistic equation*.

Field tests do not always validate the logistic model because logistic growth entails many important assumptions. The following are five of the most important:

1. The relation between a population's density and rate of increase is linear. That is, for every individual added to the population, the rate of population growth decreases slightly.
2. The effect of density on rate of increase is instantaneous.
3. The environment (and thus K) is constant.
4. All individuals reproduce equally.
5. There is no immigration or emigration.

For many laboratory cultures of small organisms, these assumptions are easily met. Thus, early tests of the logistic equation using laboratory cultures of yeast or bacteria that fed on sugar suggested that the equation was valid (Fig. 6.12). Pearl (1927),

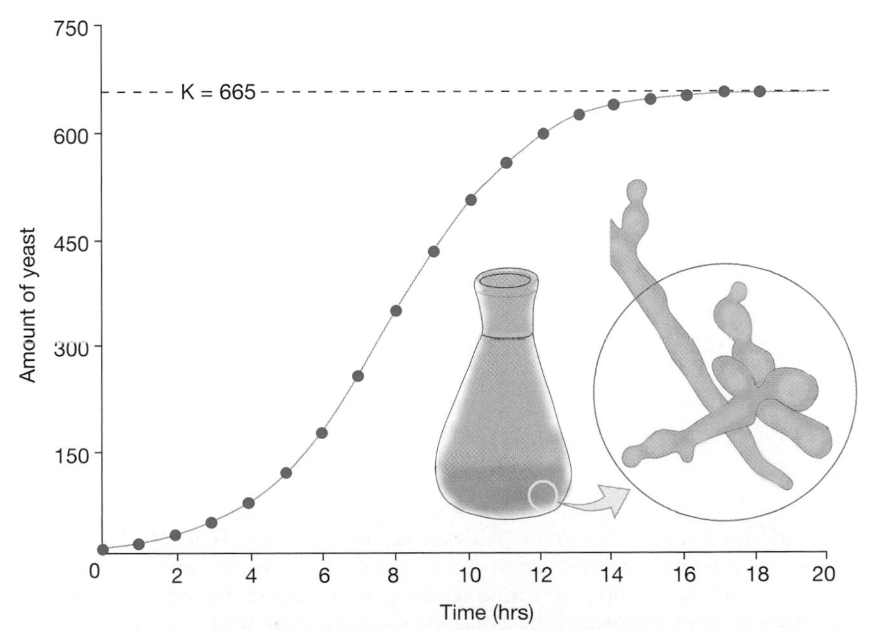

Figure 6.12 Early tests of the logistic growth curve were validated by the growth of yeast cells in laboratory cultures. These populations showed the typical *S*-shaped growth curve. *From data in Carlson (1913) reported in Pearl (1927).*

the champion of the logistic model, proclaimed it to be a universal law of population growth. Populations of slightly more complex organisms, such as *Drosophila* feeding on yeast, also fitted the logistic equation quite well. But here it was pointed out that it was not possible to maintain a uniform food level for the flies, because the yeasts themselves were growing and this was not controlled for (Sang, 1950). Thus, such experiments were not the watertight examples Pearl had proposed. Worse was to follow. Researchers now began to examine population growth using more complex species, like beetles, that fed on nonliving sources, like wheat flour. Initially, the population again seemed to fit the logistic equation. But when researchers let the experiment run longer, the size of the population seemed to drift down away from the asymptote over time or showed wild fluctuations (Fig. 6.13). For field populations of larger animals, the results were even less supportive, probably because it became very difficult to meet the five assumptions of the logistic equation. That is, contrary to those assumptions, in nature,

1. *Each* individual added to the population probably does not cause an incremental decrease in r.
2. There are often time lags, especially in species with complex life cycles. For example, in mammals, it may be months before pregnant females give birth, even when resources have been favorable for growth for months.
3. K may vary seasonally or with the climate.
4. Often, a few individuals command many matings.
5. There are few barriers preventing dispersal into or out of a population.

Thus, it is not surprising that there are few good examples of population data fitting the logistic equation. We will see how variation in just one of the components, time lags, influences whether or not, or how quickly, a population reaches an equilibrium.

The Australian physicist turned ecologist Robert May (1976) surprised ecologists by showing how variation in the parameters of simple deterministic population growth models could radically alter the behavior of population growth. He began by examining time lags. If there is a time lag of length τ between a change in the size of a population and its effect on the population growth rate, then the population growth at time t is controlled by its size at some time in the past, $t - \tau$. If we incorporate the time lag into the alternative form of logistic growth equation, viz.,

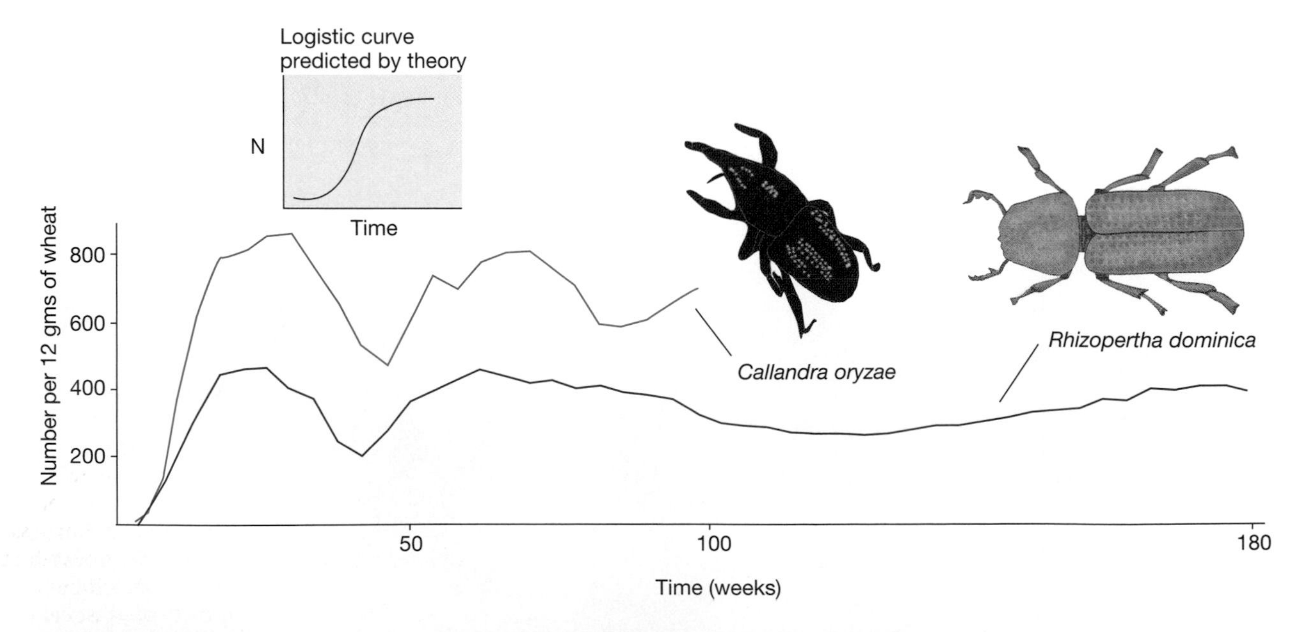

Figure 6.13 Population growth of the grain beetles *Rhizopertha dominica* and *Calandra oryzae* on a wheat medium for 3.5 years and 2 years, respectively. The generation time of these beetles is about nine weeks. The wheat medium was maintained at a constant 12 grams per container and was cleaned every two weeks by sifting. There is no tendency for either population to follow a logistic curve as theory dictates. Instead, each population oscillates quite wildly. *(Reproduced from Birch, 1953.)*

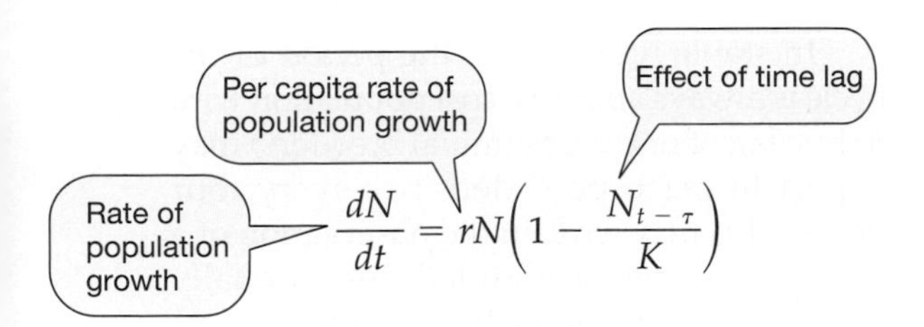

$$\frac{dN}{dt} = rN\left(1 - \frac{N_{t-\tau}}{K}\right)$$

then the time lag τ affects population growth. To see this, consider first an unlagged population, for which the alternative form of the logistic equation is

$$\frac{dN}{dt} = rN\left(1 - \frac{N_t}{K}\right).$$

Then if $r = 1.1$, $K = 1{,}000$, and $N = 900$, we have

$$\frac{dN}{dt} = 1.1 \times 900\left(1 - \frac{900}{1{,}000}\right) = 990 \times 0.1 = 99,$$

so the new population size is $900 + 99 = 999$.

If there is a time lag such that at the time the population is 900 the effects of crowding are being felt as though the population were only 800, then

$$\frac{dN}{dt} = 1.1 \times 900\left(1 - \frac{800}{1{,}000}\right) = 990 \times 0.2 = 198,$$

so the new population size is $900 + 198 = 1{,}098$. The effect of the time lag is to increase the population growth rate over that given by the ordinary logistic. Also, it is now possible for a population to overshoot its carrying capacity. Careful field observation is required to determine the time lag for a given population.

The effects of time lags on populations also depend on the response time of the population, which is inversely proportional to r, the per capita rate of population growth, so that populations with fast growth rates have short response times $1/r$. The ratio of the time lag (τ) to the response time ($1/r$), or $r\tau$, controls population growth. If $r\tau$ is small (less than 0.368), the population increases smoothly to the carrying capacity (Fig. 6.14). If $r\tau$ is large (>1.57), the population enters into a

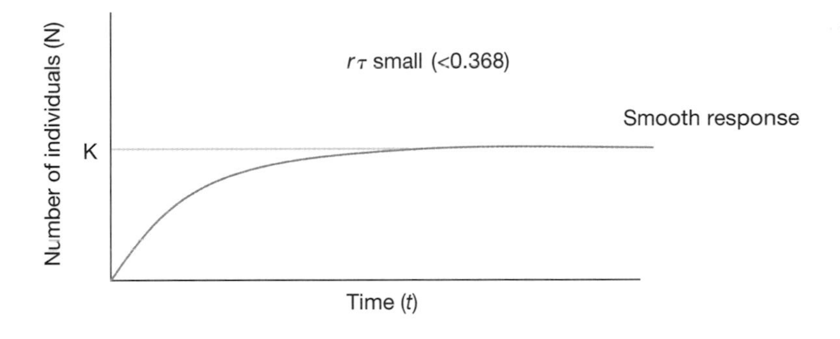

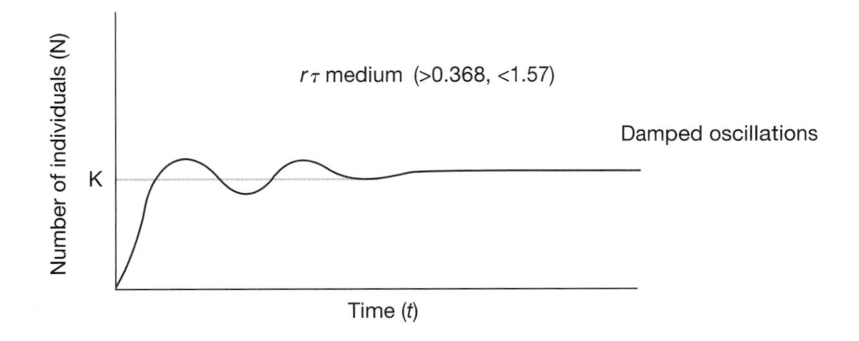

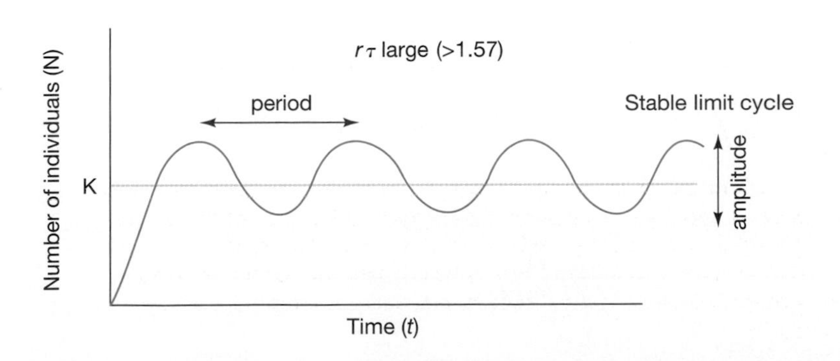

Figure 6.14 Logistic growth curves with a time lag for species with overlapping generations. Growth depends on the value of the product of the intrinsic rate of increase, r, and the time lag τ. In this figure, $r\tau$ increases from top to bottom. At a certain value of $r\tau$ called the *stable limit cycle*, population oscillations become stable and no longer converge on the carrying capacity K. (*After Gotelli, 1995.*)

stable oscillation called a *limit cycle*, rising and falling around K, but never settling on the equilibrium value. The values 0.36 and 1.57 are precise limits that are determined by the behavior of the population. We can think of a population that is just below the carrying capacity as having enough food for every individual to reproduce. After reproduction, however, few of the resulting juveniles will command enough resources to breed, and the population will fall below the carrying capacity again. In this scenario, territorial species might be less likely to undergo limit cycles, because territory holders will usually command enough resources to breed. On the other hand, invertebrate species that often scramble for resources may be more likely to undergo a limit cycle. When values of $r\tau$ are intermediate (between 0.368 and 1.57), the population undergoes oscillations that dampen with time until K is reached. It is perhaps surprising just how precise these mathematical cutoffs are.

In stable limit cycles, the period of the cycle is always about 4τ, so a population with a time lag of one year (annual breeders) may expect to reach peak densities every four years. This may explain the observation of a four-year cycle in high-latitude annually breeding mammals, such as lemmings.

For species with discrete generations, the logistic equation becomes

$$N_{t+1} = N_t + rN_t\left(1 - \frac{N_t}{K}\right).$$

Although we know that $r\tau$ controls population growth, with discrete generations the time lag is always 1.0; thus, the value of r alone controls the dynamics of the population. With discrete generations, if r is greater than 2.0, the population generally reaches K smoothly. At values of r between 2 and 2.449, the population enters a stable two-point limit cycle, with sharp "peaks" and "valleys" rather than smooth ones (Fig. 6.15). We

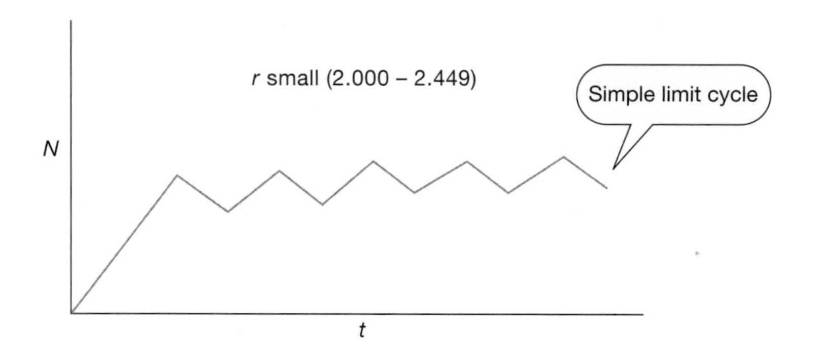

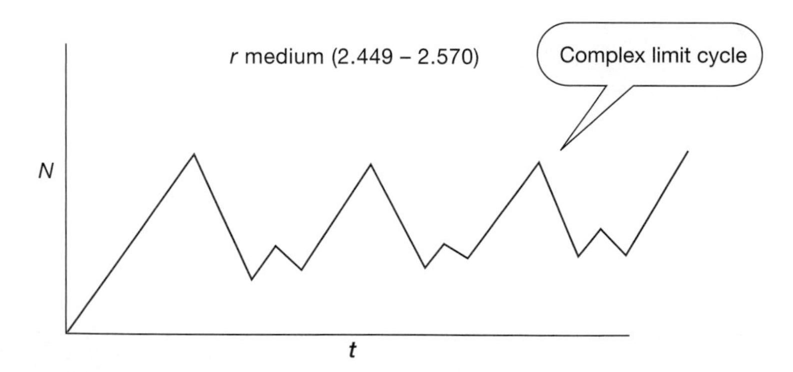

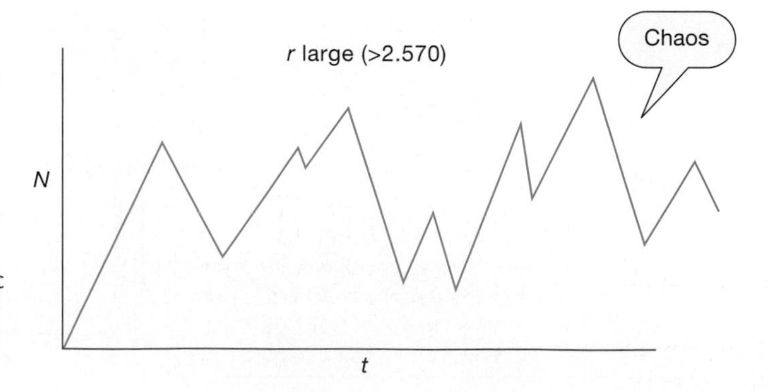

Figure 6.15 Limit cycles and chaos for species with discrete generations. The discrete logistic growth curve is affected by the size of r. Values of r increase from top to bottom in this figure, and the number of points in the limit cycle also increases. In the end, a chaotic pattern results that is difficult to distinguish from randomness. (*After Gotelli, 1995.*)

can imagine that with continuously breeding species there is more chance for these rough edges to be "smoothed" out, but with annual breeders there is less chance for this to happen. Between r values of 2.449 and 2.570, more complex limit cycles result. One illustrated in the figure has two distinct "peaks" and "valleys" before it starts to repeat. At values of r larger than 2.57, the limit cycles break down, and the population grows in complex, non-repeating patterns often known as "chaos." It is important to note that chaos doesn't mean random, although random population growth and chaotic growth certainly look similar. The point is that with chaos, the same chaotic population growth patterns would repeat every time the model is run, as long as the specified values of r and K are the same. With a "real" data set, it is sometimes hard to know whether a population is behaving in a chaotic or a random manner. Perhaps more importantly, it is easy to see how variation in just one of the assumptions of the logistic growth equation—time lags in the effect of population size on the rate of population increase—can completely disrupt a population from following a nice clear sigmoid, or S-shaped, growth curve. Changes in the other parameters that are assumed to be held constant in this equation (e.g., introducing dispersal, a mating structure, annual changes in K, and a nonlinear relationship of r to population density) add even more variation. We will not model these changes here, but you can imagine how each alone, and especially in tandem with variation in the other assumptions, can move observed population growth away from the logistic model.

6.4 Stochastic Models

It may be too simplistic to expect organisms in real life to behave exactly like numbers in an equation. Part of the reason is the genetic variability between individuals that causes some females to produce more offspring, on average, than others, or some animals to be more resistant to climatic stress, pressure from predators, and other factors. Part of the reason is environmental variation itself. Given this variability, termed *stochasticity*, can we ever hope to predict population processes accurately or to incorporate such variability into a model?

Stochastic models of population growth are based largely on probability theory. Rather than exactly two offspring, one might assume that each female has a 0.5 probability of giving birth to two offspring, a 0.25 chance of producing three progeny, and a 0.25 chance of producing one.

In a stochastic model, a coin can be flipped to mimic the probability of the outcome. Suppose a head and a tail (in either order) indicates two offspring, two tails imply one offspring, and two heads mean three offspring. Here are the results I got one night:

	Outcome of Trial			
Mother	1	2	3	4
	(number of offspring)			
1	2	3	3	2
2	3	1	1	1
3	3	1	2	2
4	1	1	3	3
5	3	1	1	1
Total population in next generation if parents die after reproduction	12	7	10	9

Some of the outcomes are above the expected value of 10, and some are below. If the tosses are continued for, say, 50 trials, a frequency histogram can be constructed (Fig. 6.16). This histogram gives the likelihood (proportion) of the observations yielding a certain final population size. The most likely outcome is a final size of 10, just as in the deterministic model. For the geometric deterministic model discussed in the previous section, if $R_0 = 2$ and $N = 5$, then

$$\begin{aligned} N_{t+1} &= R_0 N_t \\ &= 2 \times 5 = 10. \end{aligned}$$

Stochastic models of population growth incorporate the chance effects of genetic variability and extrinsic factors like climate on population dynamics.

However, with the stochastic model, other outcomes are possible; for example, all five mothers could have three offspring, for a total population of 15. This is the maximum possible number of offspring, and the likelihood of it happening is small—hence the low proportion of observations. Stochastic models can also be developed for continuous growth. Again, such a model is best explained by referring to the corresponding continuous equation, viz.,

$$\frac{dN}{dt} = rN = (b - d)\,N,$$

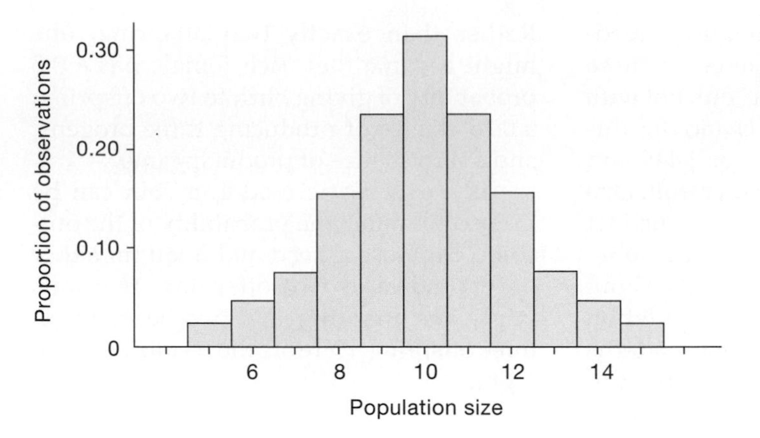

Figure 6.16 Stochastic frequency distribution for the size of a female population after one generation, beginning with five females. In this case, the probability of having two female offspring is 0.5, the probability of having three female offspring is 0.25, and the probability of having one female offspring is 0.25.

and if $b = 0.5$ and $d = 0$ (often true in new populations), then, integrating both sides, we obtain

$$\int \frac{dN}{dt} = \int rN,$$

$$\int \frac{dN}{N} = \int rdt,$$

$$ln\ N = rt$$

$$e^{lnN} = e^{rt}$$

and $N_{(t)} = N_0 e^{rt}$.
Letting $N_0 = 10$, we get

$$N_{(t)} = 10 \times e^{0.5}$$
$$= 10 \times 1.649 = 16.49,$$

so for one generation, $N_{(t)} = 16.49$.

With a stochastic model, the probability of one individual not reproducing in one time interval must be calculated as $e^{-b} = 0.6065$, and likewise, the probability of one individual reproducing once in one time interval is

$$1 - e^{-b} = 0.3935.$$

Then, for $N = 10$, the chance that no females will reproduce in one time interval is $0.6065^{10} = 0.0067$, or about 1 in 149. Similar calculations for other combinations, such as the probability that nine females will reproduce and one will not or that eight females will and two will not eventually produce a frequency histogram similar to Fig. 6.16. Similar histograms can be drawn for each generation. To get the path of population growth, we can stack these histograms on their ends (Fig. 6.17). The most likely trajectory is for the growth curve to follow the thickest part of the histogram, with the highest proportion of observations. But the actual path could connect any two bars in adjacent histograms. With stochastic models, population growth could vary dramatically, although there is still a most likely trajectory for it to follow.

If death occurs in the population, there is a chance that the population will become extinct (Pielou, 1969), according to the equation

$$\text{Probability of extinction} = \left(\frac{d}{b}\right)^{N_0}$$

Thus, if $b = 0.75$, $d = 0.25$, and $N_0 = 10$, then

$$\left(\frac{d}{b}\right)^{N_0} = 0.0000169,$$

whereas if birthrates are lower and death rates are higher (e.g., $b = 0.55$, $d = 0.45$), and if $N_0 = 10$, then the probability of extinction is 0.1344.

Thus, we conclude that, the larger the initial population and the greater the value of $b - d$, the more resistant to extinction the population becomes. In reality, $b - d$ is often zero, so $d/b = 1.0$ as time reaches millions of years and, in the limit, approaches infinity. In other words, the probability of extinction is unity, meaning that extinction is a certainty for a population that exists over a long enough time span and is likely to occur more quickly for a small population. Fischer, Simon, and Vincent (1969) maintained that probably 25 percent of the species of birds and mammals that have become extinct since 1600 may have died off "naturally," possibly because of stochastic variation due to a small population, and not because something killed the last remaining individuals. They also suggest that about 30% of birds and 15% of mammals currently are endangered because of such natural causes. These kinds of stochastic effects are particularly important when the conservation of small populations of rare species is at issue. For example, Schaffer and Samson (1985) have predicted that if the effective

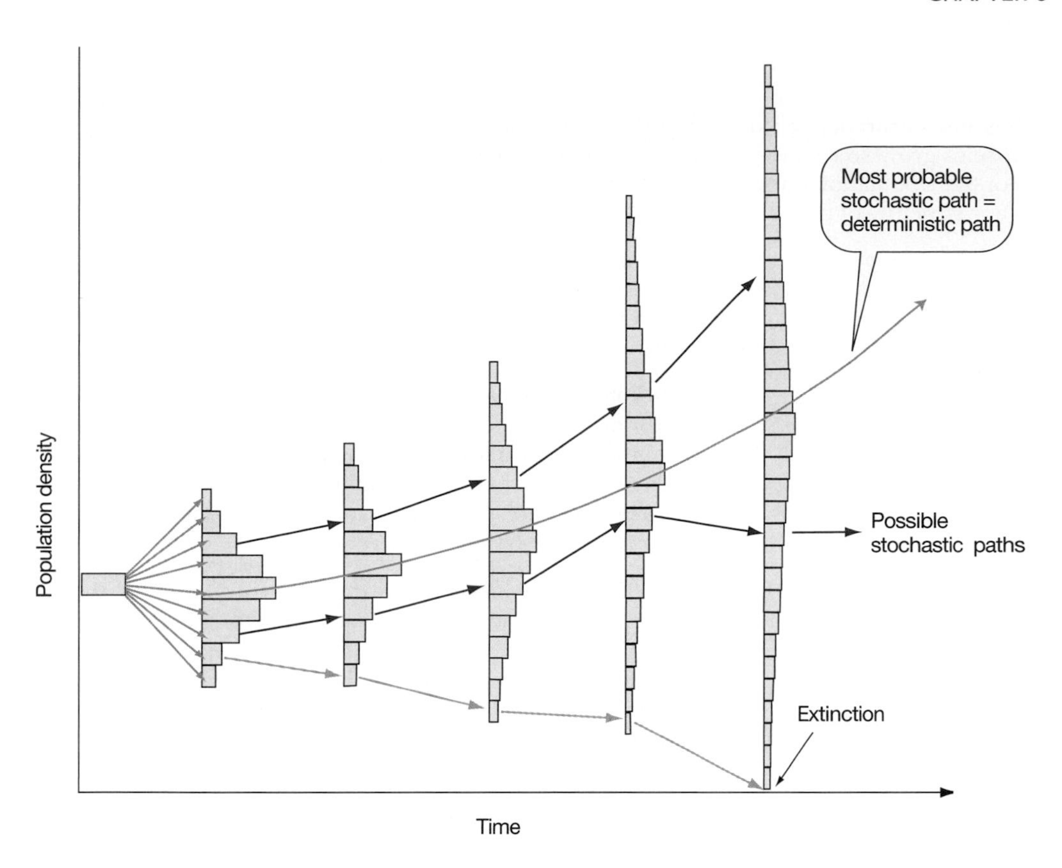

Figure 6.17 Stochastic model of geometric population growth. In effect, the figure is made up of frequency histograms, like those shown in Fig. 6.16, laid on their sides. With time, the range of possible population densities gets wider, and the probability of each population density gets smaller. Stochastic growth could occur along many paths, even one leading to extinction, but the most probable is equivalent to deterministic growth.

population size, or the number of individuals that mate within a population, is 50 for grizzly bears, demographic stochasticity alone would cause extinction once every 114 years, on average. A population model of the spotted owl (*Strix occidentalis*) suggests that demographic stochasticity is likely to extinguish local populations over the short term of decades (Simberloff, 1986b).

Stochastic models introduce biological variation into calculations of population growth and are much more likely to represent what is happening in the field. The price paid is complicated mathematics. Stochastic models become more salient as populations get smaller and so are important in examining conservation. For a population that is in the millions, as are many pest populations, deterministic models will do.

Although some readers may have by now despaired of ever producing a robust model of population growth that incorporates all the necessary factors, including time lags, variation in carrying capacity, and stochasticity, some closing generalizations can be made about population growth models. First, models provide simplified ideas about how nature operates because they are developed from basic principles about what should be happening in nature. Second, models help us figure out what things we should be measuring in nature and, perhaps, where we need to go. However, there are several dangers inherent in models. One is that they may lose their utility if they are made complex or specific. Another is that they are difficult to test if they are too complex. Finally, models that are overly simplistic cannot capture the reality of nature. To sum up, determining how populations grow is vital to applied ecology. Often, the first step is to use life tables to determine the age structure of a population. Knowing the age-specific fertility allows us to determine fecundity schedules and to calculate R_0, the net reproductive rate of the population.

Knowing the net reproductive rate allows us to construct models of how populations will grow. Where resources are not limiting, growth is *J* shaped, or geometric, but where resources are limiting, the population reaches an upper asymptote, or carrying capacity, and growth is *S* shaped.

Summary

An understanding of population growth is at the heart of applied ecology. We need to know how fast (or slow) populations of endangered species grow so that we can make adequate plans for their survival. We need to know how fast fish or tree populations grow so that we can sustainably harvest them. First, we need information on the age structure of the population:

1. Life tables provide information on how a population is structured by age. For longer lived organisms, time-specific life tables provide a snapshot of a population's age structure at a given time. Such tables often contain information on the number of organisms dying at any given age interval, the proportion surviving, and the mean expectation of life. Data in most columns can be calculated from any other column.

2. For organisms with shorter life spans—especially those completed within a year—age-specific life tables, which follow one cohort or one generation, are used.

3. Three types of survivorship curve can be recognized: Type I, wherein most individuals die after they have reached sexual maturity (exemplified by mammals); Type II, in which there is an almost linear rate of loss (exemplified by some plants and birds); and Type III, according to which most individuals are lost as juveniles (exemplified by invertebrates, fish, frogs, and plants).

4. With information on survivorship and age-specific fecundity, the net reproductive rate R_0 can be calculated and population growth determined.

5. Population growth may be examined with the use of deterministic models or stochastic models. In deterministic models, rates of population increase are given precisely determined values. In stochastic models, variation around a value is allowed, as would be the case in nature. For small populations, stochastic events are important, and stochastic models, although more difficult to develop than deterministic ones, are therefore often used in conservation theory. For large populations, deterministic models give good approximations of population growth.

6. In both deterministic and stochastic population growth, when environments are favorable and resources are not limiting, populations may undergo geometric or exponential growth, which results in *J*-shaped population growth curves. For species with one generation per year, $N_{t+1} = R_0 N_t$, and for species with overlapping generations, $dN/dt = rN$.

7. In the real world, population growth often has limits—perhaps the exhaustion of space or resources or a limitation on growth from the accumulation of waste products. As a result, most population growth curves are logistic and *S* shaped, with an upper level, or carrying capacity *K*, representing the total number of individuals or the total biomass that can exist in an area. Logistic growth is described by the equation $dN/dt = rN(K - N)/K$.

8. Population growth curves do not always have a smooth trajectory. Instead, they may oscillate about the carrying capacity or exhibit what appear to be random fluctuations. Such fluctuations may be the result of time lags and may in turn result in chaotic growth curves—curves that look as if they are generated by random fluctuations, but are actually the result of precise time lags and precise rates of reproduction.

In sum, population models are useful because they tell us how populations grow. Deterministic models usually suffice for populations in the thousands or millions, like pest insects or fish. Stochastic models are more useful in conservation ecology, which studies the more variable growth of small populations.

Discussion Questions

1. What are the most common limits on the population growth of organisms—weather, resources, pollutants, natural enemies, disease, or social factors? Pick some local plants and animals, and make some educated guesses.

2. What do you think will limit human population growth? Do you believe that the carrying capacity of the earth for humans has still to be reached, or have we overshot it? (Remember, time lags can cause the carrying capacity to be exceeded.) If fossil fuels are used up, could the human population fall to very low levels?

3. Compare the reproductive output of a family line whose mothers always have twin girls at age 20 versus a family line whose mothers have triplet girls at age 33. What conclusions can you draw? (*Hint*: Compare the two family lines' reproductive output after 100 and 200 years.) Should all couples stop reproducing after having two children?

4. In what circumstances might a population be expected to grow in a geometric fashion and in what circumstances in a logistic fashion?

5. Visit the oldest cemetery in your area, and construct an age-specific life table for some well-represented cohorts. Can you detect differences among different cohorts? If you visited a cemetery with a different ethnic mix, would you detect any differences? Now use the obituary column of a local newspaper to construct a time-specific life table. Do you see any differences between the time-specific and age-specific data? Can you get data from other countries via other newspapers in your library?

CHAPTER 7

Physical Environment

Road Map

1. Physical variables commonly limit the abundance of plants and animals. These variables may be divided into those which plants and animals use as resources, such as soil nutrients, CO_2, and water, and those which they cannot use, but nonetheless are critical to their survival, such as temperature, wind, pH, and salt concentrations.
2. Species are commonly limited by temperature extremes, such as high heat or freezing conditions. Wind modifies temperatures (ergo the wind chill index) and is a critical factor in the mortality of trees. Salt concentrations are important in aquatic and terrestrial habitats, and pH can also affect environmental suitability, particularly in aquatic habitats. Global warming is caused mainly by increased atmospheric CO_2 levels from the burning of fossil fuels. The resultant changes in global temperatures could alter the distribution patterns of plants and animals.
3. The physical environment commonly limits population abundance of plants and animals, as well as their geographic distributions.
4. The physical environment can change the composition of species that occur in an area (the community).

The local distribution patterns and abundance of many species are limited by certain physical, or abiotic, factors of the environment, such as temperature, moisture, light, pH, salinity, and water currents. Justus Liebig's "law of the minimum," coined by him in 1842, states that the distribution of a species will be controlled by that environmental factor for which the organism has the narrowest range of tolerance.

Some species are tolerant of a wide range of environmental conditions, others of only a narrow range, but each species usually functions best only over a limited part of the gradient, termed the species' *optimal range* (Fig. 7.1). It must also be remembered that part of a preferred optimal range may already be occupied by a competitively superior species: In the field, species may not occupy their full ranges, as measured in the laboratory in terms of abiotic factors, because of competition with other organisms (Fig. 7.2). (See also Chapter 9.)

The most common abiotic factors that limit the distribution of terrestrial organisms are temperature and moisture.

7.1 Physical Variables

Temperature.

Environmental temperature is an important factor in the distribution of organisms, because of its effect on biological processes and because of the inability of most organisms to regulate their body temperature precisely. For instance, many organisms, such as mollusks and organisms that form coral, secrete a calcium carbonate shell. Shell formation and coral deposition are greatly accelerated at high temperatures and are suppressed in cold water. It is therefore not surprising that coral reefs are most abundant in warmer waters and show a good

Low temperatures impose significant ecological constraints by limiting the amount of time in which physiological activity is possible.

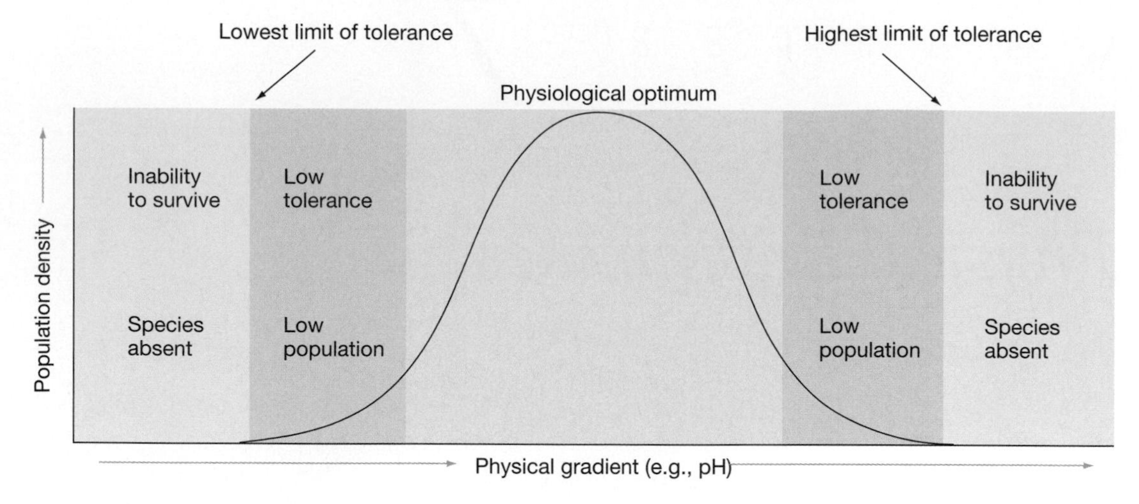

Figure 7.1 Organismal distribution along a physical gradient, such as pH.

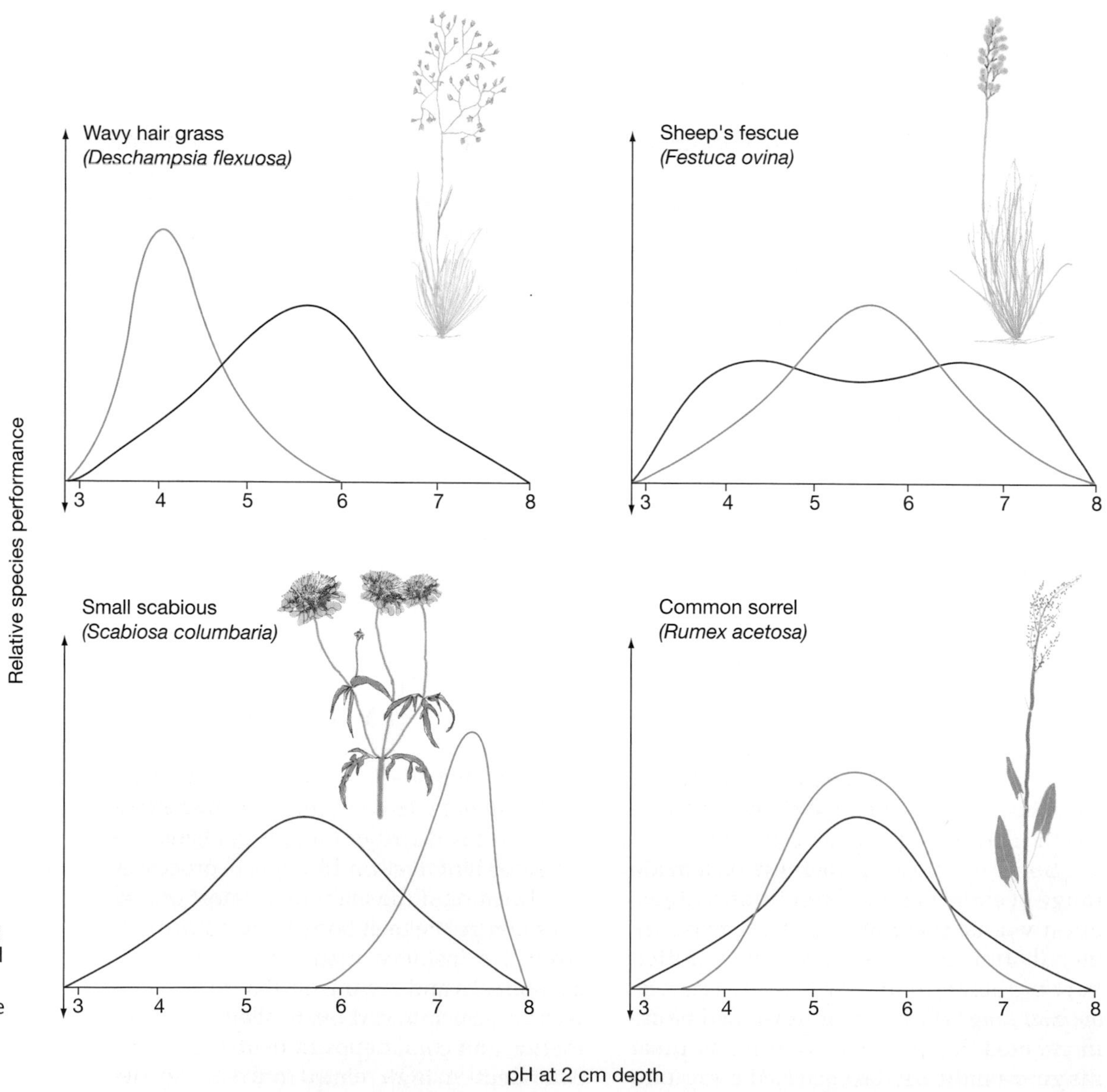

Figure 7.2 The difference between the distributions of four plant species growing in the field (ecological optimum curve) and under noncompetitive conditions in controlled laboratory plots (physiological optimum curve). *(Reproduced from Collinson, 1977.)*

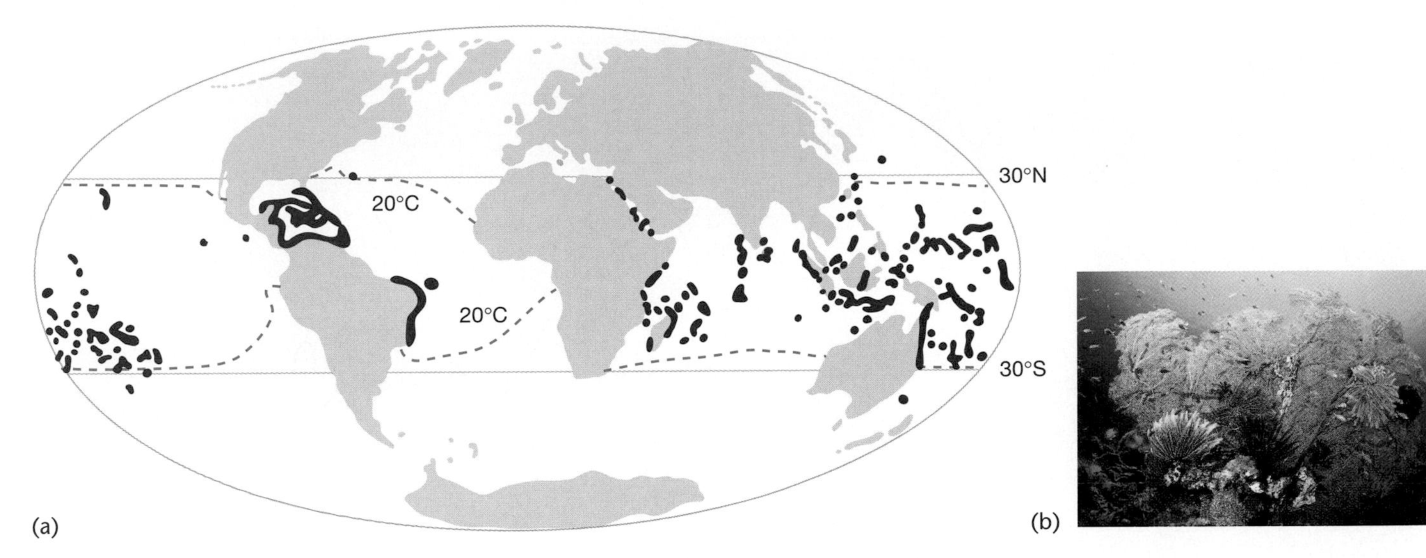

Figure 7.3 (a) The world distribution of coral reefs closely matches the 20°C isotherm for the coldest month of the year (dashed line). (*Reproduced from Clark, 1954.*) (b) Coral reef from the Pacific Ocean.

correspondence with the 20°C isotherm for the coldest month of the year (Fig. 7.3).

Few organisms can maintain a sufficiently high metabolic rate below freezing temperatures. Some organisms, of course, have extraordinary adaptations to enable them to live in very cold temperatures. Penguins can survive the coldest regions. Mammals and birds, so-called warm-blooded organisms, or endotherms, that generate heat within their own bodies, can cope with temperature extremes better than many organisms, because they can regulate their own temperature. But even mammals and birds function best within certain environmental temperature ranges. Ectotherms (cold-blooded organisms), by contrast, rely on external sources of heat and include other animals as well as plants.

In plants, temperature is particularly important because cells may rupture if the water they contain freezes. Under colder conditions, water must be bound up in a chemical form such that it cannot change to damaging ice. Usually, the water within the cells does not freeze, but the extracellular water does. This draws water out from the cell, damaging its osmoregulatory machinery in much the same way as drought or salinity does. Injury by frost is probably the single most important factor limiting plant distribution. As an example, the saguaro cactus can easily withstand frost for one night, as long as it thaws in the day, but it is killed if temperatures remain below freezing for 36 hours. In Arizona, the limit of the cactus's distribution corresponds to a line joining places where the temperature occasionally remains below freezing for more than one night. Another feature of freezing that kills plants is that air bubbles formed when xylem freezes can cause cavities in the xylem system and disrupt water transport in the plant. Although cavities are refilled in temperate, woody plants, freezing can prove fatal in some shrubs that are native to warmer areas. For example, minimum temperatures between −16°C and −20°C eliminate conductance of water in the evergreen desert shrub *Larrea tridentata* in the U.S. southwest, so the distribution of this shrub is limited to areas that usually do not get that cold (Fig. 7.4). Similarly, many tropical plants are sensitive to "chilling" and may even be damaged by only slightly cooler temperatures, even above freezing. The exact mechanism is unknown, but may be related to the breakdown of membrane permeability and the leakage of ions such as calcium (Minorsky, 1985).

The range limits of even warm-blooded animals may correlate with temperature, too. For example, the eastern phoebe has a northern winter range limit that coincides well with the −4°C isotherm of average minimum January temperatures (Root, 1988). Such limits may be related to the energy demands associated with cold temperatures. Colder temperatures mean higher metabolic costs that are in turn dependent on high

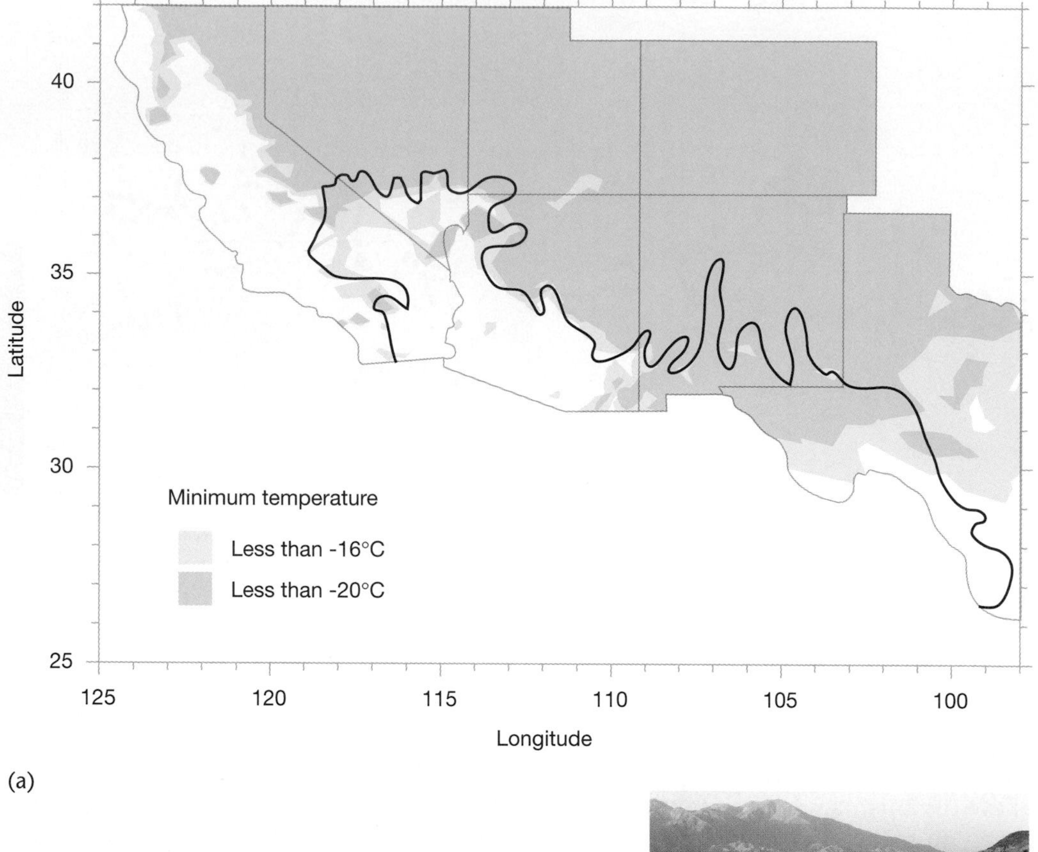

Figure 7.4 (a) The distribution of the evergreen shrub *Larrea tridentata* in the desert southwest corresponds to minimum temperatures between −16°C and −20°C because, between these temperatures, xylem freezes and disrupts water transport. *(Reproduced from Pockman and Sperry, 1997.)* (b) *Larrea tridentata* is common in the desert southwest, such as here in Death Valley National Park, California.

feeding rates. A limit appears at which the small eastern phoebe cannot feed fast enough to keep warm at such low temperatures. The distribution of the vampire bat (*Desmodus rotundus*) is from central Mexico to northern Argentina. The northern limit of its range in Mexico parallels the 10°C minimal isotherm of January because of the bat's poor capacity for thermal regulation below that temperature (Fig. 7.5). Thus, theoretically, vampire bats could live in Florida from Tampa southward, because it is warm enough. However, their natural range is confined to continental Central and South America—unless, of course, someone introduces the animal to south Florida.

We should bear in mind that most often it is not the mean average temperature that limits species, but the frequency of occasional extremes—like freezes for the saguaro cactus. Agriculturalists know this only too well: It is the frequency and strength of the periodic freezes that limits the northern distribution of oranges in Florida and the southern distribution of coffee in Brazil, not average temperatures for the coldest months. Often, experimentally moving organisms outside their normal range is a useful way to establish what controls the natural limits. Unfortunately, physical extremes, such as severe freezes or floods, may be apparent only in isolated years, so we may have to wait many years to get useful information on whether extremes are limiting in the field.

Despite the correlation between the distribution of some species and temperature, we need to be cautious about correlating species distributions with temperature maps. The temperatures measured for constructing maps are not always those which the organisms experience. In nature, an organism may choose to lie in the sun or hide in the shade and, even in a single day, may experience a baking midday sun and a freez-

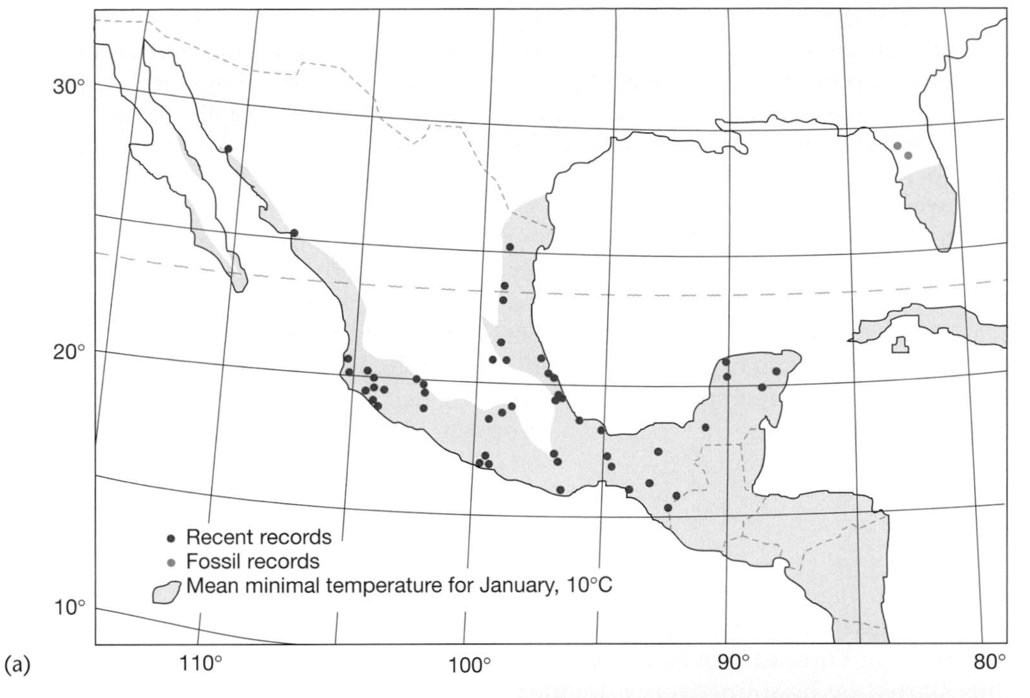

Figure 7.5 (a) The northern distribution of the vampire bat *Desmodus rotundus* in Mexico is limited to areas warmer than the 10°C January isotherm because of the animal's poor capacity to thermoregulate. Its southern limit in Argentina and Chile is limited by the same isotherm. *(After McNab, 1973.)* In theory, the bat should be able to exist in extreme southern Texas or in south Florida if introduced there. (Indeed, fossil records of *D. rotundus* exist in Florida. (b) The vampire bat. *(Zig Leszczynski.)*

ing night. Moreover, temperature varies from place to place on a far finer scale than will usually concern a geographer, but such local conditions will be important for a particular species. The winter sun, shining on a cold day, can heat the south-facing side of a tree (and the habitable cracks and crevices within it) to as high as 30°C, and the air temperature in a patch of vegetation can vary by 10°C over a vertical distance of 2.6 m from the soil surface to the top of the canopy.

These local variations in temperature create small-scale microhabitats that may be very important to an organism's distribution. For example, the rufous grasshopper (*Gomphocerripus rufus*) is distributed widely in Europe, but reaches its northern limit in Great Britain only 150 km from the south coast, where it is restricted to steep south-facing, and therefore relatively sun-drenched and warm, grassy slopes. For some organisms, such as insects, there is a linear relationship between temperature and development. If the temperature threshold for the development of a grasshopper is 16°C, it might develop in, say, 17.5 days at 20°C or 5 days at 30°C. The absolute length of time is not important; rather, a combination of time and temperature, called *degree-days*, determines development. Thus, 17.5 days × 4 degrees above the threshold = 70 degree-days, and 5 days × 14 degrees = 70 degree-days also.

High temperatures.

High temperatures impose ecological constraints as severe as low temperatures do on organisms.

Relatively few species can survive temperatures more than a few degrees above their metabolic optimum. High temperature is critical because the proteins of most organisms denature (i.e., are destroyed) at temperatures above 45°C. Organisms effectively cool themselves through water loss by evaporation, but they soon dry out if an abundant supply of water is not available.

Figure 7.6 Thermophilic giant tubeworms growing at 8,000 feet depth around deep sea vents in the Galápagos rift.

Some of the life history stages of an organism that are most resistant to heat are the resting spores of fungi, cysts of nematodes, and seeds of plants. Dry wheat grains can withstand temperatures of 90°C for short periods of up to 10 minutes. Natural hot springs are home to *Thermus aquaticus*, which grows at temperatures of 67°C, and some thermophilic organisms collected from deep-sea vents (Fig. 7.6) have been cultured at a temperature of 100°C, much higher than that originally thought to place limits on life. However, relatively few species can survive temperatures more than a few degrees above their metabolic optimum.

Many organisms are even adapted to withstand the high temperatures of fires.

Fires.

The ultimate high temperatures that organisms face are brought about by fire. Before the arrival of Europeans in North America, fires started by lightning were a frequent and regular occurrence, for example, in the pine forests of what is now the southeastern United States. Because they were so frequent, these fires consumed leaf litter, dead twigs and branches, and undergrowth before they accumulated in great quantities. As a result, no single fire burned hot enough or long enough in one place to damage large trees—each fire swept by quickly and at a relatively low temperature. Over time, the dominant plant species of these areas came to depend both directly and indirectly on frequent, low-intensity fires for their existence. The jack pine, *Pinus banksiana*, and the longleaf pine, *P. palustris*, have *serotinous* cones, meaning that they remain sealed by resin until the heat of a fire melts them open and releases their seeds. Therefore, these trees depend directly on fire for their reproductive success. Much of the rest of the fire-adapted vegetation would be supplanted by other species if fires did not suppress those species periodically. For example, in the midwestern United States, deciduous forest would take over prairies in the absence of periodic fires (Collins and Wallace, 1990).

Global warming may change the distribution and abundance of species in the future.

Management practices that attempt to maintain forests in their natural state by preventing forest fires often have exactly the opposite result. First, trees like the jack pine simply stop reproducing in the absence of fire. Second, species like the longleaf pine and wire grass, which depend on fire to suppress their competitors, are soon replaced by species characteristic of other communities. Finally, when a fire does occur, fuel has had a much longer period in which to accumulate on the forest floor, and the result is an inferno—a fire so large and so hot that it consumes seeds, seedlings, and adult trees, native and competitor alike (Fig. 7.7).

Management practices that prevent fire arise from the mistaken assumption that forests evolved in the virtual absence of fire. Lightning-caused fires are particularly frequent in the southeastern United States, but even in other areas many more fires arise naturally than most people believe. For example, in the western half of the United States, nearly half of the yearly average of over 10 thousand fires is thought to be started by lightning (Brown and Davis, 1973).

Because so many species are limited in their distribution patterns by fires or by global temperatures in general, scientists are concerned that if global temperatures change, then some species will be driven to extinction or their geographic range will change and the geographic location of many centers of agriculture and forestry will be altered. This concern is manifested in the fear of global warming.

Global warming.

The scientific debate on global warming centers on two issues: How fast is global warming occurring and how much do human-made "greenhouse" gases contribute to such warming? Political and economic debate centers upon what economic tools are available to control the anthropogenic emissions of greenhouse gases and how vigorously they should be applied.

Increased global warming is caused by something known as the *greenhouse effect*, brought about by the ability of the atmosphere to be selective in its response to different types of radiation. The atmosphere readily transmits solar radiation, which is shortwave, allowing about 50 percent of it to pass through unaltered to heat the Earth's surface. The energy absorbed by the Earth is radiated back into the atmosphere, but this terrestrial radiation is long-wave infrared, and instead of being transmitted through the atmosphere, much of it is absorbed by clouds, causing the temperature of the atmosphere to rise. A large amount of the energy absorbed in the atmosphere is then returned to the Earth's surface, causing its temperature to rise also. This mechanism is considered to operate in a manner similar to the way in which a greenhouse works, allowing sunlight in, but trapping the resulting heat inside—hence, the use of the term "greenhouse effect."

(a)

(b)

Figure 7.7 (a) When fires burn in a natural cycle, the leaf litter does not have much time to accumulate, and the fire burns with a moderate heat. *(Photo by Florida Park Service.)* (b) When fires are suppressed, much litter accumulates, and any fires that do ignite, say, from a lightning strike quickly get out of control and burn high in the forest canopy, killing mature trees. *(Photo by U.S. Department of Agriculture.)*

Without some type of greenhouse effect, global temperatures would be much lower than they are, perhaps averaging only −17°C, compared to the existing average of +15°C. The greenhouse effect helps explain both the hot temperatures of Venus, which is blanketed in CO_2, and the frigid conditions on Mars, which has very little atmosphere.

The greenhouse effect is an important characteristic of the atmosphere, yet it is made possible by a group of gases, including water vapor, that together make up less than 1 percent of the total volume of the atmosphere. There are about 20 of these greenhouse gases. After water vapor, the 4 most important are carbon dioxide, methane, chlorofluorocarbons, and nitrogen oxide (Table 7.1). All of these gases have increased in atmospheric concentration since industrial times (Fig. 7.8), and the rate of increase is accelerating. The most important of all the greenhouse gases is carbon dioxide, which has a lower global warming potential per unit of gas (relative absorption) than any of the others, but is far more common in the atmosphere.

The amount of greenhouse gases in the atmosphere is greatly influenced by natural

TABLE 7.1 Atmospheric concentrations, relative absorption, and sources for the major greenhouse gases and their contribution to global warming. Relative absorption is the warming potential per unit of gas.

	CO_2 Carbon Dioxide	CH_4 Methane	N_2O Nitrous Oxide	CFCs Chlorofluorocarbons
Radiative absorption per ppm of increase	1	32	150	>10,000
Atmospheric concentration in 1991	335 ppmv[1]	1.72 ppbv[2]	310 ppbv	0.28–0.48 ppbv
Contribution (%) to global warming	50	19	4	15
% from natural sources	20–30	70–90	90–100	0
Major anthropogenic sources	Fossil fuel use, 75%; deforestation and shifting cultivation, 25%.	Rice paddies, 17%; ruminants, 22%; landfill sites, 8%; biomass burning, 11%; coal mining + gas exploitation, 28%; animal wastes, 7%; sewage treatment, 7%.	Cultivated soils, 52%; fossil-fuel use, 25%; mobile sources, 11%; industrial sources, 12%.	Industrially manufactured (as aerosol propellants, foam insulators, etc.)

Notes:
[1]ppmv = parts per million by volume.
[2]ppbv = parts per billion by volume.

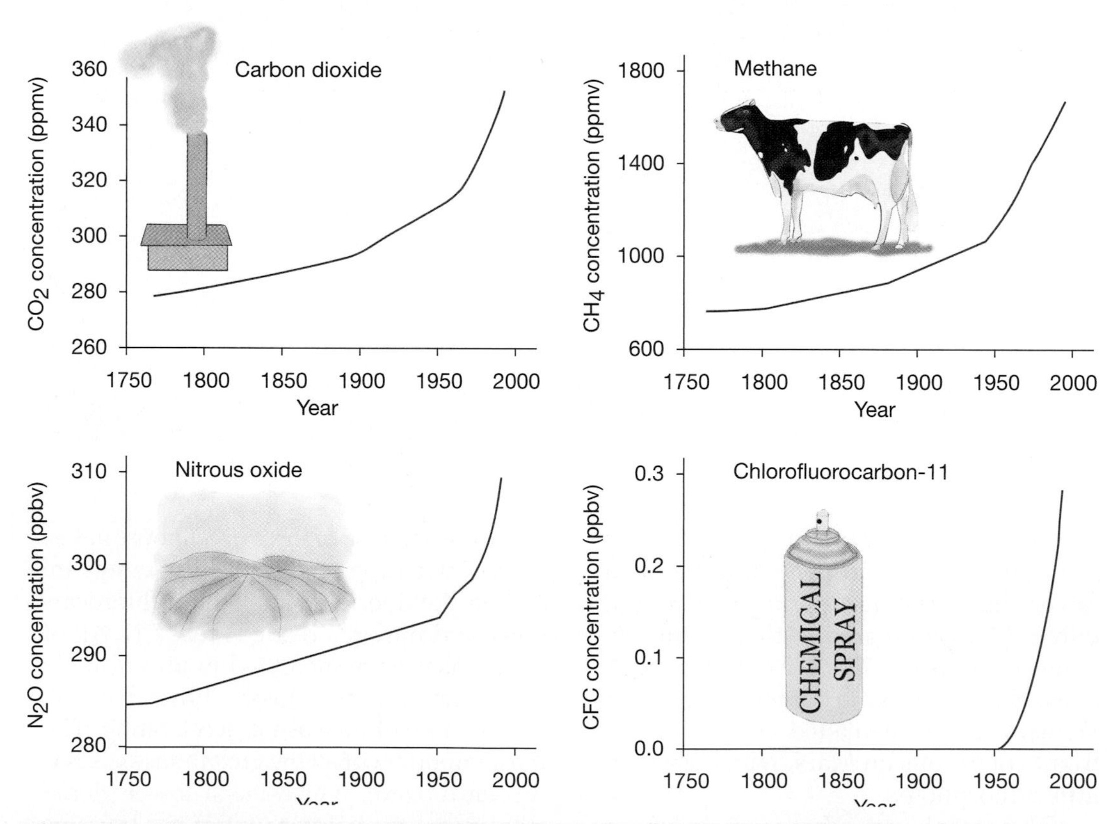

Figure 7.8 Changes since the middle of the 18th century in the atmospheric concentration of carbon dioxide, methane, nitrous oxide, and chlorofluorocarbons (CFCs). Over the past few decades, there has been a very large increase in the atmospheric concentration of CFCs, which did not exist in the atmosphere before the 1930s.

causes. For instance, nearly two-thirds of the nitrous oxide in the atmosphere comes from natural soils and the oceans, and one-third of the methane comes from bogs, swamps, and the action of termites. Volcanoes can emit huge quantities of dust and carbon, as well as sulfur, into the atmosphere. However, the steady increase in greenhouse gases over the past 200 years cannot easily be explained by natural causes.

There is little doubt that greenhouse gases are influenced by human activities. Paramount among these is the mining and burning of fossil fuels, which accounts for 75 percent of the increase in CO_2 emissions, about 39 percent of methane output, and 36 percent of nitrous oxide emissions—close to half of all the greenhouse gas emissions. The alteration of land use via deforestation, rice paddies, domestic animals, and agricultural soils accounts for about another 25 percent of the atmospheric increase in CO_2. Thus, anthropogenic effects cause nearly 75 percent of the increase in greenhouse gases, and natural effects cause only 25 percent.

Because of the huge social upheaval that would ensue if the use of fossil fuels and the production of cement were decreased, it has not yet been possible to reduce CO_2 output. Chlorofluorocarbon (CFC) production, however, has been cut dramatically, largely because of another problem: CFCs lower the concentration of atmospheric ozone. Replacing CFCs with other chemicals has not caused undue economic hardship. Replacing fossil fuels, by contrast, is a very sticky issue: Politicians are loath to do so unless the evidence is strong and the economic damage to their home states can be minimized.

What is the evidence for global warming? Temperature readings have been available for a large number of locations since the middle of the 19th century. From these readings, it is possible to construct records of the Earth's mean global surface temperature. Typically, the records show an irregular global warming amounting to about 0.5°C since 1880 (Fig. 7.9). But there are several problems with the records. For example, although the number of recording stations may be large, their geographic distribution is not truly global. The majority of stations lie in populated areas of the Northern Hemisphere, particularly in Europe and North America. Some stations may have experienced substantial warming due to changes in land use and population density—the "urban-heat-island effect." This term refers to the fact that temperatures within cities are generally higher than those in rural areas. The average urban temperature can be higher than the average surrounding

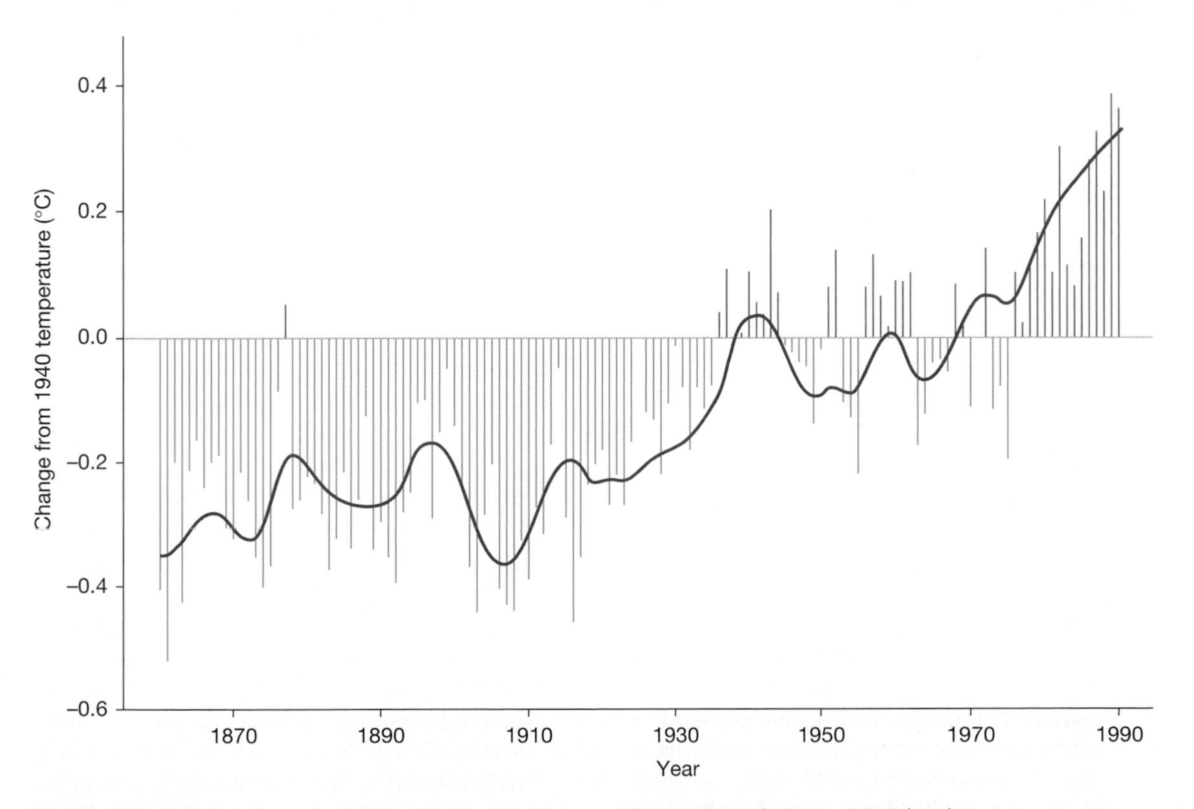

Figure 7.9 Estimated mean global surface temperature, 1860–1990, relative to 1940 levels.

rural temperature by 1 to 2°C, and summer rainfall also increases. Despite these problems, some global warming has definitely occurred. We need to know how much warming will occur in the future.

Computer models and predictions.

It is difficult to design accurate computer models of global warming which include all the mechanisms that influence the degree of warming. There are simply too many variables to incorporate into such models (Kerr, 1997; Trenberth, 1997). For instance, higher temperatures associated with an intensified greenhouse effect would bring about more evaporation from the Earth's surface, which could lead to increased cloudiness as the rising water vapor condensed. The clouds in turn would reduce the amount of solar radiation reaching the surface and therefore cause a temperature reduction, which might moderate the increase caused by the greenhouse effect. This is a negative-feedback mechanism that could prevent global warming from escalating. There are also positive-feedback mechanisms that augment global warming. The colder northern waters of the world's oceans, for example, act as an important sink for CO_2, but their ability to absorb gas decreases as temperatures rise. Global warming would reduce the ability of the oceans to act as a sink. Instead of being absorbed by the oceans, CO_2 would remain in the atmosphere, thereby adding to the greenhouse effect. Another positive-feedback mechanism is the melting of permafrost, resulting in more bogs and swamps and the production of more methane. Neither positive- nor negative-feedback mechanisms have adequately been dealt with in most models of global warming. As a result, the models tend to show more warming than has actually occurred.

The April 1996 report from the U.N.'s Intergovernmental Panel on Climate Change (IPCC), its first full report on global warming over the previous five years, was finally able to suggest a reason for the perplexing lack of fit of the models to the data: Although the chief influence on climate was the buildup of CO_2 and other greenhouse gases in the atmosphere, in many parts of the world, including much of Europe and North America, warming was masked by another form of pollution from burning fossil fuel—aerosols of sulfates and soot, which form a thin haze that reduces solar heating. This finding emphasizes the difficulty of modeling a complex interacting system such as global climate change.

The environmental impact of global warming.

It is clear that a knowledge of the speed and extent of global warming is critical. To predict the effect of global warming, most scientists focus on the year 2100, when the concentrations of CO_2 will have doubled—that is, CO_2 concentrations will have reached about 700 parts per million (ppm), compared to the late-20th-century level of 350 ppm. At that time, it is argued, global temperatures will be about 1 to 3.5°C (about 6°F) warmer than at present and will increase a further 0.3°C each decade. This increase in heat might not seem like much, but it is comparable to the warming that ended the last ice age! Much depends on the rate of greenhouse gas emissions. If the rate accelerates, the warming could be severe; if it slows down, the warming will not be so catastrophic.

If we accept the typical scenario of gradual global warming, we can ask, "What are the environmental consequences on natural and human-made ecosystems"?

Natural ecosystems.

The main drawback associated with global warming is the profound changes that will be produced in communities of organisms other than humans. Some authors point out that many species can adapt to slight changes in their environment. However, evolution is usually assumed to be irrelevant in a global-warming scenario, because it is thought that most species cannot evolve significantly or rapidly enough to counter the ensuing changes in climate. The anticipated changes in global climate are expected to occur at a rate that will probably be too rapid to be tracked by evolutionary processes such as natural selection. It is also not likely that all species can simply disperse and move north or south into the newly created climatic regions that will be suitable for them.

Many tree species take hundreds or even thousands of years to disperse substantially via their progeny. The tree with one of the fastest-known dispersal rates is spruce, which expanded its range an average of only 20 km every 10 years, beginning about 9 thousand years ago. Because plants can "move" only slowly, the resultant changes in floras will have dramatic effects on animals, both on the large biogeographic scale and on the local community scale. Herbivores cannot disperse without the foliage that they feed on being present first. As an example, paleobotanist Margaret Davis of the University of Minnesota showed that in the event of a doubling of CO_2 levels, sugar maple, presently distributed throughout all of the eastern United States and southeastern Canada, would die back in all areas except northern Maine, northern New Brunswick, and southern Quebec (Fig. 7.10). Of course, this change in the tree's distribution could be offset by the creation of new favorable habitats in central Quebec. However, most scientists believe that the climatic zones would shift toward the poles faster than trees could migrate via progeny; hence, extinctions would be increased.

Even more damaging changes could occur on some isolated islands where migration in a northerly or southerly direction is not possible, thus dooming some species to certain extinction. This has led some people into a debate over whether humans need to remove some species from the field and hold them in captive-breeding programs for reintroduction later when the climate reverts to "normal." Alternatively, wild individuals may be translocated into potentially more hospitable environments. Some changes in species distributions may already have occurred. Camille Parmesson and colleagues (1999) examined the current and historical ranges of 35 non-migrating European butterflies. Sixty-three percent had ranges that shifted to the north by 35–240 km in this century, while only 3% shifted to the south. Potential climate-induced changes in the ranges of parasites have disease centers worried ("See Applied Ecology: Diseases and Global Climate Change").

Another drawback associated with global warming is changes in rainfall patterns. Although average global precipitation should increase (Fig. 7.11), a decrease would

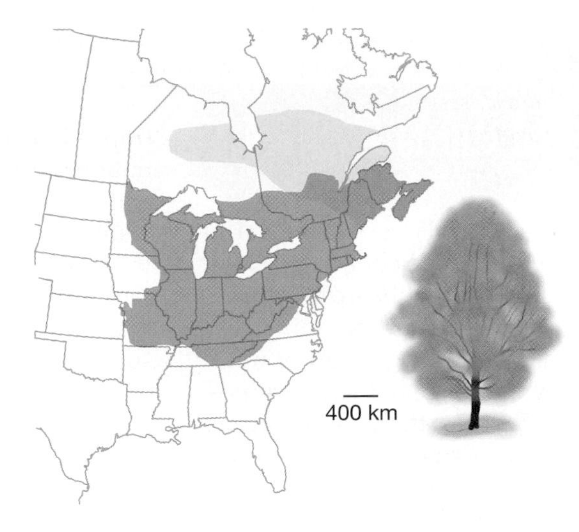

Figure 7.10 The present geographic range of sugar maple (blue shading) and its potentially suitable range under doubled CO_2 levels (yellow shading) in North America. Green shading indicates the region of overlap, which is the only area where beech would be found before it spread to new suitable areas. *(Modified from Davis and Zabinski, 1992.)*

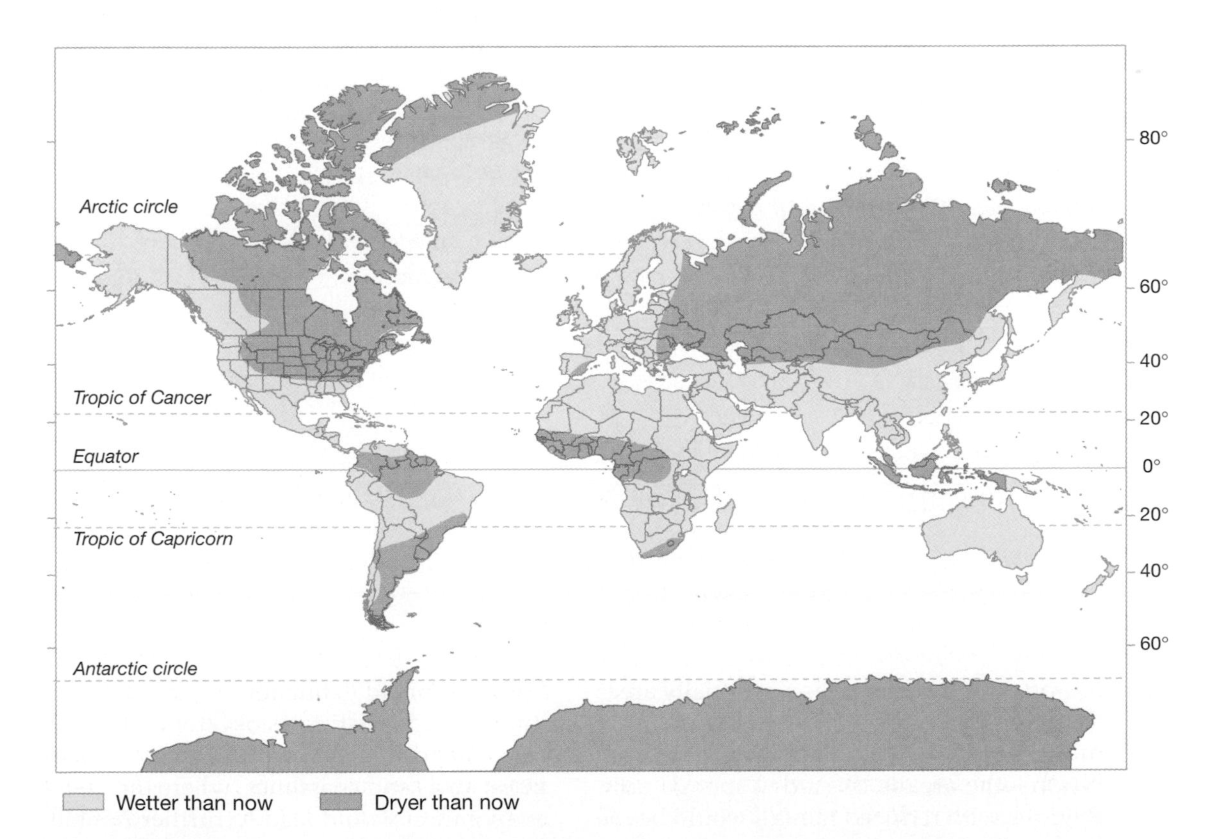

Figure 7.11 Predicted changes in precipitation patterns caused by global warming. Mid-continent regions will probably be drier than now, but coastal areas and much of the tropics could be wetter.

Applied Ecology

Diseases and Global Climate Change

One of the main concerns voiced over global warming is that the delicate balance between diseases, their vectors, and humans might be upset as tropical climates that are so hospitable to spawning and spreading diseases move poleward. The spread of infectious diseases is controlled by the range of their vectors—mosquitoes and other insects. Because these insects are all ectothermal organisms, they are susceptible to subtle changes in temperature. Increases in temperatures mean increases in the activity and ranges of the vectors. Data on recent trends support this observation: A 1°C increase in the average temperature in Rwanda in 1987 was accompanied by a 337% rise in the incidence of malaria that year as mosquitoes moved into mountainous areas in which they had previously been absent. Also, *Aedes aegypti*, a mosquito that carries dengue and yellow fever, has extended its range high into the mountain areas of such diverse countries as Colombia, India, and Kenya.

Although global warming is expected to deliver its most deadly punch via parasites in the tropical and subtropical areas of the world (see Table 1), where over 600 million people are affected (and 2 million die), the United States is not immune: A 1993 outbreak of the hantavirus respiratory illness in the southwest killed 27 people and was linked to an El Niño event. An explosion in the deer mouse population (the viral host) followed from an El Niño that precipitated heavy rains and increased the food supply of the mice.

A computer model by a Dutch public-health team proposed that an average global temperature increase of 3°C in the next century could result in 50 to 80 million new cases of malaria per year. However, infectious-disease specialists point out that public-health measures will outweigh the effects of climate. To illustrate this point, epidemiologists note that the U.S. southeast has the mosquito vector of dengue fever, a disease that is present in the Caribbean. Summer temperatures in the southeastern United States are hotter than in the Caribbean, but dengue is absent in the United States because health measures such as mosquito control and vaccination programs are in place. Nonetheless, we should remember that wildlife is not afforded the same degree of protection against mosquitoes and may be more vulnerable to new diseases.

TABLE 1 Major tropical diseases likely to spread with global warming, as assessed by the World Health Organization.

Disease	Vector	Population at Risk (Millions)	Prevalence of Infection	Present Distribution	Likelihood of Altered Distribution with Warming
Malaria	Mosquito	2,100	250 million	(sub)tropics	+++
Schistosomiasis	Water snail	600	200 million	(sub)tropics	++
Filariasis	Mosquito	900	90 million	(sub)tropics	+
Onchocerciasis (river blindness)	Black fly	90	18 million	Africa and Latin America	+
African trypanosomiasis (sleeping sickness)	Tsetse fly	50	25,000 new cases/year	Tropical Africa	+
Dengue	Mosquito	Estimates unavailable	Estimates unavailable	Tropics	++
Yellow fever	Mosquito	Estimates unavailable	Estimates unavailable	Tropical South America and Africa	+

Note: + = likely; ++ = very likely; +++ = highly likely.

be apparent in some areas—especially areas that are already dry. A rise in temperature in those regions would not increase evaporation—the area is like a dry sponge now. Regions with reduced rainfall would be the midcontinental areas of America and Asia. More droughts there would probably cause the widespread extinction of plant and animal life in existing ecosystems. On the other hand, rainfall would increase over moist areas and coastlines, where increased evaporation would lead to further rainfall. Thus, the world's current grain-producing areas would become drier than they are now following a global warming. Corn yields would be reduced in the midwestern plains of the United States, and a major increase in the frequency and severity of drought would lead to more frequent crop failure in the wheat-growing areas of Canada and Russia. This would be partly offset by increased rainfall in some tropical countries, which could lead to a boost in rice yields. Some of the grain-growing areas in Australia would have increased precipitation and higher temperatures. Thus, the pattern of the world's grain trade, which depends heavily on the annual North American surplus, would probably be disrupted and take some time to adjust to the changing conditions. Furthermore, although more rainfall could fall in many areas, it is not at all certain that this would result in increased moisture in the soil, because higher temperatures could increase evaporation, and increased plant growth could result in higher rates of transpiration, requiring more moisture from the soil.

Wind.

Because hot air rises and rising air masses push other air aside, temperature gradients can create winds. Wind can be important because it amplifies the effects of environmental temperature on organisms by increasing heat loss by evaporation and convection (the wind-chill factor). Wind also contributes to water loss in organisms by increasing the rate of evaporation in animals and transpiration in plants. In addition, wind aids in the pollination of many plants by blowing pollen from flower to flower, and wind disperses plant seeds.

Wind can be an important mortality factor as well (Fig. 7.12). Fifteen million trees in the south of England perished on October 16, 1987, in the wake of a mighty storm. Records suggest that such winds had not hit the region for at least 300 years, so this may have been a rare event. Others suggest that such severe weather will become more common because the world's climate is changing. Five of England's biggest freezes have come since 1978. The frequency of disastrous hurricanes in the South Pacific, especially over Fiji, has increased from one every 12 years to one every 7 years, and six such storms were recorded between 1981 and 1985 alone.

Winds can also modify wave action in the sea. In the rocky intertidal zone, algae such as fucoids and kelps survive repeated pounding by a combination of holdfasts and flexible structures. The animals of the zone have powerful organic glues and muscular feet to hold them in place. The current can also be critical in streams where, in turbulent upstream regions, plants and animals are in danger of being washed away. Here, rock-encrusting forms like algae and mosses grow, and animals have flattened bodies to hide under crevices.

Wind can amplify temperature gradients and be an important mortality factor in its own right.

Figure 7.12 This huge live oak tree, *Quercus virginiana*, was felled by strong winds in North Florida. Thus, besides modifying temperatures, wind can be an important mortality factor. *(Photo by Peter Stiling.)*

Salt.

Salt concentrations can affect water uptake.

For organisms in both terrestrial and aquatic habitats, the concentration of salts in the soil or in water is important, because higher concentrations increase osmotic resistance to water uptake. On land, this resistance occurs in arid regions, where water evaporates and crystalline salt accumulates. Osmotic resistance can be of critical importance to agriculture, in which continued watering in arid environments greatly increases the salt concentration and reduces crop yields.

Water availability limits the distribution and abundance of many species.

Along the salt marshes of seacoasts, salt concentrations are high, and the vegetation consists largely of halophytes, species that can tolerate higher salt concentrations in their cell sap than nonhalophytes can. Many plants also have ways of overcoming the excess salt (Adam, 1990). For example, plants such as mangrove and *Spartina* grasses have salt glands that excrete salt to the surface of the leaves, where it forms tiny white crystals (Fig. 7.13).

Windborne salt can affect the distribution of plants on sand dunes, too. On Atlantic coasts of the United States, sea oats (*Uniola paniculata*) can withstand high atmospheric salt and thus inhabit areas near the shoreline, while another dune grass, *Andropogon littoralis*, is less tolerant and occurs in protected areas back from the shoreline.

pH.

Acidity or alkalinity of the environment affects plant and animal distributions.

The pH of soil or water can exert a powerful influence on the distributions of species. Only a minority of organisms can grow below pH 4.5, as is evidenced in the decrease in the number of species in acid waters. Lake trout disappear from lakes in Ontario and the eastern United States when the pH falls below about 5.2. Although this pH level does not affect the survival of the adult fish, it affects the survival of juveniles (Mills *et al.*, 1987).

For plants, most roots are damaged below pH 3 or above pH 9. Plants have been broadly classified as *calciphobes* (growing only in acid soils), *calciphiles* growing only in basic soils), and *neutrophiles* (tolerant of either condition). Calciphobes, such as rhododendrons and azaleas, can live in soils with a pH of 4.0 or less. Calciphiles, such as alfalfa (*Medicago sativa*), blazing star (*Chamaelirium luteum*), and southern red cedar (*Juniperus silicicola*), are restricted to soils of high pH, mainly because they are susceptible to acidity. Chalk and limestone, which give grasslands a high pH (producing so-called high-lime soils), carry a much richer flora (and associated fauna) than do acid grasslands.

Figure 7.13 Special salt glands in *Spartina leaves* exude salt, enabling this grass to exist in saline intertidal conditions.

Water.

Water has an important effect on the ecology of terrestrial organisms. Protoplasm is 85 to 90 percent water, and without moisture there can be no life. The distribution patterns of many plants are limited by available water. Some, for example, the water tupelo tree in the United States, do best when completely flooded and are thus predominant only in swamps. For most plants, the limiting amount of moisture is much lower. In cold climates, water can be present but locked up as permafrost and, therefore, unavailable—a frost-drought situation. Alpine timberlines may be affected by frost drought.

Animals face problems of water balance, too, and the distribution of organisms can be strongly affected by desiccation. However, many animals can move away from dry or intolerable environments. Desert animals are usually small and can hide underground in the heat of the day. This brings home the point that some organisms have two strategies for coping with extremes—tolerance or avoidance. For plants, of course, avoidance is difficult. Larger animals too cannot avoid environmental extremes so easily, and because most depend ultimately on plants as food their distributions are intrinsically linked to those of their food sources. The distributional boundary of the red kangaroo, *Macropus rufus*, in Australia coincides with the 400 mm rainfall contour because the kangaroos are dependent on arid-zone grasses that grow only where there is 400 mm or greater of rainfall.

The importance of water in limiting animal population density was brought home by an extraordinary El Niño event (an irregular increase in water temperature in the eastern Pacific Ocean) in 1982–1983, when the rainfall on Isla Genovesa, Galapagos, increased from its normal 100 to 150 mm during the rainy season to 2,400 mm from November 1982 through July 1983. Plants responded with prodigious growth, and Darwin's finches bred up to eight times rather than their normal maximum of three (Grant and Grant, 1987). Thus, abundance as well as distribution can be severely affected by water.

Figure 7.14 This thrips insect is a pest of corn. Understanding what limits thrips densities is therefore of value to agriculture.

7.2 Physical Factors and Species Abundance

Not only does climate play an important role in the global distribution of species, but climate can affect the densities of species, too. The most famous proponents of this idea were the Australian entomologists Davidson, Andrewartha, and Birch, who studied the densities of small insects (<1 mm) called *thrips* on rosebushes (Fig. 7.14). The thrips feed by rasping the leaf surface and lapping up the exudates. The insects were actually pests of fruit orchards, and money was provided to study their biology. But the thrips also fed on rosebushes, which were more convenient to the campus where the biologists worked. The insects were counted every day (except Sundays and holidays) for 81 consecutive months! Like many insects, the thrips underwent large fluctuations in density. Davidson, Andrewartha, and Birch found that 78 percent of the variation in population maxima was accounted for by variations in weather (Andrewartha and Birch, 1954). They used a statistical technique called multiple regression to develop the following equation to predict the logarithm of the number of thrips in the spring of each year:

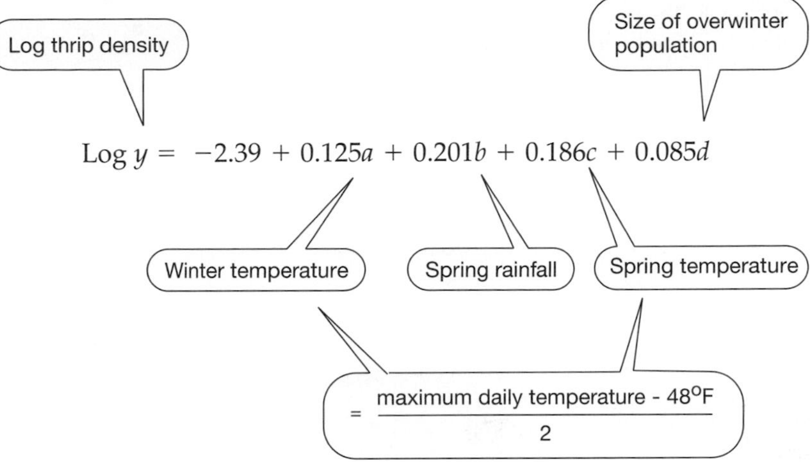

This equation did a great job of predicting the thrips density in any one year (Fig. 7.15). Andrewartha and Birch concluded that weather was the overriding influence on the density of thrips, and there was little additional variation to be explained by any other variable. It is evident that rainfall could be very important to these tiny insects, easily dislodging them from the leaves on which they fed and creating huge day-to-day fluctuations in their densities.

The physical environment can affect species abundance as well as distribution patterns.

Similar correlations between the physical environment and population density were found for other organisms. Francis (1970)

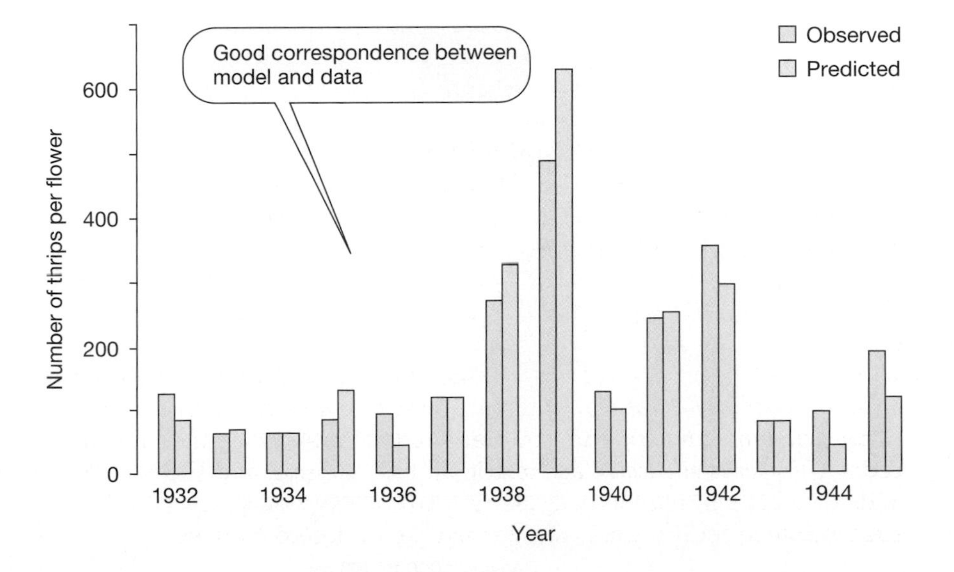

Figure 7.15 Comparison of means of observed annual population densities of thrips (*Thrips imaginis*) with densities predicted by a model based on temperature, rainfall, and numbers of overwintering thrips. The means agree well, showing that thrips densities are well predicted by abiotic factors. *(Data from Davidson and Andrewartha, 1948b).*

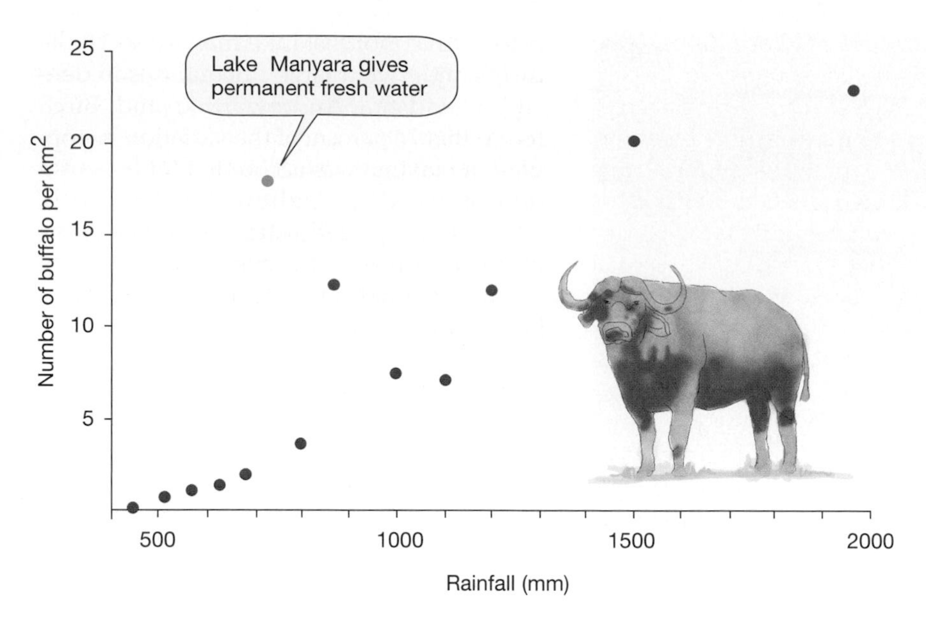

Figure 7.16 Buffalo density, in numbers km^{-2} of transect, is very much dependent on grass availability which itself is dependent on annual rainfall. The main exception is where there is permanent water, such as Lake Manyara (blue dot) where greater water availability leads to higher grass growth and buffalo densities. *(Reproduced from Sinclair, 1977).*

developed an equation that explained 99% of the variation in California quail density:

$$y = 0.929H + 0.021A - 0.120C - 0.975.$$

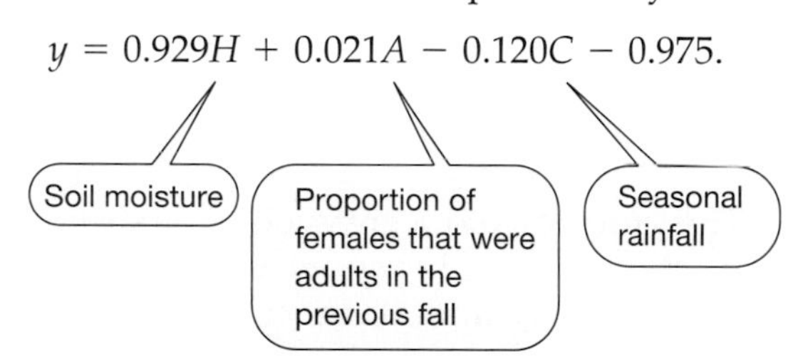

In this case, soil moisture was the most important factor, explaining 83.1% of the variance. It acted by influencing the density of insects that the quail fed on.

A similar phenomenon regulates the number of buffalo in Africa. In the Serengeti area, grass productivity is related to the amount of rainfall the previous month. Since buffalo density is governed by the availability of food, there is a tight correlation between buffalo density and rainfall (Fig. 7.16). The only exception occurs in the vicinity of lakes where groundwater promotes plant growth—for example, at an area called Lake Manyara. Not surprisingly, the densities of many plants are also limited by the availability of water: Woddell, Mooney, and Hill (1969) showed a strong correlation between creosote bush density in the Mojave Desert and rainfall (Fig. 7.17).

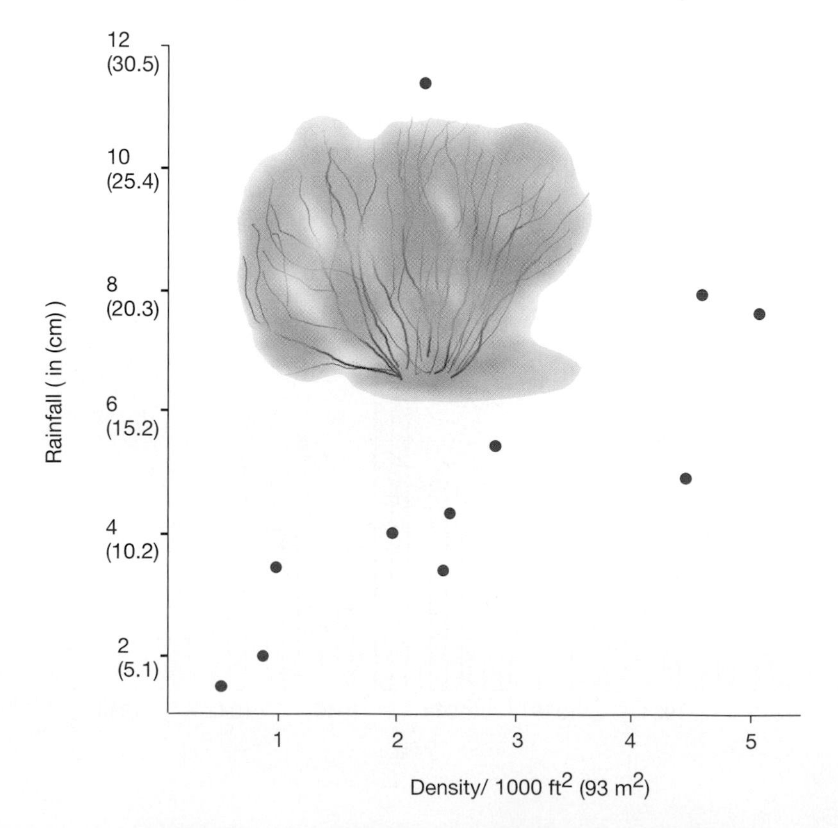

Figure 7.17 The density of *Larrea divaricata* (creosote bush) in the U.S. Southwest is governed by rainfall. Data are from 11 sites in the southwestern United States. *(Reproduced from Woodell et al., 1969.)*

In 1976, J. T. Enright, from the Scripps Institute of Oceanography in San Diego, California, suggested that the limiting effect of climate on density may be more common than we think. He argued that many ecologists are preoccupied with the effects of competitors, predators, and parasites on populations and that the role of climate is often overlooked. Biogeographers have described whole arrays of species or communities that are restricted to certain climatic zones. Why should we deny that weather and climate can profoundly influence the densities of species within their range? Often, a species' abundance declines toward the boundaries of its range and is correlated with environmental variables. Of course, in the center of a range, climatic variables may be less likely to cause observed density differences—if there are any. As we see repeatedly in the study of ecology, the issue of scale arises. Over small scales, biotic variables may be important; over larger scales, climate may be more important.

7.3 Physical Factors and Numbers of Species

Physical factors not only affect the distribution patterns and abundances of species; they also play an important part in species richness—the number of species in any given locale. David Currie and Viviane Paquin of the University of Ottawa (1987) showed that, of all the environmental factors for which data were available, evapotranspiration rates (strongly correlated with rainfall and temperature) are the best predictors of tree species richness in North America (Fig. 7.18).

In 1967, Robert Whittaker, a plant ecologist from the University of California at Irvine, first formalized the concept that a community is governed by physical variables. Whittaker invited us to consider an environmental gradient, such as a long, even, uninterrupted slope of a mountain. He considered four hypotheses purporting to explain the distribution patterns of plants and animals on the gradient (Fig. 7.19):

1. Competing species, including dominant plants, exclude one another along sharp boundaries. Other species evolve toward a close association (perhaps mutualism) with the dominants. There thus develop distinct zones along the gradient, each zone having its own assemblage of species, giving way at a sharp boundary to another assemblage of species (a community).
2. Competing species exclude one another along sharp boundaries, but do not become organized into groups with parallel distributions.

Physical factors may determine how many species occur in a community.

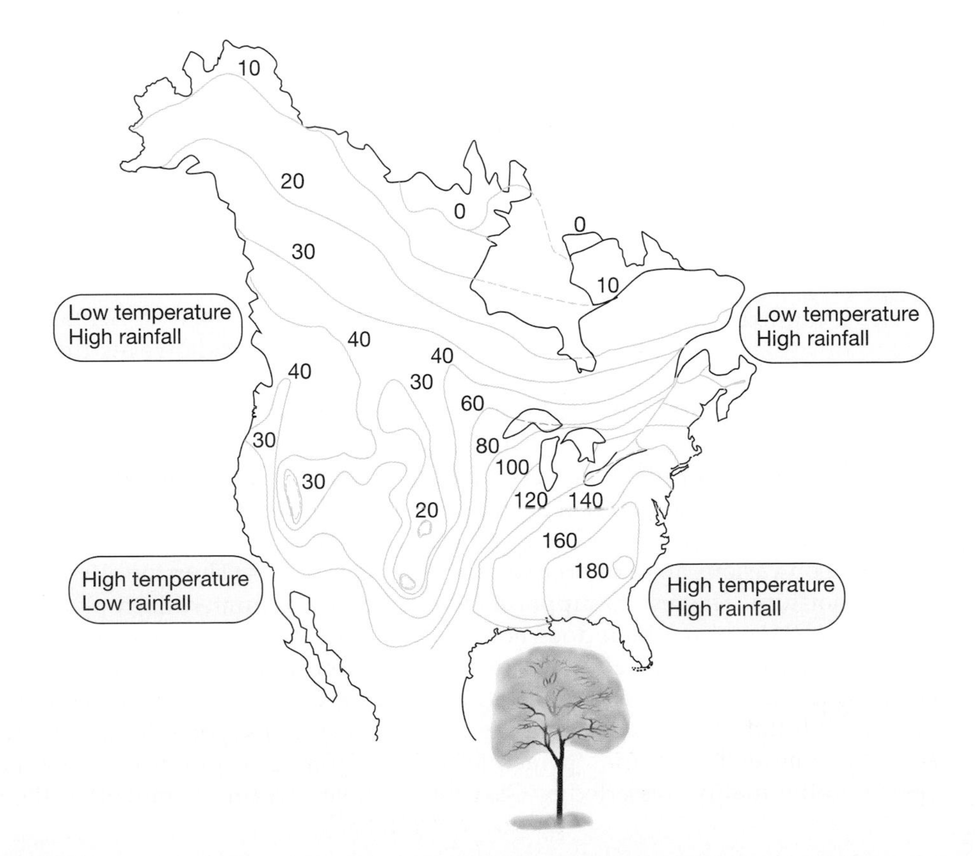

Figure 7.18 Tree species richness in Canada and the United States. Contours connect points with the same approximate number of species per quadrant. The number of species is highest in the southeast, where evapotranspiration rates, linked to rainfall and temperature, are highest. *(Reproduced from Currie and Paquin, 1987.)*

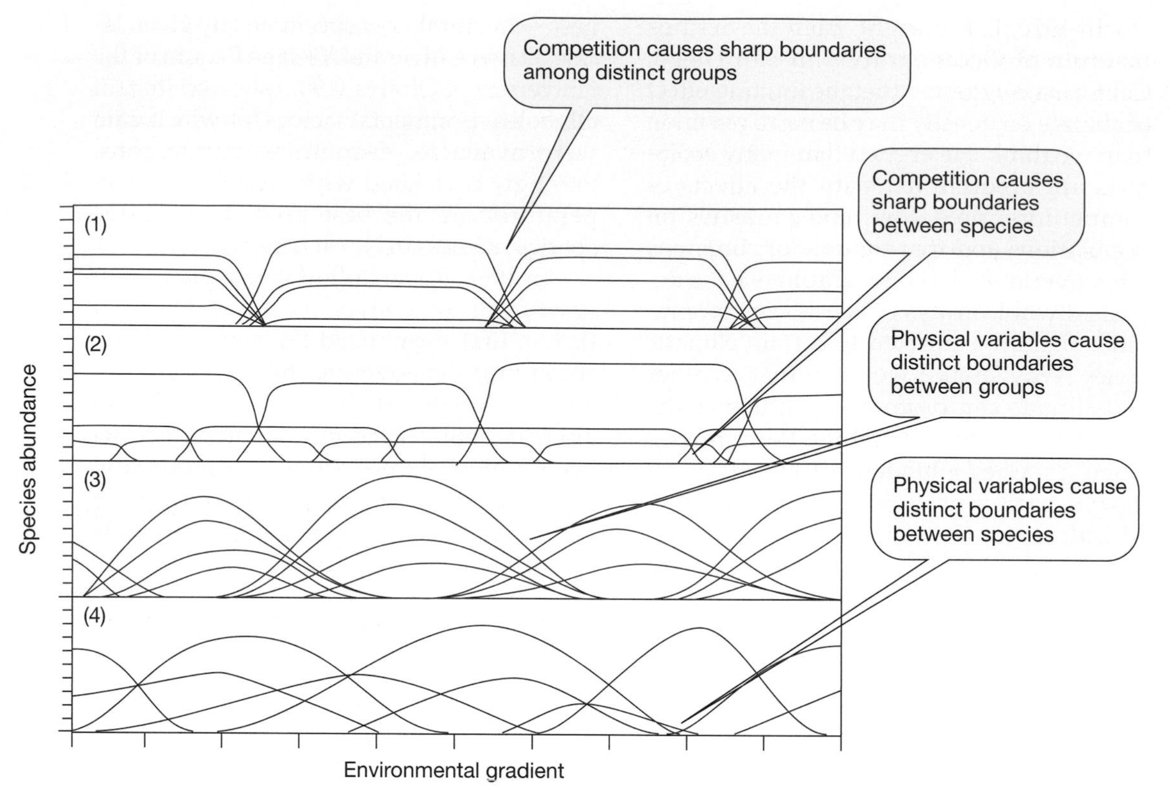

Figure 7.19 Four hypotheses explaining how species populations might relate to one another along an environmental gradient. Each curve in each part of the figure represents one species population and the way it might be distributed along the environmental gradient. *(Reproduced from Whittaker, 1970.)*

3. Competition does not, for the most part, result in sharp boundaries between species. The adaptation of species to similar physical variables will, however, result in the appearance of groups of species with similar distributions.
4. Competition does not usually produce sharp boundaries between species populations, and the adaptation of species to similar physical variables does not produce well-defined groups of species with similar distributions. The centers and boundaries of species populations are scattered along the environmental gradient, and communities are not easily recognized.

To test these hypotheses, Whittaker (1967) examined the vegetation in various mountain ranges of the western United States. Samples of plant populations were taken along an elevation gradient from the tops of the mountains down to the bases, along with data on various abiotic variables, such as soil moisture. The results supported hypothesis 4, that competition does not produce sharp boundaries and adaptation to conditions does not result in defined groups. Whittaker concluded that his observations agreed with the "principle of species individuality" (asserted by Gleason in 1926 and discussed further in the section on community ecology). That is, each species was distributed in its own way, according to its own genetic, physiological, and life-cycle characteristics. The broad overlap and scattered centers of species populations along a gradient implied that most communities intergrade continuously along environmental gradients, rather than forming distinct, clearly separated zones. Thus, Whittaker's observations showed that the composition of species at any one point in the environment was determined largely by physical factors.

Nowhere has the link between physical factors and species richness been better demonstrated, however, than on long-term grassland fertilizer trial plots at Rothamstead Experimental Farm in England, established in 1856 and monitored continuously to the present day. The unfertilized plots supported some 60 species of higher plants, including all of the species of plants found on the fertilized plots (Thurston, 1969). Species diversity on the unfertilized plots was high, and no species was clearly dominant. The vegetation was short and primary productivity low. On plots that received applications of phosphorus, potassium, sodium, and magnesium, but no nitrogen, legumes such as clover became dominant at the expense of other species (because legumes can fix their own nitrogen). The addition of nitrogen discouraged legumes, reduced their growth, and encouraged grasses. The addition of nutrients to grassland enhanced the growth of a few highly competitive and lush-growing grasses at the expense of creeping and rosette-type species. Thus, species composition changed under different nutrient regimes, and species richness was much reduced.

In sum, physical variables such as pH, wind, light, salt, and especially temperature can profoundly affect species distributions. Perhaps more importantly than this, however, variables affect species abundance and the number of species in a community.

Summary

Extremes of the physical environment, such as hard freezes or prolonged droughts, have long been known to greatly affect the distribution and abundance of plants and animals.

1. The local distribution patterns of most species are limited by certain physical, or abiotic, factors of the environment, such as temperature, moisture, light, pH, soil, salinity, and water currents.

2. One of the most important environmental variables in terms of its influence on plant and animal distribution patterns is temperature. Temperature resistance in plants, especially resistance to frost, is critical to their distributions. Resistance to high temperatures and to fires is also important. Increasing CO_2 output from the burning of fossil fuels and increasing deforestation may cause global CO_2 levels to increase so much, that the world's climate might change slightly (so-called global warming). This will profoundly influence the distribution pattern of many species.

3. Wind can amplify temperature gradients and affect species distributions directly.

4. Both soil salt concentrations and pH can affect water uptake in plants and are important for some aquatic organisms.

5. While physical factors limit where species occur, they also have a strong influence on the population densities of species as well. Studies of insects show how the weather can govern densities, while the amount of rainfall is critical to the densities of many plants and the herbivores that graze on them.

6. Because the physical environment can limit how species are distributed in nature, it can profoundly influence which species are present in a community.

The physical environment is extremely important in limiting not only where species occur, but also their abundance wherever they do occur.

Discussion Questions

1. Why do you think there is a "tree line" on the sides of many of the Rocky Mountains beyond which no trees will grow? Is it because of low temperatures, low water availability, or high wind? How would you devise an experiment to find out?

2. Do you think an estuary would be a stressful habitat to live in? Explain. Are there any disadvantages to living in an estuary? (Think in terms of salinity, food availability, and competition.)

3. Are you interested in maintaining buffalo populations in the United States? Why is fire so important to achieving this goal?

4. What is the value of laboratory experiments in attempting to identify limiting factors to explain the observed distribution patterns of plants and animals in the field?

CHAPTER 8

Competition and Coexistence

Road Map

1. Competition in nature takes a number of forms. Intraspecific competition is competition within the same species, and interspecific competition is competition between different species.
2. Intraspecific competition is common in nature and in plants is often described by a 3/2 thinning law.
3. Interspecific competition is also common, and laboratory experiments have shown how outcomes can be changed by variations in the physical environment or by the presence of other species, such as parasites or predators, that affect one species, but not the other.
4. Competition exists among about 55% to 75% of species and the most common mechanism, is over use of the same resource.
5. Mathematical models called Lotka–Volterra models predict four outcomes of competition: One species is eliminated, the other is eliminated, both species coexist, or either species is eliminated, depending on starting conditions.
6. Competing species can sometimes coexist if they partition resources between them.

Sarah Durant (1998) performed a four-year study in Africa, mapping the distribution of three predator species—lion, hyena, and cheetah—in the Serengeti National Park in Tanzania. She showed that cheetahs (Fig. 8.1) were outcompeted by both lions and hyenas, both of which killed cheetah cubs and stole cheetah kills. In addition, lions and hyenas could outcompete cheetahs by taking more of the common prey of all three species, Thompson's gazelle. Durant showed that under normal conditions cheetahs survived by preferring areas with relatively low concentrations of gazelles. These areas were less likely to attract attention from other predators, and the gazelles there were less likely to become alerted to a hunting cheetah. The ability of the highly mobile cheetahs to move long distances in search of such low-density gazelle herds helps facilitate this lifestyle. Durant went on to suggest that many other species, not only those in Africa, might "fit in" wherever they could and be widespread and relatively rare, complementing the distribution of their principal competitors. Again in Africa, the wild dog is a similar example. Competition refugia may therefore be important in promoting the coexistence of species in many systems. Of course, experimental manipulations removing lions and hyenas and observing increases in cheetahs would be the best evidence of competition, but on a large scale this is difficult, if not impossible, to do. Demonstrating competition, as we shall see, is by no means an easy task.

Figure 8.1 Cheetah and fresh kill of Thompson's gazelle in Massai Mara Game Reserve, Kenya. Cheetahs occur in regions of relatively low gazelle densities where lions are rare.

8.1 Species Interactions

In addition to being subject to influences from the abiotic environment, populations can be very much affected by the other species with which they interact. These biotic interactions can take a variety of forms and are summarized in Table 8.1. **Herbivory**, **predation**, and **parasitism** have the same general effect: a positive influence on one population and a negative one on the other. **Batesian mimicry**, the mimicry of a nonpalatable species by a palatable one, falls under the same banner. (See Chapter 10.) Competition affects both species negatively. **Amensalism** is a one-sided competition in which one species, *a*, has a negative effect on the other, *b*, but the reverse is not true (i.e., *b* has no effect on *a*). **Neutralism** is the coexistence of noninteracting species. It seems likely that most species in a community indirectly affect most other species at least somewhat, so neutralism is probably quite rare, but no one has quantified its occurrence. **Mutualism** and **commensalism** are less commonly discussed in ecology, but can be tied together under the banner of **symbiosis**. In a symbiotic relationship, the partners live in intimate association with one another. Although symbiotic relationships are often thought of as beneficial to both species, this is not always the case: the only condition required is that the two species be associated intimately with each other. Thus, parasitism can be considered a symbiotic relationship, too (Saffo, 1992). Nevertheless, the effects of mutualism and commensalism are different from those of parasitism and so are discussed separately, in Chapter 9. Basically, this is because parasitism is harmful to at least one of the species, but mutualism and commensalism are not. Mullerian mimicry, the convergence to a common color to reinforce unpalatableness, which we will discuss further in Chapter 10, can also be regarded as mutualism, since it is beneficial to both species.

Species interactions include competition, predation, parasitism, herbivory, mutualism, and commensalism.

Many different types of competition can occur in nature (Fig. 8.2). Competition may be **intraspecific** (between individuals of the same species) or **interspecific** (between individuals of different species). Competition can also be characterized as **resource** competition or **interference** competition. In resource competition, organisms compete directly for the limiting resource, each obtaining as much as it can. For example, when fly maggots compete in a mouse carcass, few individuals can command enough of the resource to survive or reproduce. Such competition is often evident between invertebrates. In interference competition, individuals harm one another directly by physical force. Often, this force is ritualized into threatening behavior associated with

TABLE 8.1 Summary of the types of interaction between species. + = positive effect; 0 = no effect; − = negative effect. *Examples of a symbiotic relationship, in which the participants live in intimate association with one another. (Mimicry explained in Chapter 10.)

Nature of Interaction	Species 1	Species 2
Mutualism,* Mullerian mimicry	+	+
Commensalism*	+	0
Predation, herbivory, Batesian mimicry, and parasitism*	+	−
Amensalism	−	0
Competition	−	−
Neutralism	0	0

Intraspecific competition between individuals of the same species.
Resource competition: each caterpillar chews as much leaf as it can.
Interference competition: Each caterpillar physically intimidates the other.
Interspecific competition between different species.
Aphid sucking leaf sap
Caterpillar chewing leaf

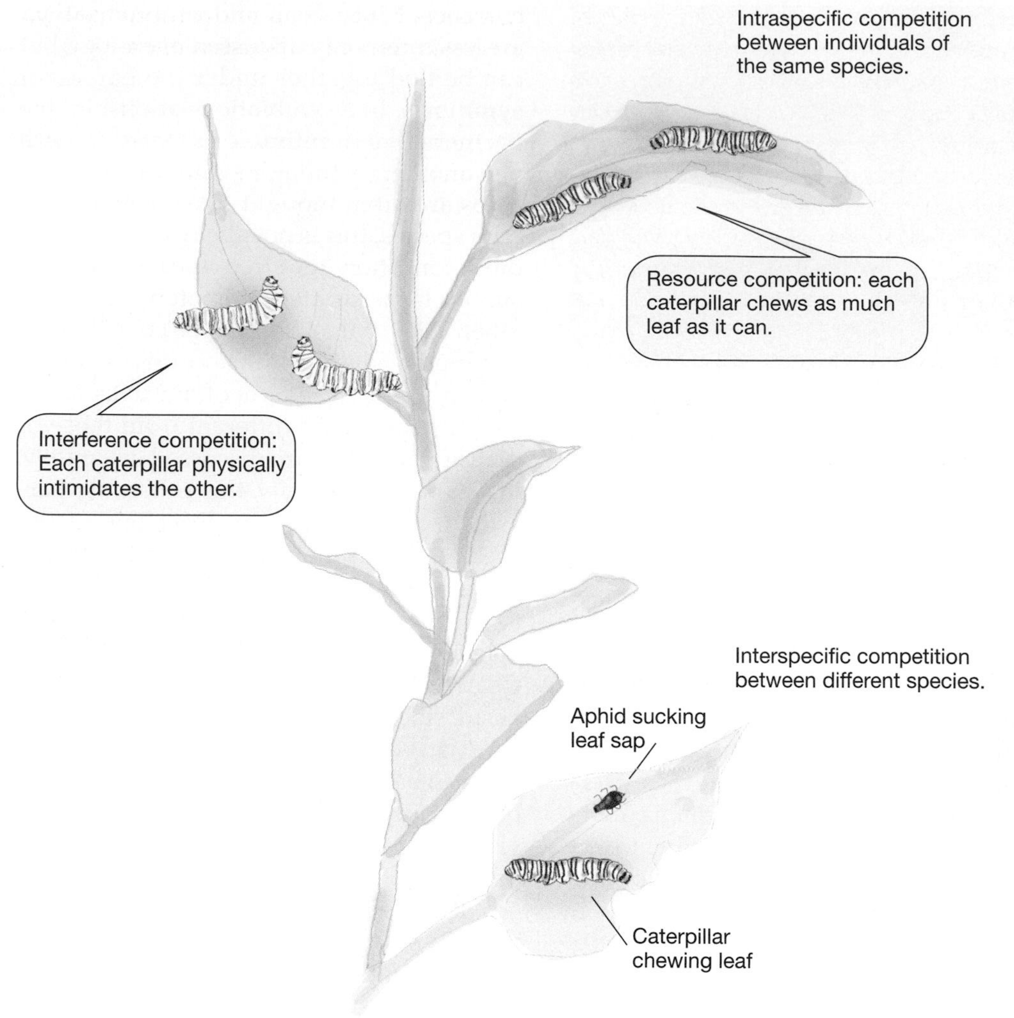

Figure 8.2 The different types of competition in nature.

territories. In these cases, strong individuals survive and take the best territories, and weaker ones perish or at best survive under suboptimal conditions. Such behavior is most common in vertebrates. Finally, competition between species is not always equal. In fact, in many cases one species has a strong effect on another, but the reverse interaction is negligible. Thus, many competitive interactions are a $-/0$ relationship, often called *amensalism*, rather than the classical $-/-$ relationship. John Lawton and Mike Hassell (1981) found that in insects, asymmetrical competition was at least twice as common as symmetrical competition.

8.2 Intraspecific Competition

Intraspecific competition between individuals of the same species is frequent in nature and in plants may be described by the 3/2 self-thinning rule.

Competition in plants is a little different than it is in many animals. Many plants are **clones**, so if some stems die, others of the same individual live on. For this reason, competition in plants is usually examined by studying the change in biomass of competitors, rather than their numbers. Being rooted in the ground, plants may suffer severe competition for water, nutrients, or light, because they can't "pick up their roots" and escape from competitors. In 1963, Yoda described competition between plants by a 3/2-power rule, sometimes called *Yoda's law* or the *self-thinning rule.* This rule describes the rate at which the biomass of individual plants increases as the number of plant competitors decreases. Mathematically,

Log weight = log plant density + a constant

$$\text{Log } w = -\frac{3}{2}(\log N) + \log c\,,$$

or

$$w = cN^{-3/2}\,,$$

where w is the mean plant weight, N is the plant density (per m^2), and c is a constant. As plants grow, they require more resources. If resources remain constant, then some plants die or are outcompeted, while the survivors continue to grow. White (1980) showed how information from 31 data sets supported the 3/2 thinning law (Fig. 8.3). Plant weight increases faster than density decreases, and a steady increase in biomass results. Of course, this increase is not infinite, and the slope changes to –1 when the maximum growth of species in the environment has been reached.

The self-thinning law has been argued to be a logical result of how plants grow in relation to the availability of light. The leaf area index L (the area of leaf per unit area of land) is related to the number of surviving plants, N, and their mean leaf area λ by the formula

$$L = \lambda N = \text{constant}.$$

The mean leaf area λ is related to other linear measurements, such as the diameter of the stem, D, by the quadratic equation

$$\lambda = aD^2$$

where a is a constant. The mean plant weight w is related to D by the cubic equation

$$w = bD^3$$

where b is another constant. When all these equations are put together, we have

$$w = b\left(\frac{L}{a}\right)^{3/2} N^{-3/2}.$$

This 3/2-power equation could explain why a large number of thinning lines obey a −3/2 thinning law. However, more recently, Weller (1987, 1991) showed that many plants obey different thinning laws. He found that only 24 of 63 data sets fit the −3/2 law. Perhaps species that did not conform to the law are limited by resources other than light.

What do *animals* do? Many may also thin their numbers, but because they move around, it is more difficult to tell. However, some sessile species, such as barnacles or mussels on a rock face, do show self-thinning slopes of about −1.5 (Hughes and Griffiths, 1988), so perhaps the −3/2 law has other, as yet unknown, derivations.

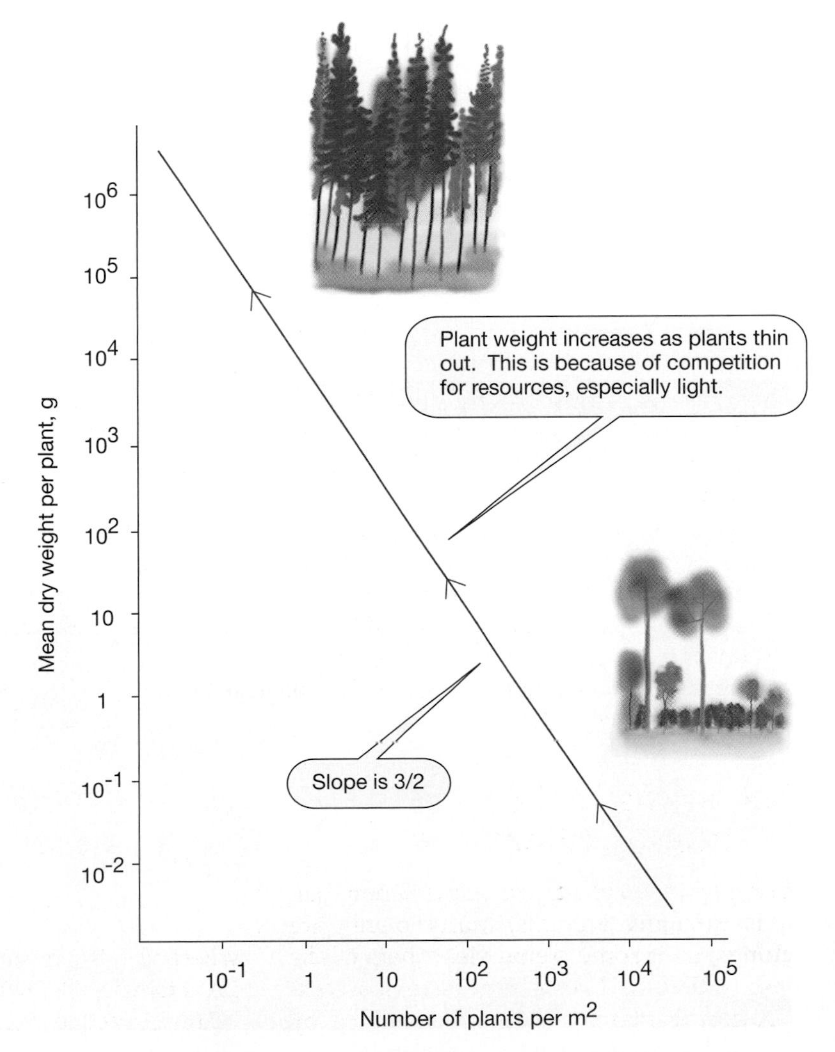

Figure 8.3 Self-thinning in plants as a result of intraspecific competition. The line represents an average based on 31 data sets in White (1980). Arrows indicate the direction of self-thinning.

8.3 Interspecific Competition: Laboratory Experiments

Laboratory experiments show how the outcome of interspecific competition can vary with changes in the abiotic and biotic environment.

As mentioned in Chapter 1, with experimental methods we can conduct both field and laboratory experiments, removing individuals of one or the other competing species, to investigate the strength and frequency of competition. Each type of experiment has its advantages. Field experiments are useful because, in nature, organisms can interact with all the other organisms in their environment and natural variation in the abiotic environment is factored in. In laboratory experiments, the investigator can control all the important factors and vary them systematically. Our first example of interspecific competition focuses on a laboratory experiment.

In the late 1940s, Thomas Park and his students at the University of Chicago began a series of experiments examining competition between two flour beetles: *Tribolium confusum* (Fig. 8.4) and *Tribolium castaneum*.

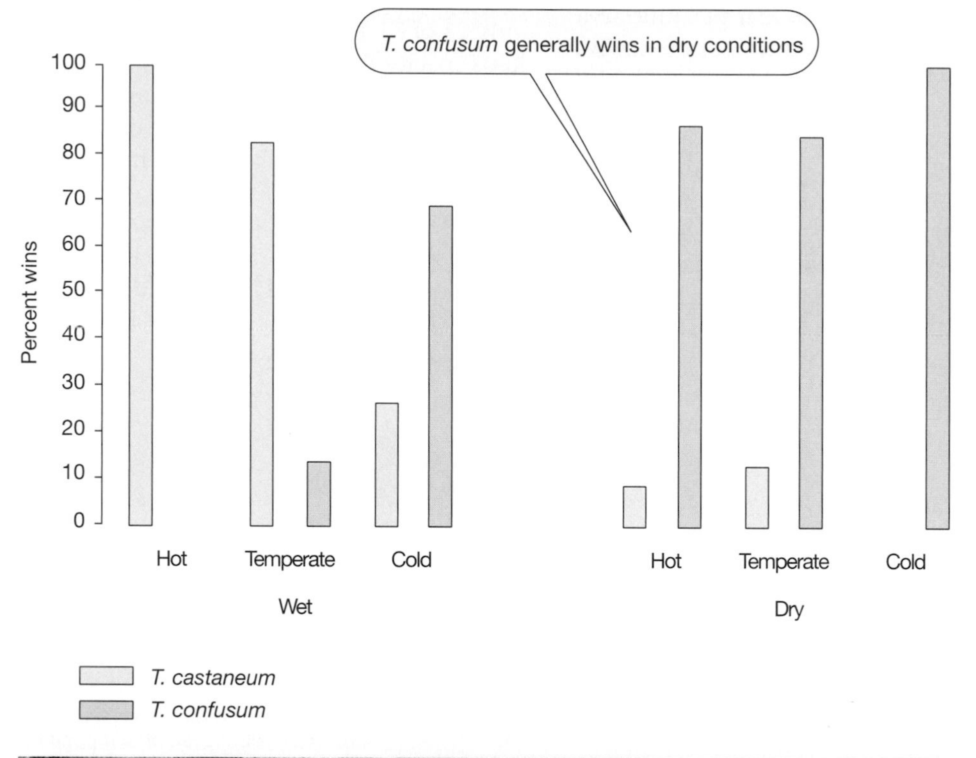

Figure 8.4 Results of competition between the flour beetles *Tribolium castaneum* and *T. confusum*. Each species usually performs better in a given habitat; for example, *T. confusum* does better in dry conditions. *(From data in Park, 1954.)*

These beetles were well suited for study in the laboratory: Large colonies could be grown in relatively small containers containing a dry food medium. Thus, many replicates of each experiment were possible. Competition experiments were conducted by putting the same number of beetles of each species into a container and seeing which species was still alive and which was extinct after a series of generations, usually two to three years later. In initial experiments, the beetle cultures were infested with a sporozoan parasite, *Adelina,* that killed some beetles, particularly individuals of *T. castaneum.* In these early experiments (Park, 1948), *T. confusum* won in 66 out of 74 replicates (89%), because it was more resistant to the parasite. Later, *Adelina* was removed, and *T. castaneum* won in 12 out of 18 replicates (67%). So the presence of a parasite was shown to alter the outcome of competition. Almost as important, with or without the parasite, there was no absolute victor; sometimes *T. confusum* won and sometimes *T. castaneum* won, so some stochasticity was evident.

Park then began to vary the abiotic environment and found that competitive ability was also greatly influenced by climate: Each species was a better competitor in a different microclimate (Fig. 8.4). Generally, *T. confusum* did better in dry conditions and *T. castaneum* in moist conditions. However, *T. confusum* also won in a cold moist climate. Once again, some stochasticity was evident, and the victor was not always absolute. (See bars in figure for cold and wet, temperate and wet, hot and dry, and temperate and dry climates.) Later, it was found that the mechanism of competition was largely predation on eggs and pupae by larvae and adults. In general, the species were mutually antagonistic; that is, they ate more eggs and pupae of the other species than they did of their own.

Next, Park varied the predatory tendencies of the beetles by selecting different strains; he obtained different results according to the strain of each beetle he used (Fig. 8.5). Strains CI and CIV of *T. confusum* usually won against all strains of *T. castaneum,* whereas strains CII and CIII always lost. Thus, it was the strain of *T. confusum,* not *T. castaneum,* that set the outcome. In sum, Park's important series of experiments illustrated that the results of competition could vary as a function of at least four factors: temperature, moisture, parasites, and genetic strains. Park's experiments also showed that stochasticity occurred even under controlled laboratory conditions. Consequently, the results of competition in nature are probably even more unpredictable, because of the vastly greater number of variables involved.

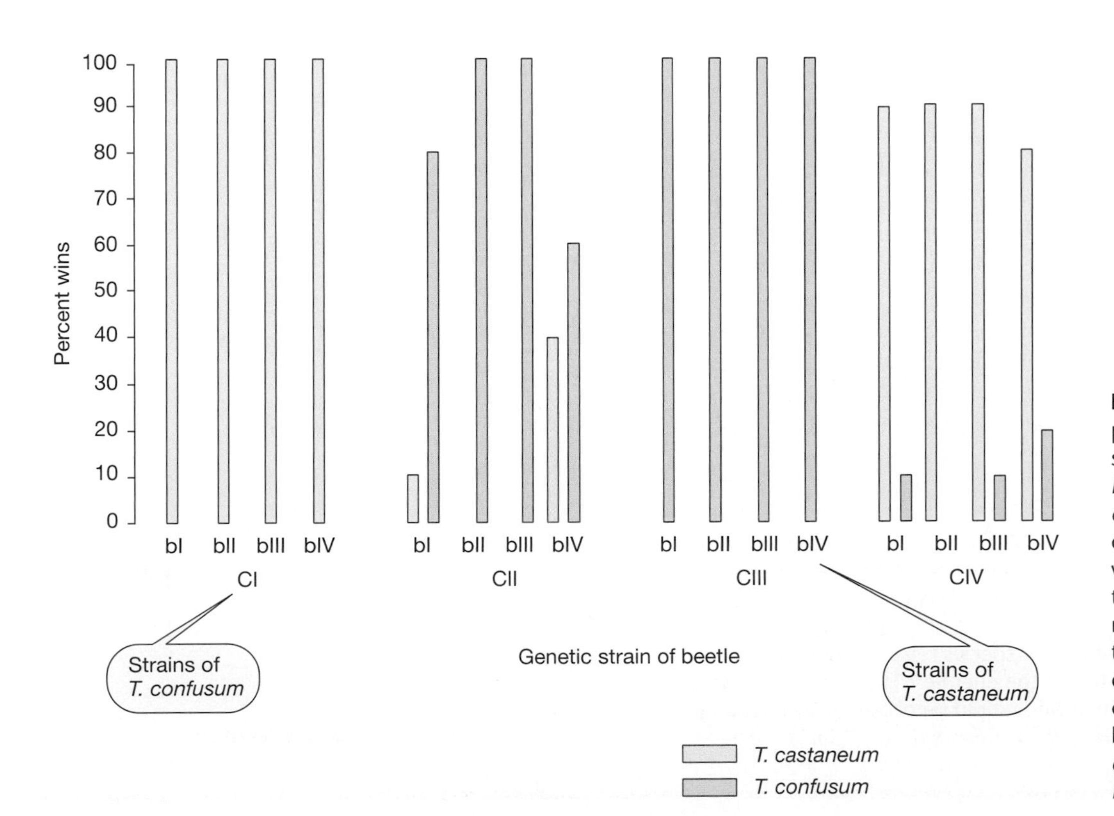

Figure 8.5 Results of competition between different strains of flour beetles *Tribolium castaneum* and *T. confusum*. Strains CI and CIV of *T. confusum* nearly always won in their trials; thus, it is the strain of *T. confusum* rather than *T. castaneum* that sets the outcome. However, once again, some stochasticity was evident. (See bars for CII versus bIV. *(After data in Park, Leslie, and Metz, 1964.)*

In nature, interspecific competition may vary among habitats and areas.

Interspecific competition: Natural systems.

What of systems in nature, where far more variability exists? One view holds that competition in nature is rare because, by now, of all pairs of potential competitors, one competitor has displaced the other (the "ghost of competition past"). An alternative view holds that competition is a common enough force in nature to be a major factor influencing evolution. Yet a third alternative is that predation and other factors hold populations below competitive levels.

The question is important in many applied situations. For example, in biological-control campaigns, it is vital to know whether releasing one natural enemy against a pest is likely to be more effective than the release of many enemies, because competition among enemies might reduce their overall effectiveness. (See "Applied Ecology Box: Is the Release of Multiple Species of Biocontrol Agents Beneficial?")

The most direct method of assessing the importance of competition is to remove individuals of species A and to measure the response of species B. Often, however, such

Applied Ecology

Is the Release of Multiple Species of Biological Control Agents Beneficial?

It is widely recognized that the control of pests is of paramount importance to agriculture. Biological control is seen by many as an alternative that is preferable to chemical control. To some, biological control is best practiced by releasing a variety of enemies against each pest and seeing which one does the best job. But is this the best strategy? It is possible that by releasing many enemies at once, competition for pest prey would be intense, lessening the effectiveness of each biological control agent. Biocontrol worker Les Ehler and his student R. W. Hall (1982) compared the establishment rates (i.e., the proportion of releases that established a viable population) of enemies released against different taxa of pest: sap-sucking bugs (Homoptera), butterflies and moths (Lepidoptera), and beetles (Coleoptera) (Fig. 1).

Wherever biocontrol had failed, the establishment rate dropped with increasing numbers of species released, indicating that competition may be important in biocontrol releases and may contribute to the failure of some campaigns. Even when biocontrol was deemed successful, the establishment rate of single-species releases was significantly greater than the rate for the simultaneous release of two or more species of natural enemies (76% versus 50%). Of course, the data do not *prove* that competition is occurring (Keller, 1984); a demonstration of competition is much more convincing when appropriate experiments are done. However, the data are consistent with the idea that competition can occur in biocontrol releases.

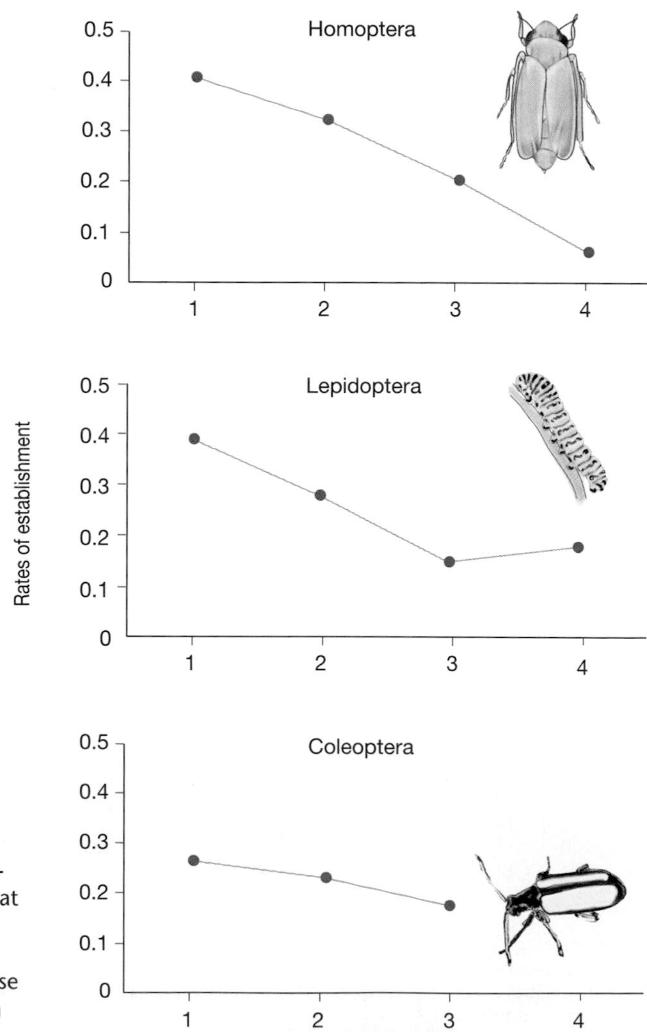

Figure 1 In cases where biocontrol was not completely successful, the establishment rate (the percentage of releases that established viable populations) of natural enemies released against Homoptera, Lepidoptera, and Coleoptera pests dropped as the number of species released increased, because of competition between the enemies. Thus, in many cases, a greater establishment rate would be obtained by releasing single species. *(After data in Ehler and Hall, 1982.)*

manipulations are difficult to make outside the laboratory. If individuals of species *A* are removed from an area, what's to stop them from migrating back into the area? If cages are used to stop species *A* from immigrating into an area, species *B* is prevented from emigrating from the area, and its numbers may reach unnaturally high levels. (This is the *cage effect* or *Krebs effect*, so called because Charles Krebs had shown that caged small mammal populations reached unnaturally high densities, changing their behavior.)

One of the best examples of competition in nature over a wide scale involves parasitic wasps. Pest control is big business in the United States, and for some biological control projects much labor is available, in the form of trained "scouts," to survey an area for evidence of successfully reproducing released enemies. Thus, when three species of parasitic wasps of the genus *Aphytis* were introduced into southern California to help control the red scale (*Aonidiella aurantii*), an insect pest of orange trees, an unprecedented amount of information was summarized by the longtime University of California biocontrol worker Paul Debach (Debach and Sundby, 1963). The first parasitic wasp released, *Aphytis chrysomphali,* was introduced accidentally from the Mediterranean in 1900 and became widely distributed in Southern California. In 1948, *A. lignanensis* was introduced from south China and began to replace *A. chrysomphali* in many areas, such as Santa Barbara County and Orange County (Fig. 8.6). Still, the

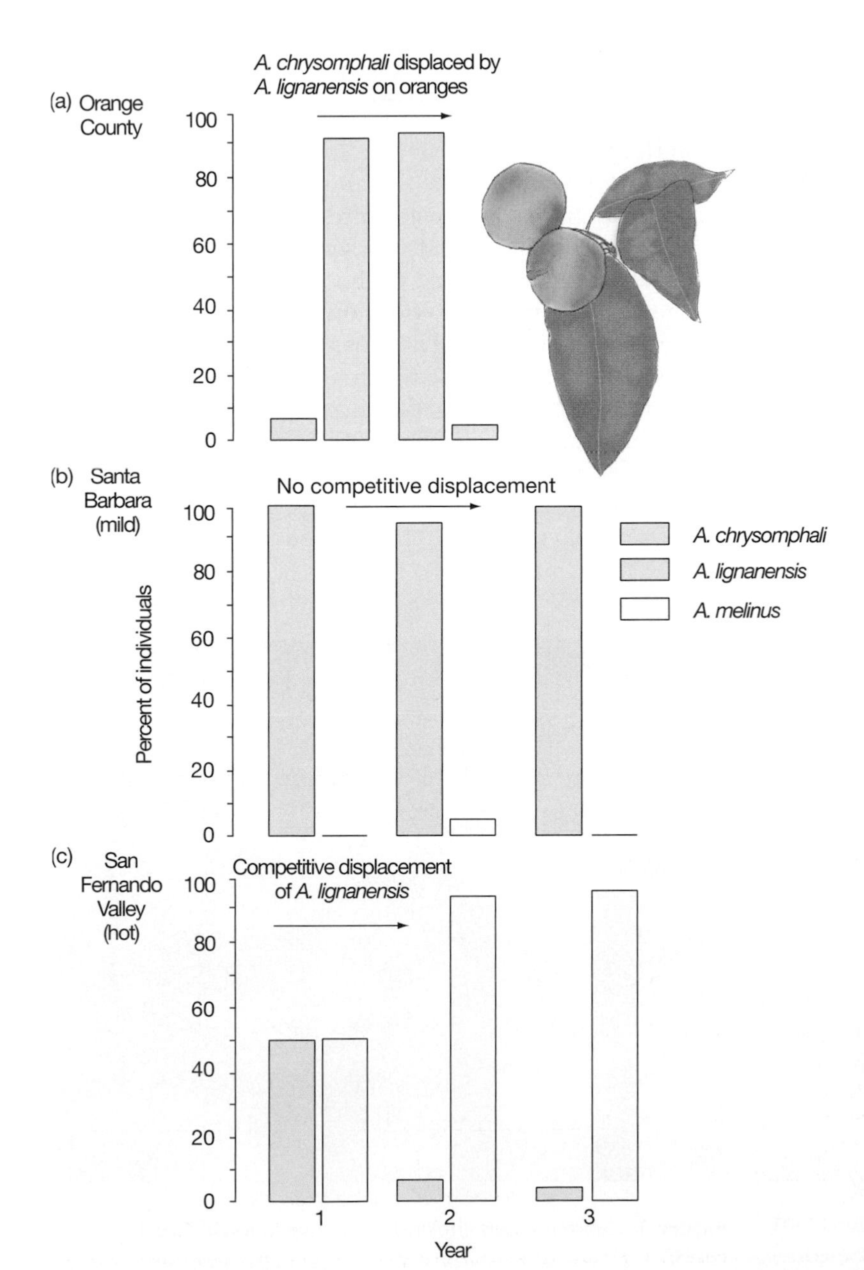

Figure 8.6 Interspecific competition: replacement of one parasite by another in the orange groves of southern California. (a) *A. chrysomphali* was replaced by *A. lignanensis* in the late 1950s as the dominant parasite of scale insects on oranges. (b) and (c) *A. lignanensis* was later replaced by *A. melinus* in hotter interior areas such as the San Fernando Valley, but not in milder areas like Santa Barbara. *(Based on data in Debach and Sundby, 1963.)*

scale pest was not completely controlled, especially in hotter areas, so, in 1956–1957 another species, *A. melinus,* was imported from India. *A. melinus* immediately displaced *A. lignanensis* from the hotter interior areas such as the San Fernando Valley. However, in the milder maritime habitat of coastal Santa Barbara, *A. lignanensis* was still dominant. As Park had noted, climate can alter a competitive outcome.

The mechanism by which competitive displacement occurred was that the female *A. melinus* could use smaller scale insects as hosts and could lay a higher proportion of "female" eggs in them than the other two species. Thus, *A. melinus* preempted most scales as hosts before the other parasites could use them. The high proportion of "female" eggs it laid gave *A. melinus* a competitive edge because its populations could grow faster. There are probably very few other examples of such well-documented competitive displacement in nature. However, the caveat here is that all species were exotic and may have been expected to be more competitive than native species that evolved together over millions of years. Nevertheless, this example underscores the point that the study of competition is important in real-world situations. For instance, an understanding of competition between native and exotic species and what the likely outcomes will be is especially important as we are faced with a growing number of invading exotic pest species. (See "Applied Ecology Box: The Effects of Exotic Competitors on Native Fauna.")

Applied Ecology

The Effects of Exotic Competitors on Native Fauna

Worldwide, humans have succeeded in establishing at least 330 species of nonnative birds and mammals in more than 15 hundred separate cases. In many instances, exotics compete with natives for resources, often with serious consequences. In New Zealand, there is a large overlap in food choice among introduced red deer, feral goats, and the native kokako bird, and the competition has threatened the kokako with extinction.

Introductions are a problem in aquatic environments, too. The introduction of the cichlid fish *Cichla ocellaris* (a native of the Amazon) into Gatun Lake, Panama, is thought to have led to the elimination of six of the eight previously common fish species within five years (Zaret and Paine, 1973). Similarly, after construction of the Welland Canal linking the Atlantic Ocean with the Great Lakes was completed, much of the native fish fauna was displaced by the alewife (*Alosa pseudoharengus*) through competition for food (Aron and Smith, 1971). Also, in the United States and Canada, introduced zebra mussels (Fig. 1) threaten to swamp out many native invertebrates in the Great Lakes. Accidentally introduced from the Caspian Sea region in ships' ballast water around 1986, the zebra mussel filters so much algae from the water column, that little is left for other species, especially native mussels. These kidney-bean-sized invaders also encrust the shells of native invertebrates, so that they cannot easily open their shells and breathe. In California, 48 of 137 species of freshwater fish are nonnative. Of these, 26 have been well studied and 24 are known to have a negative impact on native fish (Herbold and Moyle, 1986).

In continental situations, competitive interactions from both introduced birds and introduced mammals are common (found in 28% and 29%, respectively, of case studies examined; Ebenhard, 1988). Mammals are particularly serious competitors in some instances, especially in Australia, where the introduced rabbit is a serious competitor for burrows with an animal called the boodie, the only burrowing kangaroo, and with the common bandicoot. Among the predators, the dingo (feral dog) has been thought to have excluded the thylacine (a native marsupial wolflike animal) from mainland Australia. Goats and sheep probably compete with a range of kangaroo species, especially the brush-tailed rock wallaby and larger species, such as the red kangaroo and the western gray kangaroo. In Europe, the American beaver has been introduced into Finland, where

Figure 1 Zebra mussels choking out native fauna in Great Lakes.

it excludes the European beaver, with the result that the two species now have separate geographic distributions.

On islands, the consequences of exotic invasion are even more dramatic. In Hawaii, a dense prickly shrub called Koster's curse, introduced from the American tropics for erosion control, now infests more than 40,000 hectares on Oahu alone. The fire tree, an import from the Azores and the Canary Islands, is found in all the major islands. In both cases, native vegetation cannot grow under the exotics' dense canopies. The banana poka, an Andean vine imported as an ornamental, suffocates forests, causing mature trees to topple. At least 877 exotic plant species occur in Hawaii, and other islands of the world—especially the Galápagos and the West Indies—are also affected by exotic species.

Invasive plants are a problem on continents as well as islands. In the U.S. northeast, purple loosestrife is an intensive competitor, choking out species in native American wetlands. In Florida alone, three invasives—Brazilian pepper, Australian pine and, especially, punk tree (*Melaleuca*)—are suffocating native vegetation in the Everglades. Predicting which type of exotic species will have the greatest effect and in which type of native habitat is important, but as of yet, we seem to have had little success. One author found serious disagreements among three reviews since 1995 that sought to determine the common features of Britain's invasive plants (Williamson, 1999). One review found that large seeds favored invasions, one found that small seeds did, and the third found that size didn't matter.

Introduced insects may be useful in biological control, but species introduced accidentally can quickly get out of hand. One of the worst cases concerns the imported red fire ant (*Solenopsis invicta*), accidentally introduced into the United States at Mobile, Alabama, in 1918 in a ship's ballast—in this case, soil from Argentina. Since 1918, these ants have spread rapidly throughout the southeastern United States, and recently they became established in California. Although their northward expansion may be limited by their inability to overwinter in colder environments, they may yet spread along the western states to Canada. Fire ants have been found to outcompete many U.S. species of native ants and to alter the appearance of old field habitats with their prominent nests. In the contiguous 48 states, over 1,500 insect species have been introduced (Sailer, 1983), and other competitive effects are common, though not yet as well documented as those of the fire ant. Exotic species of all types are clearly altering native communities and ecosystem functions worldwide (Bright, 1998). A better understanding of how species outcompete one another is a valuable tool in the continuing fight against exotics.

8.4 The Frequency of Competition

Syntheses of experimental results show that competition occurs very frequently in nature.

Despite the difficulties of demonstrating competition, it is valuable to know how frequent it is in native communities. As previously discussed, the most rigorous method of demonstrating competition is by experimentation: Remove one species and see what happens to the other. Because of shortages of time and money, most ecologists have been able to do this only over small scales in the field. Nevertheless, it is logical that reviews of competition focus on these field experimental studies, because such experiments usually involve native systems. In 1983, Joe Connell, an ecologist at the University of California at Santa Barbara (see "Profiles" box) reviewed 72 studies on active competition that were reported in the literature. Competition was found in 55 percent of 215 species surveyed. Connell suggested that this percentage was reasonable if one took the following view: Imagine a set of resources with, say, four species distributed along it—for example, a series of different-sized grains fed upon by an ant, a beetle, a mouse, and a bird. Then, if only adjacent species competed, competitive effects would be expected in just three out of the six pairs of species (50 percent; see Fig. 8.7). Of course, the mathematics would be different according to the number of species on the axis. If there were only three species along the axis, we would expect competition in two of the three pairs, or 66% of the time. For any given pair of adjacent species that used the entire set of resources, competition would be expected, and indeed, in studies of single pairs of species, Connell found that competition was almost always reported (90 percent), whereas in studies involving more species, the frequency was only 50 percent.

In a parallel, but independent, review of 150 field experiments, Tom Schoener (1983) reported competition in 75 percent of the species studied. We can attribute the difference in Connell's and Schoener's studies to slightly different samples, methods of analysis, and even predispositions of the authors. Connell, perhaps believing predation to be a more important force in nature, was more rigorous in what he accepted as a satisfactory experiment. For example, Hairston (1989) points out that at least one of the experiments accepted as evidence of competition by Schoener did not meet the necessary requirements of experimental design. However, Connell may have been *too* restrictive

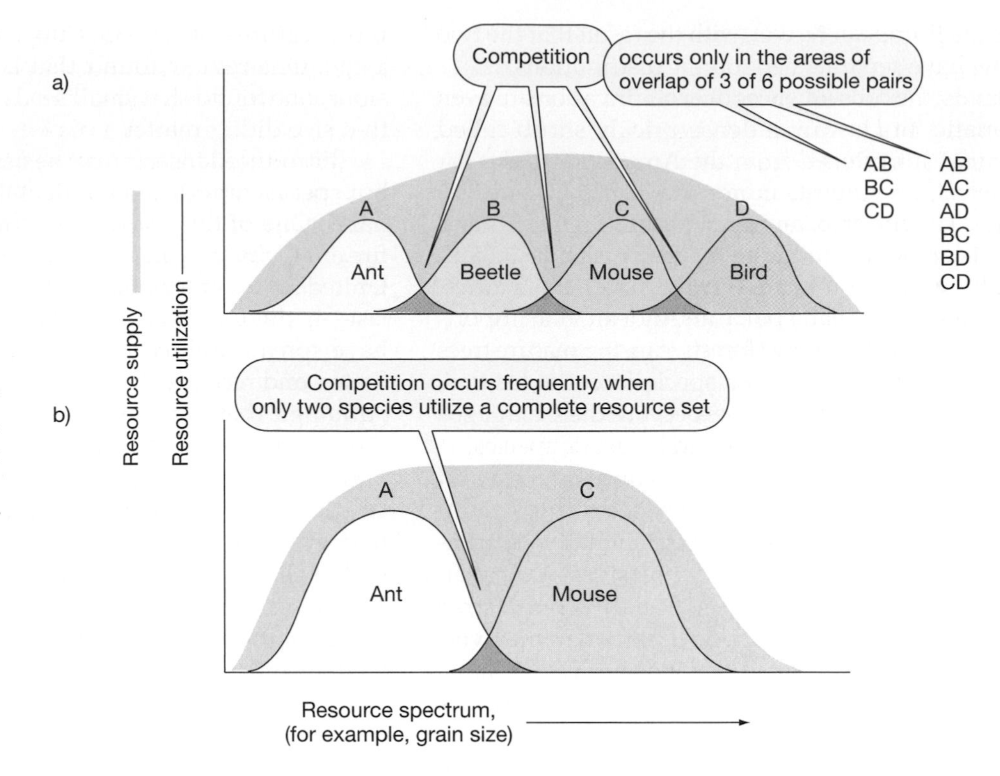

Figure 8.7 (a) Supply and utilization curves of four species, *A*, *B*, *C*, and *D*, along a resource gradient. If competition occurs only between species with adjacent resource utilization curves, then competition will be expected between *A* and *B*, *B* and *C*, and *C* and *D*—three of the six possible pairings. (b) When only two species utilize a set of resources, competition would nearly always be expected between them.

Profiles

Joe Connell, University of California, Santa Barbara

It's hard to know whom one should credit with helping in any of the successes that occur in one's career, but I'm sure that the nuns in my first eight elementary school years had a lot to do with teaching me the discipline to get a job done. I certainly learned that I'd better always have my homework done (and done correctly) when I got to school in the morning!

I had a hard time deciding what to do with my life. I lived in a small town where the only professions I saw practiced were doctor, lawyer, and engineer. So I began college in an engineering school. Only after I got in the army (during World War II) did I meet people who had studied to be biologists, usually wildlife managers. Since I had always loved watching birds and identifying trees, it was a revelation to discover that somebody would actually pay you to do that! I've never worried about the size of my paycheck, because I always reckoned that since doing ecology was such fun, one shouldn't really be getting paid for it.

In the war I was trained in meteorology, and as soon as I got out of the army I finished a B.S. degree in meteorology at the University of Chicago. Then I finally decided what to do with my life: become a biologist. I went to U. C. Berkeley, where my advisor was A. C. Leopold. He advised me to go for an M.A. degree and to do my research on a potential game animal that was very common, but completely unstudied: the brush rabbit. This choice had a great influence on my future research career: I learned that one shouldn't choose a research species solely because it had never been studied or a topic solely because it might some day be a benefit to mankind. The reason this animal had never been studied soon became clear: It was very difficult to capture, and even when I learned how (I caught one 19 times!), my results were pretty dull—I knew where I had caught them, but not much more.

It was discouraging, dull research, and at the finish I decided that this wasn't for me and became a high school teacher. This was rewarding, hard work, and I would probably have stayed with it, but I was motivated to go back to college when the government announced that any veteran who was on a G.I. Bill scholarship must use it up immediately or lose it. So I decided to go back to the university, but in a foreign country this time. I went to the University of Glasgow, in Scotland, because I had met the professor of zoology, C. M. Yonge, when he visited Berkeley. (Here I'd like to acknowledge the G.I. Bill, which gave me four years of college and the opportunity to become an ecologist; it may be the best "entitlement" our government ever provided its citizens.)

Having learned the hard way what *not* to choose as a study animal, I remembered a wonderful paper by E. S.

Deevey in the *Quarterly Review of Biology* in 1947 that I had had to review in a graduate seminar. In it, he described a 1938 paper by H. Hatton in an obscure (to Americans) French journal. Hatton's paper was in French, but I slogged through it and discovered the idea of controlled field experiments, which Hatton had conceived and carried out in 1929–1932 on three marine intertidal animals and three algae. I decided that I would study one of the same species he did—a barnacle—and since Hatton had concentrated on physical factors, I would concentrate on biological ones, namely, the effects of population density and predators, using field experiments, as his paper had taught me. (Research is usually regarded as the sole alternative to teaching by university faculty, but some of the most important teaching is done through one's papers.) So Deevey and Hatton were my early role models, both through their publications; I met Deevey later, and I went to France to try to find and thank Hatton, but couldn't; his professor thought he might have died in the war.

One aside about doing a Ph.D.: My professor wisely advised me not to try to spread my Ph.D. research too widely, but to confine it to the effects of density and predators. However, I saw what looked like a neat interaction between two competing species of barnacles, so, without telling him, I went ahead and did a field experiment on competition. I didn't include it in my final Ph.D. thesis, so he never found out about my disobedience! It turned out to be one of my most-quoted papers, so my advice is to be kind to your professor, but don't always take his advice.

I find dreaming up new ideas and getting and analyzing data easy and fun, but writing is very difficult for me. I have to force myself to stop reading a mystery novel and get down to writing; here is where the training in discipline by my elementary school nuns has its effect. (After what I put them through, they just have to be in heaven!) I usually write 5 to 10 (sometimes more) drafts of a paper before I'm satisfied, and I inflict many of these drafts on my long-suffering friends to criticize.

However, sometimes data analyses were the hardest part, particularly in two papers in which I reviewed and analyzed the data from papers written by others: one in 1983 for *American Naturalist* and one in 1997 in *Coral Reefs*. Here the problem was to understand the data from the author's published tables or figures. In a few cases, I had to write the author and ask her or him to clear up my difficulty. In two cases this resulted in the author discovering errors in the published results, so the only accurate analyses are in my papers! In one paper I simply couldn't figure out the results, so I had to use only a part of the data. I also discovered (in the 1983 paper) that papers containing only negative results apparently seldom get published.

I've not been embroiled in much controversy, although my 1983 review paper was stimulated by a particularly acrimonious criticism that I felt I had to answer. Critical comments are usually highly constructive and are routinely given by reviewers of grant proposals or papers one has submitted for publication. I certainly need them, since I seem to be unable to recognize my own shortcomings. Having colleagues who don't hesitate to point them out is essential, and I've been very lucky in this; in our group, we all realize that it's better to have one's friends, rather than one's rivals, correct your errors!

I was very lucky that my career coincided with the golden age of funding support for science, 1950 to 1980. When competition for grants got heavy in the 1980s, I had already established a track record, so I continued to be able to get National Science Foundation funding. I got turned down several times on first drafts of a proposal, but usually managed to rewrite it and get the funding the second time. Young scientists starting out now have a much tougher time than I did when I got my first grant in 1959. Luck and chance play a part in ecology and in one's life, and I've been very fortunate.

---

in accepting studies that appeared only in a narrow set of journals and rejecting articles that reported the response of only one species at one time and one place (Ferson *et al.*, 1986), therefore finding too few examples of competition. This scenario shows us how easy it is for respected scientists working on a similar problem to come to different conclusions because of different methodologies.

In reality, both reviews may well overrepresent the actual frequency of competition in nature, because of a couple of flaws commonly, and perhaps unavoidably, perpetrated by scientists as they go about their investigations:

1. "Positive" results demonstrating a phenomenon (here, competition) may tend to be more readily accepted into the literature than "negative" results, which do not demonstrate patterns indistinguishable from randomness.
2. Scientists do not study systems at random; those interested in competition may well choose to work with a system in which competition may be more likely to occur.

On the other hand, the reviews may fail to reveal the true importance of competition in evolutionary and ecological time, because

1. By now, most organisms have evolved to escape competition and the lack of fitness it may confer, and
2. Competition may occur only in certain "crunch" years in which resources are scarce (Weins, 1977), but

even this infrequent competition is severe enough to structure a community. If only one in five years is a crunch year and a researcher does experiments in any of the other four, competition may go undetected.

Assuming these tendencies cancel each other, we can accept the overall conclusion that competition is frequent in nature. Furthermore, some general patterns are evident from Connell's and Schoener's work if one assumes that they did not suffer any taxonomic biases in reporting the frequency of competition (Fig. 8.8). Plants showed a high degree of competition, perhaps because they are rooted in the ground and cannot easily escape competition's ravages. Marine organisms tended to compete more than terrestrial ones, perhaps because many of the species studied lived in the intertidal zone, where space on the faces of rocks is limited. Vertebrates can undergo both resource and interference competition, a fact that would explain why the frequency of competition is high in that subphylum compared with invertebrates. Around the time that Connell and Schoener were conducting their reviews, it was argued that many invertebrates—especially insects, of which there are numerous species—may not be common enough to compete. Lawton and McNeill (1979) had suggested a few years earlier that insect herbivores often "lie between the devil and the deep blue sea"—that is, between a huge array of predators and parasites on the one hand and abundant, but low-quality, food on the other. As a result, they could scarcely become common enough to compete. Insect herbivores also rarely undergo the direct, physical conflict of interference competition.

Competition can occur via many mechanisms, including fighting and the use of needed resources.

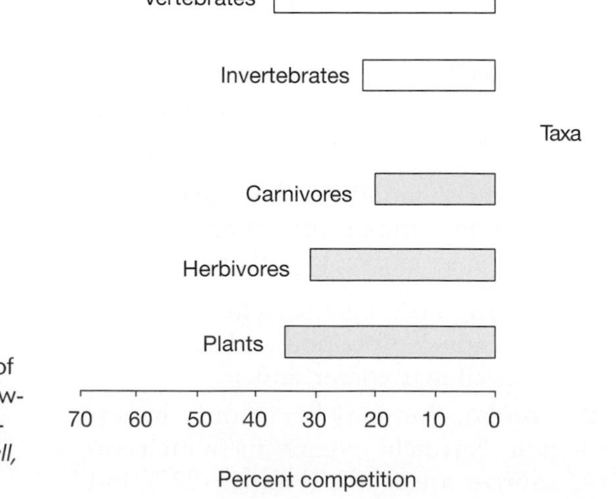

However, a later review by insect ecologist Bob Denno and his colleagues (Denno, McClure, and Ott, 1995) suggested that, in the mid-1980s to early 1990s, competition in insects was actually found more frequently than in previous decades, so now the pendulum has swung the other way. Denno's group's examination of 193 pairwise species interactions in insects showed interspecific competition in 76 percent of the cases. Why the difference? Insect ecologists had perhaps underestimated the ability of plants to respond chemically and physiologically to herbivore attack, including how far such effects could spread through the plant's transport system and how long they would last, discouraging feeding by other species. That is, herbivores occupying different parts of the plant may still compete by exploiting a common resource such as phloem sap. Aphids feeding on roots compete with other aphid species that form leaf galls without ever coming directly into contact with them (Moran and Whitham, 1990). Likewise, even herbivores that are separated in time are potential competitors if they induce a long-lasting response in the plant. For example, larvae of the winter moth (*Operophtera brumata*) feeding on young oak leaves have severe adverse effects on leaf miners that appear later in the season because the moths change the chemistry of the leaves, making them less nutritious for the leaf miners (West, 1985). If competition is frequent between insects, it is likely an even more pervasive force in nature than Schoener or Connell had imagined, since the class Insecta showed the lowest frequency of competition in their reviews.

By what mechanism does competition most commonly operate? Schoener (1983) divided the mechanisms into six kinds:

1. *Consumptive or exploitative:* using resources.
2. *Preemptive:* using space.
3. *Overgrowth:* one species growing over another and blocking light or depriving the other of some other resource.
4. *Chemical:* through production of toxins.

Figure 8.8 Percentage of experimental studies showing interspecific competition. *(From data in Connell, 1983.)*

5. *Territorial:* fighting or exhibiting other behavior in defense of space.
6. *Encounter:* transient interactions directly over specific resources.

Consumptive competition is by far the most common form of competition, occurring in 71/188 = 37.8 percent of cases (Table 8.2). Some of Schoener's findings about the mechanisms of competition are easy to interpret: Preemptive and overgrowth competition appear among sessile space users, primarily terrestrial plants and marine macrophytes and animals living on hard substrates; territorial and encounter competition occur among actively moving animals, especially birds and mammals; and chemical competition occurs primarily among terrestrial plants, because toxins tend to become too diluted in aquatic systems.

A final tidbit gleaned from both Connell's and Schoener's studies is that, most often, only one member of a pair of species responded to the addition or removal of individuals of the other species. The logic here is that such asymmetric competition should be expected—the superior competitor will probably be more strongly limited by some other factor, perhaps environmental tolerances or predators. Usually, the larger organism has the competitive advantage (Persson, 1985). Asymmetric competition is often called *amensalism* and may be particularly important in plants, wherein one species might secrete chemicals from its roots which inhibit the growth of other plants that do not secrete such chemicals. For example, in southern California chaparral, grassland shrubs such as the aromatic *Salvia leucophylla* and *Artemisia california* are often separated from adjacent grassland by bare sand 1 to 2 meters wide. Volatile terpenes released from the leaves of the aromatic shrub inhibit the growth of nearby grasses (Muller, 1966). Some plants, such as the black walnut (*Juglans nigra*), produce similar chemicals (here, juglone) from their roots. The chemicals leach into the soil, killing neighboring roots (Massey, 1925) in a phenomenon termed **allelopathy,** the action of penicillin among microorganisms of which is a classic case. In many cases, such allelopathic chemicals are toxic to some competitors, but not others, and may even have other purposes. For instance, on the tropical coral reefs of Guam, Robert Thatcher and his colleagues (1998) showed how a sesquiterpene chemical from a sponge, *Dysidea,* caused necrosis on a neighboring sponge, *Cacospongia*. However, the chemical also deterred predation by a spongivorous fish, illustrating the multiple ecological roles that some chemicals play.

About a decade ago, Jessica Gurevitch of the State University of New York at Stony Brook, Long Island, updated Connell's and Schoener's reviews by conducting a metaanalysis of competition in field experiments (Gurevitch *et al.,* 1992). This type of analysis uses statistical techniques to compare the strength of effects of competition in various studies, in contrast to the "vote-counting" procedures of both Connell and Schoener, wherein studies are simply counted "yes" or "no" according to whether they show competition or not. Gurevitch's metaanalysis

TABLE 8.2 Mechanisms of interspecific competition in experimental field studies. *(After Schoener, 1985.)* Data represent number of cases found in a literature review of 188 separate studies of competition. Consumptive competition is by far the most common form of competition, occurring in 37.8% of cases.

	Mechanism						
Group	Consumptive	Preemptive	Overgrowth	Chemical	Territorial	Encounter	Unknown
Freshwater							
Plants	0	0	1	1	0	0	0
Animals	13	1	0	1	1	5	2
Marine							
Plants	0	6	4	1	0	0	0
Animals	9	10	6	0	7	6	0
Terrestrial							
Plants	28	3	11	7	0	1	9
Animals	21	1	0	1	11	15	6
Total	71	21	22	11	19	27	17

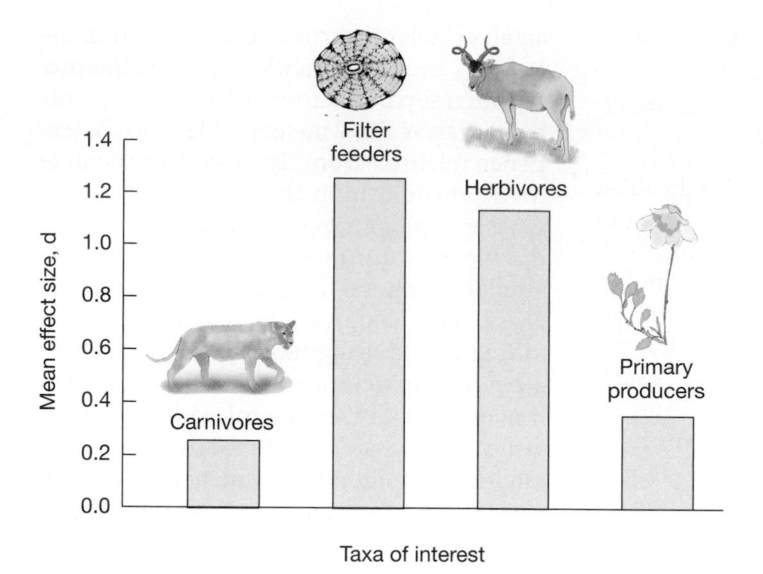

Figure 8.9 Mean size of the effect of competition on biomass for carnivores, filter feeders, herbivores, and primary producers in manipulative field experiments, as revealed by a metaanalysis. Mean effect sizes of 0.8 or greater are considered large (i.e., the experimental group mean is 80% of a standard deviation greater than that of a control group), 0.5 is a medium effect size, and 0.2 is small. (*Reproduced from Gurevitch et al., 1992.*)

The effect of competition on the growth of a population can be predicted by mathematical models.

examined field competition experiments on 93 species in a wide variety of habitats and revealed strong effects of competition on biomass. (Effects of competition on population density were not investigated.) Surprisingly, primary producers (plants) and carnivores did not show such strong effects of competition as did filter feeders and herbivores (Fig. 8.9). Competition still had a medium-sized effect on plants—much less, though, than on herbivores and filter feeders. The filter feeders were primarily barnacles and other intertidal species that inhabited rock faces where space was limited. The herbivores included vertebrates and invertebrates.

Gurevitch's review found no differences in the effects of competition in terrestrial, freshwater, or marine systems, for plants or carnivores, or in high-productivity (fields, prairies, meadows) versus low-productivity systems (deserts, arctic tundra). The latter finding is interesting in view of the fact that botanists (e.g., Grime, 1979) have argued that competition is relatively unimportant for plants in unproductive environments because plant biomass, and therefore resource depletion, is low in such environments, which are dominated by stress-tolerant species. However, others (e.g., David Tilman, 1988, of the University of Minnesota) have argued that competition occurs across all productivity gradients, but that the resources concerned may differ. Thus, in unproductive environments competition is primarily for belowground resources (i.e., nutrients and water), while in productive environments competition is primarily for light. No consensus on the "Grime–Tilman debate" has yet emerged, but Gurevitch's result supports Tilman.

8.5 Modeling Competition

Ecologists like to find shortcuts to predict when and where ecological factors will be important, and competition is no exception. Often, mathematical models are designed to provide such answers. A theoretical basis for competition was first described in the 1920s by two people working independently: Alfred J. Lotka (1925), an actuary who worked for the Metropolitan Life Insurance Company in the United States, and Vito Volterra (1926) in Italy. Lotka–Volterra competition models are based on the logistic equation of population growth—the S-shaped curve we discussed in Chapter 6.

The growth rate of populations of two species coexisting independently (i.e., not interacting with each other) can be described with the use of the logistic growth equations. As before, r is the per capita rate of population growth, N is the population size, and K is the carrying capacity. The subscripts in these equations refer to species, so r_1 is the per capita population growth rate of species 1 and r_2 is the per capita population growth rate of species 2. The two equations are

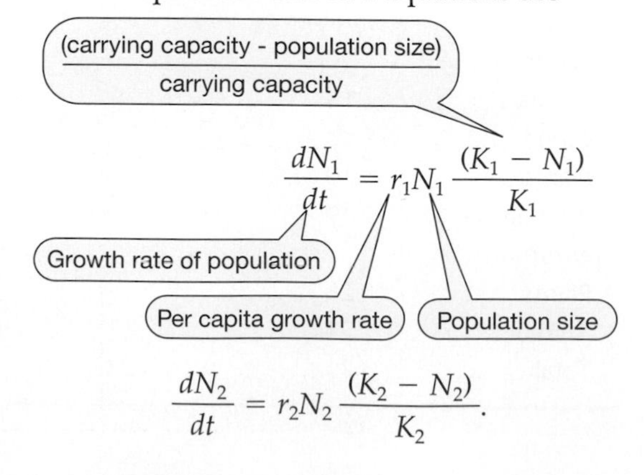

$$\frac{dN_1}{dt} = r_1 N_1 \frac{(K_1 - N_1)}{K_1}$$

$$\frac{dN_2}{dt} = r_2 N_2 \frac{(K_2 - N_2)}{K_2}.$$

In these single-species logistic equations, the growth of a population is affected by the population size relative to its carrying capacity. The quantity $K - N$ defines how far below (or above) the carrying capacity a population is, and it can easily be modified to account for the presence of a second competing species. To do this, it is necessary to define a "conversion factor" that quantifies the per capita competitive effect of species 2 on species 1 and vice versa. For example, if two individuals of species 2 take up the same amount of resources as one individual of species 1, then the per capita competitive effect of species 2 on species 1 is 0.5 (Fig. 8.10). The conversion factor is defined by the term α or β, in which

α = per capita competitive effect of species 2 on species 1

and

β = per capita competitive effect of species 1 on species 2.

The conversion factor allows the logistic equation to be modified to take into account the effects of either species on the growth rate of the other:

$$\frac{dN_1}{dt} = r_1 N_1 \frac{(K_1 - N_1 - \alpha N_2)}{K_1};$$

$$\frac{dN_2}{dt} = r_2 N_2 \frac{(K_2 - N_2 - \beta N_1)}{K_2}.$$

The population size relative to the carrying capacity is now defined by the abundance of both species, and the growth rate of each population is determined by per capita population growth rate (r), the carrying capacity for that species (K), and the abundances of species 1 (N_1) and species 2 (N_2). These equations can be used to describe the joint dynamics of the two populations (i.e., how the abundances of both species changes together).

Relationships such as these can be expressed graphically (Fig. 8.11). The increase in N_1 continues to the carrying capacity of the environment K_1 in the absence of species 2. Alternatively, the whole environment may be filled with individuals of species 2, and N_1 cannot increase because K_2 has been reached. In this case, the equivalent number of individuals of K_2 would be K_1/α. These are the two extremes marked on the axes. Between these extremes are many combinations of N_2 and N_1 at which no further increase in N_1 is possible. These points fall on the diagonal $dN_1/dt = 0$, which is often called the *zero-growth isocline*. The increase in N_2 can be represented by a similar diagram. Combining the two figures and adding the arrows vectorially illustrates what happens when the species co-occur (Fig. 8.12). Essentially, there are four possible outcomes: Species 1 becomes extinct; species 2 becomes extinct; either species 1 or species 2 becomes extinct, depending on the initial densities; and the two species coexist.

One must ask whether these mathematical formulations represent real biological systems. One of the earliest tests of the equations was performed in 1932 by a Russian microbiologist who studied competition between two species of yeast: *Saccharomyces cervisiae* and *Schizosaccharomyces kefir* (Gause, 1932). The advantage of using these species was that large colonies could be grown rapidly in the laboratory. Yeast densities were assessed at different hours by taking a subsample of the medium and counting the number of cells of each species. Gause combined data from two separate experiments

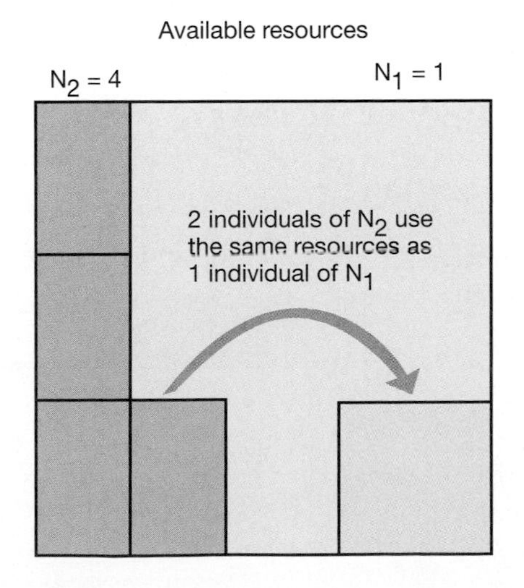

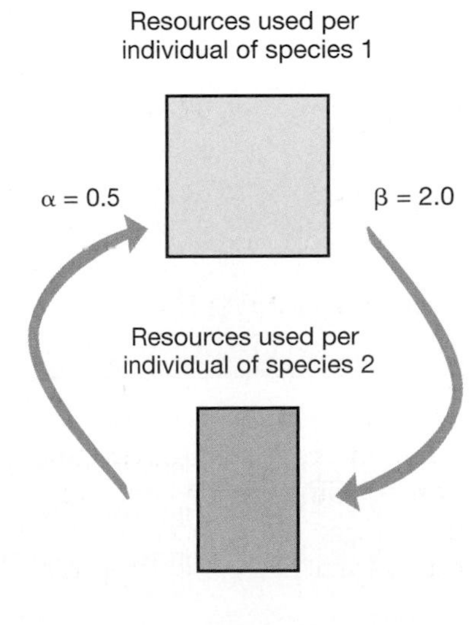

Figure 8.10 Conceptualization of "conversion factors" α and β in Lotka–Volterra competition equations. α is the per capita competitive effect of species 2 on species 1, and β is the per capita competitive effect of species 1 on species 2.

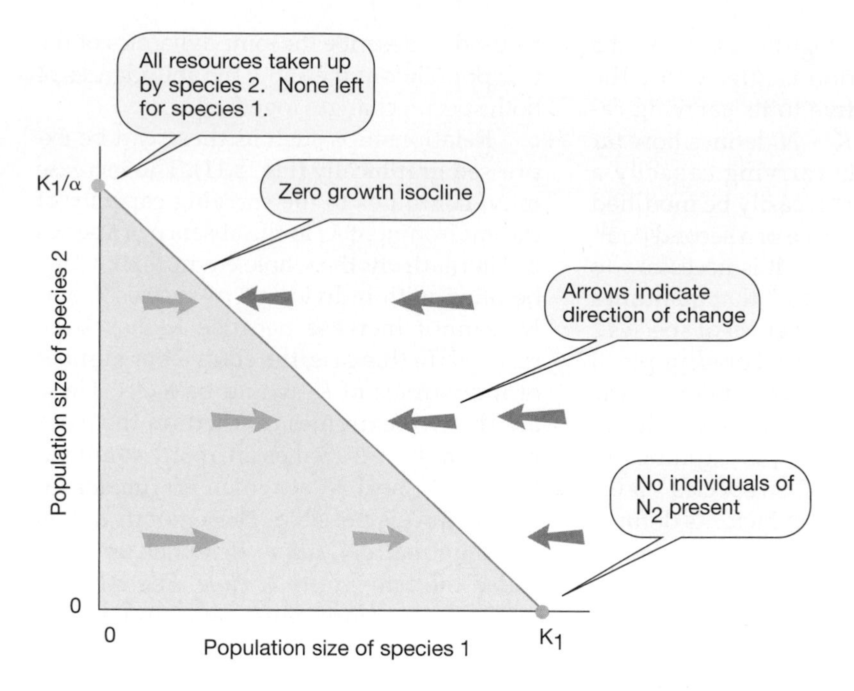

Figure 8.11 Changes in the population size of species 1 when competing with species 2. Populations in the shaded area will increase and will come to equilibrium at some point on the diagonal line. At any point on the diagonal, there is no further population growth (i.e., $dN/dt = 0$). Populations in the unshaded area will decrease in size until they reach the diagonal line.

to get enough points for his graphs (Fig. 8.13). Alone, each species grew according to the logistic curve; the asymptote reached was a function of the concentration of ethyl alcohol, a by-product of sugar breakdown under anaerobic conditions. Ethyl alcohol can kill new yeast buds just after they separate from the mother cell. In cultures where the two yeasts grew together, population densities were lower than they were under single-species conditions (Fig. 8.13). Gause observed that $\alpha = 3.15$ and $\beta = 0.44$; that is, 1

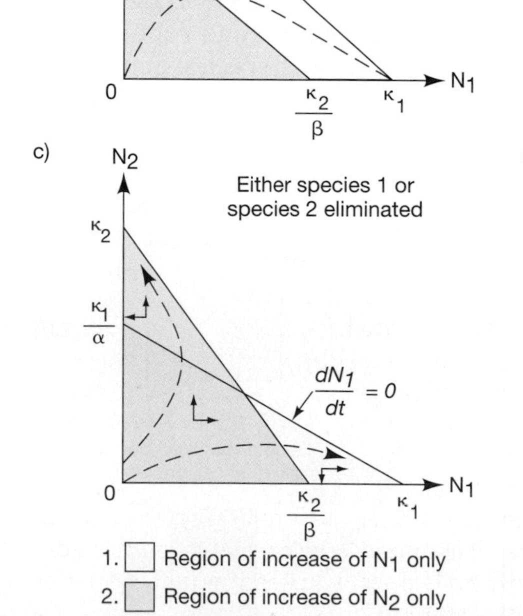

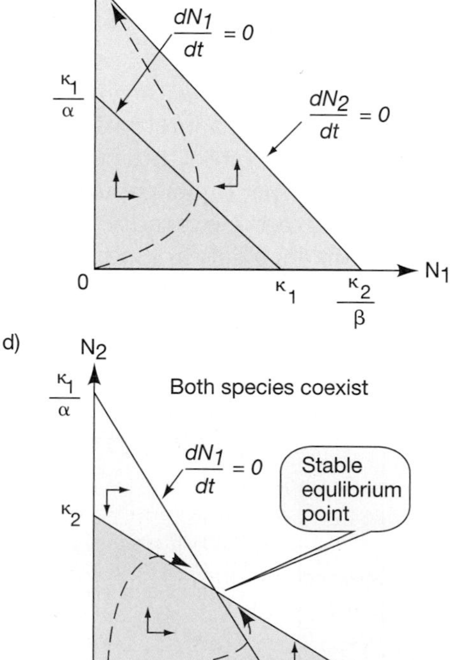

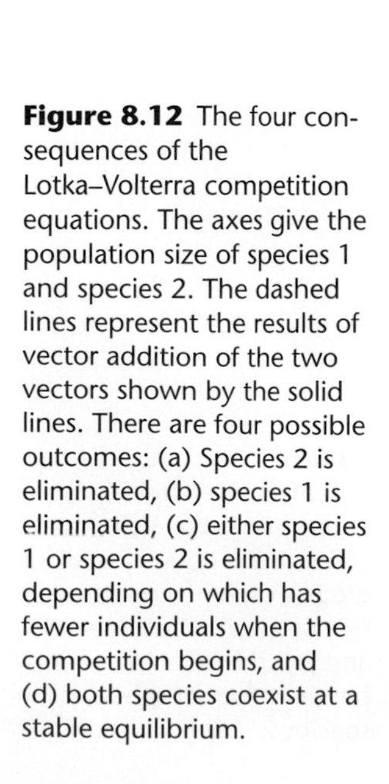

Figure 8.12 The four consequences of the Lotka–Volterra competition equations. The axes give the population size of species 1 and species 2. The dashed lines represent the results of vector addition of the two vectors shown by the solid lines. There are four possible outcomes: (a) Species 2 is eliminated, (b) species 1 is eliminated, (c) either species 1 or species 2 is eliminated, depending on which has fewer individuals when the competition begins, and (d) both species coexist at a stable equilibrium.

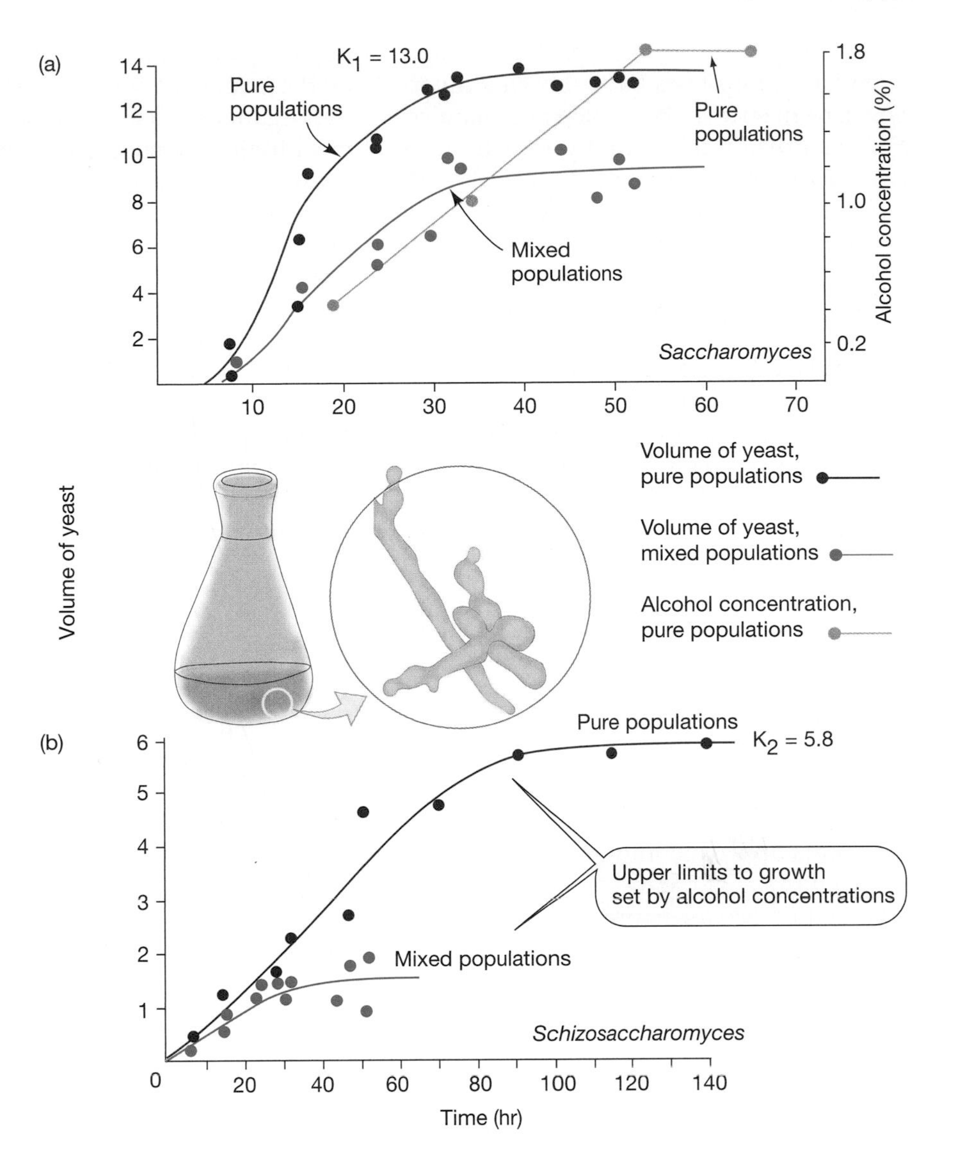

Figure 8.13 (a) The population growth of the yeast *Saccharomyces* is greater in pure cultures than in mixed cultures with *Schizosaccharomyces*, although the logistic growth curve is followed in both cases. The upper limit, or carrying capacity, is set by the alcohol concentration, a by-product that increases as the colonies grow. Eventually, the high alcohol concentration kills new cells. (b) Population growth of the yeast *Schizosaccharomyces* in pure cultures and in mixed cultures with *Saccharomyces*. *(Reproduced from Gause, 1932.)* Upper limits to growth are again set by alcohol concentrations, but Gause did not provide these values.

volume of *Saccharomyces* = 3.15 volumes of *Schizosaccharomyces*, and 1 volume of *Schizosaccharomyces* = 0.44 volume of *Saccharomyces*. Because alcohol is the limiting factor, Gause argued that he could also determine α and β by measuring the alcohol production of the two yeasts, which turned out to be 0.113 percent EtOH/cc yeast for *Saccharomyces* and 0.247 percent for *Schizosaccharomyces*. Thus, $\alpha = 0.247/0.113 = 2.18$, and $\beta = 0.113/0.247 = 0.46$. The values of α and β obtained from the Lotka–Volterra equations were indeed in general agreement with those obtained independently by a physiological method. Gause attributed the difference in the α values, 2.18 vs. 3.15, to be due to the effect of waste products other than alcohol on *Saccharomyces*. (It is interesting that in this situation population growth was limited by pollution of the environment by alcohol, not limiting resources. Many people think the same will be true for human populations.)

Park and his coworkers (Neyman, *et al.*, 1958) showed that, under certain conditions (24°C, 70% relative humidity), the competition between *Tribolium confusum* and *T. castaneum* also obeyed Lotka–Volterra principles, in that the outcome could be influenced by the initial densities of each species, as in Fig. 8.12. When the critical densities were nearly equal, the outcome was uncertain, whereas when one species had a clear numerical advantage at the start, it was assured of winning.

How realistic are the Lotka–Volterra equations in field populations? They are based on the Verhulst–Pearl logistic model, which we discussed in Chapter 6, and so suffer from the same deficiencies. That is, the maximal rate of increase, the competition coefficients, and the carrying capacity are all assumed to be constant, and there are no time lags, so every individual of species 1 and species 2 has the same effect on population growth. In reality, field tests of the equations have rarely been performed, and even laboratory tests have shown a divergence from strict Lotka–Volterra dynamics. For example, in laboratory populations of two species of competing *Drosophila* flies, *D. melanogaster* and *D. simulans*, $K_1 = K_2 = 1.0$ and $\alpha = \beta$, so theoretically, the species could not coexist. Yet coexist they did (Miller, 1964)! Francisco Ayala (1969, 1972) suggested that competition coefficients in *Drosophila* could vary with population densities, causing isoclines to be nonlinear, or bent, as illustrated in Fig. 8.14. Under such conditions, coexistence might be possible, whereas under strict Lotka–Volterra assumptions, it would not. Nonlinear relationships between density and competition coefficients have been found in other organisms, such as frogs (Smith-Gill and Gill, 1978), and nonlinearity may be a common phenomenon (Schoener, 1976).

A drawback to the Lotka–Volterra model of competition is that the mechanisms which drive the competitive process are not specified. Tilman (1982, 1987) criticized this approach and emphasized that we need to know the mechanism by which competition occurs, in order to predict outcomes more reliably. Tilman's alternative to the Lotka–Volterra model was the R^* (R star) concept, in which we need to know the dependence of an organism's growth rate on the availability of resources. For Tilman, the organism was usually a plant and the most limiting resource was nitrogen availability. As resource levels rise, so do plant growth rates (Fig. 8.15). However, plant biomass can also be lost, through the actions of herbivores, pathogens, or physical processes. At a certain resource level, commonly called R^*, gains equal losses, and the plant population is at equilibrium. At resource levels beyond R^*, the population grows. At resource levels below R^*, the population becomes extinct. The level of the resource itself, say, nitrogen, depends on both the amount consumed by the plants and the supply rate, as provided by decomposition of organic matter. Eventually, the plant population increases until the consumption rate equals the supply rate, at which point the resource concentration is at equilibrium with a value of R^*. The plant species with the lowest R^* should displace all the others from the habitat, driving the resource concentration to its own value of R^* (Fig. 8.15c).

Theoretically, for one limiting resource, the single species with the lowest R^* should displace all the others from the habitat. Usually, of course, many plant species coexist in an area, perhaps because the area has a number of limiting resources, such as nitro-

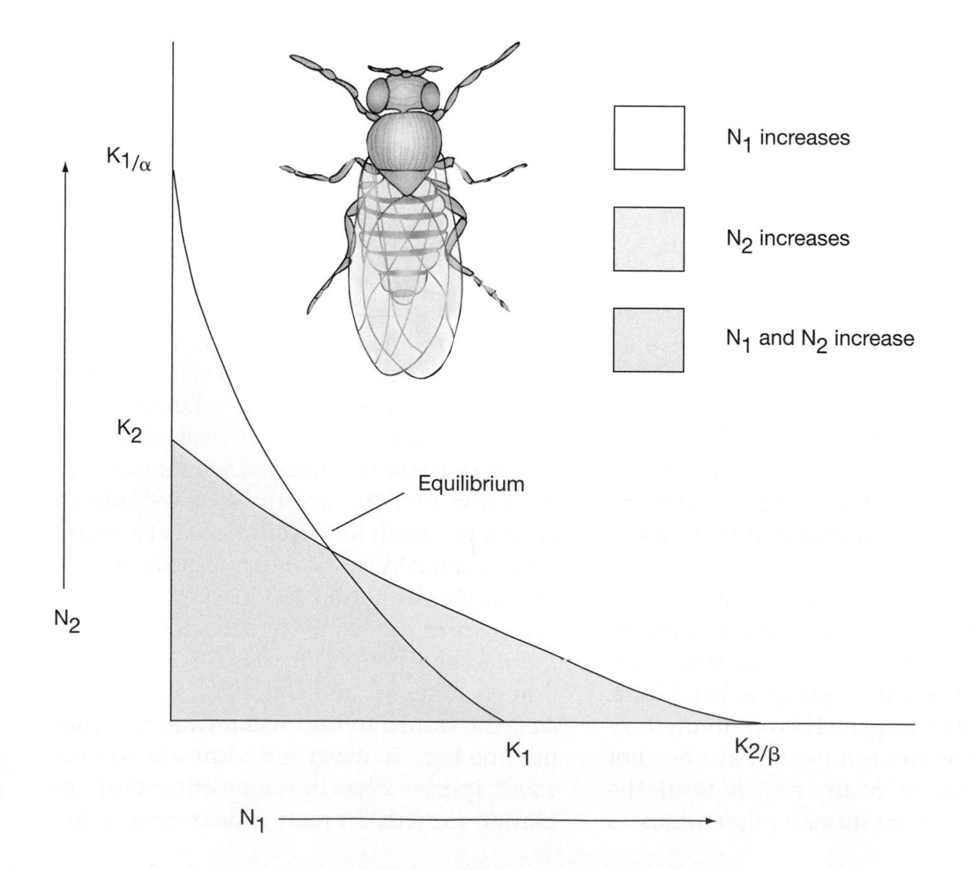

FIGURE 8.14 Nonlinear Lotka–Volterra isoclines, which might result if α and β changed as a function of density. Such isoclines have been observed in laboratory experiments with *Drosophila* flies.

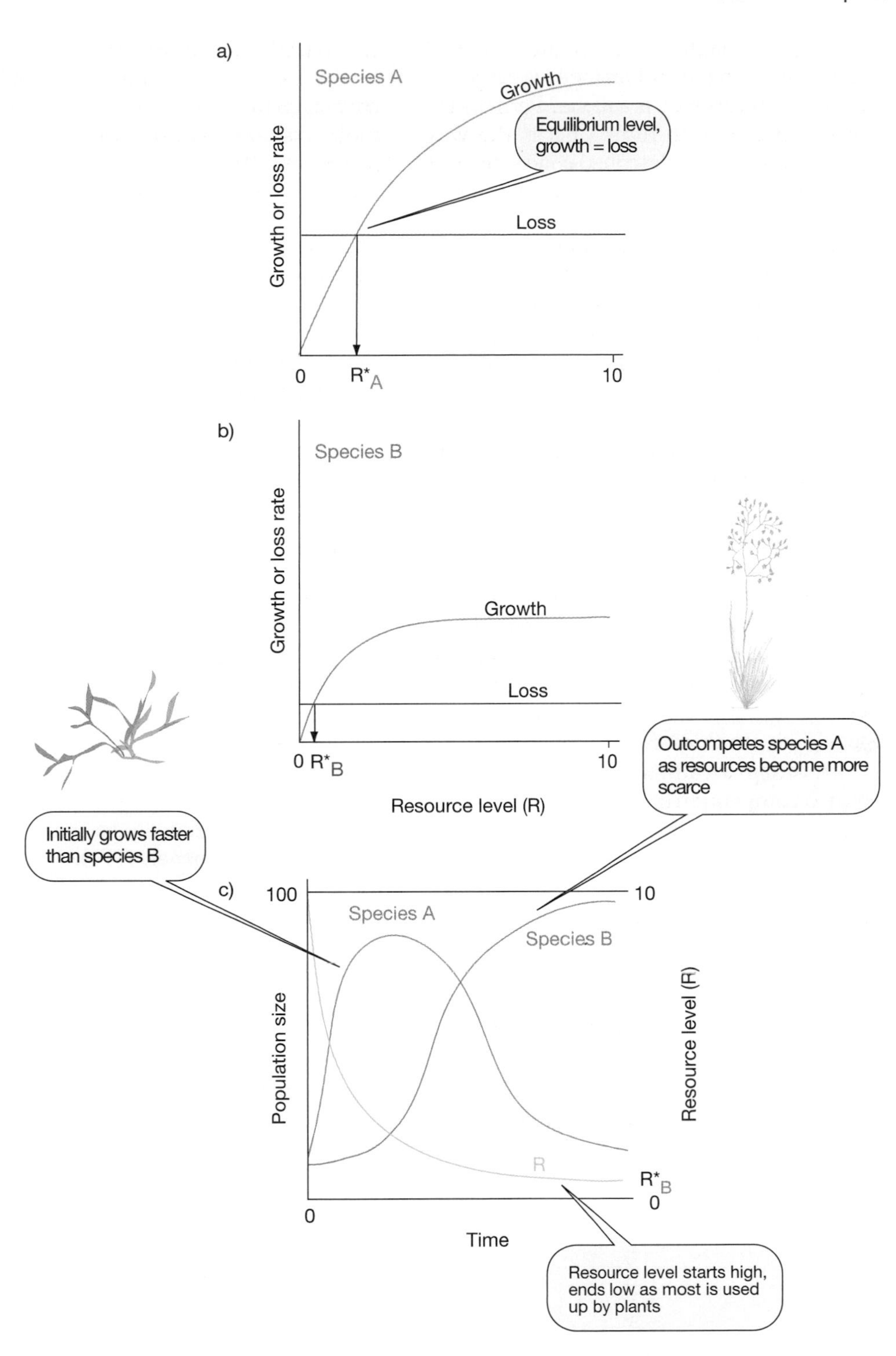

FIGURE 8.15 Tilman's R^* (R star) concept of competition between two species A and B, based on their resource utilization curves. (a) Resource-dependent growth and loss for species A. The resource concentration at which growth equals loss is R_A^*. (b) Resource-dependent growth and loss for species B and its R_B^*. (c) Dynamics of resource competition between species A and B. Note that species B, which has the lower R^*, displaces species A and drives the resource concentration down to R_B^*. *(Reproduced from Tilman, 1997.)*

gen, potassium, phosphorus, and other minerals, as well as resources like water and light. If these resources fluctuate in time and place, even more species may coexist in the same area, because no one species can always outcompete the others. Coexistence may also be possible because plants and other organisms live in a spatially variable habitat, so that even if one species is continually outcompeted in one area, it wins out in another area, and seeds from the second area constantly reach the first, allowing that species to persist there. In yet another mechanism, herbivores may continually chew down the vigorously growing competitors, so that they are never quite able to outcompete their neighbors and drive them to extinction.

8.6 Coexistence of Species

How dissimilar do species have to be in order for them to coexist?

In 1934, Gause wrote, "As a result of competition two similar species scarcely ever occupy similar niches, but displace each other in such a manner that each takes possession of certain peculiar kinds of food and modes of life in which it has an advantage over its competitor." What exactly is a niche? Perhaps the most useful way to think of it is the way Joseph Grinnell first described a niche in 1918, as a subdivision of a habitat that contains an organisms' dietary needs, its temperature, moisture, pH, and other requirements. On the other hand, British ecologist Charles Elton (1927) and G. Evelyn Hutchinson from Yale (1958) viewed an organism's niche as its role within the community, such as a small-seed feeder. Despite these operational differences, Gause was essentially saying that two species with similar requirements couldn't live together in the same place. Garrett Hardin (1960) recommended that "Gause's hypothesis" be known as the *competitive exclusion principle*, according to which direct competitors cannot coexist, just as the Lotka–Volterra equations suggested. But, as anyone who has ever spent time in the field knows, we see myriads of competing species apparently coexisting side by side all the time. How can this be?

In Britain, ornithologist David Lack examined competition and coexistence in about 40 pairs of British passerine, or perching, birds (Fig. 8.16). As a group, these birds had fairly similar lifestyles. Most were separated along some axis, habitat being the usual one. However, in about half a dozen cases, no obvious separation was noticed. Similar separations in feeding position within the tree canopy and in nesting dates were found by Robert MacArthur (1958) among five species of U.S. warblers in New England. Indeed, birds have been a frequent taxon for coexistence studies; Darwin's finches in the Galápagos (Grant, 1986) and terns on Christmas Island (Ashmole, 1968) come readily to mind. In a more wide-ranging review of over 80 natural communities, including slime molds, mollusks, insects, and other arthropods, Schoener (1974) provided the following ranks for resource partitioning:

One hypothesis explaining the coexistence of competing species states that morphological differences between competing species may allow them to coexist.

1. Macrohabitat = 55%;
2. Food type = 40%;
3. Time of day or year = 5%.

Thus, Schoener was in broad agreement with Lack that habitat was of most importance in separating species. However, some other interesting patterns were evident in his analysis: Predators partitioned resources by diurnal differences in time of activity more often than other groups did, and segregation by food type was more important for animals feeding on large food items relative to their own size than for animals that fed on relatively smaller items.

What about the species that do not appear to live in different habitats or feed in different seasons or times of day? How similar can these competing species be and still live together? This question has received more attention in ecology than any other single topic (Schoener, 1974).

In a seminal paper entitled "Homage to Santa Rosalia, or why are there so many kinds of animals?" G. Evelyn Hutchinson (1959) looked at size differences—particularly in the feeding apparatus—between congeneric species when they were **sympatric** (occurring together) and when they were **allopatric** (occurring alone) (Table 8.3). He gathered together what today appear to be scant data: six data sets from four different sources.

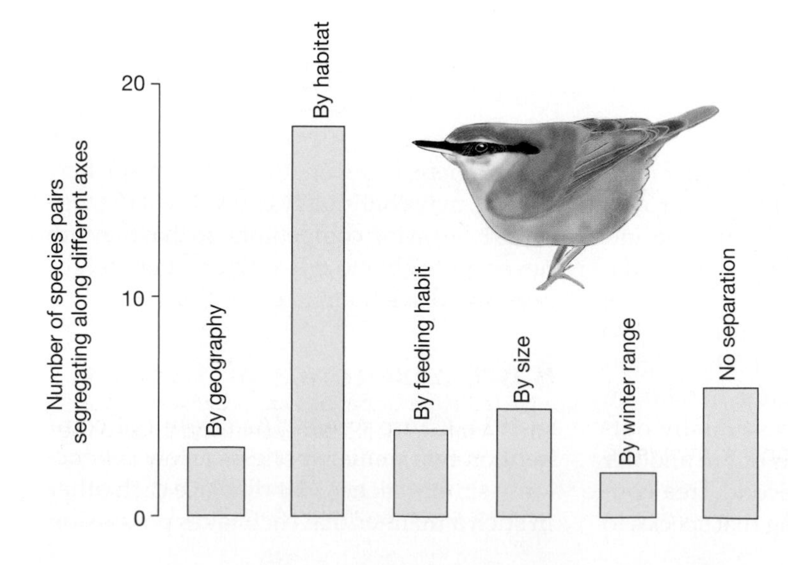

FIGURE 8.16 Type of separation among 40 species of British passerine birds. Most segregation is by habitat. In about 15% of cases, no obvious segregation was observed. *(From data in Lack, 1944.)*

TABLE 8.3 Ratio of larger to smaller dimension of feeding apparatus (culmen in birds) between congeneric sympatric and allopatric species. (*From Hutchinson, 1959*). Figures for *Camarhynchus parvulus, C. psittacula,* and *C. pallidus* are averages from three islands. Data are all of the evidence mustered by Hutchinson.

Animals and Character	Species	Measurement (mm) when Sympatric	Measurement (mm) when Allopatric	Ratio when Sympatric	Ratio when Allopatric
Weasels (skull)	*Mustela nivalis*	39.3	42.9	1.28	1.07
	M. erminea	50.4	46.0		
Mice (skull)	*Apodemus sylvaticus*	24.8	25.6	1.09	1.04
	A. flavicollis	27.0	26.7		
Nuthatches (beak)	*Sitta tephronota*	29.0	25.5	1.24	1.02
	S. neumayer	23.5	26.0		
Galápagos (i) Finches (beak)	*Geospiza fortis*	12.0	10.5	1.43	1.13
	G. fuliginosa	8.4	9.3		
Galápagos (ii) Finches (beak)	*Camarhynchus parvulus*	7.3	7.5	1.27	1.37
	C. psittacula	9.3	10.3		
Galápagos (iii) Finches (beak)	*C. psittacula*	8.9	10.3	1.35	1.09
	C. pallidus	12.0	11.2		

The ratios between the characteristics studied when the species were sympatric ranged between 1.09 and 1.43, and Hutchinson tentatively argued that the mean value of 1.28, which has subsequently been named *Hutchinson's ratio*, could be used as an indication of the amount of difference necessary to permit coexistence at the same trophic level, but in different niches. (Some authors extrapolated Hutchinson's data and concluded that ratios of over 2.0 in different types of measurement (e.g., of biomass or volume) indicated sufficient differences to permit coexistence, since mass varies to the third power of length and $1.3^3 = 2.2$.) Despite its appeal, Hutchinson's idea came under heavy fire for the following reasons:

1. In a large series of 28 studies by different authors purporting to support the hypothesis, the statistical analysis was not appropriate. Further tests on these data sets showed no more differences between species than would occur by chance alone (Simberloff and Boecklen, 1981). There was no evidence of constant size ratios between competing species for most of the species.
2. Size-ratio differences have too loosely been asserted to represent the "ghost of competition past" (Connell, 1980) when, in fact, they could have evolved for other reasons.
3. Biological significance cannot always be attached to ratios, particularly those of structures not used to gather food: Ratios of 1:1.3 have been found to occur between members of sets of kitchen skillets, musical recorders, and children's bicycles (Horn and May, 1977) (Fig. 8.17). Maiorana (1978) argued that a 1:1.3 ratio may simply reflect something about our perceptual abilities.

Needless to say, followers of Hutchinson's ideas have been swift to rebut some of these arguments. Losos, Naeem, and Colwell (1989) argued that the Simberloff–Boecklen statistics were deficient in statistical power; in other words, they erred too strongly on the side of accepting the hypothesis that competition had no effect when such a hypothesis was false. Doubtless, variations of

FIGURE 8.17 Should biological significance be attached to particular size ratios between organisms? A ratio of 1:1.3 has been found to occur between members of sets of kitchen knives, skillets, musical recorders, and children's bicycles. *(Leonard Lessin, Peter Arnold, Inc.)*

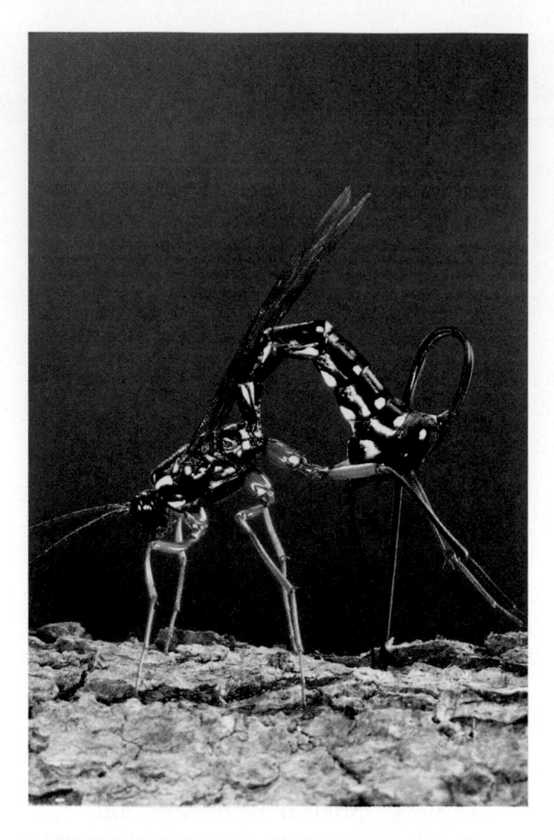

FIGURE 8.18 This parasitoid has taken up the "Derrick" position with the extended ovipositor, to lay its eggs in wood feeding larvae inside the log.

Species may coexist by partitioning resources to avoid competition.

these arguments will continue to be tossed back and forth. It is noteworthy, however, that even Daniel Simberloff, one of the strongest voices against Hutchinson's idea of size differences among competing species, subsequently published articles that support the idea of character displacement in mammals. (See, for example, Dyan and Simberloff, 1994.)

Some authors have pointed out that a number of natural experiments support the validity of Hutchinson's ratio (Price, 1970). In the primeval forests of Canada, for example, four indigenous parasitoids attacked the wood-boring insect larva of *Tremex columba* (Fig. 8.18). Each had a different-length ovipositor (egg-laying apparatus) and laid an egg only when the ovipositor was fully extended. The ovipositor can be regarded as a food-provisioning apparatus for the larva; each species laid eggs in *Tremex* cocoons at a different depth in logs. In the 1950s, a fifth species, *Pleolophus basizonus*, was introduced into the area in an attempt to control another pest, the European sawfly. The length of *P. basizonus's* ovipositor was intermediate between those of two of the existing species. Even before the introduction of *P. basizonus*, the first three species were tightly packed, but still maintained the minimum separation of about 1:1.1 noted by Hutchinson. However, when *P. basizonus* was introduced, strong competition ensued and two native species were displaced.

Besides separating by size, species may differ in how they use a particular set of resources, such as food or space. Consider three species normally distributed on a set of resources (Fig. 8.19), where K is the resource availability or carrying capacity, d is

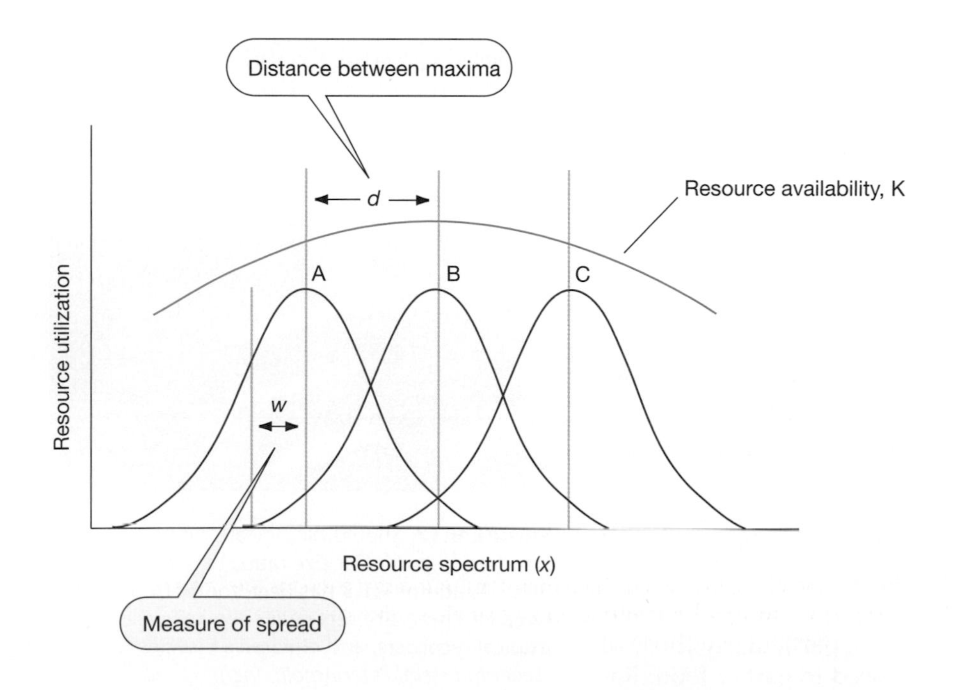

FIGURE 8.19 Resource partitioning. Three species—*A*, *B*, and *C*—with similar, normal resource utilization curves utilize a resource supply *K*. Variables: d = distance apart of means, w = standard deviation of resource utilization, and d/w = resource separation ratio.

the distance between maximum values of the abundance of different species, and w represents one standard deviation, approximately 68 percent of the area on one side of the resource distribution curve of a species. Because d measures how far apart species maxima are and w measures how spread out species are, the combination of d and w can define the amount of overlap in resource use and therefore the amount of competition. It has been argued that if $d/w > 1$, then species cannot coexist; if $1 < d/w < 3$, there will be some interaction between species; and if $d/w > 3$, then species coexist harmoniously (Southwood, 1978). One problem is that species abundances often are not normally distributed or have a discontinuous distribution. Another is that species may separate out along more than one resource set. For instance, consider three species distributed along two niche dimensions (Fig. 8.20). Overlap may be broad along the first dimension, yet narrow over the second dimension, so that niches overlap minimally or not at all in the two dimensions.

Resources can have a discontinuous distribution or occur in distinct units, like leaves on a shrub. Consider two insect

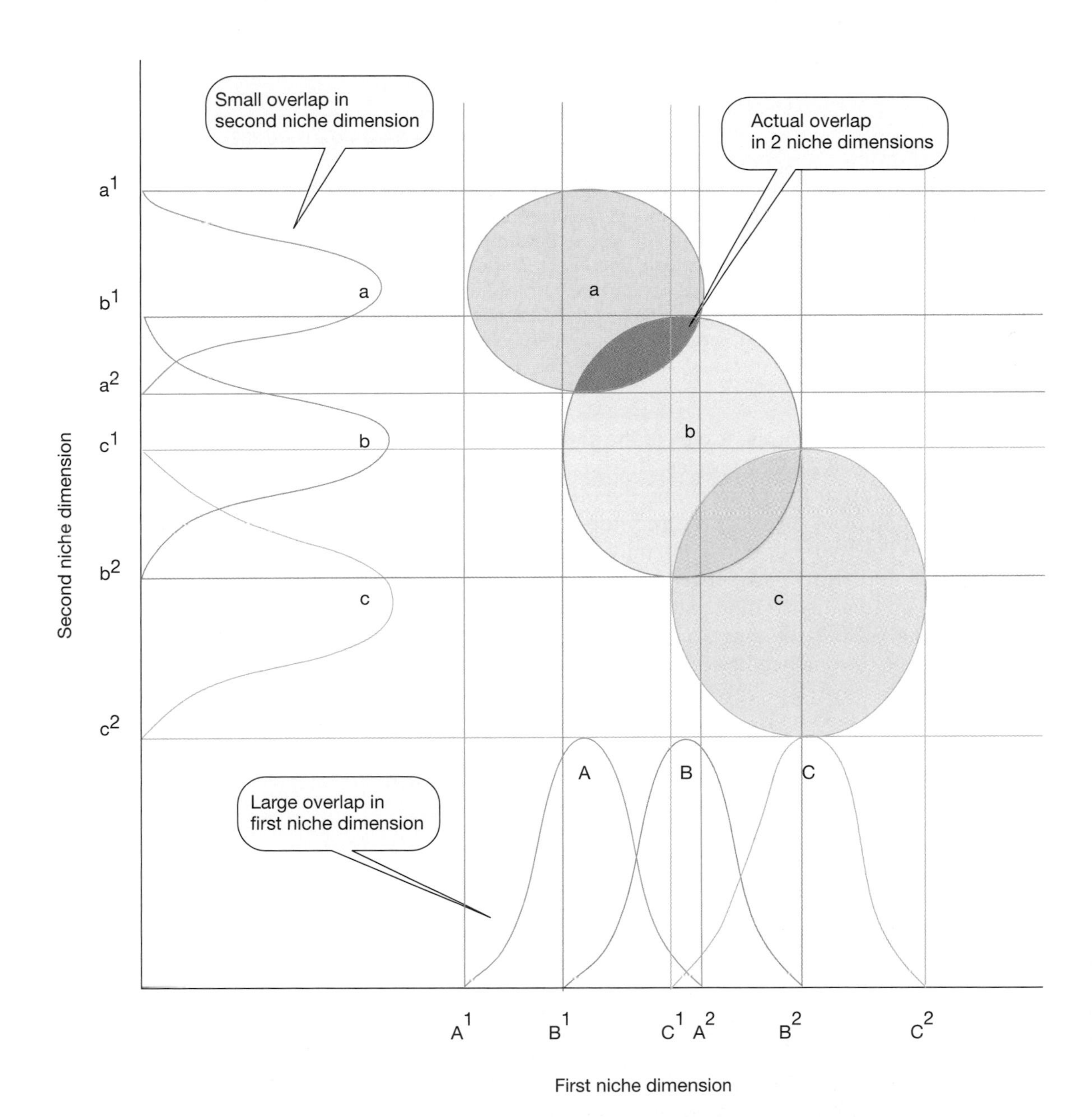

FIGURE 8.20 Division of resources along two niche dimensions. Although overlaps are broad along the first niche dimension, overlap is minimized when two niche dimensions are utilized (hatched area).

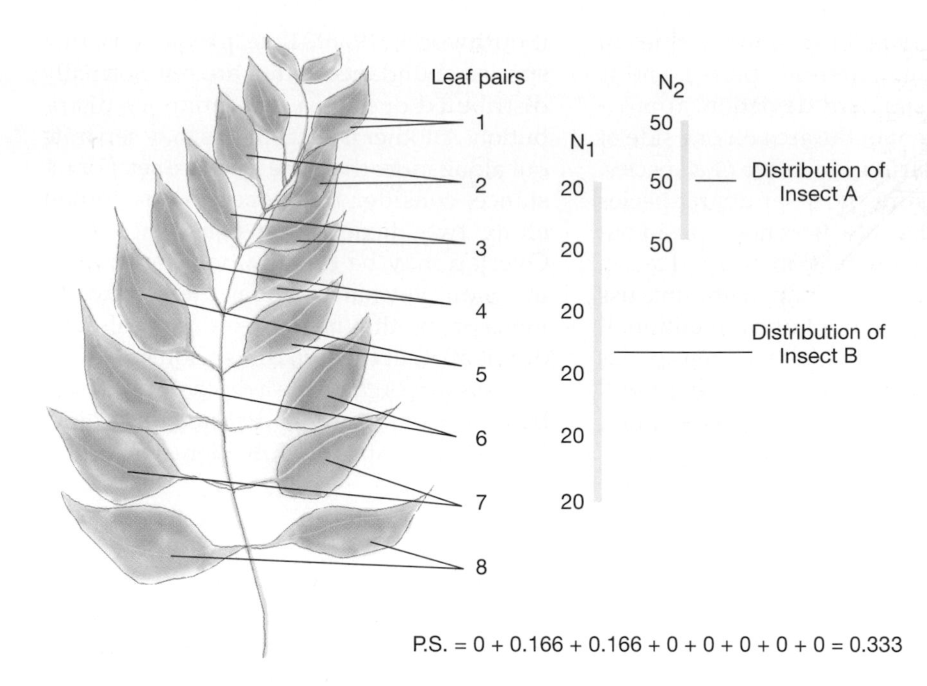

FIGURE 8.21 Resource utilization relationships of two insect species *A* and *B* when both feed on a discontinuously distributed resource such as leaves on a shrub. There are eight pairs of leaves, distributed from top to bottom, on a plant. Insect *A* is distributed only on the top three leaf pairs and insect *B* on pairs 2–7. There are 150 individuals of species *A* and 120 of species *B*. Thus, each of leaf pairs 1–3 contains a third (0.33) of the population of species *A*, and each of leaf pairs 2–7 contains a sixth (0.166) of the population of species *B*. PS = proportional similarity.

species that feed on the leaves of the shrub shown in Fig. 8.21. One species, *A*, feeds toward the top of the plant, while the other, *B*, feeds all over, except at the very top and bottom. The two species overlap only on the second and third pairs of leaves. We can represent this distribution diagrammatically in Fig. 8.22. Is the overlap small enough to permit coexistence between the insects? The niche overlap between two insect species that feed on the shrub can be measured by a quantity known as the *proportional similarity* (PS), given by

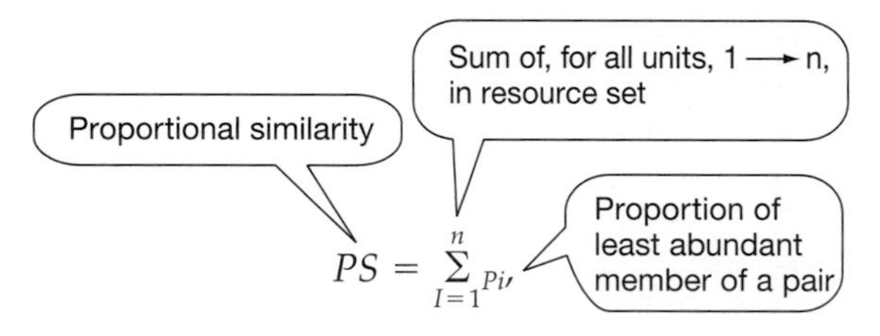

where p_i is the proportion of the least abundant species of the pair in the *i*th unit of a set of resources with *n* units. In Fig. 8.22, the proportional similarity for each of the eight units of resources is the sum of 0, 0, 0, 0, 0, 0.166, 0.166, and 0 = 0.333.

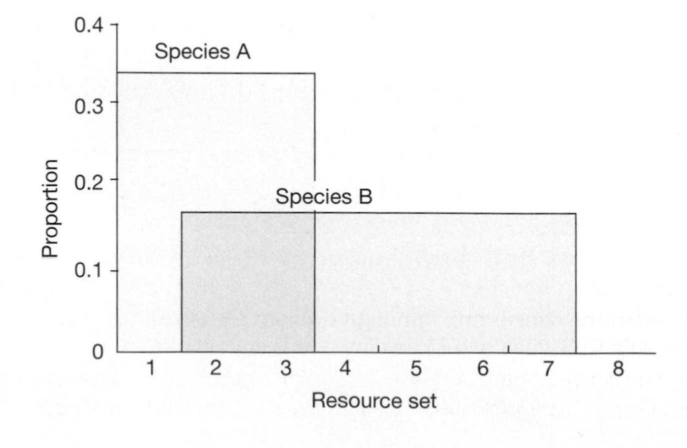

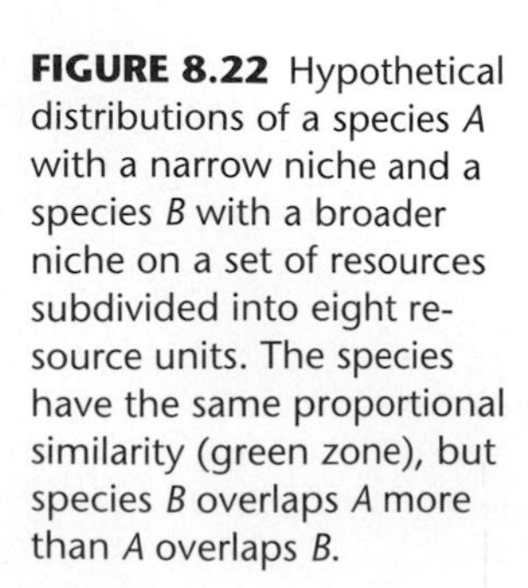

FIGURE 8.22 Hypothetical distributions of a species *A* with a narrow niche and a species *B* with a broader niche on a set of resources subdivided into eight resource units. The species have the same proportional similarity (green zone), but species *B* overlaps *A* more than *A* overlaps *B*.

In accordance with Hutchinson's ideas (i.e., there is a difference in resource use of about 30%), PS values of less than 0.70 have been taken to indicate coexistence, those greater than 0.70 competitive exclusion. However, species may differ not only along one resource axis, but also along many, such as prey size and prey taxa. For two resource axes, proportional similarity indices may be combined—proportional similarity values of 0.8 and 0.6 on two axes multiply to give an overall PS of 0.48. Theoretically, coexistence would be permitted in cases of combined PS values of $0.7 \times 0.7 = 0.49$ or less. Again, such analyses become more complex and more subject to error as new axes are included.

It is worth noting that in some situations competing species have been found to differ hardly at all in morphology and still are found together (Laurie, Mustart, and Cowling, 1997). Two reasons for this seeming anomaly have been proposed. First, in the presence of high levels of predation, competitively dominant species may be selected by predators over less abundant prey. In this situation, good competitors will probably never be able to eliminate poor competitors totally if predation occurs. (We address such predator-mediated coexistence again in Chapter 15 as a way to explain the high diversity of some communities.) Second, many real populations in nature exist, not in closed systems, but in open areas where migration is possible and they may intermingle with other populations. Hal Caswell (1978) theorized that such immigration into an area where a species is competitively inferior from an area in which it does well will often be sufficient to maintain reasonable population sizes for both competitors for an indefinite time.

Nature affords us many examples which show that species do compete with one another. However, such competition does not always lead to the displacement of one species by another: Species can partition up shared resources in various ways to promote coexistence—and even if they don't, species may still coexist if predators preferentially eat the dominant competitor or if migration of the inferior competitor into an area maintains its density there.

Summary

Competition is often thought of as a frequent phenomenon in nature, with one species always displacing another.

1. Competition may be intraspecific (between individuals of the same species) or interspecific (between individuals of different species). It may also be viewed as resource competition (for a limiting resource) or interference competition (in which individuals compete directly with one another).

2. Intraspecific competition between plants may be described by a 3/2 self-thinning rule.

3. Thomas Park's laboratory studies of competition between flour beetles showed that the outcome of competition could be changed by environmental conditions, by the presence or absence of natural enemies, and by the genetic strain of the competitors involved.

4. Experimental studies show that in nature competition occurs between different types of organisms over a broad scale. Often, such studies focus on exotic organisms; hence, their generalization to natural systems is questionable. Competition between exotics and native organisms is important because it has serious consequences for natural systems.

5. Reviews of the frequency of competition in nature indicate that competition between species is found in 55% to 75% of the species involved. Often, competition is asymmetric, with one species being a better competitor than another.

6. Schoener (1985) recognized at least six mechanisms of competition: consumptive (exploitative), preemptive, overgrowth, chemical, territorial, and encounter. Consumptive competition is by far the most common, occurring in at least 37.8 percent of cases.

7. The first models of how competing species interacted were originated by Lotka (1925) in the United States and Volterra (1926) in Italy. In a two-species interaction, four outcomes are possible: species 1 becomes extinct; species 2 becomes extinct; either species 1 or species 2 becomes extinct, depending on the starting conditions; or both species coexist. These models give no indication of the mechanism involved. For that reason, Tilman proposed his R^* concept, whereby the species that could most efficiently use a limiting resource would win.

8. It has frequently been argued that competition is minimized and that species can coexist if they utilize different resources. What is the amount by which competitors can overlap in resource utilization, but still coexist? Hutchinson (1959) suggested that a ratio of 1:1.3 in the morphological sizes of the organisms' feeding apparatus was necessary. Other authors have studied resource use more directly and have suggested that d/w values greater than unity or proportional similarity values of no more than 70 percent are necessary to permit coexistence.

Competition has been shown to be frequent and strong in nature. However, competing species do not always drive each other to extinction: Competitors can coexist if they partition their resource use.

Discussion Questions

1. Which type of competition would you expect to be the more important in nature, intraspecific or interspecific? Does intraspecific competition have to be weak in gregarious species? If interspecific competition between species can cause them to partition resources or to have different morphological characteristics, why wouldn't intraspecific competition cause even greater morphological displacement or resource partitioning between individuals of the same species?

2. Much native vegetation in the Florida Everglades is being lost. Could this be due to climate change brought about by global warming or the influence of exotic invaders? Design an experiment to test the idea that exotic trees, such as Brazilian pepper, Australian pine, or punk tree are causing the decline. Would you perform your experiment in the greenhouse or in the field? Why?

3. In question 2, how could you determine the mechanism of competition? How could you differentiate among competition for light, water, or nutrients?

4. In trying to understand how species compete, what advantages and disadvantages are there in field observations, field experiments, laboratory experiments, and mathematical models?

5. Using fully labeled graphs, explain the Lotka–Volterra approach to competition theory. What predictive power does the Lotka–Volterra model have? How is Tilman's R^* concept an improvement? What other improvements might you suggest?

CHAPTER 9

Mutualism

Road Map

1. Mutualism is an association between two species that benefits both. Pollination is a good example of mutualism, with the plant benefiting from pollen dispersal and the pollinator obtaining a meal of nectar.
2. In seed dispersal mutualism, the disperser often acquires a meal of fruit, while the plant gets its seed dispersed.
3. Through a variety of other types of mutualism, species are better able to secure resources together than alone or are able to defend themselves against natural enemies.
4. Mutualism is more difficult to model mathematically than are other population phenomena, because the models tend to result in runaway densities of both species.
5. Mutualism between species can spill over to affect whole communities, with effects on species other than those immediately involved.
6. Commensalism is an association between two species that benefits only one of them, with the other species remaining unaffected. An example is mammals transporting barbed seeds in their fur to new germination sites.

Mutualism and **commensalism** are often thought of as symbiotic relationships. However, strictly speaking, **symbiosis** is just a close association between species, so parasitism is a symbiotic relationship, too. The unique feature of mutualism is that both species benefit. For example, in mutualistic pollination systems, both plant and pollinator (an insect, a bird, or a bat) benefit, one usually by a nectar meal and the other by the transfer of pollen. In commensalisms, one species benefits and the other is unaffected. For example, in some forms of seed dispersal, barbed seeds are transported to new germination sites in the fur of mammals. The seeds benefit, but the mammals are usually unaffected. Humans have entered into mutualistic relationships with many species. In Africa, people often follow a bird known as a honeyguide. The bird leads the human to a beehive, which the human rips open for the honey inside, and the bird dines on the scraps of bees' nest and bee larvae. In this case, the human is replacing the honey badger, or ratel, which may perform the same function. The mutualistic association of humans with domestic animals or crops has permitted the most far-reaching ecological changes on Earth. (See "Applied Ecology box: Humans in Mutualistic Relationships.")

9.1 Plant–Pollinator Mutualism

Because pollination and seed-dispersal mutualism is the most frequent kind of mutualism in nature, studies of that phenomenon are more copious than those of other forms of mutualism (Fig. 9.1). The University of Arizona's Judith Bronstein, who has spent a

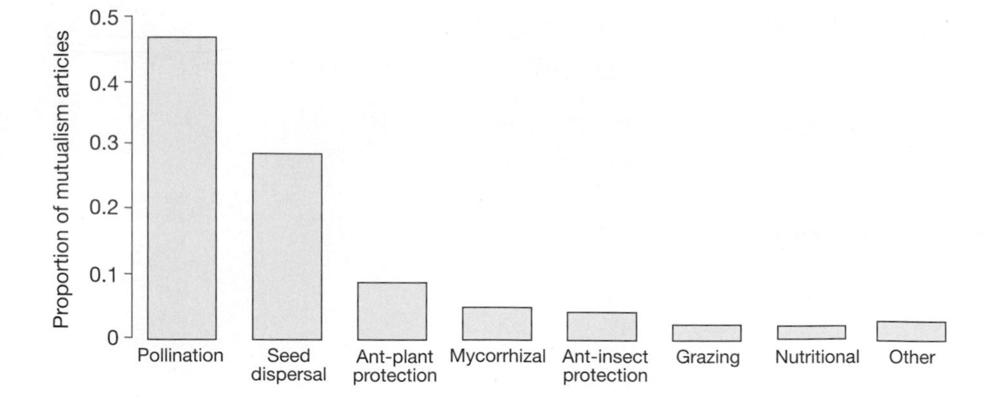

Figure 9.1 Frequency of articles on different forms of mutualism published in 675 scientific and 10 major ecology journals. *(Reproduced from Bronstein, 1998.)*

Applied Ecology

Humans in Mutualistic Relationships

The mutualisms with perhaps the most important implications are those associated with human agriculture (Figure 1). Most introduced crops remain dependent on people for their survival, requiring that water be added or that competing weeds be removed. Most of the world's crops (whose origins are shown in Table 1) have been moved only recently. Introduced animals such as sheep, probably derived from Asiatic mouflon (*Ovis orientalis*), and cows, from the European aurochs, are also recently domesticated. The population size of humans is increased in the presence of these crops and domesticated animals, and likewise, populations of the crops and domesticated animals would be reduced in the absence of humans. Fully 35 percent of the world's land area is given over to crops and grazing pastures (Table 2), and the head of livestock in the world runs into the billions (Table 3). Is there a food shortage? Not globally: The steady increase in the human population is testament to that.

Figure 1 Modern agriculture illustrates one of the most far-reaching kinds of mutualism. These radishes, near Perrydale, Oregon, could not exist without human help, and at the same time, crops sustain human populations. Such mutualisms between humans and crops have radically altered the ecological landscape. Forests have been replaced by agricultural fields. (*Bob Pool, Tom Stack, & Associates*)

Unfortunately, the side effects of this mutualism on natural systems are immense:

1. *Pollution of bodies of water*. Runoff water often contains high levels of nitrogen and phosphorus from fertilizers, as well as residual pesticides and manure, all of which contaminate streams. All these pollutants can affect communities of aquatic plants and animals.
2. *Loss of topsoil*. Some 4 billion tons of topsoil are washed into U.S. waterways each year. Topsoil can also be whipped away by wind.

TABLE 1 Geographic origins of some important crops.

Crop	Area of Origin
Apple	Western Asia
Banana	Malaysia
Carrot	Eastern Mediterranean
Coffee	Ethiopia
Cucumber	Egypt, India
Eggplant	India
Lettuce	Egypt
Orange	Indochina
Peach	China
Peanut	Central and South America
Pineapple	Amazonia
Potato	Andes
Rhubarb	Russia
Rice	Asia
Sugarcane	India
Tea	India
Tomato	Mexico
Wheat	Nile Valley

TABLE 2 Agricultural and grazing lands, in thousands of hectares.

	Continental Area	Cropland	%	Pastures	%
World	1,3078,873	1,476,483	11.3	3,170,822	24.2
Africa	2,964,595	194,869	6.2	788,841	26.6
North and Central America	2,242,075	274,626	12.2	367,062	16.4
South America	1,781,851	140,638	7.9	458,364	25.7
Asia	2,757,252	454,253	16.5	644,669	23.4
Europe	487,067	139,625	28.7	84,260	17.3
Oceania	850,967	50,285	5.9	453,026	53.2
Former U.S.S.R.	2,240,220	232,187	10.4	374,600	16.7

3. *Depletion of water supplies*. About 70% of the fresh water used by humans is expended on irrigation. Most of this water comes from "fossil" aquifers—underground water resources—at rates much in excess of natural recharge rates. In Africa, the Middle East, and China, groundwater levels can fall by as much as 1 meter per year.
4. *Salting of the land*. Salting can result from continued irrigation for agriculture, as was discovered in California in the 1970s. Even "fresh" water contains minute quantities of dissolved salts, and as the water evaporates or is used by plants, the salt is left behind to accumulate in the soil. Mesopotamia, probably the world's oldest irrigated area, on the plains of the Tigris and Euphrates rivers, had a fertility that was legendary throughout the Old World. The area now exhibits some of the world's lowest crop yields, because salt has accumulated in the soil, interfering with crop growth.
5. *Desertification*. Overgrazing by livestock can result in desertification. The cattle population in the Sahel, the region immediately south of the Sahara desert in Africa, grew fivefold in the period from 1940 to 1968, increasing the likelihood of desertification in this region.
6. *Severe loss of wildlife*. Agriculture can cause wildlife to die off, because deforestation, drainage of swamps, and the removal of hedgerows reduce wildlife habitat.

TABLE 3 Numbers of seven species of food livestock in the world in 1992.

No. of Livestock	(Millions)
Turkeys	259
Goats	574
Ducks	580
Pigs	864
Sheep	1,138
Cattle	1,284
Chickens	11,279

Clearly, crop and domestic animal mutualisms with humans are vital to sustaining human populations, but the costs to land and other organisms can be enormous.

career studying mutualism (see "Profile" box), demonstrated that over 45 percent of all studies of mutualism are of pollination systems (Bronstein, 1998). This may be partly because of the great diversity of apparently tight, **coevolved**, and interesting systems to study.

There is strong selective pressure for plants to develop intimate relationships with their pollinators. Up to 37 percent of the photosynthate that the milkweed (*Asclepias*) produces is used to produce nectar (Southwick, 1984). Some of the tightest mutualisms known involve plant–pollinator systems, such as the figs (*Ficus* sp.) and their pollinating fig wasps (Fig. 9.2; Janzen, 1979a). More than 900 species of *Ficus* exist, and virtually every one must be pollinated by its own species of agaonid wasp. The fig that we actually eat has an enclosed inflorescence containing many flowers. A female wasp enters through a small opening, pollinates the flowers, lays eggs in their ovaries, and then dies. The progeny develop in tiny protuberances called galls and hatch inside the fig. Males hatch first, locate female wasps within their galls, and thrust their abdomens inside the galls to mate with the females. The males then die, without ever having left the fig. Females collect pollen from the fig and then leave to search out new figs in which to lay their progeny. Both fig and wasp benefit—the flower is pollinated, and fruits form and so do seeds, although some are eaten by the developing wasp. Clearly, then, there is a necessary balance between mutualism and overexploitation. A similar mutualistic relationship occurs between yucca plants and

Interactions between plants and their pollinators are commonly, although not always, mutualistic.

Figure 9.2 *Blastophaga psenes*, a tiny fig wasp, which crawls inside the captifig in California to lay her eggs

yucca moths (Addicott, 1986). Both mutualisms are highly coevolved. The distribution of each species of yucca or fig is controlled by the availability of its pollinator and vice versa. In the late 19th century, Smyrna figs were introduced into California, but they failed to produce fruit until the proper wasps were introduced to pollinate them.

It is interesting that such highly coevolved systems arose, because, on a superficial examination, the needs of the plants and those of the pollinators seem to conflict. From the plant's perspective, an ideal pollinator would move quickly among individual plants, but retain a high fidelity to a plant species, thus ensuring that little pollen is wasted as it is inadvertently brushed onto the pistils of other plants. The plant should provide just enough nectar to attract a pollinator's visit. From the pollinator's perspective, it would probably be best to be a generalist and to obtain nectar and pollen from flowers in a small area, thus minimizing the energy spent on flights between patches. This seeming conflict of interests casts doubt on whether such relationships are truly mutualistic in nature or whether both species are actually "trying to win" in an evolutionary arms race. One way in which the plants encourage fidelity in the pollinator's species is by sequential flowering of different plant species through the year and by synchronous flowering within a species. Mutualistic interactions are evolutionarily stable only when both interacting species possess mechanisms to prevent excessive exploitation. For example, in the case of the yucca moths, flower abortion results if too many moth eggs are laid (Pellmyr and Huth, 1994).

There are cases where both flower and pollinator try to "cheat." In the bogs of Maine, for example, the grass-pink orchid *(Calopogon pulchellus)* produces no nectar, but mimics the nectar-producing rose pogonia *(Pogonia ophioglossoides)* and is therefore still visited by bees. Bee orchids have even gone so far as to mimic female bees; males pick up and transfer pollen while trying to copulate with the flowers. So effective are the stimuli of flowers of the orchid genus *Ophrys*, that male bees prefer to mate with them even in the presence of real female bees! Conversely, some *Bombus* species "cheat" by biting through the petals at the base of flowers and robbing the plants of their nectar without entering through the tunnel of the corolla and picking up pollen.

Finally, it is interesting to speculate about why ants, usually the most abundant insects in a given area, are so rarely involved in pollination. One reason might be that the subterranean nesting behavior of many ants exposes them to a wide range of pathogenic fungi and other dangerous microorganisms, to which they respond by producing large amounts of antibiotics via symbiotic bacteria. These antibiotics inhibit pollen function. Indeed, Peakall, Beattie, and James (1987) demonstrated that ants without metapleural glands, and therefore with lower levels of antibiotics, do successfully pollinate orchids in Australia.

9.2 Seed Dispersal

Plant seeds and fruits provide highly nutritious rewards to animals that disperse them.

Mutualistic relations are highly prevalent in the seed dispersal systems of plants. Studies of these systems also are highly prevalent in the mutualism literature, to the tune of almost 30 percent of all studies of mutualism. In the tropics, some fruits are dispersed by birds that are strictly frugivorous. These fruits provide a balanced diet of proteins, fats, and vitamins. In return for this juicy meal, the birds unwittingly disperse the enclosed seeds, which pass unharmed through the animals' digestive tract. Some plants, instead of producing highly nutritious fruit to attract an efficient disperser, simply produce abundant mediocre fruit in the hope that some of it will be eaten by generalists. Fruits taken by birds and mammals often have attractive colors—red, yellow, black, or blue; those dispersed by nocturnal bats are not brightly colored, but instead give off a pungent odor to attract the bats. In contrast, because birds do not have a keen sense of smell, fruits eaten by them are generally odorless.

In general, seed–disperser relationships are not as obligately mutual as are

Profiles

Judith Bronstein,
University of Arizona

A glance at this textbook should be enough to convince you that much of ecology deals with antagonistic interactions between organisms, such as competition, predation, parasitism, and herbivory. There is no question that in nature, mutually beneficial interactions within and between species are also very common. Ecologists only rarely study them, though. At the moment, our understanding of cooperation mostly exists through a set of well-known case studies. The goal of my research is to help move the study of cooperation away from the usual storytelling approach and toward the kind of rigorous hypothesis testing that ecologists have long used to study antagonisms.

I first became interested in community ecology and species interactions through my undergraduate courses at Brown University. As a beginning graduate student at the University of Michigan, a summer spent assisting my professor Beverly Rathcke in her field studies of plant/pollinator communities helped to focus my interests on mutualisms (cooperative interactions between species). The pivotal experience that shaped my future research came when I had the opportunity to take a semester-long course in tropical ecology in Costa Rica. These wonderful courses are offered to graduate students several times each year through the Organization for Tropical Studies, a group of universities that cooperate to provide tropical research and educational opportunities.

The instructors of our course introduced us to some of the most spectacular of the stories that have shaped our limited understanding of mutualism. The one that really piqued my interest was the fig pollination mutualism. There are about eight hundred species of figs in the tropics, each of which is pollinated by a different species of fig wasp. These tiny wasps (one of my mentors jokes that they are "about the size of cosmic dust") deliberately pollinate figs, then lay eggs in some of the fig's flowers; their offspring feed on many of the fig's seeds. What intrigued me was that for the fig trees, fig wasps are simultaneously beneficial (since adults are pollinators) and harmful (since the pollinators' offspring are seed predators). I wondered, how did these trees end up with such destructive insects as their only possible pollinators? Is there anything the trees could do to lower the damage that the wasps inflict, without killing them off entirely (since if they did, the trees would never reproduce again)? More generally, how can this mutualism persist over the long term, in light of the apparent conflict between the partners? None of the answers to these questions were known. My graduate research eventually focused on measuring those conflicts and some of their ecological consequences in one Costa Rican fig/fig wasp mutualism.

It would have been easy to have used my findings simply to add more facts to the large database on these odd interactions. Instead, my professors pushed me, and I pushed myself, to take the tougher but much more scientifically significant route of looking beyond the specifics of my study system to address big, general questions about mutualism. But what were those big questions? To my surprise, I discovered that the study of mutualism was so new that researchers were barely beginning to suggest what they were. I chose to focus on the hypothesis that in all mutualisms, not just the fig/fig wasp interaction, there are serious conflicts between the partners. (The basic idea is that mutualism is a kind of "reciprocal parasitism": each partner is out to do the best it can, by obtaining what it needs from its mutualist at the lowest possible cost to itself.) Ultimately, by taking the risk of "thinking big" in my graduate research, I found myself playing a major role in developing the very young field of mutualism studies. If instead I had been content to see my goal as simply to discover more facts about figs, I'm sure that today I would be an unemployed fig specialist rather than a professor.

More recently, the challenge has become to convince other ecologists that my chosen specialty is an important and interesting one. Many suggestions have been offered over the years for why ecologists have studied mutualism so little, especially compared to competition and predation. Some have attributed it to Western male scientists' supposed biases toward seeing aggression in nature. I tend to disbelieve this (though there does seem to be a trend for mutualism to be studied by women). Others argue that cooperation is simply not a biological phenomenon important enough to deserve much attention. I strongly disagree. But, as a scientist, it is not sufficient for me to simply believe that this view is wrong. My responsibility is to convince others that I am right by conducting good ecological studies that critically test my ideas.

I am currently associate professor of ecology and evolutionary biology at the University of Arizona. Like all university professors, I must find a way to balance research with my equally important (and enjoyable!) role as a teacher and mentor of graduate and undergraduate students. In my case, I must also set aside large blocks of time to spend with my husband and new son. Trying to balance these conflicting demands on my time has certainly proved to be the biggest challenge of my career. My primary role models these days are the increasing numbers of professionals, especially in university positions, who are successfully managing the delicate balancing act between family and career. I would like to think that the days are over when having both a thriving career and a happy home life are thought to be mutually exclusive for women.

Mutualism can take a variety of forms in addition to the well-studied pollination and dispersal mutualisms.

Some mutualisms increase the availability of resources to both partners.

plant–pollinator systems, because seed dispersal is performed by more generalist agents. Nevertheless, a wide array of adaptations exists; one has only to look at the impressive specialization of parrot beaks—powerful and sharp to crack and peel fruits—to see that the mutualistic relationship between plant and seed disperser is strong in this case. Some bizarre strategies also exist; for example, in the floodplains of the Amazon, fish have evolved that eat fruits and seeds and disperse the latter (Fig. 9.3).

"Cheating" can occur in seed dispersal systems, too. Microbes readily attack a plant's fruit without dispersing it. Janzen (1979b) suggested that microbes deliberately cause fruit and other resources like carcasses to rot by manufacturing ethanol and rendering the medium distasteful to vertebrate consumers, thereby reserving it for themselves. Animals eating such food are selected against because they become drunk and are easy victims for predators. As a countermeasure, some vertebrates have "learned" evolutionarily to tolerate these microorganisms and even to use them in their own guts to digest food. Alternatively, vertebrates often possess the enzyme alcohol dehydrogenase to break down alcohol.

Finally, it is interesting to speculate why seed dispersal is so advantageous to plants. Presumably, competition with the parent plant is avoided, but on the other hand, there is no guarantee that the seed will fall into an optimal habitat. The colonization hypothesis (Hubbell, 1979) suggests that because environmental conditions for seed germination constantly shift, the location of the parent plant is not always a good predictor of seedling success. More intriguing is the directed-dispersal hypothesis, which intimates that some dispersers distribute seeds into optimal sites. This is evident in the case of some ants, which take seeds back to their nests. Finally, the predator escape hypothesis (Janzen, 1971) suggests that seed predators congregate under parent trees to feed on the bonanza of fallen fruit, so that well-dispersed seeds suffer less predation.

Figure 9.3 A berry-eating *Piranha* fish from the Amazon. Here fish are major seed dispersers.

9.3 A Variety of Mutualisms

A variety of mutualisms exist wherein the partners benefit by protection from natural enemies or by access to new resources. Some mutualisms decrease the susceptibility of one partner to parasites. For example, on coral reefs, "cleaner" fish nibble parasites and dead skin (which might otherwise cause disease) from their "customer" fish at specific cleaning stations. On reefs cleared of their "cleaner" fish, "customer" fish developed skin diseases, and their populations declined in less than two weeks (Limbaugh, 1961).

Mutualisms and resources.

Leaf-cutting ants, primarily of the Neotropical group Attini, of which there are about 210 species, enter into a mutualism with a fungus. A typical colony of about 9 million ants has the collective biomass of a cow and cuts the equivalent of a cow's daily requirement of fresh vegetation. But instead of consuming it directly, the ants chew it into a pulp that is stored underground to grow special fungal crops. The fungus in turn produces specialized structures known as gongylidia, which serve as food for the ants. In this way, the ants circumvent the chemical defenses of the leaves, which are actually digested by the fungus. Leaf-cutting ants are major herbivores in forest communities (Fig. 9.4) and may harvest 17% of the total leaf production. Interestingly, another mutualistic player has just been identified in the system (Currie *et al.*, 1999): A parasitic fungus, *Escovopsis*, can infest the ant gardens, destroying the fungi growing inside. Normally, however, a mutualistic streptomyces bacteria on the bodies of the ants produces antibiotics that keep *Escovopsis* in check.

Nitrogen is vital to plant growth and hence to animal growth, but it is something of an enigma that most plant and animal species cannot fix atmospheric nitrogen. That is, they cannot convert atmospheric nitrogen into a useful form such as nitrate or nitrite. Instead, fixation is carried out mostly in the soil via bacteria and archaebacteria. Many of these microbes live in the roots

Figure 9.4 (upper) Leaf-cutting ants *Atta cephalotes*, in South America, chew up leaves and (lower) cultivate fungus gardens underground.

Figure 9.5 These root nodules on the broad bean, *Vicia faba*, house the nitrogen-fixing bacteria *Rhizobium* sp.

of plants, in a tight mutualism with their hosts. Any excess of fixed nitrogen that the bacteria do not use is available to the plants. Perhaps the tightest of these mutualisms occurs between *Rhizobia* bacteria and leguminous plants, which have special root nodules that contain the bacteria (Fig. 9.5). Once the legumes have increased the local soil nitrogen content, other plants can invade, and these then outcompete the legumes, thus killing the goose that laid the golden egg. After a while, however, the local soil nitrogen may decrease, again favoring the entry of leguminous species. In this way, the legumes are continually appearing and disappearing in plant communities (Cain *et al.*, 1995).

One of the oldest ideas about mutualisms is that they are more common under harsh physical conditions, when neighbors buffer one another from limiting physical stresses (Callaway, 1997). For example, a strong facultative mutualism exists between two plants—the leguminous shrub *Retama sphaerocarpa* and the understory plant *Marrubium vulgare*—in a semiarid region in southeastern Spain (Pugnaire, Haase, and Puigolefabregas, 1996). In the almost desertlike conditions, the interaction between the two species is indirect: Nitrogen and moisture in the soil are much improved in the islands of fertility that result from that interaction. *Retama* shades *Marrubium*, providing a favorable microclimate, while *Marrubium* somehow enhances the availability of water to *Retama*. Nutrient cycling is probably increased due to the accumulation of litter, and as a result, both species have a greater leaf area, leaf mass, and shoot mass, as well as more flowers and a higher leaf nitrogen content when they grow together than when they grow alone (Fig. 9.6).

Mutualism and protection from natural enemies.

There are a number of well-studied cases of insects providing protection to herbivores of plants.

In terrestrial systems, one of the most commonly observed mutualisms is that between ants and aphids. Aphids are fairly helpless creatures, easy prey to marauding ants. Yet, in general, ants tend to farm aphids like so many cattle (Fig. 9.7). The aphids feed on plant sap and have to suck a sufficient amount of it to get their required nutrients. They also excrete a lot of fluid, and some of the sugars remain in the excreta, which is called honeydew. The ants drink the honeydew and, in return, protect the aphids from an array of predators [e.g., syrphid fly larvae, coccinellid (ladybug) larvae, and neuroptera (lacewing) larvae], parasites like braconid wasps, and other competing insects. The ants attack these aphid enemies vigorously. (Incidentally, this example illustrates

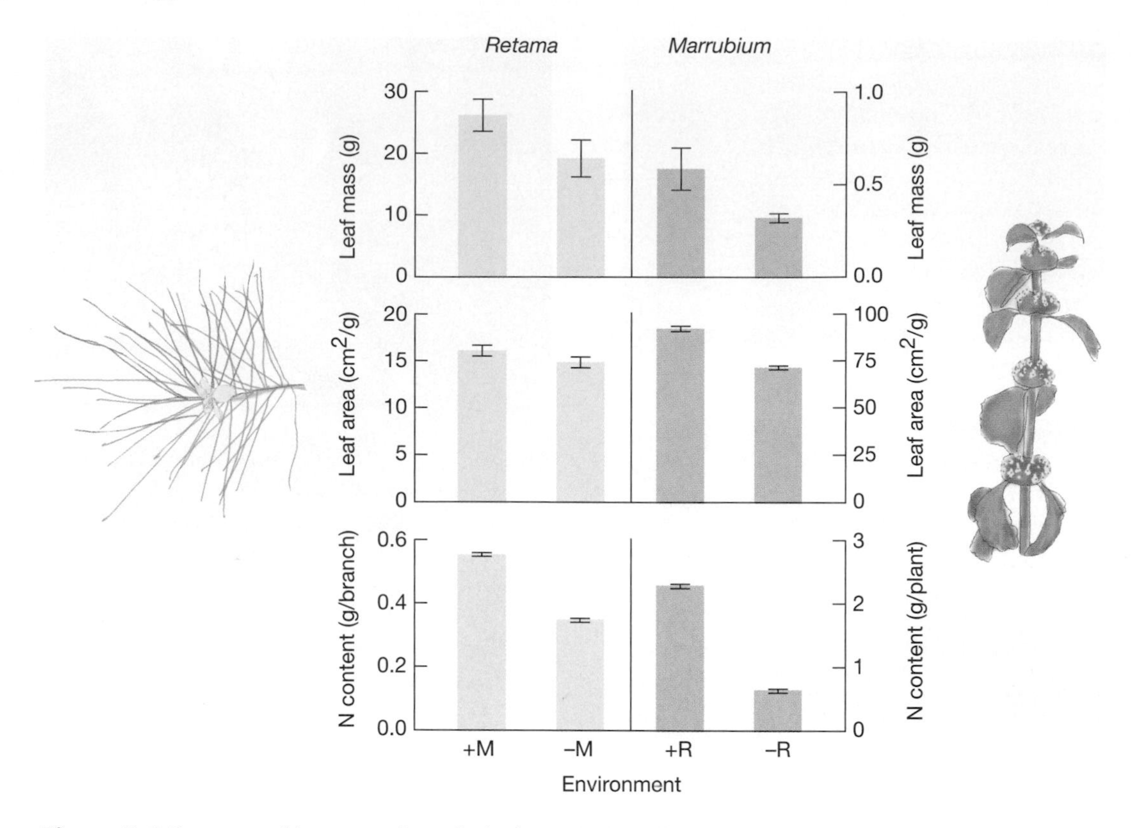

Figure 9.6 Dry mass of leaves per branch, leaf area (LA), and nitrogen content in *Retama sphaerocarpa* (*R*) and *Marrubium vulgare* (*M*) growing in association (+*R*, +*M*) or alone (−*R*, −*M*) at Rambla Honda, Spain. Data represent means; error bars represent 1 standard error. *(After Pugnaire et al., 1996.)*

another common feature of mutualism: Often, a third (or a fourth) species is involved in the relationship, such as a competitor or predator of one of the mutualists.)

Mutualism and herbivory.

There are numerous examples of insects reducing herbivory on plants.

While in some cases ants may provide protection to herbivores of plants, thus entering into a mutualism with the herbivores, in other cases ants protect the plant from herbivores, thereby entering into a mutualistic relationship with the plant itself. One of the most famous cases involves acacia trees in Central America, which provide food and nesting sites for ants inside large thorns. In return, the ants bite and discourage both insect and vertebrate herbivores from feeding on the acacia, and they also trim away foliage from competing plants and kill neighboring plant shoots. The ants' actions ensure better light, water, and nutrient supplies for the acacia. Among the most famous cases is bull-horn acacia (*Acacia collinsii,* Fig. 9.8), which has large hornlike thorns filled with a soft pithy interior that is easy for the ants to hollow out. The ants, usually *Pseudomyrmex ferruginea*, make nests inside the thorns. In return, the acacia provides two forms of food to the ants: beltian bodies and extrafloral nectaries.

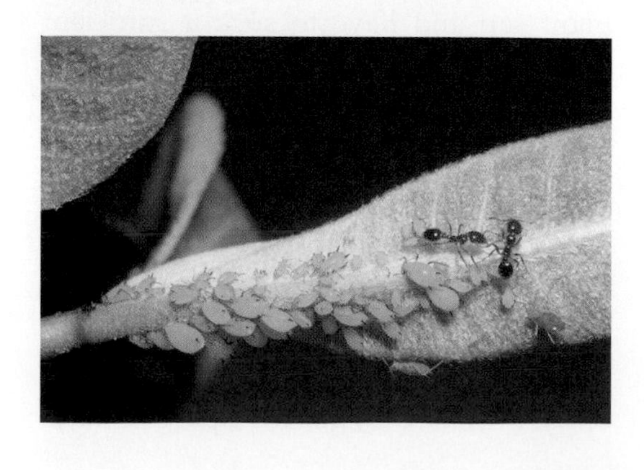

Figure 9.7 Ants tending aphids on a milkweed plant. The aphids secrete a sugar-rich substance called honeydew, which the ants eat. In return, the ants protect the aphids from marauding natural enemies. *(Robinson, Photo Researchers, Inc.)*

Beltian bodies are small protein-rich food nodules at the end of the plant's leaflets. They are named after the tropical naturalist Thomas Belt (1874), who was the first to suggest that ants may indeed protect acacia plants. The extrafloral nectaries are rich in carbohydrates and are located away from the flowers, at the base of the leaves. In this case neither species can live without the other, though this is not always the case. Acacia species that are not defended by ants have toxic compounds in their foliage to protect against herbivores. Dan Janzen

(1966) who has spent years studying tropical systems, showed that removing the ants allowed the numbers of insect herbivores to increase nearly eightfold. In turn, shoot growth was reduced sevenfold. The ants patrolled the vegetation 24 hours per day, and in return, the acacia retained its leaves year-round. (Acacia that are not defended by ants normally drop their leaves in the dry season.) Ant–plant protection mutualisms such as this have been found frequently, especially in the tropics. Fonseca and Ganade (1996) recorded a density of 377 myrmecophytic plants per hectare in a Brazilian rain forest, and Rico-Gray (1993) found 312 ant–plant associations at a single coastal site in Mexico. Even in these mutualisms, evidence of cheating has been found (Yu and Pierce, 1997). One species of mutualistic ant, *Allmomerus demeravae*, removes herbivores from leaves, but actually castrates its host plant, *Cordia nodosa*, by destroying the plant's flowers. As a result, vegetative growth increases, providing additional resources for ants in the form of stem swellings called *domatia*, in which the ants live. The system is stabilized only by the occurrence of three *Azteca* ant species that do not castrate their host and that are truly mutualistic. The presence of *Azteca*, a true mutualist, allows the host plant to survive.

Fungi may also enter into a mutualistic relationship with plants to reduce vertebrate herbivory. Some islands in the Scottish Hebrides support feral populations of Soay sheep (*Ovis aries*). On Hirta, an island in the St. Kilda archipelago, Soay sheep have grazed the native grass intensively for over 60 years (Fig. 9.9). The population exhibits fluctuations and varies from between 700 and 1,600 animals, crashing every three to five years with up to 60% mortality. Starvation is partly responsible, and increased parasitism may also contribute to the Soay's demise. However, a 1998 study by Dawn Bazely and her associates showed how a defensive mutualism may be the chief culprit. The main forage of the sheep is the grass *Festuca rubra*, which harbors an endophyte—the fungus *Acremonium*—inside its blades. The fungus synthesizes toxic alkaloids—chemicals that function as an antiherbivore defense; in return, the fungus gains access to a food source within the plant's leaves. Bazely and colleagues showed that the frequency of infection with the fungus positively correlated with grazing pressure. On nearby Dun Island, the infection

Figure 9.8 Thorns on *Acacia collinsii* in Paloverde National Park in Costa Rica are shaped like bull's horns. They provide homes for the ants *Pseudomyrmex ferruginea* which in turn deter herbivores from feeding on the plant.

rate was much lower, because the island was not grazed. The mutualism could almost be seen as an induced defense: Heavier grazing led to greater concentrations of endophytic hyphae within plants, and the levels of the alkaloid, ergovaline, were higher. The fungi were in highest concentrations in the basal regions of the plants. Thus, as the number of sheep increased and the height of the pasture decreased, the sheep would encounter higher concentrations of the alkaloids, which would reduce their numbers. Studies like Bazely's are important not only in their own right, but also because of the widespread losses to the livestock industry due to fungal-infested fescue, especially in the United States. An understanding of the fescue–*Acremonium* mutualisms may enable livestock losses to be minimized.

Obligate mutualism.

In some cases, the mutualistic relationship is so tight that neither participant could exist without the other. This kind of *obligate mutualism* is shared in by many lichens—combinations of algae, which provide the photosynthate, and fungi, which provide a safe habitat for the algae. The "lichenized" fungi have a thin layer of algal cells within their bodies that forms only 3 to 10 percent of the weight of the lichen. Of the 50 thousand or so species of fungi, 25 percent are "lichenized"

In obligate mutualism, the interaction is necessary to the existence of both partners.

Figure 9.9 Soay sheep, *Ovis ovies*, live wild only on the island of St Kilda in the Scottish Hebrides. They are the closest living relative to the early domestic sheep.

forms, which occur in deserts, in alpine regions, and across a wide range of habitats. "Nonlichenized" fungi are usually restricted to being parasites of plants or animals or to being involved in decomposition.

Many ruminants shelter symbiotic bacteria in their guts. The bacteria break down plant tissue to provide energy for their hosts. (Mammals could not otherwise digest cellulose.) Many aphids as well harbor special bacteria that synthesize necessary amino acids for their hosts. Likewise, the roots of most higher plants (except the Brassicaeceae) are actually a mutualistic association of fungus and root tissue—the mycorrhizae. The fungi require soluble carbohydrates from their host as a source of carbon (up to 40 percent of the photosynthate produced by the host), and they supply mineral resources, which they are able to extract efficiently from the soil, to the host. Most plant species are susceptible to colonization by mycorrhizal fungi, although some (e.g., many grasses) are unresponsive.

Among the many other interesting examples of obligate mutualisms are the numerous deep-sea fishes that harbor bioluminescent bacteria in specialized organs. The organs provide a safe haven for the bacteria, while at the same time, the fish take advantage of the light-emission properties of the bacteria to communicate with other fish. Corals are obligate mutualists, too. These animal polyps contain unicellular algae, dinoflagellates of the genus *Symbiodinium*. Rowan (1998) showed that each species of coral is capable of hosting two or more of a great variety of symbiont species. During adverse conditions, such as high temperatures, the mutualism between the two organisms can break down, resulting in the phenomenon of coral bleaching, in which the dinoflagellates die. Global warming is thought to have killed some corals in just such a manner. The most extreme view of mutualism is the theory of *endosymbiosis*, the idea that some cell organelles, such as the mitochondria or chloroplasts found in eukaryotes (higher organisms), actually represent the remnants of symbiotic prokaryotic organisms that have forever been incorporated into the eukaryotes (Margulis, 1976).

9.4 Modeling Mutualism

There are few appropriate mathematical models for mutualistic interactions.

Mutualistic relationships can be modeled with equations similar to the Lotka–Volterra competition equations. For facultative mutualisms, we could incorporate the positive effect of one species on the other by changing the signs of the competition coefficients such that

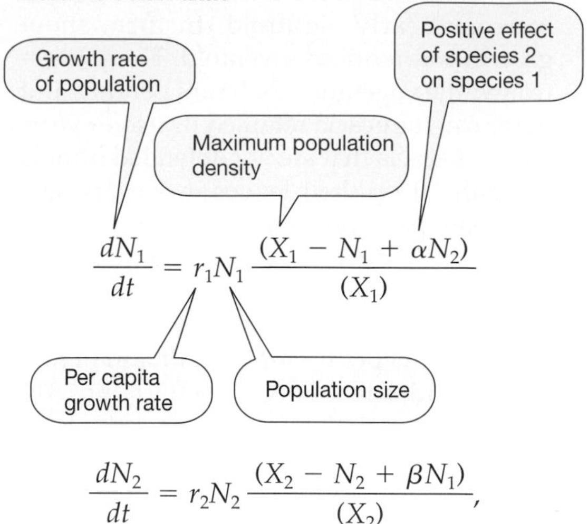

$$\frac{dN_2}{dt} = r_2N_2\frac{(X_2 - N_2 + \beta N_1)}{(X_2)},$$

where α and β are the positive effects of species 2 on species 1 and of species 1 on species 2, respectively. In this situation (Fig. 9.10), the maximal population density when each species is alone is represented by X, not K, because such densities no longer represent real carrying capacities. The reason is that, although there is a carrying capacity K while a mutualist is alone, the presence of its fellow mutualist can increase the carrying capacity. Such modifications can lead to stable solutions if the isoclines cross (Fig. 9.10a). The presence of another mutualist can also lead to unrealistic solutions, in which both populations increase to unlimited size (Fig. 9.10b). A solution to this paradox is to reduce the mutualism coefficients α and β as the populations grow (requiring a modification to the foregoing equations). Such a dependence on density would stabilize the population (Fig. 9.10c). However, some authors have suggested that instability is an accurate finding: Few obligate mutualisms in nature are stable in the face of environmental change.

For obligate mutualisms, different equations are needed (Wolin, 1995). Without going into the mathematics, a graphical approach is instructive (Fig. 9.10d–f). Some individuals of the mutualist N_2 must already be present for N_1 to grow (and vice versa), so the isoclines for each species are displaced, and the origins start partway along their respective axes. Where the isoclines cross, two outcomes are possible. First, the population trajectories may converge on an unstable equilibrium point. Second, from there, at low densities extinction occurs, and at high densities the population "explodes." Where the isoclines do not cross, the same unstable runaway dynamics as in facultative mutualism occur. Finally, as shown

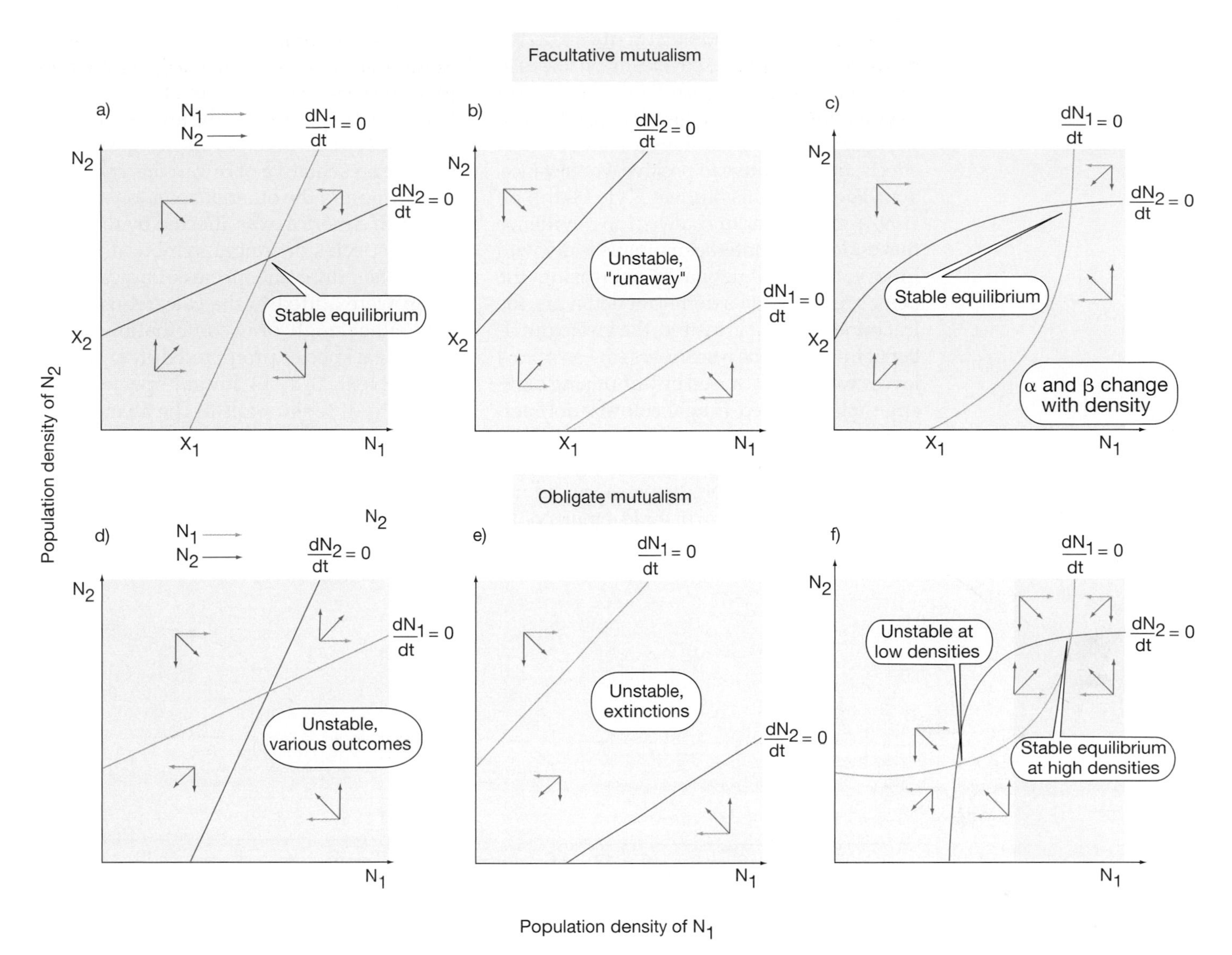

Figure 9.10 Graphical models of facultative (a–c) and obligate (d–f) mutualism. (a) and (b) are based on modified Lotka–Volterra equations. If the isoclines cross (a), then a stable equilibrium is possible. X_1 and X_2 represent the maximum carrying capacities in the absence of the mutualist. If the isoclines fail to cross (b), then runaway, or unstable, dynamics result. Modifications to the Lotka–Volterra equations such that α and β change with density (c) also result in a stable equilibrium. In obligate mutualisms, some mutualists must already be present before the other mutualist can grow, a precondition that results in more unstable outcomes (d–f).

in Fig. 9.4f, where α and β are reduced at high densities, two equilibrium points are possible—one stable (at high densities) and one unstable (at low densities). Models of facultative mutualism are, in general, more stable than those of obligate mutualism.

9.5 Mutualisms and Community Process

Although most studies of mutualism focus primarily on the two species directly involved, mutualisms can have strong indirect effects on other members of a community. For example, mycorrhizal fungi can affect herbivore loads on their host plants. As a case in point, mycorrhizal mutualists found on the roots of pinyon pines (*Pinus edulis*) can positively or negatively affect the density of the needle scale insect *Matsucoccus acalyptus* on the pine by either improving the plant's vigor or increasing the plant's investment in antiherbivore defenses. The relationship between systemic fungal endophytes—fungi that live inside leaves—and vascular plants, usually thought to be a parasitic infection, has also been viewed as mutualism. The fungi are thought to aid their hosts in defending against herbivory. For example, as mentioned previously fungal endophytes of fescue grasses (*Festuca* sp.)

Mutualisms can have communitywide effects.

produce toxic alkaloids that reduce grazing by herbivores and cause widespread losses to the U.S. livestock industry. The few studies of interactions of endophytes with insect herbivores of woody plants show a wide range of effects, from negative to positive. For instance, Stan Faeth and his student Kyle Hammon (1997a, b) did not find any strong evidence that endophytes interact mutualistically with Emory oak in Arizona by increasing the plant's resistance to a dominant herbivore, the leaf miner *Cameraria*. Instead, the greater number of infections the researchers saw on mined leaves were likely caused by leaf-mining damage that enhanced fungal colonization and penetration of the fungus into the oak leaves. An experimental infestation of fungal endophytes caused by spore suspensions did not affect either the survival of the leaf miner or its size at pupation.

Van der Heijden *et al.* (1998) have argued that mycorrhizal fungi determine plant species diversity. Using field plots and laboratory experiments, the authors showed how the floristic diversity of two grasslands, one a representative of calcareous grassland in Europe and the other of abandoned fields in North America, was affected by the number of species of fungal symbionts in the soil. When different species of mycorrhizal fungi were added to the calcareous grassland either singly or in combination, different fungal species promoted different plant hosts. Next, 0 to 14 fungal species were added to different plots of the abandoned-field communities. As the number of fungal species per system was increased, so the collective biomass of shoots and roots increased, as did the diversity of the plant community (Fig. 9.11). The mechanisms

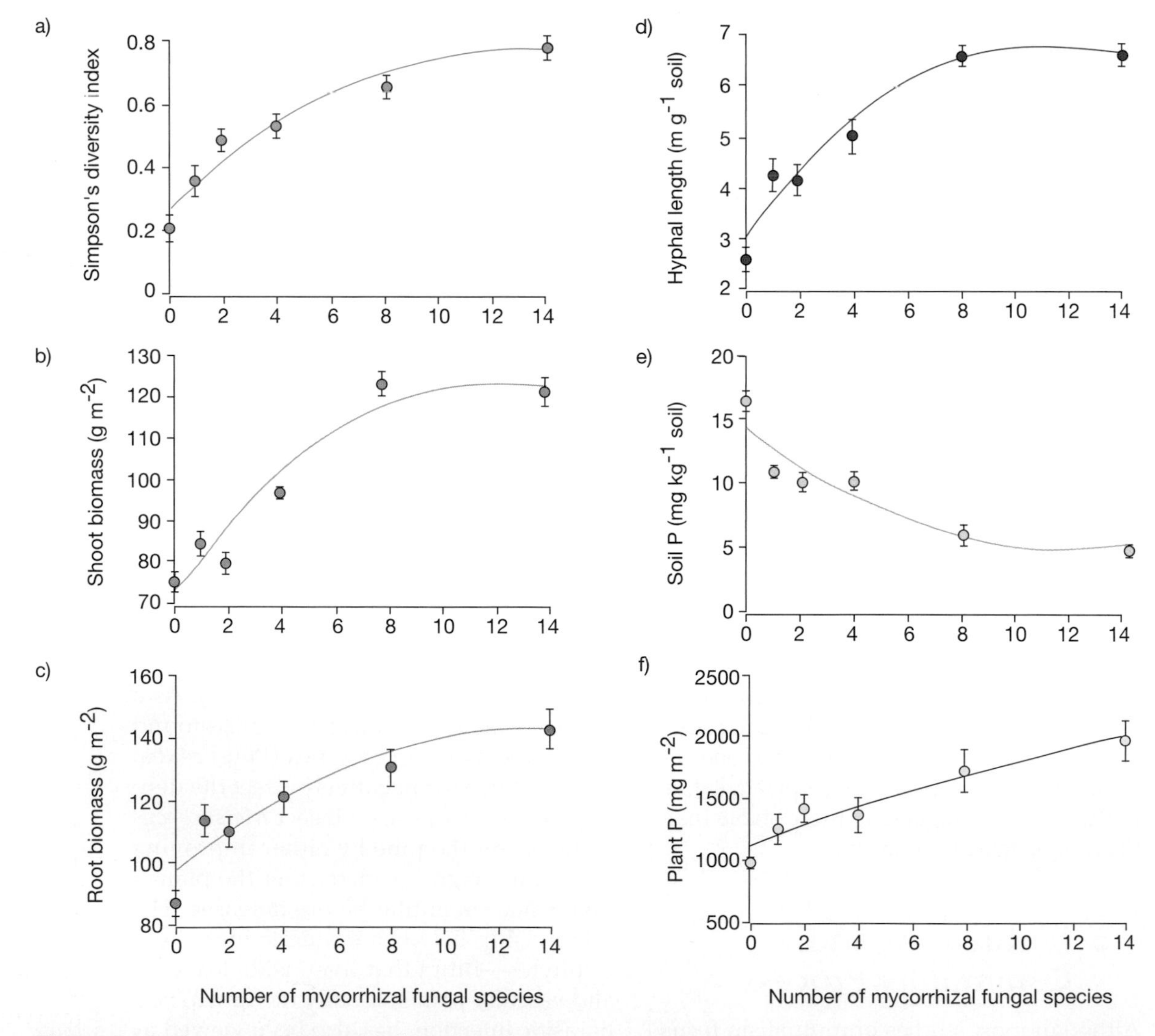

Figure 9.11 The effect of mycorrhizal species richness on communities simulating North American old-field ecosystems. As the number of mychorrhizal species in a system increased, so did (a) Simpson's diversity index, a measure of species richness, (b) shoot biomass, (c) root biomass, and (d) the length of external mycorrhizal hyphae in soil, whereas (e) the soil phosphorus concentration decreased and, consequently, (f) the total plant phosphorus content increased. Circles represent means shown with standard errors. *(Reproduced from van der Heijden et al., 1998.)*

were thought to include increased hyphal length when fungal diversity increased and hence greater uptake of phosphorus from the soil into the plants. In this experimental system, plant diversity reached a maximum when eight fungal species were added. (The result, however, may be constrained by the reconstructed nature of the communities; under field conditions, it is harder to predict what the minimum number of fungal symbionts might be to ensure maximal plant diversity.)

Microbial interactions can also *reduce* plant diversity if they promote the growth of invasive species. University of Indiana researchers Keith Clay and Jenny Holay (1999) showed how an invasive grass, tall fescue, when sown with endophyte-infested seed in experimental plots in Indiana, quickly outcompeted native grasses, reducing the number of species of native plants. Tall fescue has been widely planted for forage, turf, and soil conservation, but it is an aggressive invader of natural communities. Any control program may thus have to consider methods of reducing soil endophyte loads, as well as direct plant control measures.

9.6 Commensalism

In commensalism, one partner receives a benefit while the other is unaffected.

In commensal relationships, one member derives a benefit while the other is unaffected. An example is an orchid or some other "air plant" growing in the fork of a tropical tree: The tree gains nothing, but the orchid gains a place to live. Cattle egrets are commensal with cattle, which stir up insect prey for the birds. This creates more food for the cattle egret in fewer steps and less time (Fig. 9.12). (By contrast, some birds, such as oxpeckers, which ride on the backs of buffalo, are mutualists: The bird gets a meal, while the mammal is freed of pests. In a similar fashion, some birds pick the teeth of crocodiles, getting a meal in return.)

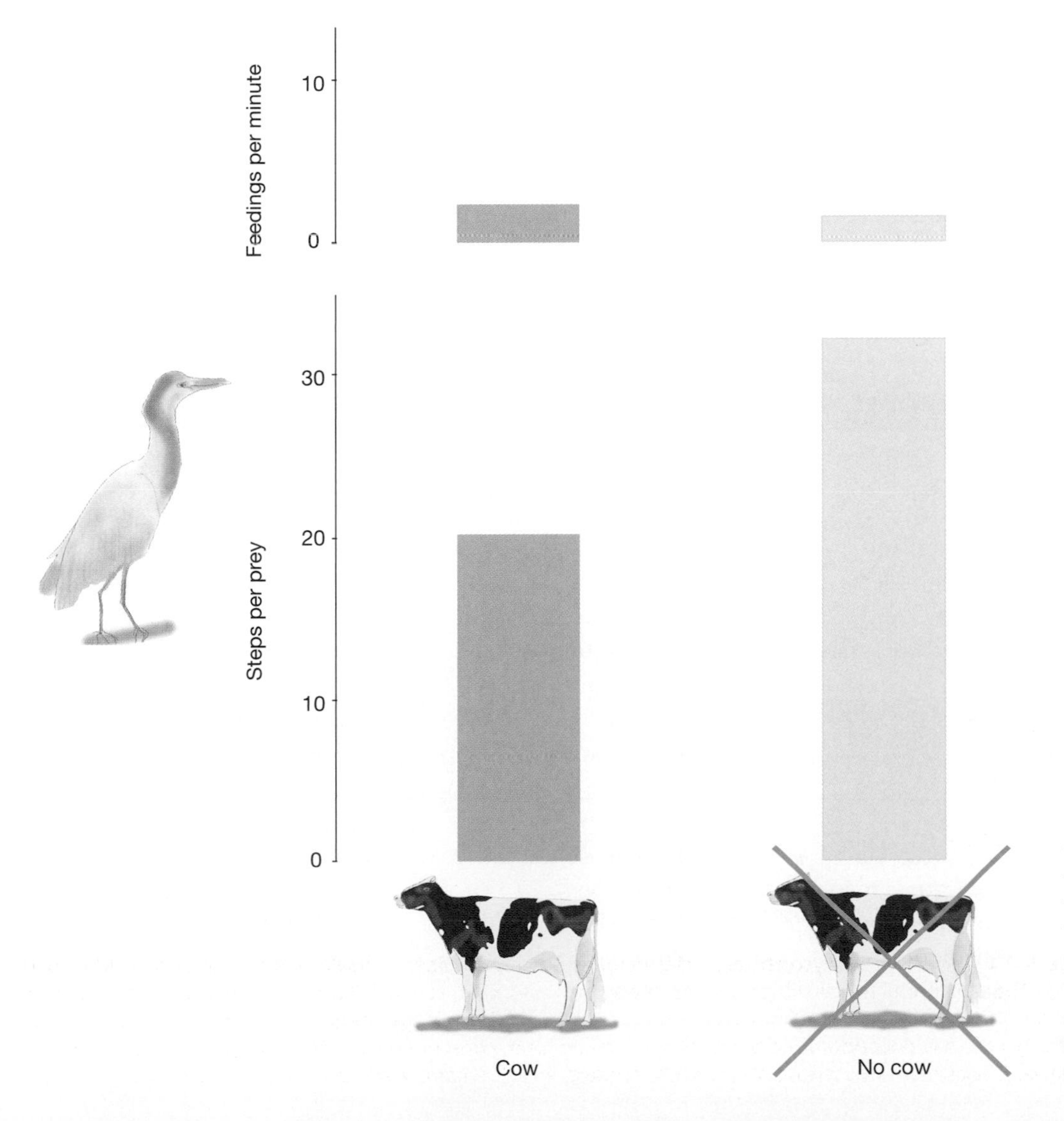

Figure 9.12 Feeding rates of cattle egrets when feeding alone and when associated with cattle. Associations with cattle lead to more feedings per minute and fewer steps per item of prey. The egret's rate of feedings per step when associated with cattle (0.129) is much higher than when the bird feeds alone (0.051). *(After data in Heatwole, 1965.)*

Another example of commensalism is *phoresy*, in which the association involves the passive and more temporary transport of one organism by another, as in the transfer of flower-inhabiting mites from bloom to bloom in the nares (nostrils) of hummingbirds (Colwell, 1973). The flowers the mites inhabit live only a short while before dying, so the tiny mites are confronted with the problem of relocating to distant flowers. They solve this problem by scuttling into the nares of visiting hummingbirds and hitching a ride to the next flower. Presumably, the hummingbirds are unaffected.

Some of the most numerous examples of commensalism are provided by plant mechanisms of seed dispersal. Many plants "cheat" their potential mutualistic seed-dispersal agents out of a meal by developing seeds with barbs or hooks. The seeds lodge in the animals' fur rather than their stomachs. In these cases, the plants receive free seed dispersal, and the animals receive nothing. This type of relationship is fairly common: Most hikers have been plagued by burrs and sticktights. However, sometimes these barbed seeds cause great discomfort to the animal in whose fur they become entangled. For instance, fruits of the genus *Pisonia* (a cabbage tree), which grows in the Pacific region, are so sticky that they cling to bird feathers. On some islands, birds and reptiles can become so entangled with *Pisonia* fruits that they die. This relationship may have crossed the line from commensalism to an antagonistic relationship such as parasitism. Similar examples exist elsewhere. For example, live oak trees in the U.S. southeast are unaffected by low levels of the epiphyte Spanish moss, but high levels can cause boughs to break. These examples serve to illustrate how hard it sometimes is to precisely demarcate when a relationship crosses the line from one category to another. Nevertheless, commensalism is probably very common in nature. (See Table 9.1.) Mark Bertness, of Brown University, and his colleagues (Bertness and Leonard, 1997) have detected many commensal interactions involving organisms on rock faces in intertidal habitats and between plants in intertidal salt marshes in New England. Many of these relationships may be regarded as facilitation, wherein one species facilitates the presence of the other.

Bertness and Hacker (1994) described an interesting relationship in the salt marshes of New England between the perennial shrub Marsh elder (*Iva frutescens*) and black grass (*Juncus gerardi*). The authors experimentally removed *Juncus* from around some *Iva* plants and found that soil salinities doubled and soil oxygen levels decreased. The photosynthetic rate of *Iva* went down, and 14 months later, the *Iva* were dead. Clearly, *Juncus* neighbors have strong positive effects on adult marsh elders, reducing the salinity of the soil and enabling *Iva* to survive at places in the salt marsh where, alone, it would die. Hacker and Bertness (1996)

TABLE 9.1 Some examples of commensalism.

Provider	Resource Provided and Mechanism Employed	Beneficiary
Pitcher plant	Habitat of water pitcher	Mosquitoes
Yellow-bellied sapsucker	Drills holes to collect sap	Hummingbirds and other feeders
Shark	Food scraps	*Remora* fish
Marine burrowing worms	Habitat (burrow)	Crabs
Cattle	Stirs up insect food	Cattle egrets
Trees	Habitat	Orchids or other air plants (epiphytes)
Sloth	Habitat in hair	Green algae
Turtle	Carapace habitat	*Baiscladia* green algae
Whale	Habitat, body surface	Barnacles
Sea cucumber	Habitat, cloacal cavity	Pearl fish
Portuguese man-of-war	Protection	Certain fish
Polar bear	Food scraps	Arctic fox
Anemone	Protection	Clown fish
Sea urchin	Protection	Clingfish

went on to show again how plant commensalisms could have communitywide ramifications. Population growth rates of aphids on *Iva* in the absence of *Juncus* were so low that the aphids were unable to produce enough offspring to replace themselves. In addition, the number of ladybird beetle predators of the aphids decreased on *Iva* in the presence of *Juncus*, so the mutualism had an effect even two trophic levels removed from the plants themselves.

Commensal interactions are increasingly being found to be common in plant communities. Ragan Callaway (1995) documented 169 cases of positive interactions among plants. Many mechanisms were responsible. Prominent among them was a mechanism of resource modification in which one species positively affected the availability of resources such as light, water, or nutrients. Another mechanism was substrate modification, whereby one species might accumulate soil around its roots that the other could use. Callaway (1998) went on to suggest that most of the facilitations were species specific. In other words, the positive effects of one plant on another were not simply due to physical effects that could be caused by inanimate objects, like rocks, that provide shade. Beneficiaries were often found associated with particular benefactor species. This finding has vast implications for plant communities, because it means that many such communities are not random groups of species, but are highly interdependent assemblages.

Plainly, a wide variety of mutualisms and commensalisms exists in nature, and there are many reasons that associations of species are mutually beneficial, ranging from increased access to resources to protection from natural enemies. Such associations are usually of greater importance to both species when harsh environmental conditions or high levels of predation or parasitism prevail. In favorable environments, species can make a successful living on their own.

Summary

Mutualism, commensalism, and parasitism are all symbiotic relationships in which the partners live in close association with one another. However, in mutualism and commensalism, neither species has a negative effect on the other, so discussions of these phenomena are usually treated together.

1. In mutualisms, both species benefit. A common example is pollination, whereby plants benefit from the transfer of pollen and the pollinator, often an insect, gains a meal of nectar. In seed dispersal systems, the plant provides a fruit meal for birds and mammals and, in turn, benefits from the dispersal of its seeds into new areas.

2. In some mutualisms, the partners either help each other obtain resources or protect each other from enemies.

3. Obligate mutualisms are mutualisms in which the species cannot live apart. Some examples are lichens, a mutualism between fungi and algae; mycorrhizae, an association between fungae and plant roots; and the symbiotic bacteria in the guts of ruminants.

4. Mathematical models of mutualisms are hard to construct, because they often have the unlikely result of runaway population densities of both species. However, models of facultative mutualisms are generally more stable than those of obligate mutualisms.

5. In commensalisms, one member derives a benefit while the other is unaffected. One of the most common examples of such a phenomenon is phoresy, the passive transport of one organism by another.

Mutualism and commensalism are frequent and important in nature. Their effects can reach beyond the immediate species involved and affect the lives of many other species in communities.

Discussion Questions

1. Where might mutualisms be more common, in temperate areas or in the more stable tropics? Why?

2. In your perusal of the literature, which type of mutualism listed in the text seems to be the most common? Are mutualisms more common in one particular kind of habitat (e.g., freshwater, marine, or terrestrial)? If so, why do you think it is so?

3. Boucher, James, and Kresler (1984) suggested that mutualisms are of four varieties: nutritional, protective, energetic, and transport. Define and give examples of each one.

4. A flock of birds with different species or mixed herds of animals on the African plains could represent facultative mutualisms. Discuss the advantages and disadvantages of such groups.

C H A P T E R 10

Predation

Road Map

1. A variety of antipredator adaptations, such as aposematic coloration, camouflage, polymorphisms, and chemical defenses, suggests that predation is important in nature.
2. Predator–prey models can explain many outcomes of the interactions between predators and prey.
3. Field data suggest that predators have a large impact on their prey populations.
4. Experiments involving the removal or introduction of exotic predators afford some of the best data on the effects of predators on their prey.
5. Field experiments involving manipulations of native predators also show predation to be a strong force in nature.

In the early 1990s, the U.S. Fish and Wildlife Service wanted to reintroduce wolves into Yellowstone National Park and to stabilize their numbers in Montana and Minnesota, the only states other than Alaska to possess viable populations of the animal (Fig. 10.1). Cattle ranchers were fearful, however, that the wolves would decimate their herds. The question was, do wolves have the ability to decimate herds of cattle, or do they take only sick and dying animals that contribute little to the production of the herd? This is a fundamental question that we shall attempt to answer in this chapter. Another important question is, are predators tied to the health of their main prey population, or can they effectively switch prey? Again, the answer to this question has important ramifications for predator recovery schemes. If wolves can switch among different prey species, reintroductions have a much greater chance of reestablishing self-sustaining populations. (By the way, wolves were reintroduced to Yellowstone in 1995, with no major effects on cattle.)

In our discussions of predator–prey relationships, several types of predation can be recognized. The traditional view of predation is **carnivory**: A predator feeds on a herbivore, with a positive effect for one species and a negative effect for the other. However, the same effects apply to herbivory and par-

Figure 10.1 Wolf reintroduction into Yellowstone National Park, Wyoming, involved much planning and preparation, but hopefully it will result in a stable population of wolves in this state.

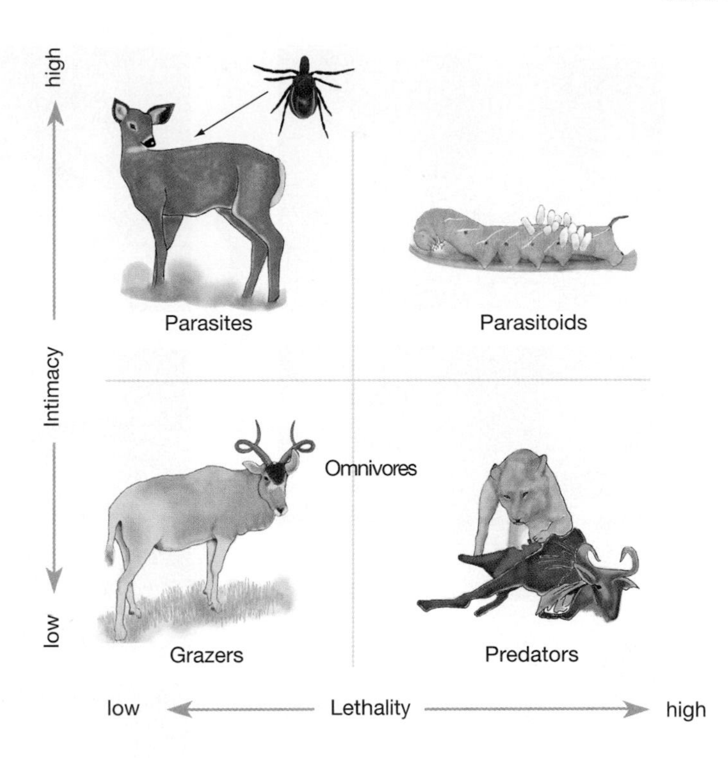

Figure 10.2 Possible interactions between populations. Lethality represents the probability that a trophic interaction results in the death of the organism being consumed. Intimacy represents the closeness and duration of the relationship between the individual consumer and the organism it consumes. Omnivores feed on both plants and animals; cannibalism may be exhibited by both predators and parasitoids.

asitism. Nevertheless, although herbivory and parasitism can be viewed in the broad context of predation, each has special characteristics that tend to set it apart, and each of those subjects will be discussed separately in subsequent chapters. Herbivory, for example, is often nonlethal predation on plants, whereas in the animal world, predation generally means death for the prey. Parasitism differs from predation because one individual prey is commonly utilized for the development of more than one parasite, as, for example, tapeworms in humans. In omnivores, individuals can feed on both plants and animals. (For example, bears feed on both berries and fish.) All these categories of predation can be classified according to how lethal they are for the prey and how closely predator and prey associate (Fig. 10.2).

10.1 Antipredator Adaptations

In response to predation, many prey species have developed strategies to avoid being eaten, including the following:

- *Aposematic, or warning, coloration.* Advertises an unpalatable taste. Lincoln Brower, an entomologist at the University of Florida, and his coworkers (1968) showed how inexperienced blue jays took monarch butterflies, suffered a violent vomiting reaction, and learned to associate the striking orange-and-black barred appearance of the monarch with a noxious reaction. The caterpillar of this butterfly gleans the poison from milkweed, its poisonous host plant. Many other species of animals, especially invertebrates, are also aposematically colored—ladybird beetles come immediately to mind. Caterpillars of many Lepidoptera are bright and conspicuous, too, because being noxious is their main line of defense—their soft bodies make them otherwise highly vulnerable to predators. Many tropical frogs have toxic skin poisons and advertise their lethality to predators by bright warning coloration (Fig. 10.3a).
- *Camouflage,* the blending in of an organism with the background color of its habitat is a common method of avoiding detection by predators. For example, many grasshoppers are green and blend in perfectly with the foliage on which they feed (Fig. 10.3b). Even the veins of leaves are often mimicked on grasshoppers' wings. Stick insects mimic branches and twigs with their long, slender bodies. In most cases, these animals stay perfectly still when threatened, because movement alerts a

The variety of strategies that organisms have evolved to avoid being eaten suggests that predation may be a strong selective force.

Figure 10.3 (a) Aposematic or warning coloration. *Dendrobates histronicus*, a poison dart frog. (*Dr. E. R. Degginger.*) (b) If predation were not important, why would this tropical grasshopper mimic leaves? (Laval, *Animals Animals.*) (c) Batesian mimicry: a harmless hoverfly, *Epistrophe balteata*, imitating a harmful wasp. *(Vock/Okapia, Photo Researchers, Inc.)* In Müllerian mimicry, the monarch (d) and viceroy (e) butterflies share a similar body coloration and a common bad taste to predators. (f) Bluff: This frilled lizard, *Chlamydosaurus kingii*, from Australia, tries to surprise potential predators by extending the frill around its neck to appear larger. *(Uhlenhunt, Animals Animals.)*

predator. Camouflage is prevalent in the vertebrate world, too: A zebra's stripes make it blend in with its grassy background, and the sargassum fish even adopts a body shape to mimic the sargassum weed in which it is found.

- *Mimicry.* Besides using camouflage, which may be viewed as mimicry of the foliage or some other feature of the background around an animal, some animals mimic other animals. For example, some hoverflies mimic wasps. Two types of mimicry can be defined.

—In *Müllerian* mimicry, (after the German biologist, Fritz Müller, 1879), many unpalatable species converge to look the same, thus reinforcing the basic distasteful design. Wasps and some butterflies utilize this form of mimicry. Sometimes a group of **sympatric** species, often from different taxa, share a common warning pattern, a phenomenon known as a *mimicry ring*. There are many such rings in Amazonian butterflies, often involving butterflies from the Pieridae and Ithomidae families.

—*Batesian* mimicry [after the English naturalist Henry Bates (1862),] is the mimicry of an unpalatable species by a palatable one. Some of the best examples involve flies—especially hoverflies of the family Syrphidae, which are striped black and yellow to resemble stinging bees and wasps (Fig. 10.3d).

Sometimes, it is very difficult to distinguish Batesian from Müllerian mimics without a great deal of work. Until very recently, monarch butterflies, aposematically colored themselves, were considered models for mimicking viceroy butterflies (Figs. 10.3d,e). However, David Ritland, a graduate student at the University of Florida (Ritland and Brower, 1991) showed that both species were unpalatable and thus coexisted as a Müllerian mimicry complex. In theory, Batesian mimics should lose their protection when they are common, because they need plenty of distasteful models whose unpalatability maintains the validity of their warning colors. If the edible mimic became too common, then predators would learn to ignore the coloration, which would also harm the model.

- Some animals put on *displays of intimidation*. For example, a toad swallows air to make itself appear larger and frilled lizards extend their collars when intimidated, so as to create the same effect (Fig. 10.3f). Both behaviors are intended to deceive potential predators about the ease with which prey can be eaten.
- *Polymorphism* is the co-occurrence in the same population of two or more discrete forms in proportions greater than can be maintained by recurrent mutation alone. Often, this phenomenon takes the form of a color polymorphism: If a predator has a preference or search image for one color form, usually the more common one, then the prey can proliferate in the rarer form until this form itself becomes the more common. Just this type of selection of prey, carried out by a visually searching parasitoid, apparently maintained the difference between two distinct color morphs, orange and black, in some leafhopper nymphs of the genus *Eupteryx* (Stiling, 1980). In pea aphids (*Acyrthosiphon pisum*), a balanced polymorphism between green and red color morphs was maintained by the action of natural enemies. Green morphs suffered higher rates of parasitism than red morphs, whereas red morphs were more likely to be attacked by ladybird beetle predators than were green morphs. Therefore, when parasitism rates in the field were high relative to predation rates, the proportion of red morphs increased relative to green morphs, whereas the converse was true when predation rates were relatively high (Losey *et al.*, 1997). In many species, the form of the polymorphism is such that *every* individual is slightly different from all the others. This is true of brittle stars, butterflies, moths, echinoderms, and gastropods. Such a staggering variety of form may thwart predators' learning processes, a phenomenon that has been called *reflexive selection* (Owen and Whiteley, 1986).
- *Sometimes the prey is phenologically separated* from the predator. For example, fruit bats, normally nocturnal foragers, are active both day and night on some small, species-poor Pacific islands such as Fiji. The bats may be constrained elsewhere to fly only at night by the presence of predatory diurnal eagles.
- One of the classic examples of a *chemical defense* involves the bombardier beetle (*Bradinus crepitans*), as studied by Cornell University biologist Tom Eisner and coworkers (Eisner and Aneshansley, 1982). These beetles possess a reservoir of hydroquinone and hydrogen peroxide in their abdomens. When threatened, they eject the chemicals into an "explosion chamber," where they mix with a peroxidase enzyme. The subsequent release of oxygen causes the whole mixture to be violently ejected as a spray that can be directed at the beetle's attackers. Many other arthropods, such as millipedes, have chemical defenses, too, and the phenomenon is also found in vertebrates, as people who have had a close encounter with a skunk can testify.
- *Masting* is the synchronous production of many progeny by all individuals in a population to satiate predators and thereby allow some progeny to survive. Masting is more commonly discussed in relation to seed herbivory in trees, which tend to have years of unusually high seed production to satiate predators. However, a similar phenomenon is exhibited by the emergence of 17-year and 13-year periodical cicadas (*Magicada* sp.). These insects are termed *periodical* because the emergence of adults is highly synchronized to occur once every 13 or 17 years. Adult cicadas live for only a few weeks, during which time females mate and oviposit on the twigs of trees. The eggs hatch 6 to 10 weeks later, and the nymphs drop to the ground and begin a long subterranean development feeding on the contents of the xylem of roots. Because xylem is such a dilute medium, nymphal development is very slow, but there seems to be no reason that adults should not emerge after 12, 13, or 14 years. It is thought that predator satiation is maximized by the exact synchrony of emergence. Worth noting in this context is the fact that both 13 and 17 are prime numbers, so no predator on a shorter multiannual cycle could repeatedly use this resource. Williams, Smith, and Stephen (1993) studied the mortality of populations of a 13-year periodical cicada that emerged in northwestern Arkansas in 1985. Birds consumed almost all of the standing crop of cicadas when the

density was low (i.e., when the cicadas first emerge in a mast year or when they make a "mistake" and emerge in a non-mast year), but their consumption fell to only 15 to 40% when the cicadas reached peak density in a mast year. Predation then rose to near 100% as the cicada density fell again. But why have both 13- and 17-year cicadas? Rick Karban of the University of California at Davis (1997) (see "Profiles" box) has shown how the 4 extra years' growth of 17-year cicadas over 13-year ones can produce females with much heavier ovaries and more eggs. This increased fecundity more than makes up for the delay in reproduction.

Profiles

Rick Karban, University California, Davis

I was an undergraduate at Haverford College, which was an excellent place to start. Students were encouraged to participate in their education, to ask questions, and to develop important skills. Haverford offered only cell and molecular biology, and a very brief exposure to this "small stuff" convinced me that biology was not for me. Fortunately, I had the opportunity to work for the National Park Service during my summers in college, and this experience convinced me that ecology might be more to my liking. I went to Cornell for my junior year and took ecology and other whole-organism courses, reinforcing my suspicion that this "bigger stuff" was pretty interesting. Dan Janzen had recently moved to the University of Pennsylvania, which was a train ride away from Haverford. I attended an informal research seminar that he gave and was impressed. Dan was one of the most charismatic personalities I had ever witnessed. He was full of stimulating ideas and stories about how nature worked.

I decided to go to graduate school in ecology at the University of Pennsylvania. Starting grad school was exhilarating and intense. Unfortunately, my feeling of excitement regarding academia was short lived. My first surprise came following a seminar presented to the biology department by an invited speaker. I was told that my questions were naive and reflected poorly on the department. Over the next few years, it became clear to me that people around me were desperately afraid of being wrong, or worse, not being as smart as the fellow down the hall. Perceptions were given a lot of thought, and there was no room for naiveté. Questions in seminars were about showing others how smart you were, not about satisfying your curiosity.

Because I was fascinated by population-level events that were synchronous, Dan suggested that I read papers about periodical cicadas that Monte Lloyd and his associates, JoAnn White and Chris Simon, had written. A personal visit to Monte in Chicago gave me the encouragement and advice I needed to start a field project that turned into my thesis.

Grad school continued to be an ordeal. I was told I was not a thinker and that I would not have a chance of getting an academic job. At one point, I left graduate school without telling anyone at the university. I went home, talked to my parents, spent time with my best friend, and thought a lot about potential careers. I decided that I really did want to be a professional biologist in some capacity and that I would not let the poor opinions that others had of my potential defeat me. I felt determined to make my own decisions. I felt that the hardest part would be returning and explaining why I had been gone for a week. I returned, fortified by the encouragement and support of being home, to find that no one had even noticed I had been gone.

Although my personal interactions in grad school were not very rewarding, fieldwork was. Fieldwork has always been the most enjoyable part of being an ecologist for me. Field seasons during my years in graduate school were fantastic, the kind of hard work that I most enjoy.

I applied for many jobs, both in academia and elsewhere. The first interview I was granted came from the entomology department at the University of California at Davis. A job in an entomology department seemed like a long shot, since I had no formal training in that subject. I had the feeling that this was an opportunity that would not come twice. In any case, I was in the right place at the right time. Dan was, well, surprised at Davis's decision. At my thesis defense just prior to moving to Davis, I was advised not to spend much time looking back, advice that I generally try to heed.

I learned some important lessons in graduate school, although I wouldn't choose to repeat the experience. I learned that working with large, soil-dwelling insects that take 17 years to complete a generation was a lot of fun, but not a good vehicle for answering the questions I was most interested in. I haven't abandoned cicadas, but I have worked more recently with mites and thrips—small arthropods that complete a generation in only a few weeks. Most importantly, I learned that I won't let anybody sell me short again. Being a successful scientist is about daring to be naive, asking questions that other people think are stupid, doing the projects that are personally interesting regardless of whether your peers agree, and finding the confidence to keep going when you don't really possess it.

TABLE 10.1 Antipredator mechanisms in arthropods (mainly insects). *(Modified from Witz, 1989.)*

	Mechanism and Rank	Example of Mechanism	Frequency (%)
1.	Chemical	Reflexive bleeding, toxic chemicals (especially beetles)	46
2.	Fighting	Stinging (especially wasps), biting, kicking	11
3.	Crypsis	Camouflage, especially caterpillars	9
4.	Escape	Running away, flying	8
5.	Mimicry	Batesian and Müllerian	5
6.	Aposematic	Warning coloration	5
7.	Display of intimidation	Posturing	4
8.	Dilution	Masting, satiation	4
9.	Mutualism	Defense by other organism	3
10.	Armor	Spines, thorns	2
11.	Acoustic	Loud noise (e.g., made by a grasshopper)	2
12.	Feigning death	Some beetles	1

How common is each of these types of defense? University of South Florida graduate student Brian Witz (1989) surveyed 354 papers that documented antipredator mechanisms in arthropods—mainly insects (Table 10.1)—in 555 predator–prey interactions. By far the most common antipredator mechanism was a chemical defense, noted in at least 46% of the papers surveyed. (I say at least, because many categories were not mutually exclusive; for example, aposematic coloration is usually coupled with noxious chemicals.) Clearly, then, predation constitutes a great selective pressure on plant and animal populations.

Despite the impressive array of defenses, predators still manage to survive by eating individuals of their chosen prey, often by circumventing the defenses in some way. The coevolution of defense and attack can be seen as an ongoing evolutionary "arms race." According to Dawkins and Krebs (1979), the prey is always likely to be one step ahead. The reason is what they termed the "life–dinner principle." In a race between a fox and a rabbit, the rabbit is usually faster because it is running for its life, whereas the fox is running "merely" for its dinner. A fox can still reproduce even if it does not catch the rabbit. The rabbit never reproduces again if it loses the race. It has been argued that this race is run not only between predators and prey, but also between parasite and host and between plant and herbivore. In the latter case, the race often proceeds by the production of toxins by the host and detoxifying mechanisms by the predator.

10.2 Predator–Prey Models

The effects of predation on the dynamics of predator and prey populations can be described with the use of graphical models.

What effect has the predator on its prey population? The answer depends on many things, including prey and predator density and predator efficiency. Using a graphical method, Michael Rosenzweig and Robert MacArthur (1963) modeled predator–prey dynamics. First, they assumed that predators feed only on one prey species. Prey **isoclines** have a characteristic hump shape (Fig. 10.4). In the absence of predators, the maximum prey density would be K_1, the carrying capacity for the prey. The lower limit is set by a value below which individuals would probably be too rare to meet for reproduction and the prey population would plummet to extinction. Between these two values, the prey population can increase, provided that predator densities aren't too high. At low prey densities, for example, even moderate predator pressure can depress prey growth. Above the isocline, prey

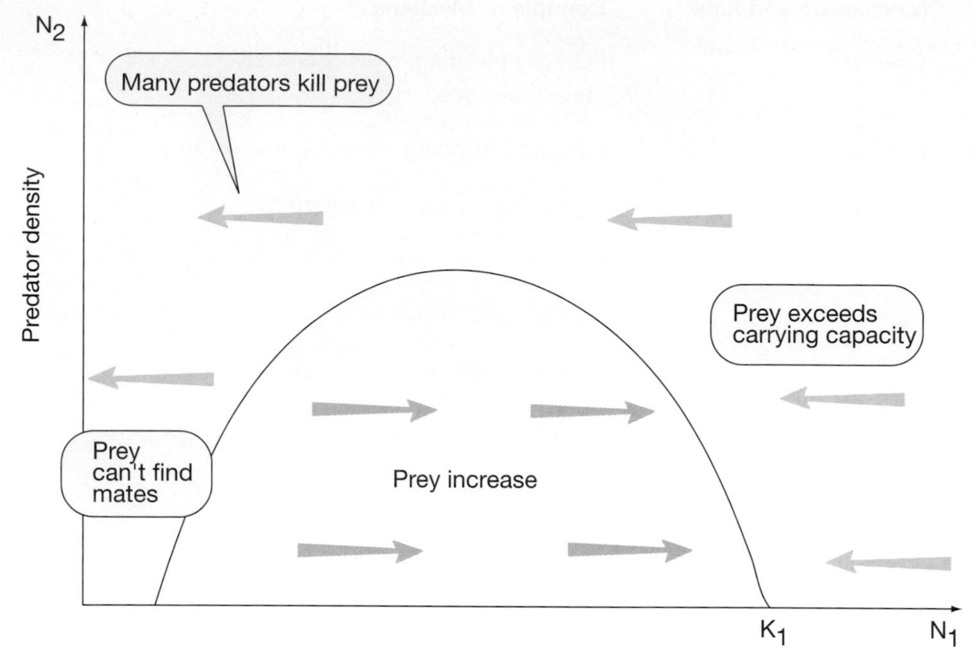

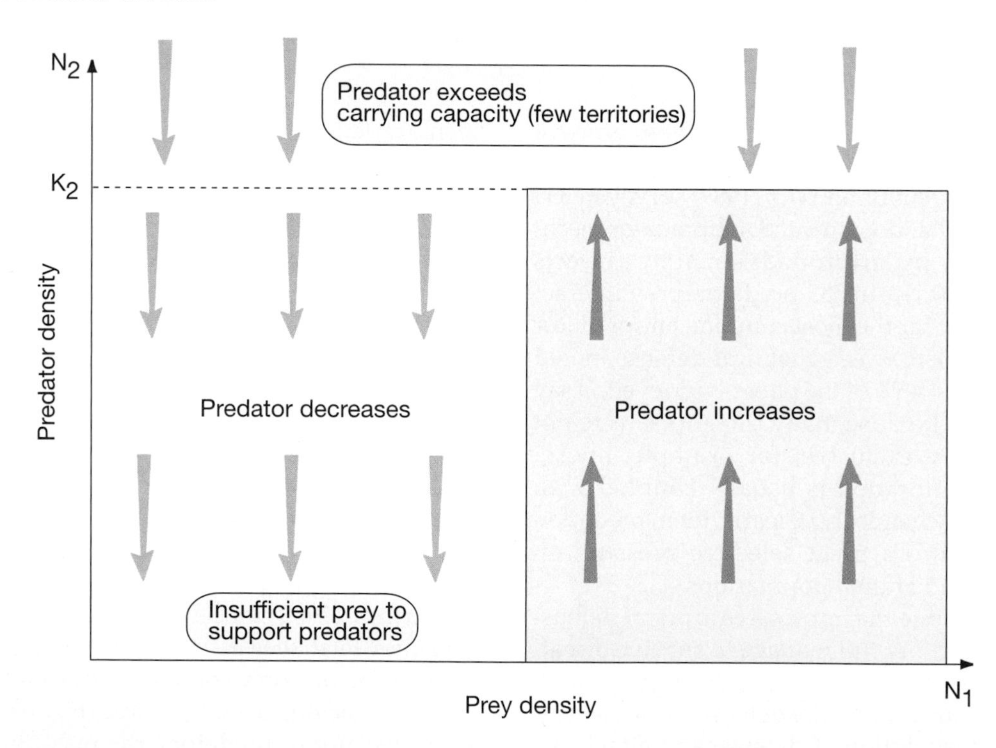

Figure 10.4 Hypothetical forms of predator and prey isoclines based on the logic of Rosenzweig and MacArthur (1963). In (i), prey populations may be limited in region *A* by too many predators, in region *B* by the carrying capacity of the environment, and in region *C* by too few conspecifics to mate with. In (ii), predator populations are limited in region *A* by the carrying capacity, which is dictated by available territories, even when prey are numerous, and in region *B* by insufficient prey to support a predator population.

decline, and below it, they increase. As long as the prey isocline has but a single peak, its exact shape doesn't alter the conclusions of the model. Next, Rosenzweig and MacArthur considered the shape of the predator isocline. First of all, there must be some threshold density above which predators can increase and below which there are too few prey to support a group of predators. (We consider a group of predators because many carnivores live in packs, but we also assume that there is no interaction among predators.) The predators increase until their carrying capacity is reached, probably as dictated by some territorial limits. If there is mutual interference or competition between predators, then even more prey are required to support a given density of prey, and the predator isocline slopes towards the right.

When we superimpose the prey and predator isoclines, we add the predator and prey trajectories (Fig. 10.5), just as we did for the competitor trajectories in the Lotka–Volterra competition model. Exactly one stable point emerges: the intersection of the lines. There are three general cases: (1) An inefficient predator requires very high densities of prey (Fig. 10.5a), and the lines of vector addition spiral inward, leading to damped oscillations of predator and prey. (2) A moderately efficient predator may lead

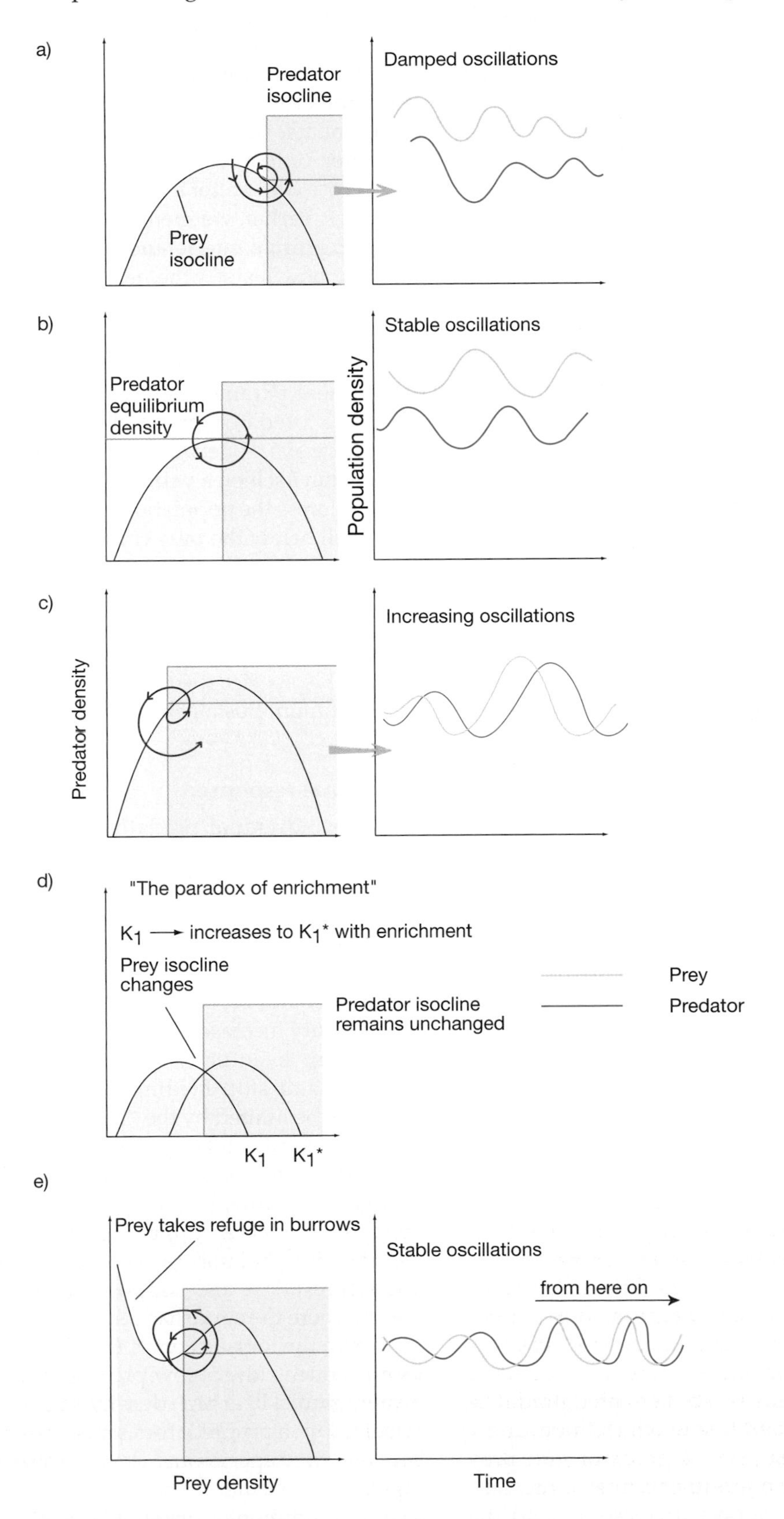

Figure 10.5 When the predator and prey isoclines of Fig. 10.4 are superimposed on one another, different interactions are possible, depending on the position of the predator isocline relative to the prey isocline. We add the predator and prey trajectories and show one resultant dotted line, just as we did for competitor trajectories in the Lotka–Volterra competition model. In panel (a), the predator is inefficient and cannot successfully exploit its prey until the prey is near the carrying capacity. In (b) the predator is moderately efficient, and the predator and prey engage in stable oscillations or cycles. In (c), the predator is highly efficient and can exploit very low densities of prey. This can result in great instability and increasing oscillations wherein one or both species run the risk of extinction. First the prey goes extinct, and then the predator does. Panel (d) represents the "paradox of enrichment." Added nutrients shift the carrying capacity of the prey to the right. Although the position of the predator isocline remains unchanged, its position relative to that of the prey isocline shifts to the left into the region of prey–predator instability. The paradox is that adding nutrients can actually destabilize an interaction. In (e), the addition of a refuge (like burrows) to which prey can retreat and where they are immune from predators effectively stabilizes the dynamics, since the predators cannot drive the prey to extinction.

to stable oscillations of predator and prey (Fig. 10.5b). Finally, (3) a highly efficient predator can exploit prey nearly down to its limiting rareness. Such a predator causes unstable, increasing oscillations between predator and prey (Fig. 10.5c). Thus, we see that a relatively simple interaction of predator and prey can produce all sorts of results, from damped to widely fluctuating oscillations, all based on how efficient the predator is.

The position of the predator isocline relative to that of the prey isocline may change if the position of the prey isocline shifts to the right or left (Rosenzweig, 1971). The latter may occur because of prey starvation (a shift to the left) or food enrichment (a shift to the right) (Fig. 10.5d). When food is particularly abundant for prey (say, nutrients are plentiful for the grasses that prey feed on), then the carrying capacity shifts from K_1 to K_1*, and the whole prey isocline changes. The position of the predator isocline is now changed from a stable position to the right of the prey isocline peak to an unstable position to its left. This situation is known as the "paradox of enrichment": Increasing nutrients or the availability of food to the prey destabilizes the system. The "paradox of enrichment" is sometimes suggested to be one reason that agricultural systems are unstable: Because the crops are dosed with fertilizers, and the "predators"—in this case the insect herbivore pests—shift to an unstable isocline, outbreaks of pests are precipitated.

The shape of the functional response can affect predator–prey interactions.

Finally, predator–prey isoclines can change dramatically with the incorporation of other factors into the scenario. For example, many fossorial (burrowing) mammals escape from predators into underground burrows, and certain passerines (songbirds) exist in well-defined territories with abundant cover. Mammals outside their burrows or birds in suboptimal habitats are often exposed to predators, whereas those in their refuges are not. The situation can be represented graphically by a practically vertical prey zero isocline at low densities (Fig. 10.5e). In other words, at low prey densities, prey can increase irrespective of predator densities, because there are enough refuges for all individuals. At higher prey densities, however, many prey exist outside the refuges and are available to predators. The refuge thus stabilizes the interaction, and a cyclic rise and fall of predator and prey occur. First, predator numbers increase when prey numbers increase beyond the number of refuges. Then, predator numbers increase and reduce prey numbers. As prey are reduced, some predators starve. Finally, prey numbers recover in the refuges, the prey population begins to expand, and the whole cycle starts again.

So, even simple predator–prey models can produce vastly different outcomes, based on tweaking just one parameter—predator efficiency—and including more components, such as refuges for the prey, can alter outcomes as well. In reality, of course, we could alter a myriad of other parameters. Earlier, we mentioned just one in passing: mutual interference among predators and how it shifts the predator isoclines to the right. Others include crowding effects on prey, limited dispersal of prey, and predators that feed on a variety of prey species. All of these parameters have the ability to stablilize predator–prey relationships in complex ways. For example, a predator that can sustain itself on a variety of prey species could depress the population of its preferred prey well below the prey carrying capacity. While some models seem to be just an abstraction of nature, we need every predictive tool possible to help us decide what harvest is sustainable for many economically valuable species (See "Applied Ecology: The Maximum-Sustainable-Yield Problem.")

Functional response.

How an individual predator responds to prey density can also affect how predators interact with their prey. The reaction is termed a *functional response* (Solomon, 1949; Fig. 10.6). There are three recognized types of functional response. In Type I, the individual predator consumes more prey as the prey density increases. In Type II, the predators can eventually become satiated (stuffed) and stop feeding. Alternatively, they may be limited by the "handling time" needed to locate and eat the prey. Handling times may also be large because of time needed to subdue, kill, and eat prey. In either case, there is an upper level or asymptote at which predators can process prey. The Type III response also supposes an asymptote, but here the curve has a sigmoid shape, similar to the logistic curve. In Type III, the feeding rate is low at low prey density, but increases quickly at high density. This sometimes happens if predators switch feeding between prey species and develop a search image for particular species once they reach a certain common or threshold density.

Applied Ecology

The Maximum-Sustainable-Yield Problem

How many deer can be shot before there is a marked effect on their population? How can people harvest part of a forest without causing long-term changes in its equilibrium numbers? How can commercial fishermen and -women prevent the catastrophic decimation of fish and the lowered economic yield of overfishing.

For a harvested population, the important measurement is the yield, expressed in terms of either weight or numbers over a particular period, to give a catch per unit effort. The catch per unit effort can then be compared year after year to determine how well a particular managed resource is doing. The maximum yield of a population is related to the maximum population increase

$$\frac{dN}{dt} = rN\frac{(K-N)}{K},$$

which occurs at the midpoint of the logistic curve (Fig. 1). Thus, the maximum yield is obtained from populations at less than maximum density, when they are constantly trying to expand into unutilized resource areas. Adjusted for fishing losses, dN/dt becomes

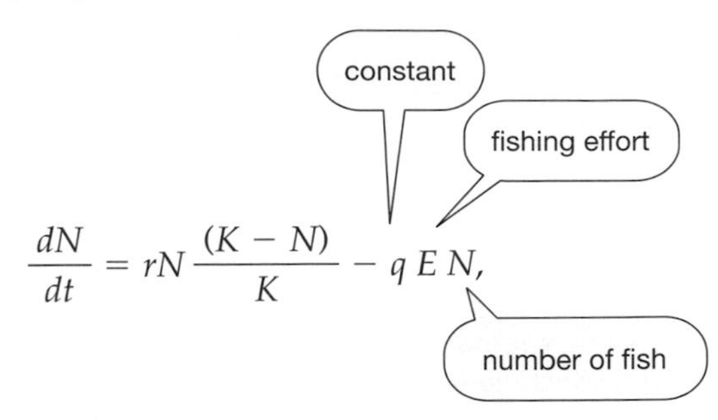

$$\frac{dN}{dt} = rN\frac{(K-N)}{K} - qEN,$$

where q is a "catch" constant per number of hours fished, E equals the amount of fishing effort (e.g., hours spent fishing), and N is the number of available fish. Then qE is the fishing mortality rate. As fishing and hence fishing losses increase, qEN can begin to affect the total catch (yield) severely.

We can calculate the maximum sustainable yield, the largest number of individuals that can be removed without causing long-term changes. This is like creaming off the interest from a bank account, but leaving the principal intact to ensure the same future interest payment in subsequent years. The maximum sustainable yield can be estimated as

$$\text{MSY} = Nr,$$

where r is the rate of natural increase and N is the average number of animals present throughout the year. The maximum value of MSY will generally be at the midpoint of the logistic curve, where the overall population growth rate is highest. Thus, if $N = 10{,}000$ and $r = 0.14$, then MSY $= 1{,}400$. However, this approach assumes a constant rate of harvesting throughout the year. If the harvest is spread out over the year, N is likely to be reduced by natural deaths. N might be reduced by, for example, 10% by natural mortality so that $N = 9{,}000$. In that case,

$$\text{MSY} = 9{,}000 \times 0.14 = 1{,}260,$$

and the overall yield will be lower if natural deaths are accounted for. The failure to account for such deaths and the failure to account for other features of populations, such as seasonal variations, fluctuating carrying capacities, and variations in rates of growth, can lead to discrepancies between theoretical and actual harvests and may be a reason that overfishing occurs.

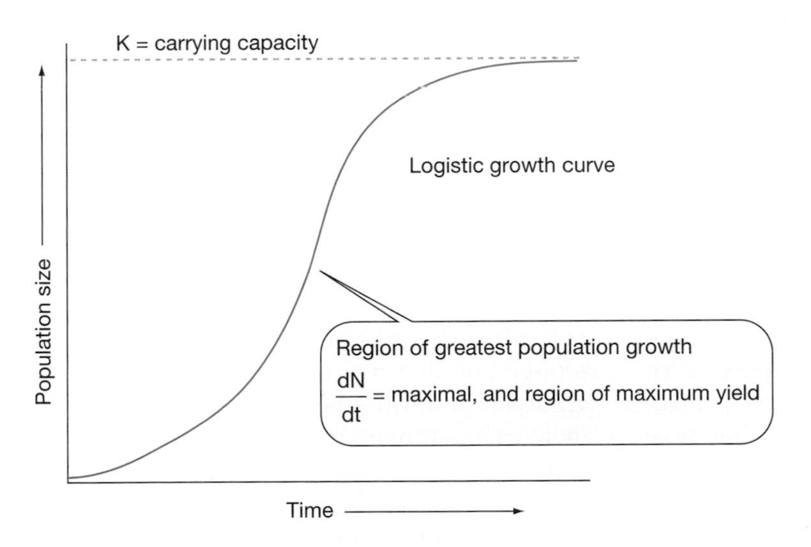

Figure 1 Maximum yield usually occurs at the midpoint of the logistic curve.

The response of predators to prey density can have important consequences for the ability of predators to control prey populations, because the proportion of a prey population that is consumed by an individual predator changes as the type of functional response changes (Fig. 10.7). The total response of predators to prey is actually determined by a numerical response as well as a functional response. While the functional response dictates how individual predators respond to prey populations, the numerical response governs how populations of predators migrate into or out of areas in response to prey densities.

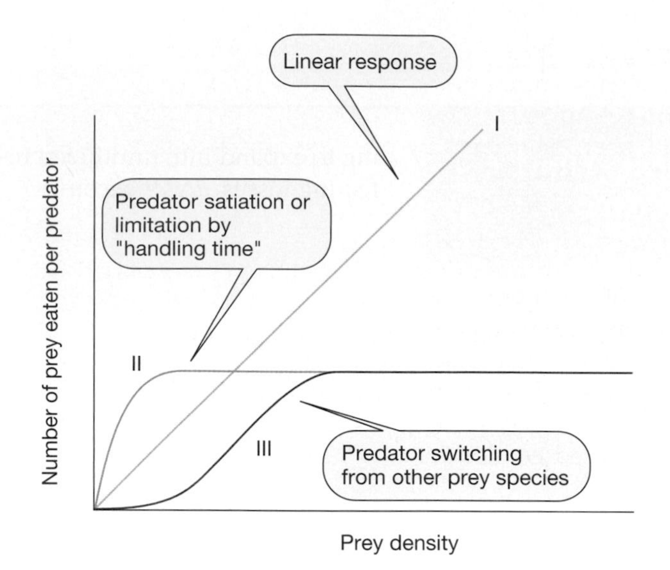

Figure 10.6 Type I, II, and III functional response curves, which govern how individual predators respond to prey density.

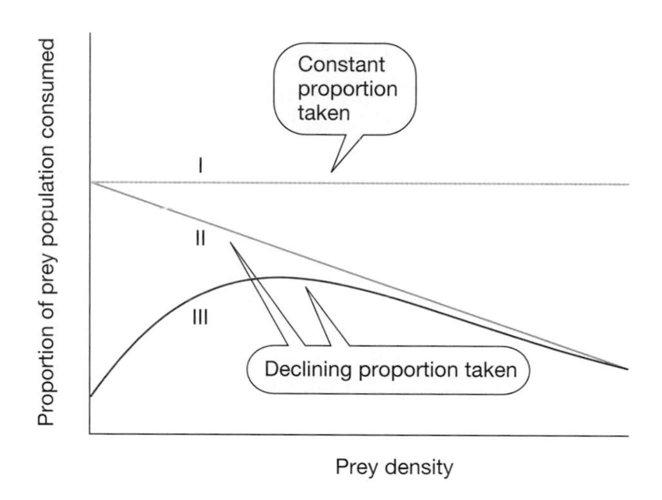

Figure 10.7 The proportion of the prey population consumed by an individual predator as a function of prey density. The three response curves I, II, and III correspond to the functional response curves of Fig. 10.6.

10.3 Field Studies of Predator–Prey Interactions

Field studies support the idea that predators can control the density of a prey population.

How do the data on predator and prey abundances in the field compare with graphical models? Do predators control populations of their prey, or do they take only such individuals that would die anyway from disease or starvation? North America has had a long history of exterminating wolves, bobcats, coyotes, and pumas under the assumption that, because predators kill to survive, they must depress the numbers of cattle or sheep as well as natural prey. While it is true that deer herds appear to increase dramatically after the removal of a predator, we cannot also control for the effect of environmental change, such as clearing forests, which has gone on concurrently.

One of the earliest professional opinions on the importance of predators controlling prey density was based on the study of the Kaibab deer herd. The Kaibab Plateau borders the Grand Canyon in northern Arizona and was declared a national park around 1900. All the big predators of the park were removed or shot by government hunters, while deer hunting itself was prohibited. There were estimates of tenfold increases in deer populations following the removal of carnivores, and these estimates were embraced by critics as evidence that exterminating predators was a bad management strategy, despite the fact that there were really no accurate censuses of deer before and after the park was established. Worse still, the critics' ideas were embraced by the textbooks of the day as sound evidence of the effects of predators, and the whole story was taken as fact. It wasn't until a careful reevaluation by Graham Caughley (1970) that it became clear that this was not the textbook example it purported to be: While predator control probably did have some influence, it could be that the cessation of deer hunting or the removal of competing sheep and cattle was enough to permit the increase in deer numbers. Clearly, what was needed was more careful documentation.

Toward that end, Paul Errington, who devoted most of his life to the study of muskrats in Iowa, undertook a more rigorous study of predation (Errington, 1946). The chief predator of the muskrat was the mink, *Mustela vison*. After 30 years of work marking and capturing muskrats and identifying muskrat remains in mink scat, Errington concluded that nearly all the muskrats eaten by mink were those doomed to die soon of other causes, especially disease. Only when muskrats wandered out of their home range (i.e., left their refuges) did the number eaten by mink go up substantially—a finding that supports Rosenzweig and MacArthur's ideas on the value of refuges in stabilizing predator–prey systems. Errington's work also supported the ideas of Murre (1944), whose work on life tables we discussed in Chapter 6 and who had argued that wolves take only weak, sick, or elderly Dall mountain sheep.

More modern data on the predators of the Serengeti plains of eastern Africa (lions, cheetahs, leopards, wild dogs, and spotted hyenas) show that these also seem to have little impact on their large-mammal prey (Bertram, 1979). Most of the prey taken are injured or senile and are likely to contribute little to future generations. In addition, many of the prey are migratory and the predators are residents of the region, so predators are more likely to be limited in numbers by prey that are resident during the dry season, when migratory ungulates are elsewhere.

Some good data sets, however, do show a strong effect of predators on prey. For many years, the moose population on Michigan's Isle Royale, a 45-mile-long island, enjoyed a wolf-free existence. Then, in 1949, during a particularly hard winter, a pair of Canadian wolves was able to walk across frozen Lake Superior and colonize the island. In 1958, wildlife biologist Durwood Allen of Purdue University began tracking wolf and moose numbers. The wolf population peaked at 50 in 1980, and then in 1981 it took a severe nosedive (Fig. 10.8). Wildlife ecologist Rolf Peterson of Michigan Technological University followed this population in the 1990s (Peterson, 1999). The wolf population continued to decline. Only four pups were born in 1992 and 1993, all to the same female in one pack. The other two packs on the island went down to just a pair of wolves each. It looked as if the wolf population was on its way to extinction, but in the late 1990s it recovered a little. As for the moose population, it increased steadily in the 1960s and 1970s, when the wolf population averaged about 24 individuals. It then declined as the wolf population increased until 1981. A record level of about 2,400 moose was reached in 1995, when the wolf

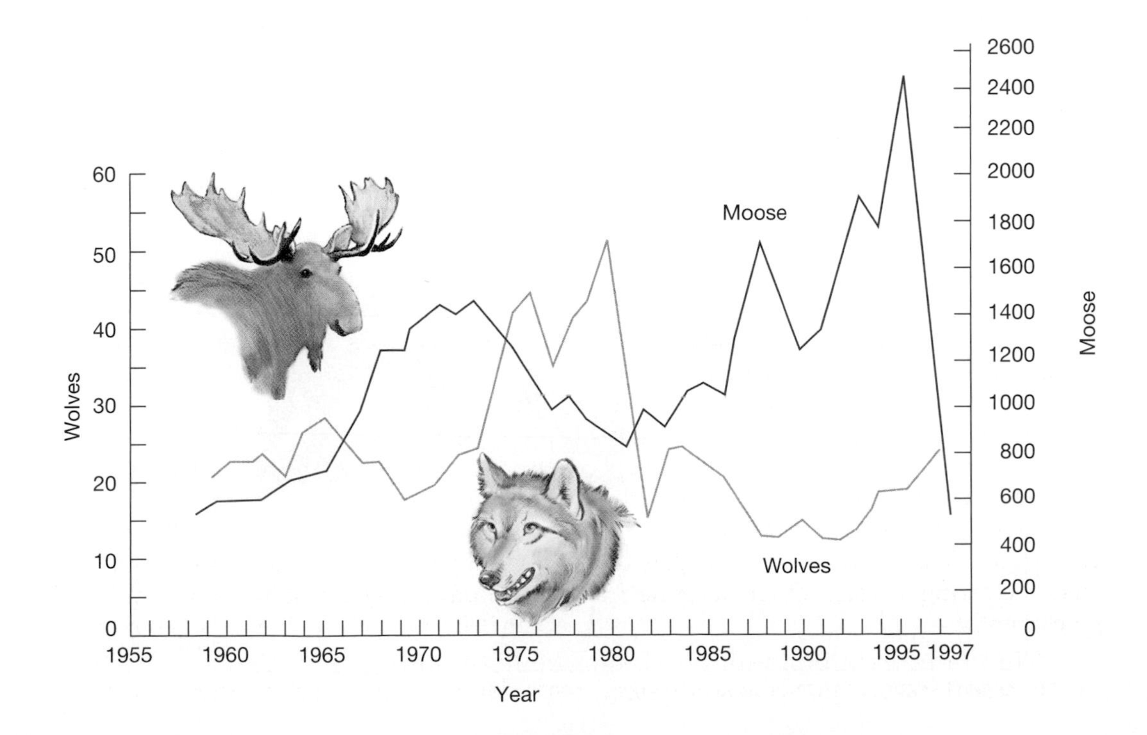

Figure 10.8 The effects of wolf predation on moose numbers on Isle Royale. As the wolf population declined, the number of moose went up. *(From data in Peterson, 1999.)*

population was low. The data from Isle Royale appear to provide good evidence that predation does have a strong effect on natural populations. However, the situation was confounded somewhat in 1996, when the moose population crashed, with 80% dying, primarily from starvation. So it looks like wolves can influence moose density, but ultimately, moose population levels are set by available food.

The study of wolves and moose on Isle Royale has important ramifications for the National Park Service on how to manage wildlife populations. In the 1980s, the emphasis was on natural regulation: Leave the system alone, and the moose will reach their equilibrium levels, based on the availability of resources. This notion strongly emphasized a "bottom-up" approach, rather than a "top-down" approach that involved significant control from the trophic level above—in this case, the wolf carnivores. Following the bottom-up approach meant that wildlife staff would no longer mimic the effects of predators by culling the herds, a policy of active management that had been advocated by others—notably, Leopold *et al.* (1943). Peterson's results showed that moose numbers in an unmanaged park are likely to be unstable and that boom-and-bust dynamics is the most probable scenario.

Why did the wolf population decrease? First of all, there was a narrow genetic base to begin with. Restriction enzyme analysis of the wolves' mitochondrial DNA turned up just a single pattern, indicating that the wolves were all descended from a single female. Thus, they had only about half the genetic variability of the mainland wolves. Second, there was evidence of a deadly canine virus in 1981, probably a result of a parvovirus outbreak in nearby Houghton, Michigan. The virus was likely carried to the island on the hiking boots of visitors (Mlot, 1993).

Another good data set concerns the Canada lynx (*Lynx canadensis*) and its prey, the snowshoe hare (*Lepus americanus*): Populations of both show dramatic cyclic oscillations every 9 to 11 years (Fig. 10.9). We have an extensive data set on this interaction because of the value of the pelts of both animals. Charles Elton analyzed the records of furs traded by trappers to the Hudson's Bay Company in Canada over a 200-year period and showed that a cycle has existed for as long as records have been kept (Elton and Nicholson, 1942). This cycle appears to be an example of an intrinsically stable predator–prey relationship, just as in Rosenzweig and MacArthur's model (Fig. 10.5b). However, Keith (1983) argued that it is winter food shortage, and not predation, that precipitates a decline in the number of hares. He showed that heavily grazed plants produce shoots with high levels of toxins, making them unpalatable to the hares. Such chemical protection remains in effect for two to three years, precipitating a further decline in hare numbers. Predators, Keith argued, simply exacerbate the population reduction. Thus, lynx cycles depend on snowshoe hare numbers and we say the cycle is donor controlled. However, the population of hares fluctuates in response to the animal's host plants.

Stuart Pimm of the University of Tennessee (1980) has argued that the importance of predation is dependent on whether the system is "donor controlled" or "predator controlled." In a donor-controlled system, the supply of prey is determined by factors other than predation, such as the food supply, so that the removal of predators has no effect on the prey population.

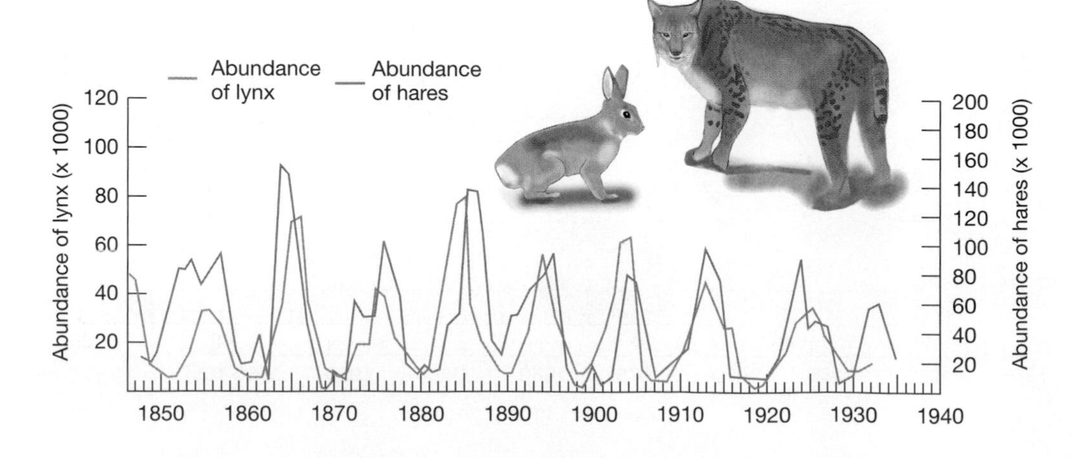

Figure 10.9 The coupled oscillations in abundance of the snowshoe hare and Canada lynx, as revealed from pelt trading records of the Hudson's Bay Company.

Obvious examples include consumers of fruits and seeds, consumers of dead animals and plants, and intertidal communities in which space plays a limiting role. By contrast, in a predator-controlled system, feeding by the predators eventually reduces the prey population and its reproductive ability. In a predator-controlled system, the removal of predators would probably result in large changes in the abundance of prey.

10.4 Introduced Predators

Perhaps the best way to find out whether predators determine the abundance of their prey is to remove predators from or add predators to a system and examine the response. One of the best examples of this kind of study involves dingo predation on kangaroos in Australia. The dingo (*Canis familiaris dingo*), an introduced species, is the largest carnivore in Australia and an important predator of imported sheep. Dingoes have been hunted intensively and poisoned in sheep country in southern and eastern Australia. Indeed, the world's longest fence extends 9,600 km to prevent the dingoes from recolonizing areas from which they have been eliminated through shooting or poisoning, providing a classic experiment in predator control. Because dingoes prey on native animals as well as imported sheep, we can examine their impact on species such as kangaroos and emus. The result of eliminating the dingoes has been a spectacular increase—one hundred sixty-six fold—in the population of red kangaroos in areas of New South Wales, over their density in south Australia, where dingoes have not been molested. Emus (*Dromaius novaehollandiae*) are over 20 times more abundant in dingo-free areas.

Experimental studies with introduced species afford an insight into the degree to which predators determine the abundance of their prey.

Dingoes are also frequent predators of feral pigs in tropical Australia (Newsome, 1990). In Cape York, in northern Queensland, there is a gross shortage of young pigs less than two years old in areas where there are dingoes. On neighboring Prince of Wales Island, where dingoes are absent, recruitment of the pigs is considerable (Fig. 10.10). This contrast suggests that predation on juveniles can inflict substantial mortality on a prey population. European foxes and feral cats are other important exotic animals in Australia. Both can do damage to domestic

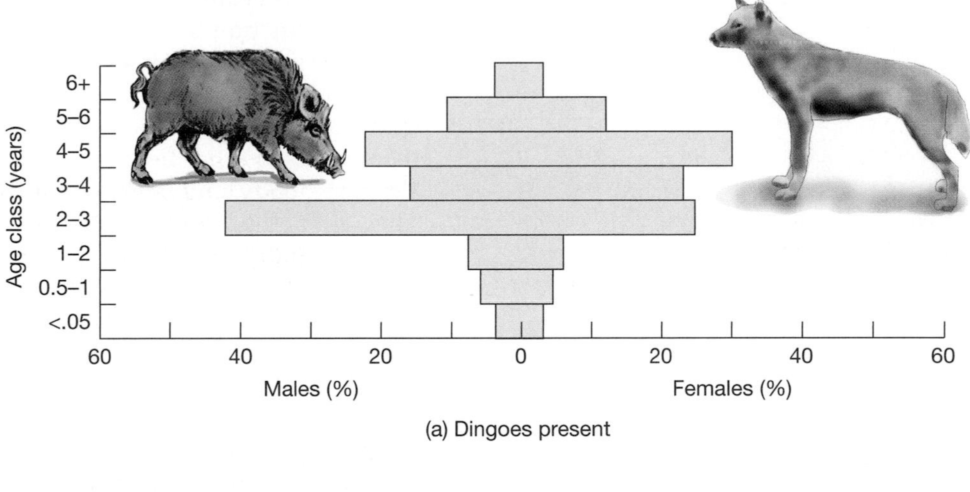

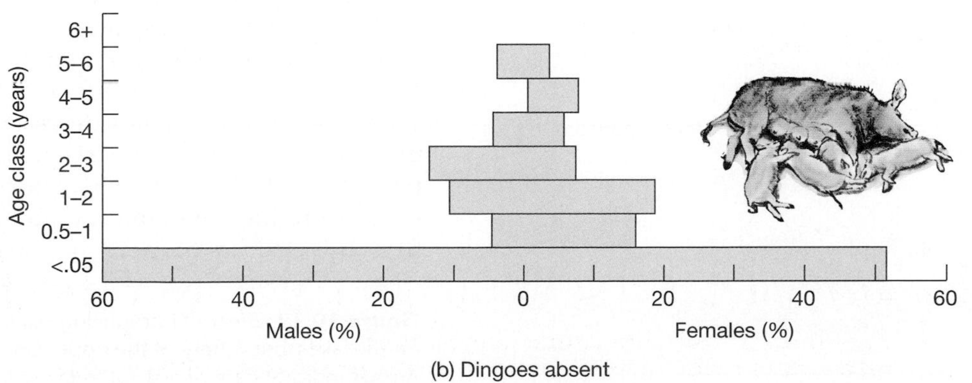

Figure 10.10 Contrasting population structures of feral pigs where dingoes are (a) present and (b) absent in tropical northern Australia. (*Reproduced from Newsome, 1990.*)

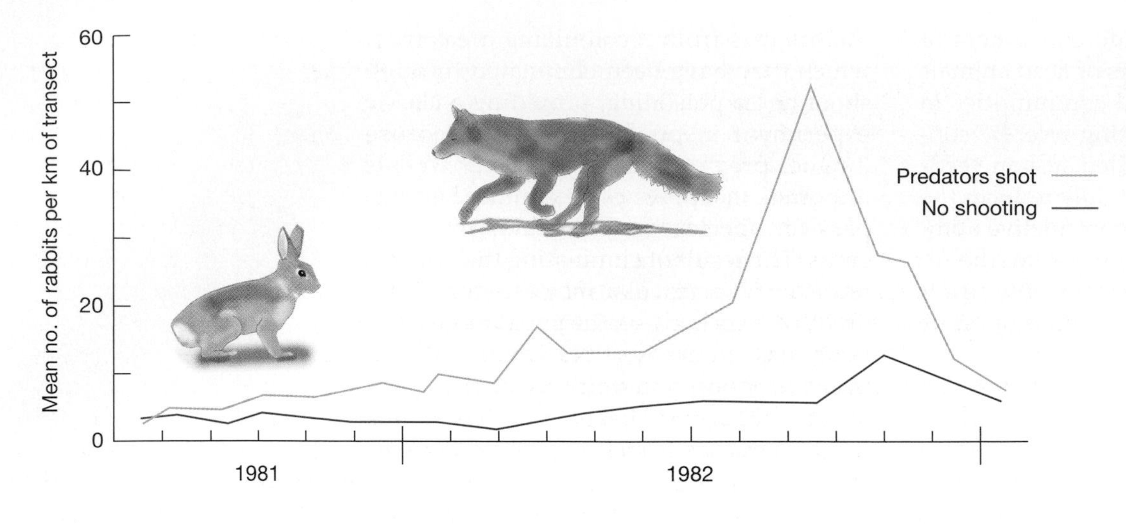

Figure 10.11 Accelerated increase in numbers of rabbits with removal of European foxes and feral cats in an Australian field experiment. The figure shows a comparison of counts of rabbits per kilometer along transects where predator populations were continually shot (blue line) or left intact (red line). *(Reproduced from Newsome et al., 1989.)*

livestock (to chickens especially) and are subject to eradication by shooting. However, in areas where these predators were shot, numbers of rabbits, also exotic in Australia, increased (Fig. 10.11). When rabbits increase, valuable rangeland may become overgrazed. Also, when dingoes are absent from an area, kangaroos can severely reduce the quality of forage for sheep and other organisms. The removal of exotic predators is a complex issue in Australia, where a mix of native and exotic herbivores also exist.

Another striking example of predation pressure has been provided by an inadvertent introduction by humans. Marine sea lampreys live on the Atlantic coast of North America and migrate into fresh water to spawn. Adult lampreys feed by attaching themselves to other fish, rasping a hole in the host's body, and sucking out its fluids. The path of the lamprey to the Great Lakes was originally blocked by Niagara Falls. When the Welland Canal was built in 1829, both boats and lampreys could pass from the St. Lawrence Seaway, through a series of locks, to Lake Erie. Surprisingly, it took nearly 100 years for the lamprey to make the journey to the upper Great Lakes. The first sea lamprey in Lake Erie was found in 1921, the first in Lake Michigan in 1936, the first in Lake Huron in 1937, and the first in Lake Superior in 1938 (Smith and Tibbles, 1987). Lake trout catches decreased dramatically following the invasion of the lamprey (Fig. 10.12). After 1951, efforts were made to reduce the lamprey population, first by electric barriers on spawning

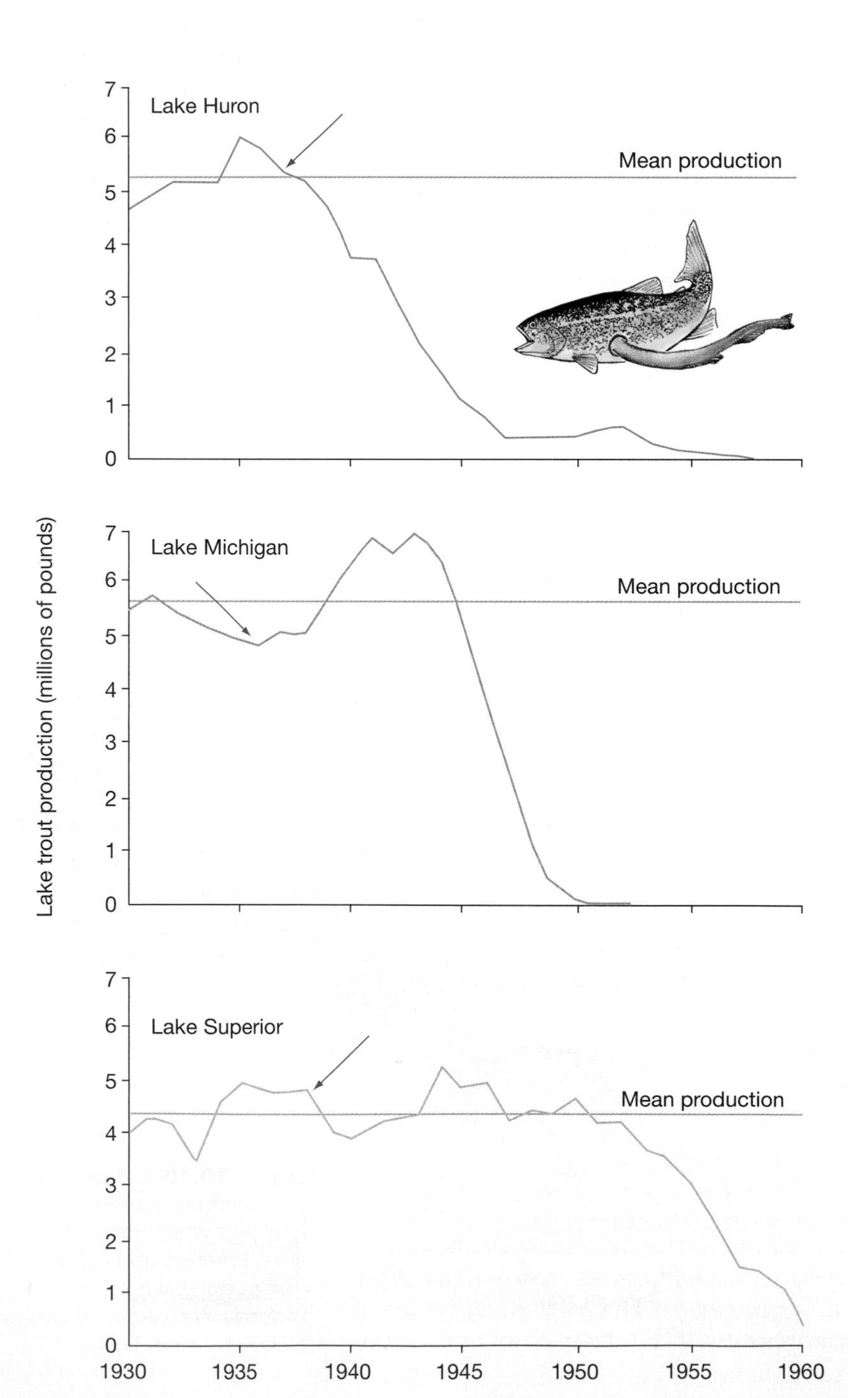

Figure 10.12 Effect of introducing sea lampreys on the lake trout fishery of the upper Great Lakes. Arrows indicate date of first catch of sea lampreys. *(Reproduced from Baldwin, 1964.)*

Applied Ecology

Humans as Predators—Whaling

The question whether whales should be exploited has been the subject of vigorous and worldwide debate since at least the 1960s. The level of popular interest in this question probably exceeds that concerning any other group of exploited animals. First of all, we have to grapple with the ethical question of whether it is wrong to kill whales at all. Following this, there are questions regarding whether any of the whale populations are sufficiently abundant to support some kind of limited harvest.

The history of Antarctic whaling in particular and whaling in general has been characterized by a progression from more valuable or more easily caught species to less attractive ones, as stocks of the original targets were depleted (Figure 1). In the Antarctic, blue whales dominated the catches through the 1930s, but by the middle 1950s few were being taken, although the species was not legally protected until 1965. As the stocks of blue whales diminished, attention was turned to the fin whale, which was originally the most abundant of all whales in the southern ocean. By the 1960s, numbers of this species had collapsed rapidly. Humpback whales, though never very numerous, were attractive because of their high oil yield and the ease with which they could be caught. Catches were never very great, but the stocks in most areas collapsed dramatically in the early 1960s. Sei whales were almost ignored by whalers until the bigger species were no longer available. Sei whales were hardly taken at all until about 1958, but then catches increased rapidly and reached a peak of about 20,000 in 1964–65. Catches declined rapidly thereafter, this time due to the introduction of limits. Then the relatively small minke whales, which were ignored in the southern ocean until 1971–72, began to be taken. Since that time, minkes make up the largest component of the southern baleen whale catch.

The story is similar in the Northern Hemisphere, although the stocks were much smaller, about 20% of that in Antarctic waters. Many countries have at one time or another attempted to establish a whaling industry. About 25 to 30 countries have done so during the present century. Before World War II, 8 to 10 countries undertook whaling, but the field was dominated by Norway and the United Kingdom, which, between them, accounted for over 95% of the catches. The United Kingdom ceased operations in 1964, and Norway did so in 1969. In 1982, perhaps because it could see the writing on the wall, the International Whaling Commission voted for a moratorium on all commercial whaling. It was proposed that commercial whaling be ended in 1985–86, a proposal that did not actually take effect until 1988.

In 1994, some countries—most notably, Iceland, Norway, and Japan—argued for a resumption of limited commercial whaling. Should we ban commercial whaling? There is no doubt that some whaling might be permissible. For example, the number of minke whales in the southern ocean was estimated to be between a half and three-quarters of a million in 1990. A take of 1,000 animals is not likely to do any harm, although taking many more than this would probably be unwise without a rigorous study of the biology of the whale itself. The good news is that following the 1982 moratorium, the populations of some whales have increased. Blue whales are thought to have quadrupled their numbers off the California coast during the 1980s, and the California gray whale population has recovered to prewhaling levels, again showing the impact that the presence or absence of a predator can have.

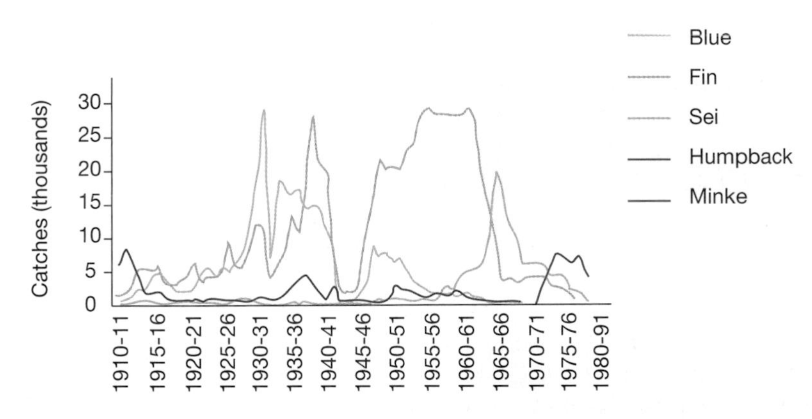

Figure 1 Sequential decline of whale populations in the Antarctic due to commercial whaling.

streams and then by a lethal chemical. As a result, lampreys became rarer, and in the 1970s and 1980s lake trout numbers rebounded substantially. However, in many cases, the trout population did not rebound to earlier levels, partly due to commercial fishing and partly due to pollution. Because lake trout were supplemented by massive restocking, any rebound in their numbers cannot be attributed solely to a reduction in the lamprey population.

One of the biggest reductions in prey in response to predation has been the systematic decline of whales of various species in response to whaling by humans (see "Applied Ecology Box: Humans as Predators—Whaling"), but this interaction cannot be regarded as a natural system.

10.5 Field Experiments with Natural Systems

Experiments with natural systems suggest that predation is an important force in nature.

Most of the examples discussed so far have focused on introduced predators. The effects of such introductions are often very strong, but again, what can we learn from such exotic systems other than that fact? We have good data on these systems because they involve species of economic value, such as sheep and fish. What of totally natural predator–prey systems in which predator and prey have evolved together for long periods? In 1903, lions were shot in Kruger National Park, South Africa, to allow numbers of large prey to increase. Shooting ceased in 1960, by which time wildebeest had increased so much, that it became necessary to cull the animal from 1965 to 1972. The drop in wildebeest numbers indicates that predators have substantial effects on their prey.

What about experiments with other taxa? Insects are small enough that whole populations can be maintained inside cages. Ted Floyd (1996) performed predator-exclusion experiments on the insects feeding on creosote bushes (*Larrea tridentata*) at the Jornada Long-Term Ecological Research site in the Chihuahuan desert of southwest New Mexico. Bird predators were excluded by using nylon mesh cages, and arthropod predators were removed by hand. In each of two years when these experiments were performed, herbivore densities became significantly higher on experimental bushes on which predators were excluded than on control bushes. As regards marine intertidal organisms, Bob Paine (1966) removed predatory starfish from sections of the Pacific Northwest Coastline and noted an increase in mussel prey. However, Paine was more interested in the effects of starfish on the number of prey species present in an area, and his work will be discussed more fully in (Chapter 15).

As for game animals, managers are continually concerned that natural predators may take individuals that might otherwise be available to hunters. The gray partridge (*Perdix perdix*), is a widespread game bird in Europe (Fig. 10.13). Over 20 million were shot annually in Britain in the 1930s, but by the mid-1980s, only 3.8 million were taken per year. Chick mortality through starvation caused by reduced insect abundance following the introduction of herbicides in the 1950s was suspected, and trials with decreased herbicide use suggested that this was indeed the case. However, it was also noted that populations of the bird were smaller in areas where there was no predator control by gamekeepers. A six-year predator-removal experiment was designed to test the effect of

Figure 10.13 A gray partridge cock crowing in Germany. A six-year predator-removal experiment showed that predation significantly affected breeding success and brood size. *(Robert Maier, Animals Animals/Earth Scenes.)*

predation in the breeding season. Foxes, crows, magpies, and jackdaws were shot with a high-powered rifle, and stoats, weasels, and rats were trapped. After the nesting period, these predators reestablished themselves. Predation control significantly increased the proportion of partridges that bred successfully and also increased the average size of their broods, so that, by August, the partridge population had increased by 75%. Incorporating the effects on breeding stocks in subsequent years led to an overall three-and-a-half-fold difference between autumn populations with and without predation control (Tapper, Potts, and Brockless, 1996).

One of the most elegant experiments on predator removal was done in Finland, where predators were thought to be driving the three- to five-year cyclic fluctuations in densities of small rodents, especially voles (Korpimaki and Norrdahl, 1998). Large-scale removals of predators were conducted in April 1992 and 1995 over 2–3 km^2 when the rodent population was just about to drop precipitously. Terrestrial predators (stoats and weasels) were reduced by trapping, and birds of prey (mainly kestrels and owls) were reduced by removing nests, nest boxes, and cavities. The result was a large increase in the rodent population by June in the experimental areas, compared with a continued decline in control areas where predators were not removed (Fig. 10.14).

Other studies suggest that bird mortality due to predation can be high. For example, O'Connor (1991) reviewed 74 studies of the nesting success of various bird species

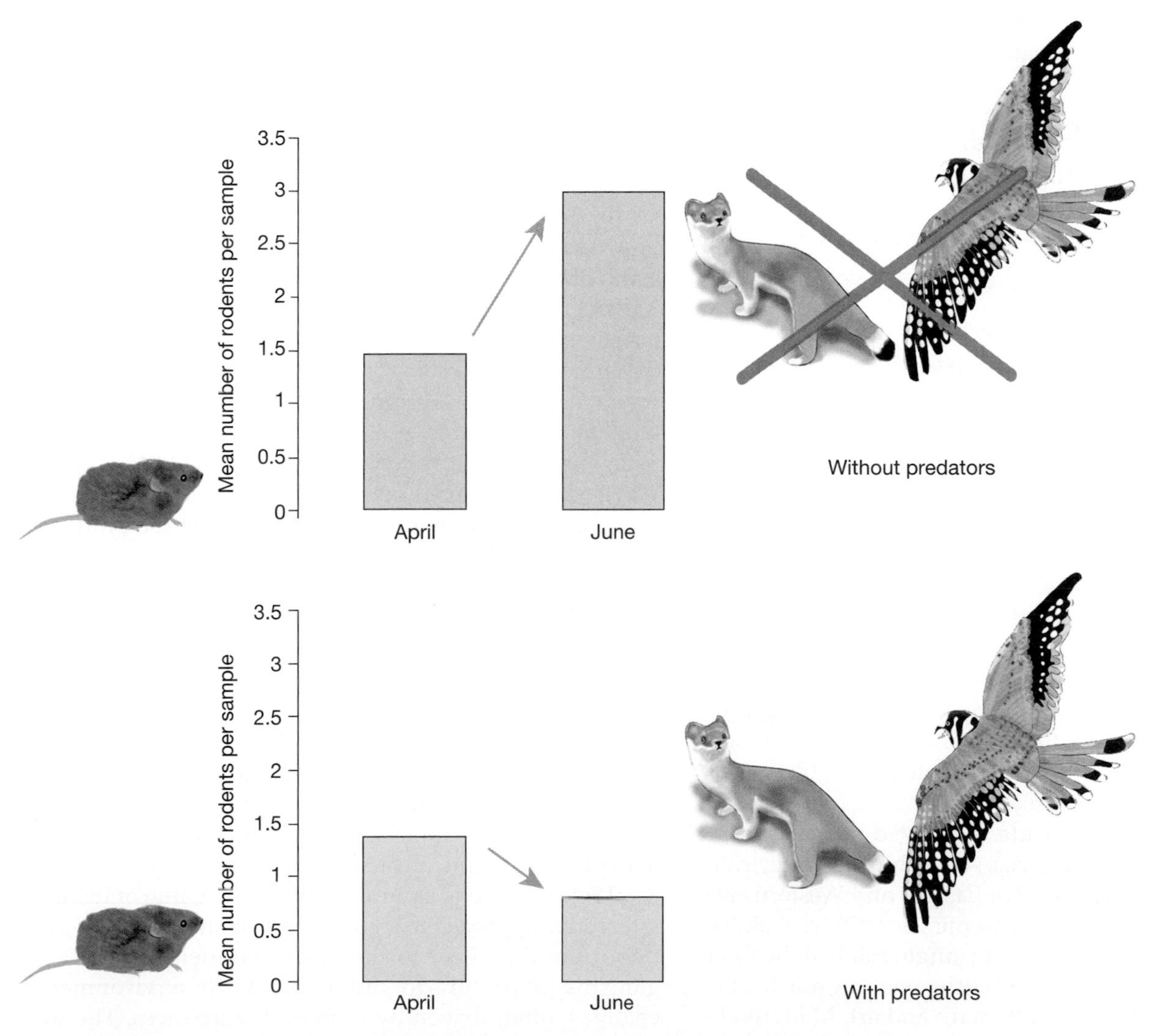

Figure 10.14 Rodent populations—especially voles—in Finland decline every three to five years. (upper) The removal of predatory stoats, weasels, and raptors reverses the decline in rodent numbers and even permits an increase in rodent density. (lower) Rodent numbers decline in the presence of normal predator densities. *(After data in Korpimaki and Norrdahl, 1998.)*

and found that one in three nests failed due to predation. Similar estimates were found in reviews by Martin (1993; 41.4% of nests lost to predation in 55 species) and by Côté and Sutherland (1997; 38.4% of nests lost to predation in 98 species). In a metaanalysis of 20 published studies of predator-removal programs, Côté and Sutherland (1997) showed an average of 75% higher hatching success in areas from which predators had been removed compared with control areas.

Andrew Sih and colleagues (see "Profiles") surveyed 20 years (1965–1984) of seven ecological journals for field experiments concerned with predation (Sih *et al.*, 1985). Their survey found 139 papers examining 1,412 comparisons. In nearly 60% of the comparisons, prey showed a significant response to predator manipulations. Nearly one half (45%) showed a large depression in prey density, of at least 50% by predators. Thus, we can conclude that in the majority of cases predators influence the abundance of their prey in the field. The variety of antipredator mechanisms discussed earlier also show how predation is important enough to cause camouflage, mimicry, and chemical defenses to evolve in prey. Taken together, these data underline how potent a force predation is in nature. At the same time, in many systems the density of prey also affects the density of their predators.

Profiles

Andrew Sih,

University of Kentucky

The theme of my career as an evolutionary ecologist is integration and synthesis. That is, I try to understand nature by using a blend of approaches in evolution, behavior, and ecology. Most of my career has focused on predator–prey interactions. As a community ecologist, I am interested in why predators have major impacts on some prey while other prey are left relatively untouched. As a behavioral ecologist, I address this community-level issue by looking at the behaviors of predators and prey. As an ethologist, I try to understand predator and prey behaviors by looking at sensory mechanisms underlying behavior. Finally, as an evolutionary biologist, I try to explain both the behaviors and their sensory basis by studying evolutionary mechanisms, including natural selection, genetics, and evolutionary histories. This integrative blend is, I think, both powerful and fun.

My integrative view of life is derived perhaps from my mixture of family influences. As a Chinese-American, I grew up immersed in a blend of Eastern and Western cultures. In addition, my parents lent me very different skills. My father is a retired engineer who taught math at the local college, while my mother's strengths were in music and art. In college, I dabbled in both math and art. Math made me comfortable with logic, theory, and modeling, while art helped to train my eyes to look at nature carefully. I joined these together when I became an ecologist whose research includes theory, field studies, and laboratory experiments.

In graduate school, I developed my lifelong interest in predator–prey interactions. My inspirations included some of the best and brightest ecologists of our time. Bill Murdoch, my major advisor, started me studying the interface between predator–prey behavior and population dynamics. Bob Warner got me excited about using the optimality framework to understand behavior. Joe Connell taught me community ecology and the value of learning by using carefully controlled experiments. Peter Abrams and Wayne Sousa showed me that 25-year-olds can write papers that shape an entire field. I drew all these influences together to mold my own scientific identity.

Perhaps my most influential work is a review paper that I wrote in 1985 with colleagues at the University of Kentucky. Our review of over 100 experimental studies showed that both predation and competition are often important factors in nature. Surprisingly, we found that in roughly one-third of all comparisons, prey did better in the *presence* of predators than in their absence. This finding helped to catalyze an interest in "indirect effects" (e.g., keystone predator effects, trophic cascades) that can help to explain this "unexpected effect."

I retain a strong optimism about the value of using an integrative approach to study nature. If anything, it is more important than ever for humans to understand how organisms cope (or do not cope) with environmental changes, often driven by human disturbances. The integrative science of ecology is critical to guiding our, hopefully, effective responses to these challenges.

Summary

Predation is often thought of as a strong selective force: nature red in tooth and claw. What is the evidence to support this viewpoint?

1. The existence of the following phenomena suggest that predation is a strong selective force in nature: aposematic coloration, camouflage, Batesian and Müllerian mimicry, intimidation displays, polymorphisms, and chemical defenses.

2. Even relatively simple graphical predator–prey models show how a variety of different outcomes is possible, from decreased predator and prey oscillation, through stable cycles, to wildly increasing and unstable oscillations. The existence of mutual interference between predators, the existence of specific territory sizes, and the ability of predators to feed on more than one type of prey make it even more difficult to predict or model how populations of predators and prey interact.

3. Large-scale field observations tend to support the idea that predators take only weak or sickly individuals and that prey populations influence predator numbers and not vice versa.

4. Accidental and deliberate introductions of exotic predators in different parts of the world have often had profound effects on populations of native prey. Although this suggests that predators can have important regulatory effects on prey, these data are not from "natural" systems and should be treated with caution.

5. Evidence from natural systems, in which both predators and prey are native, together with a 1985 review by Andrew Sih suggest that many studies which have looked for significant effects of predators on prey have found such effects.

The vast majority of the evidence suggests that predation is not a casual or unimportant force in nature, but is frequent, with often strong effects.

Discussion Questions

1. Should ranchers be concerned about the reintroduction into their vicinity of large predators like wolves or panthers? Do sea lions, otters, or dolphins decrease the stock of fish available for people who fish? Would the number of deer available for hunters be the same in the presence of large predators as it would be in the absence of them? What data would you need to collect to answer these questions?

2. What can the effects of exotic predators, such as the lamprey and the dingo, tell us about the strength of predation? What can't they tell us?

3. Which do you think more likely: that predators control populations of their prey or that prey control predator populations? Would the answer vary according to the particular system you were interested in? Give examples where you can.

4. What shortcomings do you think Rosenzweig and MacArthur's predator and prey isoclines have? What would these shortcomings mean in terms of determining how predators and prey interact? For example, do all predators live in groups? Do all prey decline to zero when there are only a few left? Why isn't the predator isocline hump shaped like the prey isocline?

5. A great many fish stocks seem to have been overfished. Examples include Peruvian anchovy, flounder, haddock, salmon, swordfish, and cod. How do you think we could prevent overfishing? Would strict enforcement of 200-mile offshore limits, tradable quotas, or fixed quotas help? What biological information do we need to have, and how can we get it when we can't even see the population in question?

CHAPTER 11

Herbivory

Road Map

1. Plants have evolved various chemical and mechanical defenses against herbivores.
2. Mathematical models predict that polyphagous herbivores have a greater effect on plants than monophagous herbivores have.
3. Herbivores remove between 15 and 18% of terrestrial plant biomass and over 50% in aquatic systems.
4. Herbivore numbers are strongly influenced by chemical and mechanical defenses of plants and by the amount of nitrogen in a plant.
5. Herbivores can change plant communities by preferentially eating dominant species, allowing other species to proliferate.

Plants have evolved a variety of chemical and physical defenses against herbivory.

Plants appear to present a luscious green world of food to any organism versatile enough to attack them and make use of their nutritive properties. But if that is so, why don't we see more plants being eaten by herbivores? After all, plants cannot even move to escape being eaten. There are three possible reasons that more plant material is not eaten. First, natural enemies, including predators and parasites, might keep herbivores below levels at which they could make full use of their resources. We will explore this possibility in more detail in Chapter 13. Second, herbivores may have evolved mechanisms of self-regulation to prevent the destruction of the host plant, perhaps ensuring food for future generations. This argument relies on group selection, however, and, as we saw in Chapter 4, that is unlikely to be the case. Third, the plant world is not as helpless as it appears: The sea of green is in fact tinted with shades of noxious chemicals and armed with defensive spines and tough cuticles.

11.1 Plant Defenses

Plants possess an array of unusual and powerful chemicals, including alkaloids (e.g., nicotine in tobacco, morphine in poppies, and caffeine in tea), mustard oils (in mustard plants), terpenoids (in peppermint), and phenylpropanes (e.g., in cinnamon and cloves) (Fig. 11.1). Two general chemical classes can be recognized: Those based on nitrogen compounds (mainly alkaloids) and those based on carbon compounds (mainly terpenoids and phenolics) (Table 11.1). The carbon nutrient balance theory suggests that plants which are limited by nitrogen invest more in carbon-based defenses and those that are limited by carbon (their growth is limited by a shortage of light or water) invest more in nitrogen-based defenses (Bryant, Chapin, and Klein, 1983). Because defensive compounds are not part of the primary metabolic pathways that plants use to obtain energy, they are referred to as *secondary chemicals*. The staggering variety of secondary chemicals in

(a)

(b)

(c)

(d)

Figure 11.1 Plants rich in secondary compounds. (a) Mustard, in the Napa Valley, California, between grapevines, contains the nitrogen compound glucosinolate. (b) Tea, on a plantation in Assam, India, contains carbon-rich tannins that bind to the gut. (c) Umbellifers, such as this Queen Anne's lace, are rich in furanocoumarins, another phenolic compound. (d) Tobacco, in eastern Kentucky, contains the alkaloid nicotine.

TABLE 11.1 The main types of secondary plant chemicals, their approximate frequency worldwide, and their effects on herbivores. Note that some substances, such as carotenoids and quinones, are not deterrents to feeding, but give color to flowers or fruits. Because they are important secondary plant chemicals, they are included here. *(Modified from Harborne, 1988.)*

Class	Approximate Number of Types	Main Distribution	Physiological Activity
Nitrogen compounds			
Alkaloids	10,000	In about 20% of angiosperms, especially in the roots, leaves, and fruits, and especially in hemlock	Toxic, bitter tasting (e.g., nicotine, strychnine, atropine of deadly nightshade)
Amines	100	Angiosperms, often in flowers	Repellent; some hallucinogenic
Nonprotein amino acids	400	Seeds of legumes (pea or bean family)	Many toxic, incorrectly assimilated into protein synthesis
Cyanogenic glycosides	30	Occasional, in fruits or leaves (e.g., locoweed)	Poisonous (cyanide), stop respiration
Glucosinolates	75	Cruciferae (mustard family)	Acrid and bitter, release isothiocyanates
Carbon Compounds			
(i) *Terpenoids*			
Monoterpenes	1,000	Widely, in essential oils	Pleasant smells (e.g., in pine resins and pollinator attractants)
Sesquiterpenes	5,000	Mainly in composites (daisy family)	Bitter and toxic, also allergenic
Diterpenoids	2,000	Especially in latex and plant resins	Some toxic
Saponins	600	Widespread	Bitter, hemolyze blood cells
Limonoids	100	Mainly in Rutaceae (citrus)	Bitter tasting
Cucurbitacins	50	Mainly in Cucurbitaceae (cucumber family)	Toxic and bitter tasting
Cardenolides	150	Especially in milkweeds and snapdragons	Bitter and toxic (e.g., cardiac glycosides)
Carotenoids	500	Always in leaves, often in flowers and fruit	Give color to flowers, fruit
(ii) *Phenolics*			
Simple phenols	200	Universal in leaves, often in other tissues	Antimicrobial
Flavonoids	4,000	Universal in angiosperms, gymnosperms, and ferns (e.g., tannins, coumarins, flavanones)	Bind to gut proteins
Quinones	800	Widespread	Give color

plants may be a testament to the staggering number of species of insect herbivores, fungi, and bacteria that feed on plants. Some insects may bypass such defenses by feeding on the phloem or xylem, which does not contain the range of defensive chemicals that appear in the foliage.

Plant defenses can be classified as quantitative or qualitative, depending on the volume of defense present in the plant.

Quantitative defenses are substances that are ingested in large amounts by the herbivore as it eats and that prevent the digestion of food. Examples are tannins (the compounds in many leaves, like tea, that give water a brown color) and resins in leaves, which may occupy 60 percent of the dry weight of the leaf. These compounds are not toxic in small doses, but they have cumulative effects. The more leaves the herbivores ingest, the more difficult it is for them to digest the leaves. Paul Feeny (1970) was the first to document such a defense by testing oaks against externally feeding caterpillars. The caterpillars were more common on oaks in the spring, when the concentrations of tannin in the leaves are the lowest (Fig. 11.2). As the season progresses and the leaves toughen, tannin concentrations increase and caterpillar feeding decreases. There are actually two main classes of tannins, each with a different biological function. *Hydrolyzable* tannins inactivate the digestive enzymes of herbivores, especially insects; condensed tannins are attached to the cellulose and fiber-bound proteins of cell walls and defend plants against microbial and fungal attack.

Qualitative defenses are, essentially, highly toxic substances, very small doses of which can kill herbivores. These compounds are present in leaves at low concentrations, 1 to 2 percent of dry weight of a leaf. Atropine, produced by the European deadly nightshade *Atropa belladonna*, is a most potent poison (Fig 11.3). The plant stores the poison in discrete glands or vacuoles in order not to poison itself. Other compounds, such as alkaloids and terpenoids, can be moved around the plant's vascular tissues. Many of these poisons have selective toxicity, so for example, they are toxic to insects and vertebrates that graze on them, but not to birds and other organisms that might disperse the fruit. Most qualitative defenses are rich in nitrogen and so are more common in nutrient-rich systems.

Qualitative and quantitative defense strategies are correlated with plant "ap-

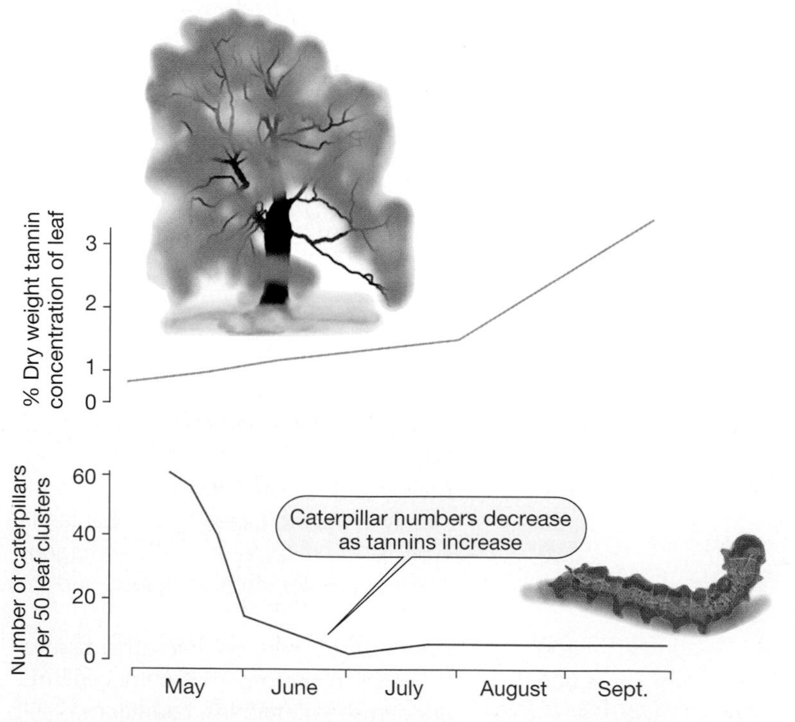

Figure 11.2 The number of butterfly and moth caterpillars per 50 clusters of oak leaves in Britain is highest in the spring, when tannin concentrations in the leaves are lowest. The numbers of caterpillars decrease in the summer as tannin concentrations increase. *(Modified from Feeny, 1970.)*

Figure 11.3 A ripe berry of deadly nightshade *Atropa belladonna* in Oxfordshire, England. An infusion of juice was formerly dropped in women's eyes, causing dilation of the pupils to produce a wide-eyed look—ergo the name *belladonna*, "beautiful lady." Interestingly, birds are immune to the poison and can safely disperse the seeds

parency" (Feeny, 1976; Rhoades and Cates, 1976). Apparent plants (e.g., oak trees) are long lived and are named thus because they are always apparent to herbivores. That is, herbivores (mainly insects) can find the plants easily. Their defenses are thought to be mainly of the quantitative kind, effective against both specialist and generalist herbivores with a long history of association with these *K*-selected plants. Unapparent plants are weeds, which are ephemeral and unavailable to herbivores for long periods. Their defenses are thought to be mainly qualitative, guarding against generalist enemies like vertebrates, which would find them only by chance. Thus, nearly all trees contain digestibility-reducing compounds, whereas weeds contain toxins. Table 11.2 illustrates some of the phenomena associated with apparent and unapparent plants. Why all the interest in secondary compounds? Partly because many are important in medicine. (See "Applied Ecology Box: Secondary Chemicals and Medicinal Uses.")

Besides chemical compounds within the leaf, a variety of other defenses is available to plants:

Mechanical defenses. Plant thorns and spines deter vertebrate herbivores, though probably not invertebrate ones. In Africa, in the presence of a large guild of vertebrate herbivores, much of the vegetation is thorny and spinose. Many Neotropical plants are also armed in this fashion, for, although now absent, large browsers were abundant on the continent until recently. In his 1992 presidential address to the British Ecological Society, Peter Grubb made some generalizations about mechanical defenses:

1. In many open sites, such as deserts, plants are primarily close to the ground and so are very spinose to protect them from grazing vertebrates.
2. Plants such as palms, with one or a few apical meristems, are also likely to protect them with spines.
3. Evergreens, such as holly, in a deciduous forest are likely to face severe herbivore pressure in the winter and so are very spinose.

TABLE 11.2 Some correlates of apparent and unapparent plants.

Apparent Plants	Unapparent Plants
1. Often in monocultures	1. Often in polycultures (patches of mixed species)
2. Large	2. Small
3. Long lived	3. Short lived
4. Chemical defenses against generalist and specialist herbivores	4. Chemical defenses against generalist herbivores that may happen upon the plant
5. Quantitative defenses, such as tannins, often > 1% of the fresh weight of a leaf	5. Qualitative defenses, such as toxins, often < 1% of the fresh weight of a leaf
6. Late successional (e.g., trees)	6. Early successional (e.g., weeds and annuals)

Applied Ecology

Secondary Chemicals and Medicinal Uses

Many medicinal drugs are derived from natural sources and make an important contribution to health care. In Western medicine, some 120 chemicals extracted in pure form from about 90 species of plants are used throughout the world.

Medicinal plants are to a large extent collected from the wild, and few are cultivated as crop plants. The United States annually imports over $20 million worth of rainforest plants for medicinal purposes. Seventy percent of pharmaceuticals now in use are derived from natural products. The cardiac stimulant digitoxin, the most widely used cardiotonic in Western medicine, is extracted directly from the foxglove plant. Even more plants are used for local remedies. An estimated 80% of people in less developed countries rely on traditional medicines for primary health care. The World Health Organization has listed over 21,000 plants used for medicinal purposes; some of these are listed in Table 1. There may be a great potential for rainforest plants to be useful in the treatment of cancer. Out of 35,000 tropical forest plants screened since 1956, the U.S. National Cancer Institute has identified over 1,400 with the potential to fight cancer, but few have proven to be commercially successful so far. The lone exception is the rosy periwinkle, native to Madagascar. Used for generations by tribal healers, this species is now utilized in the production of drugs that are effective against Hodgkin's disease and other forms of cancer. The plant yields substances called vinca alkaloids, which are difficult molecules to synthesize chemically. Indeed, it is less expensive to collect leaves of living plants and extract the valuable chemicals than to synthesize them. In addition, synthetic vincristine, used to treat childhood leukemia, is only 20% as effective as the natural product derived from the plant. The economic value of plant-related drugs is truly staggering, probably over $1 trillion globally, which in itself argues for the preservation of plant biodiversity and natural areas. In the United States, morphine and its relatives gross $600 million per annum alone.

Figure 1 Many plants have medicinal value. The cardiac stimulant digitoxin is extracted from *Digitalis purpurea*, the foxglove.

One could argue that most of the plants that are medicinally valuable have already been discovered and that we

TABLE 1 Plant species with compounds used as drugs.

Name of Plant	Name of Compound	Uses in Medicine	Country of Production or Cultivation
Atropa belladonna	Atropine	Dilate pupil of eye	Central and southern Europe, cultivated in United States, United Kingdom, eastern India, Europe, China
Cassia spp.	Danthron	Laxative	*C. senna* cultivated in Egypt
Catharanthus roseus	Vincristine	Antitumor agent	Pantropical, cultivated in United States, India, and other countries
Cinchona ledgeriana	Quinine	Antimalarial	Cultivated in Indonesia and Zaire
Datura matel	Scopolamine	Sedative	Cultivated in Asia
Ephedra sinica	Ephedrine	Chronic bronchitis	China
Digitalis	Digitoxin	Cardiotonic	Cultivated in Europe and Asia
Papaver somniferum	Codeine and morphine	Analgesic; sedative	Cultivated in Turkey, India, Burma, Thailand
Pausinystalia yohimbe	Yohimbine	Aphrodisiac	Cameroon, Nigeria, Rwanda
Rauvoilfia spp.	Reserpine	Tranquilizer	India, Bangladesh, Sri Lanka, Burma, Malaysia, Indonesia, Nepal
Silybum merianum	Silymarin	Liver disorders	Mediterranean region

are now facing diminishing returns, but relatively few plants have been screened for their chemical properties, and we have only sampled the tip of the iceberg. If we favor the logic of preserving plants of medical promise, then it could also be argued that research should focus more on tropical or desert plants. It may be that botanochemicals are most abundant in tropical and desert plants because they have evolved in response to intense competition between plants or a great need to prevent herbivory, respectively.

Regions that are known to have important concentrations of medicinal plants include Mexico and Central America, the west central region of South America (encompassing Colombia, Ecuador, and Peru), the Indian subcontinent, west Asia, and parts of northeastern Africa. However, not all medically important plants are confined to the tropics: Taxol, the new treatment for ovarian cancer, for example, is from the western yew tree in the United States.

Animals, too, yield many valuable drugs. Slow-moving, brightly colored species like marine slugs use noxious chemicals to defend themselves. A sponge yielded the adenine nucleoside of which AZT, an AIDS drug, is a synthetic derivative. Arrow poison frogs secrete vasodilators and analgesics. Leeches inject anticoagulants and anesthetics, as do ticks and vampire bats.

Repellents. Thistles produce compounds that repel certain insect larvae. Potatoes synthesize a component of an alarm pheromone released by aphids when they are attacked by a predator. The alarm pheromone causes other aphids to flee, so the potato-produced component also tends to repel the aphids.

Reproductive inhibition. Some plants, such as firs (*Abies* sp.), contain insect hormone derivatives that, if digested, prevent the metamorphosis of insect juveniles into adults. In this way, herbivory in the future is diminished by a decrease in the herbivore's reproductive output. On the other hand, the floss flower (*Ageratum*) produces a chemical mimic of the insect molting hormone ecdysone. Insect larvae that feed on the plant die when they molt prematurely.

Masting. A few tree species, notably some oaks, seem to produce more seeds in some years than others. It has been argued that the synchronous production of seeds in some years satiates herbivores, thereby permitting some seeds to survive. Nilsson and Wastljung (1987) compared seed predation on beeches *(Fagus sylvatica)* in mast and nonmast years. In mast years, 3.1% of seeds were destroyed by a boring moth; in nonmast years, the figure was 38%. Vertebrate predation of seeds was 5.7% in mast years, but 12% in nonmast years. Masting may also provide other benefits, such as enhanced pollination or seed dispersal (Kelly, 1994). If all individuals flower together, many pollinators and seed feeders are likely to be attracted, increasing the likelihood of successful seed production and dispersal. Also, weather clues may predispose plants to produce mast crops in certain years, irrespective of whether there are selective advantages to doing so. Accordingly, for each plant species, we have to be careful in attempting to determine whether masting is a predator satiation mechanism or something else.

Defensive associations. Palatable plants can gain protection against herbivores through an association with unpalatable neighbors. This was shown by Peter Hamback and his colleagues (2000) in Sweden, where the chrysomelid beetle *Galerucella calmariensis* feeds on purple loosestrife, (*Lythrum salicaria*). Sometimes the loosestrife grows on its own, and sometimes it grows in thickets of an aromatic shrub, *Myrica gale. Myrica* secretes chemicals from both its roots and its leaves, and the volatile leaf chemicals deter insects from feeding on it. The chemicals also interfere with *Galerucella* beetles searching for purple loosestrife, decreasing both the number of beetle larvae and, consequently, damage to the leaf when the plant grows in thickets (Fig. 11.4).

Of course, the opposite of associational resistance is associational susceptibility: the spilling over of herbivores from palatable neighbors. In the very next article of the journal *Ecology* in which Hamback's article was published, an article by Jennifer White and Tom Whitham (2000) from Northern Arizona University appeared, showing associational susceptibility. In northern Utah fall cankerworms (moth larvae) prefer to feed on box elder trees and are rarely found on isolated cottonwood trees of the *Populus* species. However, when cottonwoods occur *under* box elder, the cankerworms spill over and defoliate the cottonwoods (Fig. 11.5). Thus, in both associational resistance and associational susceptibility, herbivory is influenced by neighboring species.

Mutualism. Some plants defend themselves against herbivores by enlisting the

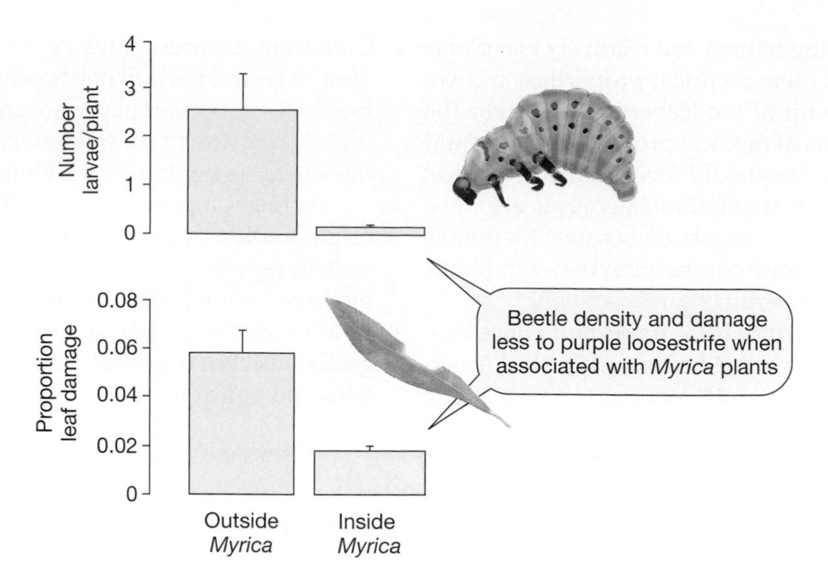

Figure 11.4 Associational resistance. The number of beetle larvae per plant and the proportion of leaves damaged on purple loosestrife is less when the loosestrife is associated with *Myrica* plants. The *Myrica* produce volatile chemicals that interfere with the beetle's ability to locate a host plant. *(Modified from Hamback et al., 2000.)*

help of other animals. Such a relationship can, of course, be seen as mutualism, as we saw in Chapter 9.

An understanding of plant defenses is of great use to agriculturalists, since the more crops are defended against pests, the higher are the yields. This line of defense, known as *host plant resistance,* may be due to physiological factors (e.g., toxic compounds within plant tissues that inhibit the

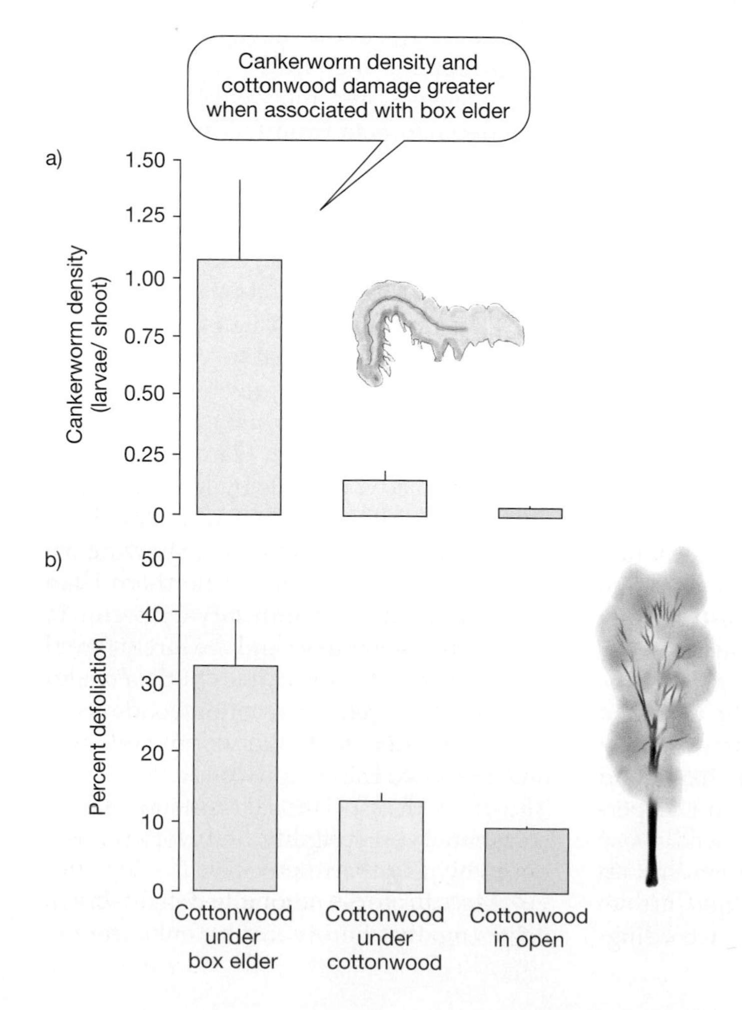

Figure 11.5 Associational susceptibility. The number of cankerworms on cottonwood plants and the damage done to leaves increase when cottonwoods are under box elder trees. The box elders are the preferred host, and their beetles spill over onto the cottonwoods when they are underneath. *(Modified from White and Whitham, 2000.)*

pest) or mechanical factors (e.g., a cuticle that is too tough for the pest to penetrate). The most serious problem associated with the development of host resistance is that it may take a long time to develop—between 10 and 15 years. Also, resistance to one pest is sometimes obtained at the cost of increasing susceptibility to other pests, and some crops can be resistant to only one pest, not to all. Finally, pest strains can appear that are able to overcome the plant's mechanisms of resistance. Circumvention of resistance by pests develops in much the same way as resistance to pesticides. Despite these problems, host resistance is a good tactic for the farmer. After the initial development of resistant varieties of plants, the cost is minimal to the grower. Perhaps more importantly, host resistance is environmentally benign, generally having few side effects on other species in the community. It is estimated that about 75% of U.S. cropland utilizes pest-resistant plants most of which are resistant to pathogens.

Of course, some herbivores can overcome these plant defenses. Certain chemicals that are toxic to generalist insects actually increase the growth rates of adapted specialist insects that can circumvent the defense or actually put it to good use in their own metabolic pathways. For example, the Cruciferae—the plant family embracing mustard, cabbage, and other, related species—contain acrid-smelling mustard oils called glucosinolates, the most important one of which is sinigrin. Cabbage butterflies (*Pieris brassicae*) preferentially feed on cabbage over other plants. In fact, if newly hatched larvae are fed an artificial diet, they do much better when sinigrin is added to it. When, upon hatching from eggs, larvae are fed cabbage leaves and are later switched to an artificial diet without sinigrin, they die rather than eat. Apparently, the secondary compound is an essential part of the diet. Similarly, the horseradish flea beetle feeds only on the horseradish plant (*Armoracea lapathifolia*), which again contains sinigrin.

This specialization of some herbivores on supposedly toxic plants has led to the notion of an evolutionary "arms race" between plants and herbivores. The idea was first formulated by Paul Ehrlich at Stanford University and Peter Raven, now director of the Missouri Botanical Garden (Erhlich and Raven, 1964). We can depict it by the following diagram:

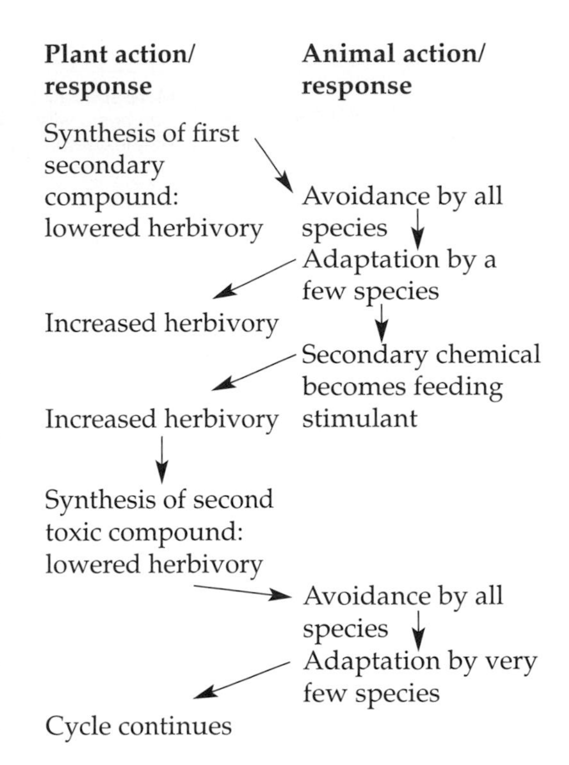

The result is a profusion of many specialized herbivores on plants. Attractive though this theory is, is it true? May Berenbaum (1981) of the University of Illinois tested the idea by examining the diversity of chemical defenses and the number of insect species in the carrot family, Umbelliferae. The simplest umbellifers were defended only by the phenolic compound hydroxycoumarin, the next group was defended by more complex linear furanocoumarins, and the most sophisticated group had complex angular furanocoumarins. Interestingly, and in support of the coevolution hypothesis of plants and herbivores, the most specialized insects, feeding only on one to three genera of umbellifers, were found on those with complex angular and linear furanocourmarins. The umbellifers without any furanocouramins were fed upon mainly by generalist polyphagous species (Fig. 11.6).

Finally, plant defenses can also have wide-ranging effects on organisms other than the intended herbivores. Interspecific differences in rates of leaf litter decomposition between plants are a consequence of differences in the antiherbivore defenses of living leaves (Grime *et al.*, 1996). The more chemically defended living leaves decompose more slowly when they abscise and fall to the forest floor, because the bacteria cannot easily degrade them. This finding suggests a critical role among leaf palatability, litter decomposition rates, and nutrient cycling in ecosystems. Only now are these

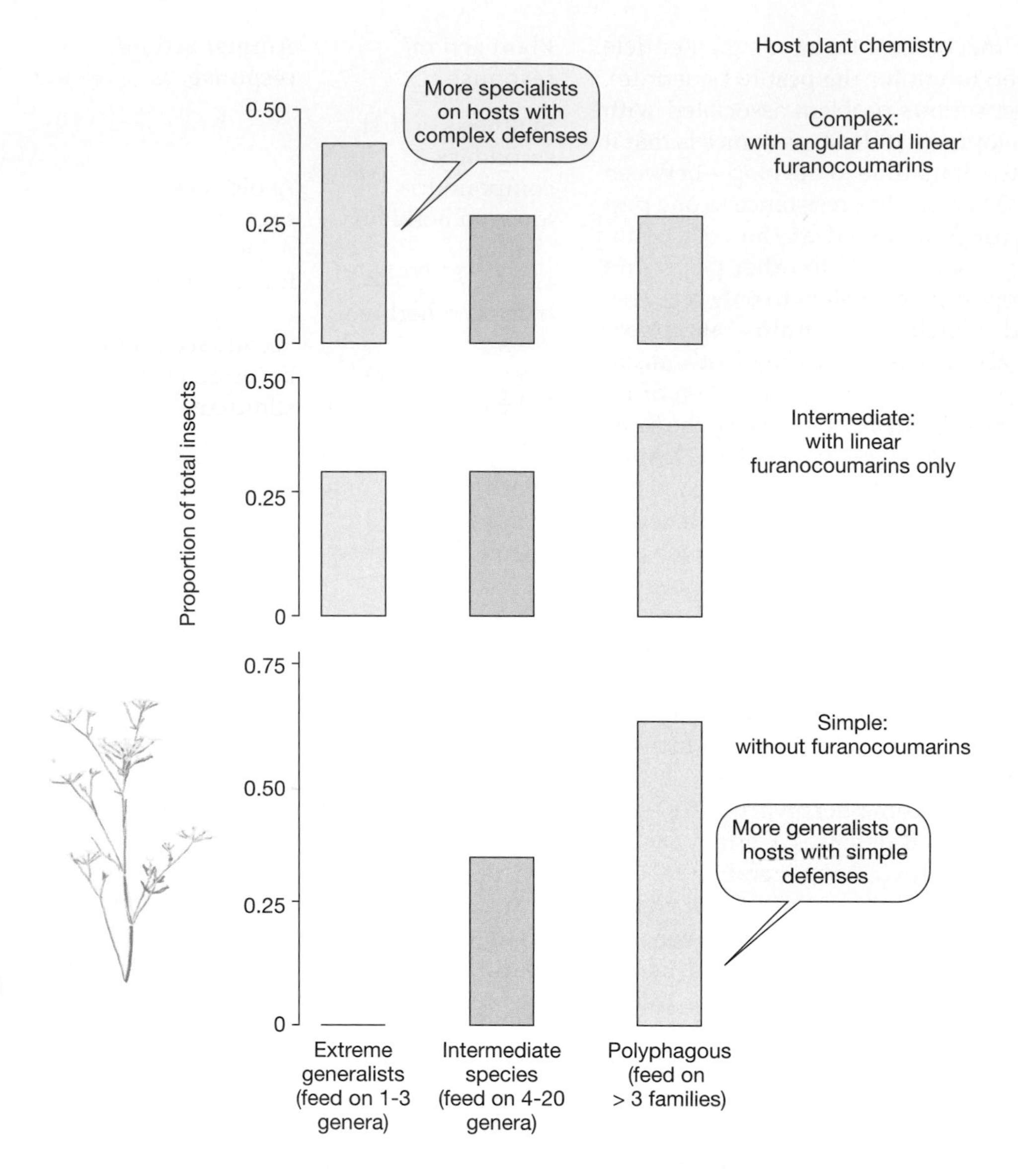

Figure 11.6 Degree of host specialization of insects feeding on umbelliferous plants according to the complexity of their secondary defenses. *(From data in Berenbaum, 1981.)*

more complicated pathways beginning to be explored.

11.2 Modeling Herbivory

In modeling herbivory, the outcomes of plant–herbivore interactions are determined by the degree of specialization of the herbivore.

Mick Crawley of Imperial College in England (1997) has devised a series of models of plant–herbivore interactions. The outcomes of these models depend to a large extent on the degree of polyphagy of the herbivore. A **monophagous** herbivore is highly specialized and feeds only on one species of host plant. By contrast, a **polyphagous** herbivore feeds on many species and so is not limited by any one. Many insect herbivores tend to specialize on one species of plant, whereas many vertebrates are generalist herbivores. Of course, there are exceptions: The panda bear specializes on bamboo and the koala bear feeds only on *Eucalyptus*, while the locust feeds on a wide variety of plants.

Let's take the case of the monophagous herbivore first. The simplest model assumes that there is an upper limit—a carrying capacity K—for a population of plants. The rate of change of a population of plants is given by

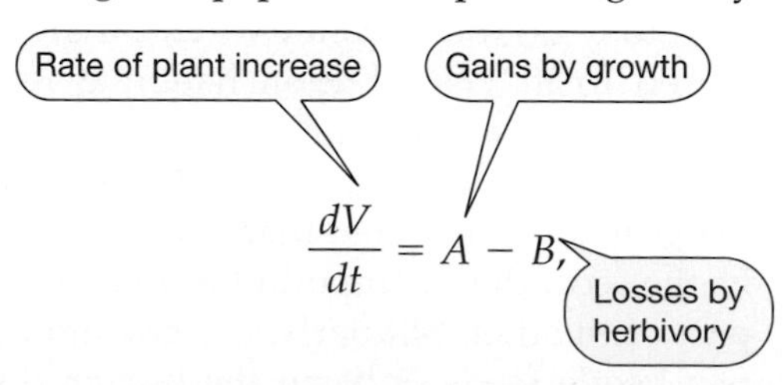

$$\frac{dV}{dt} = A - B,$$

where A denotes the gains and B the losses and V is the plant abundance, or biomass. Similarly, for herbivores,

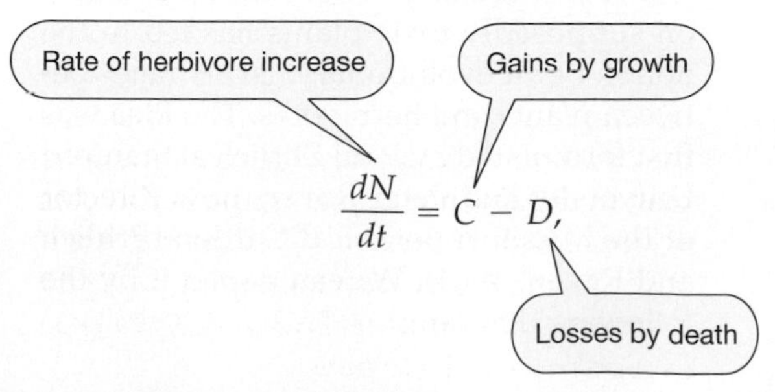

$$\frac{dN}{dt} = C - D,$$

where C and D are the gains and losses and N denotes the herbivore numbers.

It is assumed that, in the absence of herbivores, plant populations increase logistically according to the formula

$$\frac{dV}{dt} = rV\frac{(K - V)}{K}$$

where r is the plant's intrinsic rate of increase. This is the familiar logistic growth equation introduced in Chapter 6. Losses for plants, $B = bNV$, where b is the feeding rate of herbivores. The component bNV is sometimes called the *functional response*, of which, as we saw in Chapter 10, there are three recognized types. In Type I, which we are using here, the herbivore consumes more plant as the plant density increases. In this scenario, a Type I response is accurately modeled by bNV. In Type II, the herbivores can eventually become satiated (stuffed) and stop feeding. Alternatively, they may be limited by the handling time needed to locate and eat the plants. In either case, there is an upper level or asymptote at which herbivores can process plants. Type II responses are therefore modeled by a different equation that we will not use. The Type III response also supposes an asymptote, but here the curve has a sigmoid shape, similar to the logistic curve. The feeding rate is low at low plant density, but increases quickly at high density. This sometimes happens if herbivores switch feeding between plant species and develop a search image for a particular plant species once it reaches a certain threshold density, so such an assumption may be unrealistic for our specialist monophagous herbivore. The Type III response is represented by yet another equation. Thus, while we will consider a plant–insect interaction with a type I functional response by the insect herbivore, bear in mind that different functional responses can have different consequences for the ability of herbivores to control plant populations.

For the herbivores, their population growth $C = cNV$, where c describes the numerical response of the herbivores, or how much they are attracted to dense areas of plants (i.e., their degree of attraction is affected by the plant density). cNV is essentially a measure of the efficiency with which herbivores turn food into progeny. Losses for herbivores are given by $D = dN$, where d is the herbivore death rate. The assumption is that, in the absence of plants, the herbivores starve and their numbers decline exponentially. Assuming an upper carrying capacity K, and following a basic Lotka–Volterra type of logistic model, we have

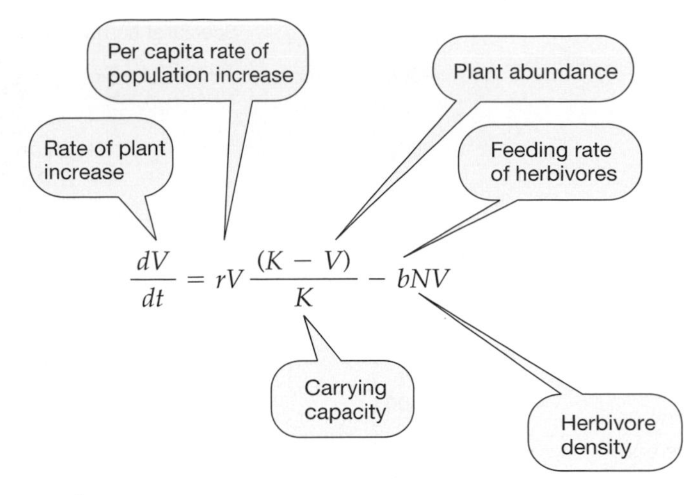

$$\frac{dV}{dt} = rV\frac{(K - V)}{K} - bNV$$

and

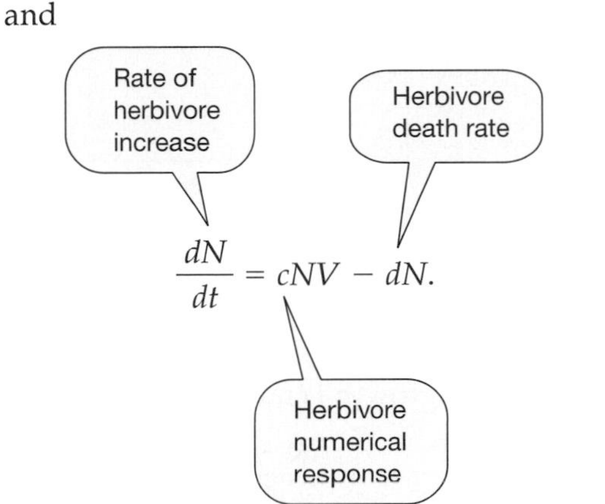

$$\frac{dN}{dt} = cNV - dN.$$

At equilibrium, both dV/dt and dN/dt are zero, so there is no population change. The equilibrium plant density V^* can then be obtained, because $A = B$ and $C = D$. We get

$$V^* = \frac{d}{c}.$$

This is a somewhat paradoxical result, because it shows that the plant equilibrium abundance has nothing to do with the plant growth rate r or carrying capacity K (Fig. 11.7a). Instead, it is determined by the herbivore death rate d and numerical response c. The higher the death rate of the herbivore, the higher the plant abundance will be. Also, the greater the herbivore numerical response (i.e., the more herbivores are drawn to an area of plant abundance), the lower the plant abundance will be. This is essentially saying that the faster the plant population grows, the more the bugs will eat them up. Of course, the model may be unrealistic because it assumes that insect herbivores are limited only by food when, in reality, they themselves may be eaten by predators when they get to high densities.

We can contrast these dynamics with those of a polyphagous vertebrate, which is taken to cause a constant level of herbivory

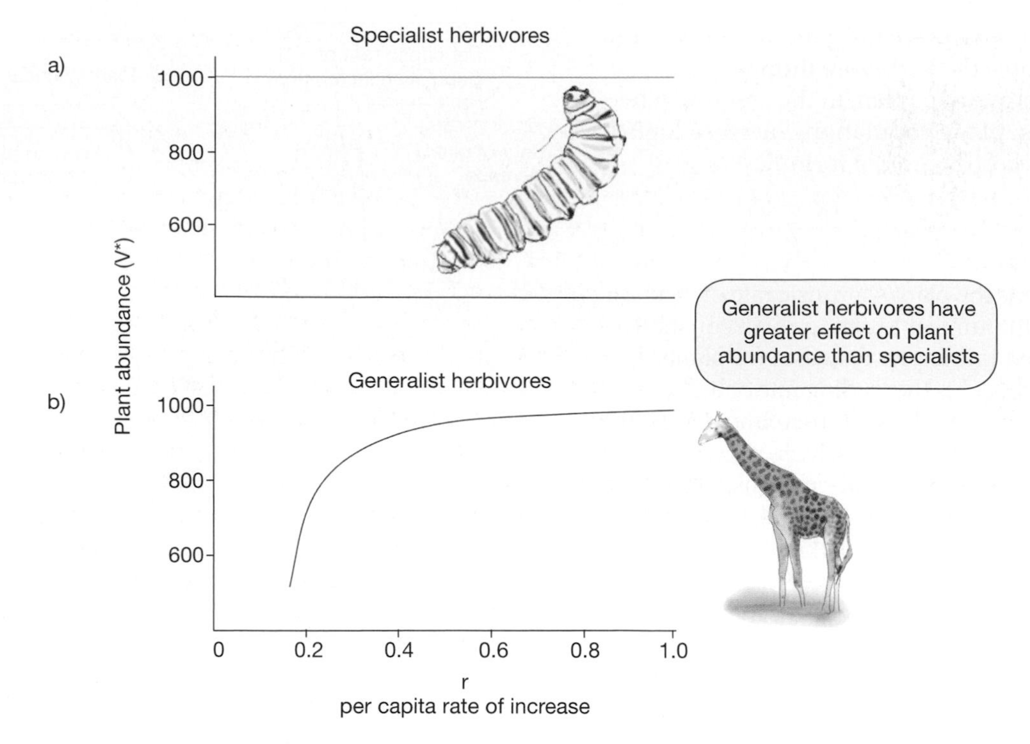

Figure 11.7 (a) The equilibrium plant abundance V^* is not affected by the intrinsic plant growth rate r when plants are attacked by monophagous herbivores. (b) With polyphagous herbivores, increases in r result in an increase in V^*. *(Modified from Crawley, 1997.)*

Herbivores can influence the growth rate, reproduction rate, and stability of plant populations.

h because its densities are not tied to any one plant. Plant growth is now represented by

$$\frac{dV}{dt} = rV\frac{(K - V)}{K} - h.$$

We can solve this equation for the equilibrium plant biomass V^*. We get

$$V^* = \frac{aK\sqrt{r^2K^2 - 4rhK}}{2r}.$$

The important point about this equation is that when the main herbivores are polyphagous vertebrates, increasing the plant reproductive rate r does increase the plant biomass (Fig. 11.7b).

There are some caveats to these models. The fact that c and b are treated as independent parameters could be misleading, because it suggests that herbivore birthrates are decoupled from the rate of consumption. This could lead to unrealistic results. For example, if c (the herbivore efficiency) is fixed, but b (the feeding rate of herbivores) goes to zero, herbivore numbers become infinitely large (Holt, 1997). Further, there are at least three different types of functional response and at least four different types of numerical response (Crawley, 1997), and we have examined only the simplest of these. Hence, it is difficult to draw generalizations from any of the models other than to say that the degree of polyphagy can radically alter our understanding of plant–animal interactions, which was the point we intended to make.

11.3 Effects of Herbivores on Plants

Despite the impressive array of defenses in their arsenals, plants do not have things all their own way in the plant–herbivore interaction. Herbivores can detoxify many poisons, mainly by two chemical pathways: oxidation and conjugation (Harborne, 1988). Oxidation, the most important of these mechanisms, occurs in mammals in the liver and in insects in the midgut. Oxidation is brought about by a group of enzymes known as mixed-function oxidases (MFOs). A secondary compound is oxidized to a corresponding alcohol. Conjugation, often the next step in detoxification, occurs when the harmful element or the compound resulting from the oxidation is united with another molecule to form one inactive and readily excreted product.

Given that herbivores can circumvent plant defenses in certain situations, what is their measured effect on plant populations in the field? An average of 7% of the leaf area was consumed in 93 cases of leaf herbivory in terrestrial systems (Pimentel, 1988). In forest systems, rates of defoliation by insects alone are generally found to lie within the 5–15% range (Landsburg and Ohmart, 1989). However, some critics have argued that these rates represent an underestimate of defoliation because they fail to account for leaf turnover. Most estimates of herbivory are made at the end of the growing season, when herbivores—especially insects—are thought to have finished feeding. This underestimates any herbivory that has removed entire leaves or caused leaves to fall. It is thought that damaged leaves are often abscised as a defense against infection (Stiling and Simberloff, 1989). Graduate student S. Zakaria (1989) [reported in Crawley (1997)] marked leaves on birch trees in spring in England and found that 17% were consumed entirely or fell prematurely by August. In an extensive literature review, Hélène Cyr and Michael Pace (1993) showed how herbivory was tied to the productivity of the plants the herbivores fed on (Fig. 11.8). In both aquatic and terrestrial systems, the more plants grew, the more the herbivores ate, a finding reminiscent of Crawley's mathematical result we discussed in the previous section. Furthermore, the amount of material found to be eaten was much higher than Pimentel's and Landsburg and Ohmart's earlier estimates. Mean values of primary productivity (leaves, stems, or roots) removed by herbivores were 18% for terrestrial plants, 30% for aquatic macrophytes, and 79% for aquatic algae! Thus, herbivory can account for the removal of large amounts of plant material. The reasons for the high values in aquatic algae are that these simple plants lack the chemically sophisticated defenses of higher terrestrial or aquatic (macrophytic) plants. Levels of primary production removed by herbivores in aquatic systems average 51%, much higher than the 18% in terrestrial systems.

Removal experiments.

Ultimately, the best way to estimate the effects of herbivory on plant populations is to remove the herbivores and examine the plants' subsequent growth and reproductive output. David Bigger and Michelle Marvier (1998), postdoctoral researchers at the University of Washington, reviewed the results of 246 experiments that compared the growth of plants that were protected and those that were unprotected from herbivores. The authors used the statistical technique of metaanalysis in generating their results. Overall, they found that the mean effect size d was 0.47, which, by convention, is considered a medium effect size ($d = 0.2$ is considered small, $d = 0.5$ medium, and $d = 0.8$ large). This means that the average plant that was protected from herbivores was significantly larger than the average exposed plant. Herbivory thus appears to be a powerful force in nature. Bigger and Marvier went on to examine which types of herbivores had the strongest effects and which types of plants were most susceptible. Just as Cyr and Pace (1993) had argued, Bigger and Marvier found that herbivory in aquatic systems was stronger than in terrestrial ones (Fig. 11.9). Also, invertebrate herbivores, such as insects, had stronger effects than vertebrate herbivores, such as mammals, at least in terrestrial systems. So, while one might consider large grazers like bison in North America or antelopes in Africa to be of considerable importance, the chances are that grasshoppers are of more importance in grasslands. In forests, invertebrate grazers such as caterpillars have greater access to many canopy leaves than vertebrates do and are likely to have greater effects than vertebrates.

Which types of plants are most affected? The effect size of herbivores was greatest in algae (Fig. 11.10), presumably because they are the least sophisticated in terms of their ability to manufacture complex secondary compounds. Grasses and shrubs were also significantly affected by

Herbivore removal experiments show significant effects of herbivores on plant growth.

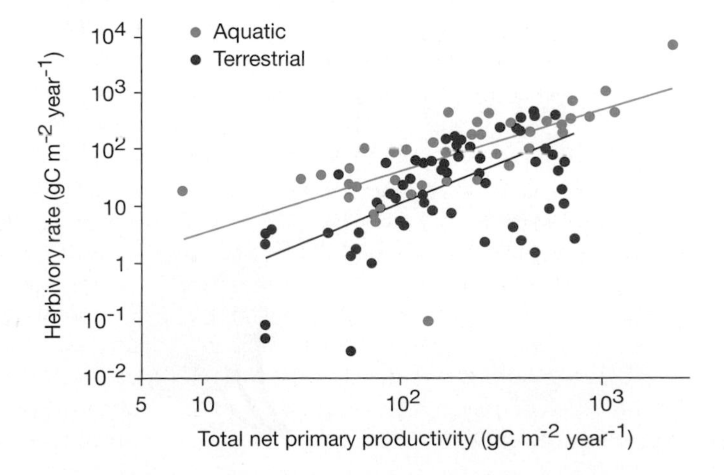

Figure 11.8 The rate of herbivory increases with plant biomass (net primary productivity) in both aquatic systems (blue circles) and terrestrial systems (red circles). *(Modified from Cyr and Pace, 1993.)*

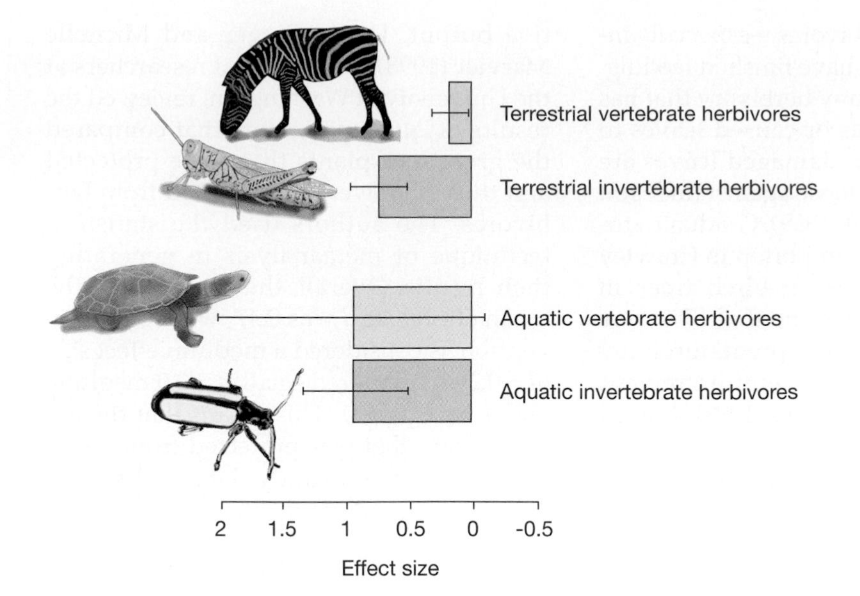

Figure 11.9 The relative impact of invertebrates (mainly insects) and vertebrates (mainly mammals) on aquatic and terrestrial vegetation. Metaanalysis shows that the effect size is much greater on aquatic plants than on terrestrial ones and that invertebrate grazers have the biggest effects. *(Reproduced from Bigger and Marvier, 1998.)*

herbivores, whereas the effect size was least for woody plants. This may be because large and long-lived trees can draw on large reserves of resources to buffer the impact of herbivores.

Biological control.

Many insects released to control exotic weeds greatly reduce weed density, demonstrating the powerful effects of herbivores.

Additional evidence on the impact of herbivores on plants comes from the biological control of weeds and from agriculture. Many weeds are invaders that were accidentally introduced to an area as seeds in ships' ballasts or in agricultural shipments. Over 50 percent of the 190 major U.S. weeds are invaders from outside the country. Many of these weeds have become separated from their native natural enemies, which were left behind in their native countries; this is one reason the weeds become so prolific. Because chemical control is expensive, many land managers have reverted to biological control, wherein the native natural enemy is reunited with the weed in its new country.

There have been many successes in the biological control of weeds by natural enemies. Klamath weed (*Hypericum perforatum*), a pest of pastureland in California, was controlled by two French beetles. A floating fern, *Salvinia molesta*, choked a lake in New Guinea and was controlled by the weevil *Cyrtobagus salvinae*, introduced from Brazil, to which the fern is native. Alligator weed was controlled in Florida's rivers by the alligatorweed flea beetle (*Agasicles hygrophila*) from South America, and hopes are high that water hyacinth can be controlled biologically, too. Finally, prickly pear cactus, (*Opuntia stricta*), an exotic pest of rangeland

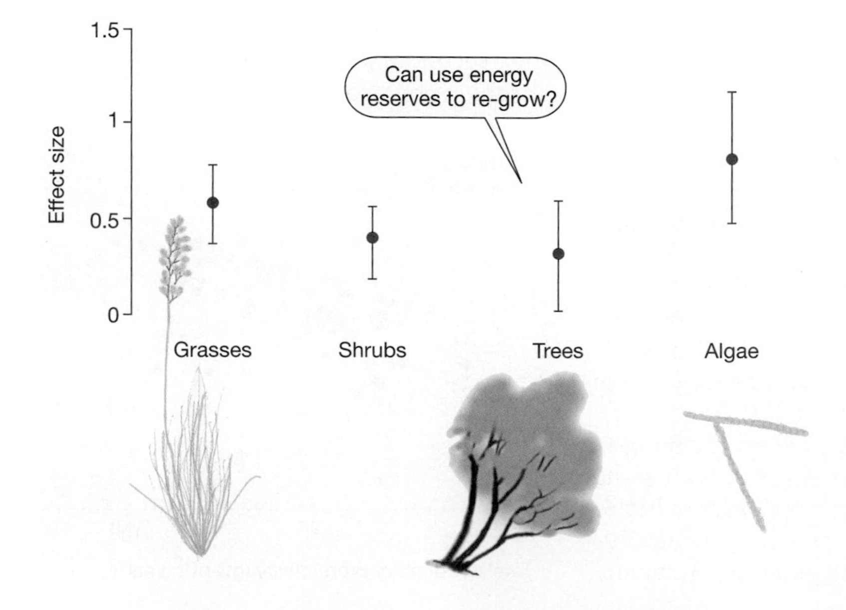

Figure 11.10 The impacts of herbivores on different plant taxa. Woody plants, perhaps having greater energy reserves in their roots, are more able to withstand grazing than are other groups of plants. *(Reproduced from Bigger and Marvier, 1998.)*

in Australia that was imported from Argentina, was controlled by introducing the cactus moth (*Cactoblastis cactorum*) from South America. This is the "poster child" of biological control, where one little moth saved hundreds of thousands of acres of valuable rangeland from cacti and allowed it to return to grassland. Thankful ranchers erected a statue in honor of the architect of the successful project, entomologist Alan Dodd (1940). Unfortunately, the moth has invaded Florida and is affecting many rare and endangered cacti, all of which are native (Fig. 11.11).

Not all biological control campaigns are successful. Large numbers of insects have been introduced to control *Lantana camara*, an introduced weed in Hawaii, but few have had any impact on the growth of the plant, though its spread is perhaps being slowed. In fact, the majority of biological control efforts are failures. In a review of 701 importations of biological weed control agents worldwide (Debach and Rosen, 1991) only 26% were rated effective, offering a noticeable reduction in pest weeds. This means that herbivores had no effect in the other 74% of the cases. Furthermore, even those cases that are successful involve plants or herbivores in exotic settings where one or both species may be changed by local environmental conditions. Therefore, the relevance of such examples to determining the effects of herbivores on plants in native systems is questionable. To further complicate matters, most biological control campaigns are conducted in the absence of the herbivores' natural enemies. Therefore, we can conclude only that biocontrol campaigns provide some evidence that, in a minority of cases, herbivores can provide strong control of plant populations when predation of the herbivores is low.

In an agricultural setting, there are many examples of huge losses of crops to herbivores, and a wide range of pest controls is used to try to reduce these losses. (See "Applied Ecology Box: Pest Control.") Even in a developed country like the United States, losses to pests and diseases amount to about 30% per year. Again, while this is circumstantial evidence that plants may even be devastated by their herbivores, most crops are exotic species, introduced from foreign countries. Most of their pests are also exotic, so large crop losses to herbivores may be because of a lack of coevolutionary history between plant and herbivore and may not reflect what happens in natural settings.

Beneficial herbivory?

In Bigger and Marvier's (1998) review of herbivory, 60 comparisons demonstrated a reduction in plant size due to natural levels of herbivory, but 10 demonstrated significant increases. How can this be? Some authors have argued that herbivory can actually be beneficial to plants (McNaughton, 1986). The rationale is that, because plants are stimulated to regrow after damage, they may end up overcompensating, growing even more than they would have had they not been damaged. Valentine *et al.* (1997) manipulated sea urchin densities inside cages over patches of turtle grass in the sea. With higher urchin densities, the turtle grass compensated for herbivory by increased recruitment of shoots. This led to a 40 percent increase in net aboveground primary production. In another example, Simberloff, Brown, and Lowrie (1978) noted that the action of isopod and other invertebrate root borers of mangroves in the Florida Keys tended to initiate new prop roots at the point of attack (Fig. 11.12). Because more prop roots lend greater stability to the mangroves against wave and storm action, root herbivory could in fact be beneficial. However, Sharon Strauss of the University of California at Davis (1988) has pointed out that very carefully designed experiments

Some plants are stimulated to regrow after damage by herbivory and may reach sizes bigger than they were before.

Figure 11.11 The effect of the exotic cactus moth, *Cactoblastis cactorum*, on *Opuntia* cacti is evidenced in this before-and-after photograph. The moth caterpillars can eat their way through the cactus pads, destroying the cacti. These photos were taken in Florida, where the moth has just become established and has begun to affect native *Opuntia* there. *(Photographs by Peter Stiling.)*

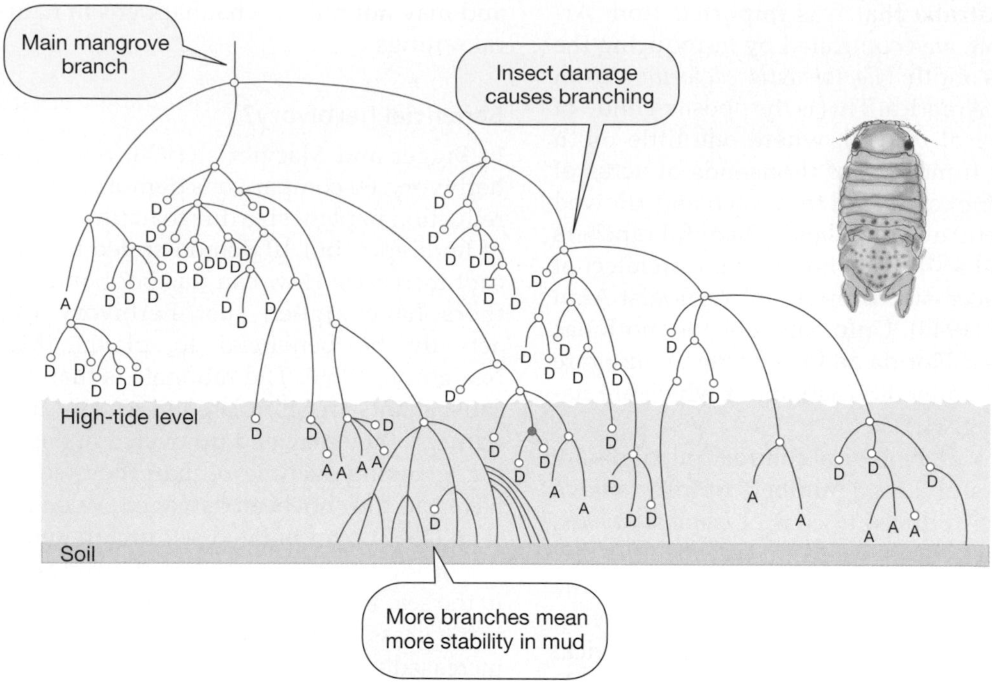

Figure 11.12 Beneficial herbivory? Branching pattern for a single *Rhizophora* mangrove root from Clam Key, Florida, following attack by herbivores. *A* = alive, *D* = dead; open square = bored by *Ecdytolopha* sp.; open circle (above water) = bored by unknown insect; open circle (below water) = bored by *Sphaeroma terebrans* (inset); shaded circle = bored by *Teredo* sp. *(Redrawn from Simberloff, Brown, and Lowrie, 1978.)*

Herbivores are strongly influenced by plant quality and chemical defenses.

involving measurements of plant size before and after herbivory are needed to address the issue of beneficial herbivory, because herbivores themselves naturally choose larger plants, which might be expected to show more growth than would plants with no herbivores (usually the controls), even after herbivory. One of the more recent studies has focused on the regrowth of scarlet gilia (*Ipomopsis aggregata*) in the San Francisco peaks area near Flagstaff, Arizona, following grazing by elk and mule deer. Ken Paige and his associates from Illinois (Gronemeyer *et al.*, 1997) actually showed not only that grazing stimulated regrowth, but also that more flowers and fruits appeared on browsed plants than on nonbrowsed control plants. Dan Doak of the University of California at Santa Cruz (1992) suggested that part of the debate over whether herbivores can be beneficial can be addressed by determining the life cycle of the plant. Short-lived species such as terrestrial annuals or aquatics have a tough time recovering from grazing, whereas long-lived perennials may be able to utilize reserves of resources to recover, just as Bigger and Marvier (1998) suggested in their review. Whether herbivory can be beneficial is a crucial issue, because it has implications for rangeland management. Many authors argue that sub-Saharan rangeland has been degraded by overgrazing, while others argue that the drastic changes took place because the area wasn't grazed enough. The latter view argues that grasses in grazed patches may be more productive, may be of better nutritional quality, and may live longer than ungrazed grasses and that leaving an area ungrazed can lead to loss of plant cover in that area.

11.4 Effects of Plants on Herbivores

According to what is known as the *nitrogen limitation theory* (for which there is much evidence), herbivores themselves select the plants that provide the most nutrition in terms of nitrogen content of the tissue. The theory holds that herbivores behave in this manner largely because of the elemental proportions of nitrogen in plants and animals. Animal tissue generally contains about 10 times as much nitrogen as plant tissue, so plants that are high in nitrogen are favored by animals. For example, red deer feed preferentially on grasses defecated upon by herring gulls (*Larus argentatus*). One study found that where the number of gull droppings increased, so did the nitrogen content of the vegetation (Iason, Duck, and Clutton-Brock, 1986). Gwen Waring and Neil Cobb of Northern Arizona University (1992) systematically reviewed the effects of nutrients and various stresses on insect performance. Plant fertilization—in particular,

Applied Ecology

Pest Control

In most parts of the world, farms specialize in growing one or a few crops, partly for reasons of economy of scale. As a result, huge monocultures of crops make ideal targets for insect pests and diseases. A wide variety of control strategies has been arrayed against them (Table 1). There is no one perfect solution; each has its advantages and disadvantages. Fifty thousand pesticides are registered for use in the United States alone. Five to six hundred million kilograms of pesticides are used each year, 70% for agriculture, 23% for forestry, and 7% for home and garden. Many of these are strong poisons—chlorinated hydrocarbons (many, like DDT, now banned), organophosphorus compounds like malathion, and carbonates such as carbofuran. Sixty percent of herbicides and 30% of insecticides are potentially oncogenic (capable of producing tumors). Residues have been found in 95% of human tissue samples, causing widespread concern and a desire to reduce pesticide use in the environment and change over to other, more environmentally benign strategies such as biological control. Even here, however, there is the potential for some environmental effects (e.g., on nontarget organisms—recall the invasion of Florida by *Cactoblastis*, attacking native cacti in the state).

TABLE 1 Herbivores can have substantial effects on crops: Witness the variety of pest control techniques arrayed against them.

Technique	Advantages	Disadvantages
Pesticides	(i) Rapid action, hours to days	(i) Nontarget effects on other species
	(ii) May degrade rapidly	(ii) May remain in the environment
	(iii) Wide range of properties, from contact poisons to systemics	(iii) Development of resistance leads to use of new pesticides and, ultimately, resistance—the pesticide treadmill
	(iv) Favorable benefit–cost ratios	(iv) May eliminate beneficial natural enemies
		(v) Risks to human health
Biological control with predators and parasitoids	(i) Even more favorable benefit–cost ratio	(i) Lack of instantaneous kill
	(ii) No environmental residues	(ii) Small reservoir of pests always needed to sustain enemies
	(iii) High specificity to certain pests	(iii) Nontarget effects on other species
	(iv) High dispersal capabilities	
	(v) Self-sustaining populations	
Growth regulators (chemicals that mimic growth hormones in insects)	(i) Environmentally benign	(i) Precise timing needed to coincide with critical time of insect life cycle
Microbial control	(i) Little toxic residue	(i) Careful timing needed
	(ii) Some specificity to pest insects	(ii) Many pathogens don't disperse well
	(iii) Low dosage requirements—may multiply on their own	(iii) Short life cycle, so short persistence
	(iv) Low likelihood of resistance	

nitrogen enhancement—had strong positive effects on the population sizes, survivorship, growth, and fecundity of almost all herbivores, including chewing organisms, sucking insects, galling insects, and phytophagous mites (Fig. 11.13). Nearly 60% of 186 studies reported positive responses by herbivores to fertilization. The proportion of positive responses was greater in cultivated versus wild plants and in herbaceous and broadleaf trees versus conifers (Fig. 11.14). Although the addition of other nutrients, such as phosphorus and potassium, can increase herbivore densities, the overall responses to these nutrients were much more variable, and positive responses were generally not as common as no responses or negative responses.

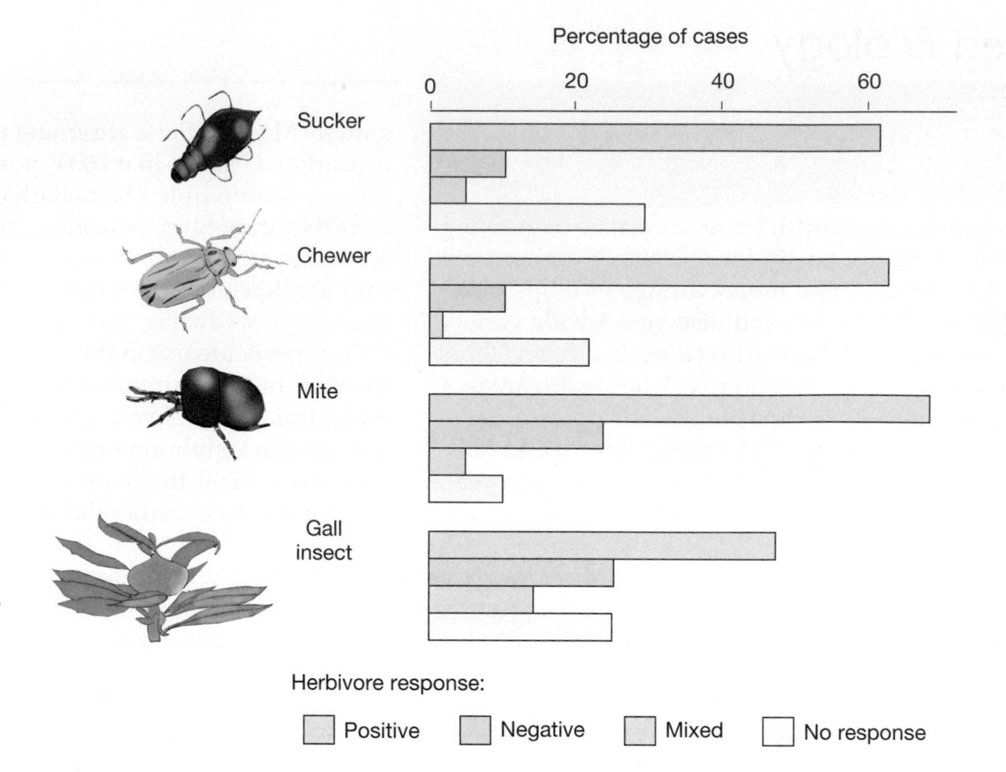

Figure 11.13 Feeding responses of herbivores to nitrogen fertilization of plants. Populations of all types of insects generally increase in size on fertilized plants. *(Reproduced from Waring and Cobb, 1992.)*

There are two variations of the nitrogen limitation theory. The first is the *stress hypothesis* (White, 1993), according to which plant stressors, such as drought, tend to increase the availability of nitrogen because many nitrogen-rich compounds (e.g., soluble amino acids) are mobilized in the phloem in response to stress. Indeed, there are many cases where drought-stressed plants accrue higher numbers of herbivores (Waring and Cobb, 1992; White, 1993). The other hypothesis is the *plant vigor hypothesis* developed by Peter Price (1991), which suggests that many herbivores select fast-growing parts of the plant and faster growing plants (such as young plants) because these are higher in nitrogen. The literature on forestry shows that many attacks by insects are on young trees. Price's own work with endophagous insects, like gallers and shoot borers, also showed higher attack rates on vigorously growing stems.

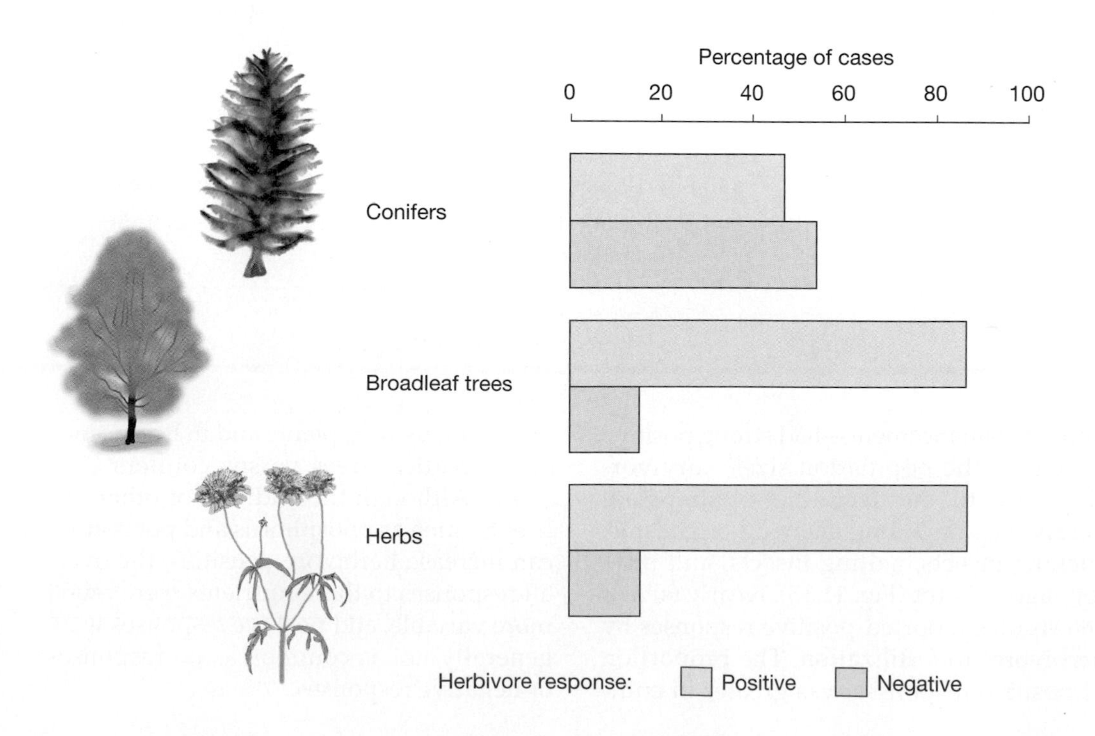

Figure 11.14 Percentage of studies in which herbivores responded positively or negatively to nitrogen fertilization in conifers, broadleaf trees, and herbaceous plants. Herbivores respond more to fertilization when on broadleaf trees and herbs. (*Reproduced from Waring and Cobb, 1992.*)

In some cases, herbivore density, is correlated with host-plant quality but observed population patterns of herbivores are more dependent on other phenomena, such as predation or parasitism. In Florida salt marshes, leaf-mining flies appear to feed primarily on the grass that is richest in nitrogen, but their population densities do not increase when the grasses are fertilized. Instead, the density of the fly is actually governed by high parasitism rates, and the flies choose primarily plants not likely to be searched by parasites; coincidentally, these plants are also the richest in nitrogen (Stiling, Brodbeck, and Strong, 1982). Cases like these may be a reason why Waring and Cobb (1992) found some species that failed to respond to increased plant nitrogen.

As a final twist on the story of how plants affect their herbivores, other mortalities may be mediated by plant quality. For example, when food quality declines, many herbivores respond simply by feeding more often or for a longer time. Either of these strategics can result in increased exposure time to predators and parasites and higher enemy-induced mortality rates—the so-called slow-growth, high-mortality hypothesis (Williams, 1999). So, although deaths of herbivores due to the depletion of food plants are witnessed infrequently (perhaps because herbivores can leave an area of poor food availability), herbivore health and the subsequent fecundity of females are strongly affected by their host plants. The relatively few examples of mass starvation due to overexploitation of plants come mainly from studies of insects that habitually undergo periodic outbreaks or from cycles of Arctic rodents.

11.5 Herbivory Affects Community Structure

Herbivores can affect plant community richness by chewing down competitive dominants.

It has been argued that, in many cases, herbivores affect community richness (numbers of species in a community), succession, and ecosystem function. Each of these topics is dealt with later on in its own chapter, so only brief mention of them, as they pertain to herbivory, is made here.

One of Darwin's own observations was that more species of grass existed in areas grazed by sheep. The idea is that competitively dominant grasses, which would normally displace many other species by outcompeting them for light or nutrients, are kept in check by heavy grazing by herbivores, permitting the coexistence of a variety of other, competitively inferior, species. A ready-made experiment to test this idea occurred in Britain in 1954. The virus disease myxomatosis was released to control rabbits. The disease was so effective that it killed nearly all the rabbits in most places. The next year, the number of plant species decreased dramatically. In the United States, bison grazing increases grassland biodiversity because the animals chew down tall C_4 grasses, allowing more light to penetrate and permitting more species of both C_4 and C_3 grasses to thrive (Fig. 11.15; Collins *et al.*, 1998). It must be remembered, however, that for this to happen, it is not sufficient for a herbivore to feed on just the dominant plant species, for if it did, the vegetation would simply become dominated by the next dominant species. Herbivores must be able to switch from one species to another, depending on its commonness. While there are many examples of herbivory increasing

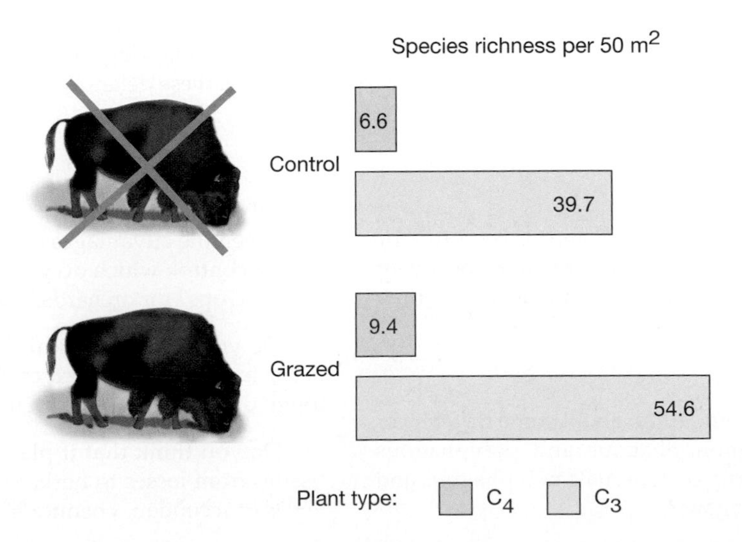

Figure 11.15 Grazing by buffalo in Kansas reduces the biomass of C_4 grasses, allowing more light to penetrate and more species of both C_3 and C_4 grasses to thrive. *(From data in Collins et al., 1998.)*

species richness, there are also many in which herbivory decreases species richness.

If herbivory can affect a plant community's composition, then it has the potential to affect its succession—how that composition changes over time. In reviewing the effects of herbivory on succession, Diane Davidson (1993) argued that herbivores retard the pace of community change from old fields to trees. The reason is that there is differentially high herbivory on old-field species such as grasses and shrubs, whose main defense against herbivory is rapid compensatory growth, therefore maintaining the field.

In sum, there is a strong evolutionary selection for plants to defend themselves and for herbivores to overcome plant defenses. Plants can affect herbivore population sizes through plant quality and chemical defenses. In turn, herbivores can affect plant populations by overcoming plant defenses. Modeling suggests that at low rates of per capita increase, specialist herbivores such as insects do not affect plant populations as much as generalist vertebrate herbivores do. However, meta-analysis shows that invertebrate herbivores have stronger effects than vertebrate herbivores.

Summary

Why is it that herbivores have not been able to eat more plant material than they have? After all, green plants are abundant worldwide.

1. A variety of plant defenses serves as a testament to the strength and frequency of herbivory in nature. There are chemical defenses, such as nicotine in tobacco and caffeine in tea, and mechanical defenses, such as spines and stinging hairs. Plants may also contain hormone mimics that disrupt insect molts. Some plants enter into mutualisms with ants that attack and remove herbivores in return for shelter and food (extrafloral nectaries).

2. Chemical defenses can be subdivided into qualitative and quantitative varieties. Quantitative defenses gradually build up inside herbivore guts and prevent food from being digested. The more foliage that is eaten, the worse the situation becomes for the herbivore. Examples are tannins and resins in leaves. Qualitative defenses are toxic compounds, such as cyanogenic compounds in leaves, which are lethal in small doses. A good example is atropine, produced by deadly nightshade.

3. Mathematical models have led some ecologists to conclude that the effect herbivores have on plants depends on whether the herbivores are monophagous or polyphagous. Polyphagous species may be more important than monophagous species in the dynamics of plants.

4. Reviews suggest that, on average, between 15 and 18% of terrestrial plant tissue is consumed by herbivores. In aquatic systems, the percentage is much higher (51%). Biological control projects have shown that many exotic weeds which have undergone population explosions in foreign countries in the absence of their native herbivores can be brought under control when the native herbivore is reunited with its host. Pest outbreaks in agriculturally important crops also suggest that herbivores can have strong effects on plants in many systems. Experiments that have removed native herbivores from native plants have shown the dramatic effects of herbivory as well.

5. Population densities of herbivores are strongly influenced by plant quality—in particular, plant nitrogen.

6. Herbivory may have substantial effects on plant communities (e.g., on species richness and on the number of plant species in a community), because herbivores can preferentially eat dominant plant species, allowing other species to grow.

The world is green probably because plants are suffused with toxic or digestibility-reducing chemicals. Herbivores are in an evolutionary arms race with plants as they try to overcome these defenses.

Discussion Questions

1. Cannabin and cocaine are secondary chemicals produced by the hemp plant (*Cannabis sativa*) and the coca plant, respectively. Can you find out the roles of these substances in the lives of their plants?

2. From a theoretical standpoint, think about some differences that might occur between monophagous and polyphagous herbivores. Why are the majority of mammals polyphagous and yet many insects are monophagous?

3. Given the advantages and disadvantages of different forms of pest control, which do you think would be more suitable for annual crops? for orchards? for forests? Why?

4. Do you think chemical defenses would be more likely to occur in lush tropical leaves that have no other defenses or in tough desert plants that can ill afford the losses to herbivores.

5. Do you think that if plants need to protect their valuable tissues from losses to herbivores, flowers and fruits have high levels of secondary chemicals? If not, why not?

CHAPTER 12

Parasitism

Road Map

1. A parasite feeds on a host, but generally does not kill it.
2. Hosts have evolved many defenses against parasites, the most important of which is the immune response.
3. Models show that the rate of spread of diseases is governed by the density of susceptibles in a population, the transmission rate of the disease and the length of life of the infected host.
4. Parasites can decrease host population sizes substantially.
5. Parasites can affect the structure of host communities by removing certain susceptible host species and allowing other, nonsusceptible species to flourish.
6. Parasitoids help in biological control by reducing the density of pests.

In the 1980s, many researchers became alarmed at what appeared to be a global die-off of frogs. Populations of many species were declining, and some species were reported to have become extinct. Associated with the declines were reports of frogs with physical deformities (Fig. 12.1). Water pollutants were investigated, but because of the wide-ranging nature of the decline, many panglobal phenomena, such as increased ultraviolet (UV) light, global warming, or windborne pollutants, were suspected. In the 1990s, the focus began to switch from pollution to a frog parasite as a cause of the decline. By 1998, *Batrachocytrium dendrobatidis*, a new (genus and) species of parasitic chytrid fungus, was being described as a lethal disease agent in frogs. It was thought that environmental deterioration, such as increased UV light, made frogs more susceptible to the pathogen, so frog decline is caused by a one–two punch: environment and **parasite**.

Figure 12.1 Infection of the western toad, *Bufo boreas*, by the trematode *Ribeioria ondatrae* causes supernumerary limbs.

12.1 Defining Parasites

The life cycle of some organisms brings into question what exactly a parasite is.

When one organism feeds off another, but does not normally kill it outright, the predatory organism is termed a parasite and the prey a **host**. Some parasites remain attached to their hosts for most of their lives; for example, tapeworms spend their entire adult life inside the host's alimentary canal. Others, such as ticks and leeches, drop off after prolonged periods of feeding. Mosquitoes remain attached for relatively short periods. Should we reserve the term *parasite* for organisms that remain in intimate contact with their hosts throughout their lives? Mosquitoes develop as larvae in a nonparasitic manner in pools of water, and the adults come into contact with hosts only for short periods. However, by the intimate-contact definition, many species of phytophagous insects can be thought of as parasitic upon their "host" plants because they spend their entire life cycle on one host plant or tree. Further, should the large ungulates of the Serengeti plains, such as wildebeest and zebra, be termed parasites? Although they feed off more than one individual host grass, the grass is not killed and will grow back later. And what about parasites of insects? Many of these develop as internal parasites of adults or immature stages of the insects. In these cases, the host almost never survives, and the term **parasitoid** is used because only one host is involved, but it is invariably killed. Even in this case, though, a further gradation between parasitoid–parasite and predator is evident, as when an egg parasitoid hatches from a host egg and has to devour several more eggs in the clutch before it is mature (Askew, 1971).

Even some flowering plants are parasitic on other plants. We can recognize two distinctions: holoparasites and hemiparasites. *Holoparasites* lack chlorophyll and are totally dependent on the other plant for all their water and nutrients. One famous holoparasite is the tropical *Rafflesia* (Fig. 12.2), which lives its entire life within the body of its host. Only the flower develops externally, and strangely, it is a massive flower, the biggest known in the world, 1 m in diameter.

Hemiparasites generally do photosynthesize, but they do not have a root system to draw water and so rely on their hosts for that function. Mistletoe is a hemiparasite. Hemiparasites usually have a broader range of hosts than do holoparasites, which may be confined to a single or a few host species. Being totally dependent on a host seems to

Figure 12.2 *Rafflesia arnoldii*, the world's biggest flower in Payakumbuh, Sumatra, Indonesia.

require greater specialization, in animals as well as plants. So external parasitic animals, like ticks and leeches, feed off a wider variety of hosts than do internal parasites, such as tapeworms and liver flukes.

We can define parasites that feed off one species or two or three closely related hosts as **monophagous**. **Polyphagous** species, by contrast, feed on many host species. Taking a different tack, May and Anderson (1979) distinguished parasites as *microparasites* (e.g., bacteria and viruses), which multiply within their hosts—usually within the cells—and *macroparasites* (e.g., schistosomes), which live in the host, but release infective juvenile stages outside the host's body. Usually, the host has a strong immunological response to microparasitic infections. For macroparasitic infections, however, the response is short lived, the infections tend to be persistent, and the hosts are subject to continual reinfection. Some of the differences between microparasites and macroparasites and between predators and parasitoids are summarized in Table 12.1.

Finally, we can distinguish *ectoparasites*, which live on the outside of the host's body (e.g., ticks and fleas) from *endoparasites*, which live inside the host's body (e.g., bacteria and tapeworms). Once again, however, problems of definition arise with regard to plant parasites, which seem to straddle both camps. For example, some parasitic plants, such as dodder, an orange, stringlike plant, exist partly outside of the host's body and partly inside. Outgrowths called *haustoria* penetrate inside the host plant to tap into nutrient supplies.

Despite these problems of definition, the biology of host–parasite relationships has a rich history of interesting, coevolved, and complex life cycle patterns. Many parasites parasitize several host species in their life cycle and are therefore faced with the problem of requiring transmission from one host species to the next. To overcome this problem, many induce changes in behavior or in the color of their host, making it more susceptible to being eaten by a second host, thus facilitating transmission of the parasite. Janice Moore of Colorado State University (see "Profiles" box) described many of these parasites in a recent review article (Moore, 1995). For example, adult lancet flukes (*Dicrocoelium dendriticum*) are about 1 cm long and live in cows. They spread their eggs in cow manure. Hungry snails eat the eggs, and the immature parasites hatch and infect the snails' intestines. The offspring make their way to the surface of the snail's body when the snail coughs them up in balls of sand. Ants then eat these slime balls loaded with flukes. The flukes migrate to the ants' heads and make the ants climb to the tops of blades of grass,

TABLE 12.1 Variations in some general life history traits along a parasite–predator continuum. There is a steady increase from microparasite through predator in body size, population growth rate, and fatality to host.

Trait	Microparasite	Macroparasite	Parasitoid	Predator
Body size	Microscopic	Smaller than hosts	Similar in size (e.g., wasp and moth)	Often larger than prey, unless social hunters
Intrinsic rate of population growth	Rapid	Faster than hosts	Comparable to hosts	Comparable to but often slightly slower than that of prey
Number per host	One host usually supports populations of several different species	One host supports a few to many individuals of different species	One host usually supports one, but can support many, individuals	Each predator eats many prey individuals
Fatality for host	Mild to severe	Usually mild	Eventually fatal	Immediately fatal
Stability of the interaction with host	Intermediate	High	Intermediate	Usually low
Ability to regulate host population	Moderate to high	Low	Moderate to high	High

Profiles

Janice Moore,
Colorado State University

I never intended that my graduate career would resemble a travelogue, but then again, there was a time when I never intended to pursue a graduate career at all, and maybe I should begin there. As an undergraduate at Rice University, I thought I might go to medical school, the *aspiration du jour* of many of my fellow students. As a consequence, I took Clark Read's parasitology course; it was rumored among the undergraduates that he was a heavyweight and good to have among your referees for medical school. (It turns out he was one of the great parasitologists of the 20th century, a fact pretty much lost on the premed contingent.)

Read taught a parasitology course that left me awestruck. He presented us with the puzzles that emerge when two organisms live in symbiosis and showed how the puzzles were nested in one another, from the dimension of the cell to the way predators find their prey. We studied animals that had no guts and that were covered with microvilli, which depended on their hosts to digest nutrients before taking those nutrients themselves. We studied animals that could live in water, in snail innards, and in cow livers, all in sequence, and that somehow knew where they were and what to do when they got there. Toward the end of the course, I told Dr. Read that I was especially interested in some stories he had told about parasites that made their hosts behave strangely, even to the point of getting eaten and thus transmitting the parasite. What a perfect amalgam of animal behavior, parasitology, and a touch of science fiction!

Read said that there weren't that many people in the United States doing that kind of work, and my best bet was to do a master's degree in behavior, get a Ph.D. in parasitology, and create my study as I went along. Ever an optimist, I rushed in, surrounded by a host of angels beside themselves with terror. I studied insect behavior at the University of Texas at Austin under the direction of Dr. Robert Barth and then entered a Ph.D. program at the Johns Hopkins University, a bastion of parasitological research. An inner-city university was something of a shock to a kid from Texas, and I was becoming tired of the fact that so few people seemed to care about the behavioral or ecological aspects of parasitism. I decided to leave Hopkins and work for a while.

Eventually, I returned to my graduate education—at the University of New Mexico. My graduate advisors, Don Duszynski and Rex Cates, were truly supportive and allowed me to do what I wanted. For once, no one was trying to get me to do "mainstream" ecology. Moreover, they literally put their money where their mouths were, with in-house grants that enabled graduate students to pursue independent lines of research. So I asked if a parasitic worm could alter the behavior of its isopod host in a way that would get it transmitted (eaten). I found that parasitized isopods behave strangely, in ways that probably increase encounters with birds, and that starlings feed their offspring more parasitized isopods than you would expect from random foraging. Talks with Dan Simberloff about the potential that parasite communities have for replication and experimental manipulation resulted in a postdoctoral collaboration that led me into the world of community ecology, with every host harboring a complete community.

My work since then has focused on various aspects of parasite-induced behavioral alterations and parasite community ecology. Looking at parasites through the eyes of an ecologist presents no small problem, especially if one focuses on the parasites themselves and not on parasites as agents of doom for what some folks see as the main event: the hosts. My position doesn't fit most job descriptions. I was happy to get a faculty position as an "invertebrate zoologist" at Colorado State University—that's a reasonable statement of what I do, if anything is. And funding can be a challenge, as it is preferred that one fit neat categories in that realm, too. Given these and other frustrations, why not simply find a neat category and fit in?

I imagine that most folks who work in the interstices of biology might have an answer similar to mine. No one is forcing us to do this. We like it. We like it a whole lot. And it allows us to fill in gaps that should never have been there in the first place, but for accidents of history and curricula. Working at the margins of ecology, behavior, neurophysiology, epidemiology . . . this gives me a world of colleagues, from vector biologists to theoreticians, that redefines the notion of intellectual diversity. Even my graduate school peregrinations were hardly the misfortune they may initially seem. The experiences and the people they brought into my life are without price.

where they are eaten by cows to start the cycle over again (Fig. 12.3).

If these life cycles seem bizarre, bear in mind that parasitism might well be the predominant lifestyle on earth. There are vast numbers of species of parasites, including those of interest to the conventional parasitologist—viruses, bacteria, protozoa, flatworms (flukes and tapeworms), thorny-headed worms (phylum Acanthocephala), nematodes, and various arthropods (ticks, mites, and fleas). There are more insects on earth than any other taxa, and at least 50% of these are parasitic. When the other large groups of parasites are considered—nematodes, fungi, viruses, and bacteria—it is clear that parasitism is a common way of life. Indeed, parasites may outnumber free-living species by four to one. Most plant and animal species harbor many parasites. For example, in the United States, leopard frogs have nematodes in their ears, filarial worms in their veins, and flukes in their bladders, kidneys, and intestines. A species of Mexican parrot carries 30 different species of mites on its feathers alone. A free-living organism that does not harbor several parasitic individuals of a number of species is a rarity (Fig. 12.4). The frequency of human infection by parasites is staggering and is testament to the importance of parasites in our life cycles. (see "Applied Ecology: Diseases and the World's Top Ten Killers.")

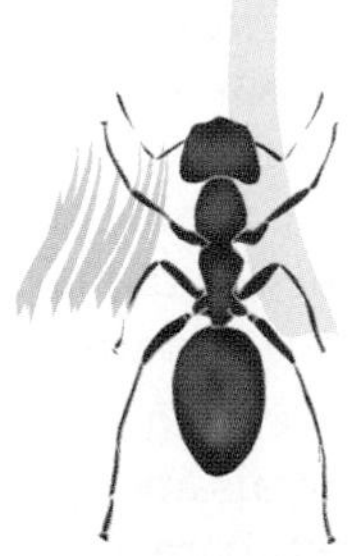

Adult flukes produce eggs inside a cow.
The eggs are passed in the cow's feces.

Ants eat the "slime balls."
Some of the flukes migrate into
the ant's brain, causing it to climb
to the tip of a blade of grass where
it can be eaten by a cow.

Snails eat the fluke eggs; later the
eggs hatch in the snail's intestine.
The eggs hatch and asexually
produce offspring.
The offspring are passed from the
snail in "slime balls."

Figure 12.3 The life history of the lancet fluke (*Dicroelium dendriticum*) is complex and involves behavioral changes in many of its intermediate hosts.

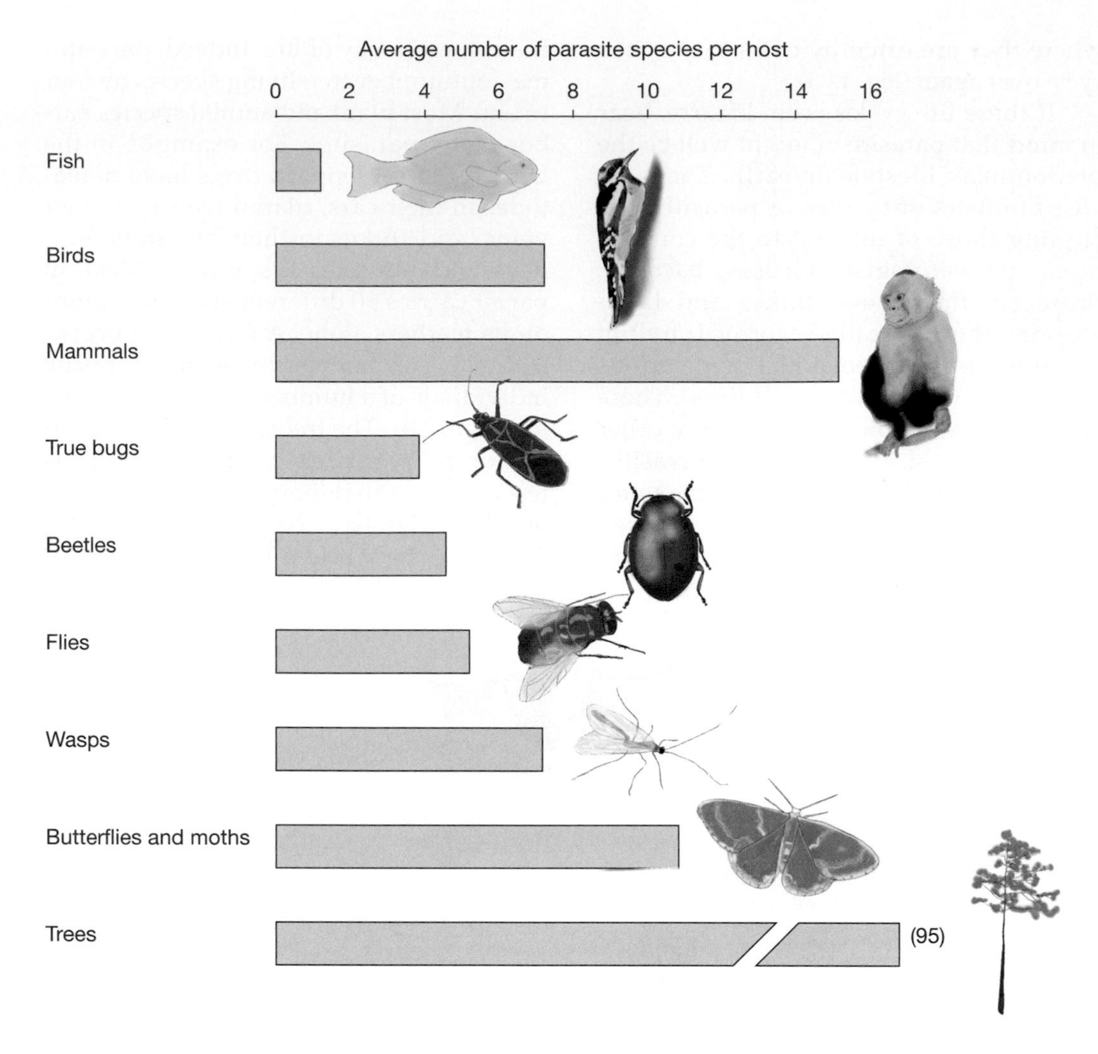

Figure 12.4 Average number of parasite species per host species in Britain. Totals for mammals, fish, and birds are only for helminth parasites and are likely an underestimate of total parasite species richness. Note that most taxa have many parasites per species.

12.2 Defense against Parasites

Hosts have developed a variety of defenses against parasite attack, including immune responses, defensive displays, and grooming behavior.

The presence of a parasite in a host usually causes a defensive reaction. If it did not, we would term the relationship commensal. The defensive reactions developed by hosts to resist parasites are impressive:

- *Cellular defense reactions*. These reactions are found particularly in insect larvae as a defense against parasitoids. The eggs of the parasitoid are rendered inviable by "encapsulating" them, or enclosing them in a tough case.
- *Immune responses in vertebrates*. These responses are the vertebrate body's defense against the parasitic microbes that cause disease in humans and other animals. Phagocytes may engulf and digest small alien bodies and encapsulate and isolate larger ones. With regard to microparasites, the host may develop a "memory" that may make it immune to reinfection. In mammals, the transmission of immunoglobulins to the young may even confer protection for more than one generation.
- *Defensive displays or maneuvers*. These actions are intended to deter parasites or to carry organisms away from them. For example, gypsy moth pupae spin violently within their cocoons to deter pupal parasites, and syrphid larvae often drop from the foliage they forage on to the ground to escape parasites.
- *Grooming and preening behavior*. This behavior, found in mammals and birds, is intended to remove ectoparasites.

Despite these defensive reactions, many hosts still die. Some parasites, such as fungi, can live saprophytically on the dead hosts; we term these parasites *necrotrophic*. Most, however, can live only on live hosts, and these are termed *biotrophic* parasites. A biotrophic parasite therefore keeps its host alive. This is why egg production and the mean weight of live parasites of many vertebrates, such as roundworms, level off when they reaches a certain level. Too many parasites would kill the host.

Applied Ecology

Diseases and the World's Top 10 Killers

If you had any doubt that parasites are important to plants and animals, a few focused figures should change your mind. First, there are five main categories of disease-causing organism that most plants and animals are susceptible to:

Class of Infectious Agent	Life History	Examples
Virus	DNA or RNA core with protein coat. No means of replication other than by host's genetic machinery.	Measles, AIDS, influenza, dengue fever, hepatitis B; 15–20% of cancers may be of viral origin. Plants are highly susceptible.
Bacterium	Cytoplasm surrounded by membrane, no nucleus.	Most abundant organisms on Earth. Animals more susceptible than plants. Tetanus, tuberculosis, diarrhea, whooping cough.
Fungus	Single-celled or multicelled long filaments called hyphae.	Yeasts, molds, mildew. Much more important for plants than animals. Dutch elm disease, potato blight, root rot. Chemicals produced to outcompete bacteria were used to produce penicillin.
Prion	Abnormal forms of proteins usually found in vertebrate brains.	Scrapie in sheep and goats and its variant, "mad cow disease" in cattle, Creutzfeldt–Jakob disease in humans.
Protist	Single-celled organisms, protozoa, with nucleus, on average 1,000 times bigger than a bacterium.	Many feed on smaller microbes; others however, are parasitic. Leishmania is transmitted to humans by sand flies and trypanosomes to humans and livestock by mosquitoes.

(Continued on next page)

Second, despite modern antibiotics and other technological advances, the number of humans killed by disease is staggering. Undoubtedly, plants and other animals suffer huge losses, too. The following are the leading causes of death by disease worldwide:

Disease	Symptoms	Worldwide Deaths, 1997 Data
Haemophilus influenzae	Influenza (respiratory infection)	3.7 million.
Mycobacterium tuberculosis	Tuberculosis	2.9 million.
Vibrio cholerae	Cholera (diarrheal disease)	2.5 million.
Human immunodeficiency virus	AIDS	2.3 million, two-thirds in sub-Saharan Africa. Thirty million people thought to be infected.
Plasmodium falciparum	Malaria	2.6 million, borne by *Anopheles* mosquitoes. May in sub-Saharan Africa. May infect 500 million people per year.
Morbillivirus	Measles	1.0 million, especially children, highest incidence in Africa.
Hepatitis B virus	Hepatitis B	0.6 million.
Bordetella pertussis	Whooping cough	0.4 million.
Clostridium tetani	Tetanus	0.3 million.
Falvirus	Dengue fever	0.15 million. Transmitted by *Aedes* mosquitoes. Asia, Latin America.

12.3 Modeling Parasitism

The effect of a parasite on its hosts is a function of the density of susceptible hosts, the transmission rate of the disease, and the fate of the infected hosts.

Models of the effect of parasites on host populations differ from models of other phenomena such as predation or herbivory, for two reasons. First, the life cycle of many parasites involves intermediate hosts. Modeling parasite transmission and the various types of host is extremely complex, involving the densities of many hosts and parasites at several different stages. Such models are outside the scope of this book. Second, with respect to diseases, models of parasite population dynamics generally describe the parasite's population growth rate by the average number of new cases of a disease that arise from each infected host, rather than by the net reproductive output of the parasite. The reason is that, for microparasites, the number of *infected hosts* is the most important factor, not the number of parasites. We can refer to the number of infected hosts as R_p, with p for parasite and R for net reproductive rate. The transmission threshold, which must be exceeded if a disease is to spread, is therefore given by the condition $R_p = 1$. For a disease to spread, R_p must be greater than 1, and for a disease to die out, R_p must be less than 1. For microparasites that are transmitted directly from host to host, R_p is influenced by

N, the density of susceptible hosts in the population;

B, the transmission rate of the disease (a quantity correlated with frequency of contact among hosts and the infectiousness of the disease); and

L, the average period over which the infected host remains infectious.

Specifically,

$$R_p = NBL.$$

Three generalizations can thus be made. First, as L, the period during which the host is infectious, increases, R_p increases. An efficient parasite therefore keeps its host alive for a long time. In the extreme, some hosts remain infectious after they are dead (e.g., some plant parasites leave a residue of resting spores on the dead plant). Second, if diseases are highly infectious (i.e., have large B's), R_p increases. Third, large populations of susceptible hosts (large N's) promote the spread of disease. By rearranging the equation, we can obtain the critical threshold density N_T, such that $R_p = 1$. N_T is an estimate of the number of susceptible hosts needed to maintain the parasite population at a constant size:

$$N_T = \frac{1}{BL}.$$

If B or L is large, N_T is small. For example, if $B = 0.1$ and $L = 0.01$ (a value considered to be large), then $N_T = 1/(0.1 \times 0.01) =$

$1/0.001 = 1{,}000$. Conversely, if B or L is small, the disease can persist only in a large population of infected hosts. For instance, If $B = 0.1$ and $L = 0.0001$, then $N_T = 100{,}000$. In humans, measles has a threshold density of about 300,000 individuals.

Cockburn (1971) cited some interesting medical and anthropological evidence to back up these ideas about transmission rates and host population size, at least in humans. Measles, rubella, smallpox, mumps, cholera, and chicken pox, for example, probably did not exist in ancient times, because the hunter–gatherer populations were small—bands of 200 to 300 persons at most. In a small population, there would be no infections like measles, which spreads rapidly and immunizes a majority of the population in one epidemic. Measles occurs endemically only in human populations larger than 500,000. In ancient times, typhoid, amoebic dysentery, pinta, trachoma, and leprosy—diseases whereby the host remained infective for long periods—were probably the most common afflictions. Malaria and schistosomiasis would also have been quite prevalent, because of the presence of outside vectors to serve as additional reservoirs. Human hookworm and roundworm can produce vast numbers of eggs per day—15,000 and 200,000, respectively—so these, too, existed in very low-density human populations. Paradoxically, civilization has increased the kinds and frequencies of diseases suffered by humans by enlarging the host pools of humans and by domesticating certain animals. Many modern diseases have arisen because of human beings' intimate association with animals and their viruses. Smallpox, for example, is similar to the cowpox virus, measles belongs to the group containing dog distemper and cattle rinderpest, and human influenza viruses are closely related to those found in hogs. AIDS is similar to a virus found in monkeys in Africa.

Many diseases appear to undergo periodic cycles. The best records are from humans and show two-year peaks for cases of measles in England (Fig. 12.5) and four-year peaks for whooping cough. Such peaks occur because host immunity is developed, which reduces the density of susceptible hosts, N. However, new births lead to fresh susceptible hosts entering the population and a concomitant increase in disease. Thus, there are parasite–host cycles just as there are predator–prey cycles. Such cycles undoubtedly occur as well in populations other than humans.

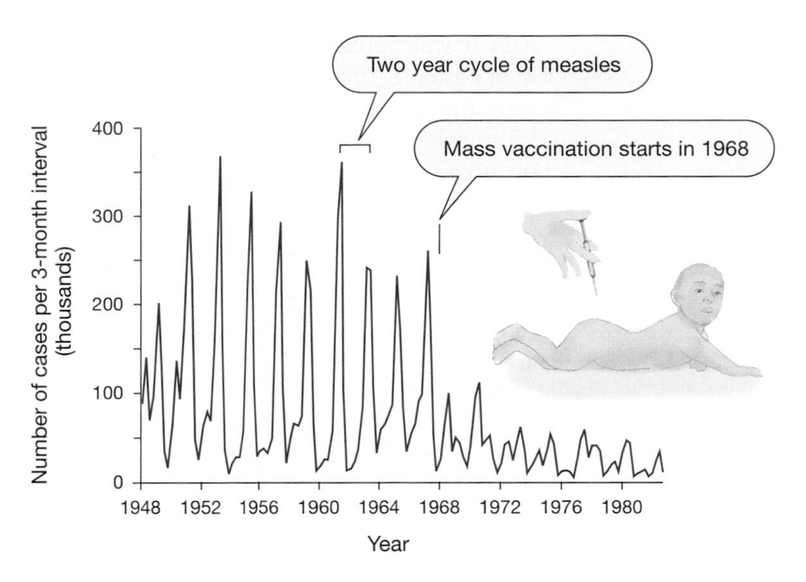

Figure 12.5 Cases of measles in England and Wales, 1948–1982. A two-year cycle is clearly evident, based upon hosts developing immunity and new susceptibles being born and entering the population. Parasite–host cycles are as common as predator–prey cycles. Note the higher incidence of measles prior to the era of mass vaccination in 1968. *(Reproduced from Anderson and May, 1991.)*

As regards parasites that are spread from one host to another by a vector (e.g., an insect), the life cycle characteristics of both host and vectors become important in the calculation of R_p. Hence, disease control measures sometimes aim directly at reducing the numbers of vectors and only indirectly at the parasite. For example, farmers use insecticides to kill aphids, which transmit viral diseases of crops, rather than direct chemicals directly at the parasite. Of course, this is not always the case: Yellow fever was eradicated in the United States by inoculation rather than the eradication of all mosquitoes.

12.4 Parasites Affect Host Populations

Parasites can dramatically decrease the size of host populations.

As with studies of predation, the best way to determine the effect of parasites on the population sizes of their hosts is to remove the parasites and reexamine the system. However, this has rarely been done with parasites, probably because of their small size and unusual life histories, which make it difficult to completely remove them from a system. Furthermore, dead hosts are often difficult to find, which means that it is not easy to determine whether a host was killed by a parasite. In addition, parasites may impair the health of their hosts rather than killing them, making the effects of parasites on host populations even more difficult to gauge. What other lines of evidence can we turn to? Inoculations, which combat disease, are known to save many human lives and to increase the length of human beings' lives. Inoculations do the same for domestic animals. Agricultural sprays reduce crop losses to disease. However, at least in humans, the effects of other factors, such as better diets and safer environments, are difficult to separate out from effects that reduce the incidence of disease in terms of their contribution to increasing longevity and population density.

Biological control seems like a promising area for studying the effect of parasites on their hosts. Brad Hawkins at the University of California at Irvine and his colleagues (1999) showed that biological control of agricultural pests, especially by parasitoids, was substantially greater in exotic, simplified, managed habitats than in natural habitats. Most often, control was exerted by a single parasitoid species, in contrast to natural systems, in which control involves a suite of generalist enemies. Thus, we cannot easily cite biocontrol projects as providing rigorous evidence that parasites are important in natural systems. What we need is evidence from populations in just those systems.

Evidence from natural populations suggests that *introduced* parasites do have substantial impacts on their hosts. In the Appalachian Mountains of North America, chestnut blight, a fungus, has virtually eliminated chestnut trees (Fig. 12.6). Most of the forests of the region consisted essentially of oak and chestnut trees. The disease was introduced from China to New York on nursery stock in about 1904, although its effects were not severe on Chinese chestnuts. In Europe and North America, Dutch elm disease has devastated elms. Twenty-five million of Britain's original 30 million elm trees were wiped out by the disease between the 1960s and the 1990s (Fig. 12.7). In Italy, canker has had severe effects on cypress.

Rinderpest, caused by a virus, has at least 47 natural artiodactyl hosts, most of which occur in Africa. The virus belongs to a class known as *morbilliviruses*, which include the viruses that cause measles and distemper. Rinderpest is spread by food or water contaminated by the dung of sick animals. Wildlife get the disease from cattle. The disease is almost always fatal in buffalo, eland, kudu, and warthogs, less so in bushpigs, giraffe, and wildebeest. Other species, such as impala, gazelle, and hippopotamuses, appear to suffer little. A major epidemic swept through Africa in the 1890s (probably from cattle imported into Somalia from India by an Italian military expedition), leaving vast areas uninhabited by certain species. More than 80% of the hoofed stock died over the entire continent. The disease traveled 5,000 km in eight years, arriving in North Africa in 1889 and reaching the Cape of Good Hope in 1897. Rinderpest was brought under control in the 1960s largely through cattle vaccination programs, but even in the 1970s distribution patterns of animals reflected the impact of the disease. For example, zebra were exterminated in an area of the Elizabeth National Park in Ugan-

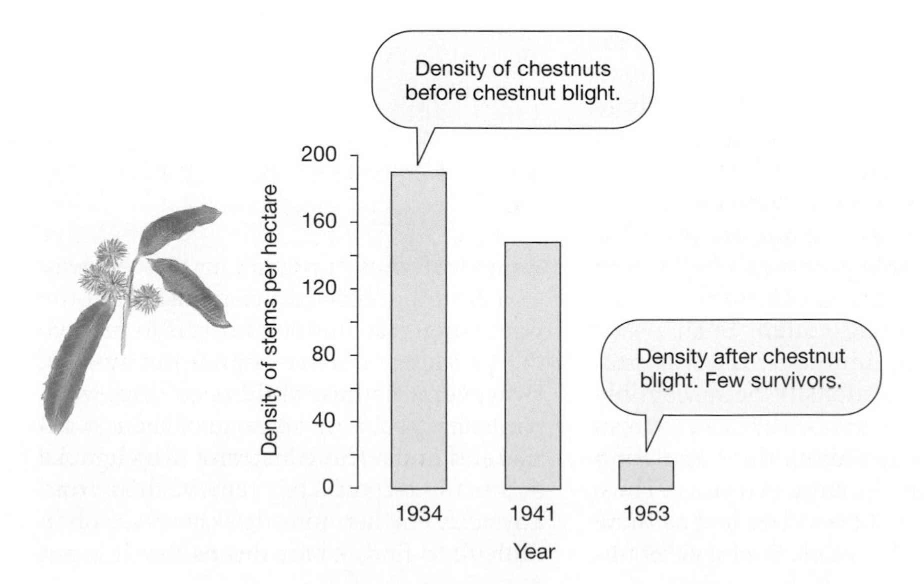

Figure 12.6 The reduction in density of American chestnuts in North Carolina forests was swift, occurring in less than 20 years. *(From data in Nelson, 1955.)*

Figure 12.7 Dutch elm disease transmitted by bark beetles, kills elm trees that it infests.

da, and they still had not recolonized the area by 1954. Because wildlife was eliminated, other parasites were affected, too. Tsetse flies became absent from large areas of Africa south of the Zambesi River. Therefore, large areas became free from trypanosomiasis (sleeping sickness), a disease borne only by the tsetse fly. One parasite, rinderpest, thus had a severe impact on the pattern of life in an area. Furthermore, in the absence of tsetse flies, humans and cattle could move in, supplanting wildlife even further. Another rinderpest outbreak occurred in Africa in the 1990s (Packer, 1997), and since 1993 some national parks in Kenya have lost 90 percent of their buffalo, antelopes, gazelle, and wildebeest. As a result, some parasitologists think that epidemics have a more severe impact on wildlife populations than does predation (Prins and Weyerhaeuser, 1987). This is a disconcerting finding for conservation biology, for it means that certain populations on small reserves could be wiped out by disease unless recolonization is encouraged.

More recent examples of the effects of introduced disease abound. In the early 1980s, a canine distemper virus caused a large decline (more than 70%) in the last remaining population of black-footed ferrets, and a few years later another outbreak exploded among seals and dolphins. In fact, over the 20-year period from 1980 to 1999, 27 mass mortalities (greater than 10% mortality within populations) have been documented in marine systems and are attributed to outbreaks of pathogens in organisms as diverse as corals, urchins, abalones, scallops, clams, seagrass, seals, dolphins, and herring (Harvell *et al.*, 1999). Terrestrial examples abound as well: In 1994, one thousand lions, one-third of the resident Serengeti National Park population, were wiped out by canine distemper virus, probably transmitted from domestic dogs.

Particularly worrisome for conservationists is the fact that many endangered animals are threatened by disease from domestic animals (Table 12.2). Guiler (1961) suggested that the demise of the thylacine (marsupial wolf) in Tasmania was because of a distemper-like disease brought about by close association with dogs. To prevent the threat of disease, some populations of endangered species have been vaccinated—for example, mountain gorillas against measles and African hunting dogs against rabies. Yet it would be very difficult to vaccinate more than a few species of wild animals against more than one or two diseases.

TABLE 12.2 The threat of disease to endangered wildlife from domestic animals.

Wildlife	Disease	From	Size of Effect
Lion	Canine distemper virus	Dogs	30% killed
Black-footed ferret	Canine distemper virus	?	70% killed
Ethiopian wolf	Rabies	Dogs	50% killed
Blanford's fox	Rabies	Dogs	Eventual extinction a possibility
Cheetah	Feline infectious peritonitis	?	Strong
Iriomole cat (Japan)	Feline immunodeficiency virus	Cats	None as yet, but threat remains strong
Serengeti wild dog	Rabies	Dogs	Extinction threatened
Mednyi Arctic fox (Aleutian Islands)	Mange	Dogs	90% killed

Data from natural systems of native hosts and parasites show that parasites exact a heavy mortality on their hosts.

Natural systems.

What of the effect of native parasites on native populations? The population dynamics of bighorn sheep in North America are dominated by a massive mortality resulting from infection by the lungworms *Protostrongylus stilesi* and *P. rushi*. These parasites predispose the animals to pathogens causing pneumonia. A fetus can become infected through the mother's placenta, and mortality in lambs can be enormous. The lungworm-pneumonia complex is regarded as one of the most influential mortality factors in many sheep populations, with infection rates of 91% and mortalities of 50 to 75% reported.

Plants, too, can be affected by parasitic plants, of which there are over 5 thousand species worldwide. For example, in Colorado pine tree plantations, mistletoe can cause 30% losses of extractable timber. *Cuscuta salina* (marsh dodder) (Fig. 12.8a) is a common and widespread plant parasite in saline conditions among marshes of North America. Steve Pennings and Ragan Callaway (1996) showed how *Cuscuta* preferentially infects the most common plant in California marshes, *Salicornia virginica*, thus

(a)

(b)

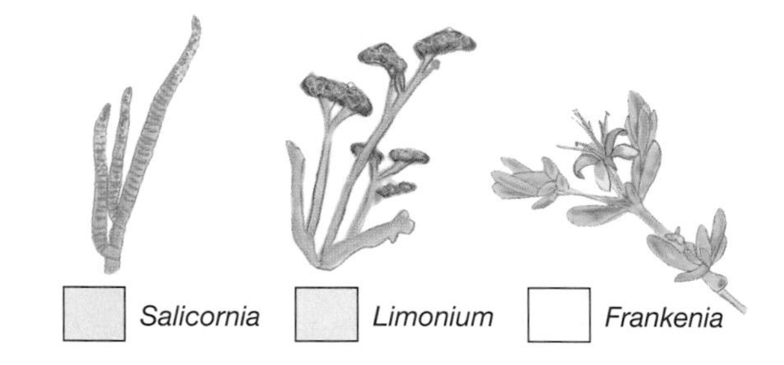

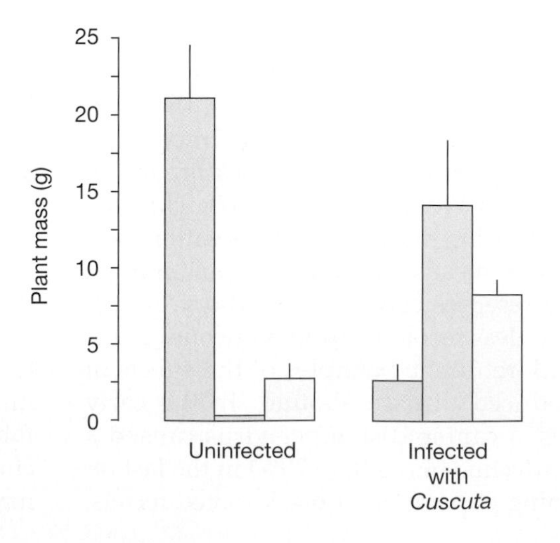

Figure 12.8 (a) Dodder, *Cuscuta spp*, can reduce the population density of many plant species, allowing other species to proliferate. Here dodder is entwined around host plants in Pennsylvania. (b) Biomass of three host plants—*Salicornia*, *Limonium*, and *Frankenia*—in California salt marshes in areas infected and uninfected by the parasite *Cuscuta*. The parasite preferentially attacks *Salicornia*, reducing its density and allowing *Limonium* and *Frankenia* to proliferate. (*Modified from Pennings and Callaway, 1996.*)

promoting the growth of two other plants, *Limonium* and *Frankenia* (Fig. 12.8b). Thus, parasitic plants can have strong effects on the relative abundance of species and, consequently, on plant community structure.

While observational data on parasite loads and host fitness provide compelling evidence of the strong effects of parasites on hosts, removing parasites in field-controlled experiments remains the ideal way to examine negative effects on host fitness. Three recent removal experiments provide evidence of such effects.

Claire Fuller and Andrew Blaustein from Oregon State University (1996) were able to compare the survivorship of parasite-infected and uninfected free-living deer mice in large outdoor enclosures. The authors found a decreased overwinter survivorship of male deer mice experimentally infected with the protozoan *Eimeria arizonensis*, compared with the survivorship of noninfected individuals. These parasites are transmitted through the ingestion of contaminated feces, making inoculation with parasite oocysts relatively easy.

Sylvie Hurtrez-Bousses and colleagues (1997) experimentally reduced the number of blowfly larvae parasites of young blue tits in nests in Corsica. The reduction was cleverly achieved by taking the nests from 145 nest boxes, removing the young, microwaving the nests to kill the parasites, and then returning the nests and chicks to the wild. Chicks from microwaved (parasite-free) nests were then compared with chicks from nonmicrowaved (control) nests, and the former were found to have greater body mass at fledging (Fig. 12.9). Perhaps more important was that complete nest failure was much higher in the control than in the treated nests (Fig. 12.9). By sucking the chicks blood, blowflies cause anemia and high mortality.

Stiling and Rossi (1997) were able to manipulate levels of parasitic infection of a gall-making fly on a coastal plant, *Borrichia frutescens*, on isolated offshore islands in Florida. To get gallfly populations with low rates of parasitism, they allowed potted plants on one island to be colonized first by gallflies. They then removed these plants before the parasitoids could find them (the low-parasitism treatment). To get gallfly populations with high rates of parasitism, they left some plants on the island longer, to allow the parasites to colonize the galls. Using these plants, the authors established replicates of both the high- and the low-parasitism treatment on other islands. Where parasitism of gallflies was high, numbers of new galls were significantly lower than where levels of parasitism were low (Fig. 12.10).

A common belief among parasitologists is that the degree of mortality the parasite inflicts on the host depends on the age of the association between the two. Older associations are thought to cause less harm and may even evolve toward commensalisms or mutualisms. For example, the myxoma virus produces a mild, nonlethal

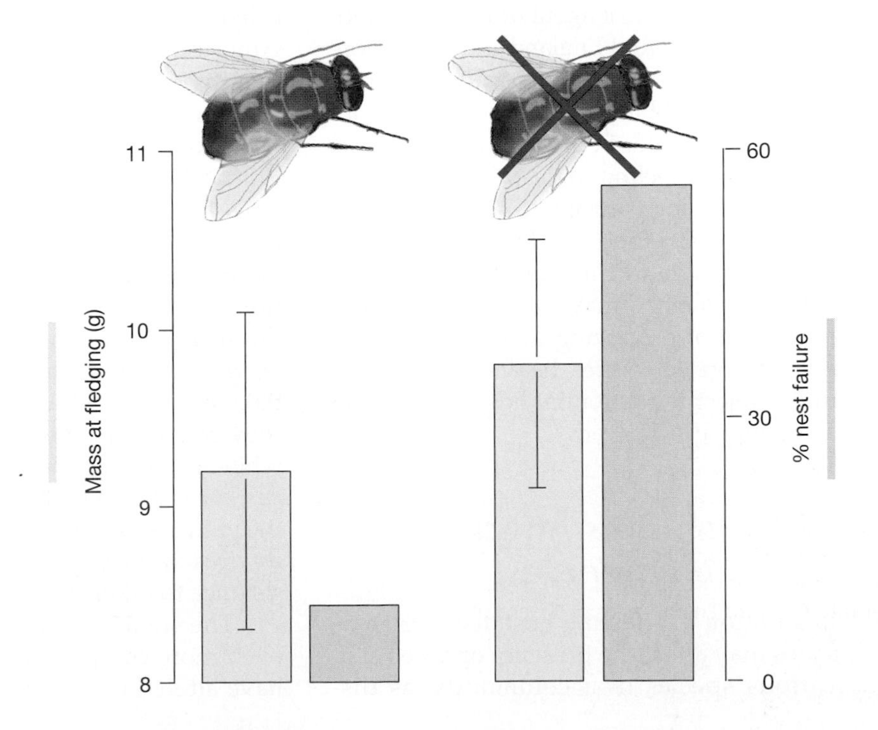

Figure 12.9 The effect of blowfly parasites on blue tit nesting success and on chick weight at fledging. *(From data in Hurtrez-Bousses et al., 1997.)*

Figure 12.10 Abundance of new insect galls of *Asphondylia borrichiae* on experimental plots during 1995 in treatments with low densities of parasitoids and high densities of parasitoids. Abundance on both types of plot tended to decrease dramatically in October and November as parasitism in both areas increased. Data are means + 1 standard error. (*Reproduced from Stiling and Rossi, 1997.*) Photo of gall on *Borrichia frutescens* is by P. Stiling.

disease in its natural host, the South America cottontail rabbit. In the 1950s, this virus was used as a biocontrol agent of European rabbits in Australia and England and killed more than 99% of the rabbits. Eventually, of course, the resistant rabbits reproduced and the population recovered. Because of this outcome, there is some feeling in the biocontrol world that introducing parasites to novel hosts may be a valuable method to help control the hosts (Hokkamen and Pimentel, 1984). Unfortunately, the method may also have the highest likelihood of producing undesirable effects on nontargeted species, because the parasite might reject the potential host and feed on something else instead.

12.5 Parasites Affect Communities

Parasites may affect whole communities by causing the demise of one species and so indirectly promoting another.

In addition to affecting host densities, parasites may affect the presence or absence of various species in a community, as discussed earlier with regard to the parasitic plant dodder in California salt marshes. Another interesting situation exists in North America as well: The usual host of the meningeal brainworm *Parelaphostrongylus tenuis* is the white-tailed deer (*Odocoileus virginianus*), which is tolerant to the infection. All other cervids and the pronghorn antelope are, however, potential hosts, and in these species the worm causes severe neurological damage, even when very small numbers of the nematode are present in the brain. This differential pathogenicity of *P. tenuis* makes the white-tailed deer a potential competitor with other cervids, because they cannot survive in the same area with whitetails. The phenomenon is known as *apparent competition*. The deleterious effects of the parasite probably include direct mortality, increased predation, and reduced resistance to other diseases.

The situation with while-tailed deer is even more complex now, because humans have altered the animal's normal distribution pattern. As northern forests were felled, the deer expanded their range from a stronghold in the eastern United States, eventually coming into contact with moose. In Maine and Nova Scotia, white-tailed deer have replaced moose as the major cervid, and they have also replaced mule deer and woodland caribou in some parts of their ranges. Whether the reintroduction of caribou into regions now occupied by white-tailed deer is possible is debatable, because of the action of the parasite (Schmitz and Nudds, 1994). So, apart from parasites directly causing mortality, competitive interactions between populations can be mediated by the action of parasites. This phenomenon is similar to that discussed in Chapter 9, in which Park (1948) found that competing populations of flour beetles had different winners in trials with and without the parasite *Adelina triboli.* A similar example exists in the Caribbean, where two *Anolis* lizard species compete on the island of St. Maarten. *Anolis gingivinus* outcompetes *A. wattsi* on most of the island, except where a malarial parasite is present. This is because the parasite attacks *A. gingivinus* and only rarely *A. wattsi* (Schall, 1992). Clearly, parasites can have direct effects on the species they parasitize and indirect effects on other species.

12.6 Parasites and Biological Control

Not all parasites are seen as detrimental by humans: Many are used as an effective line of defense against insect pests of crops. This usage has spurred the scientific community into a search for more effective biological control agents.

Only about 16 percent of classical biological control introductions attempted so far qualify as economic successes (Hall, Ehler, and Bisabri-Ershadi, 1980). Usually, organisms are released in a hit-or-miss technique. Some authors believe that this approach probably makes the best economic sense, given the high cost of research into the biology of natural enemies (van Lenteren, 1980). Just release a bunch of parasitoids and predators, and hope that one of them does the job. Others have recommended new techniques—for example, novel parasite–host associations—as the most likely avenue for control in situations where hosts have not had the opportunity to evolve complex defenses against these parasites from foreign lands (Hokkanen and Pimentel, 1984). Arguments still rage as to whether it is better to introduce one parasite at a time or many. The problem is that if more than one enemy is introduced, competition between parasites could ensue, lessening the overall level of control. As we saw when we discussed competitive interactions, a review of 548 control projects around the globe showed that the more parasites were released, the lower was the rate of their establishment (Ehler and Hall, 1982).

Other authors have tried to find what attributes make a successful biocontrol agent. Carl Huffaker and Charles Kennett from Berkeley (1969) suggested five necessary attributes of a good agent of biological control:

- General adaptability to the environment and host.
- High searching capacity.
- High rate of increase relative to the host's.
- General mobility adequate for dispersal.
- Minimal time lag effects in responding to changes in host numbers.

Stiling (1990) reviewed the methods affecting success in biological control. The factor of greatest importance was the climatic match between the control agent's locality of origin and the region where it was to be released. This finding emphasizes the value of studies in physiological ecology and shows that climatic variation is of vital importance in affecting biotic relations, as Huffaker and Kennett had suggested. The importance of climatic variation was underscored by a separate analysis of reasons for biological failures (Stiling, 1993) which found that reasons for failure related to climate (34.5%) were more common than any other type of reason, including competition or parasitism by native insects (Fig. 12.11).

Parasites and parasitoids are being used increasingly to control insect pests.

It is important to note that biological control is not the risk-free alternative to chemical control that it is often touted to be, because of effects on nontargeted species. As early as 1983, Frank Howarth, a longtime entomologist in Hawaii, lamented the reduction in numbers of native Hawaiian lepidopterans, which he thought was partly due to wasp species introduced for biological control of lepidopteran crop pests. The wasps had attacked not only the target exotic pest, but also nontarget native species. Howarth called for a more narrowly focused release effort rather than a hit-or-miss

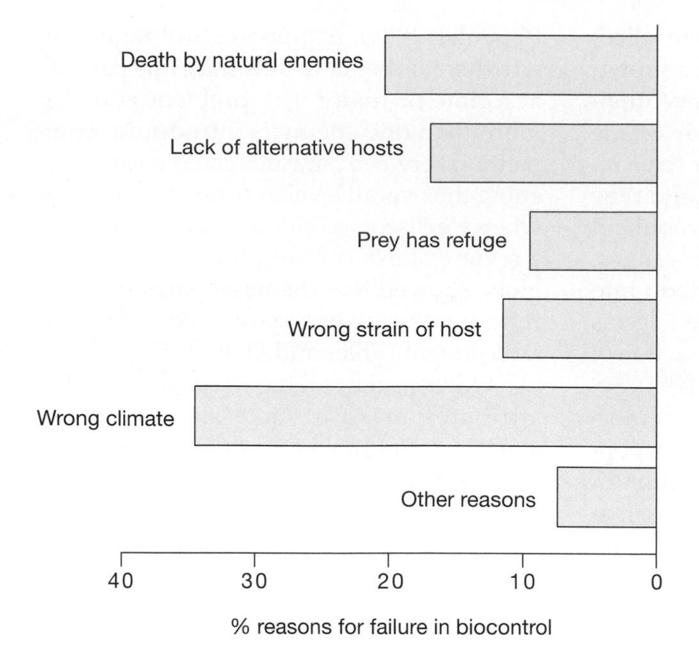

Figure 12.11 Reasons for failure in biocontrol. The most frequent reason was unsuitable climate—that is, a mismatch of the climate in the biocontrol agent's native region with the climate in the area in which it was released. *(From data in Stiling, 1993.)*

campaign—the opposite of what van Lenteren was arguing a few years earlier. Such a concern is even more important when the release of insect enemies to control weeds is considered. In this case, stringent host-specificity tests are performed to ensure that introduced insects will not turn to feeding on valuable crops, even in times of starvation. However, noncrop host plants are not tested as rigorously. Unfortunately, there are many more examples of the effects of biological control agents on nontarget species (Simberloff and Stiling, 1996), and still others surface all the time. Many biological control practitioners, feeling that they have the environmental moral high ground, are upset that they should be tarred with the same brush as the chemical control crowd. However, eventually chemicals will degrade, whereas biological control agents are essentially uncontrollable. Once released, they are forever in the environment, so some people argue that biocontrol should be employed only as a last-ditch attempt at pest control.

The release of microparasites is a subject of even more concern. Australians use viruses to control exotic pest mammals. When the British colonized Australia, they brought with them the English rabbit and the European red fox, both in the mid-1800s. The rabbits were for food, and the fox was to be hunted. Not long after these introductions, people realized that they had made a mistake: The rabbits began breeding like the proverbial rabbits, and the predatory foxes turned their attention to the native animals. So far, foxes have been implicated in the extinction of 20 species of local marsupials (Morrell, 1993). The Cooperative Center for Biological Control of Vertebrates, a government and university consortium, released a rabbit calicivirus in the Fall of 1995. The results were staggering: Rabbit numbers dropped 95 percent, and huge increases were noted in kangaroos, in other animals that competed for vegetation with rabbits, and in the vegetation itself, with a resurgence of many rare plants as well. Since the rabbits are the main item on the foxes' and feral cats' menu, a pleasant side effect was a population crash in the numbers of those species, too! However, some scientists are concerned that the fox and feral cat might now be forced to seek alternative food—the likely choices being small endangered marsupials. In addition, fears are high in some circles that the calicivirus might attack other species or evolve to do so in the future.

In the end, it is apparent that parasitism is a very important mortality factor. Animals and plants have evolved numerous adaptations against parasites, such as the vertebrate immune response, and introduced parasites can have huge effects on native populations. Parasite removal experiments have not been attempted on any regular basis because of the difficulty of obtaining and maintaining parasite-free host populations. However, the few experiments that have succeeded have shown strong effects. Clearly, parasites are as potent an ecological force as predators or herbivores, and, like those two groups, they can have dramatic effects on communities.

Summary

Are parasites important to plant and animal populations? We can't see many small parasites in the field, so how do we know that they have any effect?

1. The true definition of a parasite is problematic. Parasites may include many species that feed on plants, as well as more "traditional" parasites like tapeworms, leeches, bacteria, viruses, and parasitoids that attack animals. Parasitism is undoubtedly an extremely common way of life, with perhaps 80% of all life-forms considered parasitic.

2. The presence of various defenses against parasites, such as the immune response in vertebrates, is testament to the importance of parasitism in nature.

3. Mathematical models suggest that efficient parasites are likely to keep their hosts alive as long as possible to facilitate the transmission of the parasites to other hosts. Among the diseases that fit this description are leprosy, typhoid, and amoebic dysentery. Because human populations are now so large, diseases that were not very common, such as measles and cholera, now flourish with so many potential hosts available. Finally, the intimate association of humans with animals has allowed animal viruses to cross over to people. Smallpox in humans is derived from cowpox in cattle.

4. The huge influence of introduced diseases, such as chestnut blight and Dutch elm disease in America and rinderpest in Africa, are testament to the severe effects parasites can have on host populations, sometimes driving them close to extinction. More recent evidence from natural populations also indicates that parasites can have severe effects on host populations and host densities.

5. Parasites of insects can often be used as control mechanisms against agricultural and forestry pests. This technique is called biological control. Finding the attributes of successful biological control agents is a valuable ecological endeavor. Unfortunately, effects of biological control agents on nontarget natural populations are being increasingly found.

Parasitism is frequent in nature, with most species harboring several parasites. The effects are strong, and indeed, many rare species may now be threatened by the spread of disease from domestic animals.

Discussion Questions

1. Outline some of the main differences between microparasites, macroparasites, parasitoids, predators, and herbivores in terms of their life history strategies.

2. Why can we eradicate some diseases, such as yellow fever, through vaccinations, while we have not been able to eradicate other diseases, such as malaria?

3. Can we ever expect chemical pesticides to be replaced entirely by biological control agents? If not, why not?

4. Parasites are usually very small organisms, capable of passing through screen enclosures. With this in mind, how might you design an experiment to remove parasites from a population and compare that population's survival rates with those of an unmanipulated control population?

5. Do you think that biological control would be more successful on islands or on mainland continental areas? Against native or exotic pests? Information found in Ehler and Hall (1982), Hall, Ehler, and Bisabri-Ershadi (1980), and Debach and Rosen (1990) might help.

6. Look up and report on the life histories of four species of common parasite, such as brainworms, schistosomes, malaria, and tapeworms. Does the information you have learned give you a clue as to what the best control strategy might be?

CHAPTER 13

Evaluating the Controls on Population Size

Road Map

1. Mortalities that perturb population densities away from mean or equilibrium values are known as key factors. Among the many key factors are predation, competition, parasitism, and herbivory.
2. Mortalities that act to return population densities to mean or equilibrium levels are known as density-dependent factors. As with key factors, there are many different types of density-dependent factors.
3. Groups of populations may exist with many individuals intermixing between the groups. A collection of such groups is known as a metapopulation, and the dispersal rate, not mortality, is key to understanding its group dynamics.
4. Simple theoretical models that use only the productivity and the number of trophic levels in a system have been alleged to be able to predict when and where various mortalities are likely to be important.

The migrating wildebeest (*Connochaetes taurinus*) is the most common herbivore in Africa (Fig. 13.1). In 1993 and 1994, the wildebeest population in the Serengeti region of Tanzania declined from about its normal or equilibrium value of about 1.2 million to 0.9 million. The falloff could have been due to many factors, including increased poaching, changes in climate, renewed drought, increases in the incidence of disease, or changes in rates of predation. A 40-year study (1958–1998) by Simon Mduma and colleagues (1999) examined the reasons for the changes in the wildebeest population. Mduma found that predation plays only a minor role in reducing wildebeest numbers. Less than 3% of the adult females died from predation, so its effect on births was small. Illegal harvesting was found to kill about 20,000 animals per year, well below the number needed to depress wildebeest population sizes. The main cause of mortality, 75%, turned out to be malnutrition. Rainfall in the dry season was found to be the most important external predictor of food supply for the wildebeest. A severe drought in 1993–1994 had caused the wildebeest decline, and Mduma and his colleagues suggested that the population would rebound to predrought levels as long as poaching did not increase. Only by long-term detailed studies such as these can we hope to disentangle the effects of the multitudes of factors that act on populations.

Figure 13.1 Wildebeest herd in the Serengeti.

13.1 Comparing the Strengths of Mortality Factors

A number of factors have the potential to influence populations and perturb densities away from mean or equilibrium levels. Competition, predation, parasitism, herbivory, and mutualism are all common and powerful biotic forces. Climate can also influence populations. Which factor most commonly acts to perturb population sizes, in which communities, and at what times? The argument is as old as the oldest writings on population ecology. Reviewing these ideas is valuable because it helps us evaluate their strengths and weaknesses. In the early part of the century, when much work was being done on insect pests of economic importance, parasites were thought to be consequential, because it was argued that they could respond quickly to changes in host density and track populations of interest (Howard and Fiske, 1911). It was held that predators, being territorial, would tend to take the same numbers of prey regardless of their density and that physical factors (like cold temperatures) would have the same effect no matter what the population density either. However, in the late 1920s, ecologists became convinced of the powerful effect climate had on the birthrates and death rates of animals, especially insects (Uvarov, 1931). Later, in the 1930s, the pendulum swung back to the biotic argument when the Australian entomologist A. J. Nicholson (1933) proposed that competition was the controlling factor for populations—competition for food, for habitat, or even for space.

By the 1950s, two more Australian entomologists (Andrewartha and Birch, 1954) had argued, once again, for the importance of weather as an agent of change. Their arguments were based on years of painstakingly counting tiny thrips on roses in Adelaide, counted every day (except Sundays and holidays) for 81 consecutive months on 20 rose flowers chosen at random from along a hedge. The number of thrips varied greatly, but 78% of the variation was explained by meteorological changes, especially in temperature and rainfall.

So far, arguments for population change had been couched mainly in terms of **extrinsic** factors, such as weather, parasitism, and predation. Later, in the 1950s, Dennis Chitty (1960) advocated the idea that **intrinsic** factors could also play a role. In studies of voles, he suggested that at higher densities the animals' endocrine system becomes stressed and, at the same time, their immune system weakens, so that voles become more susceptible to disease and less able to reproduce to their full capacity.

Evaluating controls on population change involves determining the relative killing power of each type of mortality.

Different proposed causes of population change continued to go in and out of favor. In the 1970s, the prevalent view once again was that competition was of overriding importance, because many studies on vertebrates like birds and rodents showed how competitors apparently neatly divided up their habitats into discrete niches wherein each fed and lived (MacArthur, 1972). In the 1980s and 1990s, this view was again challenged, because many studies of insects indicated that competition was absent, and more emphasis was given to stochastic factors such as disturbances (Strong, Lawton, and Southwood, 1984). Although history is of little guide in telling us what the most important mortality factor is, it does give us some clues about where to look. Still, which of the variety of theories that have been advocated over the years is likely to be correct?

A comparison of mortality factors is essential to both population and community ecology theory. If we can determine which mechanism is most important, then we may be able to tell whether communities are tightly or loosely structured entities (a controversy we will examine in the next section). If biotic factors are of overriding importance, then communities may really be tightly knit entities. If abiotic forces have the most influence in determining species abundance, then community structure may be loose and ephemeral.

Mortality factors can act in either of two ways: to disturb populations away from equilibrium levels or to return them toward equilibrium. Disturbing factors are often known as **key factors**, and we will examine key-factor analysis first. Regulatory factors often act in a density-dependent fashion and will be examined later.

Key factors.

Key-factor analysis is one approach to determining the relative importance of different mortality factors on population density.

One of the most well-known techniques for empirically comparing the importance of the effect of predators, parasites, or other factors on the size of populations is key-factor analysis, a technique first developed by R. F. Morris in 1957 to compare mortality factors acting on spruce budworm outbreaks in conifer plantations in Canada. The technique was refined by the British entomologists George Varley and G. R. Gradwell at Oxford in 1960. Key-factor analysis requires detailed information on the fate of a cohort of individuals: The total mortality of a generation or some other cohort (K) is subdivided into its various causes, and the relative importance of these is compared. Consider, for example, key-factor analysis on the oak winter moth, the subject of Varley and Gradwell's work (Table 13.1) and a species probably subject to the most comprehensive key-factor analysis ever done (Fig. 13.2).

Varley and Gradwell made a detailed census of all stages of the oak winter moth on five oak tress at Wytham Wood in Berkshire, England. The adult moths emerge from pupae in the soil in December. Males have wings, but females do not. As the females crawl up the trunks of oak trees, the males mate with them. Thereafter, the females lay eggs in crevices in the bark of branches. Varley and Gradwell were able to count the number of females per tree by using "lobster-pot" traps of fabric that fastened around the tree trunk. Exactly one quarter of the tree was covered in this way, so if Varley and Gradwell multiplied their trap catch by four, they would get the total numbers of females per tree. They then calculated the canopy area of the tree—essentially, the shadow of the tree area at noon when the sun was directly overhead. They divided the number of moths caught by the canopy area to get a number per square meter. They then dissected the females they caught in the traps to get the average number of eggs laid per female moth, which turned out to be about 150. So now they knew the number of females and eggs per tree and per square meter of canopy.

Young winter moth larvae emerge from the eggs in the spring and feed on the newly developing foliage, which at this time has the lowest levels of antiherbivore chemicals such as tannins. The rewards of feeding on tannin-free leaves are high, but often larvae emerge so early that they appear before the buds burst and face starvation. To find leaves, they disperse by ballooning away on silken threads to new trees that may already have flushed. This is obviously a chancy business, and many larvae are lost; hence, the mortality from dispersal is high. This loss is known as *overwintering loss*.

Varley and Gradwell didn't actually get into the trees to count the caterpillars. They again used passive-sampling techniques to get estimates of caterpillar numbers. They relied on the fact that fully grown caterpillars leave the foliage and spin down on silken threads to pupate in the soil. By putting two metal traps on the ground under each tree, each 0.5 m^2, they could estimate the number of caterpillars per m^2. The traps filled with rainwater and the caterpillars were drowned, to be picked up and examined by scientists later. Caterpillars were dissected for parasites, so that the numbers of healthy and sick caterpillars per m^2 of canopy were determined. There was one main type of parasite: a tachnid fly called *Cyzenis* that laid its eggs on oak foliage.

TABLE 13.1 Life table for the oak winter moth, 1955–1956. *(After Varley, Gradwell, and Hassell, 1973.)*

Life History Stage	Number Alive Per m^2	Logarithm of Number Alive Per m^2	k Value
Adult female, 1955	4.39		
Eggs (= Females × 150)	658	2.82	
Larvae (after overwintering loss)	96.4	1.98	$0.84 = k_1$
Killed by *Cyzenis*	90.2	1.95	$0.03 = k_2$
Killed by other parasites	87.6	1.94	$0.01 = k_3$
Killed by microsporidian disease	83.0	1.92	$0.02 = k_4$
Pupae	83.0	1.92	
Killed by predators	28.4	1.45	$0.47 = k_5$
Killed by *Cratichneumon*	15.0	1.18	$0.27 = k_6$
Adult females, 1956 (= half of the pupae)	7.5		

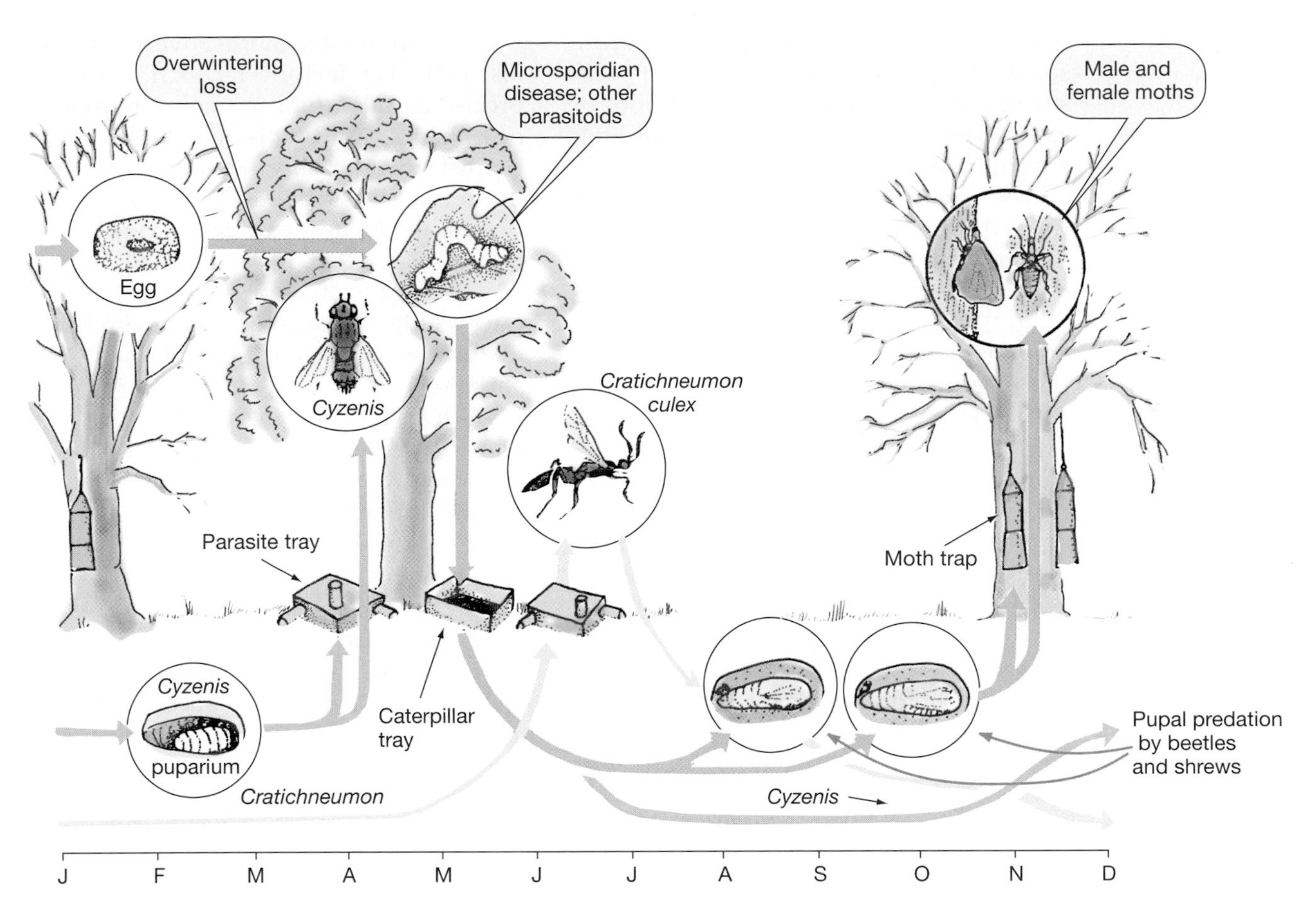

Figure 13.2 Life cycle of the oak winter moth and sampling methods. Adult, wingless, female winter moths, which climb up trees to lay their eggs, were counted in moth traps on the tree trunks; their eggs were dissected, and larvae, which occur on the leaves, were counted in the trays on the ground into which the caterpillars fall before pupation. Larvae of the parasite *Cyzenis*, which lays its eggs on the foliage, to be unwittingly devoured by the caterpillars, were counted by dissection of the fallen caterpillars. Adult *Cratichneumon*, which directly parasitize the pupae, were counted upon their emergence from the soil into the parasite traps. Pupal predation by soil-dwelling beetles and shrews was also estimated, but is not shown. (*Redrawn from Varley, 1971.*)

Some of the eggs would accidentally be ingested by feeding caterpillars, in whose body they would hatch, and the young fly larvae would feed. This is not a very successful method of parasitism, and only about 5–10% of caterpillars were infected with *Cyzenis*. Other parasites, mainly wasps, took a more active role in searching out and parasitizing caterpillars, inserting their ovipositors to lay eggs inside. However, their level of parasitism also was small (about 2–3%). A microsporidian disease infected some larvae as well, but again at a low level, about 4–5%. All this information was obtained from dissections of the drowned caterpillars in the traps.

On falling to the ground, the caterpillars pupate in the soil, but they are subject to more mortality there. First, another parasite—this time of the pupae—called *Cratichneumon* can cause substantial mortality, between 40 and 50%. Second, predators such as shrews and beetles eat lots of pupae—between 60% and 70% of all of the pupae, including parasitized ones. Once again, Varley and Gradwell used their traps to get data on the degree of mortality at the pupal stage. They inverted the trays they had used to catch caterpillars and poked a hole in the top to which they attached a small tube. They placed the inverted traps on the ground, where any moth hatching from a pupa in the ground would be caught in the tube. The difference of the number of caterpillars entering the soil and the number of adult moths emerging was due to pupal predation. The only exception was the mortality due to *Cratichneumon*, and this could be estimated from the number of that species of wasp in the traps. Also, *Cyzenis* parasitism could be estimated by adults emerging into the traps, so, for this parasite, two estimates of parasitism were available—from dissections of caterpillars and from the emergence of adult flies into traps.

At each stage, then, the researchers tracked the number of deaths and the cause of death (parasitism, predation, disease, etc.). The importance of each mortality factor (k) was estimated by calculating the amount that the factor reduced the population size, given by the formula

$$k = \log N_t - \log N_{(t+1)},$$

where N_t is the density of the population before it is subjected to the mortality factor and $N_{(t+1)}$ is the density afterward. For example, from Table 13.1, the killing power of pupal parasites of the oak winter moth is

$$k_5 = \log 28.4 - \log 15.0 = 1.45 - 1.18 = 0.27.$$

For larvae of the oak winter moth, mortality from overwintering is larger:

$$k_1 = \log 658 - \log 96.4 = 2.82 - 1.98 = 0.84.$$

The total generational mortality K can then be defined as the sum of the individual mortality factors k_x:

$$K = k_1 + k_2 + k_3 + k_4 + \cdots + k_n.$$

If this analysis is repeated for multiple generations, then graphs of K and the individual sources of mortality can be visually examined. The graphs are produced by plotting K and all its submortalities k_1 through k_x for as many years as the study lasted. The source of mortality that most closely mirrors the overall generational mortality K is then termed the key factor.

For oak winter moths, overwintering loss is the key factor (Fig. 13.3). For more precision, individual sources of mortality, or k values, can be plotted on the y-axis against total mortality, K, on the x-axis, and the key factor is then the source of mortality with the biggest correlation (the line of best fit) with K (Fig. 13.3b). However, the visual method is often preferred so that one can exclude a k that is correlated with K, but that makes only a small contribution to its variation.

What happens in other species? In plants and animals, key factors are many and vary from overwintering loss to juvenile starvation. Few generalizations can be made as to which types of key factors operate on which types of population (Table 13.2). Even for related species, such as insects, a detailed study of 48 life tables revealed no key factor of overriding importance and indicated that different species of insects are controlled by different types of key factors (Stiling, 1988).

Although key-factor analysis has been an important tool for ecologists, it has not been without criticism (Royama, 1996). First, key factors cannot always be precisely linked to specific mortality agents. Thus, overwintering mortality in the winter moth may actually be due to death of eggs or death by ballooning. The contribution of each of these to k_1, the overwintering mortality, is unknown, but it is possible that such a division would reduce the correlation between k_1 and K so much that it would disappear. Often, census data are good enough only to determine key-factor phases in life cycles—for example, juvenile mortality or

TABLE 13.2 Key factors for a variety of plants and animals. Some species have no obvious key factor. Different species tend to have different types of key factors.

Organism	Key Factor
Oak winter moth	Overwinter loss
Sand-dune annual plant	Seed mortality in soil
Colorado potato beetle	Emigration of adults
Tawny owl	Reduction in egg clutch size from maximum size
African buffalo	Juvenile mortality
Partridge	Chick mortality
Great tit	Loss of birds outside the breeding season
Broom beetle	Larval mortality on foliage
Grass-mirid insect	No obvious key factor
Cabbage-root fly	Reduction in egg production

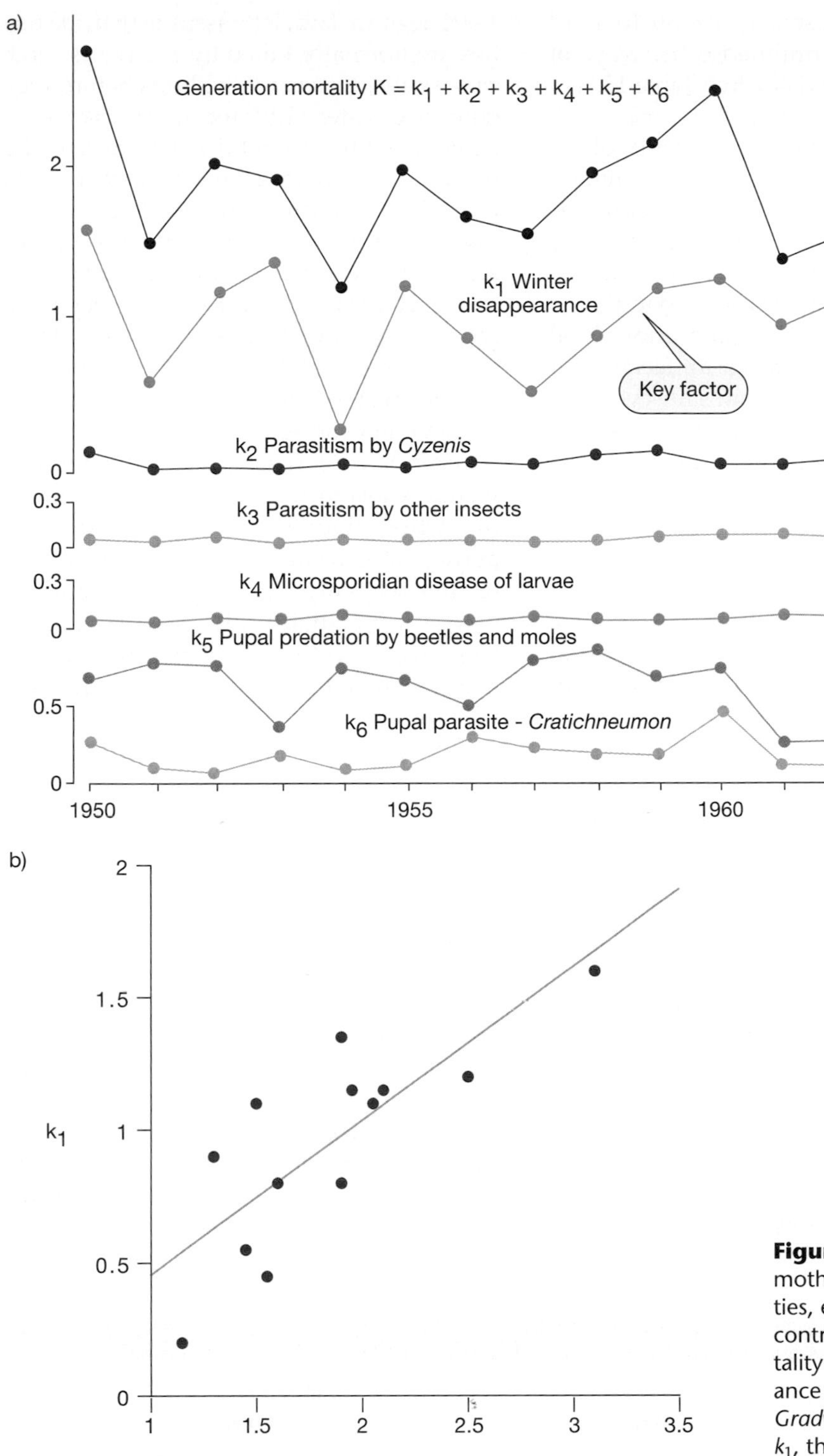

Figure 13.3 (a) Key-factor analysis of oak winter moth populations. Changes in individual mortalities, expressed as k values, show that the biggest contribution to changes in the generational mortality K comes from changes in winter disappearance of the moths. (*Redrawn from Varley, Gradwell, and Hassell, 1973.*) (b) Correlation of k_1, the overwintering loss, with K yields a straight-line relationship, indicative of a key factor.

adult mortality. (Consider the entry for African buffalo in the table.) Second, there may be intricate interactions between natural enemies, including hyperparasitoids (parasitoids of parasitoids), and such effects fail to show up in key-factor analysis. For example, if *Cyzenis* had been the key factor in the overall mortality of the winter moth, then perhaps its own abundance may have been influenced by a bacterial parasite, but the importance of such an agent would not have been detected. Third, populations can be influenced greatly by egg-bearing females that disperse into a population and that also never show up in key-factor analysis.

Perhaps for these reasons, Sir Richard Southwood (1978) outlined other ways of looking at life table data. (See Table 13.3.)

Real Mortality. This is the mortality of the population compared with the population size at the beginning of the generation. The real mortality row in the table is the only percentage row that is additive, and it is useful for comparing the role of population factors within the same generation. Real mortality is generally greater for factors that operate early in the organism's life cycle, because more individuals are actually killed, in terms of absolute numbers. So, from the table, if 200 larvae die out, of 1,000 eggs laid the real mortality at the larval stage is $200/1{,}000 \times 100 = 20\%$.

Indispensable (or irreplaceable) Mortality. This is that part of the generational mortality which would not occur should the mortality factor in question be removed from the life system, after allowance is made for the action of subsequent mortality factors. For example, consider the egg-stage mortality in Table 13.3. In Table 13.3 there are 30 adult survivors of 1,000 eggs, so 97% of the cohort dies. Now, if there is no egg mortality, 1,000 individuals enter the larval stage, at which a 40 percent mortality leaves 600 survivors to pupate; in the pupal stage, a 90 percent mortality leaves 60 survivors. Thus, now only 94% of the cohort dies. The indispensable mortality at the egg stage is then $97\% - 94\% = 3\%$. This mortality can never be made up by anything else.

Let's examine the value of protecting sea turtle nests as a conservation measure. If 1,000 eggs are laid, let's assume that 500 turtles are normally killed by predators, such as seagulls, that eat hatchlings before they enter the water (50% mortality). We'll assume that a further 400 turtles are eaten by fish predators before they can develop into reproductive adults (80% mortality). Let's say that, of the remaining 100 reproductive adults, 90 are lost to fishing nets (90% mortality). This leaves 10 turtles surviving out of the original 1,000 eggs (Fig. 13.4). How would the population fare if we protected all the nests? Then 1,000 young turtles would enter the sea, where they would be reduced 80% by fish to 200. A further 90% would be lost to nets, leaving 40. This is four times the original number surviving, but perhaps not what you'd expect from protecting so many eggs. Indispensable egg mortality is actually only 4%. We would do better by protecting the turtles from fishing nets, because then 100 turtles would survive. In fact, protecting organisms in their adult reproductive condition often yields better results than protecting the very young, especially if there are high mortalities at later stages, as in our theoretical example in Table 13.3.

Mortality–Survivor Ratio. This measure represents the increase in population that would have occurred if the factor in question had been absent. If the final population is multiplied by the mortality–survivor ratio, then the resulting value represents, in individuals, the indispensable mortality due to that factor. For example, for larvae in Table 13.3, 200 die and 300 survive, so the mortality-survivor ratio is 200:300, or 0.66.

TABLE 13.3 Various measures of mortality from life tables.

Measure	Stage: Eggs	Larvae	Pupae	Adults
n_x (number alive)	1,000	500	300	30
d_x (number dying)	500	200	270	
q_x (proportion dying)	0.5	0.4	0.9	
% real mortality	50	20	27	
% indispensable mortality	3	2	27	
mortality–survivor ratio	1.00	0.66	9.00	
logarithm of population	3.00	2.70	2.48	1.48
k values	0.30	0.22	1.00	

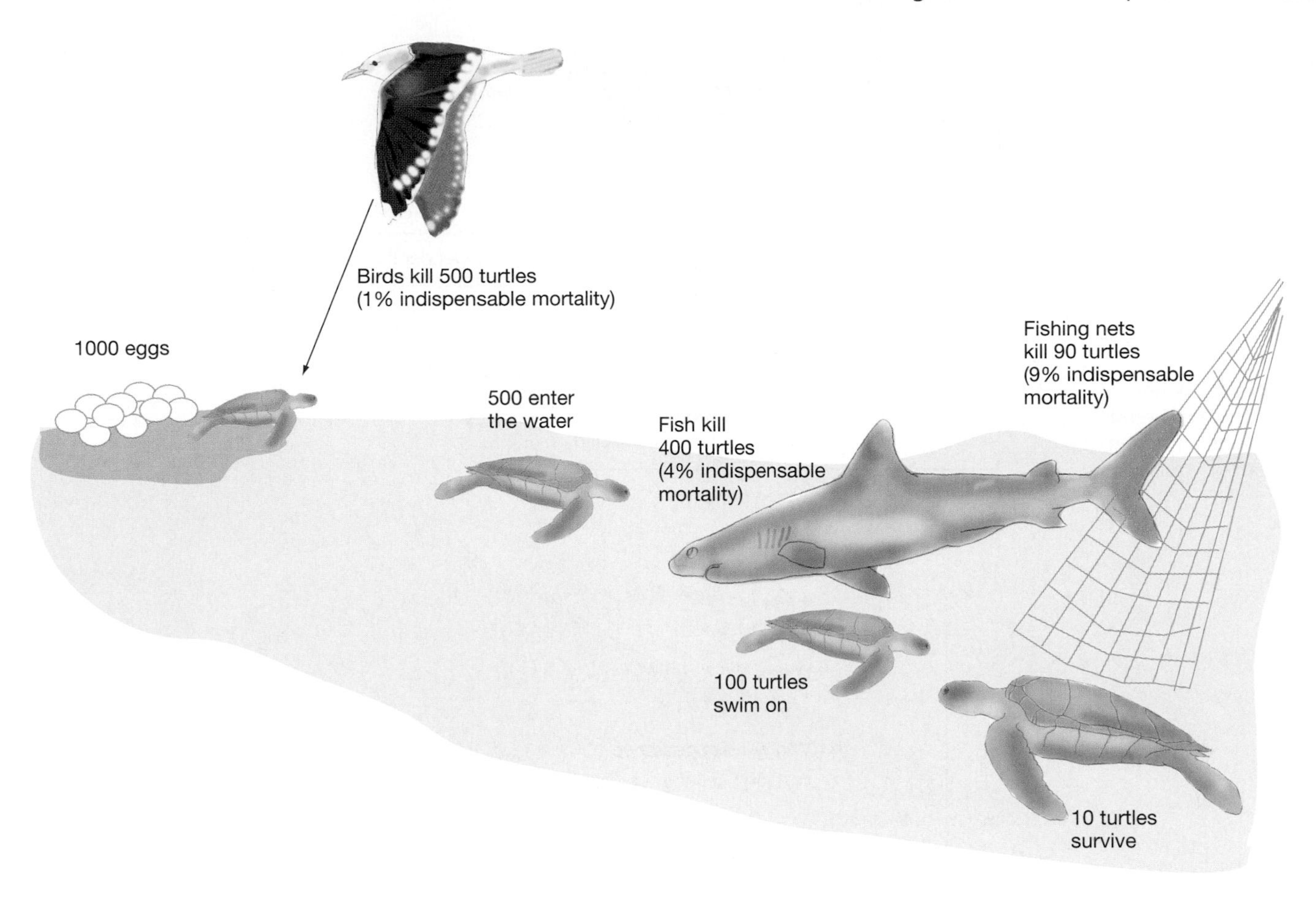

Figure 13.4 Schematic representation of hypothetical life cycle of sea turtles, illustrating the concept of dispensable mortality. Protecting eggs from predators may not do much good as protecting adults from fishing nets.

13.2 Density Dependence

Parasitism, predation, competition, and abiotic factors can all act as key factors that perturb the population densities of living organisms away from equilibrium levels. However, we often see organisms existing at virtually the same densities year after year. For instance, lake trout in Lake Michigan existed at the same densities for at least 20 years, from 1930 to 1950, before the introduction of lampreys (Fig. 10.12). Hence, there must be factors that act to stabilize population densities. Therefore, we must ask, "Which effect is most important in returning populations to equilibrium levels?" This is again a very difficult question to answer, for many reasons. First of all, it is necessary to compare the strengths of different mortality factors. Second, it is appropriate to determine which factors act in a density-dependent manner. This means that a factor must kill relatively more of a population when densities are higher and less of a population when densities are lower. Even abiotic factors could act in a density-dependent manner if, for example, there were limited refuges, like burrows, to get away from inclement weather, such as hard freezes. Determining which factors act in a density-dependent fashion has huge practical implications. Foresters, game managers, and conservation biologists are very interested in learning how to maintain populations at equilibrium levels.

Density-dependent processes may regulate population sizes around some mean value.

Density dependence can be detected when mortality, expressed as percentage, is plotted against population density (Fig. 13.5). If a positive slope results and mortality increases with density, then the mortality factor tends to have a lesser effect on sparse populations than on dense ones and is clearly acting in a density-dependent manner. Alternatively, we can detect density dependence when k values are plotted against generational mortality K and a positive slope results. In the winter moth example, predation of overwintering moth pupae in the ground (by beetles, shrews, and moles) is the density-dependent factor (Fig. 13.6). In both types of plots, factors that appear not to

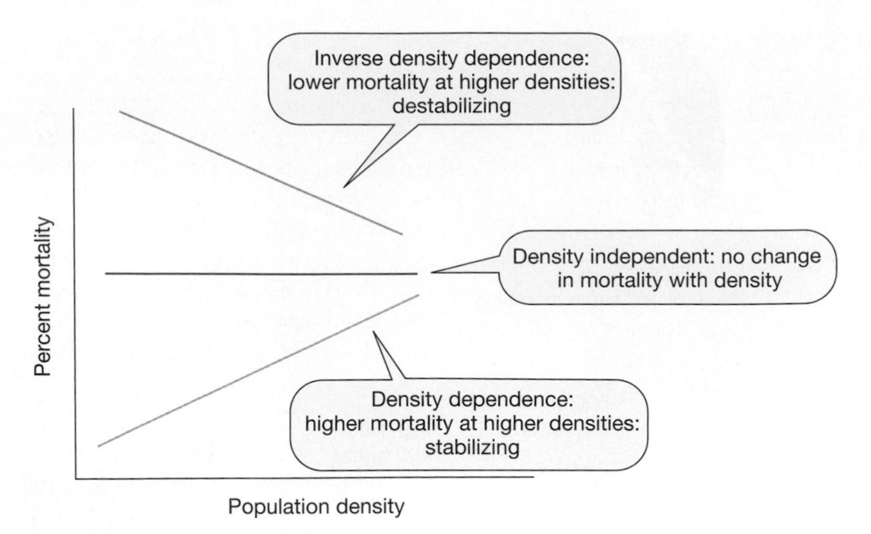

Figure 13.5 Changes in percent mortality in response to population density. A positive slope is indicative of density dependence, which will tend to stabilize population densities at mean or equilibrium densities.

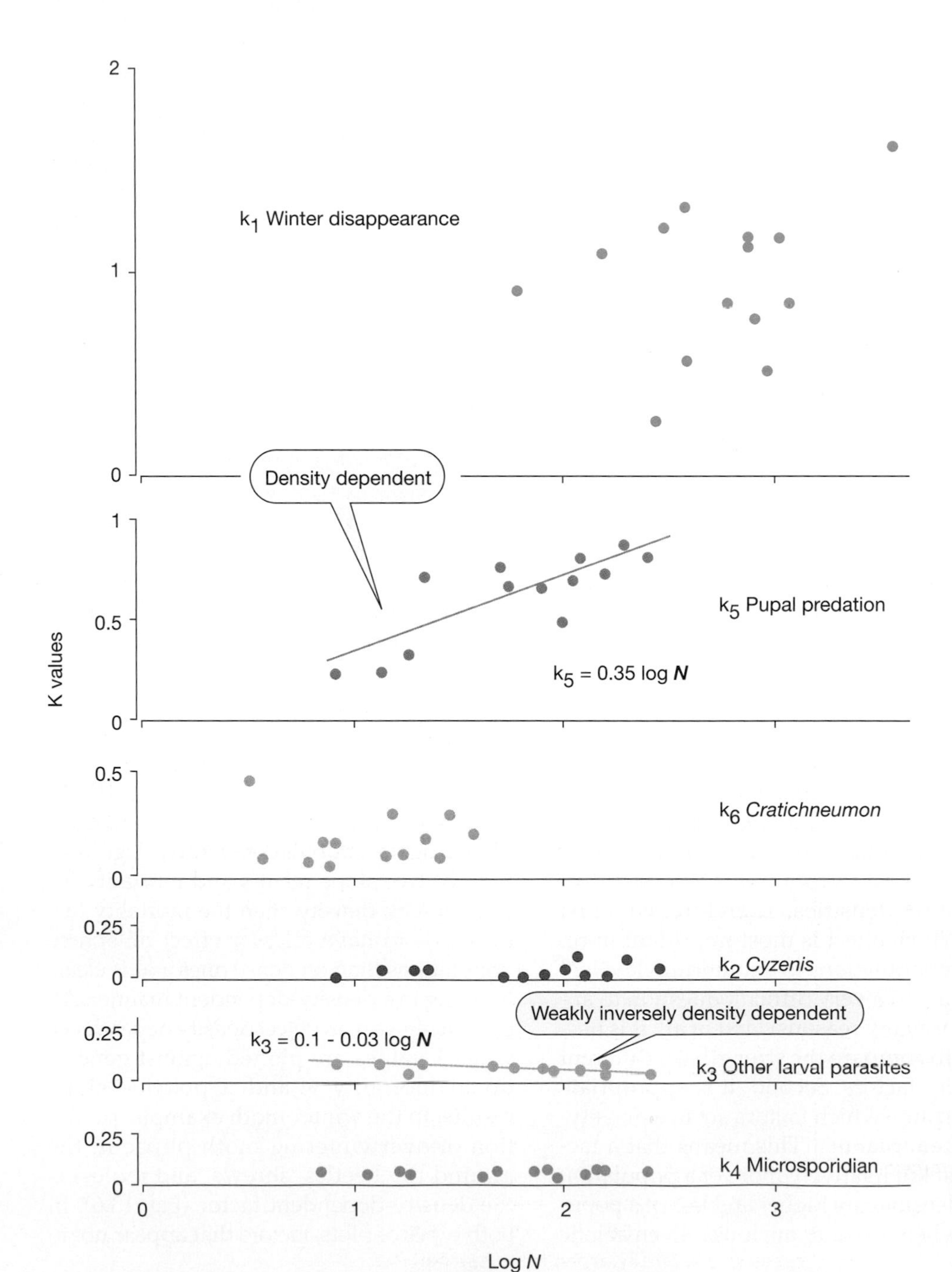

Figure 13.6 k values for the different winter moth mortalities, plotted against the population densities on which they acted. Some k values are density independent and vary considerably (k_1, k_6). Others are density independent, but are relatively constant (k_2, k_4). Still other factors may be weakly inversely density dependent (k_3) or quite strongly density dependent (k_5). (*Redrawn from Varley, Gradwell, and Hassell, 1973.*)

change with density and thus that do not contribute to population regulation, are termed *density independent*. For example, we could think of parasites that always infect about 10% of a host population, regardless of its density, as being density independent. Those sources of mortality that decrease with increasing population size are called *inversely density dependent*. For instance, if a territorial predator, like a lion, always ate the same number of wildebeest prey, regardless of wildebeest density, it would act in an inversely density dependent manner because it would take a smaller proportion of the population at higher density. Birds, being highly territorial, often act in this manner on populations of insects that are their food.

Which factors tend to act in a density-dependent manner? We can examine this question with reference to insects—the group with the most species on Earth and with perhaps the best data available to answer the question, probably because many of them are regarded as pests and hence have a large number of studies done on them. Stiling (1988) examined density dependence by any mortality factor in insect populations, using data sets for 58 species. The density-dependent factor that was most important was different for different species at different times. No single process could be regarded as a regulatory factor of overriding importance. This finding is disconcerting, because it means that generalizations about which factors are likely to act in a density-dependent manner are not easily made or are likely to be different, even for an individual taxon like the insects.

Sinclair (1989) presented a broader review of the literature, reporting on the cause of density dependence in 51 populations of insects, 82 of large mammals, and 36 of small mammals and birds (Fig. 13.7). Just as Stiling had suggested, data for insects showed a wide variety of causes of density dependence. However, there were some differences in density-dependent factors between a number of larger taxonomic groups. Food was more important for large mammals: Being bigger, they need huge amounts of food, and obtaining it is of critical importance. Space and social interactions were important for smaller mammals and birds, perhaps because they are territorial. Sinclair noted, however, that the effects of disease and parasitism had probably been grossly understudied for animals other than insects, because of the difficulty inherent in studying parasites in the field. These effects were therefore likely to be underrepresented in his table.

Sinclair also found that r-selected species with very high reproductive rates and Type III survivorship curves (insects and fish) have early juvenile (eggs and larvae) density-dependent mortality. In addition, he found that species with intermediate reproductive rates and Type II survivorship curves (birds and small mammals) have late juvenile and prebreeding regulation, while K-selected species with low reproductive rates and Type I survivorship curves (large mammals) are at

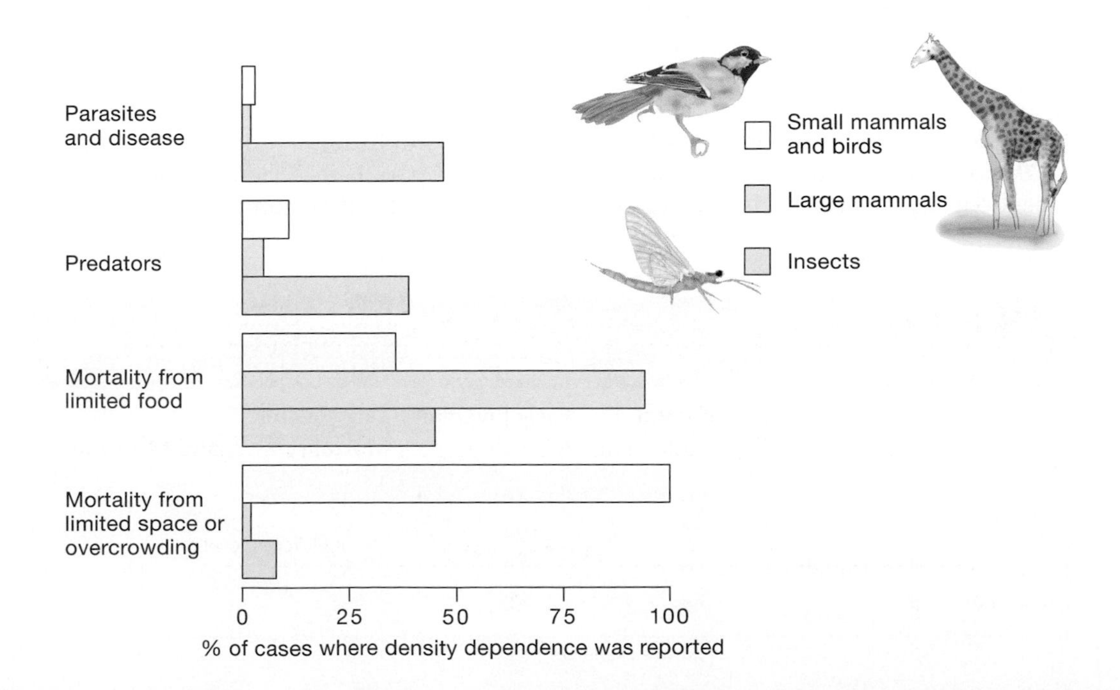

Figure 13.7 Percentage of reports of separate populations demonstrating density dependence at different life stages. *(After various sources, Sinclair, 1989.)*

least partly regulated through changes in fertility (Table 13.4).

Questions regarding which mortality factors operate in a density-dependent manner and which do not notwithstanding, Stiling (1988) noted that the overall detection of density dependence in insects was quite low—it was detected in only about half of the studies. This meant that many populations would not be tightly regulated, but would fluctuate widely and erratically in density.

There may be specific biological reasons that some mortalities do not operate in a density-dependent fashion. For example, a searching parasitoid may not always oviposit in a density-dependent manner, laying more eggs in dense concentrations of caterpillars; instead, she may save some of her eggs to oviposit elsewhere, even in a seemingly suboptimal place. The reason may be that some local catastrophe could occur in the best area, wiping out all her progeny. For example, a flock of birds could also be attracted to a dense congregation of caterpillars and eat every one. If the wasp oviposits in a few solitary caterpillars in out-of-the-way places, some progeny would survive such a catastrophe, so that behavior would be selected for. This phenomenon has become known as *spreading the risk* (den Boer, 1968, 1981).

Spreading the risk may be a good explanation for the apparently random patterns relating mortality and density that are often observed in the field. Furthermore, even if they wanted to, animals may not always behave in a density-dependent fashion in the face of a huge array of conflicting pressures. Their life history patterns may be adaptations to survival, not maximization of fitness. African hunting dogs may not prey on antelope populations in a density-dependent manner because of the presence of a pride of lions that could prey on the hunting dogs. Sinclair (1989) noted that the appeal of the density-dependent argument is that it predicts the existence of a process (density dependence) that can be measured in nature. The philosophical weakness of the argument is that it is very hard to reject; the absence of density dependence can too easily be explained away on ad hoc methodological grounds: Density dependence was not detected because the study wasn't performed correctly.

There is still no consensus on the frequency of density dependence in nature. Where it is found, it offers a control on population density. However, it is not always found, nor is it always necessary to invoke because of spreading the risk or the presence of alternative selective pressures. Where density dependence acts, there is still no simple answer as to which factors—competition, predation, parasitism, mutualism, or abiotic factors—are most likely to affect population densities. As Donald Strong (1988) summed it up, no single factor along the gamut from plant chemistry to abiotic influences to natural enemies can be ruled out, even for a minority of cases. A complex of influences participates in the regulation of most organisms. In an essay on the Internet, Jim Brown (1997) tells us that he is convinced that trying to make predictions about which factors will influence the future abundance of a plant or animal population is akin to trying to predict when and from what cause a human will die. For example, we know that age, smoking history, car accidents, and cholesterol can cause death, but because the human body is incredibly complex, we don't know how these things interact. Because communities—and, indeed, populations—are at least as complex as the human body, Brown suggests that it is difficult to predict the future abundance of species and individuals. Ecology, it might seem, is entering a period of pluralism in which simplistic, one-dimensional explanations of population phenomena are seldom sufficient. Furthermore, if populations are

TABLE 13.4 Percentage of reports demonstrating density dependence at different life stages. Density dependence can occur at more than one life stage; hence, the total number of populations is not the total of the rows. *(Reproduced from Sinclair, 1989.)*

Group	Fertility and Egg Production	Early Juvenile Mortality	Late Juvenile Mortality	Adult Mortality
Insects	30	40	28	13
Fish	6	94		
Birds and small mammals	16	19	81	16
Large mammals	73	40	1	11

linked by individuals commonly dispersing among them, then invoking density dependence as a control factor is even more problematic. If enough individuals move between populations, we can call the whole group of populations a metapopulation.

13.3 Metapopulations

A *metapopulation* is a series of small, separate populations that mutually affect one another. In this scenario, even if one individual population goes extinct, other populations survive, and they supply individuals who disperse and recolonize the patches where the population became extinct. Even if the population that supplied the colonists itself becomes extinct, it will be recolonized later. Metapopulations are viewed as sets of populations persisting in a balance between local extinction and colonization. The relevance of metapopulation theory to population biology is clear: Through metapopulations, populations could be maintained by a balance between colonization and the extinction of small patches of habitat. In this scenario, there is no mean or equilibrium level; extinctions happen at any time and density dependence is irrelevant. Persistence depends only on factors affecting extinction and colonization rates, such as interpatch distances, species dispersal abilities, and the number of patches in the metapopulation. One of the first formal mathematical treatments of metapopulation dynamics was Levins's (1969) model, which treated the rate of change in occupied patches as the colonization rate minus the extinction rate:

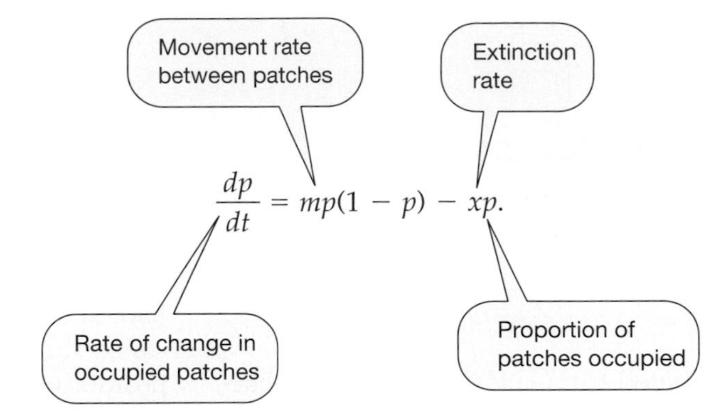

A metapopulation is a group of separate populations that are united by the migration of individuals among them.

The equation says that the rate of change in occupied patches is influenced by the rate of movement between patches, m, the number of empty patches prone to colonization $(1 - p)$ and the number of occupied patches able to provide the colonizers (p). Extinctions on the other hand, simply increase as the number of occupied patches goes up, because there are more occupied islands for extinctions to occur on. Levins's model predicts that the proportion of occupied patches will stabilize over time to $1 - (x/m)$. So if the extinction rate per period—say, a year—is 0.10, and the movement rate m is 0.4, then the proportion of occupied patches when a population is stable is $1 - (0.1/0.4) = 0.75$. This model is a good first step in examining what factor limits population sizes on islands, in terms of the rate of occupancy.

In 1991, Susan Harrison of the University of California at Davis (see Profiles box) reviewed the empirical literature and found few situations that fit the classical description of a metapopulation. More common were three related situations (Fig. 13.8):

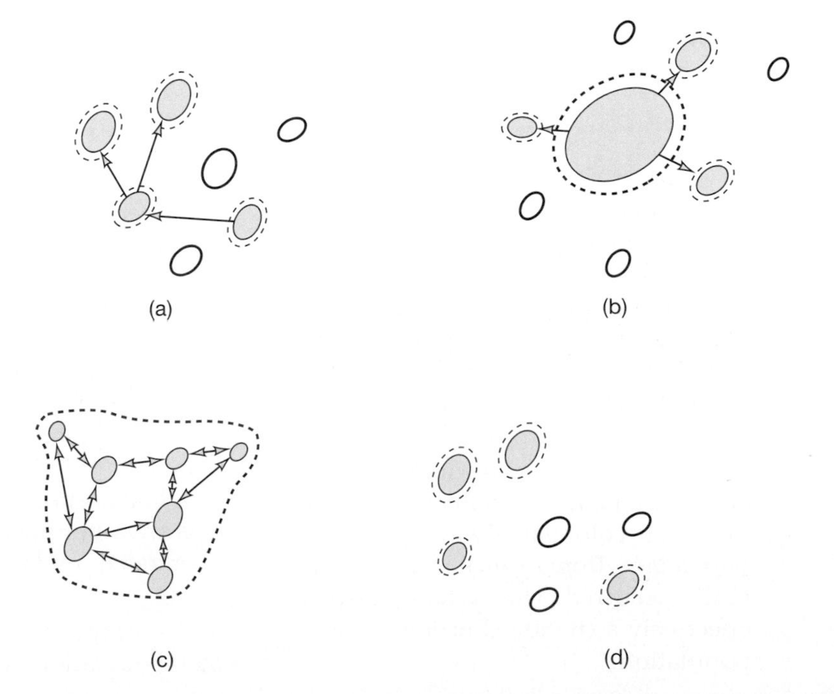

Figure 13.8 Different kinds of metapopulations. Closed circles represent patches of habitat; filled = occupied; unfilled = vacant. Dashed lines indicate the boundaries of "populations." Arrows indicate migration (colonization). (a) Classic metapopulation. (b) Core–satellite metapopulation (common). (c) Patchy population. (d) Nonequilibrium metapopulation (differs from *a* in that there is no recolonization), often occurring as part of a general regional decline. (*Modified from Harrison, 1991.*)

Profiles

Susan Harrison, University of California, Davis

When I graduated from U.C. Davis with a bachelor's degree in 1983, I'd been interested enough in nature to major in zoology and to spend my junior year in Kenya taking ecology classes. Yet it never occurred to me that I could become a research biologist, and no professor or TA ever took me aside and said, "Hey, you should consider graduate school." So I did what everyone else was doing: applied to medical school.

Then I got a summer job as Rick Karban's field and lab assistant, and Rick was great; within a week or two, he was just about letting me design his experiments. We'd go out to do fieldwork at Bodega Bay, and it was a series of revelations to me: collecting larvae and finding out what parasitoids were; using glue and twist ties to imitate the way moths damage plants; the whole idea of asking nature questions. It stretched into a year of working for Rick, Jim Quinn (who was collaborating with Rick on a study of habitat fragmentation), and Rick Grosberg, all of whom were great mentors. I went to seminars, hung out with grad students, and started some research projects. So by the time I started grad school at Davis in 1984, I already had a sense of direction. I took the Organization for Tropical Studies (OTS) course in Costa Rica, did a side project at Barro Colorado Island, and finished a master's degree in 1986. My interests really formed at this time, and I've been trying to emulate Rick's experimental finesse and Jim's grasp of theory ever since.

It seems strange by present standards that in 1986, of the various places I was interested in going for a Ph.D., Stanford, with Paul Ehrlich's group was the only one where conservation biology was respectable! Everywhere else, prospective advisors said that you should wait until you get your Ph.D (and maybe tenure) before you get involved in conservation. Paul said I could work on the conservation biology of an endangered butterfly with a fragmented habitat, so of course I did, and he also involved me in other conservation-related projects. I finished in 1989 and did a postdoc with John Lawton in England, in 1990–1991, which was great for getting a European perspective on ecology.

My present faculty position at Davis came about through the "Targets of Opportunity for Diversity" program, which allows U.C. campuses to create extra positions for women and minorities who fill particular needs. This is supposed to lead to more role models for female and minority students, without taking away from other people's chances of getting jobs. Programs like this seem essential to making ecology less of a European-American-dominated field.

In terms of significant obstacles, one that comes to mind is access to land. Metapopulation studies and studies of habitat fragmentation in general require that you survey all the patches of habitat in a large area—maybe 10 by 10 miles or more. This obviously can't be done within a single research reserve. I'm terribly jealous of the Scandinavians, whose laws say that the only thing you can't do on someone's land is camp within sight of their house. This must be one reason so many excellent metapopulation studies come from Sweden and Finland.

As far as future directions, one very significant trend I see is that nearly all the top-notch applicants to graduate school in ecology are seriously interested in conservation biology and other applied areas. Thus, I expect an increasing amount of high-quality work in these fields. Another trend is that conservation biology itself is maturing, moving away from promoting simplistic rules and principles, toward a more scientific approach that includes experiments, modeling, and considering alternative hypotheses.

---

1. Core–satellite, source–sink, or mainland–island metapopulations, in which persistence depended on the existence of one or more extinction-resistant populations, usually inhabiting large patches, which constantly supplied colonists to small peripheral patches that often became extinct.
2. Patchy populations, in which dispersal between patches or populations was so high that colonists always "rescued" populations from extinction before it actually occurred. The system was then effectively a single extinction-resistant population.
3. Nonequilibrium metapopulations, in which local extinctions occurred in the course of species' overall regional decline. Many rare species conform to this scenario, with fragmentation of their habitat reducing their population density. The failure of populations to disperse effectively eliminates a true metapopulation scenario. The conservation of species in fragmented habitats is an important area for the application of metapopulation models.

One of the most well-known studies of a metapopulation in nature was that of the metapopulation of the Bay checkerspot butterfly (*Euphydryas editha*), (Fig. 13.9); Harrison, Murphy, and Ehrlich, 1988). It consisted, in 1987, of a population of 106 adult butterflies on a 2,000-ha habitat patch called Jasper Ridge near Stanford University and nine populations of 10 to 350 adult butterflies on patches of 1 to 250 ha. Of 27 small patches of habitat in the region found to be suitable for populations to live, only those closest to the large patch were occupied. This pattern of occupancy could not be explained by differences in quality of the habitats. Instead, the distance effect appeared to indicate that the butterfly's capacity for dispersal was limited and that the large population acted as the dominant source of colonists to the small patches. From this and other evidence, it appeared that persistence in this metapopulation was relatively unaffected by population turnover on the small patches, which acted as sink populations. In 1996, the population disappeared from Jasper Ridge, after having been there since at least 1934. Whether it will ever return is uncertain.

What of other systems? Data are just now being accumulated. Nieminen (1996) invoked a metapopulation scenario for moths on small islands off the Finnish coast, and Hjermann and Ims (1996) invoked a metapopulation model to explain the presence of bush crickets on patches of habitat in Norway. There has also been much theoretical interest in the possibility that the interaction between competitors, between predators

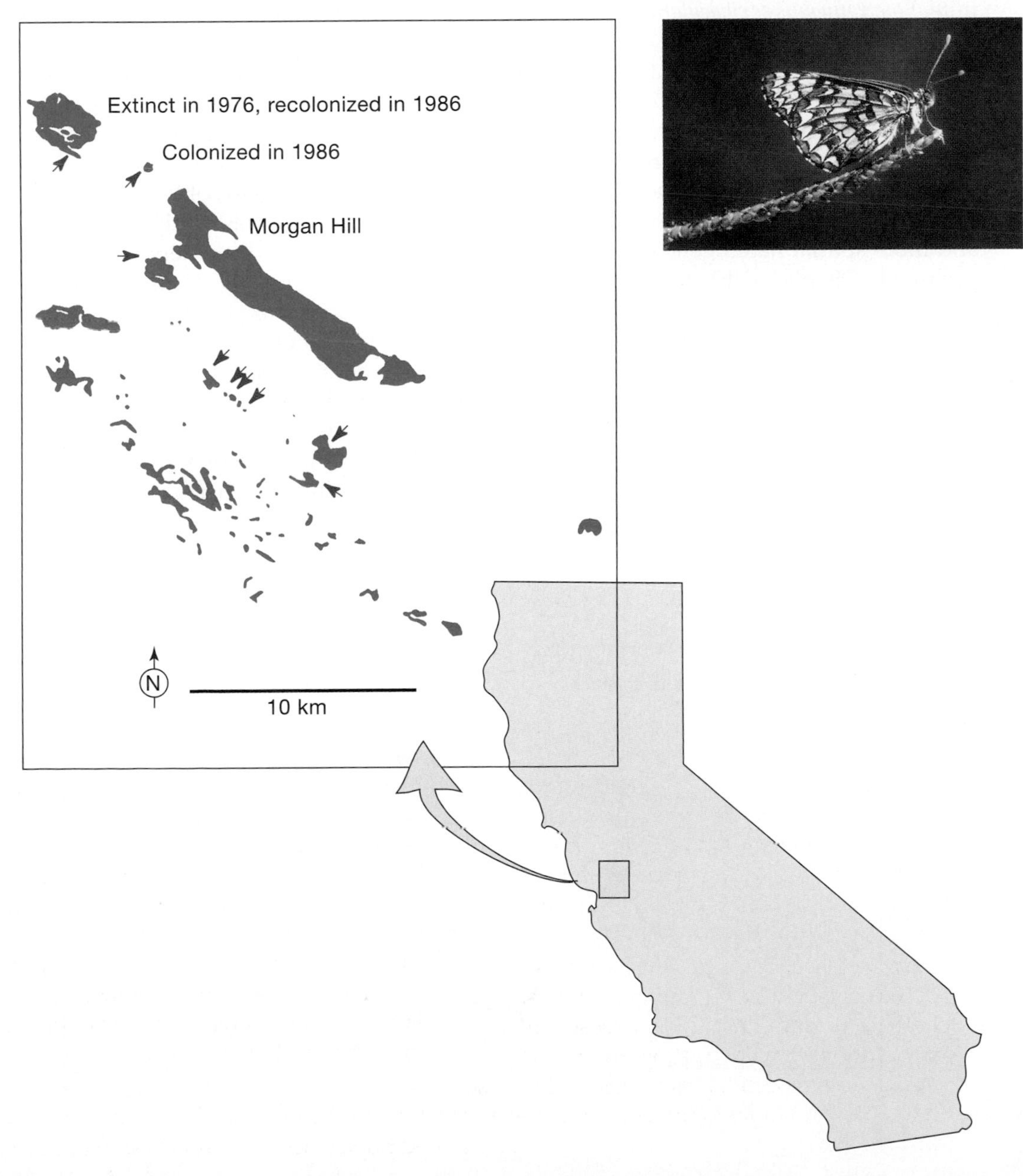

Figure 13.9 Metapopulation of the Bay checkerspot butterfly (*Euphydryas editha*) at Jasper Ridge, near San Francisco, California. The black areas represent patches of the butterfly's serpentine grassland habitat. The 2,000-ha patch labeled "Morgan Hill" supported a population on the order of 10^2 adult butterflies in 1987 and acted as the source population. The nine smaller patches labeled with arrows supported populations on the order of 10^1–10^2 butterflies in that year. Many other small patches were found to be suitable for occupation, but remained unoccupied. *(Reproduced from Harrison, 1991.) (LeRoy Simon, Visual Unlimited.)*

and prey, or between hosts and parasitoids may promote a metapopulation scenario (Reeve, 1988). For example, parasitoids may drive host patches to extinction on one patch *A*, but other patches *B*, *C*, and *D* contain other host populations that thrive in the absence of the parasitoid. Eventually, parasitoids find the hosts on patches *B*, *C*, and *D*, but by that time, patch *A* has been recolonized and the host is able to stay one step ahead of the parasitoid.

Interestingly enough, it turns out that metapopulation theory has had quite a long history, although it was not referred to by its current name. As long ago as the 1950s, Carl Huffaker was interested in the interaction of predators and prey over a fragmented landscape (Huffaker, 1958; Huffaker, Shea, and Herman, 1963). Huffaker studied a laboratory system of a phytophagous mite, *Eotetranychus sexmaculatus*, as prey and a predatory mite, *Typhlodromus occidentalis*, as predator. The prey mite infests oranges, so Huffaker used that fruit for his experiments. When the predator was introduced onto a single prey-infested orange, it completely eliminated the prey and died of starvation. Huffaker gradually introduced more and more spatial heterogeneity into his experiments. He placed 40 oranges on rectangular trays like egg cartons, but still the system eventually resulted in the extermination of the populations. Finally, Huffaker used a 252-orange universe (16 rows × 16 rows). He partially isolated each one by placing a complex arrangement of vaseline barriers in the tray, which neither mites could cross. But he facilitated the dispersal of *Eotetranychus* by inserting a number of upright sticks from which they could launch themselves on silken strands carried by air currents. The predators had only a few limited nonvaseline lines of dispersal available. The prey were thus able to keep one step (i.e., one orange) ahead of the predators. Overall, at any one time, there was a mosaic of (1) unoccupied patches, (2) patches of prey and predator heading for extinction, and (3) patches of thriving prey, and this mosaic was capable of maintaining persistent populations of both predators and prey (Fig. 13.10).

More recently, similar experiments were carried out by Marcel Holyoak and Sharon Lawler (1996), using a protist predator–prey pair—the bacterivorous ciliate *Colpidium* and the predaceous ciliate *Didinium*—in a microcosm of linked bottles. This time, too, the researchers showed how critical the dispersal rate was to the maintenance of stable systems. The dispersal rate was changed by providing (a) single containers in which the dispersal among all areas was high, and (b) more complex systems of linked containers wherein dispersal was facilitated by linking tubes between the containers. At the lowest dispersal rates, no long-term persistence of the two species was possible, because recolonization did not balance local extinctions. At the highest dispersal rates, the entire system behaved as a single large population, in which the pair could not exist. The pair persisted as predator and prey metapopulations only at intermediate dispersal rates.

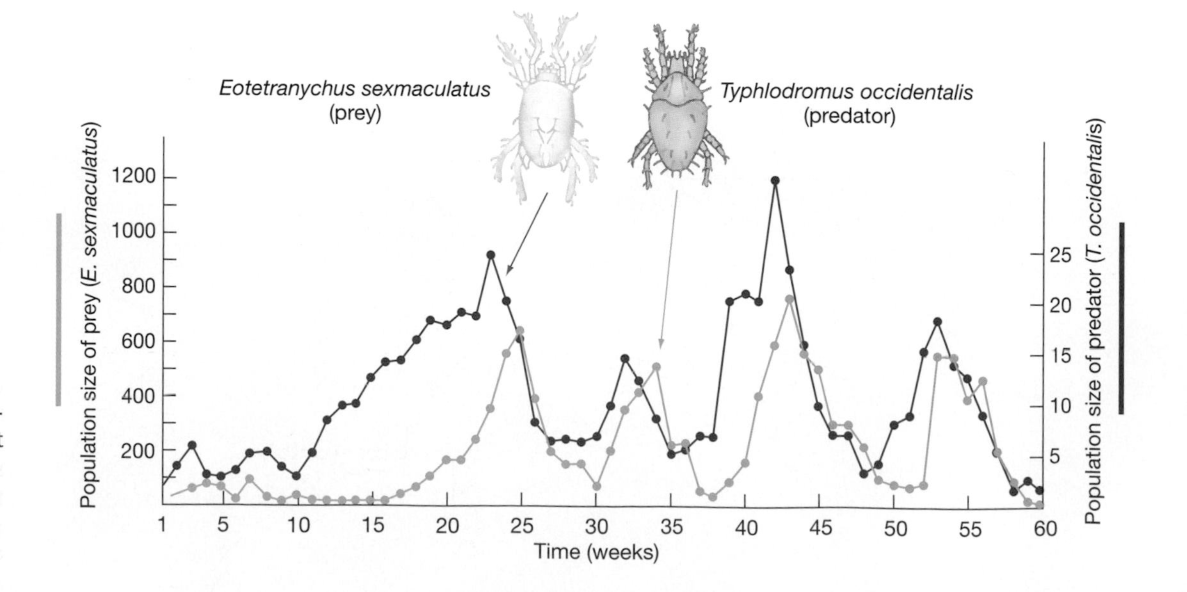

Figure 13.10
Interaction between predator and prey mites in a complex laboratory environment with a 252-orange system with sticks present to promote the dispersal of prey. *(After Huffaker et al, 1963.)*

13.4 Conceptual Models of Population Control

If the empirical data on which mortality factors are important for different taxa vary, what can theory tell us about those factors? Many different theoretical models have been proposed to describe which types of mortality factors should be most important in which systems. Some stress the importance of "bottom-up" factors, such as food, which act from the bottom of the food chain, some emphasize "top-down factors," like natural enemies, which percolate down from the top of the food chain (Fig. 13.11), and some incorporate elements of both of these ideas.

Bottom-up factors, or trophodynamics.

Among the earliest contenders for a general theory to explain population processes was Lindeman's (1942) trophic-level concept, or *trophodynamics*. This theory explained the height of the trophic pyramid by reference to a progressive attenuation of energy passing up trophic levels, from plants through herbivores, primary carnivores, and secondary carnivores. Since about 10 percent of the energy was transferred between levels (see Chapter 21), precious little energy was left for trophic levels higher than about four. Lindeman's was a true bottom-up theory based on the thermodynamic properties of energy transfer, with little use for top-down control by natural enemies. His work will be discussed in more detail in Section V, which focuses on energy transfer between populations.

Ecologists have developed conceptual models to describe the mortality factors that should be most important in different systems.

Top-down factors.

Since Lindeman's work, many different models have been proposed to incorporate the effect of natural enemies in communities. Among the first was Hairston, Smith, and Slobodkin's (1960) idea (usually referred to as HSS) that, since the Earth appears green, herbivores must have little impact on plant abundance. Hairston and colleagues suggested that this small effect is because herbivores are ordinarily limited by their predators. The implication was that plants, being abundant, endure severe competition

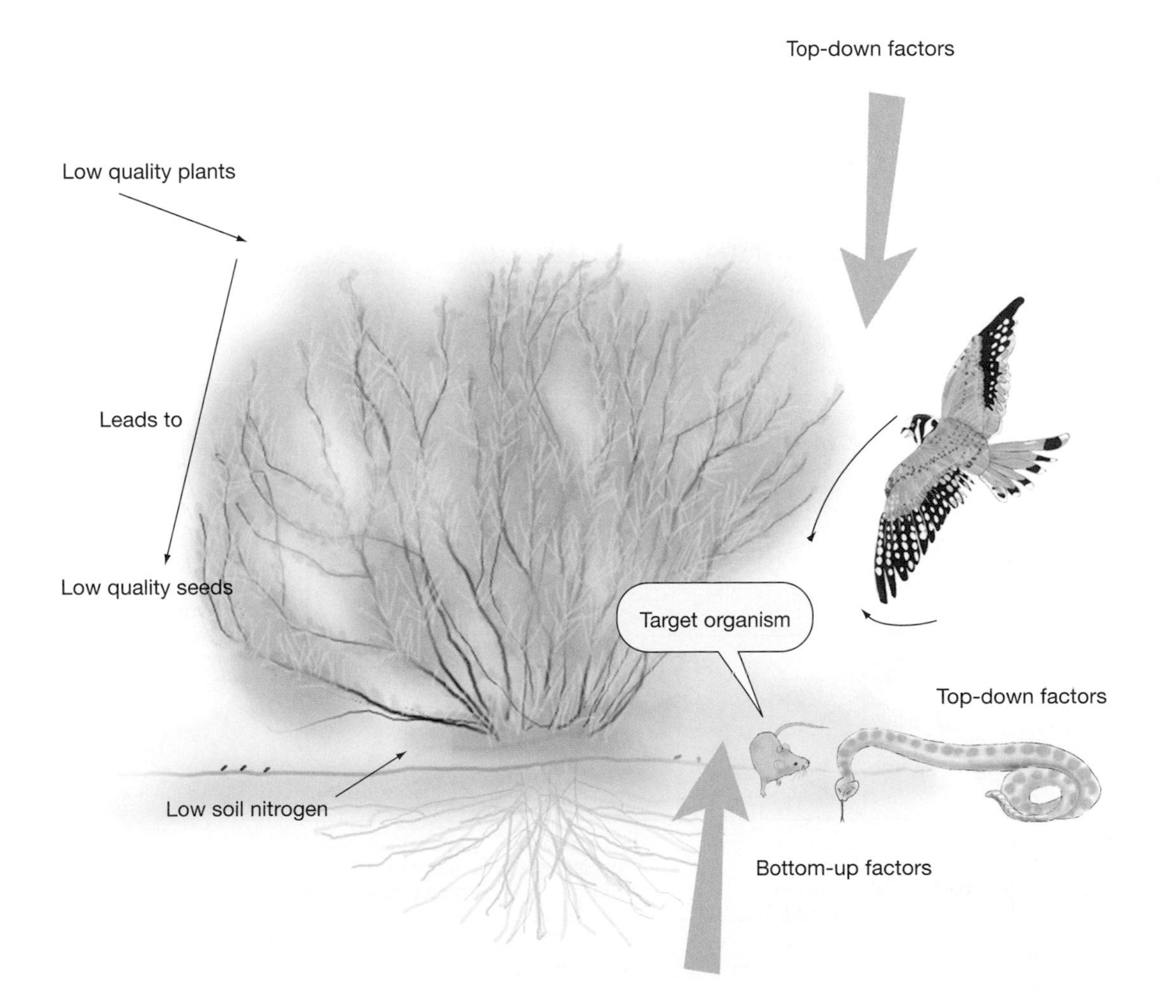

Figure 13.11
Schematic representation of top-down and bottom-up factors acting on a population of desert mice. Top-down factors include predators, like birds of prey, snakes, and parasites (not shown), and bottom-up factors include food (seed) quality and the availability of water (not shown).

for resources, but herbivores, suffering high rates of mortality from natural enemies, do not compete with each other for plant food. By contrast, natural enemies, being limited only by the availability of their prey, also compete for resources. Lauri Oksanen and colleagues (1981) extended the HSS hypothesis to a greater variety of communities. They proposed that the strength of various types of mortalities on different trophic levels varied with the type of system involved—particularly as a function of primary productivity–the amount of green plant biomass produced. They termed this idea the *ecosystem exploitation hypothesis*, or EEH (Fig. 13.12). Thus, for very simple systems with low primary productivity, such as Arctic tundra, production is so low that few plants exist. As productivity increases a little, plants become limited by resources (competition). As productivity increases still more, some herbivores can be supported, but too few to support carnivores. In the absence of carnivores, levels of herbivory can be quite high. Plant abundance then becomes limited by herbivory, not competition. The abundance of herbivores in the absence of carnivores is limited by competition for limiting plant resources. As primary productivity increases still more, carnivores can be supported, and there are three trophic levels—the HSS scenario. Finally, as productivity increases still further, secondary carnivores can be supported, and these in turn depress the numbers of carnivores, which then increases levels of herbivory and

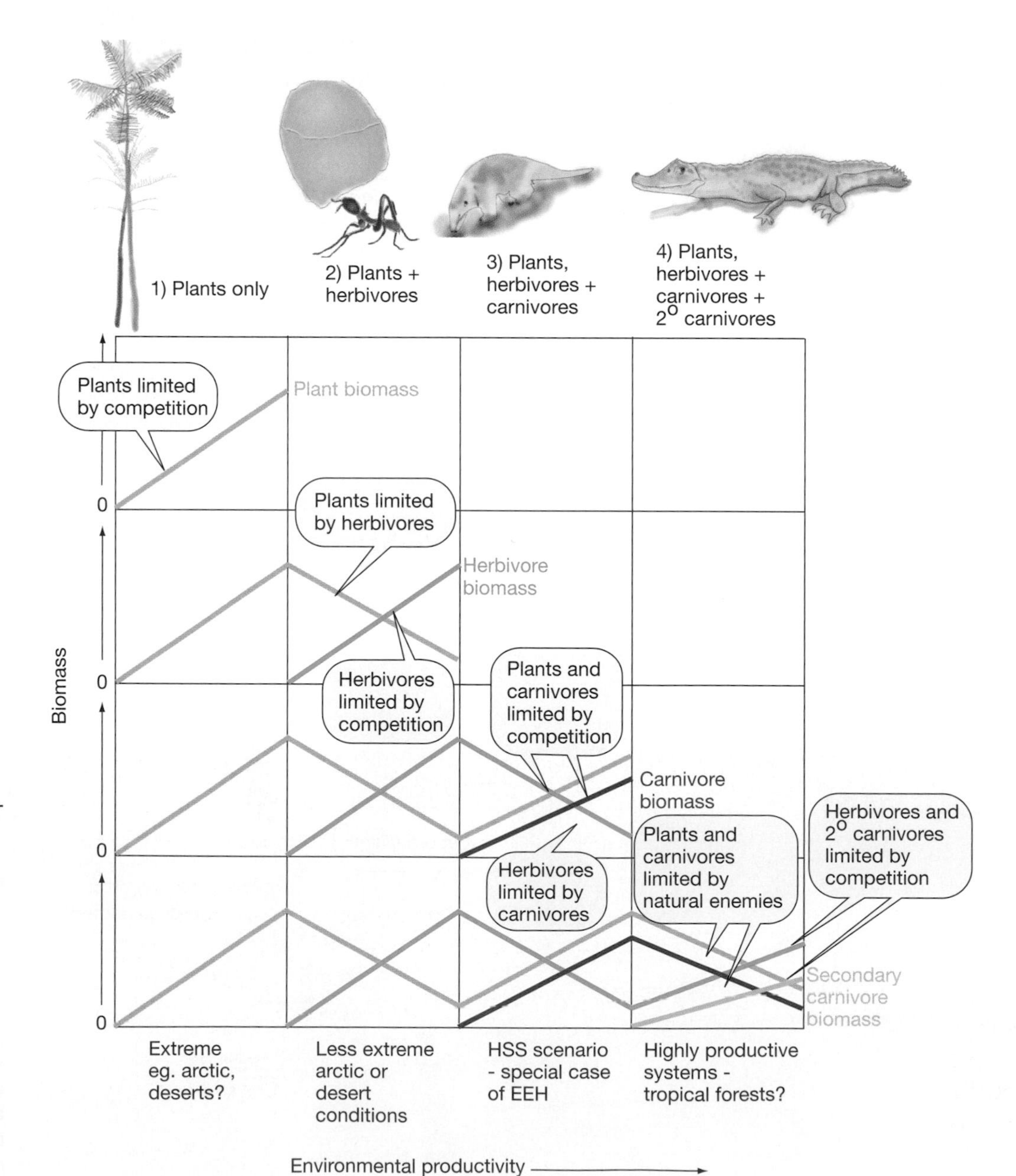

Figure 13.12
Schematic representation of the ecosystem exploitation hypothesis (EEH). The hypothesis states that the biomass of a given trophic level—plants, herbivores, carnivores, or secondary carnivores—will vary according to how many trophic levels are present and that the number of trophic levels is governed by plant productivity. In this scenario, HSS encompasses only systems with three trophic levels (plants, herbivores, and carnivores) and is a subset of the broad EEH hypothesis, which encompasses systems with varying numbers of trophic levels.

lessens competition between plants. If all this sounds confusing, note in Table 13.5 how the strength of various mortality factors varies according to trophic level. The importance of different mortalities in this scheme is related to the number of links in the food chain, or, more precisely, primary productivity.

What empirical support is there for EEH and HSS? Andrew Sih and colleagues (1985) examined 25 papers on terrestrial plants and herbivores and found that only 2 gave support to HSS. Support for EEH has come from studies of lakes and rivers, where the communities are simple and species poor relative to terrestrial communities. In these systems, carnivores greatly reduce the herbivore populations, and this has a dramatic effect on the producer organisms one or even two links down the chain from the carnivores. Such effects are known as *trophic cascades*. Why is the best support for EEH from aquatic systems? Gary Polis and Donald Strong (1996) noted that a critical element is the exceptionally high edibility, nutritiousness, and vulnerability to herbivores of algae that form the bases of aquatic food chains. Such algae lack the sophisticated chemical and even physical defenses of higher plants on land. We noted in Chapter 11 how Bigger and Marvier (1998) had shown that algae are highly susceptible to their herbivores. As a consequence, in aquatic systems, the link between primary productivity and secondary productivity is very tight.

The HSS and EEH models can both be categorized under the rubric "top-down effects," since the effects of predators are important at least some of the time. This contrasts with the bottom-up scenario, in which energy is held to control plant numbers, which in turn control herbivore densities, which in turn control carnivore numbers, and so on. Mark Hunter (see Profiles box) and Peter Price (1992) argued that bottom-up forces must logically be the most important. The removal of higher trophic levels leaves lower trophic levels present—although greatly modified. However, the removal of primary producers leaves no system at all! This argument implies that it isn't surprising that Sih found little empirical support for HSS, and EEH—trophodynamics—makes more sense. However, there is yet a third option to consider: environmental stress.

Environmental Stress.

Another option is that environmental stress, not the passage of energy or the effects of natural enemies, is the primary influence on population densities.

The environmental stress hypothesis, originated by Bruce Menge and John Sutherland (1987), postulates that the strength of various mortalities is governed by environmental stress. The hypothesis envisions that in stressful habitats, higher trophic levels have little effect because they are rare or absent, and plants are affected mainly by environmental stress. Being marine ecologists from Oregon, Menge and Sutherland couched their arguments in terms of intertidal organisms. So, for example, in really rough seas, predators would be absent in the intertidal zone, and plants would be limited by their ability to attach to the rock face. In habitats of moderate stress, there would be a little herbivory, but not enough to affect population densities. In such environments, plant densities are higher and are affected by competition. In still more benign environments, there are many herbivores, and herbivory not competition or environ-

TABLE 13.5 Effects of number of trophic levels on major mortality factors in natural systems.

→ Plant Productivity →

Taxon	Plants Only	Plants and Herbivores	Plants, Herbivores, and Carnivores	Plants, Herbivores, Carnivores, and Secondary Carnivores
Plants	Competition	Herbivory	Competition	Herbivory
Herbivores		Competition	Predation	Competition
Carnivores			Competition	Predation
Secondary carnivores				Competition

Profiles

Mark Hunter,
University of Georgia

Most ecologists whom I talk to describe one part of the job that they really love—the thing that helps them get up in the morning. It might be fieldwork, it might be writing papers, or it might be designing experiments or teaching students. For me, it's the analysis of data. I love playing with numbers and searching for the patterns that those numbers (hopefully) reveal. It's the best mix of science and detective work that I can imagine. The data might be from an experiment we've done in the field or the laboratory, or they might be long-term data collected and passed on by a kindhearted colleague. Either way, the best part of my workday is spent with numbers.

When I was growing up in the 1960s and 1970s in Scotland, there was still some gender bias in the subjects taken by high school students: Boys did physics and math, and girls did biology. Thankfully, that's changed, but it never really occurred to me to be a biologist until it was about time for me to go to university. I'd planned to study physics, but then a friend of mine lent me *King Solomon's Ring* by Konrad Lorenz. Lorenz's enthusiasm for the study of animal behavior in that book was simply infectious, and I changed my mind almost overnight. I ended up at the University of Oxford to study zoology. One of the lecturers I met there was Martin Speight, who taught courses in the ecology of insects. He was an inspiring teacher and later became a good friend, and he's probably responsible for developing my interests in both ecology and entomology.

I stayed at Oxford for my doctoral studies, working with Dick Southwood and Willy Wint on the ecology of oak-feeding insects. I began to realize that my early interest in numbers and recent interest in ecology were naturally combined in the field of population ecology. Really, I've been a population ecologist ever since. I'm fascinated by the causes and consequences of population change in ecological systems. I want to understand the mechanisms that underlie the distribution and abundance of organisms in space and time. Insects are great organisms of study for population ecologists, because they're relatively simple to manipulate experimentally: It's much easier to fit a hundred insects into a growth chamber than a hundred caribou! So I've continued to work with insects that feed on plants, hoping that some of the results of the experimental work can be applied to many more kinds of organism.

Right now, my main interest is in the relative roles of top-down, bottom-up, and lateral forces on the population ecology of herbivores. Top-down forces are the effects of natural enemies, bottom-up forces are variations in the quality and quantity of food, and lateral forces are interactions between competitors at the same trophic level. I think that all of these forces can be important to herbivores; they interact with one another and are modified by the environment in which the organisms live. It's fascinating to me to try to understand how the relative importance of these forces varies in space, varies over time, and varies from species to species. It's not an easy process and seems to require a mixture of field experiments, lab experiments, and long-term sampling data. Luckily, there are quite a few other ecologists who share this interest and who are carrying out studies in a variety of terrestrial and aquatic habitats. I'm optimistic that within the next few years we'll begin to be able to make some generalizations about these processes in a variety of environments. We are only slowly learning to deal with the complexities of this kind of research, such as interactions among ecological forces, variation in the timescales on which they operate, and lagged effects on the dynamics of herbivore populations. But we'll get there.

A few years ago, I moved to the Institute of Ecology at the University of Georgia, and Dac Crossley (now retired) has helped me to understand the importance of the ecosystem in providing a template for population-level studies. Dac has nudged me toward exploring the links between populations and ecosystems. In fact, the role of species in ecosystems may be a focus of ecological research over the next few years. Many ecologists are interested in the effects of biodiversity on ecosystem function for both theoretical and practical reasons. For example, some herbivore species appear to influence the rates of nutrient cycling and energy flow in systems. We need to understand the extent to which changes, natural and human induced, in the populations of these herbivore species will change the environment. I hope that collaboration between population ecologists and ecosystem ecologists will help to answer such questions.

mental stress, controls plant abundance. The strength of top-down factors increases as the environment becomes more benign.

As a good example of the environmental stress hypothesis in action, consider salt marshes. These are among the most productive ecosystems in the world, equivalent in productivity to tropical forests or coral reefs, but they are stressful, being inundated constantly by the tide. Here, levels of herbivory are low, as predicted by the environmental stress hypothesis. Menge and Farrell (1989) reviewed experimental studies in marine ecosystems and found that the environmental stress model was well supported. However, extensive tests have not been done in other systems and there remains the problem of defining what is a stressful habitat for some organisms.

In conclusion, we can identify key factors and density-dependent factors in many, but not all, populations. With that information, we know what perturbs densities away from mean or equilibrium levels and what returns them to those levels. This information is of great interest to fisheries personnel, foresters, and game wardens, each of whom has a vested interest in keeping populations that are under their care at equilibrium levels so that sustainable harvests can be maintained. However, not all populations have key factors or density-dependent factors. In the absence of these factors, many populations can fluctuate widely. Some populations also exist as metapopulations, wherein dispersal between populations, rather than density dependence, controls population persistence. In theory, HSS and EEH offer plausible mechanisms to estimate under which circumstances mortality factors are most important. Of course, we still need much more data to evaluate these theories properly, but at this stage at least, trophodynamics and environmental stress appear better supported than HSS or EEH.

Summary

Populations often seem to stay stable around a mean or equilibrium value. If mortality acts to perturb the population density away from a mean value in one year, the population often bounces back toward the mean in the next year. How does this happen?

1. Mortalities that perturb populations away from mean levels can be thought of as key factors and can be identified by a technique known as key-factor analysis. The key factors for plants and animals are many and varied, and there seems to be no generalization as to which key factors are important for which types of organism.

2. Factors that act so as to return populations to equilibrium levels are called density-dependent factors. Once again, there are few generalizations as to which factors (such as predators, parasites, or disease) act most frequently in a density-dependent fashion. Sometimes density dependence doesn't operate because mortality agents operate in a mode called "spreading the risk" rather than in a density-dependent mode.

3. Sometimes density dependence doesn't operate because populations exist in interdependent groups called "metapopulations." In such groups, dispersal, not mortality, is the key to understanding their dynamics.

4. Several different types of models have been proposed to describe the types of mortality factors that should be most important in different systems. Trophodynamics was among the first such models. This "bottom-up" viewpoint suggests that populations are severely limited by their food supplies. The attenuation of energy severely limits the number of trophic levels in communities. The Hairston, Smith, and Slobodkin (HSS) model suggested that, because the Earth is green with plants, herbivores must have little impact on plant abundance. This is so because predators keep herbivore numbers down. Thus, predators undergo severe competition for prey (herbivores), herbivores are limited by predators, and plants, which do not suffer much herbivory, undergo strong competition for resources. The HSS model was the jumping-off point for many other models, the most well known of which is the ecosystem exploitation hypothesis (EEH), developed by Oksanen and her associates. The EEH suggests that the frequency of competition and predation in a food web varies with productivity and the number of trophic levels in the web.

5. An alternative to both HSS and EEH suggested by Menge and Sutherland is the environmental stress hypothesis, which postulates that biotic complexity decreases with increasing stress. Thus, in polar habitats, the climate is so severe that predators and herbivores are absent, and the few plants that survive are limited by environmental stress. In benign environments, such as tropical forests, herbivory and predation are important.

At the present time, most data appear to support the trophodynamics or the environmental stress hypothesis rather than HSS or EEH. However, predicting when and where certain mortality factors will be most important, either as key factors or density-dependent factors, is a challenge for the future.

Discussion Questions

1. Contrast and compare a key factor with a density-dependent factor. Give examples whenever you can.

2. Think of some local plants, fish, birds, insects, or other animals in your area. Do they tend to have the same densities from year to year? Do you think you would have noticed if they didn't? Try to think of one example in which the abundance changed from one year to the next. What was the cause of this change? What do you think regulates the abundance of this species?

3. Do you think that density dependence has to occur in most populations because there are few examples of sigmoidal or geometric population growth?

4. What do you think is the appropriate scale on which to look for density dependence as a regulator of population size: temporal or spatial? In other words, should we follow populations through many generations to look for temporal density dependence, or is it better to look at many populations across a range of areas to see whether density-dependent processes act over a spatial scale? Are there different interpretations for these different kinds of data?

5. Are territorial animals likely to be more affected by different factors (e.g., social limitations to density) than nonterritorial animals?

6. Andrewartha and Birch (1960) found little use for the terms *density dependent* and *density independent*, because they thought that "these were neither a precise nor a useful framework within which to discuss problems of population ecology." They believed that all factors were density dependent—even the weather—because some or part of the population would always occur in "less favorable situations." For example, in a hard freeze, some organisms would survive in burrows and the rest wouldn't. The higher the population density, the lower is the percentage surviving. Do you agree, and if so, would you scrap the density-dependent-independent terminology?

S E C T I O N IV

Community Ecology

More rain in the tropics means more plants. But why are there more species in the tropics and not just more individuals of one species? Contrast the vegetation of the desert island of San Pedro Martir, in Mexico's Baja California—consisting of just one species of cactus—with the species-rich jungle vegetation of Southeast Asia. Why are there many more species in tropical forests than in temperate forests? Why are there generally a few common species and lots of rare species in a given area? Why do we see weeds and herbs in an old field gradually being replaced by shrubs and trees? Community ecology, which concerns the numbers of species in a community rather than the numbers of individuals in a population, attempts to answer these questions. Community ecology is about biodiversity.

CHAPTER 14

The Main Types of Communities

Road Map

1. In most instances, communities are less like real entities that act like superorganisms than they are individualistic associations of species adapted to the same conditions.
2. Climate causes the distribution of the major communities, or biomes, on Earth, such as rain forests and deserts.
3. Communities can be classified in many ways, but the most common is with the use of two descriptors: climate and type of vegetation (e.g., tropical rain forest or temperate grassland).
4. Some of the main differences between terrestrial communities concern the diversity of species present in them.
5. Aquatic communities are distinguished by the influence of different physical variables, such as water salinity and the strength of the current.

14.1 Are Communities More than the Organisms They Comprise?

So far, we have examined ecology in terms of the behavior of individual organisms and the growth of populations. Most populations, however, exist not on their own, but together with populations of many other species. This assemblage of many populations that live in the same place at the same time is generally known as a **community**. For example, a tropical forest community would consist not only of trees, vines, and other vegetation, but also of the insects that pollinate them, the insect and vertebrate herbivores that feed upon them, and the predators and parasites of the herbivores. Because many populations are dependent upon one another (e.g., plants depend on pollinators and pollinators depend on flowers), studying the community as a whole may provide us more insights than simply studying the individual populations.

As an ecological unit, the community is thought to be a useful object of study because it might exhibit emergent properties—properties beyond those of the individual populations that make up the community. By analogy, a cake has the emergent properties of taste and smell beyond those of its individual ingredients, and salt has emergent properties beyond those of sodium and chlorine. Indeed, life itself appears to be an emergent property. The emergent properties of communities, like the ability to withstand drought or the community's resistance to invasion by exotic species, may be related to the stability or the functioning of the community, such as its primary productivity and efficiency of utilization of soil nutrients. We shall discuss these community properties in subsequent chapters.

Some of the early scientists in the field of community ecology considered a community to be equivalent to a "superorganism," in much the same way that the body of animal is more than just a collection of organs. Indeed, the champion of this group, Frederic Clements (1874–1945), suggested that ecology was to the study of communities what physiology was to the study of individual organisms (Clements, 1905). Some modern-day ecologists still share the "superorganism" idea. For example, perhaps the ultimate extension of Clements's ideas is to regard the whole Earth, or at least the biosphere, as behaving like one giant, single living organism. In his "Gaia hypothesis," British scientist James Lovelock suggested this very thing (Lovelock, 1991).

Clements studied the stable plant communities of the Nebraska grasslands and had observed little change in these communities. He thought that the influence of animals on community structure was trivial and that communities were structured by plant formations. There is no doubt that Clements's ideas and the organismic analogy permeated many textbooks of the day. The ecological historian Robert McIntosh (1985) commented that Clementsian ecology had an orderly neatness that made it pedagogically useful and appealing. Clements's ideas were, and still are, a subject of debate among community ecologists.

Clements's ideas were challenged mostly by Henry Allen Gleason (1882–1975), who proposed an "individualistic" concept of plant association in place of Clements's organismic metaphor (Gleason, 1926). Gleason and others suggested that distinct ecological communities did not exist. Instead, species of plants were viewed as being distributed independently along gradients, so that communities could not be assigned boundaries in a nonarbitrary way. While acknowledging that some communities were fairly uniform and stable over a given region, Gleason argued that not all vegetation could be segregated into such communities. Short-term environmental changes could profoundly affect the composition of a community's flora and fauna. To Gleason and many other mid-20th-century ecologists, it was not possible to create a precisely logical classification of communities, as Clements had tried to do. By the 1950s, many plant ecologists had abandoned many of Clements's principles. In particular, Robert Whittaker's (1953, 1970) studies on the vegetational communities along elevational gradients on mountains asserted the "principle of species individuality," which stated that most communities intergrade continuously and that competition does not create distinct vegetational zones. We saw in Chapter 7 (Fig. 7.27) that the distributions of plant species do not all stop abruptly at one elevation, to be replaced by completely different plants. Thus, there are no distinct community types along mountains; rather, more often, species distributions change independently of one another. Yet it is true that the community at the top of the mountain will be different from that at the bottom, so, broadly speaking, this way of distinguishing associations of species can be useful.

The early history of community ecology focused on whether a community was just a set of associations of species or something over and above that, with a nature and properties of its own.

Communities can occur on a wide range of scales and can be nested: The tropical forest community encompasses the community living in the water-filled recesses of bromeliads, which in turn encompasses the microfaunal communities digesting the cellulose in insects' guts. Each of these entities is a viable community, depending on one's frame of reference with regard to scale (Fig. 14.1). Once a community has been identified, we can describe its type, determine its trophic structure (who eats whom), and calculate the relative biomass of its individual components. We can also count the number of species present (richness) and the abundance of each species and try to come up with an index of distribution of individuals between species (diversity).

Understanding the nature of a community is important in a wide variety of disciplines. For example, in modern tropical agriculture, the emphasis today is on the integration of pasture, livestock, and trees, a practice called **agroforestry** that maximizes the use of the land. Problems associated with particular facets of the community may develop. For example, in tropical Asia, cattle can damage young coconut trees, and the dung of cattle can serve as a breeding place for rhinoceros beetles, one of the major pests of coconut. Adjusting the species mix for optimum yields of both pasture and coconut trees is a complex problem. In conservation, land managers often strive to maximize biodiversity, so a thorough understanding of the concept of community is necessary. And in restoration ecology, practitioners are keen to know whether, following the replanting of areas with natural vegetation, the restored area will recruit animals and attain the diversity of undisturbed areas.

Much of the early work in community ecology focused on classifying groups of plants into recognizable communities. The world encompasses an enormous range of terrestrial and aquatic environments, from

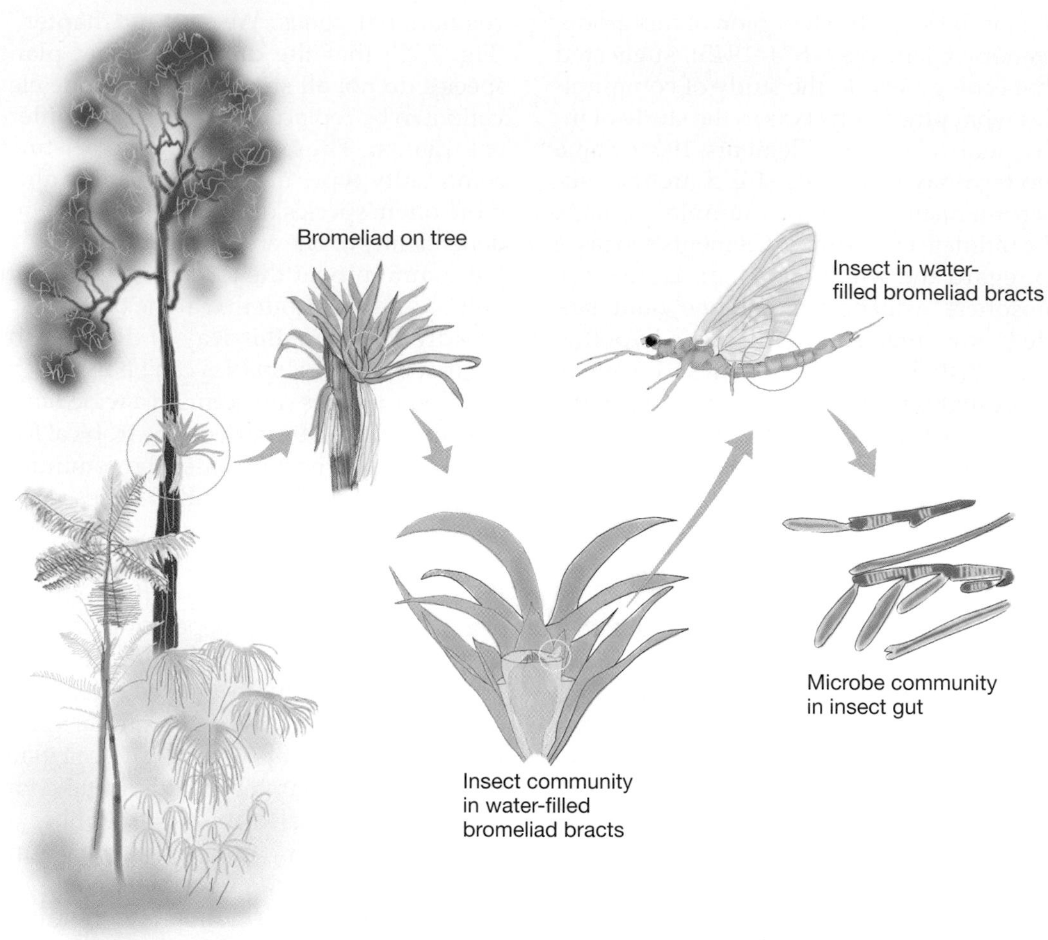

Figure 14.1 The nested nature of communities. On a large scale, we may recognize the tropical forest community and, within it, the community living in a water-filled bromeliad bract. This community consists mainly of insects, with occasional frogs. In turn, there is a separate, well-defined microbe community in each insect's gut, usually enabling the insect to digest cellulose.

polar ice caps to forests to coral reefs. The classification of this immense range of variation into a manageable system is of fundamental importance in the management and conservation of the biosphere. However, the classification of natural communities is problematic. The problems arise because these classifications are ultimately based on the assumption that the natural environment can be divided into a series of discrete, discontinuous units, rather than representing different parts of a highly variable natural continuum. As we have discussed, the latter is probably a more accurate description of the world, because species distributions often change independently of one another. Thus, community structure is only loosely valid. We recognize such broad communities as the tropical forest and tropical grasslands (savannas), but we also recognize that they often intergrade. We can best envisage communities with reference to what species are present at their centers, not at their edges. First, we need to find out what drives the formation of different communities, such as forests, grasslands, deserts, and other types of communities. The quick answer is "Climate," and we shall now examine how climate influences community structure.

Temperature differentials from the poles to the tropics create three cells of wind movement that govern the distribution of rainfall, temperature and, ultimately, community structure.

14.2 Climate and Community Structure

To understand the distribution of communities on Earth, we need to understand climate. There are substantial temperature differentials over the Earth, a large proportion of which are due to variation in the incoming solar radiation. In higher latitudes, the sun's rays hit the Earth obliquely and are thus spread out over more of the Earth's surface than they are in the equatorial regions (Fig. 14.2). More radiation is also dis-

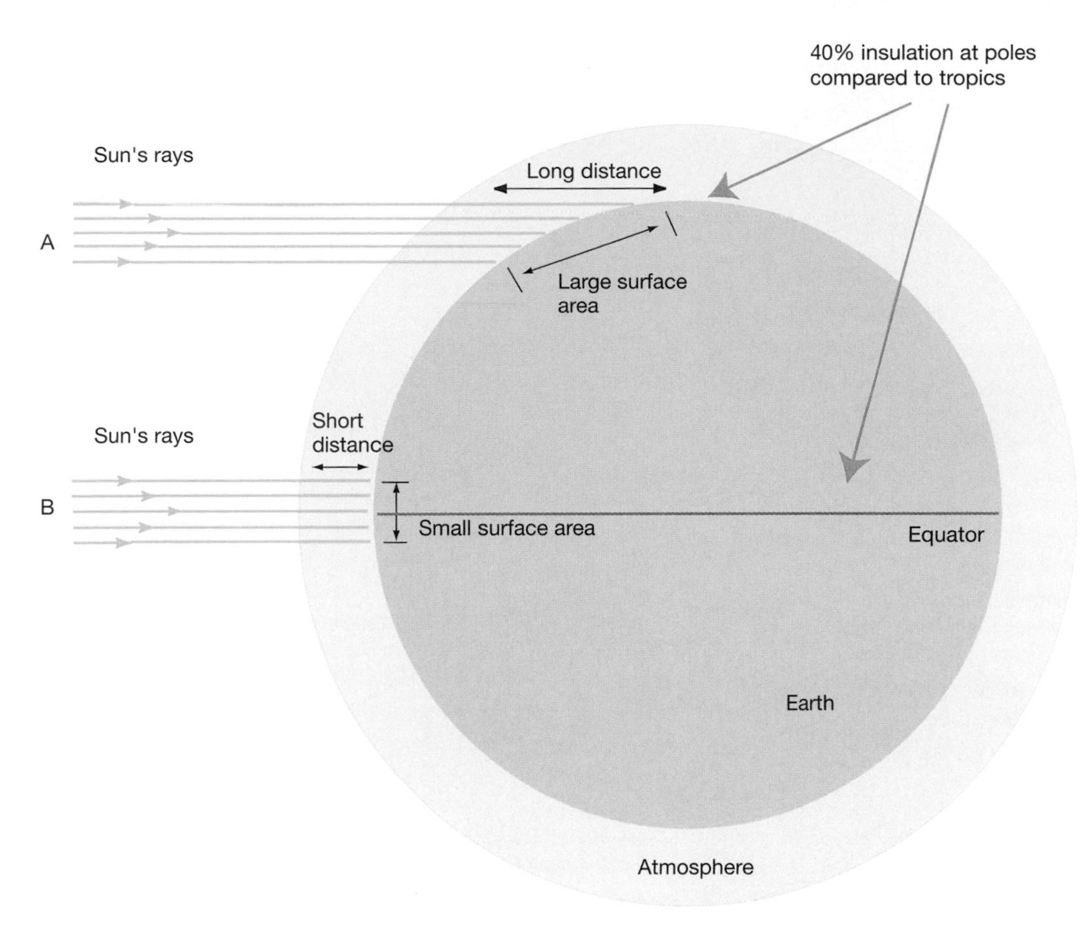

Figure 14.2 Effect of the Earth's shape and atmosphere on incoming radiation. In polar areas, the sun's rays strike the Earth in an oblique manner (A) and deliver less energy than at tropical locations (B), for two reasons: (1) because the energy is spread over a larger surface in A and (2) because it passes through a thicker layer of absorbing, scattering, and reflecting atmosphere.

persed in the higher latitudes, because the sun's rays travel a greater distance through the atmosphere. The result is a much smaller (40 percent) total annual insolation in polar latitudes than in equatorial areas. In the summer, the increased length of the day in high latitudes increases insolation, but the shorter length of the day in winter decreases the total daily insolation. The reason is that the Earth's axis of rotation is inclined at an angle of 23.5° (Fig. 14.3); so the Northern Hemisphere is treated to long summer days while the Southern Hemisphere has winter and vice versa. At the summer solstice in the Northern Hemisphere (June 22), light falls perpendicularly on the Tropic of Cancer; on December 22, it shines perpendicularly on the Tropic of Capricorn. On March 22 and September 22 (the equinoxes), the sun's ray's fall perpendicularly on the equator, and every place on Earth receives roughly the same amount of daylight. These effects do not, however, translate into a linear relationship between temperature at the Earth's surface and latitude: At the tropics, both cloudiness and rain reduce average temperature, and relatively cloud-free areas beyond this zone increase average temperature relative to insolation (Fig. 14.4).

Global temperature differentials create winds and drive the atmospheric circulation. The first contribution to a classical model of general atmospheric circulation was made by George Hadley in 1735. Hadley proposed that (1) solar energy drove winds that in turn influenced the circulation of the atmosphere and (2) the large temperature contrast between the very cold poles and the hot equator would create a thermal circulation. The warmth at the equator caused the surface equatorial air to heat up and rise vertically into the atmosphere. As the warm air rose away from its source of heat, it cooled and became less buoyant, but was unable to sink back to the surface because of the warm air behind it. Instead, it spread north and south away from the equator, eventually returning to the surface at the poles. From there, it flowed back toward the

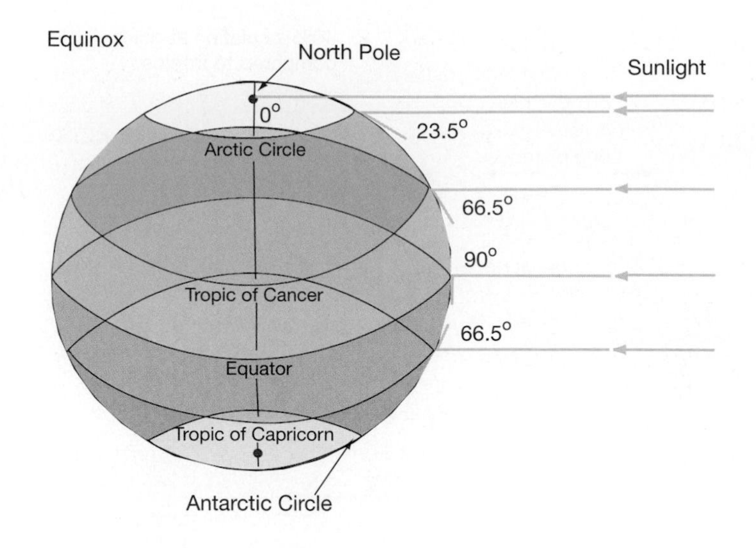

Figure 14.3 Effects of Earth's inclined axis of rotation on amount of insolation during different seasons. Earth's axis of rotation is inclined at an angle of 23.5 degrees, which causes an increasing seasonal variation in temperature and in length of the day with increasing latitude.

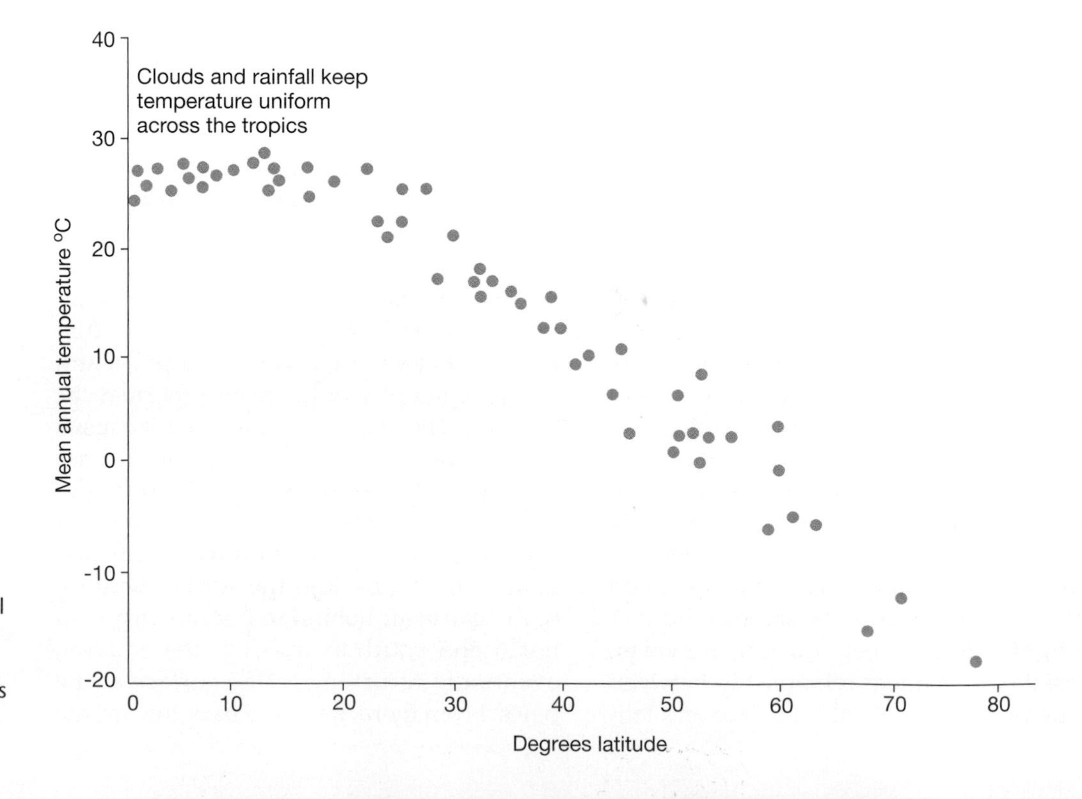

Figure 14.4 Mean annual temperature (°C) of low-elevation, mesic, continental locations on a latitudinal gradient. Note the wide band of similar temperatures at the tropics. *(Reproduced from Terborgh, 1973.)*

equator to close the circulation loop. Hadley suggested that on a nonrotating Earth, this air movement would take the form of one large convection cell in each hemisphere, as shown in Fig. 14.5.

When the effect of the Earth's rotation is added, the surface flow is deflected toward the west. This is a consequence of the Coriolis effect. In any one rotation of the Earth, any point on the equator circumscribes a greater circle than any point north or south. Thus, as the Earth rotates, the equator moves faster than any other circle of latitude. Imagine a rocket fired from the North Pole toward the equator. In the time it took the rocket to reach the equator, hours later, the target point on the equator might have moved 15° to the east. Thus, the straight line of the rocket's path effectively becomes curved toward the west. A similar phenomenon occurs with winds.

Attractive though this simple theory with one large convection cell is, we have to modify it to fit the data. In the 1920s, a three-cell circulation in each hemisphere was proposed to fit the Earth's heat balance (Fig. 14.6). With these three cells and a consideration of the Earth's rotation, the climate patterns of the Earth can be understood.

The contribution of George Hadley is still recognized, in that the most prominent of the three cells, the one nearest the equator, is called the *Hadley cell*. In the Hadley cell, the warm air rising near the equator forms towers of cumulus clouds that provide rainfall, which in turn maintains the lush vegetation of the equatorial rain forests. As the upper flow in this cell moves toward the poles, it begins to subside, or fall back to Earth, in a zone between 20° and 35° latitude. *Subsidence zones* are areas of high pressure and are the sites of the world's tropical deserts, because the subsiding air is relatively dry, having released all its moisture over the equator. Winds are generally weak and variable near the center of this zone of descending air. Subsidence zones have popularly been called the *horse latitudes*. The name is said to have been coined by Spanish sailors, who, crossing the Atlantic, were sometimes becalmed in these waters and reportedly were forced to throw horses overboard, as they could no longer water or feed them.

From the center of the horse latitudes, the surface flow splits into a pole branch and an equatorial branch. The equatorial flow is deflected by the Coriolis force and forms the reliable trade winds. In the Northern Hemisphere, the trades are from the northeast, the direction from which they provided the sail power to explore the New World; in the Southern Hemisphere, the trades are from the southeast. The trade winds from both hemispheres meet near the equator in a region that has a weak pressure gradient: the intertropical convergence zone, also called the Doldrums. Here the light winds and humid conditions provide the monotonous weather that may be the basis for the expression "in the doldrums."

In the three-cell model, the circulation between 30° and 60° latitude is just opposite that of the Hadley cell. The net surface flow is poleward, and because of the Coriolis effect, the winds have a strong westerly (flowing from west to east) component. These prevailing westerlies were known to Benjamin Franklin, perhaps the first American weather forecaster, who noted that storms migrated eastward across the colonies. In fact, the secondary peak of precipitation can

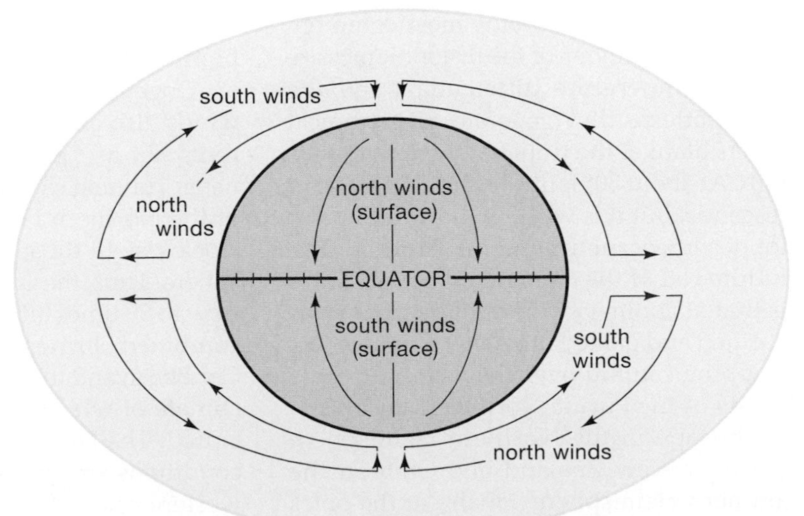

Figure 14.5 Simple convective circulation of air on a uniform, nonrotating Earth, heated at the equator and cooled at the poles, according to the scientist George Hadley in 1735. Winds are named according to the direction from which they blow, so the south wind blows from south to north.

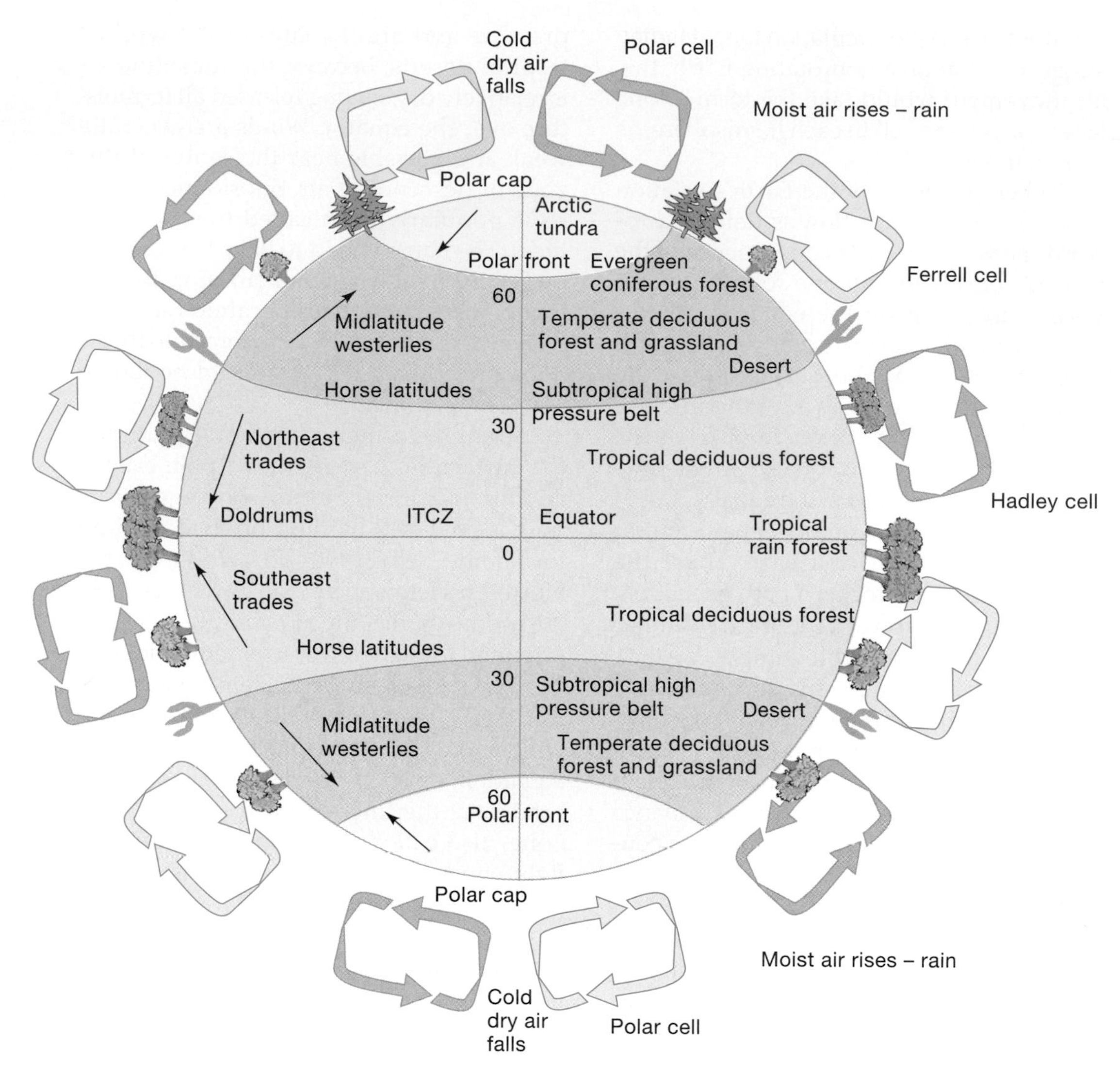

Figure 14.6 Three-cell model of the atmospheric circulation on a uniform, rotating Earth heated at the equator and cooled at the poles. The direction of airflow and the ascent and descent of air masses in three giant convection cells in each hemisphere determine Earth's general climatic zones. This uneven distribution of heat and moisture over different parts of the planet's surface leads to the forests, grasslands, and deserts that make up the planet's biomass.

come anywhere from about 45° to 60°, with between 45° and 55° being most common. So, the distributions of the major biomes are set by temperature differentials and the wind patterns they generate. Hot, tropical forests blanket the tropics, where rainfall is high. At about 30° latitude, the air cools and descends, but it is without moisture, so the hot deserts occur around that latitude. The middle cell of the circulation model shows us that at about 45°–55° latitude the air has warmed and gained moisture, so it ascends, dropping rainfall over the wet, temperate forests of the Pacific Northwest and Western Europe in the Northern Hemisphere and over New Zealand and Chile in the Southern Hemisphere. Finally, at the poles the air has cooled, and it descends, but it has little moisture left, explaining why many high-latitude regions are desertlike.

Together with the rotation of the Earth, winds also create currents. The major currents act as "pinwheels" between continents, running clockwise in the ocean basins of the Northern Hemisphere and counterclockwise in those of the Southern Hemisphere. Thus, the Gulf Stream, equivalent in flow to 50 times all the world's major rivers combined, brings warm water from the Caribbean and the U.S. coasts to Europe, the climate of which is correspondingly moderated. The Humboldt Current brings cool conditions almost to the equator along the western coast of South America (Fig. 14.7).

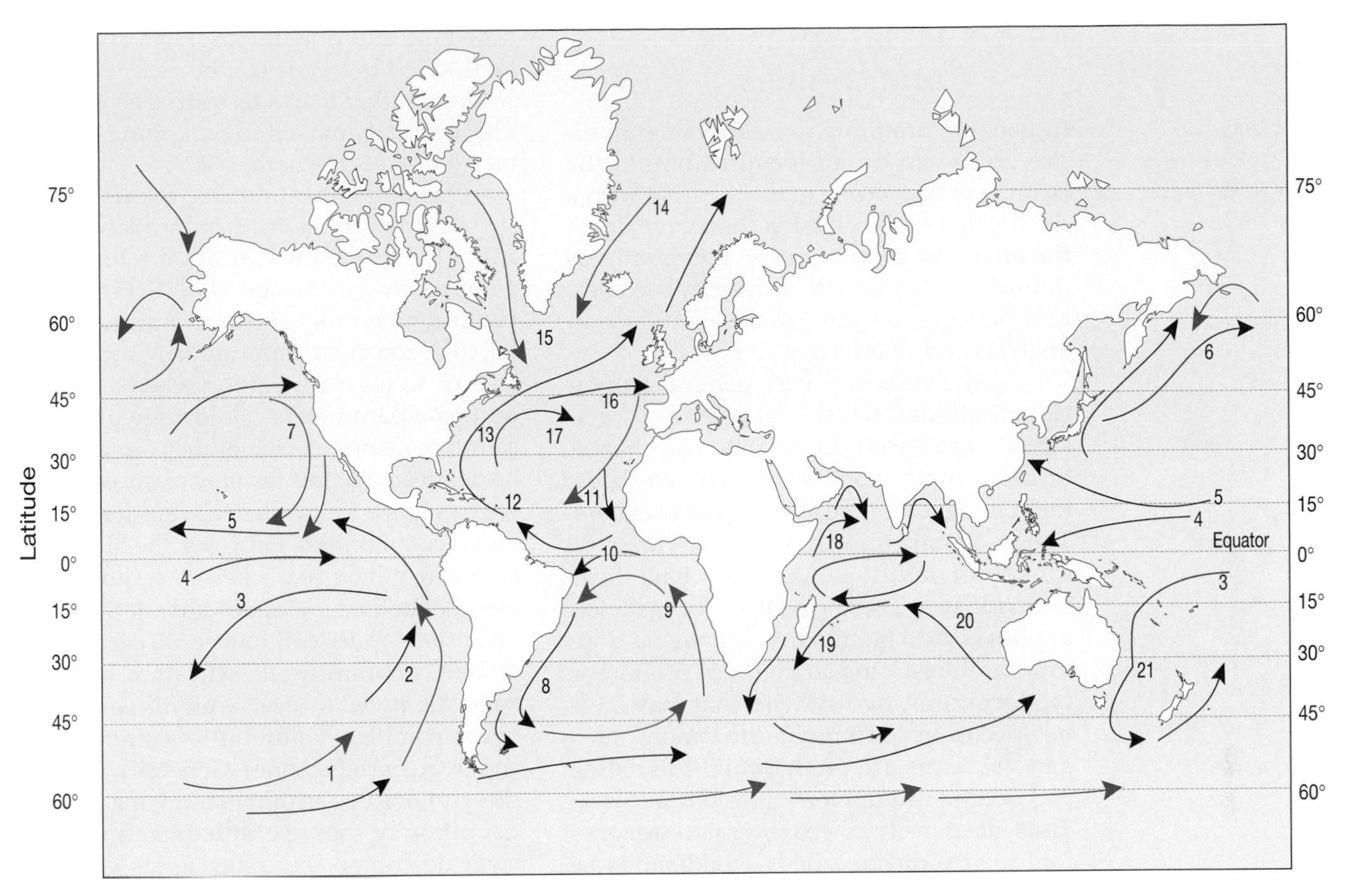

Figure 14.7 Ocean currents of the world: (1) Antarctic West Wind Drift; (2) Peru Current (Humboldt); (3) South Equatorial Current; (4) Counter-Equatorial Current; (5) North Equatorial Current; (6) Kuroshio Current; (7) California Current; (8) Brazil Current; (9) Benguela Current; (10) South Equatorial Current; (11) Guinea Current; (12) North Equatorial Current; (13) Gulf Stream; (14) Norwegian Current; (15) North Atlantic Current; (16) Canaries Drift; (17) Sargasso Sea; (18) Monsoon Drift (east in summer, west in winter); (19) Mozambique Current; (20) West Australian Current; (21) East Australian Current. Red arrows represent cold water.

Superimposed on the broad geographical temperature trends are the influences of altitude and landmass. On mountains, temperatures decrease with increasing altitude, a result of a process known as **adiabatic cooling.** The process works as follows: Increasing elevation means a decrease in air pressure. When wind is blown across the Earth's surface and up over mountains, the air expands because of the reduced pressure; as it expands, it cools, at a rate of about 10°C for every 1,000 m, as long as no water vapor or cloud formation occurs. (Adiabatic cooling is also the principle behind the function of a refrigerator—freon gas cools as it expands coming out of the compressor.) Higher elevations are cooler as well, because the less dense air allows a higher rate of heat loss by radiation back through the atmosphere. A vertical ascent of 600 m is roughly equivalent to a trek north of 1000 km. This explains why alpine vegetation, even in tropical areas, can be tundralike in appearance.

Landmass affects climate because land heats and cools more quickly than the sea does. The land surface reflects less heat than the water, allowing the surface to both warm more quickly and lose heat more quickly than does the sea. The warmed air rises, and cooler air flows in to replace it. This pattern creates the familiar onshore and offshore sea breezes in coastal areas in the morning and afternoon, respectively. The sea therefore has a moderating "maritime" effect on the temperatures of coastal regions and especially islands. The climates of coastal regions may thus differ markedly from those of their climatic zones; many such regions never experience frost, and fog is often evident. Thus, along coastal areas, we may get different vegetation patterns from those seen further inland. In fact, some areas of the United States, including Florida, would be deserts were it not for the warm waters of the Gulf of Mexico and the moisture-laden clouds that form above it.

14.3 Classification of Communities

Communities may be classified in various ways according to climatic conditions and types of vegetation.

In general, attempts to classify communities are based on an identification of the plant taxa that occur in them, along with a description of the physical characteristics of the area. The more rigidly a community is defined, the more site specific it becomes, and hence, the more limited is its use in analysis and planning.

At the extreme, very general habitat classifications ("forest," "grasslands," "wetlands") are based on the physical characteristics and appearance of an area, independently of the species that live there. These classifications cover such a wide range of conditions, that they have little heuristic use: The term *forest*, for instance, applies both to highly diverse lowland tropical rain forests and to northern coniferous monocultures, two systems that may have no species in common. Furthermore, such general terms are often difficult to define and delimit in a universally applicable way. Thus, the density of tree cover necessary before an area can be called a woodland is arbitrary. Similarly, it is difficult to define for how long and how intensely an area must be flooded before it can be classified as a wetland rather than a terrestrial ecosystem. This naturally makes any mapping of communities a problematic task.

One of the most notable efforts at constructing a major global classification system is the Life Zone Classification scheme developed by Holdridge (1967) (Fig. 14.8). Holdridge considered temperature and rainfall to be the most important environmental factors, so his classification scheme focuses on those parameters. Holdridge gave descriptive names to each of the communities he delineated in this fashion, many of which we recognize today, such as rain-forest and desert communities. However, the difference between others (e.g., desert scrub, thorn woodland, thorne steppe, and very dry forest) seems more difficult to discern. Most global community classification systems have attempted to steer a middle course between complex definitions of communities and oversimplified ones. Generally, the best descriptions use a combination of a general definition of the type of habitat and a climatic descriptor—ergo such terms as "tropical forest," "temperate grassland," and "warm

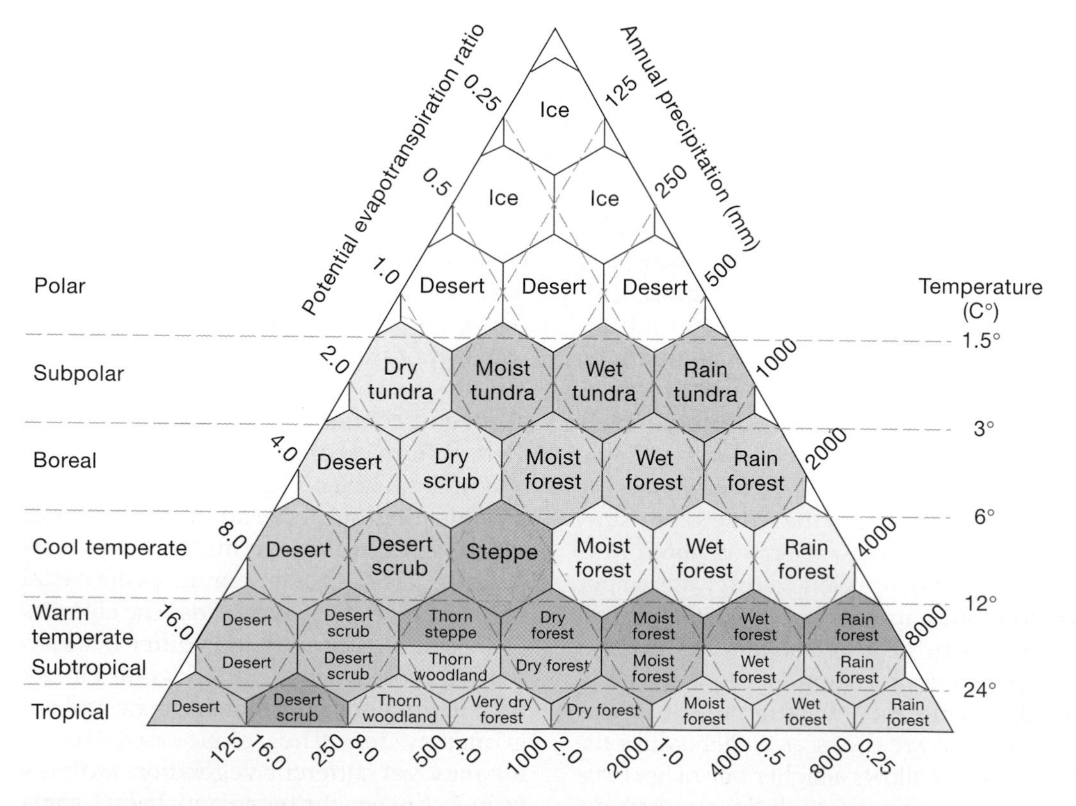

Figure 14.8 Holdridge Life Zone Classification scheme. Potential evapotranspiration is the amount of evaporation that would occur if water were limited. Annual precipitation usually occurs as rainfall, but can be snow. The biotemperature is the mean annual temperature above 0°C, based on the assumption that temperatures below freezing are equivalent to freezing for plants, because they are dormant at and below freezing.

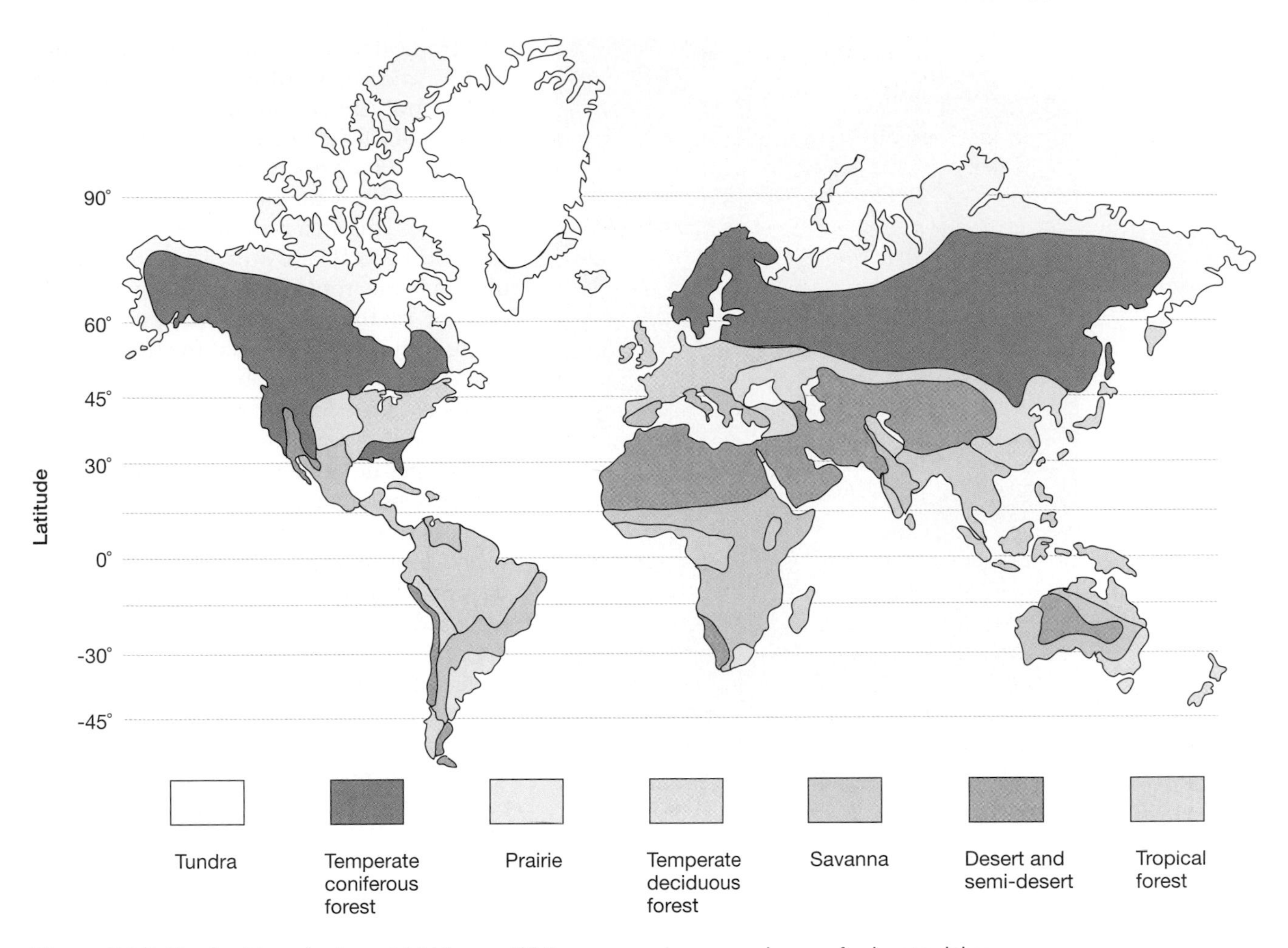

Figure 14.9 The Earth's major terrestrial biomes. White areas are ice covered or are freshwater lakes.

desert." In this scheme, we can recognize seven community descriptions:

1. Tropical forest, always deciduous
2. Temperate forest, deciduous
3. Temperate forest, coniferous (called taiga)
4. Tropical grasslands, called savanna
5. Temperate grassland, called prairie
6. Desert
7. Tundra

We can divide the world into these types of community, or **biomes** (Fig. 14.9). Why should we worry about the classification of communities? If we wish to conserve the world's biota, we need to know in which biome species richness is greatest and in which most plants and animals are threatened. (See "Applied Ecology Box: The Distribution of Threatened Species among Biomes.") However, we should be aware that there are many other classification schemes that rely on different criteria, such as the vegetation of different types of soil, and each of these is a valid way to describe communities. We should not expect there to be a single "best" classification.

14.4 Terrestrial Communities

Terrestrial communities include tropical and temperate forests, savannas, prairies, deserts, and tundra.

Tropical forests.

Tropical forests contain an astonishing diversity of plant and animal species.

Tropical forests are found in equatorial regions where annual rainfall exceeds 240 cm a year and the average temperature is more than 17°C. Thus, neither lack of water nor low temperature is a limiting factor for tree growth. Surprisingly, soils in such areas can be fairly poor, yet still support a luxuriant vegetation (Fig. 14.10). Many of the nutrients are leached out by heavy rainfall. There is no rich organic layer of old leaves, called humus, as there is in temperate systems; thus, fallen leaves are quickly broken down and nutrients are returned to the vegetation, where most of the mineral reserves are locked up. Consequently, cleared tropical forestland does not support agricultural practices well for long.

Figure 14.10 Tropical rain forest on the southwest coast of Costa Rica near Golfo Dulce. The tropical forest has distinct vertical layers that provide niches for animals living on the top of the canopy, in the mid-layers, on epiphytes, and on the forest floor. Leaves and fruits exist year round, permitting a rich array of specialist feeders. *(Gregory G. Dimijian, Photo Researchers, Inc.)*

Tropical forests cover much of northern South America, Central America, western and central Africa, southeast Asia, and various islands in the Indian and Pacific Oceans. The total land area is about three billion hectares (abbreviated ha; 1 hectare is about 2.5 acres). This area represents 23 percent of the world's total land area. The human population in the tropical forestlands is about 20 percent of the world's total population. There is much concern that the tropical forest is being affected by human activity. The diversity of species in tropical forests is staggering, often reaching more than 50 tree species per ha; indeed, the record for most tree species in an area alternates back and forth between southeast Asia and South America, as different areas are censused. The tropical biologist Alwyn Gentry (1988) recorded 283 tree species in one ha of Peruvian rain forest. Sixty-three percent of the species in a 1-ha plot were represented by a single tree each, and there were only twice as many individuals as species.

Rain forest trees are often smooth barked and have large, oval, waxy leaves narrowing to "drip tips" at the apex so that rainwater drains quickly from the fronds. Many trees have shallow roots with large buttresses for support. The tallest trees reach heights of 60 m or more and emerge above the tops of lower trees, which interdigitate to form a closed canopy. Little light pene-

Applied Ecology

The Distribution of Threatened Species among Biomes

It would be valuable to know which types of habitat contained the most threatened species so that we can focus our efforts to preserve habitat. However, such data are available only for the birds of the world and the mammals of Australia and the Americas. (See accompanying figure.) The type of habitat with the largest percentage of threatened species is clearly the tropical forest. This distribution explains why conservation biologists are so keen to conserve all the tropical forests possible.

In general, the world's threatened bird species occupy the same types of habitat as the threatened mammals. However, there are a couple of noticeable differences, the most important of which is that there are far more threatened birds than mammals on oceanic islands. Furthermore, while complete data are lacking, oceanic islands are also known to have many species of threatened plants. For example, Hawaii has one of the most distinctive and most threatened floras in the world, with 108 endemic taxa that have already gone extinct and 138 that are endangered. Other oceanic islands show similar patterns. St. Helena, in the Atlantic Ocean, has 46 endemic plant species that are threatened. Bermuda has 14 of 15 endemic plant species threatened. So tropical forests and oceanic islands appear to be good habitats for maximizing the protection of endangered vertebrates.

Another valuable statistic is the frequency of endemic species in each biome—that is, species unique to that biome. Of the top 25 "hot spots" for endemic species in the world, 15 are in tropical forests and 5 occur in Mediterranean scrub habitat (Myers *et al.*, 2000). This pattern again underlines the importance of these types of habitat to conservation biologists.

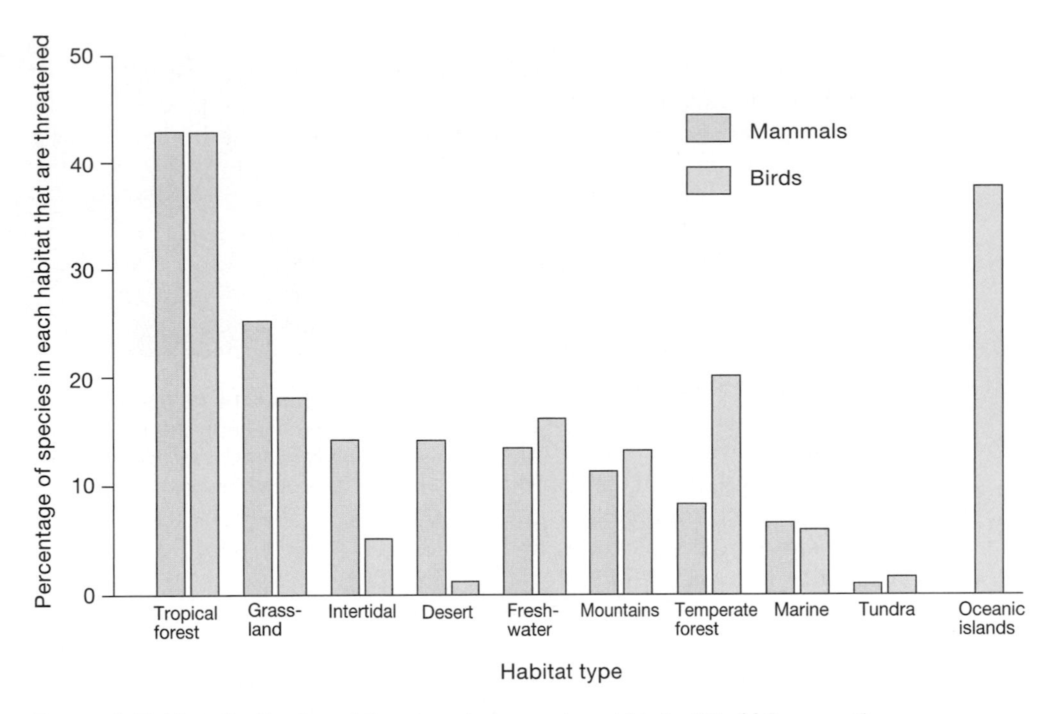

Figure 1 Habitat distribution of threatened mammals and birds. *(World Conservation Monitoring Centre, 1992.)*

trates this canopy, and the ground cover is often sparse. Tropical rain forests are also characterized by epiphytes—plants that live perched on trees and that are not rooted in the ground, although some are rooted in accumulations of organic matter in the branches. Bromeliads are common epiphytes in New World forests. Lianas (climbing vines) are also common.

Animal life in the tropical rain forests is also diverse; insects, reptiles, amphibians, and birds are well represented. Because many of the plant species are widely scattered in the forest, it is more risky for plants to rely on wind to be pollinated or to disperse their seed. This means that animals are important in pollinating flowers and dispersing fruits and seeds. Many plants rely on mutualistic interactions with animals to deliver pollen. As many butterflies can be found on the tropical island of Trinidad as occur in the entire United States—500 to 600 species. Tropical rain forests are the great reservoirs of diversity on the planet; as many as half the species of plants and animals on Earth live in them. Bright protective coloration and mimicry are rampant. Large mammals, however, are not common, although monkeys may be important herbivores.

Temperate forests.

Temperate forest is the type of forest with which many people in the United States and Europe are most familiar (Fig. 14.11). Tem-

Temperate forests occupy midlatitude regions and are dominated by a combination of evergreen and deciduous trees.

Figure 14.11 Temperate forest, like this Douglas fir forest at Cathedral Grove, Vancouver Island, British Columbia, has sparser vegetation than its tropical counterpart, with fewer layers between the forest canopy and the ground. Lower temperatures mean a slower rate of decay, so fallen leaves and branches remain on the ground longer, resulting in a richer layer of humus in the soil. *(S. J. Krasemann, Peter Arnold, Inc.)*

perate forests occur in regions where the temperature falls below freezing each winter, but not usually below −12°C, and annual rainfall is between 75 and 200 cm. Large tracts of such habitat are evident in the eastern United States, east Asia, and western Europe. Commonly, leaves are shed in the fall and reappear in the spring. In the Southern Hemisphere, evergreen *Eucalyptus* forests occur in Australia, and large stands of southern beech (*Nothofagus*) occur in southern South America, New Zealand, and Australia.

Species diversity is much lower in temperate forests than in the tropics; several tree genera may be dominant in a given locality—for example, oaks, hickories, and maples are usually dominant in the eastern United States. Many herbaceous plants flower in spring before the trees leaf out and block the light, although, even in the summer, the temperate forest is usually not as dense as the tropical forest, so there is often some ground cover. Epiphytes and lianas are few. Soils are richer because the annual fall of leaves or detritus is not as quickly decomposed. With careful agricultural practices, soil richness can be conserved, and as a result, agriculture can flourish.

Like temperate plants, temperate animals are adapted to the vagaries of the climate; for example, many mammals hibernate during the cold months. Birds migrate and insects enter diapause, a condition of dormancy usually passed as a pupa. The reptile fauna, dependent on solar radiation for heat, is relatively impoverished compared to the reptile fauna in the tropics. Mammals include squirrels, wolves, bobcats, foxes, bears, and mountain lions.

Deserts.

Deserts occupy areas that experience regular water deficits and are dominated by species tolerant of extreme aridity.

Deserts are biomes suffering from a deficit of water. They are found generally around latitudes of 30°N and 30°S, between the latitudes of tropical forests and temperate forests or grasslands. About one-third of the Earth's terrestrial surface is occupied by hot, dry desert regions. Prominent deserts include the Sahara of north Africa, the Kalahari of southern Africa, the Atacama of Chile, the Sonoran of northern Mexico and the southwest United States (shown in Fig. 14.12), the Gobi of central Asia, and the Simpson of Australia.

Deserts are characterized by two main conditions: a paucity of water (less than 30 cm per year) and usually high daytime temperatures. However, cold deserts do exist and are found west of the Rocky Mountains in the United States, in eastern Argentina, and in much of central Asia. Lacking cloud cover, all deserts quickly radiate their heat at night and become cold. The degree of aridity is reflected in the ground cover. In true deserts, plants cover 10 percent or less of the soil surface; in semiarid deserts, like thorn woodlands, plants cover 10 to 33 percent of the surface. Only rarely do deserts consist of ostensibly lifeless sand dunes, but there are such places: In some regions of the Atacama Desert of western Chile no rainfall has ever been recorded.

Figure 14.12 Sonoran Desert, Arizona. The prominent plants include the tall, columnar saguaro cactus, (*Carnegiea gigantea*), the green spraylike ocotillo (*Fouquieria splendens*), and numerous smaller cholla cacti (*Opuntia* sp.), which are present in the foreground. (*Photo by Peter Stiling.*)

Three forms of plant life are adapted to deserts: (1) annuals, which circumvent drought by growing only when there is rain; (2) succulents, such as the saguaro cactus (*Carnegiea gigantea*), and other barrel cacti of the southwestern deserts, all of which store water; and (3) desert shrubs, such as the spraylike ocotillo (*Fouquieria splendens*), which have short trunks, numerous branches, and small, thick leaves that can be shed in prolonged dry periods. As a strategy against water-seeking herbivores, many plants have spines or an aromatic smell indicative of chemical defenses, although the physical structure of the desert plants—their few leaves and sharp spines—probably also help conserve water and minimize heat absorption. The aboveground parts of perennial desert plants are more widely spaced than those of their forest counterparts, because the roots of the desert plants are longer and occupy greater areas to ensure maximum water-gathering potential (Fig. 14.13).

To conserve water, desert plants produce many small seeds, and animals that eat those seeds, such as ants, birds, and rodents, are common. Reptiles are numerous

Figure 14.13 Illustration of the regular spacing that allows desert plants to maximize their water uptake after rains.

because high temperatures permit these ectothermic animals to maintain their body temperature. Lizards and snakes are important predators of seed-eating animals. Like the plants, desert animals have also evolved many ways of conserving water, such as dry excretion (uric acid and guanine crystals), heavy wax "waterproofing" in insects, and generally crepuscular habits and the use of burrows. Some invertebrates [e.g., brine shrimp (*Artemia* sp.)] follow a strategy similar to that of annuals: Grow and reproduce when it rains, leaving eggs that can survive during drought.

Because of the large amount of sunlight available, irrigated deserts can be extremely productive for agriculture. However, large volumes of water must flow through the system, as the rapid rate of evaporation brings salts from deeper layers of the soil to the surface, where they accumulate and inhibit plant growth. Desert civilizations that harnessed the flow of such rivers as the Tigris, Euphrates, Indus, and Nile dominated early human history. Unlike tropical forests, deserts seem to be expanding under human influence, because overgrazing, faulty irrigation, and the removal of what little woody vegetation exists all speed up desertification. The Sahel region, a narrow, low-rainfall band south of the Sahara whose name is derived from the Arabic word for border, is often argued to be a case in point. The acacia tree, ubiquitous in many arid zones and useful as firewood and forage, was common around Khartoum, capital of Sudan, as recently as 1955. By 1972, the nearest trees were 90 km south of the city, indicating that the desert region had spread southward.

Grasslands dominate regions that are too wet to be deserts and too dry to be forests.

Grasslands.

Grasslands occur in the range between desert and temperate forest in which the rainfall, from 25 to 70 cm, is too low to support a forest, but is higher than necessary to support only desert life-forms (Fig. 14.14). Grasslands are sometimes divided into prairie (temperate grasslands) and savanna (tropical grasslands), according to average temperatures. Some ecologists think that, in addition to the limiting amounts of rain, fire and grazing animals prevent the establishment of trees in the grassland areas. From east to west in North America and from north to south in Asia, grasslands show differentiation along moisture gradients. In Illinois, with about 80 cm of rainfall per year, tall prairie grasses about 2 m high, such as big bluestem (*Andropogon*) and switchgrass (*Panicum*), dominate, whereas along the eastern base of the Rockies, 1,300 km to the west, where rainfall is only 40 cm annually, short-grass prairies exist, rarely exceeding 0.5 m in height and consisting of buffalo grass (*Buchloe*) and blue grama (*Bouteloua*). Similar gradients occur in South Africa (the veldt) and in Argentina and Uruguay (the pampas).

Figure 14.14 In between the areas of desert and temperate forest are the grasslands—vast areas of treeless plain such as this prairie pasture in Nebraska. Some of the largest mammalian herbivores and their predators exist on prairies and savannas. (*Grant Heilman, Grant Heilman Photography, Inc.*)

Nowadays, few original grasslands remain. Prairie soil is among the richest in the world, with 12 times the humus found in a typical forest soil. Historically, and where the grasslands remain, large mammals are the most prominent members of the fauna; examples are bison (buffalo) and pronghorn (antelope) in North America; wild horses in Eurasia; large kangaroos in Australia; and a diversity of antelopes, zebras, and rhinoceroses in Africa, as well as their associated predators (lions, leopards, cheetahs, hyenas, and coyotes). Burrowing animals, such as gophers (in North America) and mole rats (in Africa), are also common.

The taiga is the coniferous forest that dominates subarctic latitudes in North America and Eurasia.

Taiga.

North of the temperate-zone forests and grasslands lies the biome of coniferous forests, known commonly by its Russian name, *taiga* (Fig. 14.15). Most of the trees are evergreens or conifers with tough needles that may persist three to five years before being replaced by new needles. Spruces (*Picea*), firs (*Abies* and *Pseudotsuga*), and pines (*Pinus*) generally dominate, but some deciduous species, such as aspens, alders, and willows, occur along watercourses. All these species are highly tolerant to freezes and can withstand temperatures of −60°C. Many of the conifers have conical shapes to reduce the likelihood that their boughs will break from heavy loads of snow. As in tropical forests, the understory is thin because the dense, year-round canopies prevent sunlight from penetrating. Soils are poor because the fallen needles decay slowly in the cold temperatures, and organic matter often builds up in a layer of litter. The layer of needles also acidifies the soil. Snakes are rare, and there are few amphibians. Insects are strongly periodic, but may reach outbreak proportions in times of warm tem-

Figure 14.15 Taiga, the coniferous forest of North America and Eurasia, here in Quebec, Canada.

peratures. Mammals, such as bears, lynx, moose, beavers, and squirrels are heavily furred. The taiga is famous for cyclic population patterns, of which hares and lynx are well-known examples. In the Southern Hemisphere, little land area occurs at latitudes at which one would expect extensive taiga to exist. (See Fig. 14.9.)

Tundra.

The tundra is the last major terrestrial biome, occupying roughly 17 percent of the Earth's surface (Fig. 14.16). Like taiga, tundra exists only in the Northern Hemisphere (north of the taiga), because there is very little land area in the southern hemisphere at the latitude where tundra would occur. Precipitation in the tundra is generally less than 25 cm per year, and because it is often locked up as snow, it is largely unavailable to plants. Deeper water can be locked away for a large part of the year in permafrost. With so little available water, trees cannot grow. Summer temperatures are only 5°C, and even during the long summer days the ground thaws to less than 1 meter in depth. Midwinter temperatures average −32°C. Vegetation occurs in the form of fragile, slow-growing lichens, mosses, grasses, sedges, and occasional shrubs, which grow close to the ground. In some places, so little moisture falls that desert conditions prevail. Because permafrost is impenetrable, water drainage is inhibited, and water lies in shallow lakes and ponds on the surface of the Earth in the summer. The anaerobic (oxygenless) conditions of the waterlogged soil and the low temperatures slow nutrient cycling. Organic matter cannot completely decompose, and the humus often accumulates in thick layers known as peat. Animals of the arctic tundra have adapted to the cold by having good insulation. Many birds—especially shorebirds and waterfowl—migrate. The fauna is much richer in summer than in winter. Many insects spend the winter at immature stages of growth, which are more resistant to cold than the adult forms. The larger animals include herbivores such as musk oxen and caribou in North America and reindeer in Europe and Asia, as well as the smaller hares and lemmings. Common predators are arctic fox, wolves, snowy owls, polar bears near the coast, and others.

Regions that are very cold are occupied by tundra communities.

Other biomes.

Of course, not all communities fit neatly into the six major types of biome (or seven if you divide grasslands into prairie and savanna).

Other, smaller terrestrial biomes, like temperate rain forests, tropical seasonal forests, and chaparral, also exist.

Figure 14.16 Tundra, the vast treeless zone north of the taiga, here shown on the rolling hills of Polychrome Pass in Denali National Park, Alaska.

Marine biomes include the ocean's intertidal rocky and sandy shores, coral reefs, pelagic zone, and benthic zone.

As with most things ecological, there are characteristic regions where one type of biome grades into another. For example, coniferous forests occur in some temperate lowlands where one might think deciduous trees would dominate. In the eastern United States, most of New Jersey's coastal plain is sandy, nutrient-poor soil that cannot support deciduous forest. Instead, this coastal area is dominated by pine barrens, a type of scrub forest that contains grasses and low shrubs growing among open stands of pine and oak trees. Similar habitats exist in the coastal plains of North Carolina, South Carolina, Georgia, and Florida. Another distinct type of biome is *chaparral*, a Mediterranean scrub habitat adapted for fire. Chaparral is common along the coastlines of southern Europe, California, South Africa, and southwest Australia. In such terrain, rainfall may be sufficient to support large trees, but frequent fires prevent the trees from surviving long enough to grow tall.

Mountain ranges must be treated still differently. The nature of a biome is predominantly a function of climate, and on mountains the temperature decreases with increasing altitude by adiabatic cooling, as discussed earlier. Precipitation changes with altitude, too, generally increasing in desert elevations, but decreasing on the leeward sides of slopes, which are in a rain shadow. Approaching clouds usually have dumped all their moisture on the windward side. Thus, a biome may change from temperate forest through taiga and into tundra on an elevation gradient in the Rocky Mountains and even from tropical forest to tundra on the highest peaks of the Andes in South America. In the tropics, of course, daylight varies little from 12 hours per day throughout the year. So instead of an intense period of productivity, vegetation in the tropical alpine tundra exhibits slow, but steady, rates of photosynthesis and growth all year.

14.5 Aquatic Communities

Within aquatic environments, different types of biome can also be recognized: rivers, freshwater lakes, and, within saltwater oceans, the intertidal rocky shore, sandy shores, coral reefs, and the pelagic zone, or open ocean. These biomes are distinguished by differences in such parameters as salinity, oxygen content, strength of the current, and availability of light.

Marine habitats.

Marine environments are among the most extensive and uniform on Earth. Marine ecosystems are found over nearly three-quarters of the Earth's surface. Like freshwater communities, marine communities are affected by the depth at which they occur. The shallow zone where the land meets the water is called the **intertidal zone**. Beyond the intertidal zone is the **neritic zone**, the shallow regions over the continental shelves. Past the continental shelf is the open ocean, or **oceanic zone**, which may reach great depths. The open ocean is often referred to as the **pelagic zone**. At the bottom is the sea floor, or **benthic zone**. We may also recognize the stratum of water near the surface to which depth light penetrates as the **photic zone** and, below, the dark **aphotic zone**. Phytoplankton, zooplankton, and most fish species occur in the photic zone. In the aphotic zone, production by plants is virtually zero, and only a few invertebrates and luminescent fish are able to live there, feeding on the debris from the photic zone, which rains down upon them.

The intertidal zone, where the land meets the sea, is alternately submerged and exposed by the daily cycle of tides (Fig. 14.17). The resident organisms are subject to a great daily variation in the availability of seawater and in temperature. They are also battered by waves, especially during storms. In temperate areas, they may be subject to freezing in the winter or very hot temperatures in the summer. At low tides, they may be dry and vulnerable to predation by a variety of animals, including birds and mammals. High tides bring predatory fishes. Commonly, there is a vertical zonation consisting of three broad zones, most evident on rocky shores. The upper littoral is submerged only during the highest tides. The midlittoral is submerged during the highest regular tide and exposed during the lowest tide each day. Life in this biome may be quite rich, consisting of green algae, sea anemones, snails, hermit crabs, and small fishes living in tide pools. Competition for space on the faces of rocks may be intense. The lower littoral is exposed only during the lowest tide, and the diversity and richness of organisms is great. Along sandy and muddy shores, few large plants or other sessile organisms can grow, because the sand or mud is constantly shifted around by the tide. Instead, this ecosystem contains burrowing marine worms, crabs, and small isopods.

Coral reefs exist in warm tropical waters (Fig. 14.18), a conspicuous and distinct biome. Currents and waves constantly renew nutrient supplies, and sunlight penetrates to the ocean floor, allowing photosynthesis. Coral reefs are composed of organisms that secrete hard external skeletons made of calcium carbonate. The skeletons vary in shape, forming a substrate that other corals and algae grow on. An immense variety of microorganisms, invertebrates, and fish live in the coral, making the reef one of the most interesting and richest biomes on Earth. Probably 30 to 40 percent of all fish species on Earth are found on coral reefs. Prominent herbivores include snails, sea urchins, and fish, which are in turn consumed by octopuses, sea stars, and carnivorous fish.

In the pelagic zone (open ocean), nutrient concentrations are typically low, although the waters may be periodically enriched by upwellings of the ocean that carry mineral nutrients from the bottom to the surface. Pelagic waters are mostly cold, warming only near the surface, where many photosynthetic plankton grow and reproduce. Their activity counts for nearly half the photosynthetic activity on Earth. Scientists have suggested that if phytoplankton productivity were increased, much of the carbon dioxide produced by burning fossil fuels would be soaked up, and global warming would be slowed. One of the limiting factors in phytoplankton growth seems to be the availability of iron. Large experimental additions of iron to the Pacific have increased phytoplankton production (Van Scoy and Coale, 1994), raising the possibility that this should be done on even bigger scales to help reduce the atmospheric buildup of CO_2.

Zooplankton, including some worms, copepods (tiny shrimplike creatures), tiny jellyfish, and the small larvae of invertebrates and fish, graze on the phytoplankton. The pelagic zone also contains free-swimming animals, called nekton, which can move against the currents to locate food. The phytoplankton and zooplankton move with the current. The nekton include large squids, fish, sea turtles, and marine mammals that feed on either plankton or each other. Only a few of these organisms live at any great depth. Those which do may have enlarged eyes, enabling them to see in the dim light. Others have luminescent organs that attract mates and prey.

Figure 14.17 Purple sea stars (*Pisaster ochraceus*) in the midlittoral zone at Olympic National Park. There is a distinct vertical zonation on rocky shores: upper littoral (submerged only during the highest tides), midlittoral (submerged during normal tides), and lower littoral (exposed only in the lowest tides). (*Jim Zipp, Photo Researchers, Inc.*)

Freshwater.

Freshwater habitats include communities in running and still waters.

Freshwater habitats are traditionally divided into standing-water lentic habitats (from the Latin *lenis*, calm—as are lakes, ponds, and swamps) and running-water lotic habitats (*lotus*, washed—as by rivers and streams). The ecology of lentic habitats is governed largely by the unusual properties

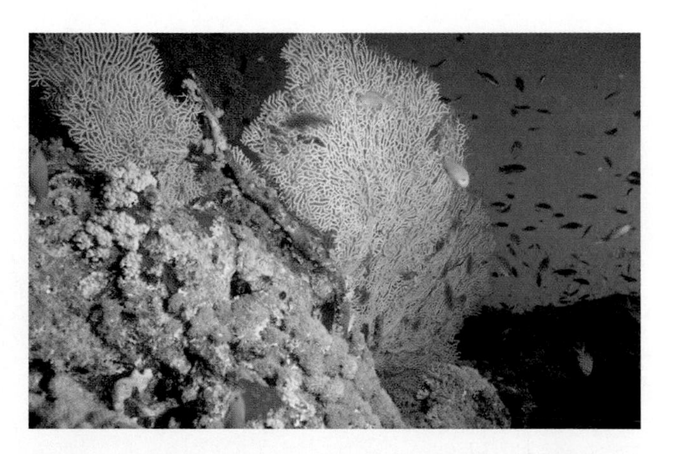

Figure 14.18 A coral reef in the Red Sea. Coral reefs have the most species-rich marine communities on Earth. (*Jeff Rotman, Peter Arnold, Inc.*)

of water. First, water is at its least dense when frozen; ice floats. From a fish's point of view, this property is advantageous, because a frozen surface insulates the rest of the lake from freezing. If ice sank, all temperate lakes would freeze solid in winter, and no fish would exist in lakes outside the tropics. Water is at its densest at 4°C. Thus, as long as no water in the lake is colder than 4°C, the warmest water is at the surface, and temperature declines with depth, although not in a linear fashion. Normally, three layers are present (Fig. 14.19). An upper layer, called the *epilimnion*, is warmed by the sun and mixed well by the wind. Below this lies the *hypolimnion*, a cool layer too far below the surface to be warmed or mixed. The transition zone between the two is known as the *thermocline*. Organisms that need oxygen cannot usually live in the hypolimnion, because its oxygen is depleted during the summer, a phenomenon known as summer stagnation.

Like the ocean, a lake has zonations based on the availability of light. The upper layer, to which light penetrates, is the *autotrophic*, or *photic*, zone. Below, in darkness, is the *profundal*, or aphotic, zone, where heterotrophs live and subsist on the rain of material from above. The depth of the photic zone depends on the availability of light and the clarity of the water. The level at which photosynthate production equals the energy used up by respiration is the lower limit of the photic zone and is known as the **compensation level** or *compensation point*. In the summer, in temperate lakes, the compensation level is usually above the thermocline.

The degree of productivity in lakes also determines their characteristic fauna and flora. The least productive lakes are termed **oligotrophic**. Such lakes generally have a low nutrient content, largely as a result of their underlying substrate and young geologic age. Young lakes have not had a chance to accumulate as many dissolved nutrients as have older ones. Oligotrophic lakes are relatively clear, and their compensation levels may lie below the thermocline. If so, photosynthesis can take place in the hypolimnion, adding oxygen. Low nutrient concentrations keep the algae and rooted plants in the epilimnion sparse, and little debris rains down upon the inhabitants of the hypolimnion. As a result, oligotrophic lakes are clear and often contain fish such as trout.

Even though few nutrients are present in oligotrophic lakes, eventually they do begin to accumulate. Sediments are deposited, and both algae and rooted vegetation begin to bloom. Organic matter accumulates on the lake bottom, the respiration of bottom dwellers increases, the water becomes more turbid, and the oxygen levels of the water go down. Fish such as trout are excluded by bass and sunfish, which thrive in warm water and at low oxygen concentrations. This process of aging and degradation is natural and is termed **eutrophication**; its end result is a **eutrophic** lake. Eutrophication, however, can be greatly speeded up by human influences, which increase nutrient concentrations through the introduction of sewage and fertilizers from agricultural runoff. This human-produced eutrophication is often termed *cultural eutrophication*. A measure of eutrophication is the *biochemical oxygen demand*, which is the difference between the production of oxygen by plants and the amount of oxygen needed for the respiration of the organisms

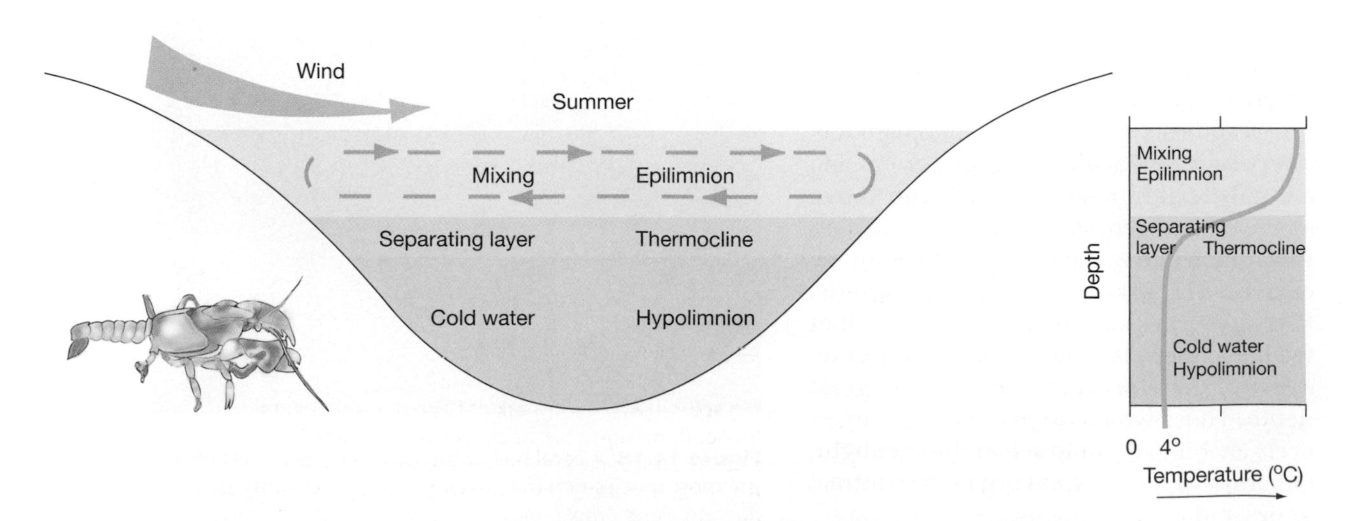

Figure 14.19 Cross section of lake stratification and profile of temperature with depth.

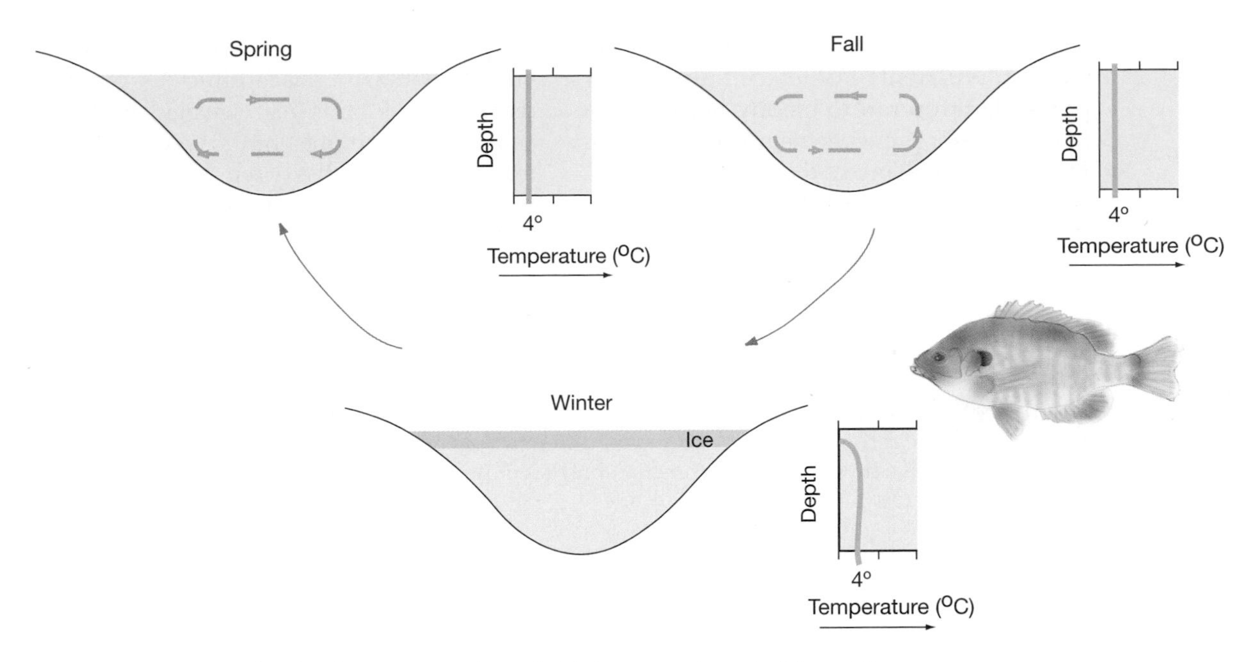

Figure 14.20 Annual cycle of a temperate lake. Cold air temperatures in fall cool the upper layers, and the dense, cold water sinks, thoroughly mixing the waters of the lake. The lake surface freezes in winter. When the ice melts in the spring, the cold water again sinks and mixes the lake waters.

in the water. Biochemical oxygen demand is normally measured in the laboratory as the number of milligrams of oxygen consumed per liter of water in five days at 20°C.

The stratified nature of temperate-zone lakes in summer does not last all year (Fig. 14.20). In the fall, the upper layers cool and, as their density increases, sink, carrying oxygen to the bottom of the lake and forcing the bottom layers upward. The water in the lake is thoroughly mixed by this action and by storms, and the thermocline disappears. In winter the surface usually freezes, and no turnover of water occurs; once again, a gradient is set up. Then, as spring returns, the ice melts and sinks, and the water temperature rises, producing another mixing, called the **spring overturn**. In contrast to temperate lakes, tropical lakes are often isothermal (that is, all at one temperature) or exhibit at most only a weak temperature gradient from top to bottom, because of the absence of seasonality. Little mixing occurs, and deep lakes are generally unproductive, with oxygen-poor, fishless lower depths, as the builders of tropical dams learn (to their chagrin!). Worse still, most water from dams is drawn off from the base of the structure, meaning that the streams below dams are much less oxygen rich than those above them.

Lotic, or running-water, habitats (Fig. 14.21) generally have a fauna and flora completely different from those of lentic waters. Plants and animals are adapted to help

Figure 14.21 North Fork of Payette River, central Idaho. Running-water habitats contain animals, such as Caddis flies, that have a dorso-ventrally flattened body and are adapted to remain in place despite a strong current. (*Frazier, Photo Researchers, Inc.*)

them remain in place despite an often strong current. Nutrients fail to accumulate and phytoplankton fail to bloom, because the flowing water prevents them from doing so. The current also mixes the water thoroughly, providing a well-aerated regime. Animals of lotic systems are therefore not well adapted to low-oxygen environments and are particularly susceptible to oxygen-reducing pollutants (e.g., sewage). Fish such as trout may be present in rivers with cool temperatures, high oxygen, and clear water. In warmer, murkier waters catfish and carp may be abundant.

Summary

A community consists of populations that live in the same place at the same time and that are dependent upon each other. Communities can be recognized over a variety of different scales, from tropical forests, to the insect communities that live in water-filled bromeliad bracts, to the microbe communities in insect guts.

1. Community ecologists have long debated whether communities are superorganisms or merely associations of species. Although most ecologists hold the latter view, communities may have emergent properties.

2. Differences in temperature between the poles and the tropics create three cells of wind movement. Together with Earth's rotation and the locations of landmasses, these cells govern the world's temperature and rainfall distribution, which in turn governs the distribution of the various types of communities around the globe.

3. Global classification schemes for communities, such as the Holdridge Life Zone Classification are based mainly on the types of vegetation that prevail in different temperature and rainfall regimes.

4. Large-scale terrestrial communities are referred to as biomes and include tropical forests, temperate deciduous forests, deserts, temperate grasslands (prairies), tropical grasslands (savannas), temperate coniferous forests (taiga), and tundra. Tropical forests cover 23% of the world's land area, have astoundingly high species diversity, and are particularly susceptible to disturbance by human activity.

5. Aquatic communities include marine and freshwater communities and vary in properties such as salinity, strength of the current, availability of light, and oxygen content. Marine communities include the intertidal rocky shore, sandy shore, coral reef, and open ocean. Freshwater communities are divided into lentic habitats (with still waters, such as lakes) and lotic habitats (with moving waters, such as rivers).

6. Tropical forests and oceanic islands have a high proportion of the world's endangered vertebrates, and tropical forest and Mediterranean scrub communities are particularly rich in endemic species. Accordingly, these types of communities are of particular interest to conservation biologists.

Global climate patterns determine biome locations on a broad scale, and local variables such as the availability of nutrients in the soil modify the distribution of these communities on a finer scale. Species richness differs dramatically from biome to biome and must be taken into account in conservation decisions.

Discussion Questions

1. Which do you think is the most meaningful scale on which to examine communities? For example, are the insects and other invertebrates that inhabit rotting logs a more tightly knit, and therefore more biologically meaningful, type of community than the temperate grassland community?

2. Do you think that the broad divisions of biomes, such as temperate forest, temperate grassland, tropical forest, and tropical grassland, are the most useful? What finer divisions can you envisage? Bailey (1989) and Heywood (1995) count a multitude of divisions, including forest meadow and semishrub deserts. When would these categories be useful?

3. Which communities or biomes do you expect to be most degraded by humans and why?

CHAPTER 15

Global Patterns in Species Richness

Road Map

1. From the tropics through temperate areas to the poles, there is a gradient in species richness, with the highest richness in the tropics and the lowest in the polar areas. Many biotic and abiotic factors can explain the changes in species richness over these large scales.
2. Some communities in similar habitats from different parts of the world, such as plants in deserts, converge in numbers of species. Others, such as lizards in deserts, do not.
3. It is estimated that there are 12.5 million species on Earth.
4. To preserve biodiversity, we can focus on saving countries with the highest numbers of species or saving areas with the highest numbers of endemics.
5. The preservation of biodiversity is important because recent experiments show that communities perform best when they have a full complement of species.

Canada has a system of about 34 parks located in nearly all of the principal biomes in the country. The parks were established for the benefit, education, and enjoyment of Canadians and are supposed to be maintained so as to leave them unchanged for future generations. One measure of change is the number of species they contain, commonly called the *species richness* of the parks. In 1999, Daniel Rivard and colleagues compared lists of species of terrestrial and aquatic vertebrates when the parks were established with lists produced from recent systematic surveys. They found that some species, including the bison, wolf, beaver, cougar, lynx, marten, fisher, turkey, caribou, and black bear, had been lost from certain parks. The missing species were mainly those hunted by humans (Fig. 15.1). At the same time, species richness *increased* in some of the parks, through an influx of species also associated with humans: rats, starlings, sparrows, and pigeons. Other species had been deliberately introduced and included largely game species, such as salmonid fish, deer, moose, and elk. If we could understand what factors influence species richness, we would be in a better position to manage parks for richness. Rivard and colleagues found that changes in species richness were strongly related to climate, particularly mean annual potential evapotranspiration. Evapotranspiration is water lost through evaporation or transpiration in plants. Sometimes evapotranspiration is limited by the amount of water that is available. Potential evaporation is water that

Figure 15.1 Canadian lynx and young; one of the flagship species Canadian parks are designed to preserve.

would be lost if it were superabundant and not limiting to plants. For example, transpiration is low in deserts because there is little water there, but potential evapotranspiration is high because it is so hot that evapotranspiration could be higher were water available. So the change in species richness in Canadian parks was highest in the warmer, southern areas. In that case, the change in species richness coincides with evapotranspiration because human activity varies with climate. The warmer areas of Canada are the areas most affected by humans. Understanding what causes species richness to change in other parts of the world is central to the mission of conservation biology.

15.1 Explanations of Species Richness Gradients

In general, the number of species in any habitat increases from polar areas through temperate areas and reaches a maximum in the tropics. For example, the species richness of North American mammals increases from Arctic Canada to the Mexican border (Fig. 15.2a), and so does the richness of birds (Fig. 15.2b). Such changes are linked to temperature.

Species richness is also increased by topographical variation—hence, the increase of birds and mammals in the west in Fig. 15.2. This pattern occurs because moun-

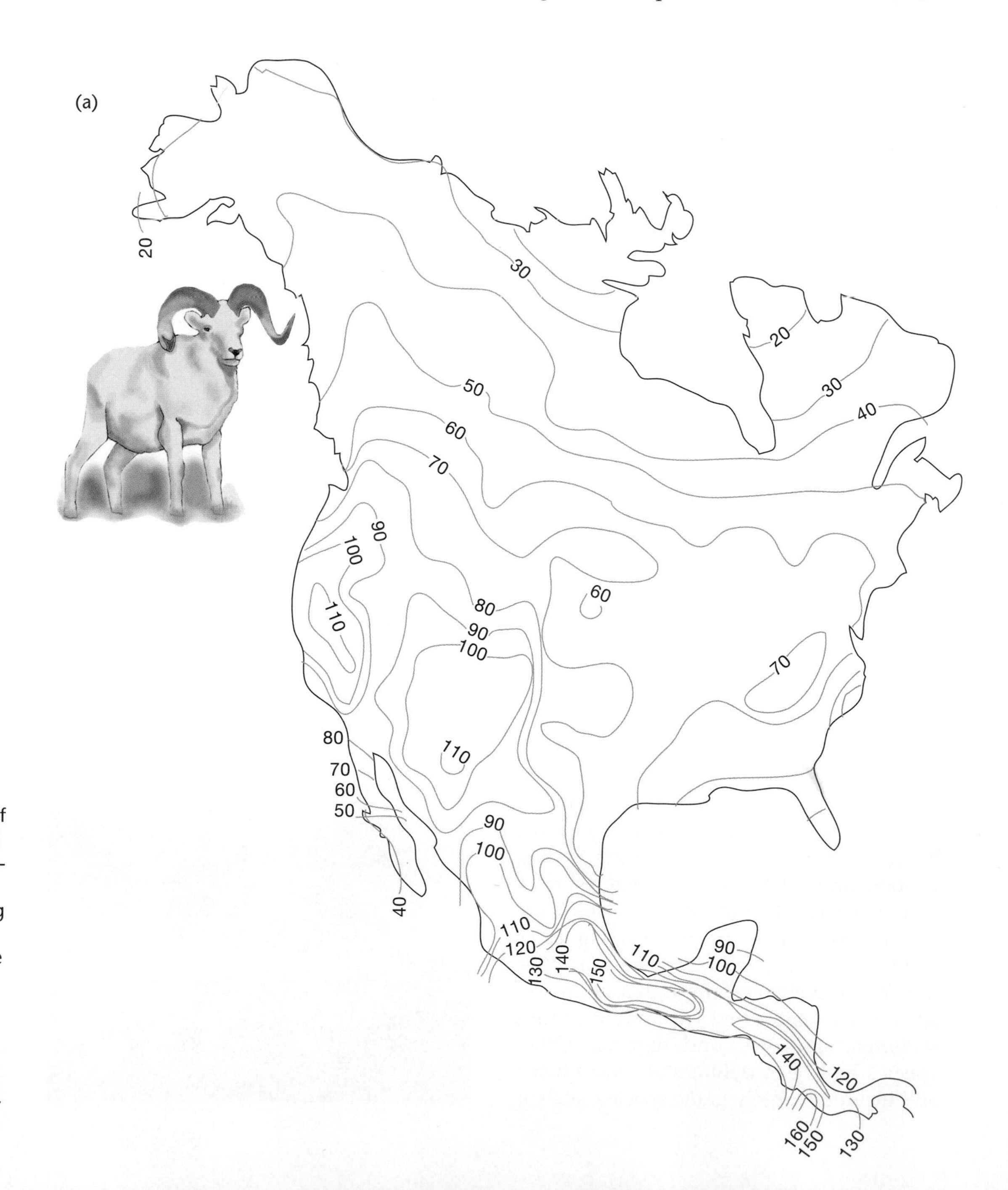

Figure 15.2 Geographic variation in species richness of North America. Contour lines show equal numbers of species of (a) mammals and (b) land birds. Note the pronounced latitudinal gradients in both groups heading south toward the tropics and the high diversity in the southwestern United States and northern Mexico, a region of great topographic relief and habitat diversity. ([a] *Redrawn from Simpson, 1964* [b] *redrawn from Cook, 1969.*)

tain ranges provide peaks, valleys, and a wide range of habitats, which increases richness. Species richness is also affected by the "peninsular effect," but in the opposite direction: The number of species falls as one progresses southward toward Florida. Note as well that there are more species of birds than mammals in any given region of the United States. Indeed, as one moves southward, the diversity of bird species increases twelvefold in the 60 degrees of latitude shown in Figure 15.2, whereas the diversity of mammal species increases only eightfold. Why this is true is not yet clear, but the trend of species richness increasing with decreasing latitude is not unique to birds and mammals, although other trends in species richness may occur in other taxa. For example, rainfall can influence the richness of plant communities.

The richness of trees in North America is not well linked to latitudinal gradients (Fig. 15.3). Trees do not grow well in deserts of the U.S. Southwest, despite decreases in latitude and increases in topographical variation. In this case, species richness is linked to rainfall levels. In order to grow, trees need high levels of moisture in the soil, and as rainfall increases toward the U.S. Southeast, so does the species richness of trees. Sorting out the effects of temperature, moisture, topography, and other variables on species

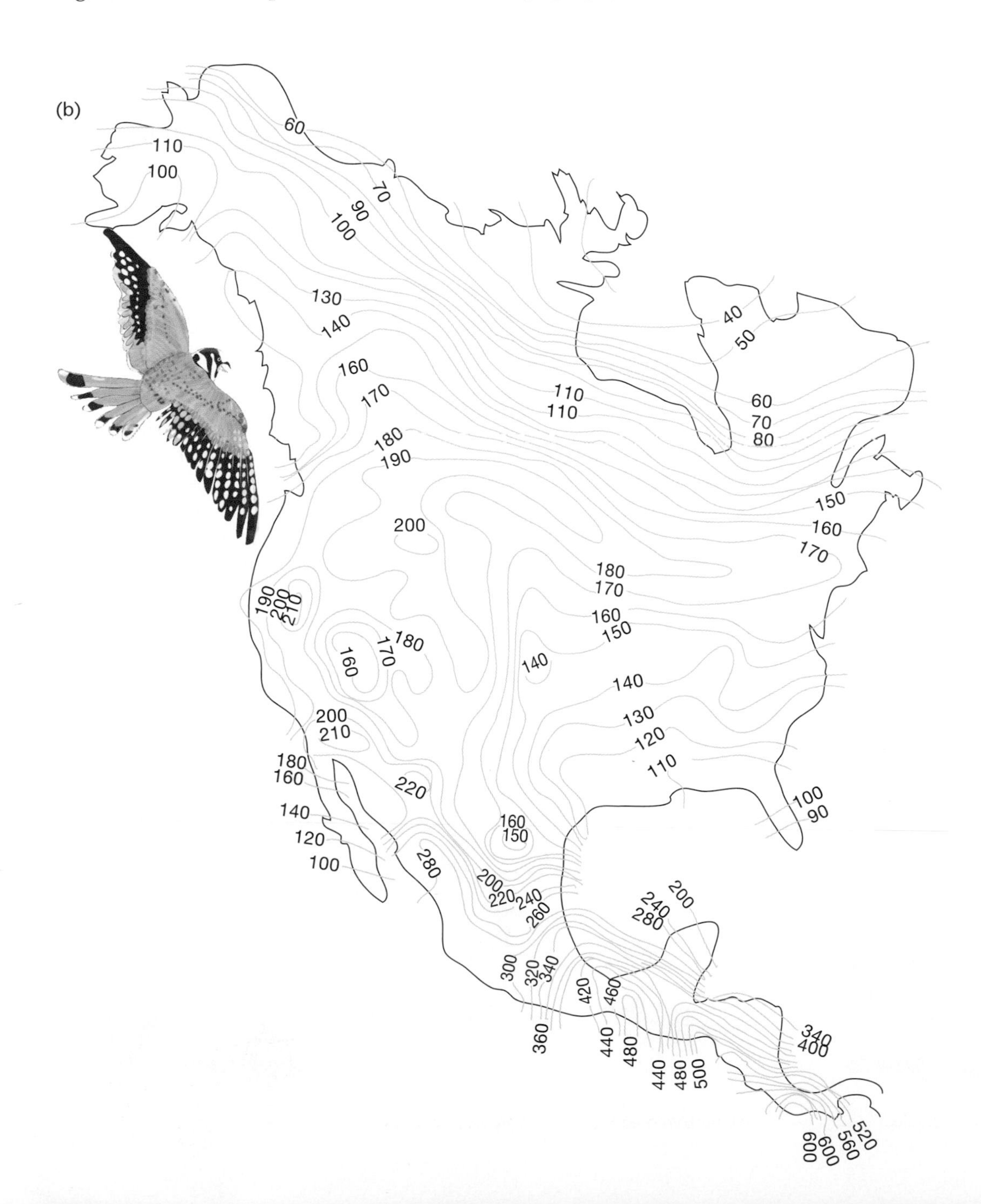

Figure 15.2 Continued

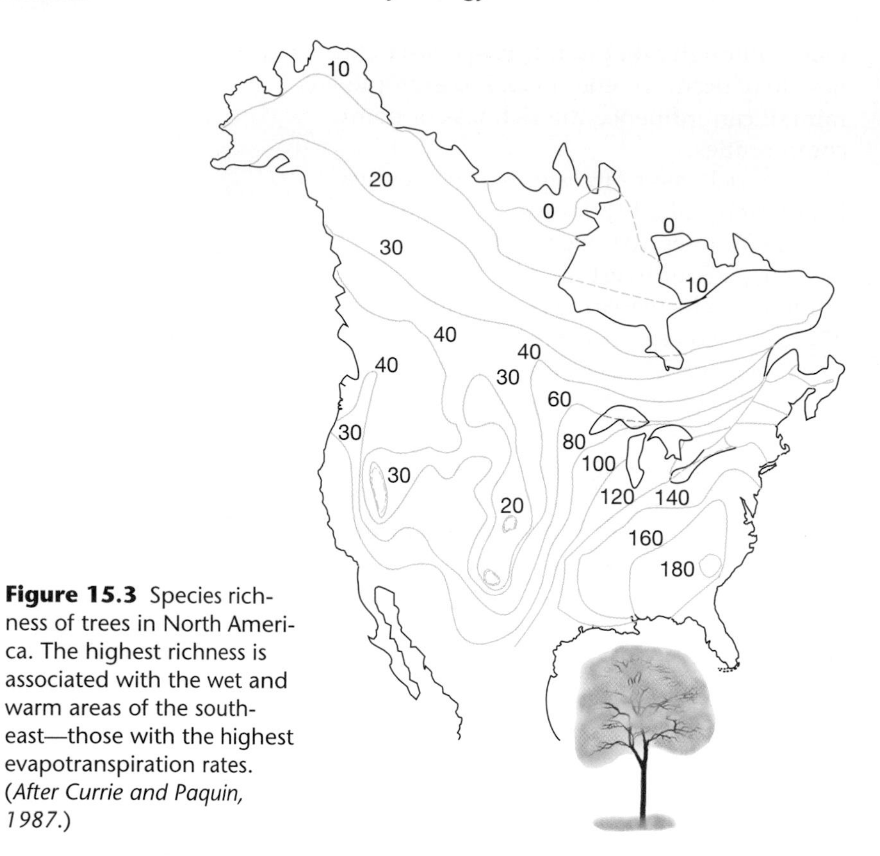

Figure 15.3 Species richness of trees in North America. The highest richness is associated with the wet and warm areas of the southeast—those with the highest evapotranspiration rates. (*After Currie and Paquin, 1987.*)

Biotic explanations for diversity gradients maintain that the increased diversity in the tropics reflects changes in interactions among species.

richness is a major task for ecology. At least 28 theories explaining temperate-to-tropical progressions have been advanced (Rohde, 1992). The importance of knowing which, if any, of these theories is correct lies in the current move to conserve biodiversity on Earth: Knowing what promotes diversity helps us preserve it. To identify the correct theory, let us examine the 8 most broadly proposed theories and the evidence in support of them. It may even be that different explanations operate for different taxa, such as birds, trees, or marine molluscs. Initially, however, it is useful to understand theories one at a time. We can divide them up into two main categories: biotic explanations and abiotic explanations.

Biotic explanations.

Spatial Heterogeneity Theory. Generally, there are more plant species in the tropics than in any other climatic zone. These greater numbers of plant species support higher numbers of herbivorous animal species and hence more carnivores. Richness of vegetation increases the numbers of herbivore species in two ways: by increasing the numbers of monophagous herbivores (those eating just one kind of plant) directly and by creating a more diverse architectural complexity, which affords more niches for herbivores to occupy. The traditional evidence to support this theory was provided by MacArthur and MacArthur (1961), who related bird species diversity to both plant species diversity and richness in the heights of foliage. More recently, Siemann and colleagues (1998) showed experimentally how insect herbivore richness was influenced by plant richness. Examining grassland plots in Minnesota with different richnesses of prairie plants, they found that insect herbivore richness increased with the number of plant species per plot. Of course, the spatial heterogeneity theory does not address the reason for the higher numbers of plant species themselves and, in that regard, remains incomplete.

Competition Theory. In 1950, the evolutionary biologist Theodosius Dobzhansky argued that, in temperate climates, natural selection operates mainly through harsh physical extremes and that species are generally r selected. In the more constant tropical temperatures, species are thought to become more K selected, to compete more keenly, and to interact more. The keen competition narrows the breadth of niches available, allowing more species to pack along the resource axes. Because competition has

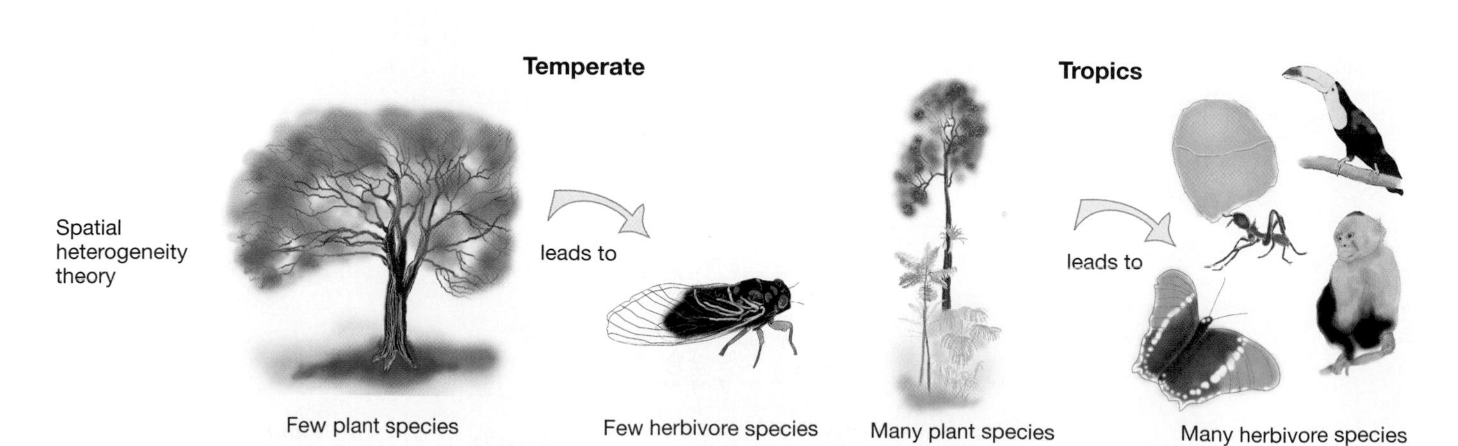

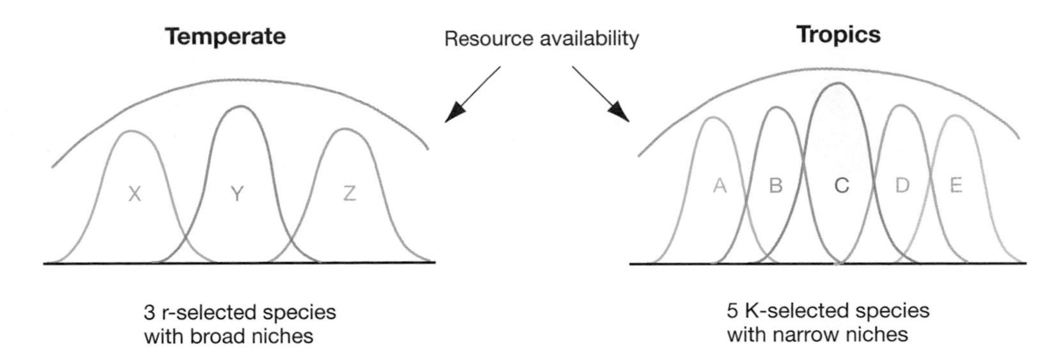

been found to be frequent and important in nature, the fact that it could explain the pole-to-equator gradients in biodiversity warrants closer examination. However, the theory has never been critically evaluated, and competition has not been measured for a sufficient variety of species groups to determine whether it in fact causes the tropical-to-polar gradients.

Predation Theory. Proposed by Robert Paine (1966) of the University of Washington at Seattle, the predation theory runs contrary to the competition hypothesis, arguing that the more numerous predators and parasites in the tropics hold populations of their prey down to such low levels that more resources remain and competition is *reduced*, allowing more species to coexist. The increased richness in turn promotes more predator species. Paine provided evidence to support the mechanism underlying this theory from studies on the intertidal communities of the U.S. northwest coast at two wave-exposed sites: Mukkaw Bay and Tatoosh Island in Washington State, where the food web was fairly constant and the starfish *Pisaster* was the top predator.

Pisaster preyed on a predatory whelk, *Thais*, and on chitons, limpets, acorn barnacles, and bivalves. *Thais* itself preyed on bivalves and acorn barnacles. After *Pisaster* was removed from a section of the shore, diversity decreased from 15 to 8 species. A bivalve, *Mytilus*, increased its numbers, crowding out other species. In unmanipulated sections of shore, *Pisaster* removed *Mytilus* and other species, preventing any one species from monopolizing the space. Removing any other single species from the system did not affect species diversity so drastically as removing *Pisaster*. For this reason, *Pisaster* was termed a **keystone species**, by analogy with the keystone that holds all the other stones in an arch in place. It is not unusual for other top predators to specialize on the most abundant prey, because they develop a search image for it. Such a phenomenon, wherein predation allows the coexistence of many prey species, was noted even by Darwin (1859), who observed more grass species coexisting

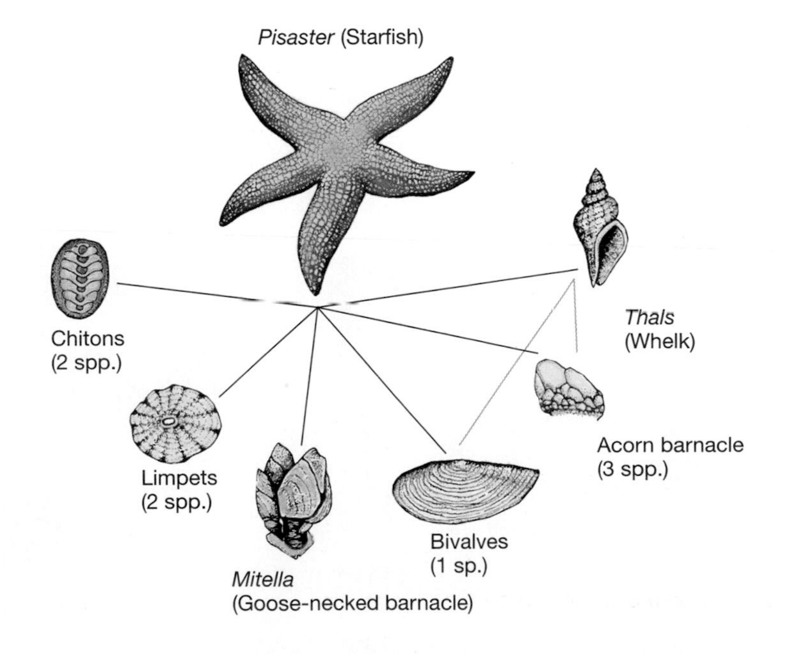

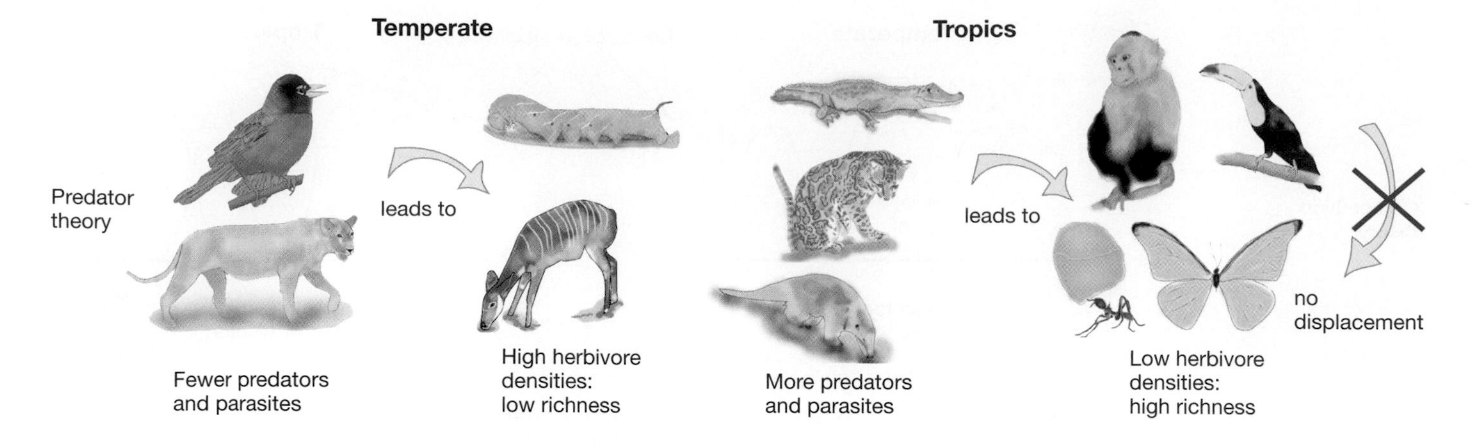

in areas grazed by sheep or rabbits than in ungrazed areas.

Of course, for such a theory to explain tropical richness, predation would have to be intense on the majority of species at all trophic levels in the tropics, and no data are available as yet to test such an idea. We would also have to explain why there are more predators and parasites in the tropics in the first place. Indeed, whether *Pisaster* in fact acts as a keystone species even on most of the shores of the Pacific coast has been brought into question (Foster, 1990).

Abiotic explanations of diversity gradients emphasize the differences in environmental characteristics that parallel changes in diversity.

Animal Pollinators Theory. In the tropics and other humid parts of the world, winds are less frequent and are of lower intensity than in temperate regions. (See Chapter 14.) The effect is accentuated by dense vegetative cover. Therefore, most tropical plants are pollinated by animals: insects, birds, and bats. Even some grasses that are typically pollinated by wind throughout most of the world are probably pollinated by insects in the tropics (Soderstrom and Calderon, 1971). Often, close associations build up between plants and specific pollinators, increasing the reproductive isolation between plant populations and, consequently increasing speciation rates. Coevolution of plants and pollinators then ensures high rates of speciation of animal pollinators as well. If mutualisms like this are common in nature, they could explain increased biodiversity in certain areas. Still, although such a theory is attractive when applied to terrestrial ecosystems, it cannot easily explain the similar diversity gradients that exist in aquatic ecosystems, where plant–pollinator relationships do not exist.

Abiotic theories.

Time Theory. Time theories argue that communities diversify with time and that temperate regions have younger communities than tropical ones just because they are younger: They have only recently recovered from glaciations and severe climatic disruption. According to this viewpoint, species that could live in temperate regions have not migrated back from the unglaciated areas into which the ice ages drove them (an ecological theory), or resident species have not yet evolved new forms to exploit vacant niches (an evolutionary theory) (Fischer, 1960).

Sanders (1968) provided evidence in support of an evolutionary time theory by

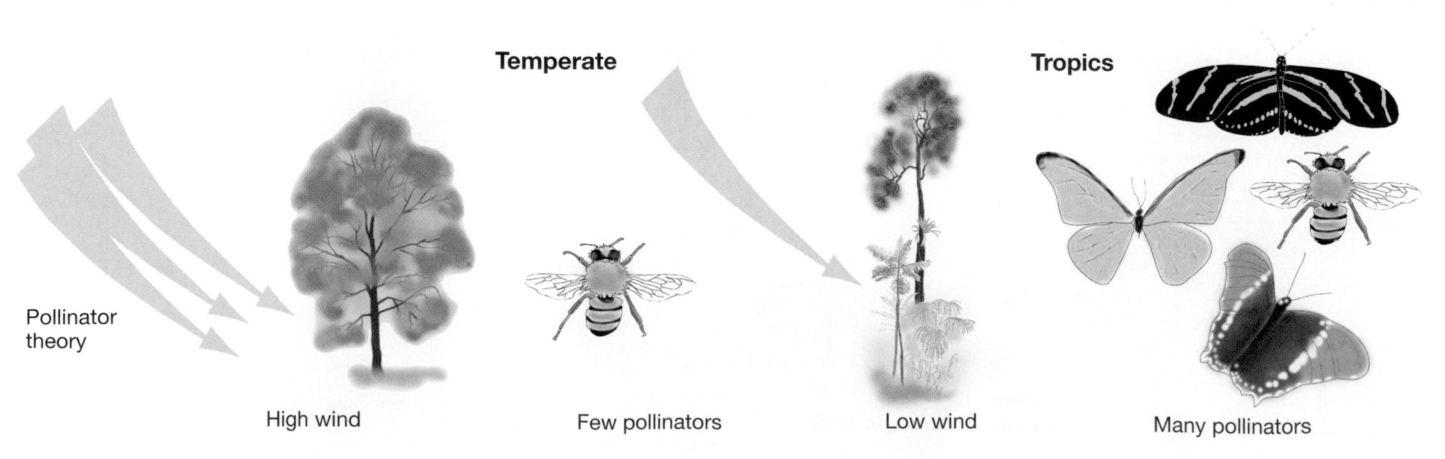

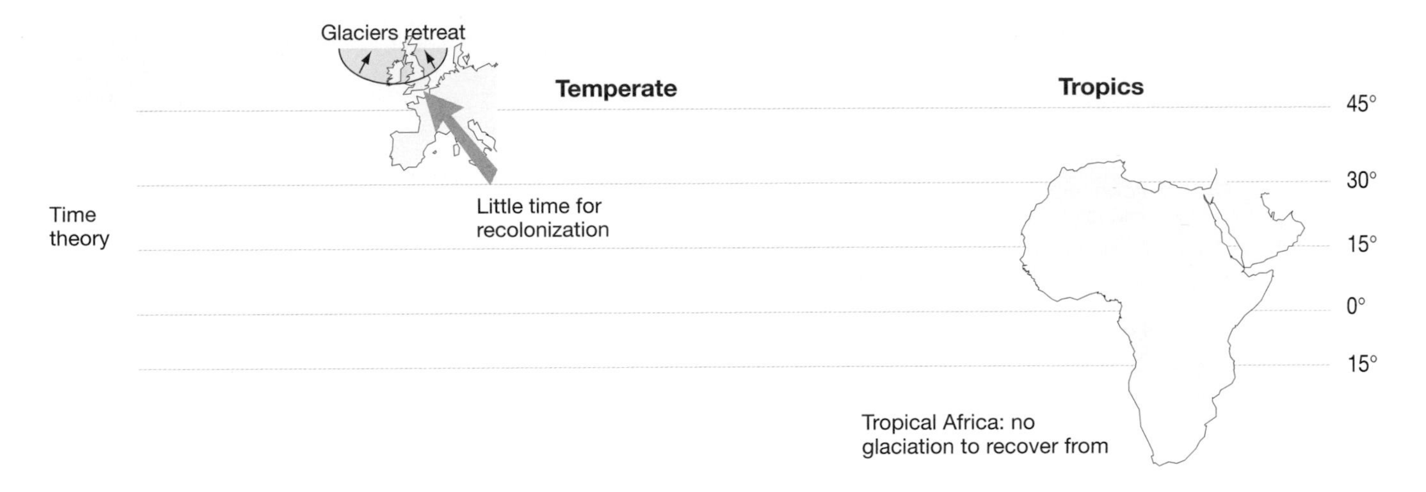

comparing bottom-dwelling (benthic) invertebrate diversity in glaciated and unglaciated Northern Hemisphere lakes that occur at similar latitudes. Lake Baikal in the former Soviet Union is an ancient unglaciated temperate lake and contains a diverse fauna, including 580 species of benthic invertebrates. Great Slave Lake, a comparable lake in glaciated north Canada, contains only 4 species in the same zone.

Another test of the time hypothesis involved insect herbivore diversity on British trees. Since the end of the last Ice Age, trees have recolonized Britain, and in the past two thousand years, humans have introduced trees as well. Sir Richard Southwood (1961) was the first to examine the number of insect species associated with each tree species, and he found good correlations of insect richness with the length of time the species of tree inhabited Britain. However, the American ecologist Donald Strong (1974) showed that insect species diversity on each tree species was better correlated with the *area* over which a tree species could be found. (See Chapter 19.) Furthermore, Strong and students Earl McCoy and Jorge Rey (1977) provided more detailed information on another system: sugarcane. They examined sugarcane pest richness, the area planted, and the dates of the plant's introduction into at least 75 regions of the world over the past three thousand years. They found no support for the time hypothesis, but good support for an area hypothesis. (See next.) Moreover, while we might not expect terrestrial species to redistribute themselves quickly following a glaciation—especially if there is a barrier like the English Channel to overcome—there seems to be no reason that marine organisms couldn't easily shift their distribution patterns during glaciations; yet the polar–tropical richness gradient still exists in marine habitats.

Area Theory. The area theory is based on the notion that larger areas contain more species because larger areas support bigger populations than smaller areas. Bigger populations result in fewer extinctions; hence, large areas support more species. In particular, large areas of climatic similarity will have greater species richness. Worldwide, the climates of the polar regions in the Northern and Southern Hemispheres are similar, and so are the climates of the temperate regions

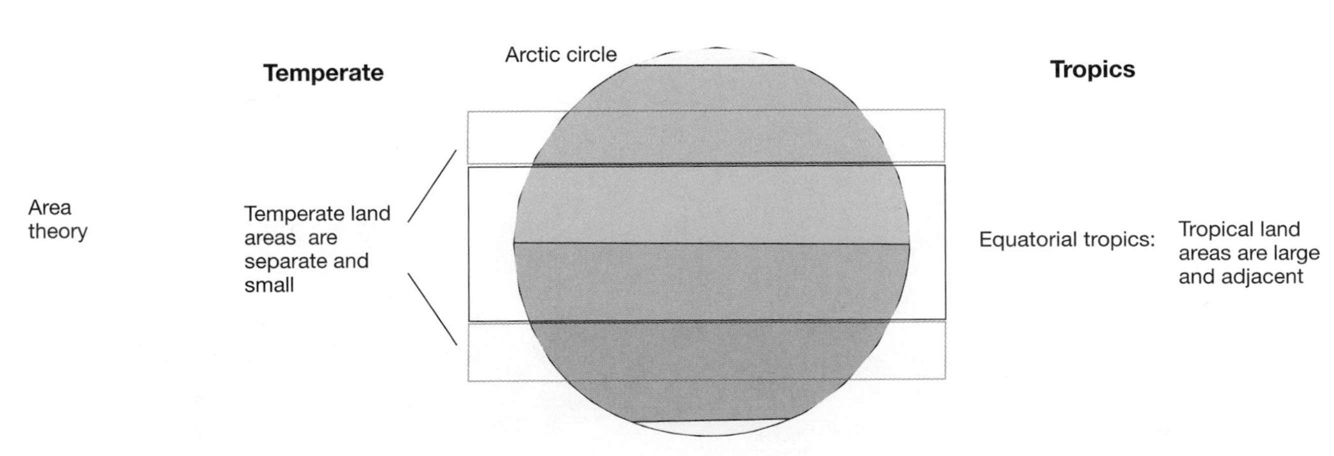

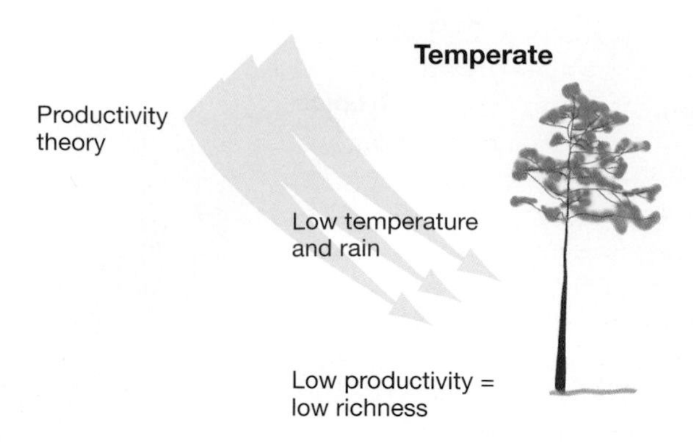

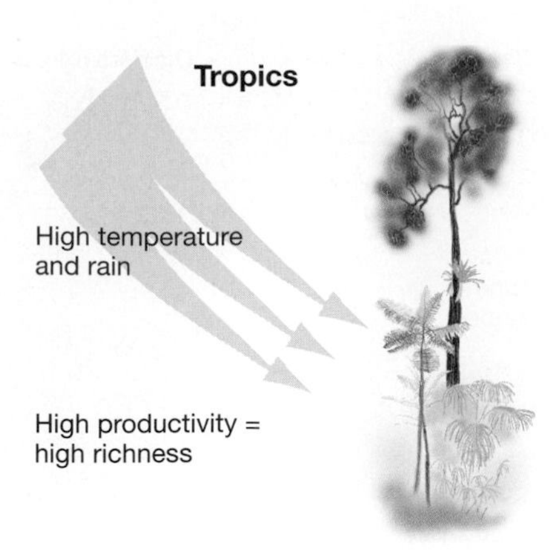

in those hemispheres. The tropical regions of the two hemispheres are adjacent, creating one large area with one climate. This is one reason the tropics may have greater species richness. The area theory, however, seems unable to account for the relatively few species in the vast contiguous landmass of Asia. Furthermore, the tundra may be the world's largest land biome, yet it is species poor. Finally, the largest marine system, the deep sea, which has the biggest volume of any habitat, has fewer species than the tropical surface waters, which have a relatively small area (Rohde, 1998).

Productivity Theory. This theory proposes that greater primary production by plants results in greater overall richness in a community; that is, a broader base on the energy pyramid permits more species in the pyramid (Wright, 1983). David Currie, of the University of Ottawa, has worked on this idea for many years. Currie and Paquin (1987) showed that the diversity of tree species in North America is best predicted by the evapotranspiration rate. (See Fig. 15.3.) Annual evapotranspiration is correlated with primary production because plants grow better where it is warm and wet. Currie (1991) later expanded his arguments to discuss how diversity in North American birds, reptiles, amphibians, and mammals was also linked to productivity. In Britain, a similar argument was made by Turner, Gratehouse, and Carey (1987), who demonstrated a correlation between the richness of British butterflies, which are ectothermal, and sunshine and temperature during the months they were on the wing, again suggesting a relationship between energy and species richness. Finally, a simple prediction from the productivity theory is that the number of resident species in seasonal habitats should change according to the seasons. Turner, Lennon, and Lawrenson (1988) showed that this is true for British birds: They found that the number of birds present in Britain in winter is less than that in summer, and the pattern is consistent with the amounts of energy present.

Once again, however, there are phenomena the theory fails to explain. Some tropical seas have low productivity, but high richness. By contrast, eutrophic lakes and lakes polluted with fertilizers have high productivity, but low richness, and so do

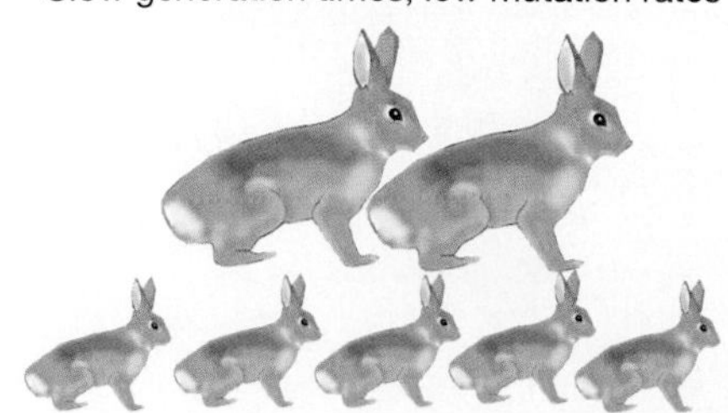

coastal salt marshes, presumably because they are stressful to many organisms. In marine ecoystems, some of the highest productivities are found in the depths of the Antarctic Ocean, which has few species. Finally, Robert Latham and Robert Ricklefs (1993) showed that, while patterns of tree diversity in North America provide evidence for the evapotranspiration theory, the pattern does not hold for broad comparisons between continents: The temperate forests of eastern Asia support substantially higher numbers of tree species (729) than climatically similar areas of North America (253) or Europe (124). The three areas have different evolutionary histories and different access to source regions from which new species might invade.

Evolutionary Speed Theory. In the view of Klaus Rohde (1995), an Australian ecologist who has worked on species richness patterns for 20 years, effective evolutionary time promotes high species richness. High effective evolutionary time is a result of two things: evolutionary speed and the geological time during which an ecosystem has existed under more or less the same conditions. The theory holds that evolutionary speed is in turn promoted by high temperatures, which foster (a) shorter generation times, (b) higher mutation rates, and (c) increased selection, leading to the fixation of favorable mutants. We know that the generation time is accelerated by temperature, but there is as yet no evidence to support hypotheses (b) and (c), even though intuitively it appears that they should be favored by higher temperatures.

The evolutionary speed theory is attractive because it predicts high species richness in the tropics, which have high temperatures and a long, uninterrupted community existence. The theory also predicts high richness in the deep sea, which has had little disturbance from human beings. In addition, the theory accounts for the fairly high levels of richness adjacent to these areas if species can migrate or "bleed over" into other areas, as they can in marine habitats or in mainland subtropical areas.

The eight theories we have just presented are neither exhaustive or mutually exclusive and can be combined in many permutations. Nevertheless, there is a strong tendency among ecologists to search for a common, single explanation of the phenomena they investigate. Unfortunately, as pointed out by Rohde (1992), some biotic explanations are insufficient. In particular, explanations that invoke increased competition, predation, or disease are secondary explanations (Table 15.1); a primary explanation is still needed to explain why these mechanisms themselves might be more or less important in certain areas. As for the abiotic explanations, there may be good correlations between various abiotic variables, such as evapotranspiration rates and species diversity, but it is not clear why increased productivity promotes diversity and not simply higher population densities of just a few species. In effect, why don't one or two species of tropical tree simply usurp all the resources?

Perhaps the most realistic way of examining causes of richness gradients is that of Robert Ricklefs and Dolph Schluter, who, in their 1993 book *Species Diversity*, suggested that different processes may act on different scales. Biotic factors may influence diversity more on a local scale and evolutionary processes on a provincial or global scale (Fig. 15.4). So, for example, we know that Paine's (1966) work on the intertidal communities of the U.S. northwest showed without doubt that predation can cause changes in diversity along local populations on small scales. On a bigger scale, however, like a region, there is no evidence to suggest that predation pressure affects species richness. On a regional scale, we know that evolutionary time—at least the time since the last glaciation—has the potential to change patterns of species richness, although it is not likely to affect local populations differentially. So, at any given point on the globe, richness may be affected by many different factors simultaneously.

TABLE 15.1 Explanations of latitudinal gradients in species richness and their flaws. *(After explanations in Rohde, 1992.)*

Explanation	Type of Flaw
Competition Predation Animal pollinators Spatial heterogeneity	Secondary explanation; primary explanation still needed
Productivity Area Time Evolutionary speed	Insufficient evidence

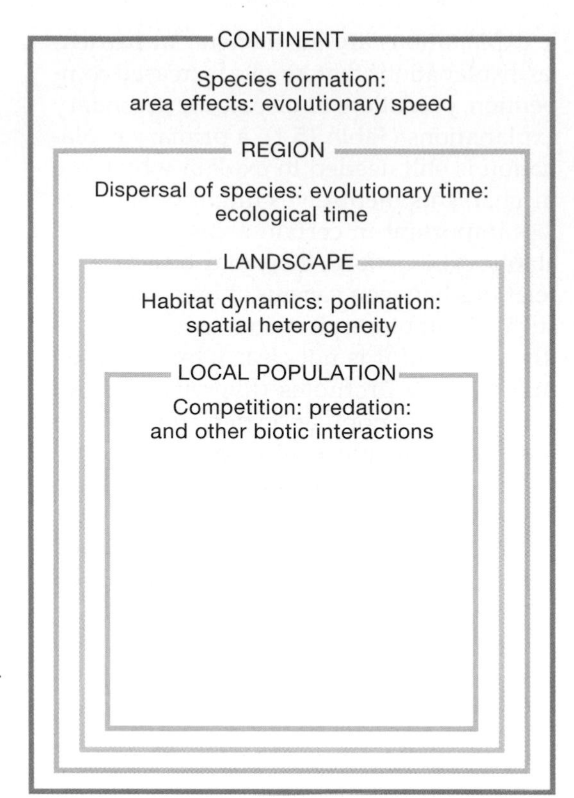

Figure 15.4 A hierarchical viewpoint of processes influencing species diversity. Each level includes all lower ones and nests within all higher ones.

15.2 Community Similarity

Similar communities from different parts of the world may converge in species richness.

Besides the generally recognized progression in species diversity from the poles to the equator, communities from different parts of the globe are similar in some respects. For example, plant biologists have long noticed the similarity of vegetation in climatically similar areas around the globe: Cacti-like plants occur in deserts of all types. Thus, if species can converge in morphology, can communities converge in species richness?

As is often the case in ecology, the evidence is equivocal. Schluter and Ricklefs (1993) provided a list of many examples of similar species diversity in similar habitats around the globe (Table 15.2a). In the same text, however, was a practically equally sized list of *dis*similar species diversities in similar habitats around the globe (Table 15.2b). They concluded that near-identical richness does not always result from similar environmental conditions.

One of the most comprehensive field studies of convergence in species richness—or rather, the absence of it—was performed by Eric Pianka (1986) at the University of Texas. Examining desert lizard species, Pianka was surprised to find quite different numbers of lizards in deserts around the world: 61 species in Australia and 22 in southern Africa, but only 14 species in North America (Fig. 15.5). Could it be that deserts are not functionally equivalent in different areas of the globe, or could there be historical evolutionary constraints also? Additional data from other habitats, such as wetlands, show that while there are fewer species of lizards in wetlands globally, Australia still has more lizard species than South Africa, whatever the habitat. This suggests that there are indeed strong evolutionary constraints in the number of lizards in southern Africa. Plant ecologist Robert Whittaker (1972), who was aware of similar patterns in vegetative communities, named the differences in richness *within* habitats (in this case, deserts) *alpha* (α) *diversity* and the difference *between* habitats *beta* (β) *diversity*. The *overall* difference in diversity between two geographical regions, called *gamma* (γ)

TABLE 15.2a Examples of nearly equal species richness in similar habitats around the globe. *(After Schluter and Ricklefs, 1993.)*

Group	Habitat	No. Species	No. Species
Plants	Desert	Arizona (250)	Argentina (250)
Plants	Semiarid	North America (70)	Australia (65)
Sea anemones	Rocky shore	Washington (11)	South Africa (11)
Ants	Desert	Arizona (25)	Argentina (25)
Saprophytic insects	Tree hole	North America (6)	Australia (6)
Fishes	Forest lakes	Wisconsin (4)	Finland (4)
Lizard	Mediterranean scrub	California (9)	Chile (8)
Birds	Desert	Arizona (57)	Argentina (61)
Birds	Mediterranean scrub	California (30)	South Africa (28)
Birds	Shrub desert	Australia (5.5)	North America (6.3)
Small mammals	Scrubland	California (7)	South Africa (6)

TABLE 15.2b Examples of highly dissimilar species richness in similar habitats around the globe. (*After Schluter and Ricklefs, 1993.*)

Organism	Habitat Type	No. Species	No. Species
Mangroves	Mangal	Malaysia (40)	West Africa (3)
Algae	Rocky shore	Washington (17)	South Africa (3)
Chitons	Rocky shore	Washington (10)	South Africa (3)
Insects	Streams	Australia (60)	North America (26)
Insects	Bracken	England (21)	North America (5)
Bees	Desert	Argentina (188)	Arizona (116)
Bees	Mediterranean scrub	California (171)	Chile (116)
Ants	Desert	Australia (37)	North America (16)
Ants	Mediterranean scrub	California (23)	Chile (14)
Amphibians	Wetlands	Zambia (22)	Australia (14)
Lizards	Desert	Australia (27)	North America (7)
Birds	Peat lands	Finland (33)	Minnesota (18)
Rodents	Desert	Arizona (16)	Argentina (5)

Figure 15.5 Desert lizards from Australia, North America, and Africa.

diversity, would then be the product of alpha and beta diversity. Thus, the difference in species richness of lizards in, say, Australia versus Southern Africa would be a combination of the numbers of different habitats in Australia and southern Africa (beta diversity) and the difference in diversity between identical habitats in each area (alpha diversity; Fig. 15.6).

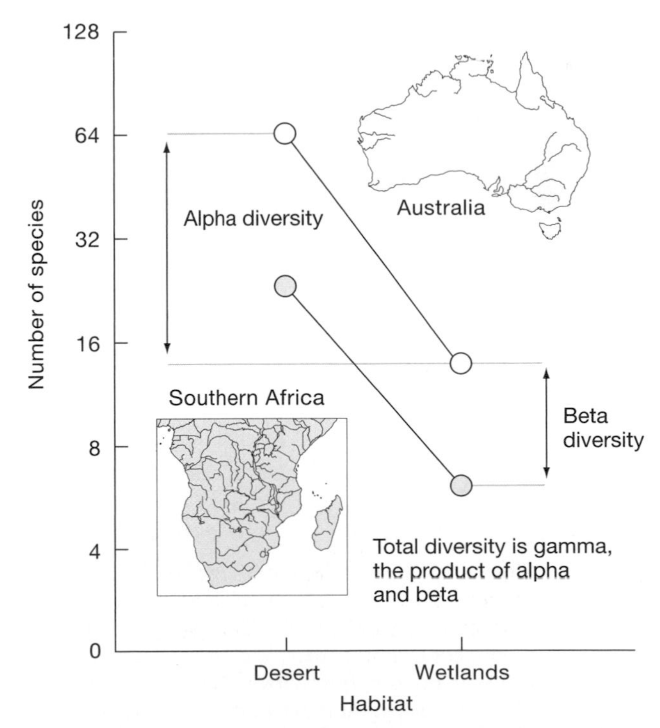

Figure 15.6 The number of lizard species in deserts and tropical wetlands of Australia and southern Africa and their relation to α, β, and γ. (*After Schluter and Ricklefs, 1993.*) Lizard communities in Australia are usually richer than those in southern Africa, regardless of the type of habitat.

Figure 15.7 Bracken fern, *Pteridium aquilinum.*

John Lawton and his colleagues (1993) have examined convergence in diversity in a different way, by examining convergence in **guilds**—the actual way species utilize a common resource. Bracken fern, *Pteridium aquilinum*, is a widespread and common native member of the flora of all the nonpolar continents (Fig. 15.7). In many places, it is regarded as a serious weed, and its herbivorous insect fauna has been thoroughly studied, sometimes with a view toward controlling the plant by biological control. Over the last 20 years, surveys of insects have been conducted in Hawaii, New Mexico, Great Britain, South Africa, Brazil, New Guinea, and Australia, usually by Lawton or one of his colleagues. The species assemblage varies remarkably, giving no evidence of taxonomic similarity in the fauna (Fig. 15.8). There are, for instance, no beetles on bracken in Britain and no hymenoptera on the plant in South Africa. Thus, insects appear to have independently colonized bracken in different parts of its range over evolutionary time. Hawaii has no confirmed bracken herbivores, while New Guinea has the richest fauna on bracken, with about 30 species. The variation in the total number of insect species exploiting bracken is partly a function of how common and widespread the plant is in each geographical region (Fig. 15.9), reminding us once again that there is a strong species–area effect.

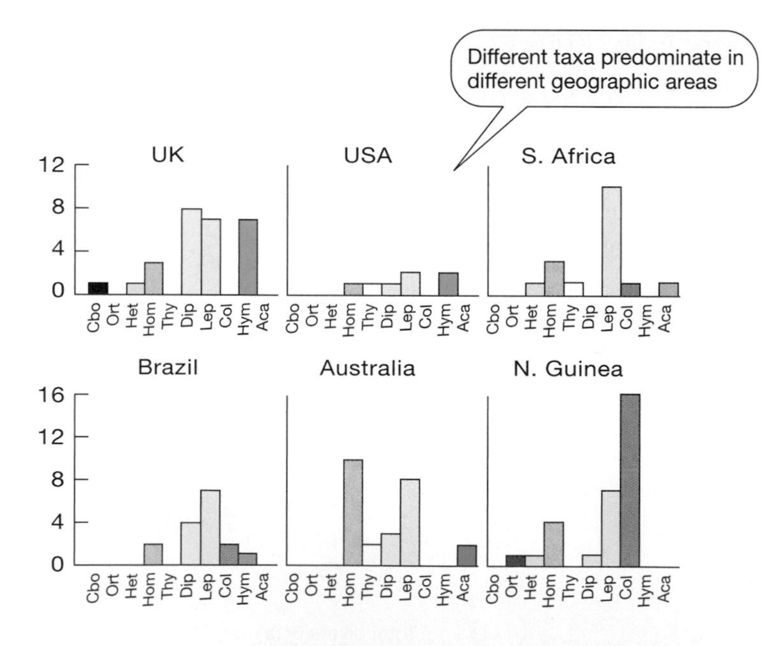

Figure 15.8 The taxonomic composition, by order, of the arthropod assemblages feeding on bracken is very different in different parts of the world. For example, there are no beetles in the United Kingdom, but many in New Guinea, that feed on bracken. Abbreviations: *Cbo*, Collembola; *Ort*, Orthoptera; *Het*, Heteroptera; *Hom*, Homoptera; *Thy*, Thysanoptera; *Dip*, Diptera; *Lep*, Lepidoptera; *Col*, Coleoptera; *Hym*, Hymenoptera; *Aca*, Acarina. (*Reproduced from Lawton et al., 1993.*)

Lawton and colleagues also asked whether biotic interactions—in particular, competition—have shaped richness. They argued that if competition were important, one might expect it to produce convergence in the ways in which bracken is partitioned among herbivores, with guilds including chewers, sap suckers (feeding on phloem or xylem), leaf miners (feeding between the leaf surfaces), and gallers (feeding inside tumorlike insect-induced growths on the plant). The plant is structurally similar everywhere, consisting of stem, leaf, and leaf veins. It turns out that the distribution of species across resources on the plant is idiosyncratic from locality to locality, with numerous vacant niches by certain types of feeders (Fig. 15.10). In other words, parts of the plant go unutilized in certain areas of the world. It does not look like there is convergence of feeding types across regions, nor does it look like competition is important such that high densities of, say, chewers of one part of the plant exclude other groups from feeding on the leaf. About the only pattern is that the leaf seems to be exploited more than the other parts, but this could be because the leaf is the softest part of the

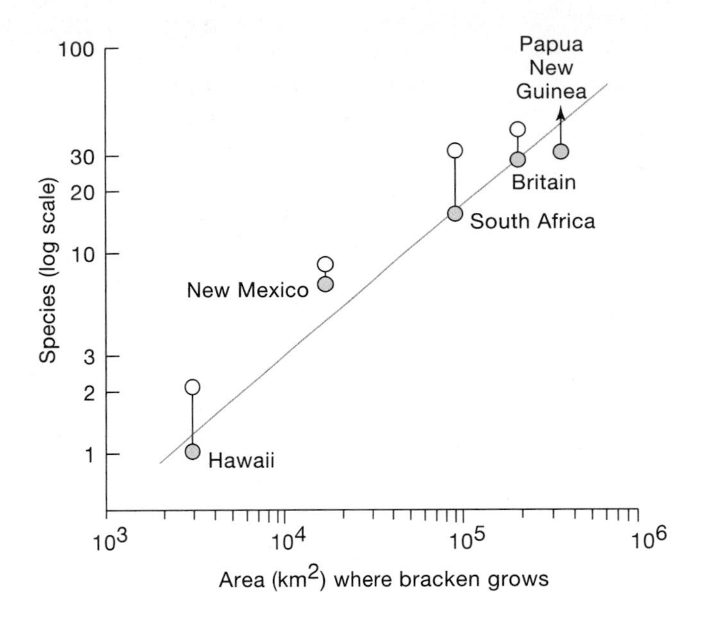

Figure 15.9 Species–area relationship for the number of species of herbivores definitely feeding on bracken (green circle) in different parts of the world. Also shown (yellow circle) are the total numbers of species feeding on bracken, including possible and occasional feeders and instances where records are uncertain. The arrow in the data for Papua New Guinea indicates that the number may be an underestimate. (*After Compton, Lawton, and Rashbrook, 1989, and Lawton, 1984.*)

plant. After 20 years of study of bracken, Lawton and colleagues (1993) summarized the main rules that determine the species richness of insects on bracken (Table 15.3). According to those researchers, the colonization of species in different parts of the world over evolutionary time has been idiosyncratic. Diversity is affected by latitudinal gradient and species–area relationships. The effects of species interactions look feeble. Finally, just as Ricklefs and Schluter argued, different processes may be important at different spatial scales in affecting community structure.

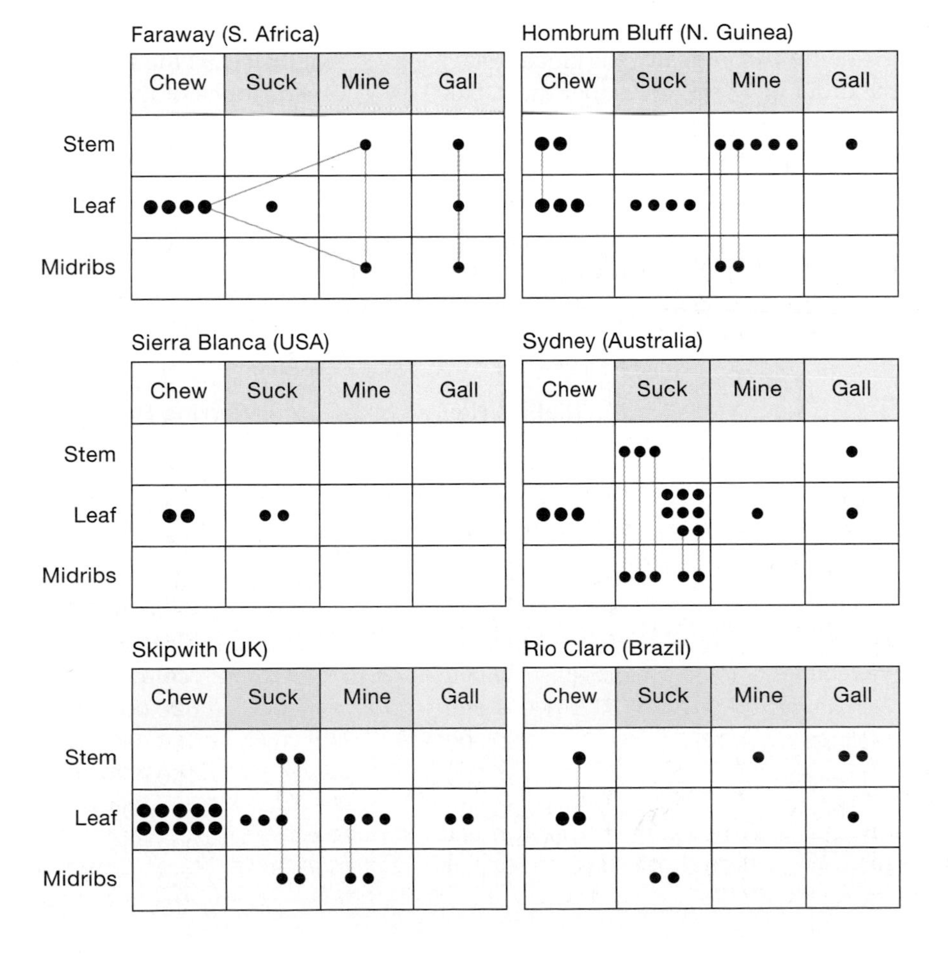

Figure 15.10 The feeding sites and feeding methods of herbivorous arthropods on bracken at sites in different parts of world. Each dot represents one species, and feeding sites of species exploiting more than one part of the plant are joined by lines. Species richness and feeding methods change dramatically among different parts of the world, but many parts of the plant remain unutilized by certain types of feeders. (*Reproduced from Lawton et al., 1993.*)

TABLE 15.3 Factors that influence the diversity of insects feeding on bracken. *(After Lawton, Lewinsohn, and Compton, 1993.)*

Factor	Scale	Indicators	Importance
History	Continental	Faunal taxonomic composition	Strong
Plant range	Continental	Regional species–area effect	Strong
Latitude	Regional	Local assemblage size, turnover	Small, but significant effects
Seasonality	Regional	Insect succession, dynamics	Variable
Habitat heterogeneity	Regional to local	Insect distribution	Variable
Patch size	Local	Local species–area effect	Weak
Community interactions	Local	Assemblage saturation, functional convergence	No detectable effects

15.3 Global Species Richness

The estimate of the total global richness of species is about 12 million.

While ecologists strive to explain latitudinal gradients and other patterns in species richness, one even larger question remains: How many species are there on Earth? Fewer than 2 million species have been classified, and best estimates suggest that about 12.25 million exist, with a maximum estimate of as much as 118 million (Table 15.4). We have no idea why, theoretically, the global total is on the order of 12 million rather than 10,000 or 10 billion (May, 1988). Also, insects represent at least two-thirds of the total number of species, but why? Could it be because of their size? For any given taxon, it is not the smallest or the biggest organisms that are the most abundant, but species intermediate between smallest and medium sized. For example, this is the case for terrestrial mammals, land birds, and freshwater fish in the United States (Fig. 15.11). At the moment we are not sure what drives this pattern. However, if we construct a similar plot of the number of species on Earth against body size, measured as length, we get the same right skew as in most other graphs of this nature (Fig. 15.12). The body size of insects falls right where the number of species is highest.

TABLE 15.4 Numbers of species in those groups of organisms likely to include in excess of 100,000 species, plus estimates for vertebrates. *(After World Conservation Monitoring Centre, 1992.)*

	Estimated Species	
	Highest Figure	**Working Figure**
Viruses	500,000 +	500,000
Bacteria	3,000,000 +	400,000
Fungi	1,500,000 +	1,000,000
Protozoans	100,000 +	200,000
Algae	10,000,000 +	200,000
Plants	500,000 +	300,000
Vertebrates	50,000 +	50,000
Nematodes	1,000,000 +	500,000
Molluscs	200,000 +	180,000
Crustaceans	150,000 +	150,000
Arachnids	1,000,000 +	750,000
Insects	100,000,000 +	8,000,000
Total	118,000,000 +	12,230,000

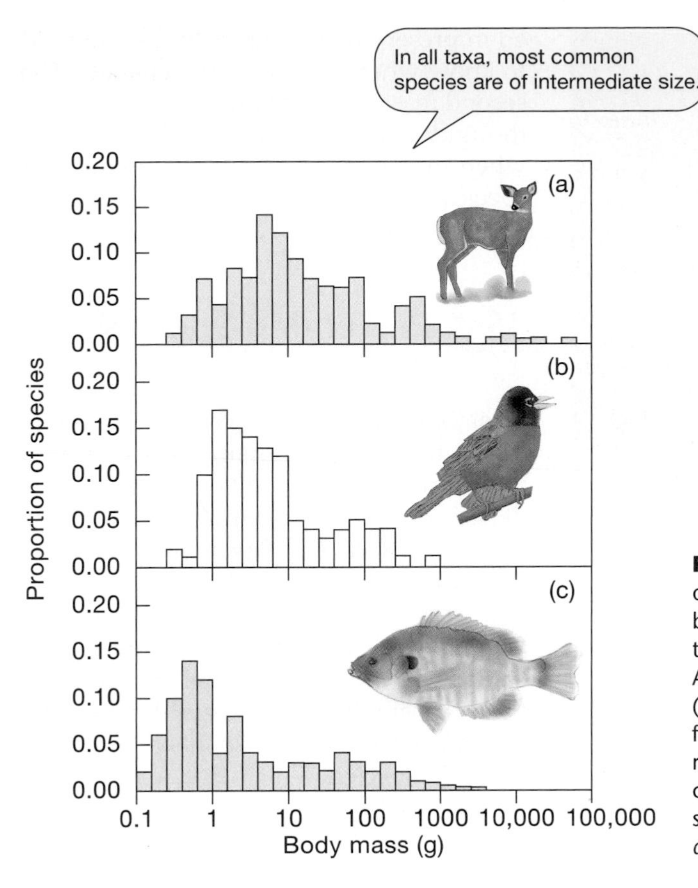

Figure 15.11 Frequency distributions of numbers of species as a function of body mass (on a logarithmic scale) for three groups of vertebrates in North America: (a) terrestrial mammals, (b) land birds, and (c) freshwater fishes. All three distributions have a right-skewed shape, and the most common species are of intermediate size. (*Reproduced from Brown, Marquet, and Taper, 1993.*)

So if we want to get an accurate determination of the number of species on Earth, we have to get a good handle on insect richness. We have to determine whether the total number of insect species on Earth is closer to 1 million (as currently described), 8 million (the likely number), or 100 million (the maximum estimated number). The estimate of the total number of species on Earth will be influenced more by insects than by any other taxon. Also, estimates of total numbers of species are not likely to change much for the best-known groups, such as plants, birds, mammals, and, indeed, vertebrates as a whole.

The high estimate of insect numbers is based largely on the work of the Smithsonian entomologist Terry Erwin (1982), who fumigated 19 trees of one species in Panama. On the strength of samples of beetles taken from the canopies of these trees, he proposed that there might be 30 million species of arthropods (principally insects) in tropical regions alone (Fig. 15.13). To arrive at this estimate, Erwin had found 1,100 species of beetles on the canopy of one species of tree in Panama. He suggested that about 160 were specific to that species of tree and that beetles represented about 40 percent of all arthropods, so there would be about 400 arthropods specific to the given tree's canopy. Erwin went on to suggest that there would be a total of about 600 insect species on the canopy and the rest of the tree combined. If, then, there are 50,000 species of tropical tree, there will be 30 million

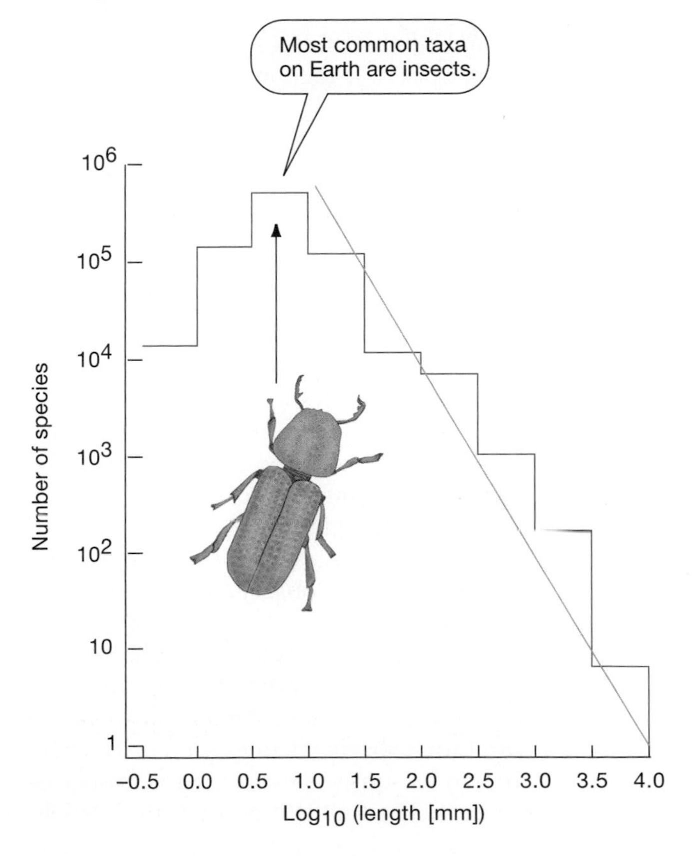

Figure 15.12 A crude estimate of the distribution of the number of species of all terrestrial animals, categorized according to their characteristic length *L*. Insects, which are of intermediate size compared to other taxa such as microbes or vertebrates, are the most species-rich taxon. (*After May, 1988.*)

Conservation of species richness emphasizes the preservation of areas of the world rich in total species or endemics.

Figure 15.13 To estimate insect richness some researchers collect all the species that are attracted to UV lights; as here in Chiriqui Province, Panama.

arthropods on tropical trees alone. Add in the insects in the soil and insects on the foliage of temperate forests and grasslands, and we could triple the estimate to 90 million or a round 100 million.

Subsequently, other entomologists (Stork, 1988; May, 1988; Thomas, 1990) pointed out many problems with Erwin's assumptions, raising doubts about the estimate of 100 million insect species. For example, many tropical insects probably utilize more than one tree species—a necessary strategy because of the distances between trees of the same species in the tropics. If each insect species utilized just two tree species, then the estimate of tropical insect species would be halved, to 15 million. Furthermore, Kevin Gaston (1991) surveyed insect taxonomists to obtain the input of experts on the likely richness in insects of different orders. He found that many insect species which were taken to be newly discovered had actually been described earlier on, but long forgotten—a problem known as synonymy. With such large numbers of insects, this is quite a common problem—and one that may continue to grow, given the ever-increasing numbers of species identified! Gaston concluded that a figure of between 5 and 10 million insect species was more probable, and this is probably the most accurate estimate so far.

In discussing species richness, we have emphasized geographic patterns and how we can explain them. We have also talked about total species richness on Earth. There are two reasons we are interested in species richness. The first is that if we are interested in preserving species richness, we need to know where on Earth it is greatest. The second reason is more pragmatic: It is often thought that communities function best when they have their full complement of species. We will now discuss each of these reasons in turn.

15.4 Preserving Species Richness

A major goal of modern conservation biology is to ensure the maintenance of high species richness, or biodiversity. In the past, most resources have been allocated to single large species called "flagship" species like the Indian rhinoceros or the American buffalo, and preservation of these species has been ensured by trying to maintain their habitat. Conservation of biodiversity differs from this single-species focus, because the aim is to conserve the habitats that contain the most species. Which areas of the globe support the most species? First, let's consider which habitats contain the most species. We find that, generally, tropical habitats have the most species. A 13.7-km^2 area of La Selva Forest Reserve in Costa Rica contains almost 1,500 plant species, more than the total number of species of plants found in Britain, which has an area of 243,500 km^2. Ecuador contains more than 1,300 bird species, or almost twice the number found in the United States and Canada combined. But there is much variation even among tropical countries. The South American and Asian tropics are much more diverse than the African tropics, and even within each of those continents some areas are more diverse than others.

One method of identifying areas for conservation is to target countries with the greatest numbers of species—the so-called megadiversity countries—a strategy favored by World Wildlife researchers such as Russell Mittermeier (1988; Mittermeier and Werner, 1990). For instance, Mittermeier and colleagues (McNeely *et. al*, 1990) used country species lists of vertebrates, plants, and butterflies to identify 12 such megadiversity countries: Mexico, Colombia, Ecuador, Peru, Brazil, Zaire, Madagascar, China, India, Malaysia, Indonesia, and Australia, which together hold up to 70 percent of the diversity in those groups of organisms. In this approach, bigger countries, because they hold more species (a topic we will take up again in Chapter 19), fare better than smaller countries. Since conservation is most often managed at the national level, the proponents of the megadiversity approach believe that it works well because big countries with large areas garner most of the available international funds, thus promoting the survival of the largest numbers of individuals of most species. Perhaps the greatest drawback of the strategy, however, is that, although these areas contain the most species, they do not necessarily contain the most *unique* species (Williams *et al.*, 1996; Reid, 1998). For example, the list of mammal species contains 344 entries for Peru and 271 for Ecuador, with 208 species common to both countries. What is needed is some measure of the uniqueness of species—the endemics that are restricted to a country.

British conservationist Norman Myers and his colleagues (2000) identified 25 "hot spots" for endemic tropical forest plants that together contained 133,149 endemic plant species (44 percent of the world's total) and 9,645 endemic vertebrate species (35 percent of the world total) in only 1.4 percent of the world's land area (Fig. 15.14, Table 15.5). South Africa's Cape Region has many endemic plant species (Fig. 15.15). Protecting these hot spots would prevent the extinction of a larger number of endemics than would protecting areas of a similar size elsewhere. Because only limited funds are available, we must make choices about the areas to protect, and endemism is an important consideration.

The next question, of course, is whether the same pattern of endemism follows for other taxa. If the pattern of endemism is the same, protecting the aforementioned hot spots will preserve a lot more than just plants. Bibby *et al.* (1992) compared data for birds, mammals, reptiles, and amphibians and showed that, at least among larger vertebrates, areas rich in endemics of one taxon are often rich in endemics of another. An examination of Table 15.5, however, reminds us that such is not always the case. For example, while Wallacea (a system of small islands of Indonesia lying between Borneo and New Guinea) is rich in endemic mammals, it

Endemics are species confined to a certain region and only that region.

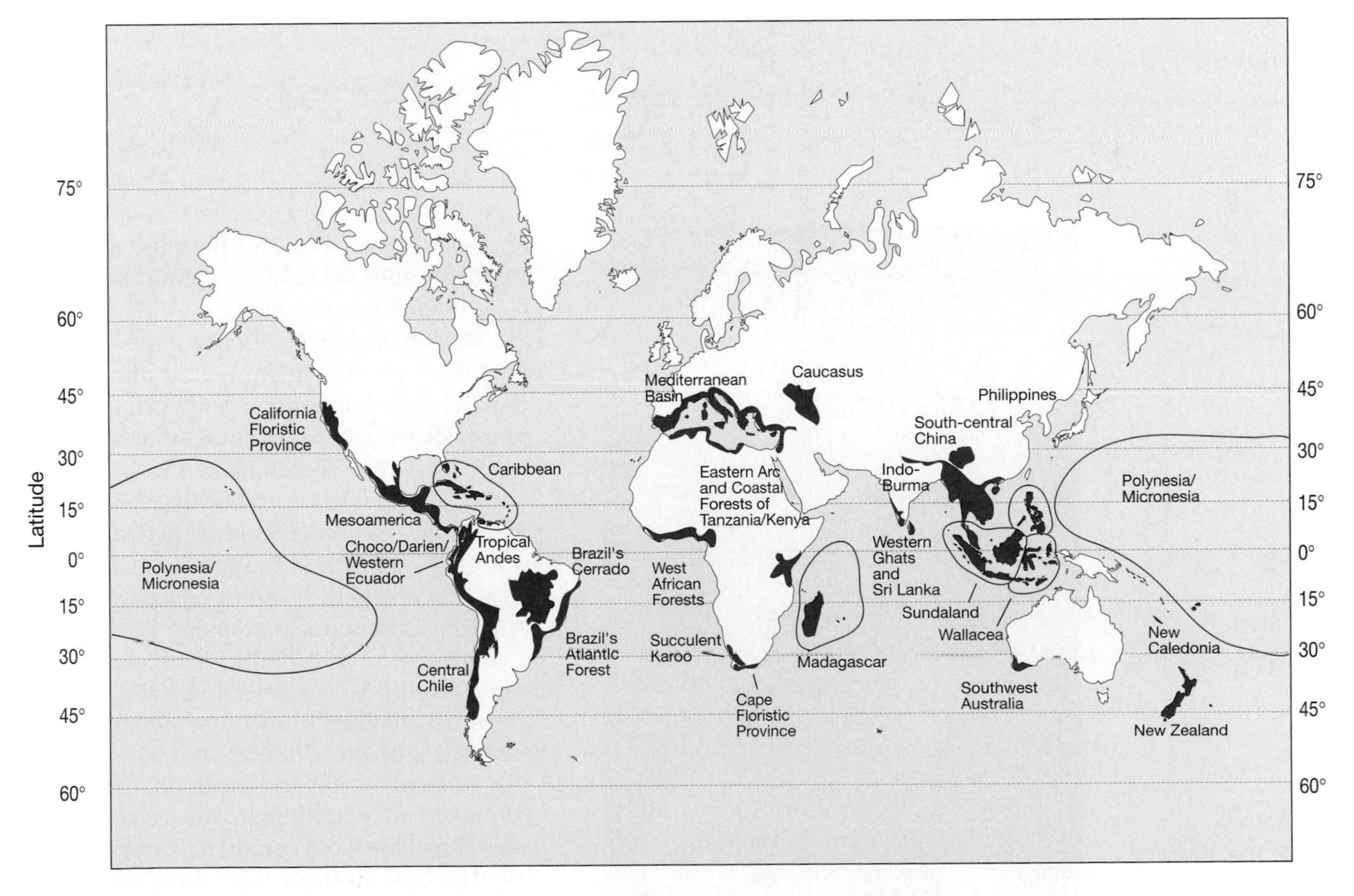

Figure 15.14 Location of endemic "hot spots" around the world, according to Myers *et al.* (2000). Actual areas of the hot spots vary between 3 and 30% of the red areas.

TABLE 15.5 Numbers of endemic species present in 25 "hot spots." *(After Myers et al., 2000.)*

Hot Spot	*Plants*	*Birds*	*Mammals*	*Reptiles*	*Amphibians*
Tropical Andes	20,000	677	68	218	604
Mesoamerica	5,000	251	210	391	307
Caribbean	7,000	148	49	418	164
Brazil's Atlantic Forest	8,000	181	73	60	253
Chocó/Darien/Western Ecuador	2,250	85	60	63	210
Brazil's Cerrado	4,400	29	19	24	45
Central Chile	1,605	4	9	34	14
California Floristic Province	2,125	8	30	16	17
Madagascar	9,704	199	84	301	187
Eastern Arc and Coastal Forests of Tanzania/Kenya	1,500	22	16	50	33
Western African Forests	2,250	90	45	46	89
Cape Floristic Province	5,682	6	9	19	19
Succulent Karoo	1,940	1	4	36	4
Mediterranean Basin	13,000	47	46	110	32
Caucasus	1,600	3	32	21	3
Sundaland	15,000	139	115	268	179
Wallacea	1,500	249	123	122	35
Philippines	5,832	183	111	159	65
Indo-Burma	7,000	140	73	201	114
South-Central China	3,500	36	75	16	51
Western Ghats/Sri Lanka	2,180	40	38	161	116
Southwest Australia	4,331	19	7	50	24
New Caledonia	2,551	22	6	56	0
New Zealand	1,865	68	3	61	4
Polynesia/Micronesia	3,334	174	9	37	3
Totals	133,149	2,821	1,314	2,938	2,572

Figure 15.15 The so-called "Fynbos" area of the Cape Floristic Province, South Africa, rich in endemic plants.

is poor in endemic amphibians, and the tropical Andes are rich in endemic birds, but poor in endemic mammals.

The downside of preserving species richness through endemism is that most areas rich in species or endemics are tropical rain forests. Scientists have argued that we need to conserve representatives of all major habitats (Woinarski, Price, and Faith, 1996). Thus, the Pampas region of South America, which is arguably the most threatened ecosystem on the continent, does not compare to the rain forests in richness or endemics, but it is a unique area that, without preservation, could disappear. By selecting habitats that are most distinct from those already preserved, many areas that are not biologically rich, but that are nonetheless threatened, may be preserved in addition to the less immediately threatened, but species rich, tropical forest. The best strategy may, then, employ a "portfolio" taking into account richness, endemism, and type of habitat.

15.5 Species Richness and Community Function

Stanford biologist Paul Ehrlich and Harvard biologist E. O. Wilson (1991) suggest that the loss of biodiversity should be of concern to everyone for at least three reasons. First, we have a moral responsibility to protect what are our only known living companions in the universe. Second, humanity has already obtained enormous benefits from biodiversity in the form of foods, medicines, and industrial products like wood and rubber and has the potential to gain many more benefits. By this reasoning, however, because it is couched more in terms of conserving individual species and not biodiversity in general, we should preserve varieties of wheat, rice, and corn, which are useful to us, and we should permit the extinction of species with little value, such as most birds. This brings us to the third reason, which focuses on the array of essential services provided by natural communities.

Natural communities provide a number of essential "services," including the maintenance of the "correct" gaseous composition of the atmosphere, which prevents global warming, and the maintenance of soil biodiversity, so that the soil can support forests and crops, decompose organic matter, and recycle nutrients. Among other community functions are the maintenance of a reservoir of natural enemies to prevent pest outbreaks and the maintenance of a reservoir of pollinators to pollinate crops and other plants. A skeptic might say that one grass or one tree can function as well as any other in helping control the hydrologic cycle, one predator is as good as another in controlling a potential pest, and one tree species is as good as another in terms of lumber production. The truth is that most organisms are adapted to specific physical and biotic environments—and substitutions are likely to prove unsatisfactory. But "likely" is the operative word here; only now are scientists starting to provide data to back up the rhetoric. Two recent experimental studies indicate that more diverse communities perform better than less diverse ones.

Shahid Naeem and colleagues (1994) used a series of 14 environmental chambers, each 1 m², in a facility termed the Ecotron, at Silwood Park, England, to replicate terrestrial communities that differed only in their biodiversity (Fig. 15.16). The communities consisted of 9, 15, and 31 species, spread across four trophic levels, with the species-poor communities being subsets of the more diverse communities (Fig. 15.17). Loss of richness was designed to be equivalent on all trophic levels: decomposers (earthworms and soil insects), primary producers (annual plants), herbivores (insects, snails, and slugs), and parasites attacking some of the herbivores. The experiment ran for just over six months, and species were added when the trophic level below them

Figure 15.16 The Ecotron facility, Silwood Park, England, is a series of 14 environmental chambers like this one, used to investigate the link between biodiversity and ecosystem function. (*Photo by Shahid Naeem, University of Minnesota.*)

Species-rich communities usually function better than species-poor communities.

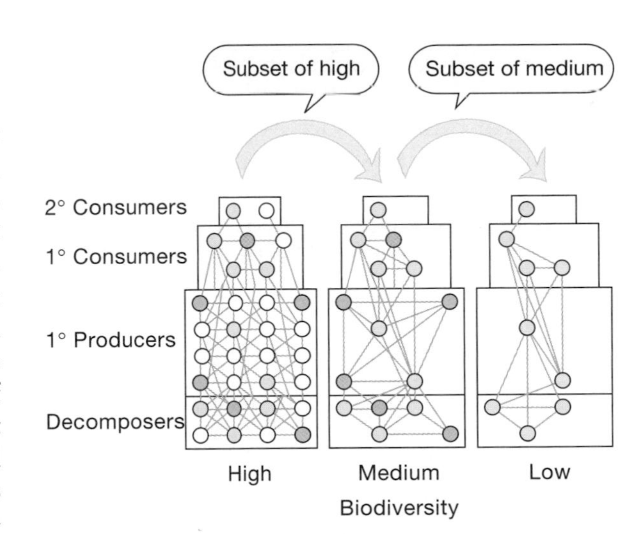

Figure 15.17 Community diagrams of the three types of model terrestrial ecosystems developed in the Ecotron. Circles represent species, and lines connecting them represent biotic interactions among the species. Note that each lower diversity community is a subset of its higher diversity counterpart and that all community types have four trophic levels. (*Reproduced from Naeem et al., 1994.*)

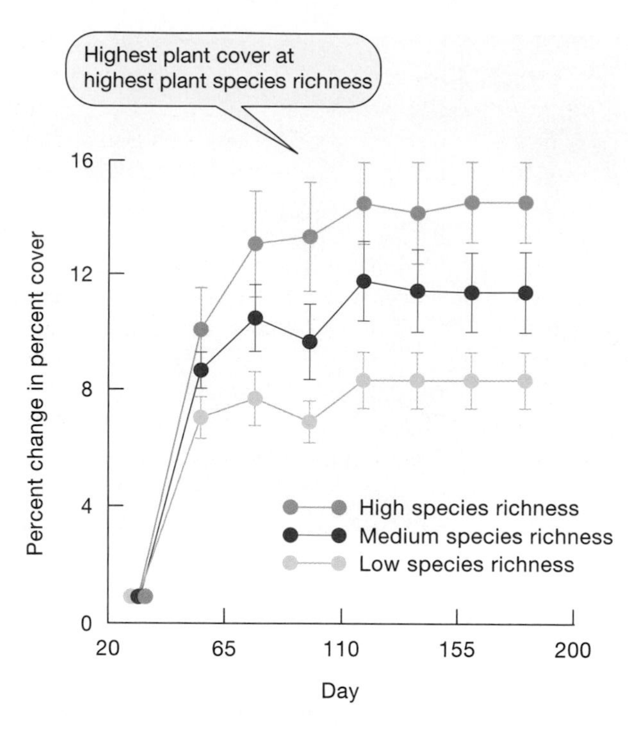

Figure 15.18 Plant productivity is linked to community species richness, as measured by the Ecotron experiments. The chart shows the percent change in vegetation cover (mean with error bar showing ±1 S.E.) from initial conditions, as determined by analysis of video images of canopies. (*After Naeem et al., 1994.*)

was established. For example, parasites were not added until herbivores were abundant. The researchers measured a whole host of variables, such as community respiration, decomposition, nutrient retention rates, and productivity. The result was that, as richness increased two- to threefold, so did community productivity (Fig. 15.18). That is, as plant richness increased, so did plant growth, measured as percent cover.

Experiments investigating the relationship of community function to species richness had never before been used to demonstrate that communities differing only in species richness differed in productivity. In the field studies conducted by previous workers, changes in many other environmental variables confounded the results as communities changed in richness. For example, in field observations, soil moisture or nitrogen levels also changed as diversity changed, and we cannot be sure that any changes in productivity that might be associated with changes in richness were not caused by the changing soil moisture or nitrogen levels. By contrast, the Ecotron experiments controlled environmental variables, and the mechanism for increased plant production was held to be an increase in light energy associated with plant canopies that intercepted sunlight as they filled three-dimensional space more completely with the addition of more plant species.

Further experiments by Naeem and colleagues (1996) focused on a single trophic level (plants) and documented changes in net aboveground primary productivity as a result of changes in plant diversity. (See "Profiles box: Shahid Naeem, University of Washington.") Naeem used species richness in replicate experiments in English weedy fields composed of from 1 to 16 species and measured plant biomass at the end of one growing season. On average, species-poor assemblages were less productive.

David Tilman and colleagues at the University of Minnesota (Tilman 1996;

Profiles

Shahid Naeem,
University of Washington

Two great things about being an ecologist are that just about anything you learn can be useful in ecology and that the work never gets boring. As a resident of New York City, I was exposed to a tremendous diversity of people and ideas, none of which had anything to do with ecology or nature. I went to the City College of New York largely because I was bored with my work as an inventory controller for a publishing firm in Manhattan. I moved to California and transferred to the University of California at Berkeley primarily because I ran out of courses to take at the City College of San Francisco. I majored in biology only because it allowed me to take a wide variety of courses, not because I had any plans to be a biologist. By my second year at Berkeley, I found the tertiary structure of the sodium ionophore in membranes of neurons to be the epitome of what the biological sciences stood for: focused, high tech, abstract, aesthetically appealing—and, ultimately, boring. So I decided I'd be an illustrator.

My interests changed dramatically, however, when, quite by chance, I began working for Dr. Robert K. Colwell, who was then a professor at Berkeley. His research centered on understanding the significance of complexity in biological communities. This struck me as totally wacky, but I found the question way cool. Even more wacky was the fact that Rob's way of approaching the question was by studying mites that rode from flower to flower on the noses of hummingbirds, an approach that required endless trips to the tropics and many adventures, including nearly losing his life to a snakebite. As a city slicker, I had no particular desire to go through similar experiences, but I was infected with the desire to understand the role of complexity in nature. So I switched my major from cell-molecular biology to organismal population biology and later obtained both a master's degree and a Ph.D. at Berkeley.

One might think that being a city person trained primarily in cell biology would be a poor start. In fact, it turned out to be ideal training for dealing with the logistical problems of international travel and fieldwork, as well as working in the laboratory whenever the need arose. I have worked in the lowland tropical rain forests of Central America and the Caribbean, in the low-elevation bogs of the Sierra Mountains, and in the mudflats of Denmark. Now I am working in the prairie grasslands of the North American Midwest. I have also worked on model ecosystems such as outdoor tanks stocked with plankton in Michigan; with assemblages of plants in greenhouses in Minnesota and England; with artificial terrestrial communities of plants, animals, and microbes in England; and, currently, with microbial microcosms in the laboratory here at Minnesota.

My research on artificial terrestrial ecosystems is a good example of where having a diversity of experiences helps a great deal. Professor John H. Lawton, Dr. Lindsey Thompson, Dr. Sharon Lawler, Richard Woodfin, and I attempted to answer our own personal wacky question: Does biodiversity affect the functioning of ecosystems? This question may seem strange, but an international panel of ecologists who met in Germany in 1992 could provide no clear answer to it (Schulze and Mooney, 1993). That is, aside from the obvious economic, moral, and aesthetic values of biodiversity, what benefits are there to having 10 to 100 million species on the planet? Does a rain forest need 300 species of trees per hectare, or will a monoculture of bananas sequester carbon dioxide and cycle nutrients just as well? Does a prairie grassland need its 200 species of plants, or can the depauperate cornfields that have replaced them do just as well in terms of nutrient cycling and energy flow?

Two books strongly influenced my approach in tackling these questions: Schlesinger's *Biogeochemistry* (1991) and De Angelis's *Dynamics of Nutrient Cycling and Food Webs* (1992). Together, these books suggested that the key to understanding the relationship between biodiversity and ecosystem functioning was understanding how variation in community structure affected biogeochemical processes. This meant that, to test the idea, we had to build a series of replicate ecosystems identical in every way except one—community structure. Building and maintaining replicate model ecosystems was difficult work. It took a tremendous amount of planning and team effort to keep our 14 model ecosystems going. Equipment failures, contamination, computer problems, and much more kept us up at all hours, often putting us in an ill humor. Indeed, fieldwork appeared easier by contrast. My experience in working with people, managing inventory, conducting research in both the laboratory and the field, and working with plants, animals, and microbes all paid off.

One particularly worrisome aspect of this research was the fact that we carried out our experiment in the "Ecotron," a controversial system of controlled environmental chambers built at Imperial College of London's Centre for Population Biology under the directorship of Professor Lawton. Nearly two million dollars went into building the Ecotron and the center that supported it. At a time when Thatcherism had ravaged the British economy and environmental research in the United Kingdom was suffering from severe shortages of funds, the Ecotron was viewed with tremendous skepticism by the British ecological community. Was this the best way to spend limited funds? Was laboratory research even appropriate for ecology (cf., for example, Carpenter, 1996; Drake *et al.*, 1996)?

When I delivered our findings to the British Ecological Society meetings in 1994, the auditorium was packed and overflowing into the hallways with ecologists wanting to see and hear about the infamous Ecotron. What we had found was that, even if ecosystems are intact, still, as biodiversity declines, ecosystems change their functioning (Naeem *et al.*, 1994; Naeem *et al.*, 1995). The audience was thrilled with our findings. Here was experimental evidence that the dramatic declines we are experiencing in global biodiversity could mean an irreversible change in the way our ecosystems function—an important reason to work toward the preservation of biodiversity. After we published our study, the Ecological Society of America honored it with the Society's Mercer Award, and the paper became the fifth most-cited research paper on global change in 1996. Professor Lawton's vision of a large-scale laboratory facility for ecological experiments proved its value, and the Ecotron continues to do exciting research in global change ecology.

There continues to be much debate over the Ecotron's findings, but I am used to such debate. My research and that of my students focuses on addressing controversies in ecology by direct field or laboratory experimentation. This approach gets us in trouble, but I guess we feel that it is better to do something that is intellectually exciting, even if it is risky, than to do something technically flawless, but boring.

As global change continues, the need for novel approaches will be critical. Ecology is fast becoming the premier science in solving environmental problems, and these problems are topping both national and international agendas. The future of ecology will be one in which groups of investigators bring all their experiences and training together and take risks to tackle complex issues by any means possible.

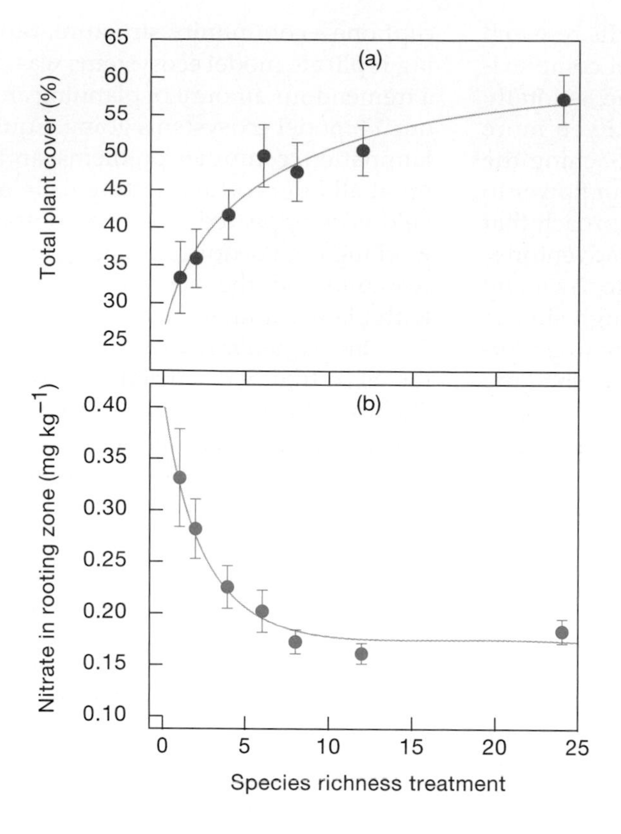

Figure 15.19 The relationship between (a) richness and community biomass and (b) uptake of nitrogen on experimental plots. With increasing species richness, biomass increases and nitrogen is used up, leaving less nitrogen in the soil and more species of plants. (*Reproduced from Tilman et al., 1996.*)

Tilman, Wedin, and Knops, 1996) took such experiments a step further by performing them in a Minnesota prairie. Earlier, Tilman and Downing (1994) had suggested that species-rich grasslands are more resistant to the ravages of drought and recover from drought more quickly than species-poor grasslands. That is, they hypothesized that diversity increases both resistance and resilience (as we will discuss further in Chapter 17). However, this older work was the result of a study of natural communities, not experimentally generated ones. The newer work directly manipulated plant diversity and measured plant production and nitrogen extraction rates from the soil. One hundred forty plots, each 3 m × 3 m and with comparable soils, were sown with seeds of 1, 2, 4, 6, 8, 12, or 24 species of prairie plants, replicated 20 times for each treatment. Exactly which species were sown into each plot was a random draw from a pool of 24 native species. Thus, any differences in results between plots with 1 or 2 species and plots with 12 or 24 species would be due not to the identity of the species used, but to the number of species per se. The results were clear: More diverse plots used nutrients more efficiently than less diverse ones and exhibited increased productivity (Fig. 15.19). Further, not only was this pattern evident in the experimental plots, but it could also be found in a sample of 30 different unmanipulated native prairies (Fig. 15.20). Since Tilman and Downing's experiments, similar results have been found in eight European countries (Hector *et al.*, 1999). The mechanism underlying the correlation between species richness and biomass was that in more diverse communities, interspecific differences in soil use among plant species allow fuller use of nitrogen, the main limiting nutrient. So, as species richness increased, more nitrogen from the soil was used up (Figs. 15.19b and 15.20b). Although we have focused on how richness affects nitrogen uptake and productivity, an increase in species number is also thought to promote other community processes, such as resistance to invasion by exotic species. (See Applied Ecology: Loss in Species Richness Weakens Community Resistance to Invasion by Exotic Species.)

Tilman argued that the relationship between productivity and diversity is not linear, but is strongest at lower levels of diversity and weaker at higher levels of diversity. Thus, diversity affects community processes only up to a point, after which it matters less. As we saw in Chapter 1 when we discussed ecosystem function, the most likely relationship between diversity and ecosystem function was that of the rivet hypothesis or the redundancy hypothesis. Both of these hypotheses suggest that the function of some species is redundant within the

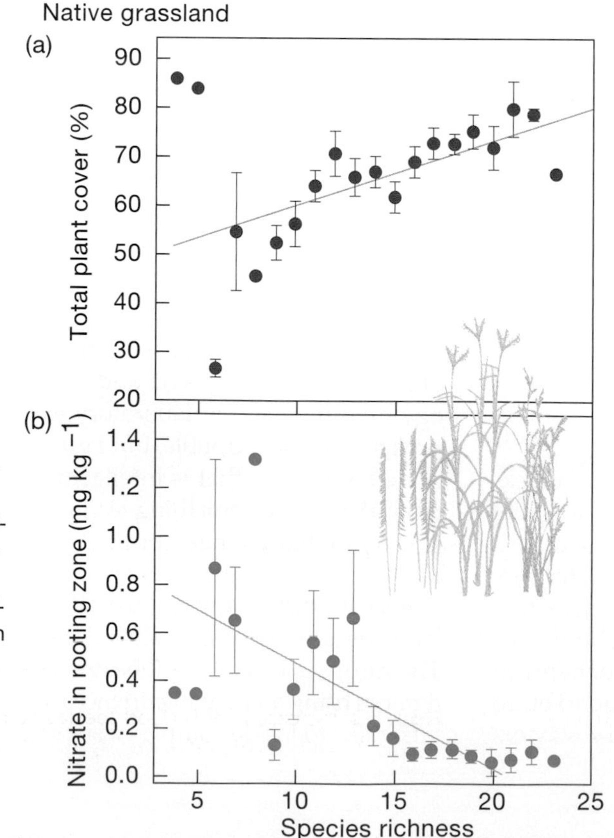

Figure 15.20 The relationship between (a) richness and community biomass and (b) uptake of nitrogen on native prairies from different plots. As seen on experimental plots, as species richness increases, biomass increases and nitrogen from the soil is used up. (*Reproduced from Tilman et al., 1996.*)

Applied Ecology

Loss in Species Richness Weakens Community Resistance to Invasion by Exotic Species

As long ago as the 1950s it was argued by Charles Elton, one of the leading ecologists of the day, that community richness would affect aspects of community function. Elton (1958) suggested that species-rich communities would be more resistant to invasion from outside organisms. The logic was that richer systems would more fully utilize limiting resources, and few would be left over for invaders to use. Elton also suggested that increased richness would reduce the severity of plant disease. He reasoned that, in epidemiology (the science of the spread of disease), transmission rates are proportional to the host species population. Most diseases are specific to one plant species; therefore, because rich communities have fewer individuals of any one host, transmission rates are low. A third hypothesis was that increased plant richness should lead to increased diversity of insect herbivores and hence increased predator and parasite richness.

Recently, Johannes Knops and colleagues (1999) provided a test of Elton's ideas: In two field experiments in Minnesota, plant richness and composition were directly manipulated. The "small biodiversity experiment" had 140 plots that were 3 m × 3 m. Each plot was weeded frequently so as to control the composition of the species that inhabited it. The "large biodiversity experiment" had 342 plots 13 m × 13 m. These plots were not weeded as carefully or as frequently, but they were large enough to examine the effects of richness on disease and insect herbivore attack rates. The small plots had either 1, 2, 4, 6, 8, 12, or 24 native grassland species seeded within them, with 20 replicates of each species. The large plots had either 1, 2, 4, 8, or 16 species. In the small plots, the frequency of invaders not planted in the plots increased with decreased plant species richness (Fig. 1), and the biomass of the invaders also declined (not shown). Also, for each plant species, the level of foliar fungal disease was negatively correlated with plant species richness (Fig. 2). Finally, arthropod species richness increased with plant species richness, albeit rather weakly (Fig. 3). Knops and coworkers showed how, 40 years previously, Elton had been right on all three counts: As species richness increases, fewer species invade the area, plants are less susceptible to disease, and insects are more diverse.

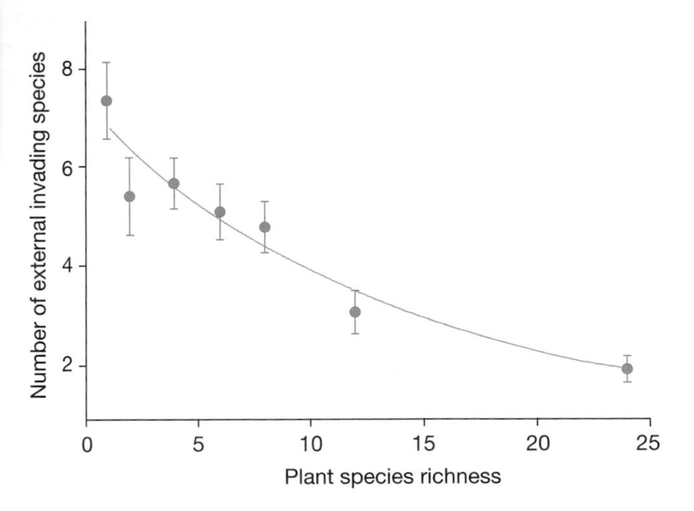

Figure 1 Decrease in susceptibility to invasion of temperate grasslands by exotic species with increase in plant species richness. *(Reproduced from Knops et al., 1999.)*

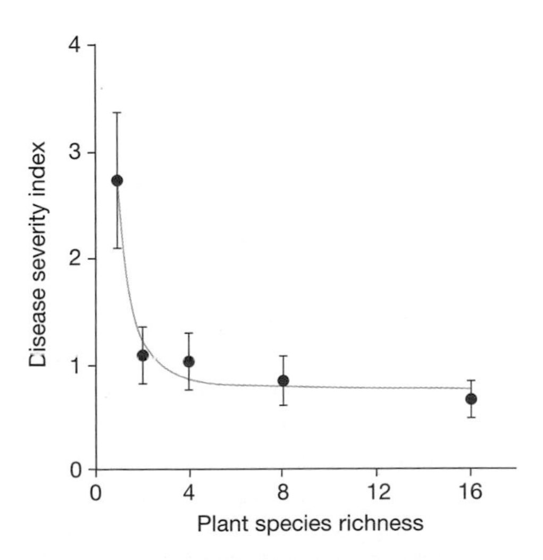

Figure 2 Decrease in disease severity index for plant species in temperate grasslands with increase in plant species richness. *(Reproduced from Knops et al., 1999).*

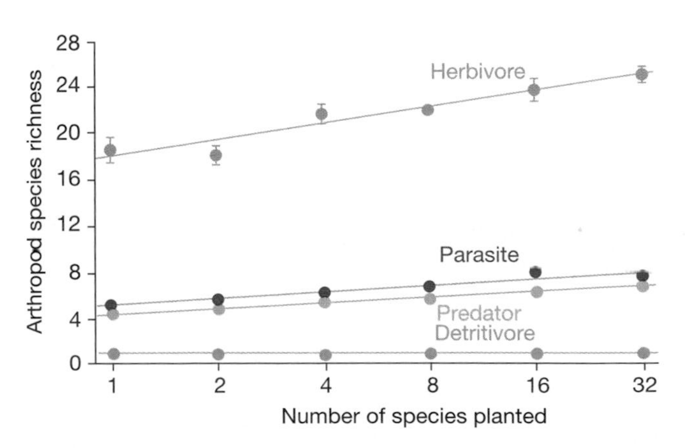

Figure 3 Increase in arthropod species richness with increase in plant species richness *(Reproduced from Knops et al., 1999).*

ecosystem, so that the loss of a particular species would not necessarily alter function of the ecosystem. When we compare temperate forests, we see that productivity is roughly the same despite different numbers of tree species present—876 in East Asia, 158 in North America, and 106 in Europe. Productivity would probably increase dramatically as tree richness increased from 0 to 40 species, but might increase little after that. In effect, diversity matters, but only up to a point.

Summary

It is well known that species richness is highest in tropical forests and other tropical locations. Explanations for this pattern are many and varied:

1. Species richness increases from the poles to the equator. Thus, tropical areas have more species than temperate areas, which in turn have more species than arctic areas. A variety of factors have been proposed to explain this phenomenon. The explanations can be grouped under two headings: biotic and abiotic. However, as pointed out by Rohde (1992), most of these explanations seem secondary or incomplete. Over different scales, from the local to the regional, the predictive power of the explanations varies considerably. For instance, biotic factors such as predation and competition may be important at a local scale, whereas evolutionary time may be important on a larger scale. Perhaps the best explanation is that evolutionary speed is higher in the tropics, thereby creating more species.

2. Besides acknowledging the progression of species richness from the poles to the equator, we can look for convergence in species richness in different habitats. For example, do deserts or wetlands have the same numbers of species? An equal number of studies say no as say yes. Species richness between areas has been examined with reference to the terms α, β, and γ diversity. α diversity exists *within* habitats, and β diversity obtains *between* habitats. The total diversity γ is the product of α and β.

3. Estimating the total richness of species on Earth is fraught with difficulty, but a figure of about 12.5 million species has been proposed.

4. A variety of methods have been proposed for managing richness or biodiversity on Earth, including paying increased attention to so-called megadiversity countries with very high species richness and to "hot spots" with high numbers of endemic species.

5. Evidence is emerging that species richness can affect ecosystem function: In general, species-rich communities perform better than species-poor communities.

Because richness is important in community function, paying attention to it helps us make wise conservation decisions.

Discussion Questions

1. Summarize the evidence for and against abiotic and biotic factors as drivers of the increase in species richness from the poles to the tropics.

2. Although most groups of organisms become more species rich toward the tropics, sometimes the reverse pattern is found, as with aphids (sap-sucking insects) (Dixon *et al.*, 1987) and parasitic insects (Owen and Owen, 1974). Why do you think this might be? If you do not know, try reading these papers.

3. To preserve biodiversity on Earth, should we focus primarily on species-rich areas, on "hot spots" of rare or endemic species, on underrepresented habitats, or on some combination thereof?

4. Some people have argued that more diverse plant communities are more efficient simply because they are more likely to contain a big important plant species than are smaller communities. This has been termed the "portfolio" effect by analogy to economic investment, wherein investors with a more diverse portfolio are more likely to select a greatly successful stock just by chance than are investors with a small portfolio. On the other hand, the portfolio effect could be regarded as a genuine ecological law. What do you think? Houston (1997) and Wardle (1999) discuss the topic fully.

5. Give an annotated list of emergent properties that communities might exhibit, such as stability, resistance to invasion, and other such properties.

CHAPTER 16

Species Diversity

Road Map

1. Diversity indices take into account numbers of individuals and numbers of species, indicating how evenly the individuals are distributed among species.
2. By graphing the abundance of individuals of all species, rank abundance plots provide us with a more accurate measure of diversity than do simple diversity indices.
3. Similarity indices indicate the proportion of species common to the communities being compared.
4. Cluster analysis permits the comparison of similarity among three or more communities.

In 1710, tin ore was discovered by Dutch colonists on the Indonesian island of Bangka. Over the centuries, the mining industry has engaged in deforestation on a massive scale, both to allow access to the tin and to use the trees as a source of fuel for the smelters. Together with slash-and-burn agriculture and the removal of large tracts of the forest by Japanese pulp mills and Javanese oil palm producers, deforestation has left only about 3% of the primary forest intact, compared with 60% remaining on Borneo, 30% on Sumatra, and 10% on Java. A study by Howard Passell of the University of New Mexico investigated whether bird diversity increased on minimally restored mining land that was simply revegetated with seedlings planted into a small hole, filled with natural topsoil collected from nearby. At first glance, the restored areas looked the same as the unrestored site; both had only 9 species of bird, compared with 16 in the nearby forests (Table 16.1). However, on further examination, the restored sites looked more diverse—at the very least, they had more individuals (Fig. 16.1). If we measure how these individuals are spread out among species, we find that the restored sites are more diverse. This measure of spread is calculated by using diversity indices. How to calculate and interpret diversity indices to compare communities is the subject of this chapter.

16.1 Diversity Indices

As we saw in Chapter 15, the simplest measure of diversity is to count the number of species in an area. The result is termed **species richness**. Although many analyses of community diversity are compared this way, one major problem with such an approach is that it does not take species abundance into account. For example, imagine two hypothetical communities, *A* and *B*, both with 100 individuals:

	Number of Individuals of Species 1	Number of Individuals of Species 2
Community A	99	1
Community B	50	50

TABLE 16.1 Abundance of bird species on unrestored tin-mining areas, in restored areas, and in unmined pristine forest areas of Indonesia. *Data from Passell (2000).*

	Unrestored Sites	Second-Year Restored Sites	Unmined
Yellow-vented bulbul	7	11	44
Olive-backed sunbird	5	10	12
Ashy tailorbird	2	3	20
Spotted dove	1	4	8
Orange-bellied flower-pecker	1	3	8
Blue-throated bee-eater	0	4	5
Cuckoo doves	2	0	4
Magpie robin	0	0	13
Less coucal	0	1	5
Plaintive cuckoo	0	0	7
Savanna nightjar	3	4	0
White-headed munia	3	4	0
Scaly-breasted munia	0	0	0
Richard's pipit	2	0	0
Red-eyed bulbul	0	0	6
Woodpecker	0	0	4
Black-shouldered kite	0	0	4
Striped titbabbler	0	0	4
Large-tailed nightjar	0	0	0
Eastern marsh harrier	0	0	1
Scarlet-backed flower pecker	0	0	1
Species richness	9	9	16
Number of individuals	26	44	146

The species richness of Community *B*, with two species each of population 50, would equal that of Community *A*, in which the populations of the same two species were 1 and 99. In actuality, though Community *B* must be considered more diverse: One would be much more likely to encounter both species there than in Community *A*. Species richness is highly susceptible to sample size: The greater the number of individuals sampled, the higher is the number of species recorded. As another example, suppose you have been charged with the task of finding the most species-rich forest in the Appalachians. In one forest, you count 100 trees and find 10 individuals of each of 10 species (community 1 in Table 16.2). In a second forest, you also find 10 species and 100 trees, but here the distribution is less equitable: Most of the trees are of one species (species *j* in the table). So community 1 is actually more diverse than community 2. To measure diversity, then, one must incorporate both abundance and species richness in measures of species diversity. There are a great many of these indices but we can divide them into two broad categories: dominance indices and information-statistic indices.

Figure 16.1 The orange-bellied flower pecker, an Indonesian tropical bird species found in the region where tin mining occurs.

TABLE 16.2 The total number of species, usually called the species richness, and the total number of individuals in two hypothetical communities, 1 and 2, are identical: Both communities have 10 species and 100 individuals. However, the individuals are more evenly distributed in community 1, so we say that community is more diverse.

Species	Individuals Per Species	
	Community 1	Community 2
a	10	5
b	10	5
c	10	5
d	10	5
e	10	5
f	10	5
g	10	5
h	10	5
i	10	5
j	10	55
Total number of individuals	100	100
Total number of species	10	10

Dominance indices.

Dominance indices are weighted toward the abundance of the commonest species. Perhaps the simplest dominance index is that due to Berger and Parker (1970), who proposed the index

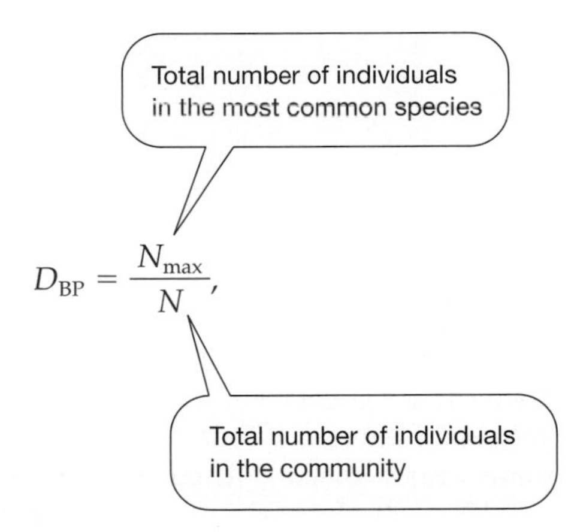

$$D_{BP} = \frac{N_{max}}{N},$$

where N_{max} is the number of individuals in the most abundant species and N is the total number of all individuals in all species. The idea is that if one species tends to dominate a community, the community will likely not be very diverse, whereas if no species dominate, the community is probably diverse. In community 1 in Table 16.2, out of the 100 trees, the most common species has only 10 individuals, so no one species dominates, and $D = 10/100 = 0.1$. For community 2, $D = 55/100 = 0.55$. Thus, while community 1 is more diverse than community 2, it has a lower value of D. Accordingly, to express greater diversity with a numerically greater value, we usually use a reciprocal form of the index such that $D_{BP} = 1/D$, so the more diverse community actually does have a higher index of diversity. In this case, D_{BP} for community 1 is $1/0.1 = 10$, and D_{BP} for community 2 is $1/0.55 = 1.82$. A huge advantage of the Berger–Parker index over others is that it is very easy to compute.

Common species influence estimates of diversity from dominance indices more heavily than do rare species.

A more widely used index than Berger–Parker's is Simpson's index (Simpson, 1949), which gives the probability that any two individuals drawn at random from an infinitely large community will belong to different species. For example, the probability that two trees picked at random from a tropical forest will be from the same species is low, because there are many different tree species in any given tropical forest. By contrast, in a Canadian temperate forest, where there are only one or two species of tree, the chances of both individuals being of the same species would be relatively high. Simpson's density index is

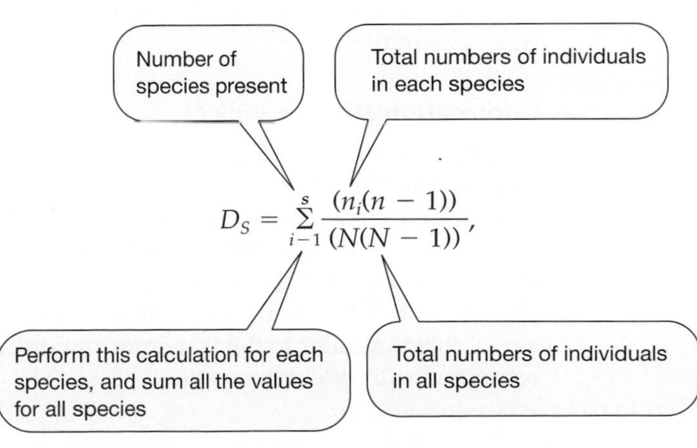

$$D_S = \sum_{i-1}^{s} \frac{(n_i(n - 1))}{(N(N - 1))},$$

where S is the number of species, n_i is the number of individuals in the ith species and N is the total number of individuals in the community. Using this formula, we see again that the higher the value of D, the lower is the diversity, so, since D and species diversity are negatively related, Simpson's index is usually expressed as $1/D$ in order to obtain increasing values of the index with increasing diversity. As an example, consider the following table:

Tree Species	Number of Individuals
1	100
2	50
3	30
4	20
5	1
Total	201

For this hypothetical data set on tree abundance and richness in a certain community, $D = (100 \times 99)/(201 \times 200) + (50 \times 49)/(201 \times 200) + ... + (1 \times 0)/(201 \times 200) = 0.338$. Then $D_S = 1/D = 1/0.388 = 2.96$.

The disadvantage of Simpson's index is that it is again heavily weighted toward the most abundant species, as are all dominance indices. Thus, the addition of many rare species of trees with one individual will fail to change the index. This property is seen to be obvious by examining the contribution of species 5 in the preceding table to the overall value of the index—it is effectively zero. As a result, Simpson's index is of limited value in conservation biology if an area has many species with just one individual. However, an advantage is that, unlike the Berger–Parker index, Simpson's uses information from a broader array of species in the community and so may be regarded as more accurate.

Information-statistic indices.

Information-statistic indices reflect both richness and evenness.

Because information-statistic indices can take into account rare species in a community, they may be more useful wherever such species exist. **Information-statistic indices** are based on the rationale that diversity in a natural system can be measured in a way that is similar to the way information contained in a code or message is measured. By analogy, if we know how to calculate the uncertainty of the next letter in a coded message, then we can use the same technique to calculate the uncertainty of the next species to be found in a community. A message consisting of *bbbbb* has a low uncertainty, because the next letter is virtually certain to be a *b*. A message consisting of *bhwzj* has a high uncertainty and hence a high index value. The higher the index value, the higher is the uncertainty of being able to tell the next letter in the sequence, or, analogously, the next species in a community. This means that the community is diverse.

Shannon index.

The Shannon index is given by

$$H_s = \Sigma p_i \ln p_i \, ,$$

where p_i is the proportion of individuals found in the ith species and "ln" denotes the natural logarithm. The following table gives an example:

Species	Abundance	p_i	$p_i \ln p_i$
1	50	0.5	−.347
2	30	0.3	−.361
3	10	0.1	−.230
4	9	0.09	−.217
5	1	0.01	−.046
Total 5	100	1.00	−1.201

Note that even the rare species with one individual, species 5, contributes some value to the Shannon index, so if an area has many rare species, their contributions would accumulate. Also, remember that the Shannon index has a minus sign in the calculation, so the index actually becomes 1.201, not −1.201. The minus sign exists solely to give us a positive index, which is vastly more appealing than negative indices. Values of the Shannon diversity index for real communities are often found to fall between 1.5 and 3.5.

Brillouin index.

A second information-statistic index, designed to reflect species abundance, is the Brillouin index

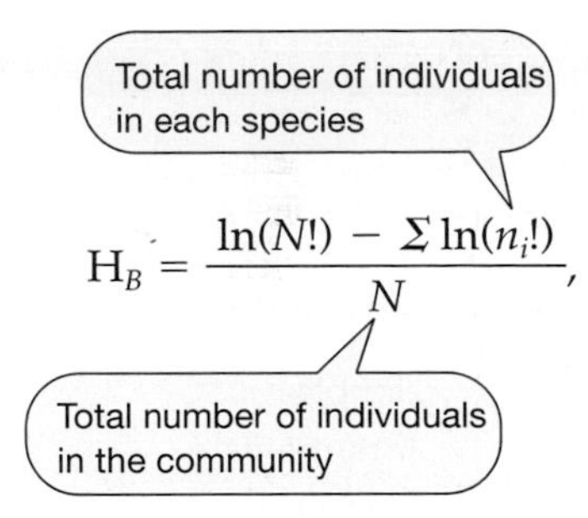

$$H_B = \frac{\ln(N!) - \Sigma \ln(n_i!)}{N},$$

where N is the total number of individuals and n_i is the number of individuals in the ith species. $N!$ is a factorial, which means that if $N = 5$, then $N! = 5 \times 4 \times 3 \times 2 \times 1 = 120$, and if $n_i = 4$, then $n_i! = 4 \times 3 \times 2 \times 1 = 24$. A worked-out example appears after the following table:

Species	Number of individuals	$\ln(n_i!)$
1	5	4.79
2	5	4.79
3	5	4.79
4	5	4.79
5	5	4.79
$S = 5$	$N = 25$	$\Sigma\ln(n_i!) = 23.95$

$$H_B = \frac{\ln(25!) - 23.95}{25} = \frac{58.00 - 23.95}{25} = 1.362.$$

So, what's the difference between the Brillouin index and the Shannon index? One major difference is that the Shannon index does not change with species abundance, provided that the number of species and their proportional abundance remain constant, while the Brillouin index does change under those circumstances. We can say that the Brillouin index discriminates between these two samples and the Shannon index does not. The Brillouin therefore has a better discriminant ability. Whether this is desirable is open to debate. For example, consider the following table and the Shannon and Brillouin indices derived from it:

Species	Individuals in Sample 1	Individuals in Sample 2
1	3	5
2	3	5
3	3	5
4	3	5
5	3	5
H_S	1.609	1.609
H_B	1.263	1.362

From a conservationist's point of view, an area rich in individuals and in species (sample 2) might be more valuable to preserve than an area that is species rich, but that has fewer individuals (sample 1), so the Brillouin index may have some real advantage over the Shannon index and may be preferable in conservation biology. A second major difference in the two indices is that the use of factorials in the equations quickly produces huge numbers that are unwieldy. This is why the Shannon index is often chosen—for its computational simplicity.

So which index should we use: dominance or information statistic, Berger–Parker, Simpson, Shannon, Brillouin, or just a simple count of species richness? Generally, values of diversity are correlated, which means that they generally increase together as diversity increases. We can work out the Berger–Parker, Simpson, Shannon, and Brillouin indices for the data on bird species diversity in the restored, unrestored, and unmined sites in Indonesia (Table 16.3). As you can see, these indices generally give the

TABLE 16.3 Calculations of the Berger–Parker, Shannon, and Brillouin indices for the data in Table 16.1 show the same general trend that restored sites are more diverse than unrestored sites. The value for Simpson's index does not indicate this trend, showing that, on occasion, different indices can give different results. Results vary according to the particular distribution of individuals among species.

Diversity Index	Unrestored Sites	Restored Sites	Unmined
Berger–Parker (D_{BP})	3.72	4.0	3.32
Simpson (D_S)	8.83	7.76	7.91
Shannon (H_S)	1.81	2.02	2.32
Brillouin (H_B)	1.62	1.88	2.16

TABLE 16.4 Comparison of the effectiveness of different diversity indices. (*After Magurran, 1988*)

Index	Discriminant Ability	Sensitivity to Sample Size	Biased toward Rare (*R*) or Dominant (*D*) Species	Calculation	Widely Used
S (species richness)	Good	High	*R*	Simple	Yes
Shannon	Moderate	Moderate	*R*	Intermediate	Yes
Brillouin	Moderate	Moderate	*R*	Complex	No
Simpson	Moderate	Low	*D*	Intermediate	Yes
Berger–Parker	Poor	Low	*D*	Simple	No

same result. Three of the four indicate that bird diversity is greater in the restored site, suggesting that restoration can produce results in a relatively short time. Thus, the choice of the "correct" diversity measure may not be as critical as might be feared and depends on how much one values isolated rare species—in which case information-statistic indices are the most valuable—and how much one thinks that total abundance of individuals is important—in which case Brillouin is best. A summary of the effectiveness of the indices examined in this section is outlined in Table 16.4.

There is no one single best diversity index. One's choice of index often depends on the question asked.

It is clear from an examination of the table that no one index is excellent across the board—with good discriminant ability, with low sensitivity to the sample size, without bias toward dominant species, and yet easy to calculate. Thus, it is not possible to recommend a single index as superior to all others. Furthermore, it is sometimes debated whether we can treat all species as equals in conservation. For example, is a rare bird equivalent to a rare earwig? If not, how can we account for that difference? One idea is to use *weighted* indices. (See "Applied Ecology" box.)

It is also important to remember that we cannot mix diversity indices when we are comparing communities. So we can't compare communities using a Shannon index for one and a Simpson's for another; we must use one index throughout.

All information-statistic indices are affected by both the number of species and their equitability or evenness. A higher number of species and a more even distribution both increase diversity. For any information-statistic index, the maximum diversity of a community is found when all species are equally abundant. We can compare a community's actual diversity, H_S, to the maximum possible diversity, H_{max}, by using a measure called *evenness*:

$$\text{Evenness} = \frac{H}{H_{max}}.$$

E is constrained between 0 and 1.0 and can be determined using any of the information-statistic diversity indices discussed. In Table 16.2, H_S for community 1 is maximal, as the individuals are equally distributed among the species. Thus, $H_S = 2.30$, $H_{max} = 2.30$, and $E = 1.0$. In community 2, $H_S = 0.329$, so $E = 0.329/2.30 = 0.143$. Generally, adding species and increasing evenness both increase species diversity (Table 16.5). Surprisingly, a community with two species can be more diverse than a community with three if the dis-

TABLE 16.5 Species diversity, measured here by H_S, the Shannon index, increases with the addition of species and with increasing evenness, measured as H_S/H_{max}.

	Relative Abundance of Species			H_S	H_{max}	Evenness
Community	Species 1	Species 2	Species 3			
1	90	10	0	0.325	0.69	0.471
2	50	50	0	0.69	0.69	1.00
3	80	10	10	0.639	1.10	0.58
4	33.3	33.3	33.3	1.10	1.10	1.00

Applied Ecology

Weighting Biodiversity Indices: The Use of Ordinal Indices

It would be good to incorporate as much biological information as possible in a diversity index especially when one is trying to use such an index as a measure of which communities to conserve. All the indices discussed so far treat all species equally. Is this right, or should more weight be given to a rare species than to a common species? Should we give more weight to a large rare predator than a small rare nematode? Is this just a question of human preference, or would the larger, rare predator be more important in the community, using up more energy or acting as a keystone species. Indices that attempt to rank or order species in importance can be thought of as *ordinal* indices. Indices that treat all species as equals are *cardinal* indices, just counting the numbers of species. All the indices discussed so far are cardinal.

Vane-Wright, Humphries, and Williams (1991) have explored how to weight "taxonomically rare" species in diversity indices (Fig. 1). Species counts to be used in a diversity index are multiplied by a weighting factor (Table 1). In a cladogram, which details the relatedness of species, those which have the most branches between the stem and the tip are set to a value of unity, and then the sister groups are given a weight or score (W) equal to the sum of all the other branch values (in the case of Fig. 1, up to 8). However, this approach may overweight the value of the taxonomically distinct species. For example, for reptiles, May (1990) pointed out that the two living species of tuatara reptiles (Fig. 2), which live only in New Zealand, would be weighted equally to the sum of all 6 thousand other species

TABLE 1

How to combine taxonomic distinctiveness in diversity indices. In community *A*, there are two species, each with 10 individuals. The unweighted Shannon index gives a value of 0.694, greater than the unweighted Shannon index for community *B*, 0.395, where the community is dominated by one species with 18 individuals and two very rare species with one individual each. If we now calculate a weighted Shannon index for community *B*, giving more weight to these rare species, we see that the Shannon index is 0.926, greater than that for a weighted index for community *A*. The weighting factor (WF) = taxonomic distinctiveness (TD), divided by total units in the taxonomic distinctiveness column (16). Therefore, WF = TD/total.

Community A

Taxonomic distinctiveness (TD)	N	p_i	$p_i \ln p_i$	Weighting factor (WF)	N x WF	Weighted p_i	$p_i \ln p_i$
1	10	0.5	−0.347	6.25	62.5	0.5	−0.347
1	10	0.5	−0.347	6.25	62.5	0.5	−0.347
2	0	0	0	12.5	0	0	0
4	0	0	0	25	0	0	0
8	0	0	0	50	0	0	0
Total = 16	20	1.0	−0.694	100	125	1.0	−0.694

Community B

Taxonomic distinctiveness (TD)	N	p_i	$p_i \ln p_i$	Weighting factor (WF)	N x WF	Weighted p_i	$p_i \ln p_i$
1	18	0.9	−0.095	6.25	112.5	0.6	−0.306
1	0	0	0	6.25	0	0	0
2	0	0	0	12.5	0	0	0
4	1	0.05	−0.150	25	25	0.133	−0.268
8	1	0.05	−0.150	50	50	0.266	−0.352
Total = 16	20	1.0	−0.395	100	187.5	1.0	−0.926

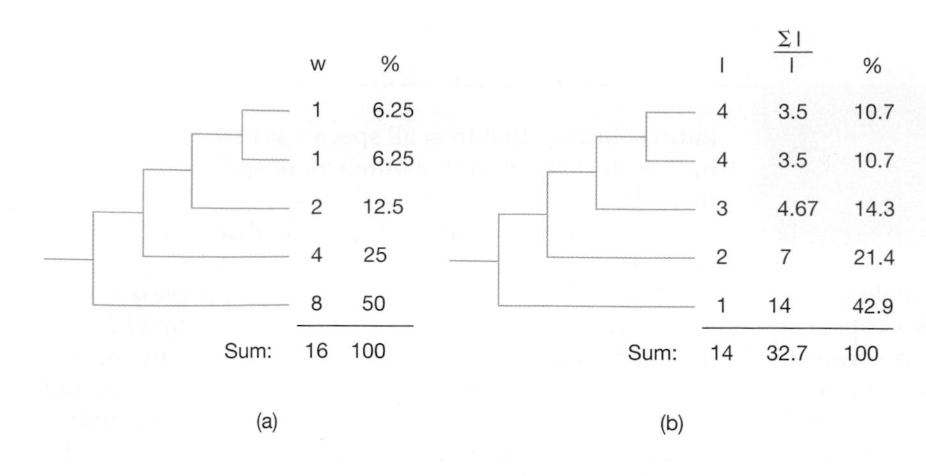

Figure 1 Two ways of measuring taxonomic distinctiveness: (a) via common ancestry, where *w* represents a weighted score; (b) measures of genetic distance apart, where *I* represents an information index.

Figure 2 *Sphenodon punctatus*, the tuatara, from New Zealand, is the most taxonomically isolated reptile in the world. How can we weight rare or unusual species in diversity indices? *(John Cancalosi, Peter Arnold, Inc.)*

of snakes and lizards. A second approach is based on an information index (I), which in turn is based on the number of branchings in the tree (Fig. 1b). The sum of the I values is then divided by the value for the individual species itself, and the contribution is then expressed as a percentage. Vane-Wright, Humphries, and Williams (1991) used this technique on the *Bombus subrircus* group of bumblebee species, which is distributed worldwide. If a simple species count were used, then maximum diversity would occur in Ecuador (10 species = 23% of world total), but if taxonomic distinctness were accounted for, maximum diversity would occur in the Gansu region of China (23% of world total).

Rank abundance diagrams visually describe the allocation of individuals to species in communities.

tribution of individuals among the two species is more even.

16.2 Rank Abundance Diagrams

Descriptions of whole communities by one statistic of diversity run the risk of losing much valuable information. A more complete picture of the distribution of species abundances in a community is gained by plotting proportional abundance (usually on a logarithmic scale) against rank abundance. The abundance of the most common species appears on the extreme left and that of the rarest species on the extreme right (Fig. 16.2). A rank abundance diagram can be drawn for the number of individuals, biomass, ground area covered (for plants), and other variables, all plotted against rank abundance. There are at least 18 different theoretical forms of rank abundance diagrams (Tokeshi, 1999), but we will consider the three most well known: the geometric series, lognormal, and broken-stick. The distribution of species abundance is most equitable in the broken-stick, less equitable in the lognormal, and less equitable still in the geometric series.

A common biological explanation for the pattern appearing in a geometric series is that the first or most dominant species to colonize a new area appropriates a fraction of the available resources and, by competitive interaction, preempts that fraction, say, 50%. The second species then preempts a similar fraction of the remaining resource, and so on with further colonists (Fig. 16.3). Fits to this model have been found for plants from a subalpine forest community (Whittaker, 1975) and for benthos in a polluted fjord (Gray, 1981). The model may best fit communities of relatively few species wherein a single environmental factor predominates. Only one species is best fit to survive in such a habitat, so that species becomes numerically dominant. Both extreme temperatures (subalpine) and pollutants limit species richness—hence the fit of such communities to the geometric series

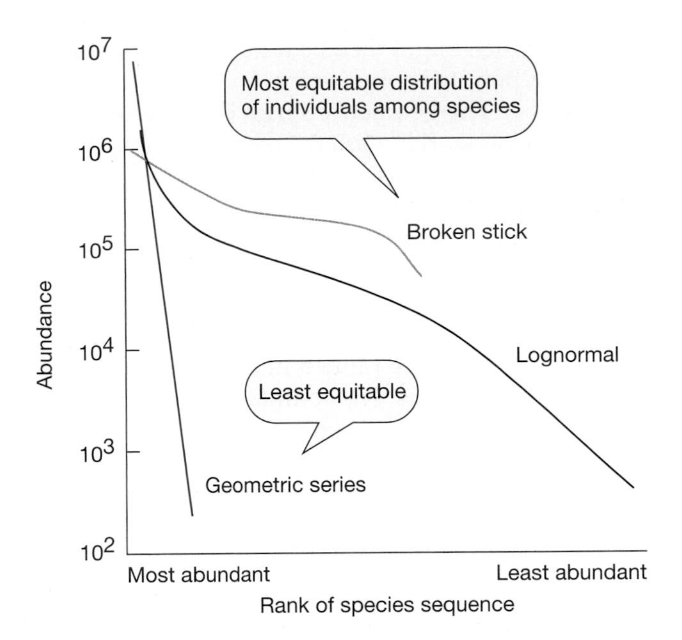
10^7
10^6
10^5
10^4
10^3
10^2
Abundance
Most equitable distribution of individuals among species
Broken stick
Lognormal
Least equitable
Geometric series
Most abundant
Least abundant
Rank of species sequence

Figure 16.2 Rank abundance plots illustrating the typical shapes of three species abundance models: geometric series, lognormal, and broken stick. In these graphs, the abundance of each species is plotted on a logarithmic scale against the species' rank, in order from the most abundant to the least abundant species. Each type of curve can tell us something about the nature of the community under investigation. The shape of the curve is specific to the distribution it represents.

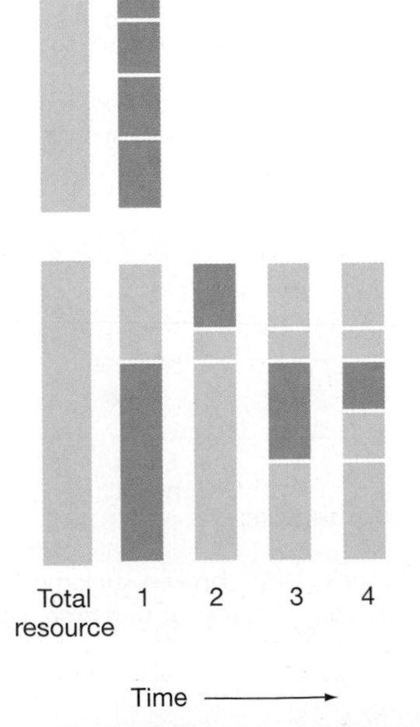
Geometric series
Each successive species preempts some fraction, here 50%, of the resource. The second species preempts a similar fraction of the remaining resource, and so on. Alternatively, the most dominant species preempts 50% of the resource, the second most dominant preempts 50% of the remainder, and so on. Shading indicates resource taken by each new species.
Broken stick
Resources are divided randomly or instantaneously between all colonists.
Lognormal
Stick is again broken sequentially, not instantaneously. Any part of the resource may be broken at each step, including those preempted by previous colonists.
Total resource
1
2
3
4
Time

Figure 16.3 Diagrammatic representation of species abundance curves and ecological explanations. Each community starts at time 1 and progresses toward time 4. The geometric series, lognormal, and broken-stick models are the most commonly discussed in the ecological literature.

which might then be used as an indicator of a polluted community.

The geometric series distribution contrasts starkly with the more equitable distribution represented by the "broken stick" model of Robert MacArthur (1957). In this model, the resources (represented by a stick) are divided at one time into segments over the whole length of the stick, implying instantaneous colonization by all species. The segments are ranked in decreasing order of their lengths. The abundance of each species is assumed to be proportional to the length of the segment associated with that species (Fig. 16.4). A few bird, fish, worm, and predatory gastropod communities were found to fit the broken-stick model (MacArthur, 1960; King, 1964), but the model is realistic on relatively few occasions (Hairston, 1969).

The lognormal distribution is a common descriptor of how individuals are apportioned to species in communities.

The lognormal distribution is more common than the broken stick for most communities that are rich in species. One ecological explanation for the lognormal distribution refers to the broken-stick model, but the stick is broken sequentially, not instantaneously. The stick is broken at random into two parts, and then any one part is chosen at random and broken again, giving three parts. One of the three is chosen at random and broken again, and so on.

The lognormal distribution owes its place in ecology to Frank Preston (1948), who obtained a normal, or bell-shaped, curve when he plotted number of species (on the y-axis) against the logarithm of species abundance (expressed as abundance classes [e.g., 1–10, 11–100, 101–1,000, and 1,001–10,000]) on the x-axis. We can see how this distribution works by translating the lognormal rank abundance plot of Fig. 16.1 into a graph of number of species versus a logarithmic scale, as shown in Fig. 16.5. Williams (1964) went on to produce many lognormal distributions for a variety of com-

(a) A stick, here 100 units long, represents a resource gradient.

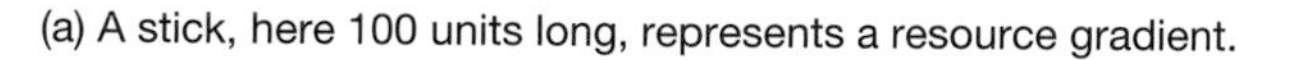

(b) For a 6-species (n) community, 5 (n–1) random throws are made at the stick.

(c) The stick is broken at each point where a throw landed. The 6 fragments of stick represent the 6 species; the length of each segment represents the fraction of the resource used by that species, and hence its abundance.

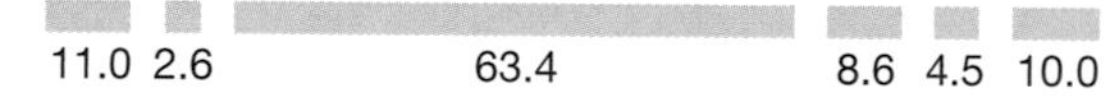

(d) In a ranked abundance plot, the 6 species are arranged in sequence of decreasing order of abundance and the abundances plotted on a log scale.

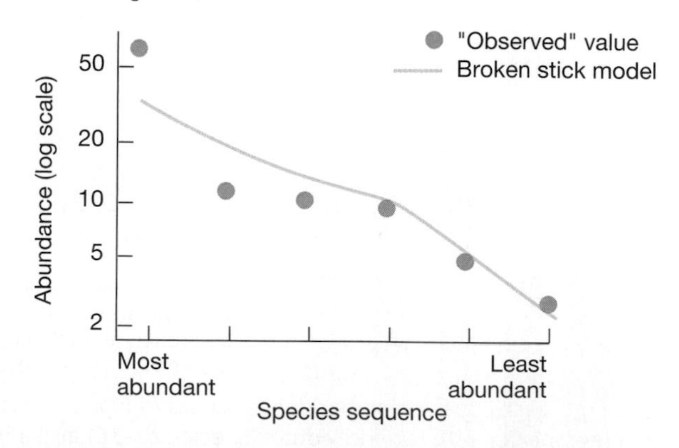

Figure 16.4 Simulation of MacArthur's (1957) broken-stick model, in which all colonizers arrive and partition resources at one time. *(Reproduced from Wilson, 1993.)*

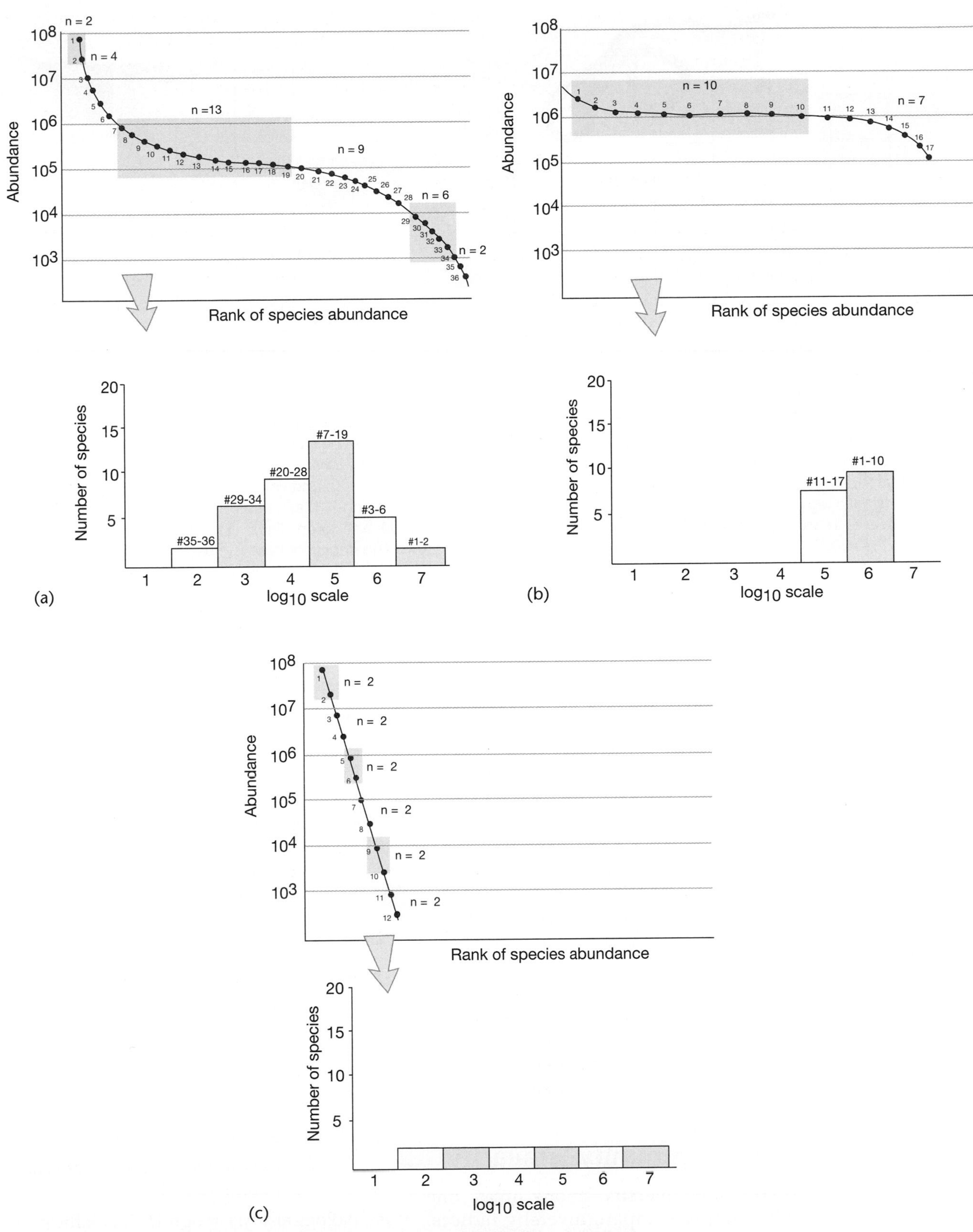

Figure 16.5 Rank abundance curves and their underlying distribution of species in abundance classes. (a) How the lognormal rank abundance curve got its name. We rank 36 species in a community in a standard rank abundance diagram. Next, each species is given a number. We then group the species in abundance classes of $\log_{10}$. The result is a "normal" distribution: a few species with few individuals (species 29–36) and a few with a lot of individuals (species 1–6), but most with an intermediate number (species 7–28). (b) The broken stick yields a much narrower bell-shaped curve, while (c) the geometric series yields a flat-line relationship between number of species and $\log_{10}$ abundance.

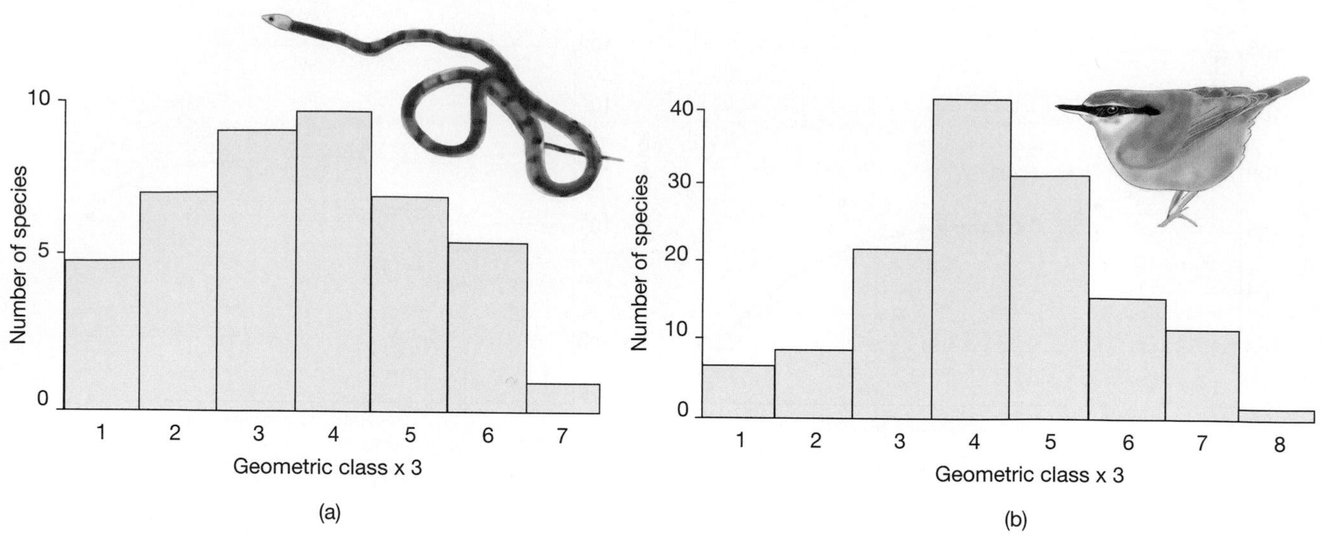

Figure 16.6 The lognormal distribution. The "normal," symmetrical bell-shaped curve is achieved by plotting the logarithm of the species abundances on the *x*-axis. A variety of bases can be used. (a) $\log_3$. Successive classes refer to treblings of numbers of individuals. Thus, in this example showing the diversity of snakes in Panama, the upper bounds of the classes are 1, 4, 13, 40, 121, 364, and 1,093 individuals. Although used widely, $\log_3$ is rarely employed today. *(Data from Williams, 1964.)* (b) $\log_{10}$. Classes in $\log_{10}$ represent increases in order of magnitude of 1, 10, 100, 1,000, 10,000, 100,000, and so on. Thus, the scale is 0–1, 2–10, 11–100, 101–1,000, 1,001–10,000, 10,001–100,000, and so on. This choice of base is most appropriate for very large data sets, such as this one: birds in Britain. *(Data from Williams, 1964.)* In all cases, the *y*-axis shows the number of species per class.

munities, using a wide array of bases (Fig. 16.6). Preston also recognized the existence of the truncated lognormal distribution, in which plots of the numbers of species (on the *y*-axis) against individuals per species on a logarithmic scale (on the *x*-axis) again followed a normal distribution, but were truncated to the left of the mode (Fig. 16.7). The truncation was explained as being due to species that were present in the habitat, but not in the sample. If larger samples were taken, more species would be obtained, and the mode would move to the right. Therefore, in determining which type of rank abundance diagram fits which community best, it is critical that the community have been adequately sampled. We could not be certain that pollution causes a geometric series unless we do exhaustive sampling.

16.3 Community Similarity

Community similarity is a measure of how many species are common among communities.

To compare diversity among areas, one could simply compare diversity indices. Another method is to compare diversity of sites directly by the use of indices called *similarity coefficients*. These coefficients usually compare the numbers of species common to all areas. In discussing community similarity, we will generally consider a simple presence–absence matrix for two areas *A* and *B*:

		Area *A*	
		No. of Species Present	**No. of Species Absent**
Area *B*	**No. of species present**	*a*	*b*
	No. of species absent	*c*	*d*

where

a = the number of species common to both sites;

b = the number of species in site *B*, but not in *A*;

c = the number of species in site *A*, but not in *B*; and

d = the number of species absent in both samples.

Is *d* biologically meaningful? Only if the pool of species is well known. If two areas both have few species despite a potentially rich pool of species nearby, then they may be more similar than some similarity coefficients would show. In this case, *d*, a measure of the negative matches, is potentially biologically meaningful. In reality, it is almost impossible to know *d*, so most similarity coefficients

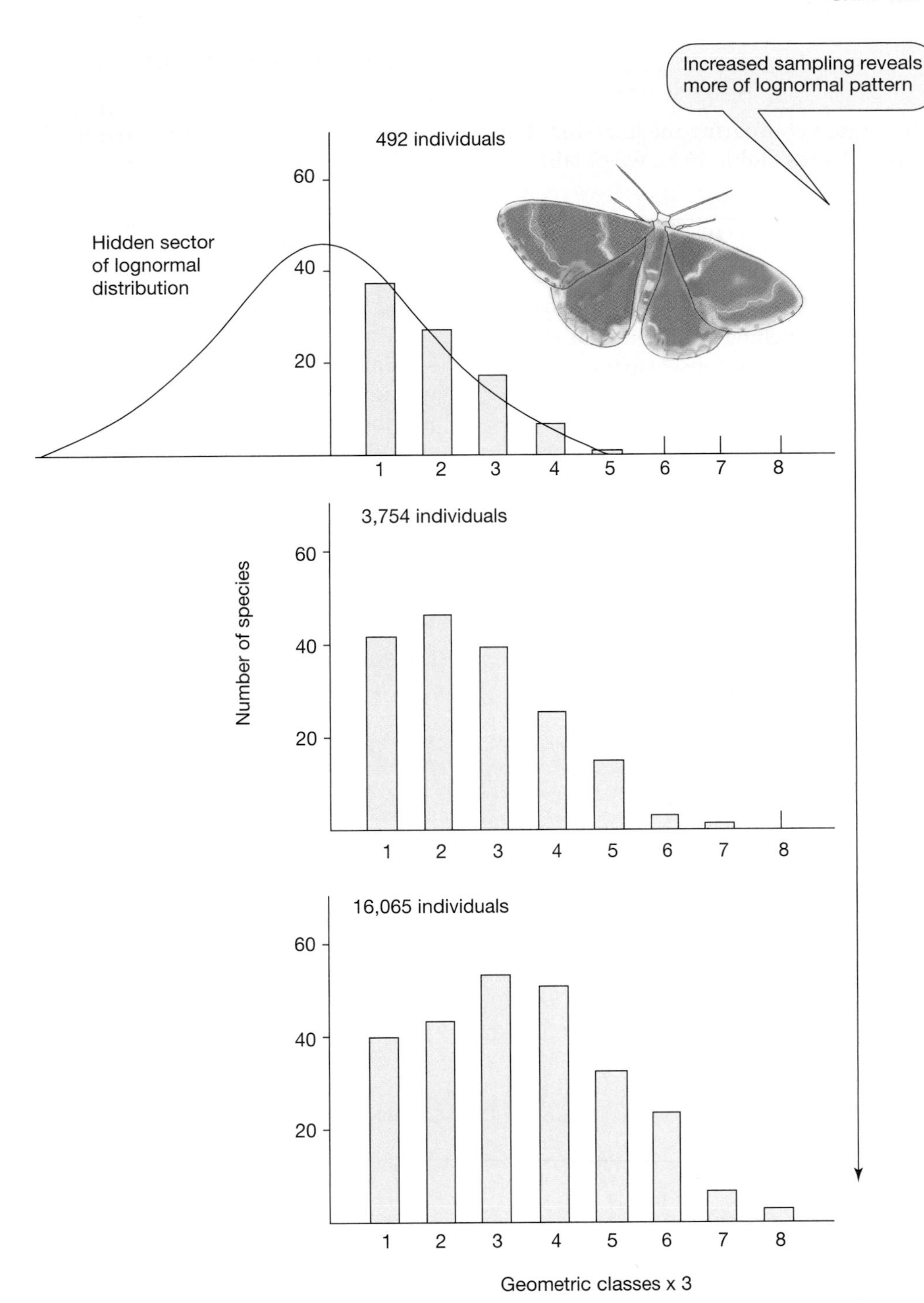

Figure 16.7 Distribution of abundance in moths captured by light traps in England. In the top figure, the true distribution of abundance is hidden behind the veil of the *y*-axis. As sampling becomes more intense, the distribution pattern moves to the right to reveal the distribution of rare species. (*Reproduced from Williams, 1964.*)

rely only on a, b, and c. Out of over 40 common similarity coefficients, we will consider the three simplest: the Jaccard index (Jaccard, 1912), the Sorensen index (Sorensen, 1948), and the simple matching coefficient.

Jaccard coefficient.

The Jaccard coefficient is calculated using the equation

$$C_J = \frac{a}{(a + b + c)}.$$

Thus, for the data outlined in Table 16.1 between the unrestored and unmined sites,

$$C_J = \frac{6}{(6 + 10 + 3)} = 0.32 .$$

Sorensen coefficient.

The Sorensen coefficient is similar to the Jaccard coefficient and uses identical variables. However, the Sorensen coefficient weights matches in species composition between the two samples (a, which we know) more heavily than mismatches (which we are less sure of because of the possibility of undersampling). Whether this is valuable or not has not yet been resolved. The formula is

Cluster analysis can determine the similarity of three or more communities.

$$C_S = \frac{2a}{(2a + b + c)}.$$

Thus, again comparing the unrestored and unmined sites (Table 16.1), we obtain

$$C_S = \frac{12}{(12 + 10 + 3)} = 0.48.$$

The simple matching coefficient C_{sm}. The simple matching coefficient makes use of the number of species absent in both areas (d):

$$C_{Sm} = \frac{a + d}{a + b + c + d}.$$

We cannot use the data in Table 16.1 without knowing the total number of species in the area. If we suppose that $d = 21$, the total number of species in all areas, then

$$C_{Sm} = \frac{6 + 21}{6 + 10 + 3 + 21} = \frac{27}{40} = 0.675.$$

16.4 Cluster Analysis

When research is being done on more than two sites, one method of comparing the sites is cluster analysis. The method starts with a table or matrix giving the similarity between each pair of sites, as measured by the use of any similarity coefficient. The two most similar sites in this matrix are combined to form a single cluster. The analysis then proceeds by successively clustering similar sites until all are combined in a single figure called a **dendrogram**. There are a number of techniques for deciding how sites should be joined into clusters and how clusters should be combined with each other, but the original method was called single-linkage clustering.

In single-linkage clustering, we start out with a matrix of similarity coefficients (Table 16.6). In this example, we seek to determine the similarity of moths captured by light

TABLE 16.6 Hypothetical matrix of similarity coefficients for moth species composition at three sites in a conifer plantation and two sites in oak woods. These same data were used to produce the dendrogram of community similarity shown in Figure 16.7.

		OAKWOOD		CONIFER		
		1	2	1	2	3
Oak	1	1.00	0.65	0.05	0.02	0.01
	2		1.00			
Conifer	1			1.00	0.47	0.4
	2				1.00	0.7
	3					1.00

STEP	ACTION	RESULT		ACTION	TREE
1	Find greatest similarity less than unity	0.7	(conifers 2–3)	Join conifers 2–3 at level 0.7	2 3
2	Find next-highest similarity	0.65	(oaks 1–2)	Join oaks 1 to 2 at level 0.65	1 2 2 3
3	Find next-highest similarity	0.47	(conifers 1–2)	Join conifer 1 to conifers 2 and 3 at 0.47 level	1 2 1 2 3
4	Find next-highest similarity	0.4	(conifers 1–3)	1 already joined to cluster containing 3; therefore, ignore	
5	Find next-highest similarity	0.05	(oak 1 to conifer 1)	Join oak 1 and 2 cluster to conifer 1, 2, and 3 cluster at 0.05 level	1 2 1 2 3

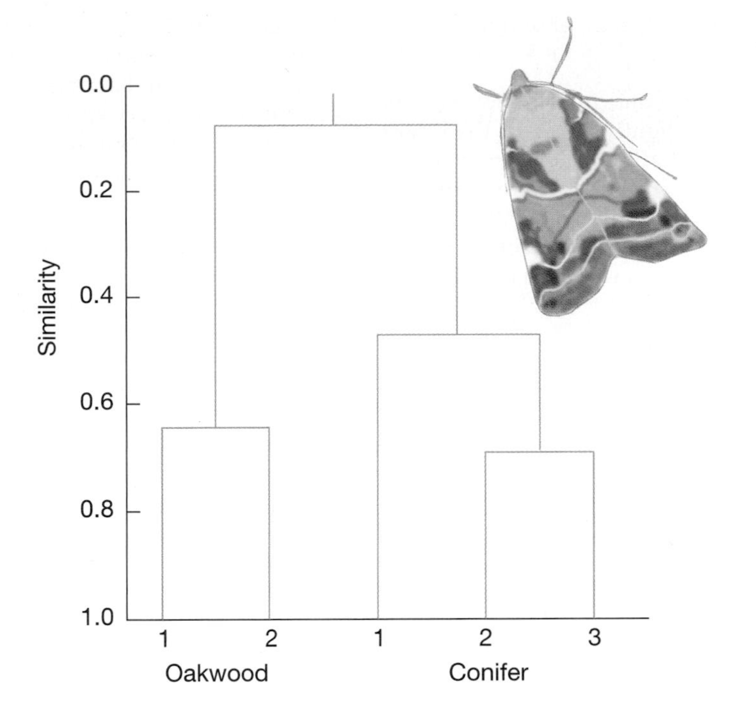

Figure 16.8 An example of a single-linkage cluster dendrogram showing the similarity between moths found at three light-trap sites in a conifer plantation and two light-trap sites in an oakwood. The dendrogram shows much greater similarity within the two woodland habitats than between them. (*Reproduced from Magurran, 1988, and based on data in Table 16.5.*)

traps in different woods. From the given matrix, we find the most similar pair of samples. In this case, the greatest similarity is between conifer sites 2 and 3, equal to 0.70. Next, we find the second most similar pair of samples, which turns to be oak sites 1 and 2, equal to 0.65, so these sites are linked together. The next-highest similarity is between conifer sites 1 and 2, at 0.47. Therefore, conifer site 1 is linked with the cluster of conifer 2 and 3, to form a three-community cluster. Finally, the last joining occurs between the oak 1 and 2 cluster and the conifer 1, 2, and 3 cluster, at similarity 0.05. Figure 16.8 shows the tree summarizing this cluster analysis.

Similarity coefficients and cluster analysis can be useful in conservation. Say we can afford to preserve only two of five tropical forests. Would you choose the most diverse? Maybe, but you certainly wouldn't pick two that are very similar, even if they are both diverse. Similarity coefficients can aid us in conserving rich, but close-to-unique, areas.

As we noted earlier, the management of biodiversity necessitates measurement—hence the interest in diversity indices and similarity coefficients. But there has always been another side to the importance of understanding diversity: It has been one of the hallowed tenets of community ecology that diversity begets stability. For land stewards, foresters, and wildlife managers, stable systems are inherently easier to manage than unstable ones. Accordingly, we seek to examine the link between diversity and stability. But before we do that, we must define what stability is.

Summary

Species richness gives a good measure of biodiversity and is the most common measure used. However, in comparing areas to make conservation decisions, the distribution of individuals among species should be taken into account. This added precision allows policymakers to more accurately focus their conservation efforts on areas that best meet their goals.

1. Biodiversity may be expressed in many ways: as species richness (numbers of species) or by various indices that take the numbers of species and the numbers of individuals of each species into account.

2. Dominance indices include Simpson's index and the Berger–Parker index. Both are easy to calculate and easy to interpret, but both are biased toward dominant or common species. Simpson's index uses more information than the Berger–Parker index and may therefore be more realistic.

3. Information-statistic indices pay more attention to rare species than dominance indices do. Two information-statistic indices are the Shannon index and the Brillouin index. The

Brillouin index is more sensitive to species abundance, but is much more difficult to calculate.

4. Each of the various dominance and information-statistic indices may give values that are slightly different than those obtained from other indices, but all of the values are generally well correlated. Each has its own particular strengths. Some are easy to calculate. Others have good discriminant ability. There is no one overall best index, and choosing the appropriate index for the appropriate situation depends on what sort of question is being asked.

5. Most diversity indices can be referred to as cardinal indices: They treat every species as equal. However, species are not always equal. Indices that attempt to weight rare species or any other type of species are known as ordinal indices.

6. Diversity indices attempt to describe whole communities with just one statistic. A more complete description of a community can be obtained by plotting the proportional abundance of every species against its rank of abundance. The result is a rank abundance diagram. The form of the rank abundance diagram can be one of at least three slightly different shapes referred to as geometric, lognormal, and broken stick.

7. To compare diversity among areas, one could simply compare diversity indices, or one could use similarity coefficients, which compare the numbers of species common to all areas. Such indices include the Jaccard, Sorensen, and simple matching coefficient indices.

8. Cluster analysis may be used to compare the similarity of two or more sites.

Diversity indices are many and varied, and each tends to give a somewhat different value for a given data set. Rank abundance diagrams give a more complete picture, but diversity indices are simpler to use. The most appropriate index in a given situation depends on the question being asked.

Discussion Questions

1. In developing ordinal indices of species diversity, what factors would you use in weighting your indices: rarity, biomass, importance in nutrient cycling, or others?

2. We discussed three types of rank abundance diagrams: geometric, broken stick, and lognormal. Yet there are many more. Think of some other ways you might "break a stick." Do you always take the biggest piece? Do you always break off the same fraction? Do your ideas make sense biologically? You can read Tokeshi (1999) for examples if you get stuck.

3. Choose an area of your campus, and work out the richness and diversity of trees and of flowering plants. Are they different? Explain.

4. Calculate similarity coefficients between the unrestored and restored sites and between the restored and unmined sites in Table 16.1. What can you conclude?

5. The following data represent numbers of butterflies of various species in logged and unlogged forests in Borneo, where $n > 50$ individuals per species (Willott *et al.*, 2000):

Butterfly Species (Family)	Unlogged Forest	Logged Forest
Eurema tiluba (Pierid)	19	72
Cirrochroa emalea (Nymphalid)	132	43
Parthenos sylvia (Nymphalid)	14	58
Ragadia makuta (Satyrid)	140	144
Idea stolli (Danaid)	45	17
Drupadia ravindra (Lycaenid)	34	17
Logania malayica (Lycaenid)	36	14
Neopithecops zalmora (Lycaenid)	79	6
Jamides para (Lycaenid)	38	37

Calculate the species diversity and evenness of the two butterfly communities.

Are your results similar to those for birds in mined and unmined areas in Indonesia given earlier in the chapter? If not, why not?

CHAPTER 17

Stability, Equilibrium, and Nonequilibrium

Road Map

1. Community stability includes resistance to change and resilience—the ability to rebound from change.
2. For a long time, ecologists have held that more diverse communities are more stable.
3. Laboratory experiments with simple communities have usually failed to show a link between diversity and stability. Field experiments with complex communities, by contrast, have shown such a link.
4. The intermediate-disturbance hypothesis suggests that the most diverse communities, such as rain forests and coral reefs, exist at intermediate levels of disturbance, due to storms or some other kind of turbulence.

In 1997, Jan Bengtsson and his colleagues (Bengtsson, Ballie, and Lawton, 1997) examined the stability of British woodland bird communities over the period 1971 to 1992 (Fig. 17.1). They studied over 18 different woodland plots, mainly in the south of England, for the whole 22-year period and for shorter periods of time. Each year, the numbers of individuals and numbers of species of bird in each plot were recorded. Most species changed little in abundance on plots over the course of the census. Bengtsson *et al.* concluded that the community of woodland birds was relatively stable in terms of both numbers of species and individuals. They argued that density-dependent population regulation acted upon most of the bird species, keeping their relative abundance similar. Each bird species probably competed with so many others that relative abundances were stabilized.

Figure 17.1 The blue-tit, *Parus caeruleus* (above) and the Robin, *Erithacus rubecula* (below), both common species of British woodland bird shown to exist in stable communities over relatively long periods.

17.1 Community Stability

Resistance to change and resilience (the ability of a community to return to equilibrium after a disturbance) are components of community stability.

A community is often seen as stable when no change can be detected in the population sizes and numbers of species over a given period. The community may then be said to be in *equilibrium*. The frame of reference for detecting change may encompass a study of a few years or, preferably, a few decades. Long-term data on other communities also show a constancy over time. For example, because bird-watching is a popular pastime, long-term data exist for other geographical areas that show a stability over time. Indeed, in Eastern Wood, Brookham Common, England, some bird communities have appeared stable for nearly 30 years (Fig. 17.2).

However, community predictability can decrease over time. In the Bengtsson study, the bird community changed very little over short time spans, but changed more the longer the community was observed. In a similar fashion, the Eastern Wood data show variation upon a careful examination. There is a slow rise in species richness from 1959 onward, so that community stability appears greater between adjacent years than between widely separated ones. Bengtsson and his colleagues cautioned that, in order to fully document stability, the temporal scale of change should be accounted for. In other words, if we wish to document community stability in a national park, we should be aware that change is more likely to happen over longer time scales than shorter ones. Environmental variability is thought to cause this increased community variability over time, with harsh winters or dry springs likely to be spread apart rather than occurring in adjacent years.

Community stability is important. A decrease in community richness over time may tell us that something is wrong. In the 1970s, many communities of birds of prey declined greatly in species richness because of the effects of DDT. This was a signal that the pesticide was finding its way into all levels of the food chain. Later in the decade, DDT was banned. Land managers are very much interested in maintaining stable communities for public enjoyment and to maximize the survival of species. Instability can lead to the extinction of certain species.

It is important to realize that just recording the number of species in an area may obscure potential changes in the community—such as the fact that some species go extinct or emigrate and others immigrate. Thus, even though the number of species in an area remains constant, some species may become extinct and others may immigrate, causing a turnover in species. Thus, a definition of stability usually presupposes a stable equilibrium for each population in the community, not just a constant number of species. It is also worth pointing out that, while a community in equilibrium for a long time is likely to be stable, a stable community need not be in equilibrium. For example, a cycle of predator and prey may exhibit a predictable periodicity over years, yet equilibrium in numbers is never reached.

Despite the apparent simplicity of the concept of stability, it is actually difficult to define. There are, in fact, many more different ways of thinking about stability than just whether the number of species stays constant over time. A community might be thought of as stable if it quickly returns to the way it was following a disturbance or if it can return from a huge disturbance like an oil spill. These different ways of thinking about stability often are given different names:

1. **Resistance:** how big a force is needed to change the community.
2. **Resilience:** the ability of the community to return to equilibrium after a perturbation. Resilience includes

   (i) *elasticity*—how quickly a community can return to equilibrium—and

   (ii) *amplitude*—how much disturbance the community can return from.

These concepts are represented diagrammatically in Fig. 17.3.

Sometimes the ability to return from high-amplitude disturbances is termed **global stability**, whereas the ability to return from only low-amplitude disturbances

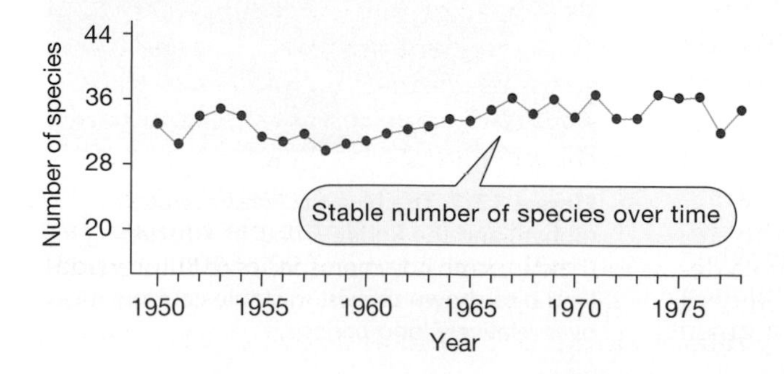

Figure 17.2 Time plots of the total number of species of breeding birds of Eastern Wood, Brookham Common, England. The near absence of change is taken to represent stability in the community. *(Modified from Williamson, 1987.)*

Resistance: force needed to change a community

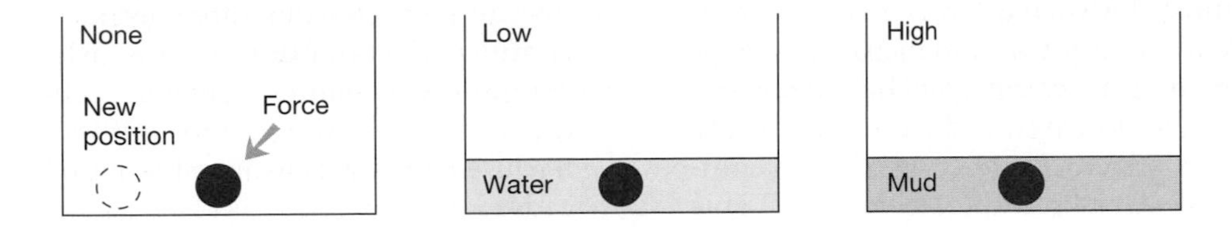

Resilience–elasticity: how quickly a community can return to equilibrium

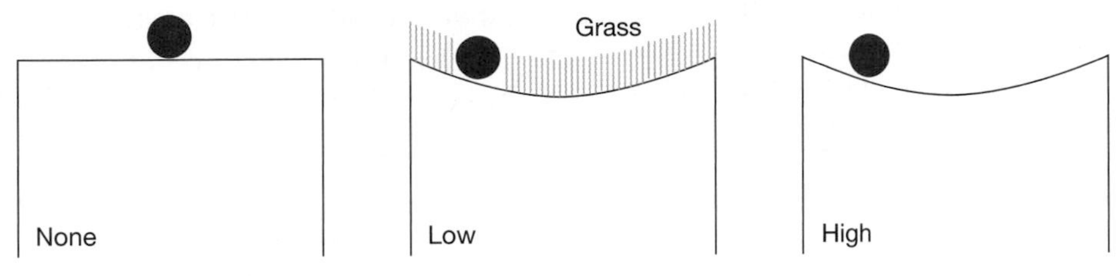

Resilience–amplitude: degree of change from which a community can return

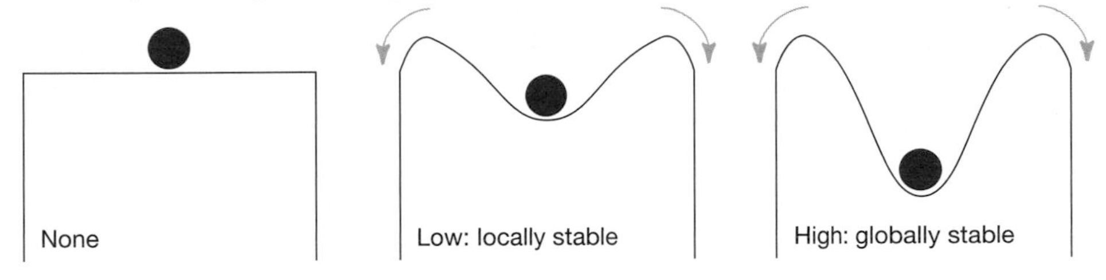

Multiple stable states: can communities exist in more than one state?

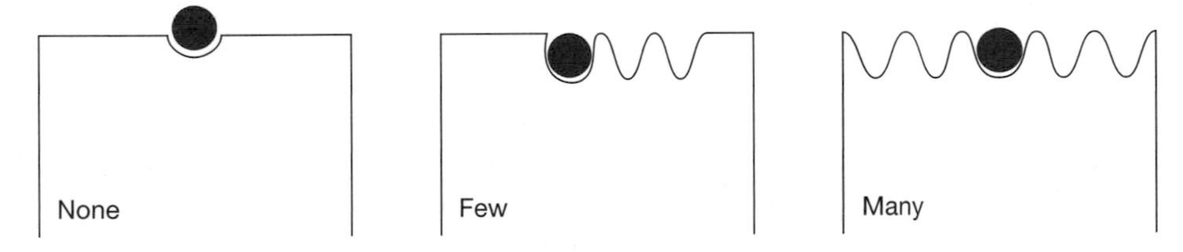

Figure 17.3 Different ways of thinking about stability. Here, the community is represented by a ball on an imaginary topographic "habitat" map. Once the ball goes "over the edge," the community becomes extinct.

is termed **local stability**. After high-amplitude disturbances, communities with only local stability would never return to their previous state.

The independent concepts of resistance and resilience are sometimes correlated and sometimes not. Lakes are often only weakly resistant, because they can concentrate pollutants from a variety of sources, and weakly resilient as well. Some truly barren deserts, on the other hand, are highly resistant and resilient, because they are so deprived of water that very little grows in them and only huge volumes of water would change things. Other deserts with more widespread vegetation may be more easily changed by disturbances such as off-road vehicles tearing up their vegetation. Rivers are not particularly resistant, but may be resilient because the fast-flowing water often cleanses them quickly.

Finally, we should ask, following a disturbance, must a community return exactly to how it was before in order for it to be labeled "stable"? If, for example, following an oil spill, populations of 90 of 100 intertidal species declined significantly and only 85 of these populations recovered, would the community itself have recovered and be deemed stable, or would it be a different community? Also, if a "recovered" community had different abundances of species than the original community had before the oil spill, would we consider the community to have changed? In other words, can communities exist in multiple stable states? This was an important question following the *Exxon Valdez* oil spill, with Exxon arguing that communities can exist in different states and the state of Alaska arguing that they cannot (See "Applied Ecology: Can Marine Communities Recover after Oil Spills?"). Joe Connell and Wayne Sousa (1983) wrestled with the idea of multiple stable states and concluded that there are few examples of such states in nature. They argued that communities occupying an area before and after a disturbance are interpreted as alternative stable states when, in fact, the physical variables at the site have been changed: Nutrient levels may have risen, levels of toxic materials may have increased, there may be greater sediment loads in lakes, or hunting or fishing efforts may have changed.

17.2 Is There a Link Between Diversity and Stability?

More diverse communities, such as tropical forests, appear to be more stable than less diverse ones, such as taiga.

The link between diversity and stability was first explicitly recognized by the English ecologist Charles Elton (1958). He suggested that disturbances in diverse communities would be cushioned by large numbers of interacting species and would not produce as drastic effects as they would on species in a small community. Thus, a predator or parasite introduced into the community could cause extinctions in a simple system, but perhaps not in a complex system, where the effects of the intruder would be buffered by interactions with more species. This suggestion is in contrast to the idea that such an invader would do even greater harm in a diverse community because there are more species it could affect. Again, land managers and conservation biologists have a vested interest in what promotes stability, because they are striving to maintain communities under their jurisdiction at stable, equilibrium levels. Elton (1958) marshaled together several lines of evidence to support his theory:

Evidence for a link between diversity and stability

1. Small, faunistically simple island communities are much more vulnerable to invading species than are species-rich continental communities, and island communities suffer greater consequences. For example, mosquitoes introduced into the Hawaiian Islands have caused many extinctions of native birds.
2. Outbreaks of pests are often found on cultivated land or land disturbed by humans, both of which are areas with few other naturally occurring species.
3. Tropical rain forests do not often have insect outbreaks like those common in less diverse temperate forests.

However, since Elton posed his original arguments, each one of his main points has been countered as follows:

Evidence against a link between diversity and stability

1. There are many examples of introduced species invading continents and assuming pest proportions, including rabbits in Australia and pigs in North America.
2. Agricultural systems may suffer from pest outbreaks not because of their simple nature, but because their individual components often have no coevolutionary history whatever, in complete contrast to the long associations evident in forest biomes (Murdoch, 1975). Natural monocultures of native plants, such as *Spartina* and *Juncus*, two salt marsh species, and of bracken fern seem to be stable.
3. Goodman (1975) argued that the stability of tropical ecosystems is a myth: There are reports of cases in which insects nearly completely defoliated Brazil nut trees and monkeys succumbed in large numbers to epidemics. Because tropical systems are not nearly as well understood as temperate ones, we simply don't know as much about tropical instability as temperate instability. It is well known, however, that locusts in tropical countries undergo huge outbreaks. In addition, rain forests, which are highly complex systems, seem particularly susceptible to human-made perturbations.

The next major step in the diversity–stability debate focused on a mathematical argument. The Australian physicist-turned- ecologist Sir Robert May (1973) (see Profile at end of

Can Marine Communities Recover after Oil Spills?

In marine systems, a common contaminant is crude oil, which can kill directly through coating and asphyxiation, a route that especially affects intertidal life, or by poisoning from contact or ingestion, as in plants and preening birds, respectively. Water-soluble fractions can be lethal to fish and invertebrates and may disrupt the body insulation of birds, resulting in their death from hypothermia.

In the United States, a particularly severe oil spill resulted from the wreck of the *Exxon Valdez* near Valdez, Alaska, March 24, 1989. The result was one of the worst oil spills in U.S. waters: Almost 11 million gallons spilled, most of it in the first 12 hours (Fig. 1). By international standards, the spill was not exceptional: Between 1967 and 1994, 39 other wrecks had spilled more oil, and several, such as the *Torrey Canyon* (1967), *Amoco Cadiz* (1978), and *Castillo de Bellver* (1983), spilled five to eight times as much.

A week after the *Exxon Valdez* spill, the resultant slick covered nearly 900 square miles. Hundreds of miles of shoreline were blanketed with oil, as much as 6 inches deep in places. Officially, 27,000 birds, 872 sea otters, and untold numbers of fish died, although the true numbers are probably higher because many dead birds and otters probably sank and were not recovered.

The effects of the oil spill carried over into the terrestrial ecosystem when bears, otters, and bald eagles feasted on the oily carrion washed up on the beach and Sitka black-tailed deer ate kelp on the beaches. Few of these animals were expected to be found dead on the beaches, because they generally return to their normal habitats before the effects become apparent. Still, the Fish and Wildlife Service found over 100 dead eagles, and most pairs in the area failed to produce young that year. The Exxon Corporation spent $3.2 billion on cleanup, an amount that, at fiscal-year 1995 rates, would support the National Science Foundation's General Ecology program for 227 years (Paine *et al.*, 1996). Two billion dollars was spent in the rescue and rehabilitation of affected wildlife, including $18.2 million on sea otters alone. Was the effort a success?

The importance of a good definition of community recovery is seen in the contrasting goals of the two major entities involved in the cleanup: Exxon and the State of Alaska. The state viewed recovery as successful if the affected communities reached the point where they would have been had no oil spill occurred. The Exxon view was that recovery occurs by the reestablishment of a healthy biological community in which the plants and animals characteristic of that community are present and functioning normally. The goal that all former plants and animals be present is probably unattainable, because we can never fully know what the community was like before the spill; not enough baseline data exist to tell us that. The goal that the former plants and animals be functioning normally pays little attention to densities and age structures of the population. However, it is clear that a good definition of a stable community is vital in courtroom situations. Unfortunately, even if populations were shown not to be fully recovered, we cannot be absolutely sure that the spill was to blame: Two months prior to the accident, in January 1989, the Alaskan port of Valdez suffered a record-setting freeze, with the coldest temperatures ever recorded; this could also have severely affected biological populations in the area.

The long-term consequences of the spill were examined by comparing nine oiled areas that were set aside and not cleaned with (a) areas that were oiled and cleaned (Fig. 2) and (b) areas that were not oiled. The percentage cover of rockweed, *Funcus gardneri*, which reaches 50 percent on unoiled areas, returned to normal values by 1991 on oiled, noncleaned sites, but not until 1992 if those sites were cleaned. Cleaning also reduced the diversity of species found in soft-sediment cores for at least two years after the spill, while oiled sites that had not been cleaned showed reductions in species abundance, but not diversity. Surprisingly, these findings suggest that the best cleaning is to leave the beach alone.

Figure 1 The oil tanker *Exxon Valdez* aground off the Alaska coast in 1989, with a boom placed around it in an attempt to minimize the amount of oil washed up on the shore.

Figure 2 Clean up on Latouche island. Cleaning efforts were slow and laborious. Steam cleaning was sometimes employed. Although the hot water removed the oil more effectively than cold, it killed the animals that lived on the shore.

chapter) argued that increasing complexity actually reduces stability in models. May assembled food webs at random, using a variety of trophic links. He argued that food webs are likely to be stable only if the inequality

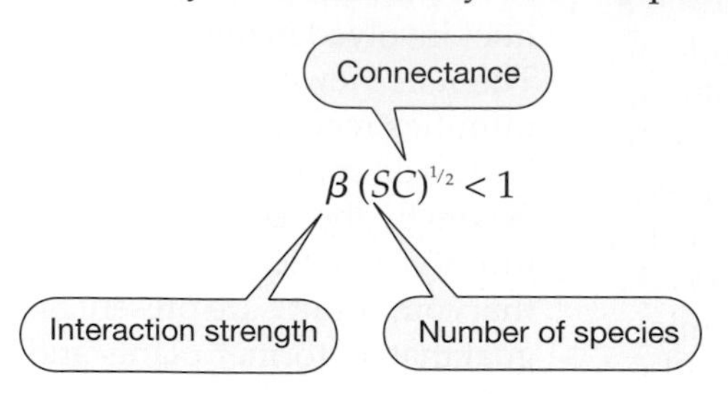

is satisfied, where β is the interaction strength between species, S is the number of species, and C is the connectance (a measure of the number of links between species in a food web, which will be discussed further in Chapter 20). Any increase in S or C would therefore increase instability. Thus, more diverse communities are more unstable than simple communities. May cautioned ecologists that if diversity causes stability in nature, it does so not as a direct consequence of the mathematics of the situation. However, some authors have noted that randomly assembled food webs often contain biologically unrealistic elements, such as predators without prey or prey feeding on predators. Mathematical analyses of food webs that are constrained to be realistic show that they are more stable than their unrealistic counterparts (Pimm, 1979). But even then, stability still declines with complexity. Dealing with randomly assembled food webs is probably not as good a method for detecting a link between diversity and stability, if one exists, as the experimental approach is.

17.3 Experimental Tests of the Diversity–Stability Hypothesis

Laboratory experiments have often failed to show a link between diversity and stability.

What might help sort out the link between stability and diversity is an experimental approach. One of the first experimental studies was that of Nelson Hairston and colleagues (1968) with microorganisms. The researchers set up various combinations of one, two, or three species of bacterial prey, one, two, or three species of *Paramecium* predator, and sometimes, one or two species of protozoan that fed on the *Paramecium*. Cultures were monitored for 20 days. When the two-trophic-level systems with two species of *Paramecium* were cultured with bacterial prey, the least abundant *Paramecium* species showed more of a tendency to become extinct with only one species of prey present (extinction rate = 68% of cultures) than with two or three prey species (extinction rates = 39% and 30%, respectively) (Fig. 17.4a). These early results supported a link between diversity and stability because more bacteria species reduce the probability of extinction. However, when the number of bacteria species was held constant at three and the number of species of *Paramecium* was increased from two to three, extinction rates increased (Fig. 17.4b). This result did not support a link between diversity and stability.

In Hairston and his colleagues' experiments, which particular species went extinct in the *Paramecium*–bacteria cultures was dependent on which species were together in culture. Thus, species were not simply interchangeable numerical units. This result again argued against a simple link between diversity and stability. Finally, when a third trophic level was added in the form of the predatory protozoan *Didinium* or *Woodruffia*, there was a further decrease in stability, because *Paramecia* were usually forced to extinction regardless of the number of their species or whether one or two predator species were present. In 50 replicates of all three trophic levels, only three contained any protozoans after 13 days. In these simple systems, diversity did not automatically lead to stability. Additional studies on small microcosms by Luckinbill (1973, 1974) showed very much the same thing. However, as Hairston would later comment, getting students interested in this sort of work was difficult, and over 20 years would elapse before anyone would try again.

In the early to mid-1990s, Sharon Lawler and Peter Morin at Rutgers University performed experiments with microorganisms to examine changes in population dynamics that occur as food chains increase in length from two to three trophic levels. In their experiments, bacteria were the prey, at trophic level 1, bactivorous protists occupied trophic level 2, and predators such as the protozoan *Didinium* that could feed on either trophic level 2 or trophic level 1 occurred at trophic level 3 (Fig. 17.5). Censuses focused on the variability of trophic level 2, and not on trophic level 1, as a measure of stability (Lawler, 1993; Lawler and Morin, 1993). Both species of protists, *Colpidium* and *Tetrahymena*, displayed remarkably constant abundances in shorter food chains, where they occupied the top trophic level (i.e., food

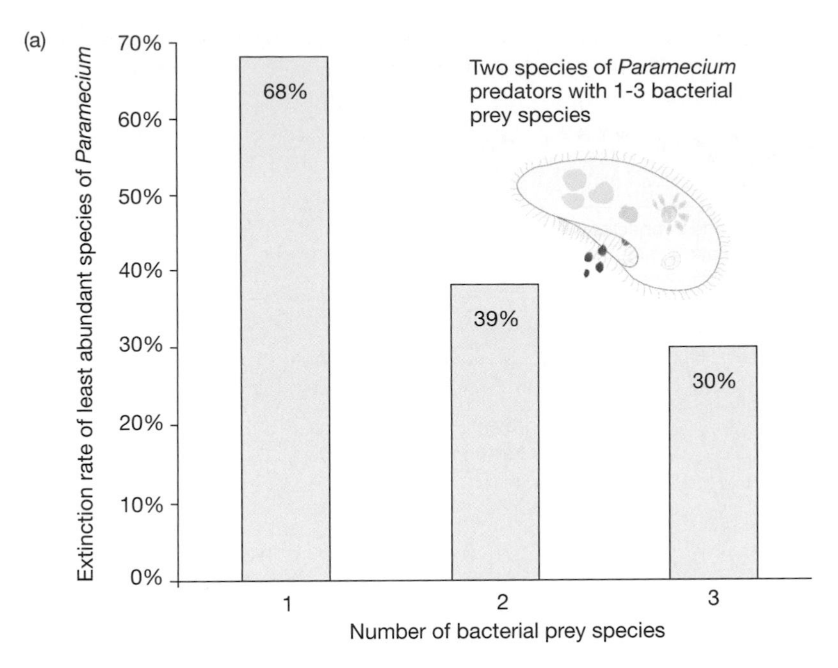

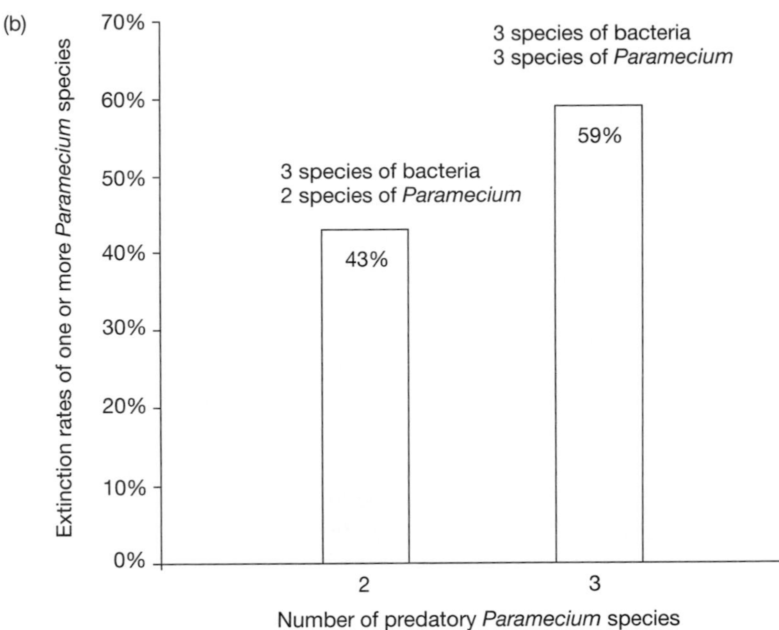

Figure 17.4 The relationship between diversity and stability as revealed by laboratory experiments involving microorganisms. (a) In simple two-trophic-level systems with two species of *Paramecium*, an increase in the number of bacterial prey species does decrease extinction rates and therefore increase stability. (b) In more complicated systems with more species of *Paramecium*, stability decreased with increased numbers of predator species. *(From data in Hairston et al., 1968.)*

(a)

(b)

Figure 17.5 A bacteria-eating protist, *Colpidium* (a), a relatively small freshwater protozoan, that feeds on bacteria shown here with an open mouth and a predatory protozoan, *Didinium* (b) shown feeding on a *Paramecium*. The *Didinium* releases a toxin which kills and paralyzes its prey.

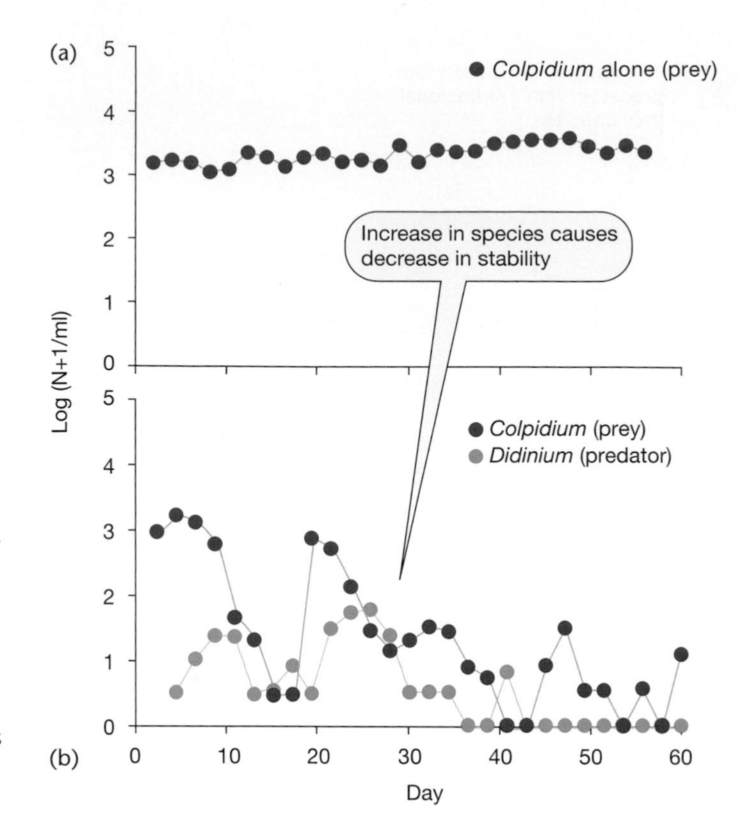

Figure 17.6 Population dynamics of *Colpidium* (trophic level 2) in (a) the absence and (b) the presence of the predator *Didinium* (trophic level 3) in experimental microcosms. *(After Morin and Lawler, 1995.)* Y axis denotes log numbers of microbes + 1 per milliliter of medium.

chains without predators on the third trophic level). The addition of predators significantly increased the variability of their prey populations, which still occupied trophic level 2 in 9 out of 13 different combinations of species of predators and protist prey (Morin and Lawler, 1995) (Fig. 17.6). The degree of reduction in stability depended on the identity of both the prey and predator species. The impact of a species that preys on *Tetrahymena* did not necessarily affect the variability of *Colpidium*. Once again, species could not always be regarded as simply interchangeable, casting doubt on mathematical models of stability that treat species as simple equivalents. However, laboratory cultures involving many species are difficult to set up and maintain, because it is virtually impossible to replicate the full array of environmental conditions, nutrients, fluctuating abiotic factors, and other trophic levels that exist in nature. To better assess diversity under natural conditions, we turn to experiments done on a larger scale in the field.

Field tests have shown a link between stability and diversity.

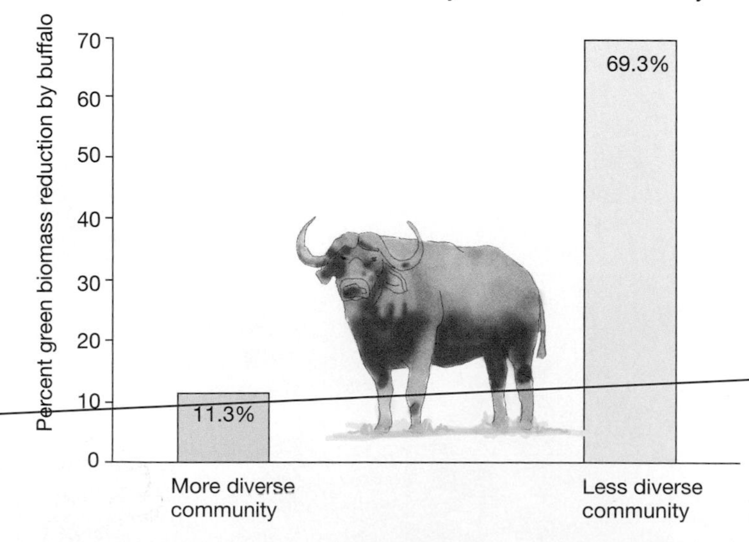

Figure 17.7 Effects of buffalo grazing on the biomass of East African savanna plant communities. In the species-rich plots, the plants not eaten by buffalo could grow more and compensate for the material eaten by the buffalo. *(After data in McNaughton, 1977.)*

Field tests of the diversity–stability hypothesis.

Among the earliest experiments to examine the relationship between diversity and stability on a large scale were those by Sam McNaughton in Africa. Working out of Syracuse University in New York, McNaughton has studied large vertebrates in Africa for over 20 years. In the 1970s, he disturbed savanna communities in the Serengeti by allowing buffalo to graze in some areas, but not others. The experiments were repeated in both species-rich and species-poor grassland communities. The plant biomass changed more in the species-poor community than it did in the species-rich community (Fig. 17.7). McNaughton argued that this difference was clear evidence of a link between diversity and stability: Although buffalo ate the same amounts in both types of grassland, the noneaten species grew more in the more diverse plots, compensating for the material removed by the buffalo. McNaughton was adamant that his data showed a clear link between stability and diversity, and he was scathing in his attack of mathematical models that had failed to show such a link. Better to trust less rigorous verbal and empirical models that appeared right, he argued, than accept more rigorous mathematical treatments that did not jibe with the data.

Since his East African studies, McNaughton has gone on to address the diversity–stability link in other systems. In 1991, together with Frank, he examined the effect of a severe drought on the composition of species in the plant community of the diverse grasslands of Yellowstone National Park (Fig. 17.8). McNaughton expressed diversity with the Shannon index H_S, and he calculated an index of resistance to change,

$$R = 1 - \sum_{i=1}^{n}\left(\frac{\Delta p_{ij}}{2}\right),$$

where $\Delta p_{ij}/2$ is the change in abundance of the ith of n species in community j between 1988, the drought year, and 1989, a normal year. The resistance to change in species composition, measured as the arcsine of the square-root-transformed version of R, was greater in more diverse communities (Fig. 17.9).

Figure 17.8 Resistance to drought, such as this drought in Yellowstone National Park in 1988, may be greater in more diverse plant communities.

Following McNaughton's work, others have taken up the challenge to address the diversity–stability hypothesis. David Tilman (1996) examined the relationships between biodiversity and stability for both population and ecosystem traits in a long-term study of grassland plots in Minnesota. His results demonstrated that biodiversity stabilizes

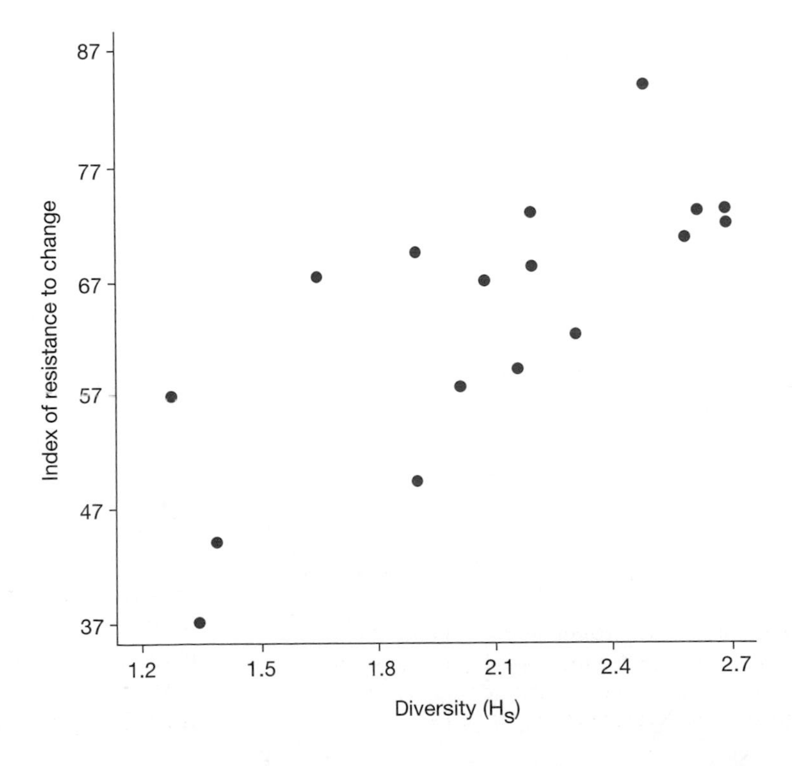

Figure 17.9 Positive relationship between an index of resistance to change in species composition and diversity, as measured by H_S. *(Modified from Frank and McNaughton, 1991.)*

community and ecosystem processes, but not population processes. The year-to-year variability in total aboveground plant community biomass was significantly lower in plots with greater plant species richness for the 11 year period of the experiment, including a drought year (Fig. 17.10a).

In contrast, the year-to-year variability in species abundance was *not* stabilized by plant species richness (Fig. 17.10b). Tilman suggested that the difference between species versus community biomass likely results from interspecific competition. When climatic variations harm some species, unharmed competitors increase. Such compensatory increases stabilize the total community biomass, but cause species abundance to be more variable. Tilman suggested that his results helped to reconcile the long-standing dispute over the diversity–stability relationship because they supported May's theoretical results concerning the effects of diversity on population stability while at the same time upholding Elton's original ideas of a diversity–stability link.

The portfolio effect suggests that the diversity–stability relationship arises out of the laws of statistical averaging.

Daniel Doak and his colleagues (1998) at the University of California at Santa Cruz explained the diversity–stability relationship in terms of what they called the "portfolio effect," or statistical averaging. The analogy they made was that prudent investors in the stock market usually diversify their investments by holding stock in a variety of companies. Diversification of holdings is advisable because the value of a variety of stocks is less variable through time than the value of a single stock. Diversification also broadens one's chances of including one stock that will perform exceptionally well, increasing the value of the portfolio. It is an unusual investor who can pick just that one successful stock, and the situation is analogous as regards diversity: Adding more species to a community results in an averaging of their fluctuations and hence less variation in total biomass. A drought occurs and some species die, but others may profit and their biomass may increase. Include more species in the community, and the chances that one of them will be populous and very productive will be good. This is actually a purely statistical argument for why diversity begets stability, and no special biological properties are invoked. Doak discovered that his statistical

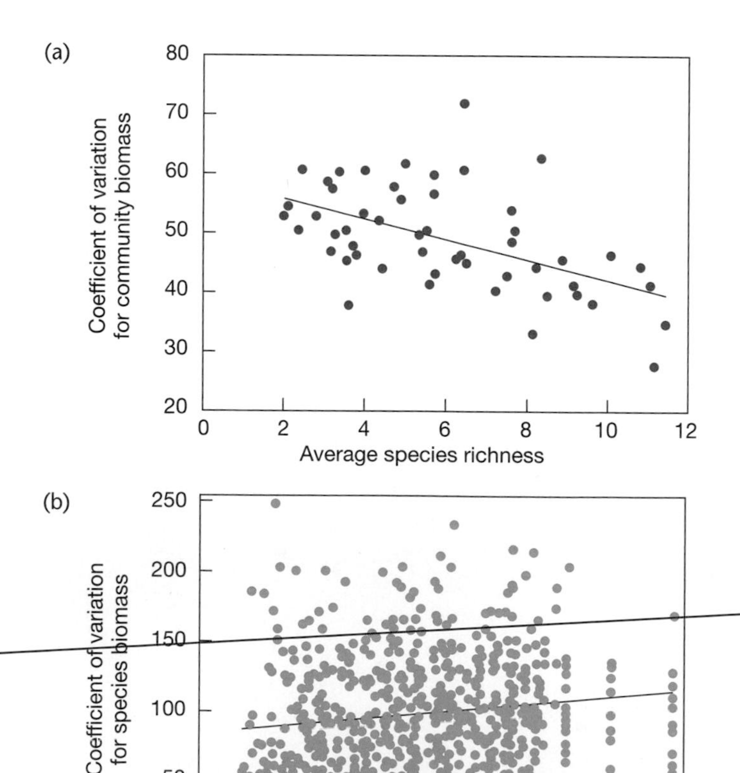

Figure 17.10 (a) Coefficient of variation in community biomass for experimental grassland plots in Minnesota. The 11 years of study showed that, as plant communities become more diverse, the community biomass becomes less variable. This finding contrasts with (b), where, for any given species, biomass becomes *more* variable as richness increases. *(After Tilman, 1996.)*

averaging effect could explain Tilman's diversity–stability results on Minnesota grasslands. Tilman responded (Tilman, Lehman, and Bristow, 1998) that whether statistical averaging or interspecific competition caused the diversity–stability effect mattered little, because the effect was just as real either way.

17.4 The Intermediate-Disturbance Hypothesis

We have so far focused on the question, "Are diverse communities more stable than species-depauperate communities?" Evidence from studies of microbial communities has failed to establish a link between diversity and stability, but large-scale field studies verified that stability is increased by diversity. What if we rephrase the question to "Are stable systems more diverse than unstable systems?" Joe Connell, who was interested in the stability of marine intertidal systems, argued (1978) that the highest local diversities are actually maintained not in stable systems, but in communities of intermediate levels of disturbance (Fig. 17.11). He reasoned that, at high levels of disturbance, only good colonists, which would be *r*-selected species, will survive, giving rise to low diversity. At low rates of disturbance, competitively dominant species will outcompete all other species, and only a few *K*-selected species will persist, again yielding low diversity. The most diverse communities lie somewhere in between. Connell argued that natural communities seem to fit into this model fairly well. Tropical rain forests and coral reefs are both examples of communities with high species diversity. Connell pointed out that coral reefs maintain their highest diversity in areas disturbed by hurricanes and that the richest tropical forests occur where disturbance by storms is common. Areas with no mature trees, called light gaps, created by mud slides allow many *r*-selected species to exist.

Interestingly, some of the richest plant communities in the southeastern United States occur on army bombing ranges, which, of course, suffer some of the highest known rates of disturbance. Seth Reice, who was interested in river communities, argued (1994) that disturbance operates in habitats of all types, and naturally disturbed areas are nearly always more diverse than undisturbed areas. Although storms are frequent in many habitats, other common disturbances include fire, freezes in subtropical habitats, and floods along rivers. Reice suggests that many communities therefore exist in nonequilibrium states in which the species composition constantly changes. This perspective contrasts with the earlier-discussed view of communities having a stable equilibrium—especially bird communities.

The most diverse systems may exist at intermediate levels of disturbance.

Wayne Sousa (1979) provided an elegant experimental verification of the intermediate-disturbance hypothesis in a marine intertidal situation. He found that small boulders, which were easily disturbed by waves, carried a mean of 1.7 sessile plant and animal species (Fig. 17.12). Most colonizing species were crushed by these frequently moving boulders. Large boulders, which were rarely moved by waves, had a mean of only 2.5 species. On such boulders, competitive dominants supplanted many other species. Intermediate-sized boulders had the most species, an average of 3.7 per boulder, because they contained a mix of

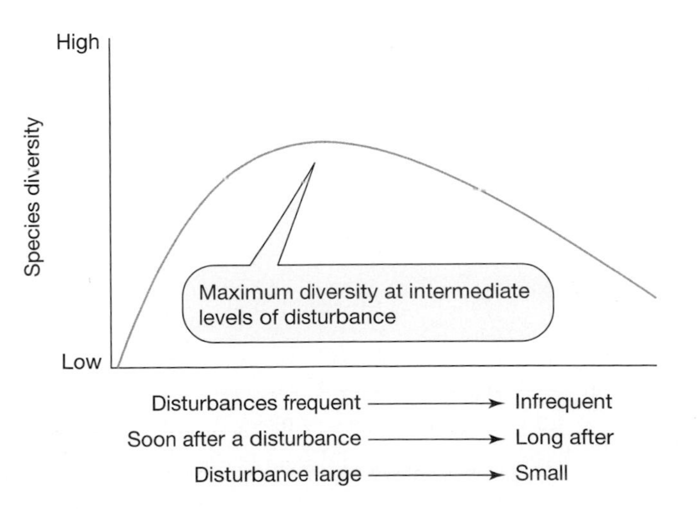

Figure 17.11 The intermediate-disturbance hypothesis of community organization. The species composition of a community is never in a state of equilibrium, and high species diversity can be maintained only at intermediate levels of disturbances, such as those produced by fires or windstorms. *(Redrawn from Connell, 1978.)*

Figure 17.12 Sessile communities along the southern California rocky shore. Small boulders get moved easily, and their fauna and flora are crushed. Large boulders are dominated by one or two species. Species richness is highest on intermediate-sized boulders that move only periodically.

r- and *K*-selected species. Sousa then cemented small boulders to the ocean floor and obtained an increase in species richness to near the value for large boulders, showing that his results derived from rock stability, not rock size.

What of other tests of the intermediate-disturbance hypothesis—especially those done on a bigger scale? There have been very few. To evaluate whether the intermediate-disturbance hypothesis applies on regional scales, Hiura (1995) examined the relationship between species diversity and a gap-formation regime of beech forest in Japan over a scale of 10° latitude. Locations that sustained an intermediate frequency of disturbance in terms of the mean windstorm interval had the highest species diversity (Fig. 17.13). However, when we examine the intermediate-disturbance hypothesis on a larger scale, we run the risk of confounding latitudinal diversity gradients, which operate on large scales, with disturbance-related phenomena, which operate on smaller scales. This means that, regardless of the frequency of windstorms, tree diversity would probably increase going south in Japan anyway because of temperature increases. So does the frequency of windstorms also increase diversity? Hiura himself noted that the most predictable model for species diversity over a 10° latitude range in Japan was a model composed of two factors: the interval between windstorms and the cumulative temperature of the growing season.

In sum, stability does seem to be linked to diversity. Although no such link was revealed in laboratory experiments, large-scale field experiments showed a definite link between the two. However, if we turn the question around, we find that the most diverse communities are not necessarily the most stable. Rather, the most diverse communities seem to exist at intermediate levels of disturbance.

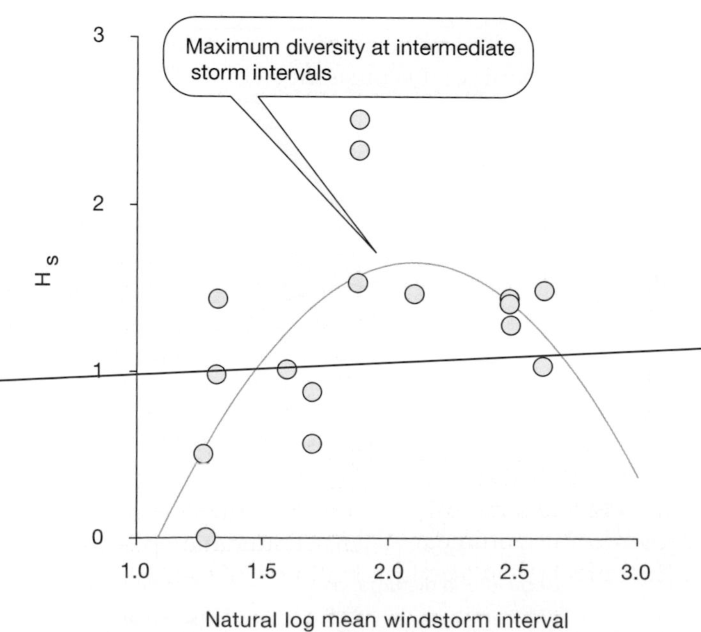

Figure 17.13 Relationship between disturbance interval [natural logarithm of mean windstorm interval (MWI)] and species diversity H_s. *(After Hiura, 1995.)*

Profiles

Sir Robert M. May,
Oxford University

I went to Sydney University to get a degree in chemical engineering, but ended up with a Ph.D. in theoretical physics. I had never even thought about a career as a researcher when I entered the university, but once I became aware that it was possible to spend my life as a teacher/researcher, being paid (albeit not a lot!) for the pleasure of trying to understand how the natural world works and communicating that to others, I seized the chance.

After a postdoc at Harvard, I returned to Sydney University, becoming a full professor at age 33. Soon after, in the late 1960s, I became active in the movement for Social Responsibility in Science in Australia. This led to my becoming interested in environmental and ecological problems. As one result, and largely by accident, I came across the "stability–complexity" question. Around 1970, the conventional wisdom—following the work of two of my heroes, Elton and Hutchinson—was that "complex" communities (those with more species and richer webs of interaction among them) were more "stable" (better able to resist or recover from disturbance, human created or natural). Comparing mathematical models of ecological communities with few species against the corresponding models with many species, I showed that there could be no such simple and general rule; all things being equal, complex systems are likely to be more dynamically fragile. This work, I believe, has refocused the subsequent agenda in the area, as people have become more careful about distinguishing the productivity of a community as a whole from the fluctuations in individual populations and, more broadly, have sought to understand the intricate patterns that particular ecosystems have woven, over evolutionary time, in ways that undercut any glib generalizations.

Around this time, I arranged a sabbatical leave from Sydney in order to work at the Plasma Physics Laboratories near Oxford in the United Kingdom and at the Institute for Advanced Study in Princeton. But, meeting people like Southwood and the younger Hassell and Lawton in the United Kingdom and Robert MacArthur and others in New Jersey, I found my interests increasingly engaged by ecological problems. This was, I think, a romantic era for ecologists. In the hands of MacArthur (my number-one hero) and others, many key questions were being framed in physics in ways that combined theory and field studies or experiments: What, if any, are the limits to similarity among coexisting competitors? How do the nonlinear feedbacks of intraspecific competition affect population dynamics? and ultimately, What are the causes and consequences of biological diversity? By great luck, I stumbled into this field at a time when a theoretical physicist could be useful. In 1973, following MacArthur's death of cancer at age 42, I moved to Princeton University and became a professor of biology. The ecology course I taught at Princeton was the first biology course I had attended since the age of 12. This says a lot about Princeton's willingness to take risks.

Many interesting projects, most of them collaborative with empirical researchers, followed. The 1973 book *Stability and Complexity in Model Ecosystems* (with a second edition in 1974) drew some of this work together. Probably the most important thing I did was, along with Jim Yorke, George Oster, and others, bring "chaos" center stage through our studies of simple deterministic, but nonlinear, models of biological populations with discrete, nonoverlapping generations (first-order difference equations). The ideas thus generated in ecological research quickly spilled over into physics, chemistry, and other sciences. In the 1980s, in collaboration with Roy Anderson, I turned to how infectious diseases can often regulate the numerical abundance or geographical distribution of populations of plants and animals. This work preadapted us to contribute to early research on the AIDS epidemic in developed and developing countries. In 1988, feeling it was time for a change, I moved to Oxford as a Royal Society research professor. A narcissistic high point of my career was when the Royal Swedish Academy of Sciences awarded me its 1996 Crafoord prize for "pioneering ecological research in theoretical analysis of the dynamics of populations, communities, and ecosystems"; this award is intended to complement the Nobel prizes, cycling on a three-year basis between mathematics, earth and space sciences, and the nonmedical areas of biology.

My current interests have mainly to do with conservation biology and—to take a different tack—theoretical immunology (in collaboration with Martin Nowak).

Through all the twists and turns of my research interests, I have had fairly consistent encouragement from funding agencies in Australia, the United States, and the United Kingdom. Entering unfamiliar fields of research, I have found many people enthusiastic about new ideas and new approaches, others interested only in ploughing their own furrows, and a few fiercely hostile. What else would you expect? Science is done by people! I hope this diversity of experience with jobs, people, and research areas has helped prepare me for my current position as chief scientific adviser to the U.K. Government and head of the U.K. Office of Science and Technology (which includes the Research Councils, roughly equivalent to the National Science Foundation plus the nonclinical part of the National Institutes of Health).

Summary

For a long time, ecologists, beginning with Charles Elton (1958), have thought that more diverse communities were more stable. This viewpoint was shattered when Robert May showed that, mathematically, more diverse communities actually reduce stability.

1. Stability can be thought of in different ways—as resistance to change or as resilience, which refers to the ability of the community to return to equilibrium after perturbation. Resilience in turn can be divided into two concepts: elasticity, which measures how quickly a community can return to a former state, and amplitude, which measures how big a disturbance it can return from.

2. A community could exist in more than one form or stable state, but the evidence for this hypothesis is weak.

3. The anecdotal evidence for a link of diversity with stability, first suggested by Elton (1958), has not held up well under scrutiny.

4. Experimental studies of communities of microgranisms have not shown a strong link of diversity with stability. Large-scale field experiments with plants have shown a link between diversity and stability.

5. In 1996, Tilman reconciled the opposing viewpoints in the diversity–stability debate by showing how communities of prairie plants exhibit stability in biomass and ecosystem function, but not in species abundance.

6. The diversity–stability link may be caused by a "portfolio," or statistical averaging, effect.

7. The older, more conventional view called the equilibrium hypothesis proposes that most communities are stable. The more modern nonequilibrium viewpoint argues that disturbances are frequent and species composition is constantly changing, so that stability is elusive.

8. The intermediate-disturbance theory suggests that the most diverse communities, such as tropical forests and coral reefs, exist at intermediate levels of disturbance.

Only in the 1990s have large-scale field experiments with plants shown that more diverse communities are more resistant to change (e.g., drought) and so are more stable. However, the diversity–stability relationship is clearly complex, and more study will no doubt reveal further details, such as how much change in diversity is needed to change community stability.

Discussion Questions

1. Classify the biomes listed in Chapter 14 (temperate and tropical grasslands and forest, deserts, tundra, rivers, lakes, estuaries, coral reefs, and the open sea) in terms of their resistance and resilience.

2. Do you think that pesticides might reduce the stability of agricultural systems? Explain how such an effect would work.

3. If the intermediate-disturbance hypothesis is correct, how does that influence the conservation movement? Should conservationists promote disturbance to maximize diversity?

4. What differences might you expect between equilibrium and nonequilibrium communities in terms of numbers of species, stochastic effects, and life history strategies of species?

5. How would you set about establishing a link between diversity and stability in nature?

6. Do you believe in the idea of multiple stable states? Does this idea help or hinder the restoration of natural habitats by polluters in terms of how far they are expected to go to restore communities?

CHAPTER 18

Succession

Road Map

1. Communities develop over time.
2. In facilitation, each succeeding species makes its habitat more favorable for subsequent species.
3. In inhibition, each species inhibits the species that try to succeed it.
4. In tolerance, a species neither facilitates nor inhibits its successors.
5. Species richness usually increases during succession.
6. Biotic processes, such as herbivory, competition, and mycorrhizae, can deflect the path of succession.

At 8:32 A.M. on May 18, 1980, Mount Saint Helens, a theretofore little-known peak in the Washington Cascades, erupted. The blast felled trees over a 600-km^2 area, and the avalanche of mud that followed destroyed everything in its path. How the area has recovered is an interesting story in succession from which ecologists gleaned much knowledge. First, ecologists learned the telltale signs that such avalanches create. Using that information, they found evidence of more than 400 prehistoric avalanches—evidence that disturbance and recovery in these mountains is frequent. Second, ecologists learned how important stochastic events, or chance, is in recovery. The ecological dogma had been that such areas recover very slowly because the land, newly covered with ash and mud, would be sterile and it would take decades for small plants to colonize the area and then be followed by bigger plants and associated fauna. The truth, however, was that some species were quick to colonize. Most seeds blown in by wind to a moist spot germinated, regardless of what species they were. Some species even benefited: The dead wood was a bonanza for both wood-boring insects and the woodpeckers and finches that feed on them. The fallen trees had been uprooted, carrying soil and seeds above the ash layer, and many of the seeds germinated quickly. Because the eruption occurred in May, many of the lakes were capped by 30 cm of frozen ice, and the organisms underneath were little affected. More than that, the eruption had created 120 new lakes, which were promptly colonized. Algae in lakes bloomed, because the shoreline fringe of the trees had been blown down, allowing more sunlight to penetrate the water. Frogs, whose tadpoles feed on the algae, also flourished. Salamanders too, thrived, especially in the new lakes. Salamanders have aquatic larval stages and metamorphose into land-dwelling adults. But how did the salamanders cross kilometers of often-barren land to reach the lakes? It turns out that they were aided by pocket gophers, which prefer meadows for their underground tunneling. When the trees were felled, the blast zone was converted into meadow habitat, and the gophers thrived. As they tunneled, they brought seeds to the

Figure 18.1 Mount Saint Helens erupted in May 1980. Surprisingly, many species, such as meadow plants, pocket gophers, and salamanders, benefited from the eruption.

surface. The seeds germinated, providing more meadow habitat. Somehow the adult salamanders found the gopher tunnels and used them to disperse to the isolated lakes. Succession, it seems, can be greatly speeded by such chance effects and indirect interactions (Fig. 18.1). Only now, with modern detailed studies of community recovery on Mount Saint Helens, are we beginning to appreciate the intricacies of succession.

Each stage in succession is called a sere, and the last or end sere, is called the climax community.

18.1 Development of Communities

Succession can be either primary, on bare ground, or secondary, starting from a more advanced beginning such as an old field.

In severe disturbances, such as volcanic eruptions, avalanches, and fire, species richness can be reduced to zero. In other disturbances, like hurricanes or floods, species richness can be greatly decreased. Species richness gradually changes as the community returns to "normal," and early community ecologists suggested that this change was predictable and orderly. They termed such changes **succession. Primary succession** occurs when plants invade an area in which no plants have grown before, such as bare ground or new lakes created by the retreat of glaciers. In primary succession on land, the plants must often build the soil, so a long time—hundreds of years—may be required for the process. Only a tiny proportion of the Earth's surface is currently undergoing primary succession.

Secondary succession can be considered a modification of the longer term primary succession. Secondary succession occurs not on bare ground, but on partially cleared land. Clearing a natural forest and farming the land for several years is an example of a severe forest disturbance that doesn't kill all of the native species. Some plants and many soil bacteria, nematodes, and insects are still present. Cessation of farming may lead to a distinct secondary succession that can be quite different from the kind of succession which develops after a natural disturbance. The ploughing and lack of plant cover and the added fertilizers, herbicides, and pesticides may have caused substantial changes in the soil.

Most of the theory regarding succession relates to primary succession, but in the 1950s the ecologist Frank Egler studied secondary succession at Aton Forest, his own private estate in Connecticut. He found that in secondary succession most species existed in the ground in the seed bank or as old roots. Therefore, the rate of root regeneration or seed generation governed the order in which species appeared. Eventually, however, larger, slower growing trees would outcompete smaller pioneer species. Egler (1954) called his ideas the *initial floristic composition model*.

The American botanist Frederic Clements is often viewed as the father of successional theory, as well as perhaps community ecology in general. His early work (Clements, 1916, 1936) emphasized succession as a deterministic phenomenon, with a community proceeding to some distinct end point or climax. Each "unit" of succession was called a **sere** or **seral stage**. The initial seral stage was known as the pioneer seral stage (Fig. 18.2). A disturbance could return later seral stages to earlier seral stages, but generally, the community headed toward climax. Succession could be governed by disturbances caused by abiotic factors, in which case the succession is called **allogenic**. Alternatively, biotic disturbances, such as herbivores eating later seral species, could dominate, in which case the succession is called **autogenic**.

A key assumption was that each invading species made the environment a little different—say, a little less salty, a little more shady, or a little more rich in soil nitrogen—so that it then became more suitable for *K*-selected species (those living in mature communities, discussed in Chapter 5), which invaded and outcompeted the earlier residents. This process, known as **facilitation**, supposedly continued until the most *K*-selected species had invaded, when the community was said to be at **climax**. A return to earlier stages in the sequence was not possible unless another disturbance intervened. The climax community for any given region was thought to be determined

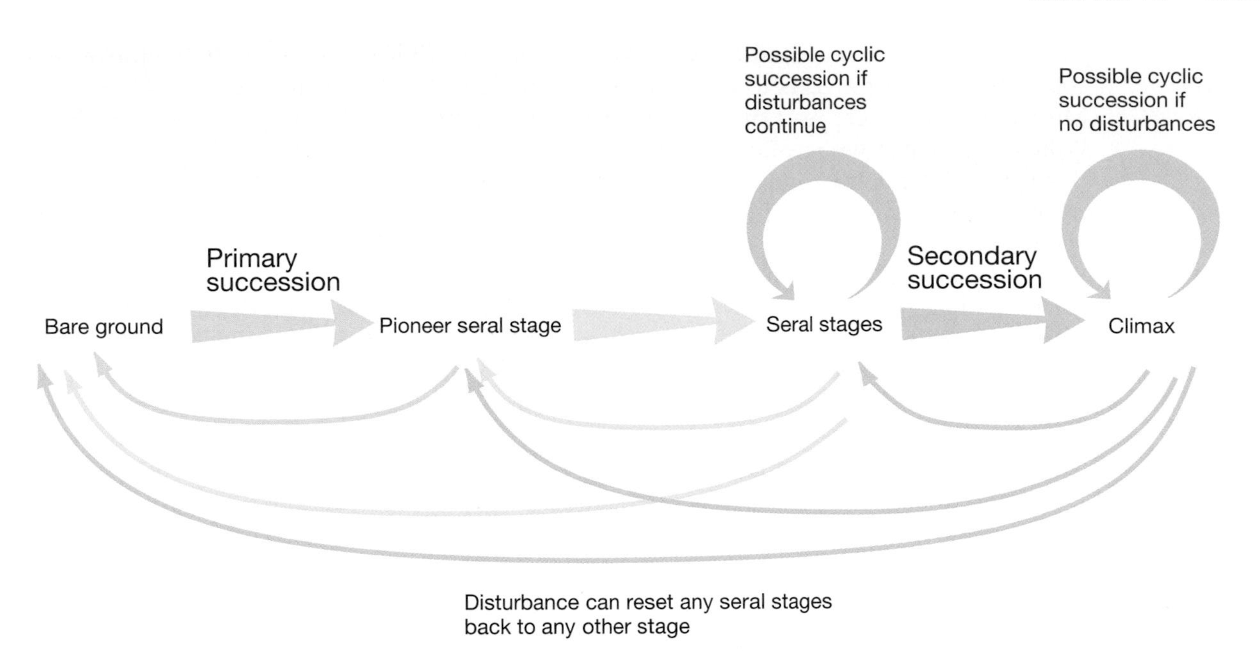

Figure 18.2 Diagrammatic representation of the different stages in succession. Secondary succession can occur from any stage after the pioneer seral stage.

by climate and soil conditions. Clements worked in the stable prairie communities of the Midwest, where dominant species rarely change; hence, he was predisposed toward the view he held. But do other data fit such a model?

18.2 Facilitation

Initially, succession following the retreat of Alaskan glaciers was argued to fit the Clementsian pattern of facilitation (Cooper, 1923; Crocker and Major, 1955). Over the past 200 years, the glaciers in the Northern Hemisphere have undergone dramatic retreats, up to 100 km in some instances (Fig. 18.3). The area is one of the few for which we can trace the chronology of physical change. In 1794, Captain George Vancouver visited the inlet known as Glacier Bay and made notes on the positions of the glaciers. In 1879, John Muir revisited the area and found that the ice described by Vancouver had retreated, and Muir could paddle up Glacier Bay. Muir noted the absence of forests in the upper part of the bay, but he also observed that the region had many old stumps, indicative of a forest that had been smothered by an advancing glacier centuries earlier. Ring counts of the oldest living trees in different areas gave rough approximations of when the forests formed in each area of Glacier Bay. This allowed the age of the

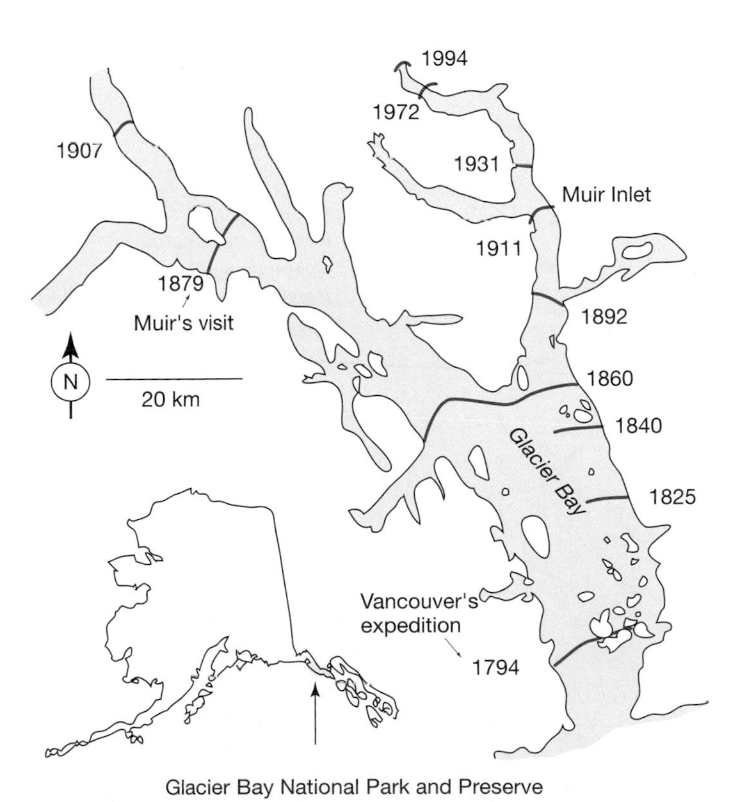

Figure 18.3 Glacier Bay fjord complex of southeastern Alaska, showing the degree of recession of the ice since 1760. Plants have colonized the bare areas left by the retreating glaciers, and ecologists have studied the resultant patterns in succession. *(After Fastie, 1995, and others.)*

Facilitation assumes that each invading species makes the habitat a little more favorable for succeeding species.

forest when the ice retreated between Vancouver's and Muir's visits to be estimated. Muir's accounts were read by the ecologist William S. Cooper, who realized that Glacier Bay was a natural environment in which to study succession empirically. Cooper had been a student of Henry Chandler Cowles, who had examined succession in Lake Michigan sand dunes; hence, Cooper was primed to work on succession at Glacier Bay. So began a long tradition of academic work in the area: From about 1910, direct measurements on ice retreats have been made periodically.

There is a basic ecological sequence of succession in Glacier Bay (Fig. 18.4). As glaciers retreat, they leave tills and moraines, which are deposits of stones and pulverized rock, respectively, that serve as soil brought forward by the ice. In Alaska, the bare soil has a low nitrogen content with little organic matter. In the early stages, it is first colonized by a black crust of blue-green algae, lichens, liverworts, horsetail (*Equisetum varietum*), and the occasional river beauty (*Epilobium latifolium*). The blue-green algae are nitrogen fixers and the soil nitrogen increases a little, but litter fall and soil depth are still minimal. At this stage, there may be a few seeds and seedlings of willows (*Salix*), *Dryas* (a dwarf shrub of the rose family), alders, and spruce, but they are rare in the community. After about 40 years, the nitrogen-fixing *Dryas drummondi* (mountain avens) comes to dominate the landscape. Soil nitrogen has increased, as has soil depth and litter fall.

At about 60 years, the nitrogen-fixing alder forms dense, close thickets. Nitrogen-fixing bacteria that live mutualistically in the

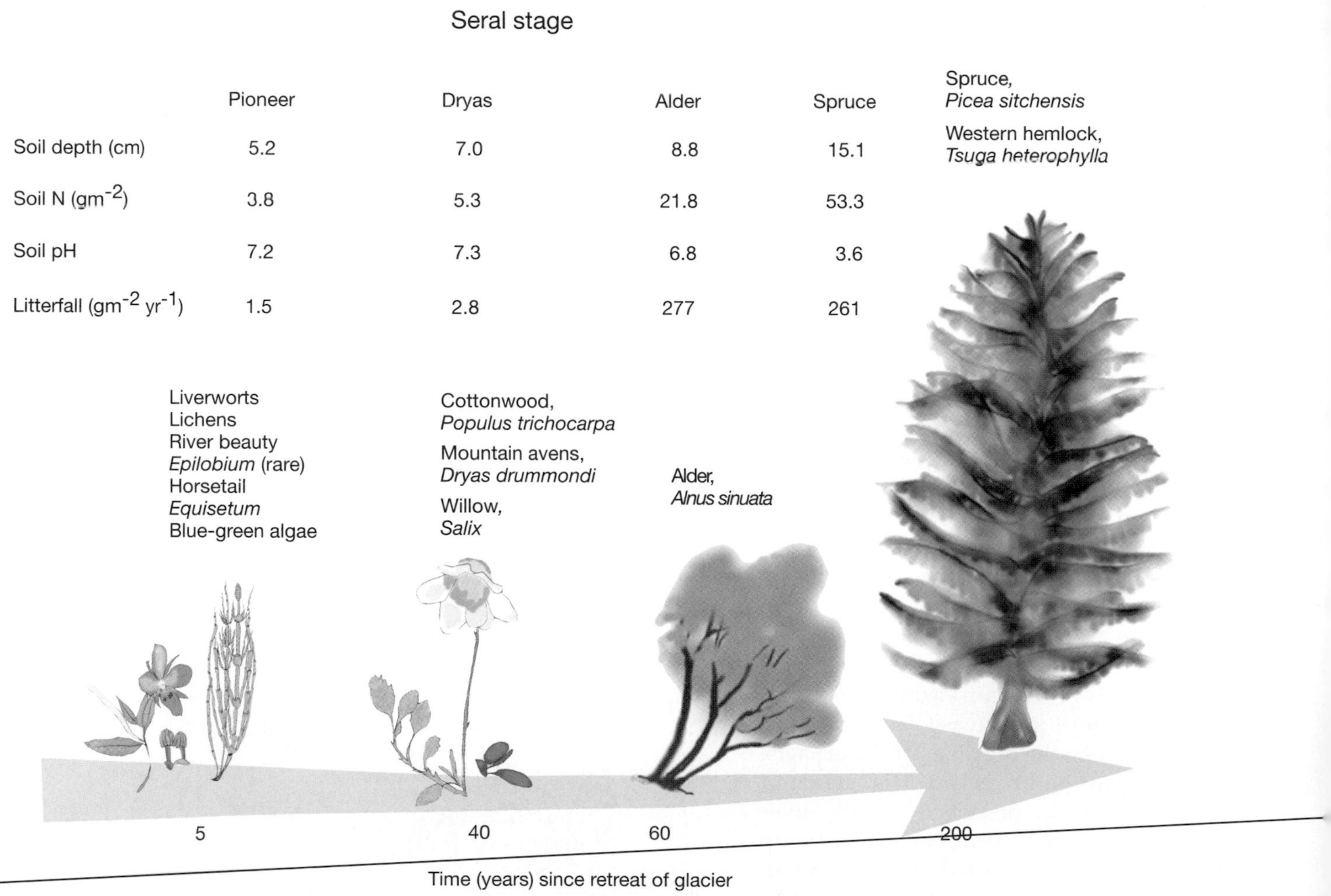

Seral stage

	Pioneer	Dryas	Alder	Spruce
Soil depth (cm)	5.2	7.0	8.8	15.1
Soil N (gm^{-2})	3.8	5.3	21.8	53.3
Soil pH	7.2	7.3	6.8	3.6
Litterfall ($gm^{-2}\ yr^{-1}$)	1.5	2.8	277	261

Figure 18.4 Diagrammatic representation of succession at Glacier Bay, Alaska. The first species to colonize the bare earth following retreat of the glaciers are small species, such as blue-green algae, mosses, and horsetails. They increase the soil nitrogen a little, but they produce little litter and the pH of the soil remains unchanged. After about 40–50 years, the community is dominated by *Dryas*, a dwarf shrub, together with scattered alder, *Alnus sinuata*. *Alnus* is a nitrogen fixer, and soil nitrogen increases dramatically as does the litter that is produced. After about 50 years, the community is dominated by alder, but some pine trees, chiefly *Populus* and *Picea*, invade. Soon these trees grow taller than the alder and shade it, so that after 200 years the forest is dominated by pines.

roots of nitrogen fixers convert nitrogen from the air into a biologically useful form. The excess nitrogen fixed by these bacteria and not used by the tree accumulates in the soil. The level of nitrogen in the soil increases dramatically, as does litter fall. After about 75–100 years, the spruce trees begin to overtop the alders, shading them out. Litter fall is still high, and the large volume of needles turns the soil acidic, so that its pH is low. The shade causes competitive exclusion of many of the original understory species, and only mosses carpet the ground. At this stage, seedlings of western hemlock (*Tsuga heterophylla*) and mountain hemlock (*T. mertensiana*) may also occur, and after 200 years a mixed spruce–hemlock forest results (Fig. 18.5).

How generally does facilitation occur in nature? Interestingly, other studies conducted in the interior of Alaska on the Tanana River show a pattern similar to that characterizing facilitation in Glacier Bay (Walker, 1995). On well-drained slopes in Alaska, spruce and hemlock usually constitute the climax community. However, in poorly drained soil, the climax community is different. Here, the forest is invaded by *Sphagnum* mosses, which accumulate water and further acidify the soil. Most trees die from lack of oxygen to the roots. Sometimes, a few individuals of lodgepole pine (*Pinus contorta*) are found, because they can tolerate poor aeration. In these conditions, the climax community is called a *muskeg bog*.

At first, ecologists suggested that each species facilitated the entry into the community of the next species. For example, the blue-green algae are nitrogen fixers and increase soil nitrogen levels, allowing other species to enter the community. However, Stuart Chapin from the University of Alaska, who has studied succession in the Glacier Bay region of Alaska for many years, showed that facilitation occurs only during part of the process (Chapin *et al.*, 1994). Because *Dryas* and alders can fix atmospheric nitrogen, the nitrogen content of the soil increases markedly in these seres. The increase in soil nitrogen facilitates the invasion of spruce trees. On the other hand, competition is important, because the taller alders shade out the shorter *Dryas*. Again, competition is

Figure 18.5 Succession at Glacier Bay, Alaska. (a) First comes a mix of herbaceous species, including dwarf fireweed. (b) Then *Dryas drummondi* appears, followed by (c) alder trees (*Alnus sinuata*), which fix nitrogen and facilitate the entry of (d) spruce trees. *(Tom Bean, DRK Photo.)*

TABLE 18.1 The decomposition of oak leaves is facilitated by large organisms, such as earthworms, as revealed by the decomposition experiments of Edwards and Heath (1963).

Bag Mesh Size (mm)	Fauna that Could Enter Bags	Percent Oak Leaves Gone in Nine Months
7.0 (large)	All	93
0.5 (medium)	Small invertebrates, microorganisms	38
0.003 (small)	Microorganisms only	0

evident after about 50 years, when the dense stands of alder begin to be shaded out by Sitka spruce. Facilitation, originally thought to fuel the entire sequence of succession, was important only in the establishment of the first three seres; competition was important in later phases of succession. Thus, one should always keep in mind that in classical facilitation only half of each interaction is facilitative: early species facilitate later ones, but later species outcompete earlier ones.

What other evidence is there of facilitation? The decomposition of plant material, such as logs, also incorporates elements of facilitation. A now classic experiment by Edwards and Heath (1963) demonstrated this phenomenon. They put oak leaves in nylon bags in the soil and examined decomposition rates. By varying the mesh size of the bags, they could vary the sizes of the decomposers entering them. Large-meshed bags permitted access by all soil dwellers, including earthworms, nematodes, collembolans, and other insects. Medium-mesh bags allowed only small invertebrates and microorganisms, while small-mesh bags allowed only microorganisms. Edwards and Heath found that the larger the mesh, the more complete was the decomposition. In the small-meshed bags, microorganisms alone were unable to decompose the leaves (Table 18.1). Although microorganisms are very important in decay, they cannot begin their work until the size of the particles they ingest is reduced by larger organisms. In the soil, earthworms are most important in the initial decay process. Thus, on a small scale, facilitation occurs in the decomposition of plant material.

Inhibition implies that early colonists prevent later arrivals from replacing them.

Other examples of facilitation have been found in terrestrial and marine systems. The nitrogen content of heathland soil in Western Europe increases from about 1 to 13 g of nitrogen per square meter over time spans of about 50 years (Berendse, Schmitz, and de-Visse, 1994). Such changes are accompanied by a replacement of the dwarf shrubs *Calluna vielgoris* and *Erica tetalix* by the grasses *Molinia caerulea* and *Deschampsia flexuosa*. The experimental addition of *Calluna* litter or nitrogen fertilizer had the same result: an increase in the biomass of the grasses and a reduction in the biomass of *Calluna*. Thus, *Calluna* litter actually alters the environment, enriching the soil and hastening invasion by other species.

Although soils do not develop in marine systems, facilitation may still be encountered when one species enhances the quality of settling and establishment sites for another. Dean and Hurd (1980) put out experimental test plates at different times of the year in Delaware Bay and examined colonization. Some facilitation was found, in that hydroids enhanced the settlement of tunicates and both facilitated the settlement of mussels, which were the dominant species in the community. Also, Teresa Turner (1983) found facilitation in the marine intertidal zone off the Oregon coast. Here, the presence of some algae facilitated the succession of surfgrass (*Phyllospadix scouleri*), because the barbed surfgrass seeds became attached to the algal species and could then grow.

The ultimate form of facilitation is enablement, whereby species *B* cannot invade unless species *A* has already invaded first. If this type of phenomenon occurs, the community is often thought to obey certain "assembly rules" (Drake, 1991). The species in a community can be thought of as pieces in a jigsaw that fit together only in a certain order.

18.3 Inhibition

Many types of succession do not show elements of facilitation at all. Another view is that possession of space is all important: Who gets there first determines community structure. In such communities, the principal mechanism affecting successional change is the **inhibition** of subsequent colonists. For example, Facelli and Facelli (1993) removed the litter of *Setaria faberii*, an early successional species in old fields of

New Jersey, and noted that the removal increased the biomass of a later species, *Erigeron annuis*. They concluded that the release of phytotoxic compounds from decomposing *Setaria* litter or physical obstruction by the litter may contribute to the inhibition of *Erigeron*. Without the litter, *Erigeron* became dominant and reduced the biomass of *Setaria* (Fig. 18.6).

Inhibition has been seen as the dominant method of succession in the marine intertidal zone, where space is limited. Here, early successional species are at a great advantage in maintaining possession of valuable space. Wayne Sousa (1979), of the University of California, experimentally examined how inhibition worked by scraping rock faces clean or putting out fresh boulders or concrete blocks. The first colonists of the area he worked in were the "weedy" species, the green algae *Ulva* and *Enteromorpha*. Much later, these were replaced, first by large brown algae and finally by the red alga *Gigartina canaliculata*. By removing *Ulva* from the substrate, Sousa was able to show how both *Gigartina* and brown algae were able to colonize more quickly (Fig. 18.7). But does this ever occur in nature, and if so, how? First, *Ulva* is subject to herbivory by the crab *Pachygrapus* and is also susceptible to drying out, and both of these disturbances favor the invasion of other species. Second, the middle species—the brown alga—is commonly overgrown by epiphytes. The last species, *Gigartina* is not as susceptible to either of these disturbances, so both favor its appearance in the community.

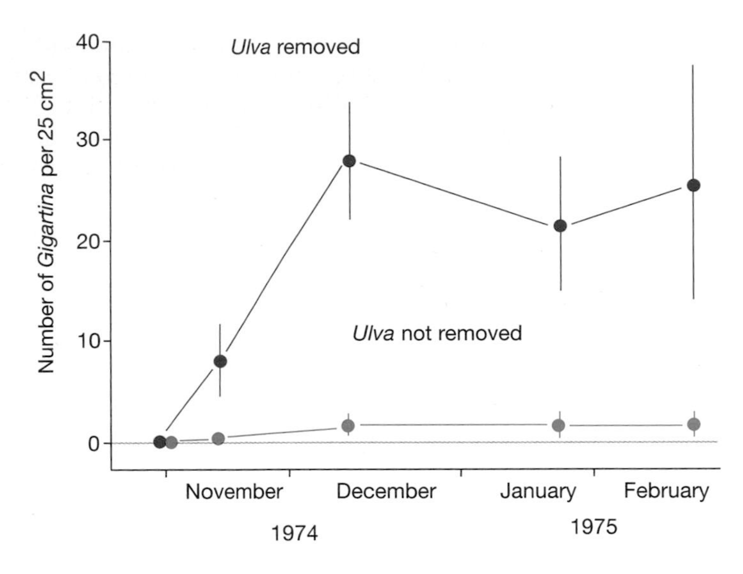

Figure 18.7 Removing the green alga *Ulva* from intertidal rock faces allowed colonization by the brown alga *Gigartina*. *(After Sousa, 1979.)*

(a)

(b)

Figure 18.6 (a) *Erigeron annuis*, daisy fleabane. Litter from (b) giant foxtail, *Setaria faberii*, prevents *Erigeron* from colonizing an area.

Many scientists (e.g., Tilman, 1982) believe inhibition to be the main process determining the rate and direction of succession. "Possession is nine-tenths of the law." But if this is true, what are the factors that cause a dominant to become susceptible to invasion by its successor?

Van der Putten, Van Dijk, and Peters (1993) suggested that in terrestrial communities species-specific soil pathogens might turn off inhibition so that new species could invade. In Europe, sand dunes are built around marram grass (*Ammophila arenaria*), which is followed in sequence by fescue (*Festuca rubra*) and sand sedge (*Carex arenoria*), with sea couch (*Elymus athericus*) at the landward edge. Van der Putten and coworkers carried out reciprocal transplant experiments using sterilized and unsterilized dune sand. Two-week-old seedlings of all four species, raised from local seed, were planted into a replicated series of pots containing sand collected from beneath each of the species. In half of the pots, the soil was sterilized by gamma radiation. The investigators demonstrated that the biomass production of each plant species was reduced by the soilborne diseases of its successors, but not of its predecessors. Once a successor gains a foothold, its associated pathogens help it outcompete its predecessor. Burial by new sand presumably provides marram grass with relatively sterile sand in which roots can develop and escape the ravages of pathogens. Thus, inhibition can be mediated through compounds in the litter, through soilborne pathogens, or merely by holding space.

18.4 Tolerance and Other Patterns of Succession

Tolerance suggests that early colonists neither facilitate nor inhibit later colonists.

The huge differences in the two mechanisms of succession in nature—facilitation and inhibition—prompted Joe Connell and Ralph Slatyer (1977) to view the different models as extremes on a continuum. They termed the classical view of succession proposed by Clements, which posits an orderly progression to a predictable climax community, the *facilitation model*, because each species makes the environment more suitable for the next. For the type of succession in which competition between species was great, they coined the term *inhibition model*, because colonists tend to prevent subsequent colonization by other species. In this model, succession depends on who arrives first. Succession proceeds as colonists die, but it is not orderly. As a sort of an in-between model, Connell and Slatyer proposed a third concept, which they termed the **tolerance model**. In this model, any species can start the succession, but the eventual climax community is reached in a somewhat orderly fashion.

Connell and Slatyer thought the best evidence for the tolerance model came from Egler's (1954) work on floral succession. Egler showed that, in succession in many flower communities, there is a tendency for most species to be present at the outset, as buried seeds or roots. Whichever species germinated first or grew from roots would start the succession sequence. But because it relies on buried seeds or roots, Egler's model works only for secondary succession, whereas the facilitation, inhibition, and tolerance models have most often been discussed with reference to primary succession. All three models predict that the most likely early colonists will be r-selected species—weeds in many cases. The key distinction between the models is in how succession proceeds. In the classical facilitation model, species replacement is facilitated by previous colonists; in the inhibition model, it is inhibited by the action of previous colonists; and in the tolerance model, it is unaffected (Fig. 18.8).

In 1987, John Lawton proposed a fourth model, which he termed **random colonization**. In this model, there is no facilitation or competition, and succession proceeds by chance alone, governed by who arrives first and happens to be present when favorable conditions prevail. Although we can think of few studies that have invoked the random-colonization model, succession on Mount Saint Helens, discussed at the beginning of the chapter, seemed to show that chance does play an important role.

It is worth reiterating that neither facilitation nor inhibition is usually the primary driver of succession in all seral stages. Thus, facilitation may be important early and inhibition important later, or vice versa. As a good example of this, consider another classic study, that of secondary succession in the Piedmont Plateau of North Carolina, done by Henry Oosting (1942) and his student Catherine Keever. In the Piedmont, much old-growth forest was felled for agriculture, especially tobacco farms, and then the fields were often abandoned and new forest was felled. There was a continuum of ages in the fields of the region on which to study succession. Indeed, succession was more precise in this system than perhaps in any other known. In most communities, there is a gradual overlapping of species over relatively long periods of time. In the Piedmont, the changes were relatively quick, resulting in distinct combinations of species.

One year after the fields were abandoned there were 35 species recorded (though not on every plot), all annual or perennial herbaceous species. Two species dominated: crabgrass (*Digitaria sanguinalis*) and horseweed (*Erigeron canadensis*), together regarded as the pioneer dominants. Surprisingly, in the second year, the same species were present, but now aster (*Aster ericoides*), which was not present the previous year, was a dominant, along with rag-

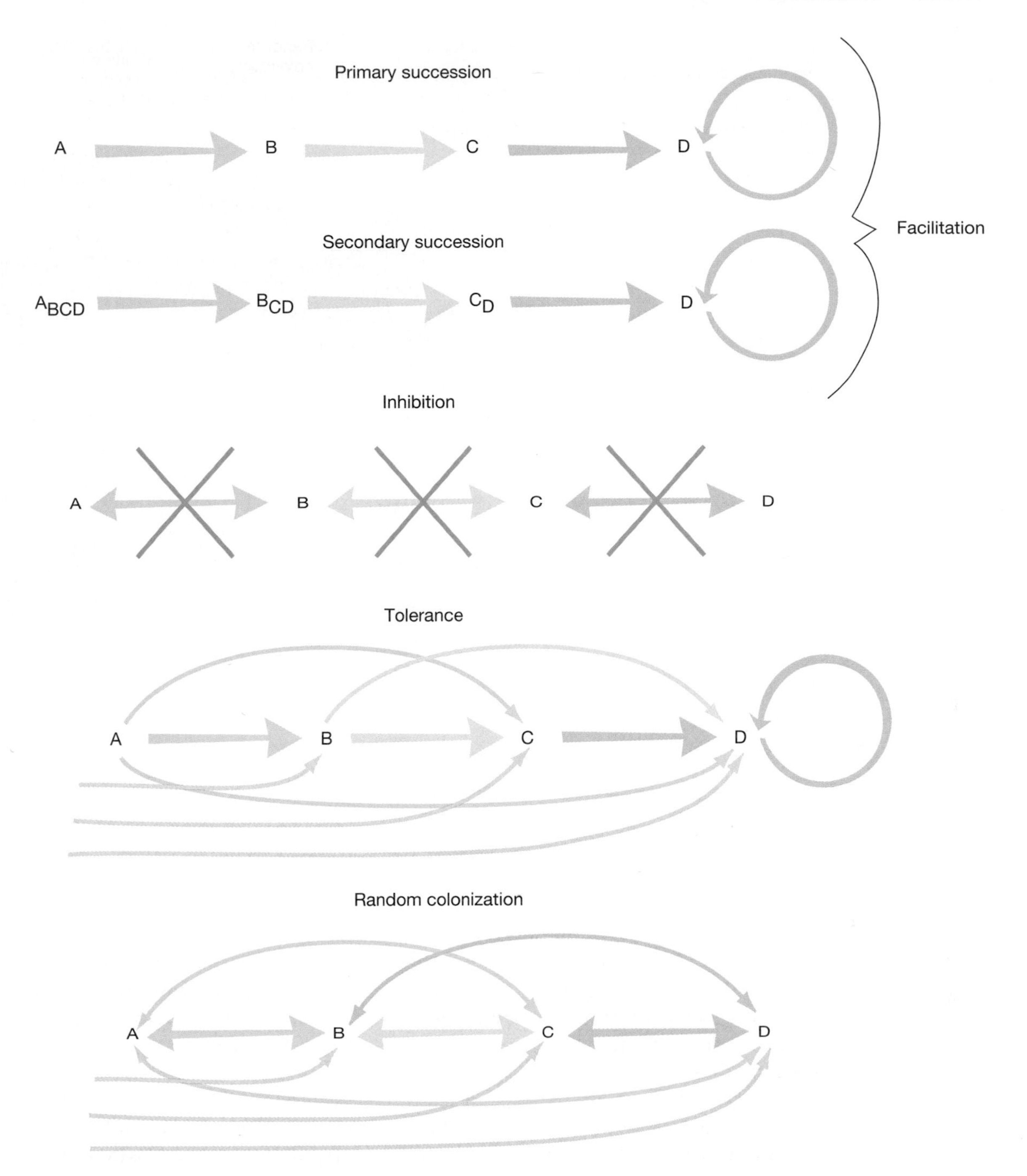

Figure 18.8 Four models of succession. Four species are represented by *A*, *B*, *C*, and *D*. An arrow indicates "is replaced by." The facilitation model is the classic model of succession. In the inhibition model, all replacements are possible, and much depends on who gets to the area first. The tolerance model is represented by a competitive hierarchy in which later species can out-compete earlier species, but can also invade the area in their absence. The random colonization model suggests that chance processes play the biggest role in succession.

weed (*Ambrosia artemisifolia*), which was only of minor importance in the first year. Twenty-six new species were added in the second year. In the third year, species richness declined dramatically, because yet a third species, broom sedge (*Andropogon virginicus*), a grass, became dominant and remained so for several years. During this time, seeds of pines, predominantly loblolly (*Pinus taeda*), and some hardwood trees arrived via wind dispersal. The pines became established in the fifth year and formed a closed canopy by the tenth year. However, pine seedlings have a hard time germinating under a pine canopy, so the hardwood understory thrived. By the time

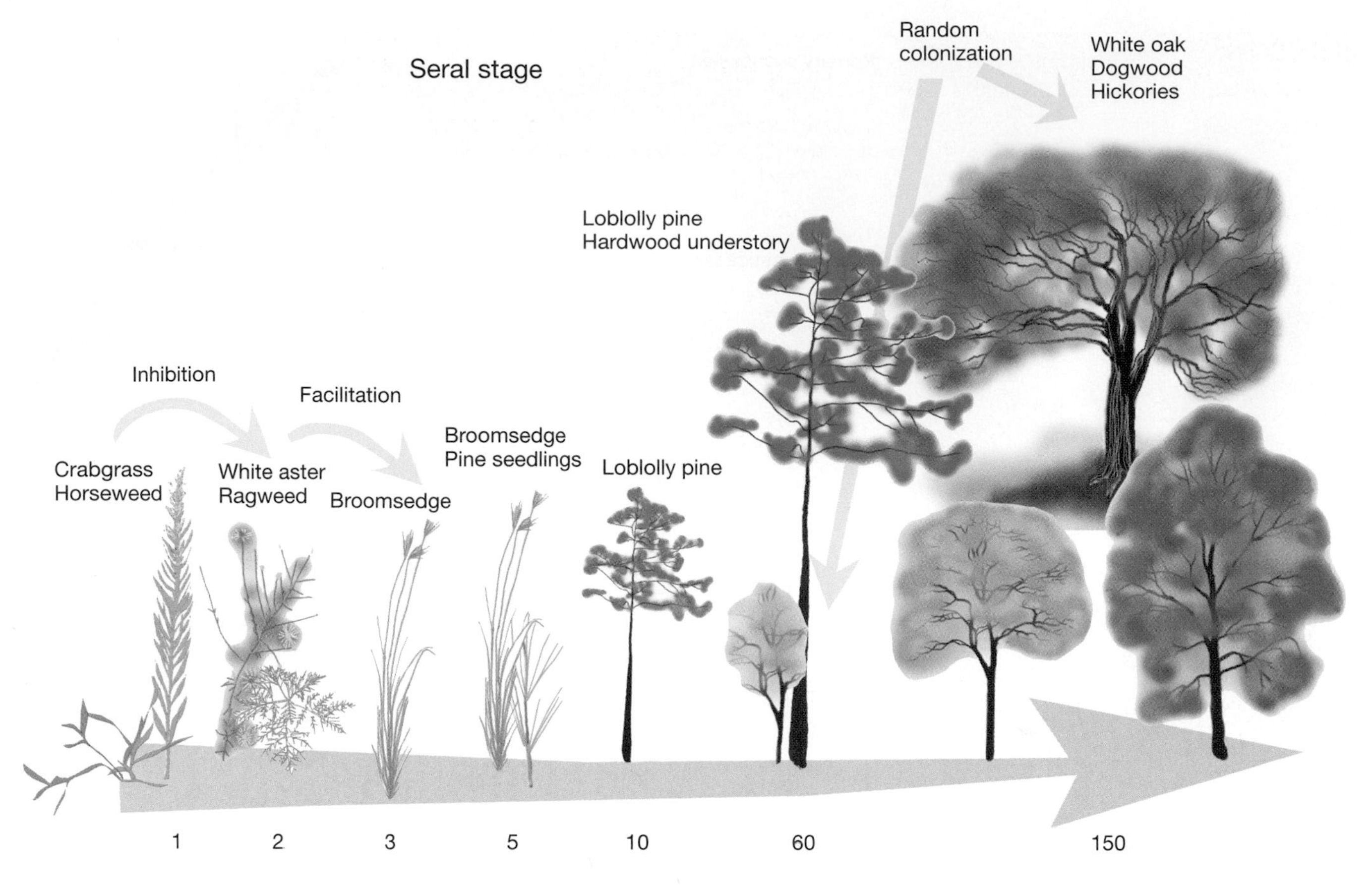

Figure 18.9 Secondary succession in the Piedmont region of North Carolina following the abandonment of tobacco fields. Inhibition, facilitation, and random colonization are all important at various stages.

100 years had passed, most stands had as many hardwoods as pines, and by 200 years only scattered pines remained.

Catherine Keever (1950) showed that in the early seres rotting horseweed inhibits the growth of aster. On the other hand, aster stimulated the growth of its successor, broom sedge, because aster roots increase the amount of organic matter in the soil. Within the space of a few years, first inhibition and then facilitation played primary roles in succession in these old fields. In 1991, Diane DeSteven showed that the identity of hardwood species that became established under the pine depended on the amount of wind-dispersed seeds, often called "seed rain." Therefore, in the establishment of hardwoods, the random-colonization model appears to be supported. So, in postagriculture secondary succession in the North Carolina piedmont, each seral stage had a different mechanism or path to succession (Fig. 18.9). This pattern of different mechanisms dominating at different successional stages is probably frequently seen in community succession (Callaway *et al.*, 1996).

Species richness of both plants and animals is low in the early stages of succession, but often rises quickly in later seres.

Understanding how succession proceeds is not merely of academic interest: Restoration ecology, the attempt to restore communities following devastating disturbances, has emerged as an important practical discipline. To succeed in this endeavor, understanding succession on "repaired" communities is of vital importance. See "Applied Ecology: Restoration Ecology."

18.5 Patterns in Species Richness during Succession

Some of the most important general trends in succession have been summarized by the renowned University of Georgia ecologist Eugene Odum (1969) and Harvard biologist Fahriki Bazzaz (1979) (Table 18.2). As we have already seen, species in early seral stages are often dispersed by the wind, and the seeds can live a long time, maximizing their chances of successful colonization. In addition, plants in the early stages of succession acquire nutrients quickly, grow fast, and

TABLE 18.2 Some characteristics of early and late seral stages in succession. *From Odum (1969) and Bazzaz (1979).*

Attribute	Early Stages	Later Stages
Seed dispersal	Good	Poor
Plant efficiency at low light	Low	High
Resource (nutrient) acquisition	Fast	Slow
Biomass	Small	Large
Stability	Low	High
Diversity	Low	High
Species life history	*r*	*K*
Seed dispersal vector	Wind	Animals
Seed longevity	Long	Short

Applied Ecology

Restoration Ecology

Despite the best efforts of conservationists, some species or communities can become extinct in a particular area following the loss of their habitat. If there are other populations of the species elsewhere and other remnants of the habitat, it may be possible to take young animals and seeds of plants from these areas and use them to restore the damaged area. How best to do this is the province of restoration ecology.

Restoration ecologist is in its infancy, with many of the techniques yet to be well worked out and refined. For example, following the wreck of the oil tanker *Torrey Canyon* in 1967, some cleanup methods appear to have caused more damage to the indigenous biota than the oil itself. A natural equilibrium was restored more rapidly to areas where no intervention occurred than to those in which cleanup processes were used. The careless use of suction devices, scrapers, oil dispersants, and the like may cause more stress to the community than the material of the spill itself, thus retarding succession and the community's return to equilibrium.

The foundation of an economical and successful restoration program is a clear understanding of the environment and the plants, animals, and people involved with it. A restoration program should proceed via the following steps:

1. *The knowledge of why a species or community disappeared in the first place.* This is obvious in the case of activities such as oil spills or mining, but the causes of an individual species extinction are not so clear. For example, no one knew why the nene became extinct in Hawaii. Thus, the release of 1,244 of these geese back into Hawaii over a 16-year period failed. The goose could not establish a self-sustaining population.
2. *An understanding of the natural history of similar ecosystems.* The native vegetation, the soil characteristics, and the animals occupying comparable undisturbed native ecosystems should be determined, gathering as much information as possible on the interactions among plants, animals, and humans. When this information is available, a draft plan for restoration can be developed.
3. *Test plots.* Plots should be established to evaluate the strategies for restoration that appear promising. During this time, a seed-collection program should be initiated and seed nurseries should be established if needed to increase seed stocks.
4. *Soil preparation.* Preparing the soil for reseeding may include techniques such as ploughing, to break up the surface of the soil and make it rough, and sterilization. Sometimes, the soil is sterilized by chemicals. Another important sterilization technique is solarization, which involves moistening the soil, covering it with transparent plastic, and letting the sun heat the soil to a high temperature, thereby killing both the seeds of weeds and many pathogens.
5. *Revegetation.* The essential elements of a minimum-cost restoration effort are the introduction of appropriate seeds and related symbionts to sites that provide suitable soil and moisture conditions for rapid root growth and the establishment of plants. Weed control helps slow-growing native plants to compete. Controlled burning at the time weed species are most vulnerable can also help reduce weeds.
6. *Advanced techniques.* Other restoration elements that may be of value include pest control (cages or fencing to protect plants from vertebrate herbivores, or chemical sprays to kill insects), irrigation programs, and fertilizer. Fertilizer should be used with care, because it may increase shoot rather than root growth, enhance competition by weeds, and make plants more palatable to pests. Mulching and composting can also provide many benefits.
7. *The reintroduction of animal components.* In most cases, animals are assumed to be able to find their own

way back into restored habitats. Less commonly, as in the reclamation of areas mined for phosphate in Florida, species such as gopher tortoises and Florida mice are actively returned to form a nucleus for new populations.

The careful consideration of the preceding elements may accelerate the pace of succession on restored land and promote the return of a climax community, thereby achieving the goal of helping to restore the community to its undisturbed state.

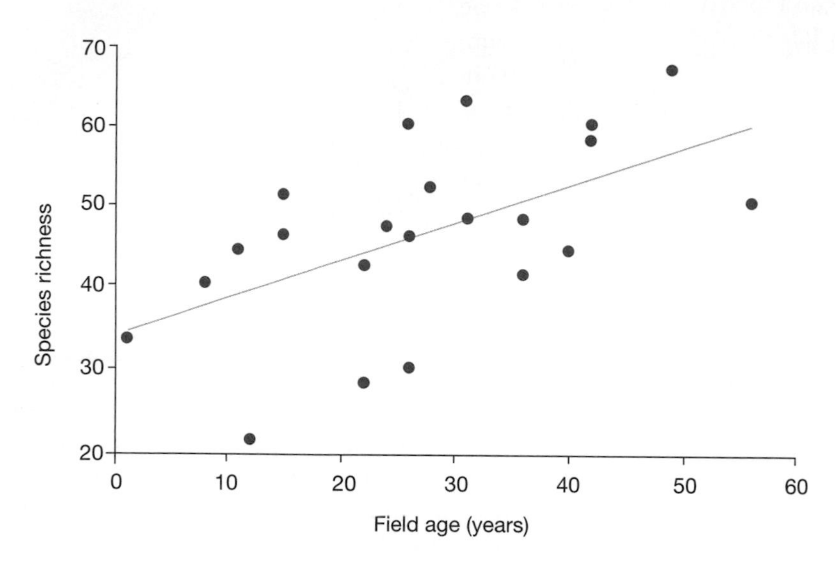

Figure 18.10 Increase in plant species richness with age of the field in Minnesota old fields. *(Reproduced from Inouye et al., 1987.)*

reach maturity at low biomass. Species richness and community stability also change in somewhat predictable ways through the stages of succession.

Plant species richness generally increases as succession proceeds. This pattern was well illustrated by Richard Inouye and colleagues (1987), who studied species richness in old fields in Minnesota (Fig. 18.10). The older the field, the greater was the number of species, because there had been more time for species to colonize. Of course, such an increase in species richness can't go on forever.

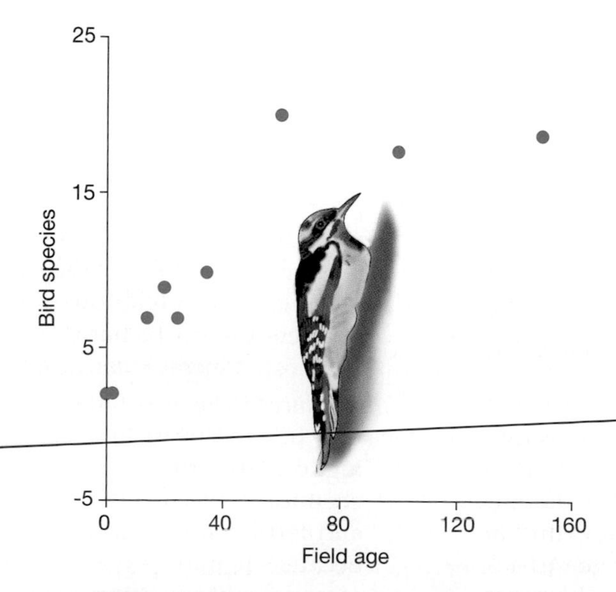

Figure 18.11 Increases in bird species richness on different aged old fields in Georgia during the nesting season. *(From data in Johnston and Odum, 1956.)*

So far, succession has been seen as a largely botanical affair, probably because plants commonly provide 99% of the biomass and structure of communities. Most of the animals are simple followers of plant succession. We can see this by revisiting the Piedmont, where Johnston and Odum (1956) examined the richness of bird communities in areas ranging from grassland to mature oak–hickory forest (Fig. 18.11). In grasslands there were only about 2 bird species, in grass shrub 7, in 35-year old pine forests 10, and in old pine forests of 60–100 years 18–20 species. In the mature climax oak–hickory forests there were about 19 species. This increase in bird species probably occurs because, as more plants invade an area, there are a greater variety of insect herbivores that feed on them, and these insect herbivores in turn support a greater variety of bird species. Increased plant height also provides more niches for perching and nesting birds. Once again, however, as with plant species, animal species richness cannot increase indefinitely.

In fact, in climax communities, both plant and herbivore richness can *decrease*. In England, Val Brown and her colleague G. Edwards-Jones (Edwards-Jones and Brown, 1993) showed that plant species richness in areas of different age increased only in the middle seres. At climax, the community became dominated by a birch woodland, so plant species richness dropped precipitously to two species of birch, *Betula pendula* and *B. pubescens* (Fig. 18.12a). Such climax communities are highly stable. Insect richness also decreased dramatically in climax communities, because the communities were dominated by birch trees, so only birch-feeding insects were present (Fig. 18.12b).

In some instances, animals play a more active role and can influence the course of succession. We have already seen how crab herbivory influences succession on marine intertidal boulders because crabs eat *Ulva*, the early colonist, and the removal of *Ulva* allows other species, such as the brown alga, to enter the zone. In Africa, elephants are prolific browsers of trees, keeping the savannas relatively open. Although fires can destroy trees, too, the exclusion of elephants can result in a more than threefold increase in tree density, changing the appearance of a region from a grassland to a more closed woodland community (Deshmukh, 1986).

In marine systems, fish grazing may deflect the path of succession. Hixon and Brostoff (1996) performed some elegant manipulations of fish grazing in Hawaii. In the absence of grazing, algal succession proceeded through three seres in one year: early dominance by green and brown filaments, a midsuccessional stage of thin, red filaments, and a late stage of coarsely branched, thick filaments. Where grazing was intense, all erect algae were replaced by a low-biomass, low-diversity assemblage of crusts and prostrate blue-green mats.

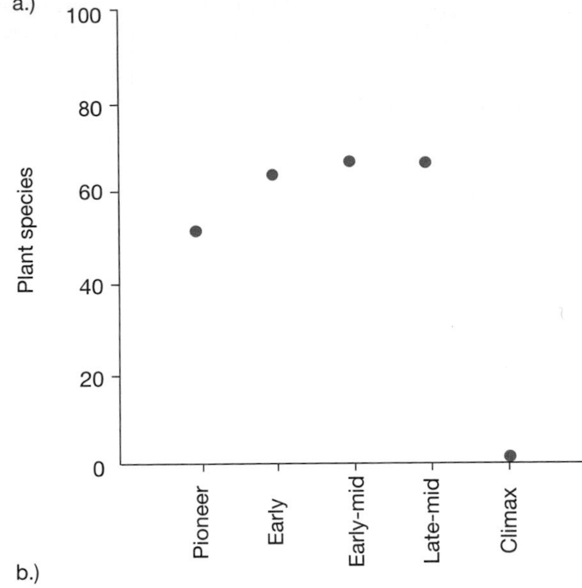

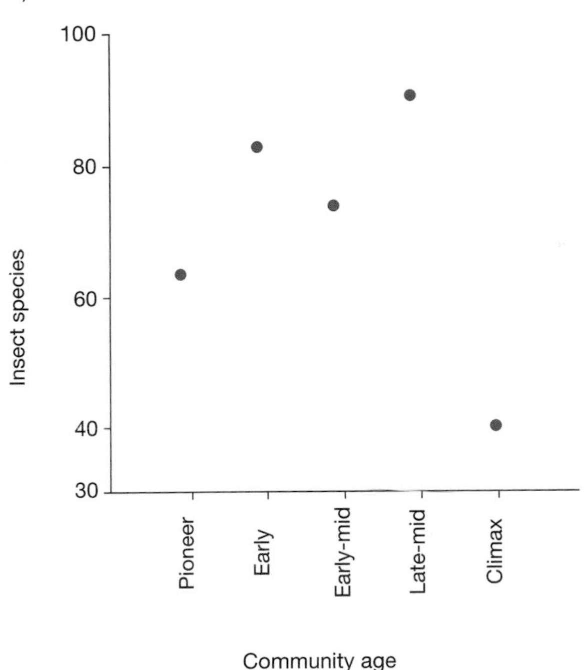

Figure 18.12 Species richness changes in (a) plants and (b) their insect herbivores in seres of different ages in England. *(From data in Edwards-Jones and Brown, 1993.)*

18.6 Biotic Interactions and Succession

Besides herbivory, myriad other biotic processes have the potential to affect how succession proceeds. Ten years after the facilitation, inhibition, and tolerance models were proposed, Walker and Chapin (1987) suggested how some of these biotic processes might affect succession (Fig. 18.13). We can summarize Walker and Chapin's main conclusions as follows:

1. Mechanisms promoting seed dispersal are more important in primary succession, whereas buried seeds and surviving vegetative propagules are more important in secondary succession. Many species produce a high number of small seeds to promote their dispersal by the wind.
2. Stochastic variation is more important in severe and low-resource environments. For example, fire is important in dry environments, flooding affects the colonization of riverbanks, and sea levels are important in dune communities.
3. Facilitation is more important in severe environments such as deserts, in primary succession, and in the early stages of community development, where low levels of nutrients, shade, and water may prevail. Mutualisms may be important in these same circumstances.
4. Competition is probably more widespread than facilitation (see also Tilman, 1985), especially in more favorable environments.
5. Maximum potential growth is particularly important in favorable environments, in which resources are sufficient to promote rapid growth.
6. Differences in longevity among plant species are more important in older communities, because the longer a plant lives, the better chance it has of outgrowing its competitors.

Biotic processes such as herbivory, competition, mutualism, and other processes can affect the course of both primary and secondary succession.

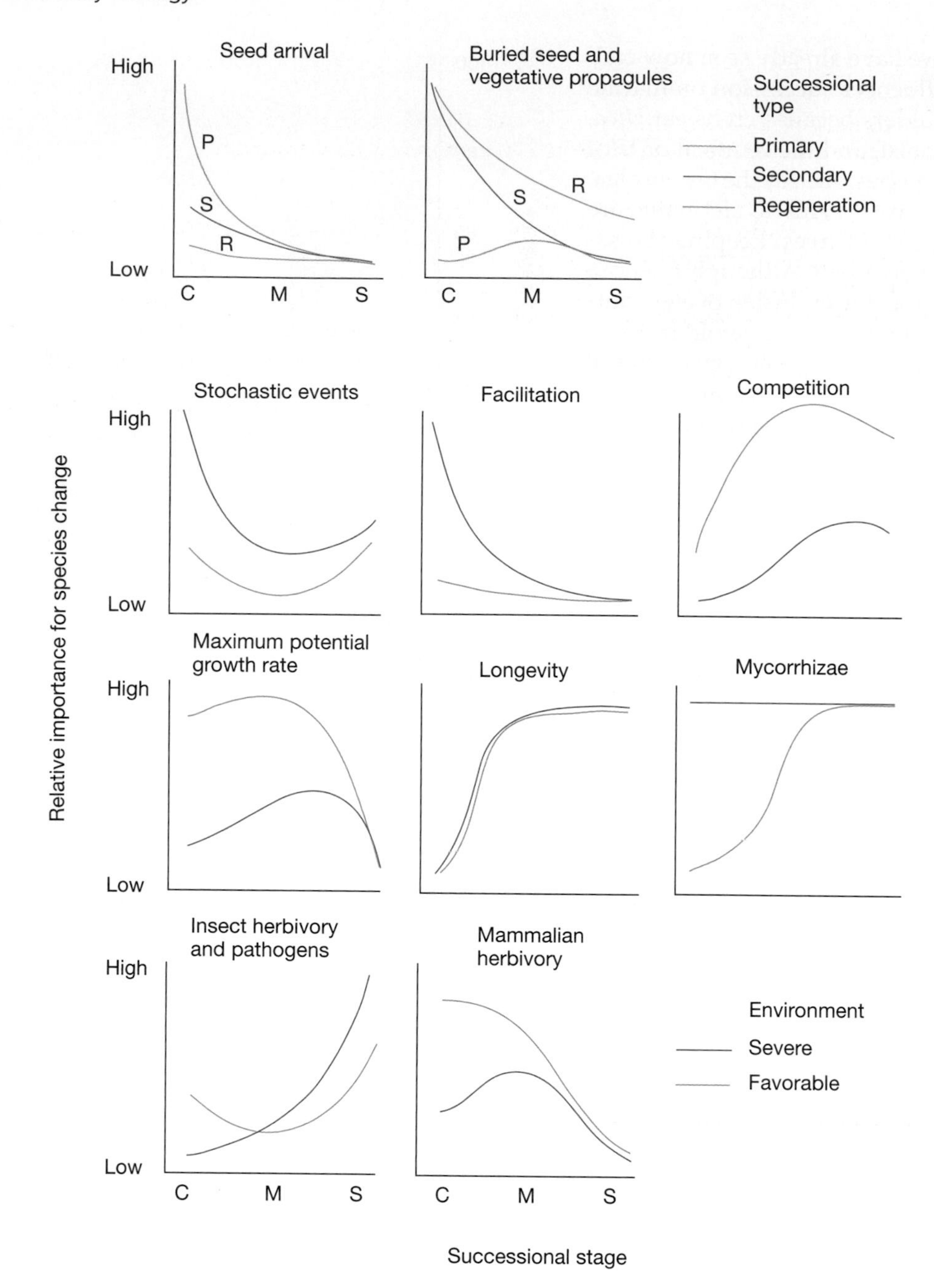

Figure 18.13 Influence of abiotic and biotic factors in succession in different seres: C = colonization, M = maturation, and S = senescence. *(Reproduced from Walker and Chapin, 1987.)*

7. Mycorrhizae may be especially important in severe environments, where plants need the extra help to survive.
8. Insect herbivores and pathogens are more important on mature vegetation, because large species are much easier to locate and there is more room on them for the insects to grow to high densities.
9. Mammalian herbivores are more important in early and midsuccessional communities, because the smaller plants of earlier stages can be completely devoured by large herbivores.

Walker and Chapin's conclusion was that succession is complex and is driven by many processes acting simultaneously in any given situation. Thus, the effects of such factors as competition, seed arrival time, insect and mammalian herbivory, and stochastic events vary in importance according to the stage of succession—early, middle, or late. For example, competition is not very important to colonizing species in the early stages of succession, because few competitors are present. However, in mature communities, competition can be an important force. Similarly, stochastic events like fire can devastate early successional communities, but may have less effect on species change in mature communities, because species such as large trees are able to withstand periodic burns. In essence, succession is a multifaceted ecological phenomenon, and many factors, both biotic and abiotic, can affect its path.

Summary

Succession concerns the development of pioneer communities that appear soon after a disturbance to mature communities that change little in the absence of a disturbance.

1. Primary succession occurs when species invade an area in which no organisms are present, such as land unearthed after receding glaciers. Secondary succession is the change in species composition following a change in land usage, such as the reversion of old fields to forests after agriculture has stopped.

2. Early theories of succession viewed the entire process as facilitative, in which each species makes the environment more suitable for the next.

3. Later work on succession revealed the existence of inhibition, wherein early colonists actually prevented colonization by other species.

4. In 1977, Connell and Slatyer recognized the existence of a third type of succession, which they termed tolerance, that, in essence, was intermediate between the other two. In this model, any species can start the succession, but the eventual climax community is reached in a somewhat orderly fashion.

5. In 1987, John Lawton proposed the random-colonization model of succession, which posits that there is no facilitation or competition and that succession proceeds by chance alone.

6. In succession, early communities are often species poor and later communities are species rich. However, this is not always the case.

7. Recently, it has been recognized that the path of succession can be modified by herbivory, disease, and other factors. Thus, there is no single universal cause of succession, which is a multifaceted process.

Successional processes may usually be viewed along a continuum from facilitation to inhibition. Along this continuum, many biotic and abiotic factors can change the path of succession.

Discussion Questions

1. If agriculture on once-virgin tropical forest areas were to stop, how could we speed up the process of secondary succession so that the forests returned?

2. Which types of communities (marine versus freshwater versus terrestrial, forest versus prairie, temperate versus tropical, or primary versus secondary) do you think would support which model of succession (facilitation, tolerance, or inhibition)? For example, in marine intertidal habitats, nitrogen levels of the soil are not important, but competition for space on rocks is vital, so whoever arrives first might have an advantage and inhibition might be common.

3. How does the existence of the climax stage in many communities affect the diversity–stability debate? If diversity is not maximum at the climax, does it mean that climax communities are less stable than others?

C H A P T E R 19

Island Biogeography

Road Map

1. Island biogeography theory suggests that the number of species on an island is affected by the size of the island and by the proximity of the island to a source pool of colonists. The theory also suggests that species continually emigrate to islands and go extinct, so that species turnover occurs.
2. There is much evidence to support the hypothesis that, as the size of an oceanic island or a continental habitat island increases, so does the species richness.
3. There is also evidence to suggest that more distant islands have fewer species.
4. Although species turnover on islands occurs, it does not appear to be common.

A massive volcanic explosion in 1883 on the island of Krakatau destroyed two-thirds of the island, originally 11 km long and covered in tropical rain forest. Life on the remainder, an island known as Rakata, and the two neighboring islands, Sertung and Panjang, was eradicated, suffocated by tens of meters of volcanic ash. In 1930, another island, Anak Krakatau (Krakatau's child), formed in the center of the three remaining islands, the product of more volcanic eruptions. Recolonization on these islands has been studied by a series of scientists.

Nine months after the 1883 eruption, the first reported colonist of Rakata was a spider spinning its web. By 1896, there were 11 species of ferns and 15 of flowering plants. Sixteen of these species were dispersed by the wind, and another eight by the sea. Colonization of Rakata was greatly affected by how well plants were able to disperse. The early plant community was dominated largely by grasses, but 25 years later it was given over to *Cyrtandra* bushes and then, in the 1920s, to a forest of *Neonauclea* trees. The insect fauna changed, too, as their host plants came and went. Further investigations showed that wind- or sea-dispersed plants were able to colonize the island much more readily than plants dispersed by animals (birds) (Fig. 19.1), but by 1929, over 40 years after the original eruption, animal-dispersed plants became as common as sea-dispersed ones, probably because birds had become more abundant on the island. The subsequent development of the floras and faunas of Sertung, Panjang, Rakata, and Anak Krakatau varied, based partly on the size of the islands and partly on subsequent eruptions, which affected some islands more than others. Succession in the archipelago was affected by the distance of the island from a source of colonists, by the ability of the various organisms to disperse, and by the area of the island. Only much later was a holistic theory of succession on islands developed.

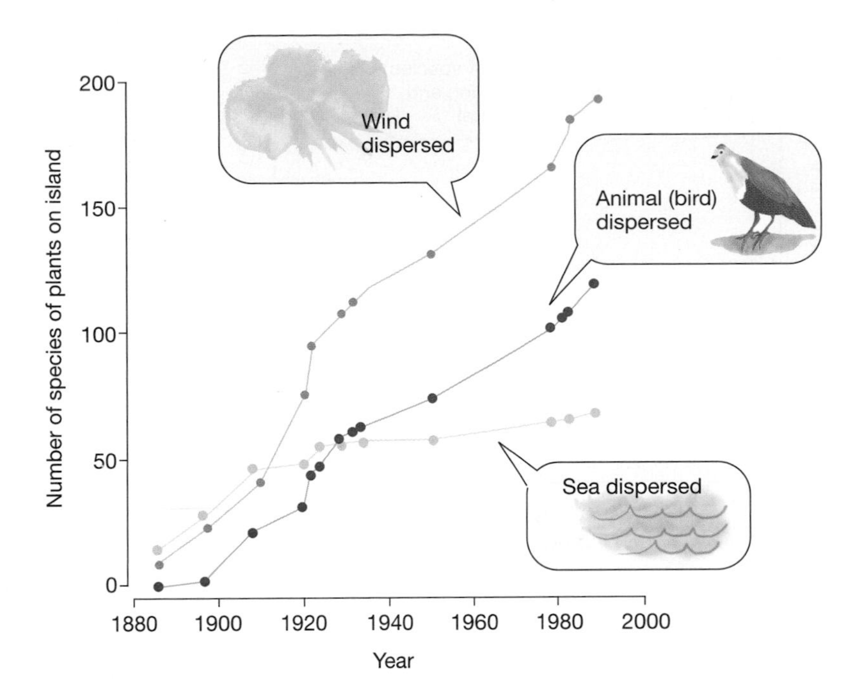

Figure 19.1 The mode of dispersal of plants, whether by wind, animal (mainly birds), or sea, made a difference in the colonization rates of Rakata, following the eruption of Krakatau in 1883. (*From data in Whittaker et al., 1992.*)

19.1 Theory of Island Biogeography

The scientists who first formally developed a comprehensive theory to explain the influence of the size of an island and its distance from the mainland on succession on the island were Robert MacArthur and E. O. Wilson (1963, 1967). Their theory was called the "equilibrium theory of insular zoogeography." MacArthur had studied mathematics as an undergraduate and had received a master's degree in the subject from Brown University. He switched to ecology only when he was pursuing a Ph.D. at Yale. It was understandable that his approach to ecology began with a mathematical or graphical model of the way nature worked. From these ideas, he generated predictions that could be tested by collecting data from the field. This approach is often known as the hypothetico-deductive method. Wilson, on the other hand, was a meticulous field biologist who had been interested in island biology for quite a while. Together, they made a great team, and together, they suggested the following ideas:

1. The number of species on an island, S, tends toward an equilibrium number $\hat{S}$.
2. $\hat{S}$ is the result of a balance between the rate of immigration and the rate of extinction. The rate of immigration of new species is highest when no species are present on the island, so that each species which invades the island is a new species. As species accumulate, many subsequent immigrants no longer represent new species. The rate of extinction is low at the time of first colonization, because few species are present and many have large populations. With many species, the population diminishes, so the probability of extinction by chance alone goes up. Thus, species may arrive and become extinct, but the number of species on the island remains the same (Fig. 19.2). In actuality, MacArthur and Wilson reasoned that both the immigration and extinction lines would be curved (Fig. 19.3). First, species arrive on islands at different rates. Some organisms, including the good fliers and good swimmers, are more mobile than others and will arrive quickly. Later, the poorer dispersers arrive. These include the nonfliers and poor swimmers, some of whom may arrive on floating logs or other debris. This pattern causes the immigration curve to start off steep, but get progressively shallower. As regards the extinction curve, extinctions rise at accelerating rates, because, as later species arrive, competition increases and more species are likely to go extinct. Also, the r-selected species, which arrive first, are generally not good competitors, whereas the later arriving species are generally more K-selected and hence better competitors. Therefore, later arriving species usually outcompete earlier arriving ones.

The colonization of islands following a disturbance has been described with the theory of island biogeography, which predicts that the equilibrium number of species on an island should be positively correlated with island size and inversely correlated with distance to the mainland.

e.g. previous page

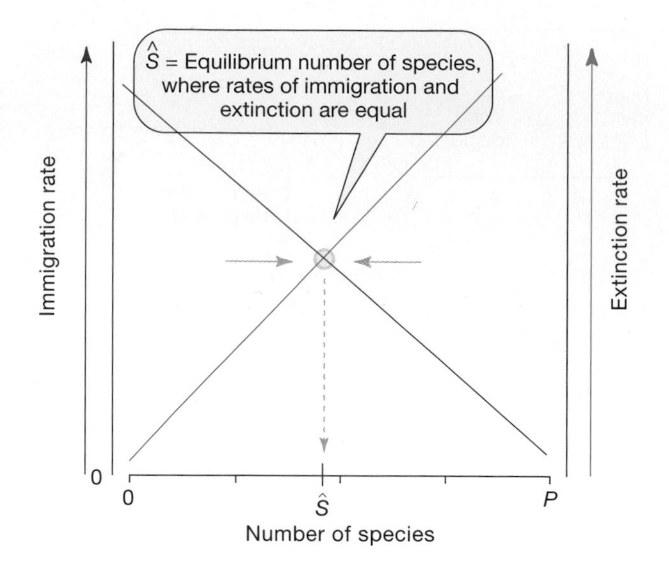

Figure 19.2 The interaction between the immigration rate and the extinction rate to produce an equilibrium number of species on an island, $\hat{S}$, which varies from 0 species to *P* species, the total number in the species pool of colonists. The basis of the MacArthur–Wilson island biogeography model is the assumption that all species have the same immigration and extinction rates and do not affect each other. The species richness on an island will then tend toward a steady state that is less than the richness of the source pool.

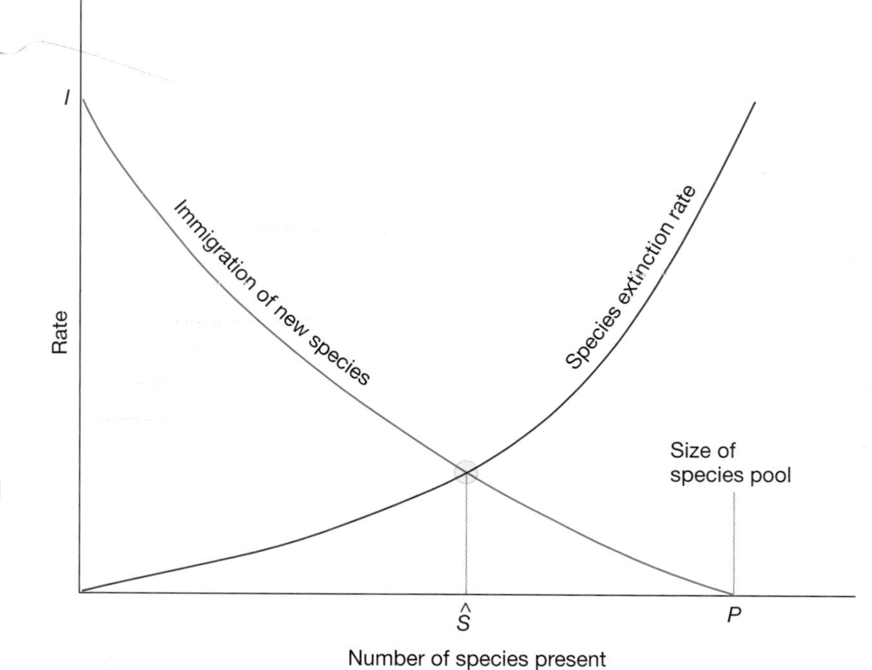

Figure 19.3 In MacArthur and Wilson's model of island biogeography, the rates are curved because species add to each other's extinction rates and because some species immigrate more readily than others.

3. $\hat{S}$ is determined only by the island's area and position, which influence the rates of immigration and extinction. MacArthur and Wilson argued that extinction rates were greater on smaller islands and immigration rates greater on islands near source pools (Fig. 19.4). Extinction by chance alone was more likely on small islands because population sizes would be smaller there and more susceptible to extinction. Immigration rates are greater on islands closer to shore or to other source pools because species don't have as far to travel.
4. Equilibrium is dynamic; hence, following colonization of an island, we should see the numbers of species on the island stay constant through time, although the extinction of some species and the immigration of new species change the identities of the species on the island. Thus, we will see a turnover of species.

Since its original exposition, MacArthur and Wilson's theory of island biogeography has undergone three major modifications:

1. *The target effect (Whitehead and Jones, 1969).* The rate of immigration depends on an island's size because larger islands present larger "targets" for colonists than do smaller islands. This idea assumes that if species dispersed like arrows shot from a bow, more would hit a bigger target (a bigger island).

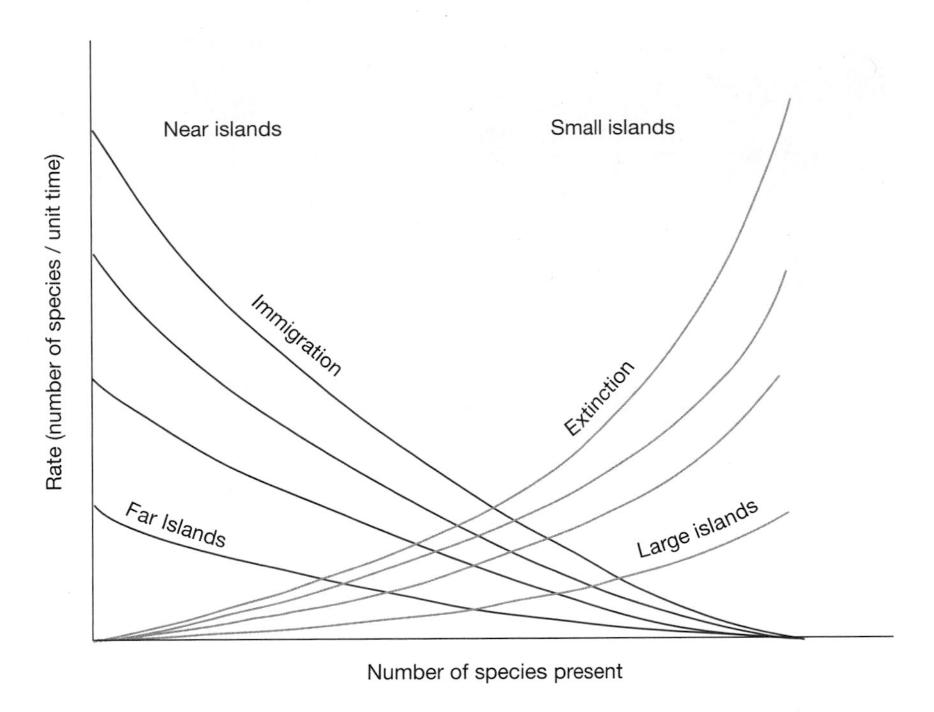

Figure 19.4 Equilibrium models of biotas of islands with varying distances from the principal source and of varying size. An increase in distance (*near* to *far*) lowers the immigration curve; an increase in area (*small* to *large*) lowers the extinction curve. (*Redrawn from MacArthur and Wilson, 1963.*)

2. *The rescue effect*. Jim Brown and Astrid Kodric-Brown (1977) proposed that the distance from an island to a source pool of potential colonists affects both the rate of extinction and the rate of immigration. This is because the immigration of individuals of taxa already resident on the island slows the rate of extinction of those taxa by keeping population sizes higher than they would be in the absence of immigration. Also, even if a species were not able to be supported on an island and were going extinct, we might not detect its extinction because of continued immigration. In this view, we are unable to differentiate residents from transients of the same species that just arrived. Thus, although the basic MacArthur–Wilson model described only two mechanisms—the effect of *area on extinction* and the effect of *distance on immigration*—the target and rescue effects extend and effectively complete the model by including the effect of *area on immigration* and the effect of *distance on extinction*, respectively (Fig. 19.5).
3. *The concept of an island*. Patches of particular habitat on continents are viewed as islands in a sea of other, unsuitable habitat. Thus, patches of grassland in the Craters of the Moon National Monument in Idaho are surrounded by extensive lava flows (Fig. 19.6). For the mice that inhabit the grassland, the lava flows may as well be a real ocean because they cannot survive in such inhospitable areas. Similarly, freshwater species that live in lakes behave as if they were in islands surrounded by a "sea" of inhospitable dry land.

Island biogeography theory can be applied to continent patches as well as true islands.

University of Pennsylvania ecologist Dan Janzen (1968, 1973) extended the habitats-as-

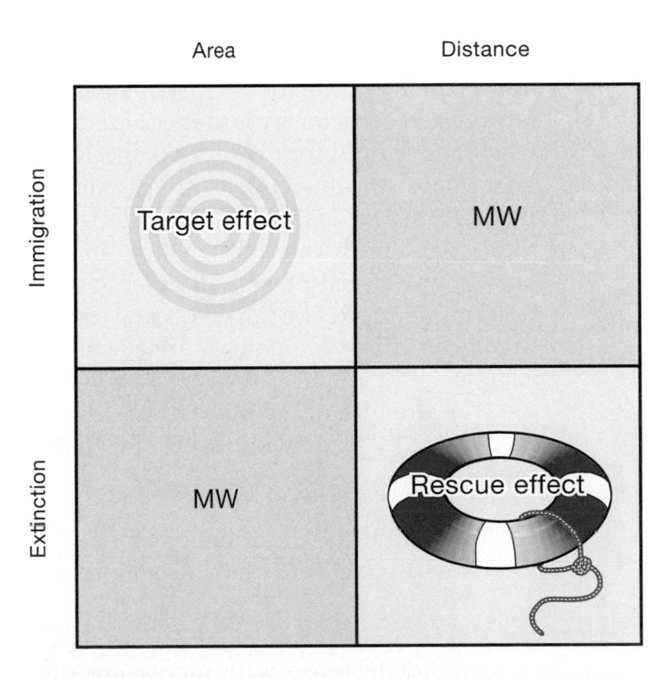

Figure 19.5 The effects of area and distance on the immigration and extinction of species on islands, as originally proposed by MacArthur and Wilson (MW), and as modified by Whitehead and Jones (target effect) and Brown and Kodric-Brown (rescue effect). (*Modified from Gotelli, 1995.*)

Figure 19.6 Idaho's "craters of the moon" and Minnesota's 10,000 lakes. For animals that live there, each habitat can be seen as an island in a sea of inhospitable terrain.

Many studies support the prediction that the number of species should increase with the size of the island.

islands concept by proposing that individual host-plant species could be islands to their associated herbivore fauna, which was adapted only to feed on that particular type of vegetation. For example, many insects have sophisticated biochemical machinery that allows them to detoxify certain plant tissues. However, the machinery is so specialized that it will work only for one or two species of host plant. All other plants are essentially inedible to the insect. As example is the monarch butterfly caterpillar, which can feed only on milkweed plants. All other vegetation might as well be "open ocean." So for monarchs, milkweed patches are effectively islands.

The strength of the MacArthur–Wilson theory was that it generated some falsifiable predictions:

> Prediction 1. The number of species should increase with increasing island size.
>
> Prediction 2. The number of species should decrease with increasing distance of the island from the source pool.
>
> Prediction 3. The turnover of species should be considerable. The number of species on an island might remain the same, but the identities of those species should change as new species arrive at the island and others become extinct.

We shall deal with these predictions one by one and see how they are supported by the data.

19.2 Species–Area Effects

Oceanic islands.

The West Indies has traditionally been a fertile hunting ground for ecologists studying island biogeography. This is because the islands are very well known, in terms of both their physical geography and their biota. Generations of naturalists have enjoyed compiling detailed lists of the plants and animals of the West Indian islands. Furthermore, the Lesser Antilles, from Anguilla in the north to Grenada in the south, enjoy a similar climate, are surrounded by deep water, and have no historical connections to the mainland (Fig. 19.7). Robert Ricklefs and Irby Lovette (1999) summarized the available data on the richness of species of four faunal groups—birds, bats, reptiles and amphibians, and lepidopterans (butterflies)—over 19 islands that varied in area over two orders of magnitude (13–1,510 km^2). In each case, there was a significant relationship between area and richness of species (Fig. 19.8). Note that these relationships are traditionally plotted on a double logarithmic scale, a so-called log–log plot, in which the horizontal axis is the logarithm to the base 10 of the area and the vertical axis is the logarithm to the base 10 of the number of species. This is done so that a straight-line relationship is obtained. A linear plot of the area vs. the number of species would be very difficult to produce, because of the great range of area and richness of species involved. Logarithmic scales condense this variation to manageable limits.

Note that these plots are not isolated instances. Frank Preston (1962) provided many other examples that showed a species–area relationship: birds of the East Indies, beetles on West Indian islands; ants in Melanesia; vertebrates on islands in Lake Michigan; and land plants in the Galápagos. At least 17 other examples from islands were cited in a review by Jim Quinn and Susan Harrison (1988).

We can represent the relationship of species richness to area with the equation

$$S = cA^z,$$

or, in logarithmic form,

Species richness

A constant

$$\log S = \log c + z \log A$$

Slope

Area

where S denotes the number of species, c is a constant measuring the number of species per unit area (i.e., per unit of forest, grassland, or desert), and A represents the area. The constant c is likely to be higher for more productive habitats.

In this relationship, a z value of 0.301 means that, as we multiply the area by 10, we double the richness of species. Among those studies for which the species–area relationship was viewed as proven, one of the early topics of discussion was the significance of variations in the value of z. A high value of z indicates steep increases in the number of species as the size of an island increases, whereas low values indicate much smaller differences in numbers of species among islands. We shall return to this topic a little later.

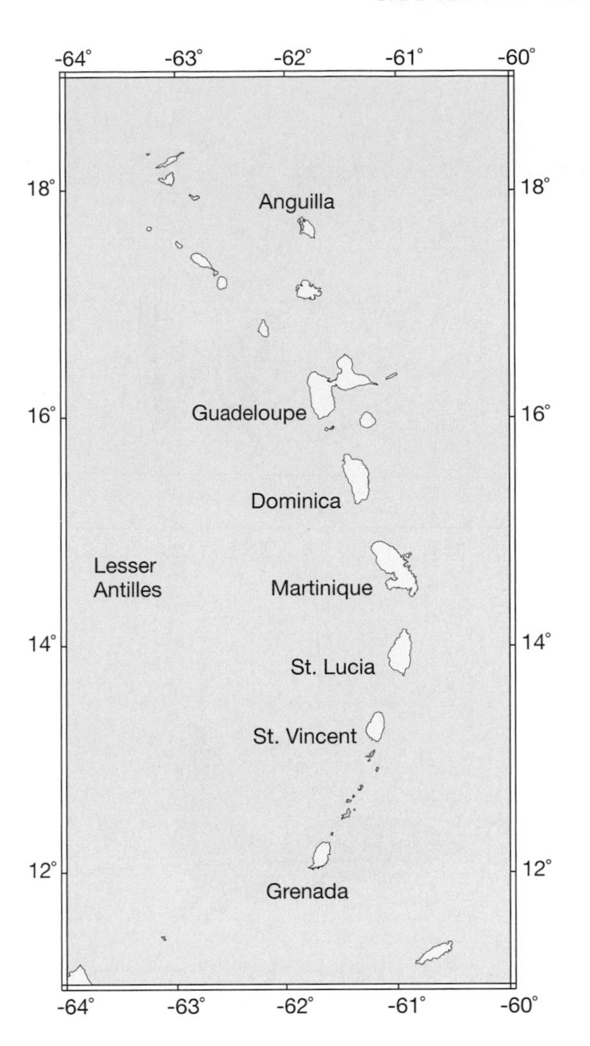

Figure 19.7 The Lesser Antilles extend from Grenada in the south to Anguilla in the north and have been much studied in island biogeography.

Habitat "islands."

While at the University of Arizona, James Brown (1978) studied the distribution of boreal mammals and birds in the mountain ranges of the Great Basin in the United States. The mountain ranges are essentially isolated from one another, and the mammalian fauna is a relict community of a bygone age when rainfall was higher and the boreal habitats in the range were contiguous. Each mountaintop is essentially a forest island in a sea of desert (Fig. 19.9). Brown found a significant relationship between species and area for both mammals and birds. Data from eight studies of other habitat isolates showed similar species–area relationships (Quinn and Harrison, 1988), proving that this relationship, as seen on a log–log plot, holds in many instances.

Of particular interest is the value of z in the equation $S = cA^z$, where z is the slope of the line. Values of z obtained from continental areas are often 0.15 to 0.25, lower than those from truly insular situations, for which z is often 0.20 to 0.40. This means that, as larger areas are sampled, fewer new species are added on continents than on

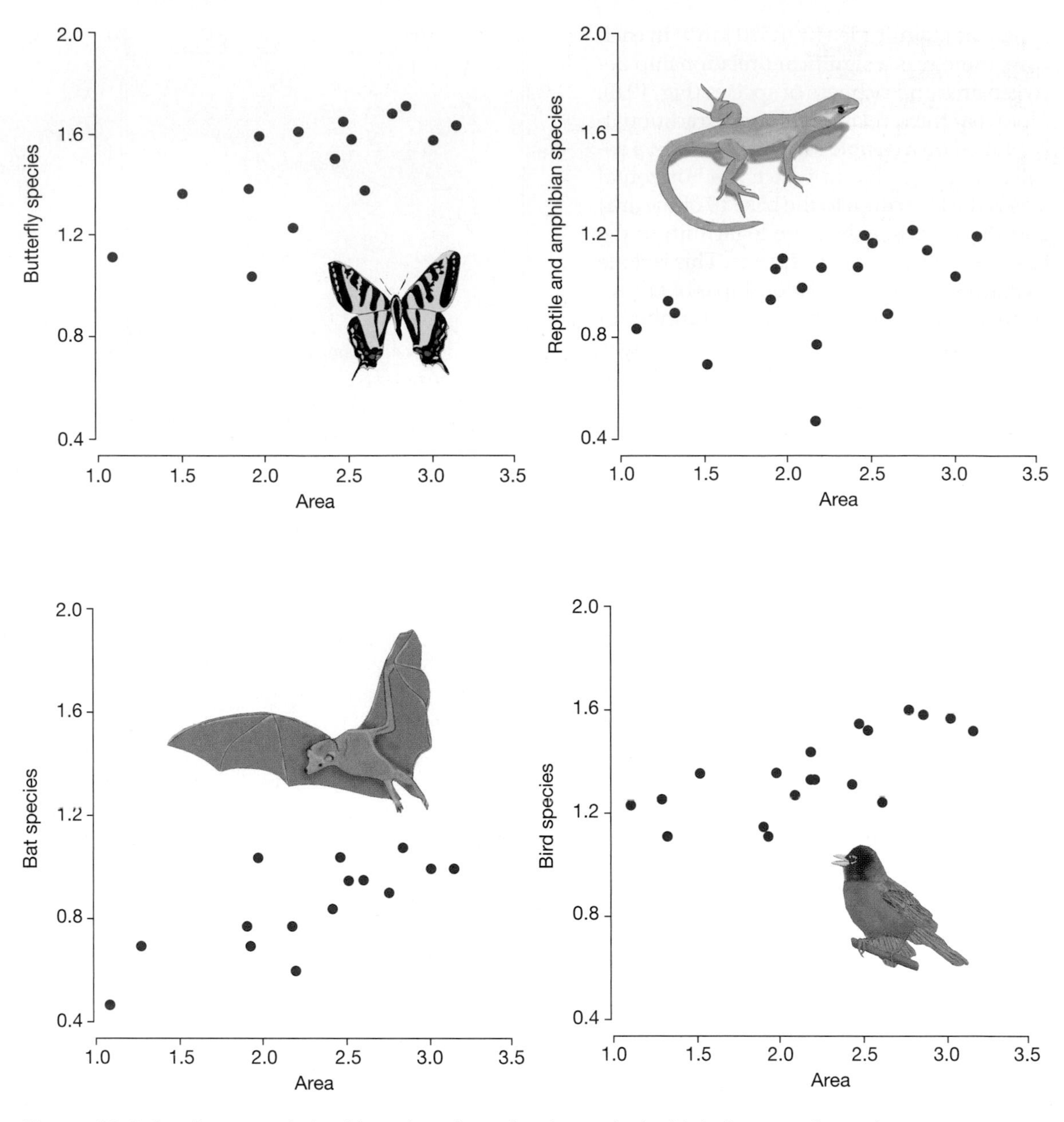

Figure 19.8 Species–area relationships, plotted on a log–log scale, for birds, bats, reptiles and amphibians, and butterflies, for islands of the Lesser Antilles. (*From data in Ricklefs and Lovette, 1999.*)

islands. The biological explanation for this phenomenon is that continental areas contain some transient species from adjacent habitats because, on continents, some species simply rest up in patches while passing through an area. There are proportionately more transients passing through smaller areas, raising the apparent number of species in those areas, so the slope of the species–area curve is shallow for continents. Islands, by contrast, are actual isolates with reduced migration rates, because species cannot rest up in the ocean between the islands (otherwise they would drown), so the number of transients in an area is minimal.

The species–area relationship for birds on Brown's mountaintops had a slope of 0.165, and the slope for mammals was 0.326 (Fig. 19.9), more like that found on islands. The reason is that there is no mammalian migration between mountaintops because mammals would have to walk down the mountain, across the valley, and up the next mountain. In this situation, mountaintops behave as true islands. In contrast, birds disperse more than mammals, because they can simply fly between mountaintops, and the value of z in their case is much more in line with a mainland type of relationship.

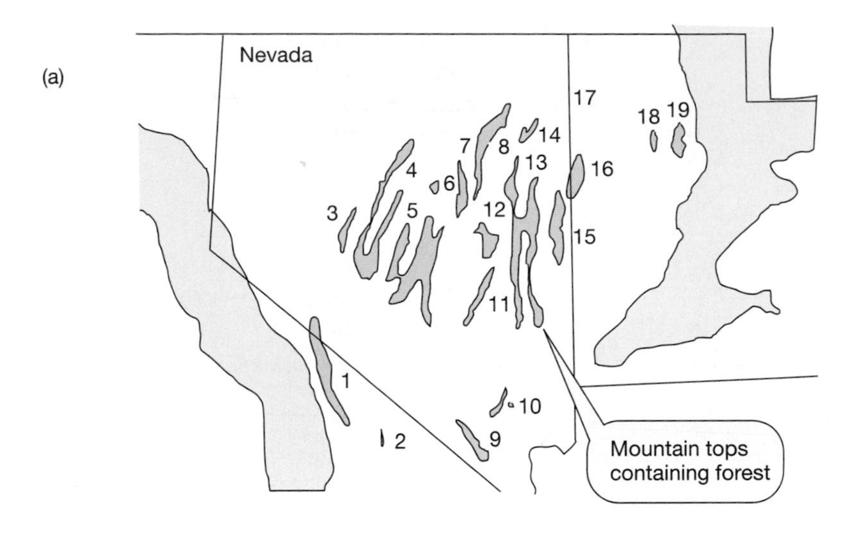

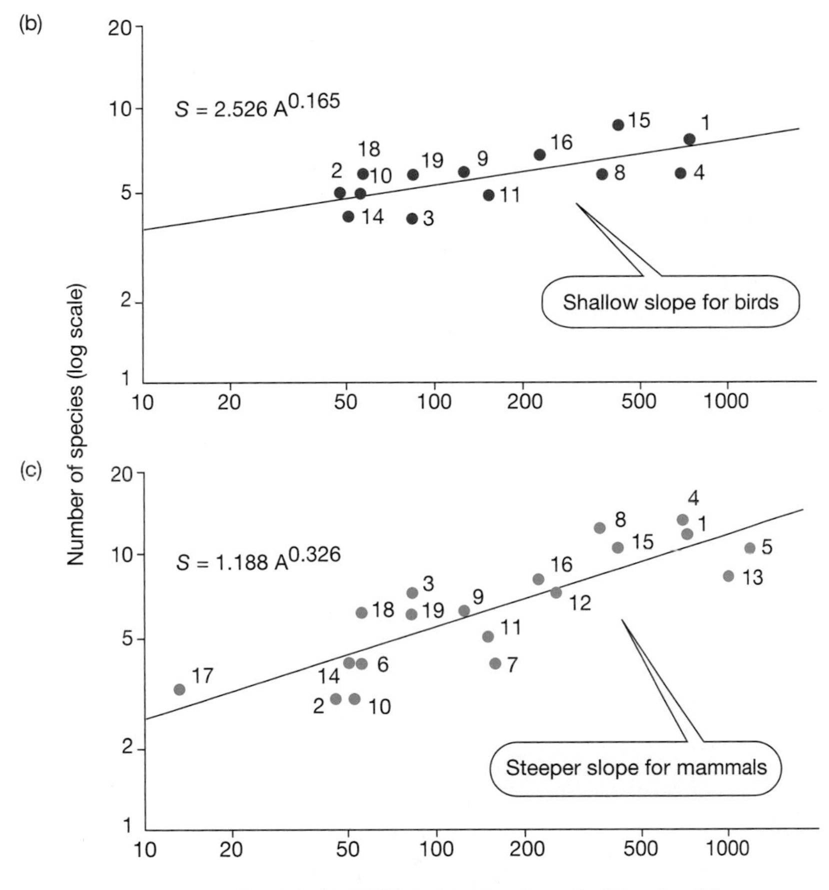

Figure 19.9 Island biogeography applied to mountaintops. (a) Map of the Great Basin region of the western United States, showing the isolated mountain ranges between the Rocky Mountains on the east (*right*) and the Sierra Nevada on the west (*left*). (b) Species–area relationship for the resident boreal birds of the mountaintops in the Great Basin. (c) Species–area relationship for the boreal mammalian species. Numbers refer to areas on the map. (*Redrawn from Brown, 1978.*)

Species as islands.

Janzen's idea that species of host plants act as islands in a sea of other vegetation for the herbivores that eat from the plants were elaborated by Donald Strong (1974), who found a species–area relationship between the geographical area of distribution of British tree species and the number of insect herbivore species (Fig. 19.10). Strong's work was made possible because detailed surveys of the distribution of trees in Britain were available on maps. The entire island of

For insects, individual trees may often act as islands in a sea of other vegetation on which they cannot feed.

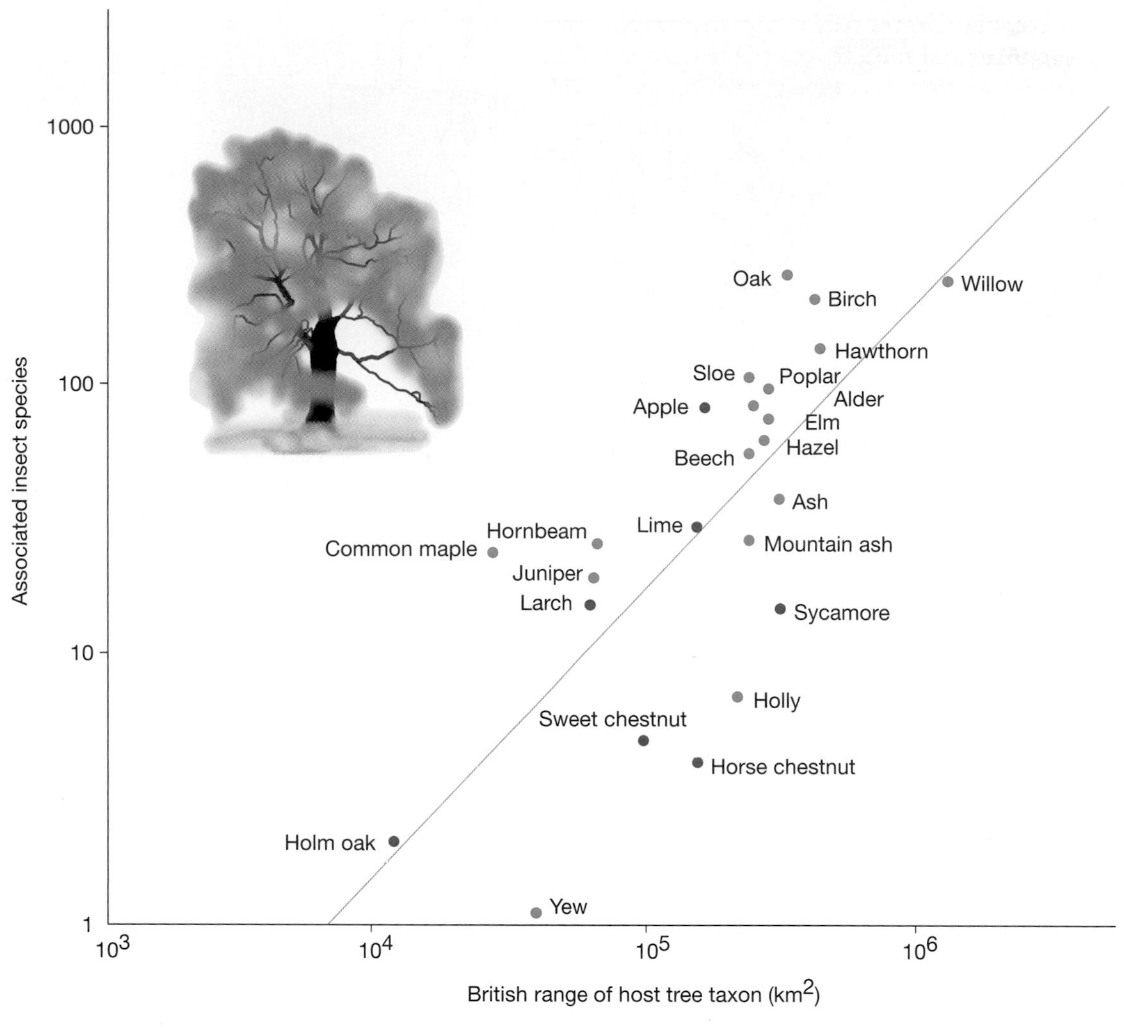

Figure 19.10 The species–area relationship for the insects associated with British trees. Red dots indicate introduced taxa; blue dots indicate British natives. (*Modified from Strong, 1974.*)

Great Britain was divided up into 10-km^2 grids, and every time a species of tree was found in an area, a dot was made on the map. By counting up the number of dots, Strong could work out the area the tree occupied in Britain. The second piece of information, the number of associated insect herbivores per tree species, came from detailed lists of insect herbivores on each species of tree made over the years by generations of naturalists. Such an undertaking was made possible by the relatively small size of Britain, the low number of tree species and herbivores present, and the large number of naturalists on the island. The task would be too onerous in a large geographical area like the United States. Since Strong's original work, many other workers have documented species–area relationships for insect herbivores on host-plant species. For example, Paul Opler (1974) found a similar relationship between California oaks and their leaf miners, species of tiny moths whose larvae live in blotchlike blisters within the surfaces of the leaf.

The studies we have just discussed show that MacArthur and Wilson's prediction that species richness increases with area is borne out in observation. However, the exact mechanism may not always be that extinction through chance is more likely on smaller islands because populations on those islands are smaller, as MacArthur and Wilson envisaged. Hart and Horwitz (1991) provide at least five reasons for a species–area relationship:

1. Extinction rates are greater on small islands (the original MacArthur–Wilson hypothesis).
2. The passive effect of increased sampling effort in bigger areas increases the number of rare species found. Bigger areas contain more individuals. The more time we sample these bigger areas, the greater are our chances of collecting new, rare species. We saw a similar

effect, in Chapter 16, with increased sampling revealing more of the lognormal distribution as rare species become apparent.

3. Speciation may be more likely in bigger areas, an explanation also given for the greater species richness in the tropics, as discussed in Chapter 15.
4. Larger areas contain more "core" areas, which are less affected by disturbances such as wind and temperature variations that affect perimeter areas. Also, the perimeter areas contain more species that are sensitive to these disturbances.
5. The species–area relationship may more likely be the result of an increased diversity of habitats on large islands than just an increase in area relationships.

Larger areas often contain a greater diversity of habitats—different soil conditions, slopes, elevations, salinities, pH, and the like—than do smaller islands. A large island in the West Indies, for example, is more likely to contain mountain habitats than is a smaller island. In fact, this is probably the biggest factor behind the species–area relationship.

In Australia, Barry Fox (1983) investigated the relationship between species, area, and habitat diversity in Australian mammals. He classified habitats into seven broad types and showed that larger areas do include more types of habitat (Fig. 19.11a). He also showed that the number of mammalian species is well predicted by area (Fig. 19.11b). However, species richness was better predicted from the number of habitats than from area (Fig. 19.11c).

William Tonn and John Magnuson (1982) showed much the same thing. They recorded habitat variables and fish species from 18 shallow lakes in Wisconsin. The area of a lake predicts the number of species in it, but so does the lake's habitat diversity. The problem is that area and habitat correlate so closely that either variable alone can predict diversity.

We can also return to the study of species–area relationships done by Ricklefs and Lovette (1999) in the Lesser Antilles. These researchers were able to examine the diversity of habitats on the islands by quantifying the amount of the five major types of habitat present: coastal mangrove forest, dry scrub woodlands, secondary rain forest (recovering after logging), primary rain forest,

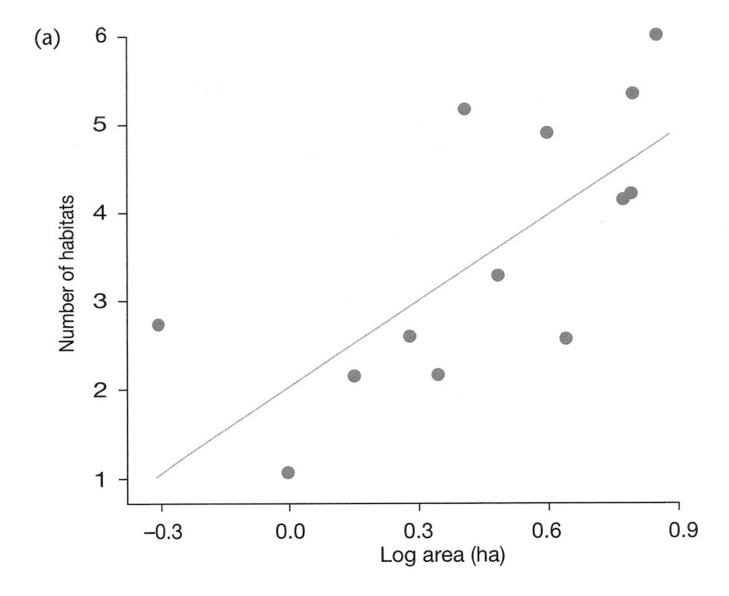

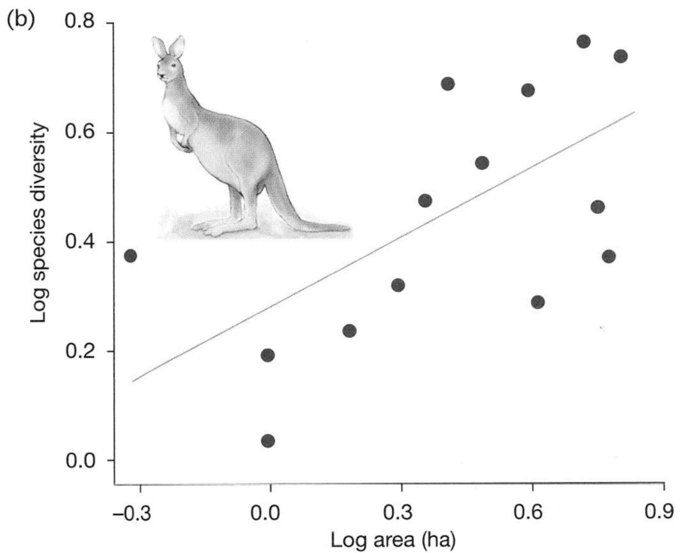

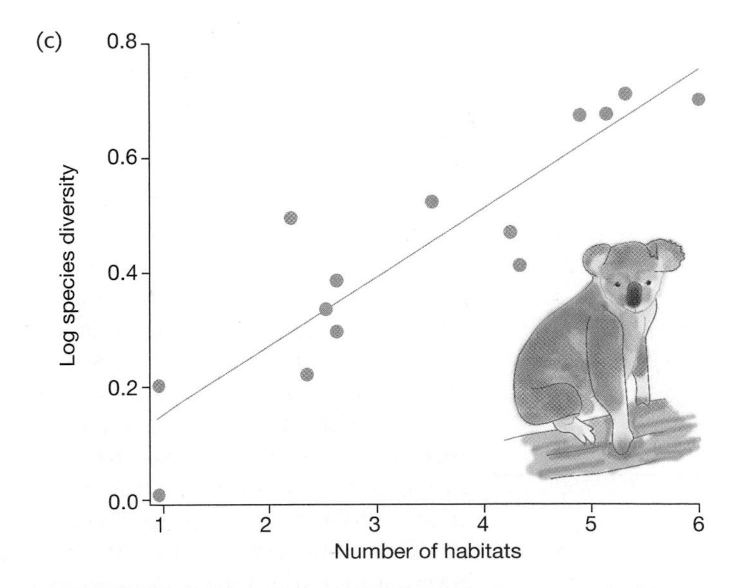

Figure 19.11 In Australia, (a) larger areas have more habitats, and (b) larger areas have more mammal species in the southeast. However, (c) the number of habitats better explains the number of mammalian species, as is seen by the tighter clustering of points around the line. *(Modified from Fox, 1983.)*

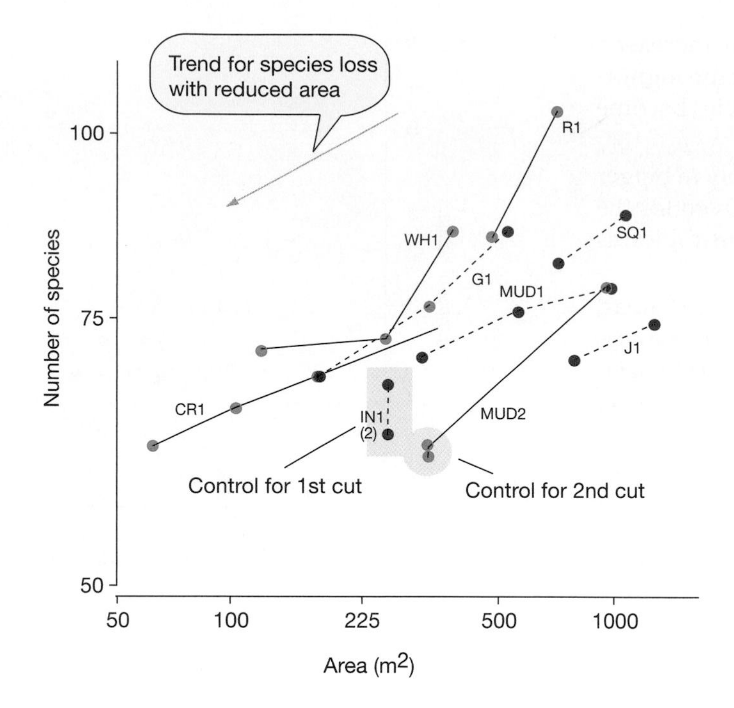

Figure 19.12 Reduction in numbers of species on mangrove islands with area reduced by chain-saw. Note that on four islands the area was reduced more than once. (*After Simberloff, 1978.*)

and elfin woodland at high elevations, where the high altitude produces a dwarf, moss-covered forest. Ricklefs and Lovette used Simpson's index on the five types of vegetation to measure habitat diversity, which they once again showed to have a slightly greater effect on species richness than did area.

The best way to study the effect of area alone on the richness of species is to choose a habitat that doesn't change as you sample bigger islands of it. Dan Simberloff (1976a,b) (a student of E. O. Wilson) did just that. We shall discuss his experiments more fully in the next section; suffice it to say here that Simberloff studied islands of pure mangroves, of varying size in the Florida Keys. The islands consisted of clusters of mangrove trees emerging directly from the water; there was no soil above the water level. Simberloff neatly created his own reduced-area islands by taking a chain-saw and felling trees to make the island smaller. The felled material was hauled away in a barge. One could probably not get a permit to do this sort of experiment today, as mangroves now are legally protected. Simberloff painstakingly crawled on his hands and knees to collect every species that fed on the islands—mainly very small insect species both before and after the chain-sawing. Although the chain-saw treatment did nothing to the quality of habitats on the remaining island area, the reduction in area itself did cause a diminution in the richness of invertebrate species. Simberloff reduced the area of eight islands and left one as a control. Seven months later, after enough time for insects to become reestablished at equilibrium, he censused the islands again. Insect densities had dropped on all eight experimental islands, but not significantly on the control island (Fig. 19.12). He then further reduced the areas of four of the islands (WH1, CR1, MUD1, and G1) and left one island (MUD2) as a control. A year later, a final census revealed that species richness had again dropped on the four islands, but not in MUD2. The results clearly indicated that area affects the number of species.

19.3 The Effect of Distance on Island Immigration

Many studies have shown a decrease in immigration rates of species onto islands as the distance from a source pool increases.

MacArthur and Wilson didn't spend as much time marshaling evidence for the effect of the distance of an island from a source pool as they did for the effect of area on species richness. Their best evidence came from a study of the numbers of land and freshwater bird species on four groups of islands: the Moluccas, Melanesia, Micronesia, and Polynesia. They divided these islands up into three subgroups: those far away (>3,200 km) from a mainland source (New Guinea) those near the source (<800 km), and those of intermediate distance. The groups contained many islands that were the same size and that differed only in their distance from the source pool. MacArthur and Wilson's study showed that the closer

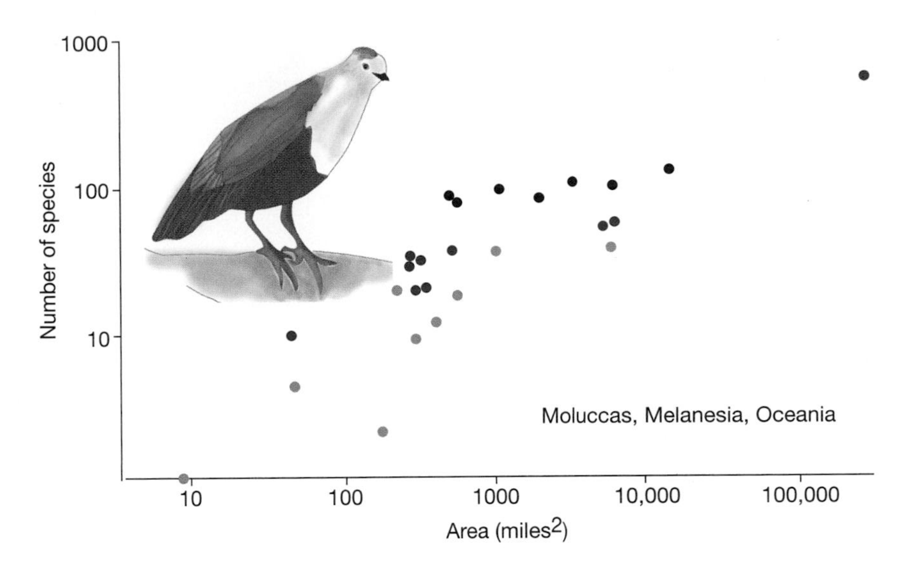

Figure 19.13 The numbers of land and freshwater bird species on various islands in the Moluccas, Melanesia, Micronesia, and Polynesia. Near islands, less than 500 miles from New Guinea, are represented by black dots, far islands (>2,000 miles) by blue dots, and intermediate islands by red dots. The relationship between area and number of species is clear, and there is also a distinct effect of distance, with nearer islands supporting more species (*Reproduced from MacArthur and Wilson, 1963.*)

islands had more species than the intermediate sized islands, which in turn had more species than the distant islands (Fig. 19.13).

Jared Diamond (1972), working out of UCLA in California, expressed the relationship between distance and number of species in a different way. He tabulated the land birds on islands close to the source area of New Guinea (Fig. 19.14) and assumed that these islands had 100 percent of the available birds. He then documented the drop-off in species richness on islands with increasing distance from New Guinea (Fig. 19.15) and expressed the richness of bird species as a proportion of the number of bird species found on New Guinea, a figure he called the *degree of saturation*. Again, there were strong declines in the number of species with increasing distance. Thus, although most of the studies testing the distance–immigration relationship dealt with birds, it does seem that MacArthur and Wilson's prediction of species richness falling off with increasing distance from the source pool is well substantiated.

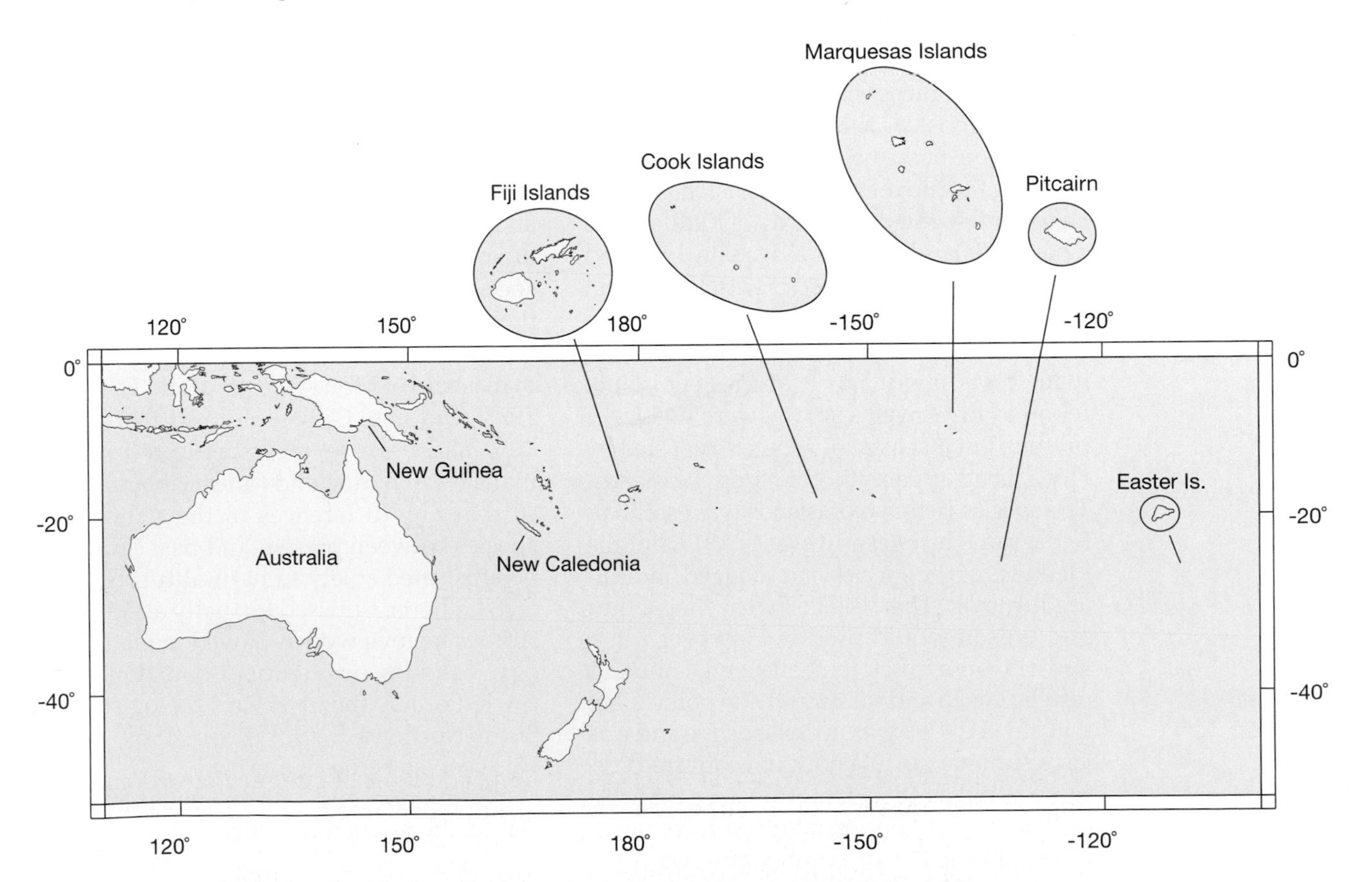

Figure 19.14 Map of Australia, New Guinea, and Pacific Islands, from New Caledonia to Fiji, Cook Islands, Marquesas, Pitcairn, and Easter Island.

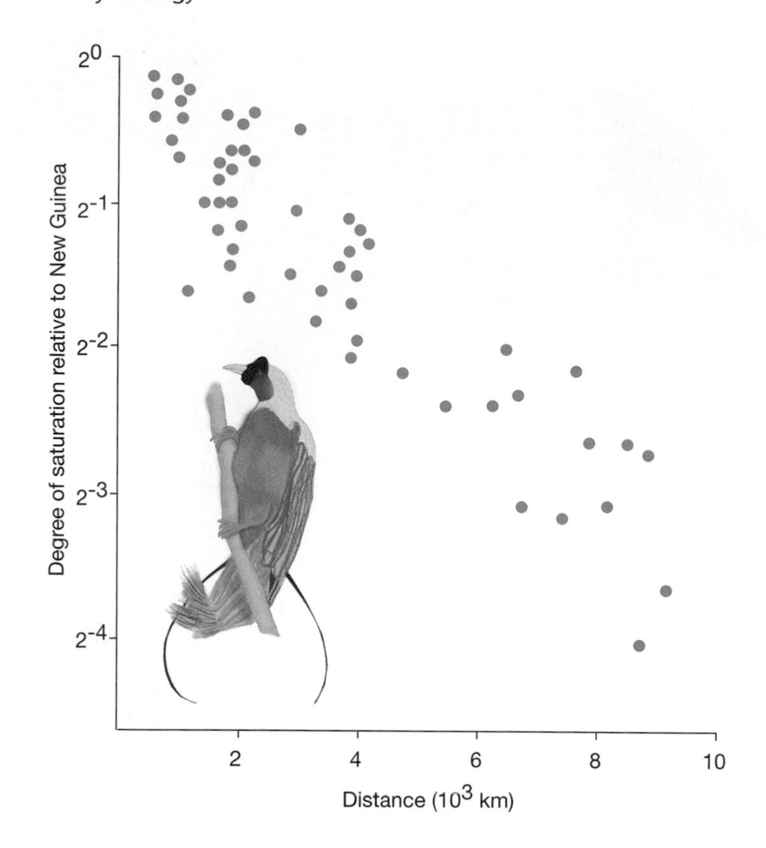

Figure 19.15 The numbers of lowland bird species on islands with increasing distance from the source pool (New Guinea), expressed as the percentage of the number of lowland bird species found on New Guinea. (*After Diamond, 1972.*)

19.4 Species Turnover

There are few estimates of the rates of colonization, extinction, and turnover on islands. Most evidence suggests trivial rates of turnover involving transient species.

Francis Gilbert (1980), a British ecologist at the University of Nottingham, managed to find 25 investigations carried out in order to demonstrate turnover, but as Gilbert went on to show, most of them suffered from fatal flaws. Some of the best-known work includes studies of turnover of birds on islands. Jared Diamond studied the birds of California's Channel Islands National Park in 1968 and compared his lists of species with those of an earlier study published in 1917 by A. B. Howell (Diamond, 1969). Diamond reported that many birds, about 5 to 10 species per island, reported in Howell's survey were no longer present, but just as many species *not* listed by Howell had apparently colonized the islands. The species richness of islands was essentially the same in 1969 as it was in 1917, but the identities of the species had changed, indicating turnover. This was just what MacArthur and Wilson would predict. However, other authors were quick to challenge Diamond's results. Lynch and Johnson (1974) pointed out that Howell's list was not exhaustive and was just a summary of all known breeding records, some as old as 1860. Unfortunately, the actual 1968 lists of species involved have never been published, so one cannot try to verify the accuracy of the comparison.

In many cases, comparing current lists with old lists is about the best we can do. However, because old lists are often generated in an anecdotal fashion, without regard to scientific rigor, it is not easy to accept them at face value. In fact, Diamond went on to carry out similar investigations on islands near New Guinea, but once again, the earlier surveys included no extensive field notes, and Diamond was forced to compare his lists of species with earlier ones. However, if a species had been seen, but not collected, then the comparisons would be invalid. A similar study was done on birds of the West Indian island of Mona (Terborgh and Faaborgh 1973), but here Gilbert showed that the island had been severely disrupted by humans with mining and forestry plantations, so that any differences in the richness of species between present and past could not be attributed solely to natural turnover. In fact, Gilbert dismissed virtually all 25 of the studies he reviewed as flawed in methodology, statistics, or quality of data. Only one carried out by then Harvard biologists Dan Simberloff and E. O. Wilson (1969, 1970), was seen as being of merit.

Simberloff and Wilson censused small (11 to 25 m in diameter) red mangrove (*Rhizophora mangle*) islands in the Florida Keys for all terrestrial arthropods. They then

Figure 19.16 Experimental defaunation of a mangrove islet in the Florida Keys. (a) Construction of a scaffold frame. (b) Installation of a large tent into which insecticide was introduced, killing all arthropods. Commercial pest control operators from Miami were hired to perform the fumigation. The tent and scaffold were removed after defaunation, and recolonization was monitored. *(Photos by Daniel Simberloff, Florida State University.)*

fumigated what they dubbed their "experimental islands" with methyl bromide to kill all arthropods on them (Fig. 19.16). Periodically thereafter, they censused all the islands for several years. After 250 days, most of the islands had similar numbers of arthropod species as they had to begin with (Fig. 19.17), just as MacArthur–Wilson theory suggested. Simberloff and Wilson observed colonization and extinction, and the data did indicate that colonization rates during the first 150 days were higher on nearer islands than on far islands—again, as MacArthur–Wilson theory predicted. However, calculated rates of turnover were found to be very low, 1.5 extinctions per year. Thus, Simberloff (1976b) later suggested that the data afforded only weak support for the turnover idea of MacArthur–Wilson theory and that very few other studies seemed to support the theory either. When Simberloff looked again at his earliest data, he found that most of the same species returned to the islands he had fumigated. This indicates the existence of biological processes that shape the final community structure the same way every time an island is recolonized, a finding that is contrary to the theory of island biogeography, which treats

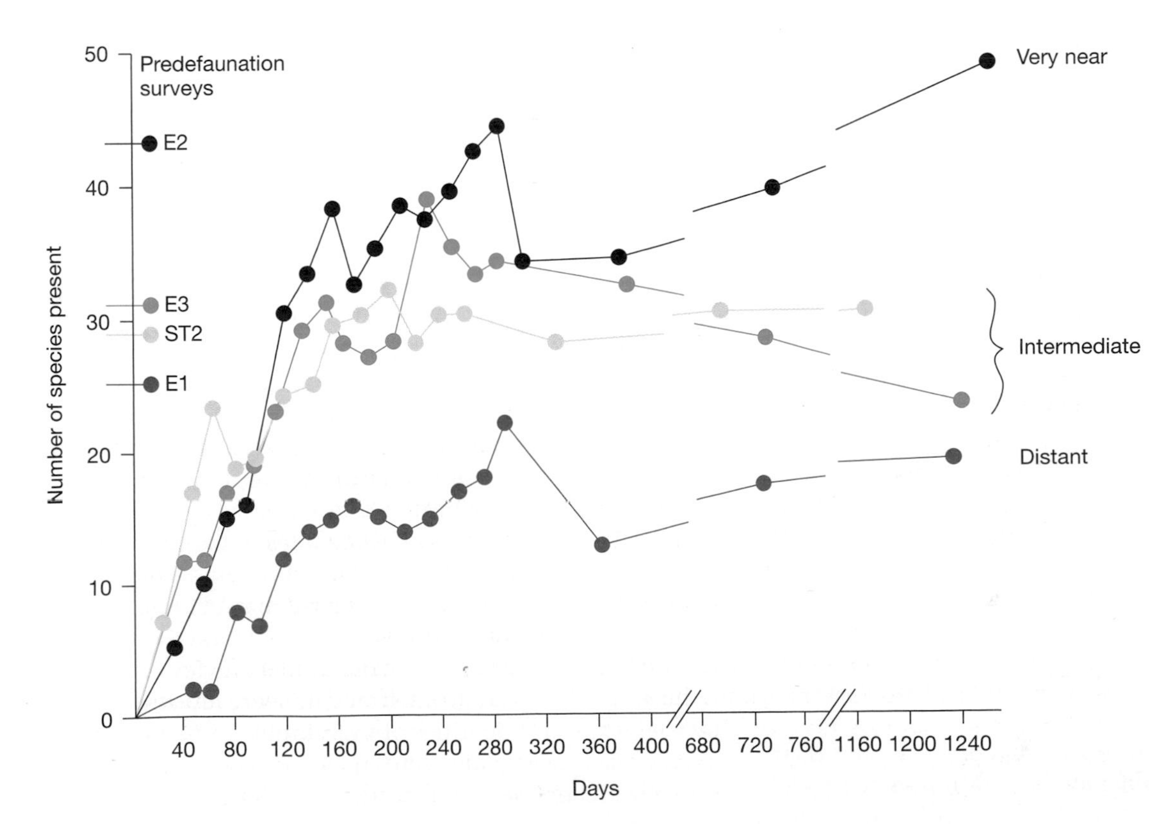

Figure 19.17 Colonization curves for arthropods on mangrove islands near Sugarloaf Key, south Florida, after defaunation by fumigation. Predefaunation species numbers are on the *x*-axis. Descriptions refer to distances of islands from mainland. *(Redrawn from Simberloff, 1978.)*

the dynamics of different colonizing species as equivalent, with community properties essentially unimportant. As Simberloff (1978b) concluded, turnover probably involves only a subset of transient or unimportant species, with the more important species becoming permanent after colonization.

And what of more recent studies? Jorge Rey (1981, 1983), a student at Florida State University with Donald Strong, defaunated islands of pure salt marsh cordgrass, *Spartina alterniflora*, in north Florida. Turnover was estimated at about 0.14 species per island per week, but rates of colonization and extinction could not be shown to be related to the distances of islands from a source pool. In addition, later extensive and detailed studies of herbivory on *Spartina* revealed only about 30 herbivores and their parasites associated with the plant (Stiling, Brodbeck, and Strong 1982; Stiling and Strong 1983), indicating that some of Rey's 56 recorded species may have been transients and not subject to the MacArthur–Wilson theory. Transients are species that use islands only as a resting place while they are passing through an area; the MacArthur–Wilson theory deals only with resident species that feed and reproduce on the island.

Abbott and Black (1980) measured the known extinctions and immigrations of plants on 40 islands off the coast of western Australia. The islands were censused sporadically (one to four times between 1956 and 1978). Extinctions and immigrations tended to match each other, but again, turnover was low, with most islands showing turnover rates of between zero and two species per census (<0.1 species per year). Morrison (1997) found low turnover on 77 cays in the Bahamas, which he surveyed annually over a four-year period. Most of the observed turnover (usually less than 1 percent per year or less than one species per year) was due to immigrants that never became established. The take-home message from most of these studies is that recorded rates of turnover are low, which gives little support to this part of the MacArthur–Wilson theory.

In sum, MacArthur and Wilson's theories on island biogeography have stimulated much research. There do seem to be strong effects of area and distance on species richness. However, turnover appears low, which suggests that succession on most islands is a fairly deterministic process.

Finally, island biogeography theory is not just an academic issue: It is thought to be useful in the continuing debate over the design of wildlife preserves, particularly in the question of whether planners should design many small preserves or a few large ones, given the caveat that the total area preserved would be the same. Island biogeography theory suggests that fewer extinctions will occur in larger parks, so we should strive to make our parks as large as possible, but the species–habitat relationship suggests that maximizing habitat diversity is more important (see Applied Ecology Box: The Theory of National Park Design).

Applied Ecology

The Theory of National Park Design

Since the early 1970s, there has been much discussion about the shape, design, and management of nature reserves. The debate on reserve design has centered on island biogeographic theory, which suggests that large parks hold more species than small parks. The International Union for Conservation of Nature and Natural Resources (IUCN), a major worldwide conservation group, stated that refuge design criteria and management practice should be in accord with the equilibrium theory of island biogeography (IUCN, 1980). However, our current knowledge of theory suggests that this recommendation may be on shaky ground.

It is widely agreed that the ideal strategy would be to have lots of large refuges. Of course, large refuges cost a lot of money, so many countries have to settle for preserving small areas of land rather than large ones. The debate then becomes, should single large areas or, instead, several small ones be preserved? (This is the so-called SLOSS debate, for "single large or several small.") Single large preserves may buffer populations against extinctions, because they contain large populations of species. However, many studies suggest that multiple small sites contain more species, because a series of small sites is likely to have a broader variety of habitats contained within it than is one large site. Hence, in practical terms, it may actually be better to have several small sites.

Jim Quinn and Susan Harrison (1988) reviewed species richness for 15 island groups; all data were reported island by island species identities of individuals were confirmed, and at least six islands were included in each survey the two researchers took. The resultant species–area relationships for vertebrates, land plants, and insects showed that groups of small islands generally harbor more species than do comparable areas composed of one or a few large islands. Also, faunas were shown to be richer in collections of small national parks than in the larger parks. Apparently, having more types of habitat outweighs the importance of area, so several small preserves seem to be better than one large preserve in maintaining species diversity. This finding has an important bearing on future land purchases for conservation purposes.

Summary

The theory of island biogeography suggests that species richness on islands is affected primarily by the size and distance of an island from a source pool. Once equilibrium in species richness is reached, species will come and go, but richness will remain the same.

1. Island biogeography theory predicts that the equilibrium number $\hat{S}$ for species on an island is determined by a balance between immigration of species onto that island and extinction of species already there.

2. The theory suggests that the number of species, $\hat{S}$, is determined by an island's size and position relative to a source pool of colonists. Extinction should increase on small islands, because of their smaller populations, and immigration should decrease on far islands, because colonists have a difficult time reaching distant places. Indeed, many studies show that species richness does increase with island size and that distance does affect colonization rates of islands.

3. Island biogeography theory also suggests that there is much turnover on islands as new species arrive and old ones become extinct. There is little evidence, however, to support this prediction. Most turnover that has been documented suggests that rates of turnover are low and center mainly on transient species.

4. Island biogeography theory may be applied to "habitat islands" as well as real islands. In the relationship between species richness and area, the slope of the line may be steeper for true islands than habitat islands and steeper for poor dispersers like mammals than for good dispersers like birds.

Island biogeography theory has been useful in stimulating empirical studies to address whether area and distance affect species richness. Succession on islands, whether real or continental, such as on Mt. St. Helens after it erupted in 1980, is better understood with reference to distance and area effects.

Discussion Questions

1. How can island biogeography be important for conservation biologists? Should we place reserves close together to facilitate colonization among them? Should we connect reserves via corridors?

2. How does logging national forests, particularly in the western United States and Canada, affect species richness in the fragments of forests that remain? How would you design a logging policy if you were a forester?

3. The MacArthur–Wilson theory suggests a linear relationship between area and species richness. Can you see any limit to the relationship?

4. Preston (1962) gave the following data for island size and richness of bird species in the East Indies:

Island	No. of Bird Species	Area (square miles)
New Guinea	540	312,000
Borneo	420	290,000
Philippines	368	144,000
Celebes	220	70,000
Java	337	48,000
Ceylon (now Sri Lanka)	232	25,000
Palawan	111	4,500
Flores	143	8,870
Timar	137	18,000
Sumba	108	4,600

Construct a species area curve for these data. Which island is richer in bird species than it should be, based on area alone, and which is depauperate (has fewer species than expected)? Why does Ceylon (now known as Sri Lanka) have more bird species than expected?

5. The following is a map of the Galápagos Islands and a list of their areas and numbers of plant species. Construct a species–area curve. Pinzon and Santa Fe have about the same area, but different numbers of species. Why is this?

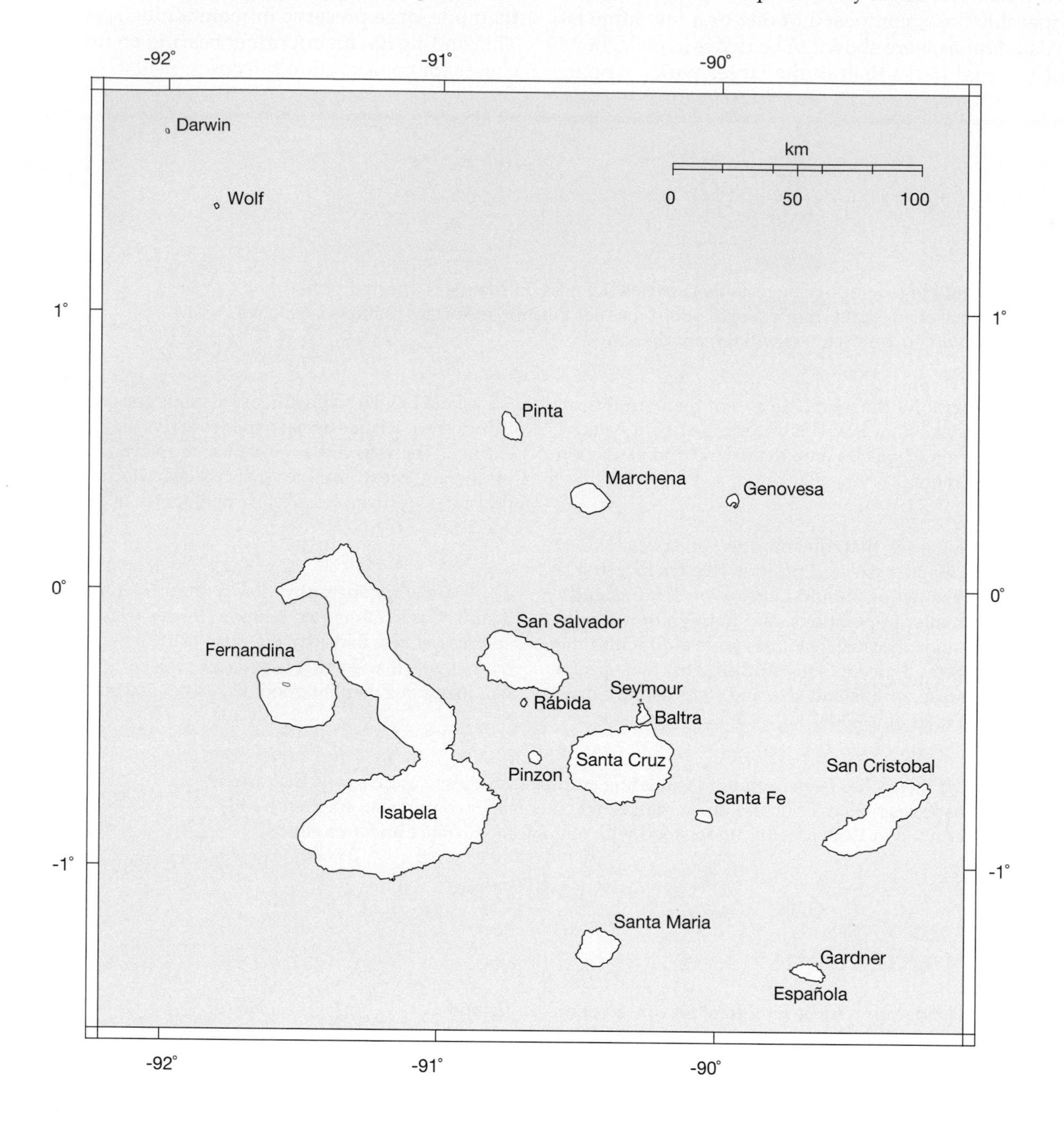

No.	Name	Area (Sq. Mile)	Species	No.	Name	Area (Sq. Mile)	Species
1	Isabela	2,249	325	10	Seymour	1	52
2	Santa Maria	64	319	11	Santa Fe	7.5	48
3	San Cristobal	195	306	12	Gardner	0.2	48
4	San Salvador	203	224	13	Marchena	45	47
5	Santa Cruz	389	193	14	Rábida	1.9	42
6	Pinta	20	119	15	Genovesa	4.4	22
7	Pinzon	7.1	103	16	Wolf	1.8	14
8	Fernandina	245	80	17	Darwin	0.9	7
9	Española	18	79				

SECTION V

Ecosystems Ecology

Three links in the food chain: Seven-spot ladybird larva eating blackbean aphids feeding on a plant leaf. How much energy is passed from one link to another? How many links are there in food chains? Ecosystems ecology focuses on the flow of energy and nutrients through food chains and attempts to answer questions like this. *(Nigel Cattlin, Photo Researchers, Inc.)*

CHAPTER 20

Trophic Structure

Road Map

1. Food webs connect all species that feed on one another in an ecosystem.
2. Generally, the connectance remains constant between different food webs, but the mean chain length increases as the number of species in the web increases.
3. Problems such as imperfectly known links between species, the taxonomy of species, and the amount of prey eaten by each species can profoundly affect food web patterns.
4. Ecological guilds are groups of species that feed on the same resources in the same way, and the species involved are likely to be competing for resources.
5. Keystone species have an effect on an ecosystem out of all proportion to their biomass.

In 1935, the British plant ecologist A. G. Tansley coined the term *ecosystem*, which he took to include not only the biotic community of organisms in an area, but also the abiotic environment around that community. Ecosystem ecology concerns the movement of energy and materials through communities, and like the concept of a community, the ecosystem concept can be applied at any scale: A drop of water inhabited by protozoa is an ecosystem, and a lake and its biota are another. Most ecosystems can never really be regarded as having definite boundaries. Even in a clearly defined pond ecosystem (Fig. 20.1), waterfowl may be moving in and out. The big advantage of studying ecosystem ecology is the common currency of energy or nutrients, which allows the biology of communities and populations to be compared between and within trophic levels.

20.1 Food Web Complexity

Many ecosystems can be considered to be large food webs.

Few ecosystems are so simple as to be characterized by a single unbranched **food chain** in which one plant species is eaten by one herbivore species that is in turn fed upon by one species of carnivore. Usually, many herbivore species feed on the same plant species. For example, one may find tens to hundreds of insect species, as well as several vertebrate grazers, all feeding on one species of tree. Also, many species of herbivores eat several different plant species, not just a single one. Such branching of food chains occurs at other trophic levels as well. For instance, frogs eat many different types of insect, and owls may eat herbivores such as field mice and also predatory organisms like snakes. It is more correct, then, to draw relationships between these plants and animals not as a simple chain, but as a more elaborate interwoven food web, as illustrated in Fig. 20.2.

The first measure of the relative complexity of a food web can be described by a measure known as the *chain length*, which denotes the average number of links between trophic levels. So, if there are three trophic levels—plant, herbivore, and carnivore—there are two links between the three species, and the chain length is two. The second measure is ***connectance***:

$$\text{connectance} = \frac{\text{actual number of links}}{\text{potential number of links}}.$$

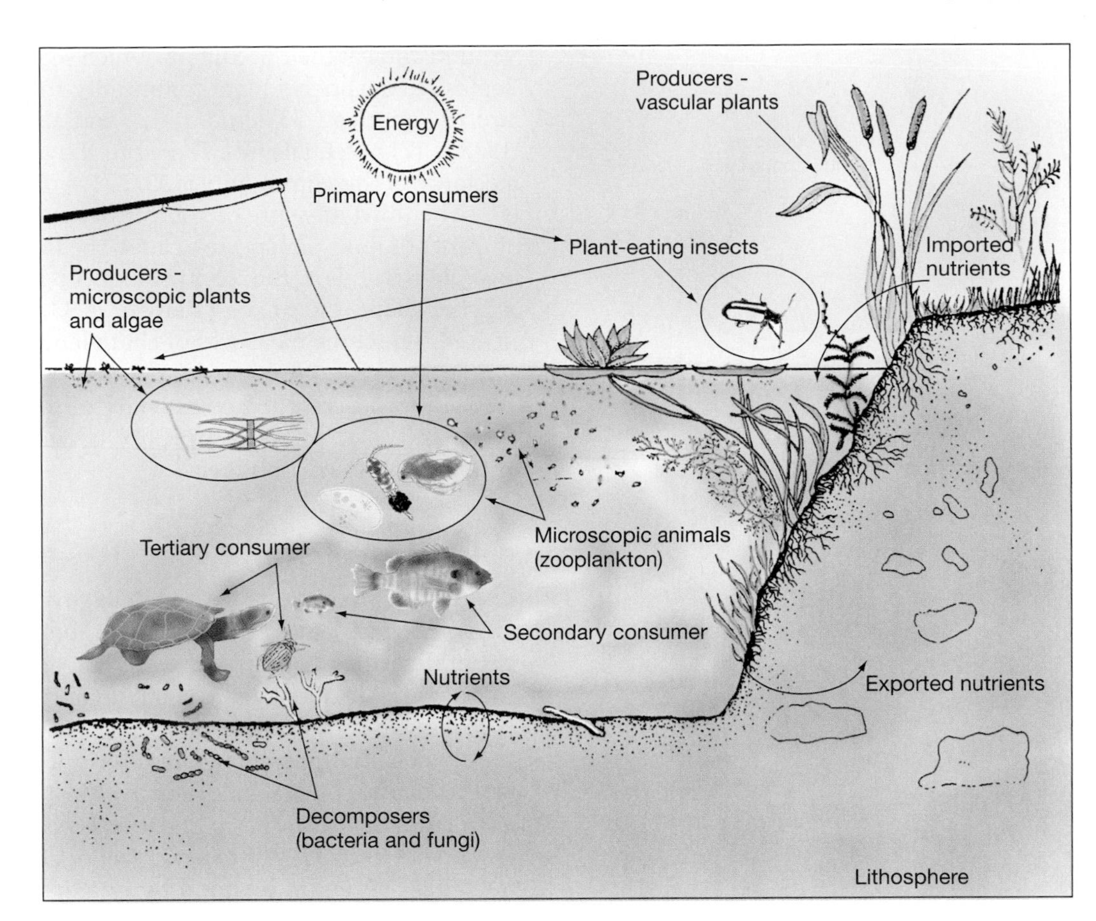

Figure 20.1 Food web in a pond ecosystem. Each trophic level is occupied by a different species. Microscopic plants or rooted plants are the primary producers. Microscopic animals (zooplankton) or insects are the primary consumers or herbivores. Small fish feed on the insects and zooplankton and can be considered secondary consumers. Finally, larger fish or, in this case turtles, feed on the smaller fish and constitute the tertiary consumers.

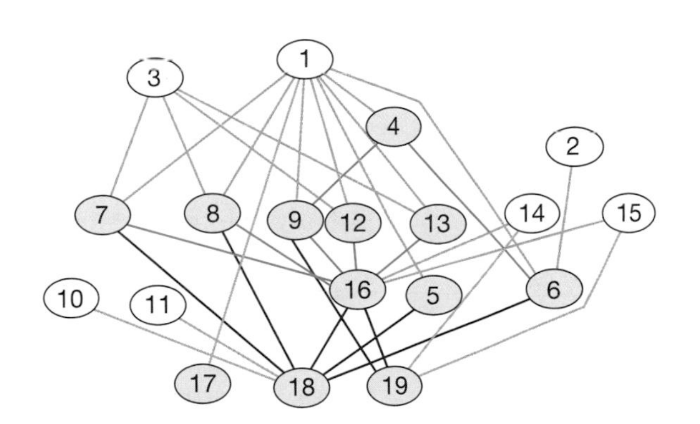

Key: (1) *Misumenops nepenthicola*; (2) *Encyrtid* (near *Trachinaephagus*); (3) *Toxorhynchies klossi*; (4) *Lestodiplosis syringopais; (5) Megaselia sp. (? nepenthina); (6) Endonepenthia schuitemakeri; (7) Triperoides tenax; (8) T. bambusa; (9) Dasyhelea nepenthicola; (10) Nepenthosyrphus sp.; (11) Pierretia urceola; (12) Culex curtipalpis; (13) C. lucaris; (14) Anotidae* sp. 1; (15) *Anotidae* sp. 2; (16) bacteria and protozoa; (17) live insects; (18) recently drowned insects; (19) older organic debris.

Number of top predators is 7: species 1, 2, 3, 10, 11, 14, 15.

Number of basal species is 3: species 17, 18, 19.

Number of intermediate species is 9: the rest.

Number of linkages:

intermediate to top 15;
basal to top 5;
intermediate to intermediate 7;
basal to intermediate 7.

(a)

(b)

Figure 20.2 (a) A simple food web of the insects in the pitcher plant *Nepenthes albomarginata* in West Malaysia. Each line represents a trophic linkage; predators are higher in the figure than their prey. (*After Beaver, 1985.*) (b) Pitcher plant, *Nepenthes macfarlanei*, in Montane Forest at Bintang, Malaysia. Pitcher plants like this have their own self-contained ecosystems, like those represented in (a), which lend themselves well to food web analysis. (*Fletcher & Baylis, Photo Researchers, Inc.*)

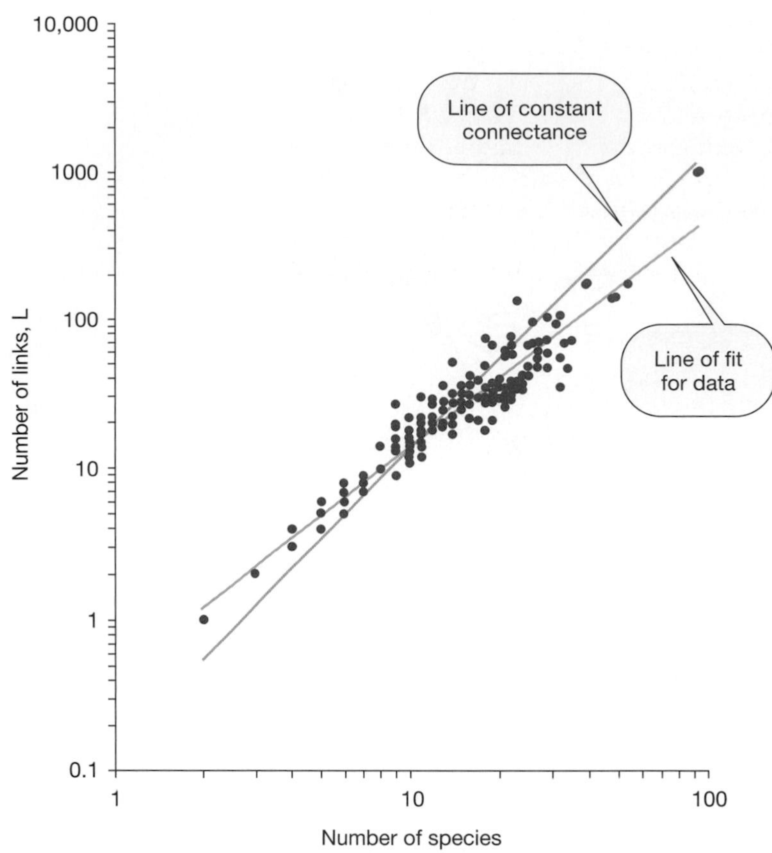

Figure 20.3 Food web connectance. The gray line gives the line of constant connectance such that the number of links in a food web goes up uniformly with the number of species in the web. The red line gives the line of best fit for data from 175 actual food webs. Note that the scale on both axes is logarithmic. *(Modified from Martinez, 1992.)*

Patterns in food webs focus on connectance values, chain lengths, and linkage densities.

In an ecosystem of n species, the number of potential links (assuming that links cannot go in both directions) is

$$N = \frac{n(n-1)}{2}.$$

This means that each species can eat any other species, but "prey species" cannot turn around and eat their predators, which seems logical. In the pitcher plant community, there are 19 "species," 33 actual links, and $19 \times 18/2 = 171$ potential links. Therefore, the connectance of this community is $33/171 = 0.19$.

The third measure of complexity is the number of links per species, called the ***linkage density***. For the community in the pitcher plant, the linkage density is $33/19 = 1.74$. The three measures of chain length, connectance, and linkage density are widely used to describe the complexity of food webs and to compare complexity between different types of food webs.

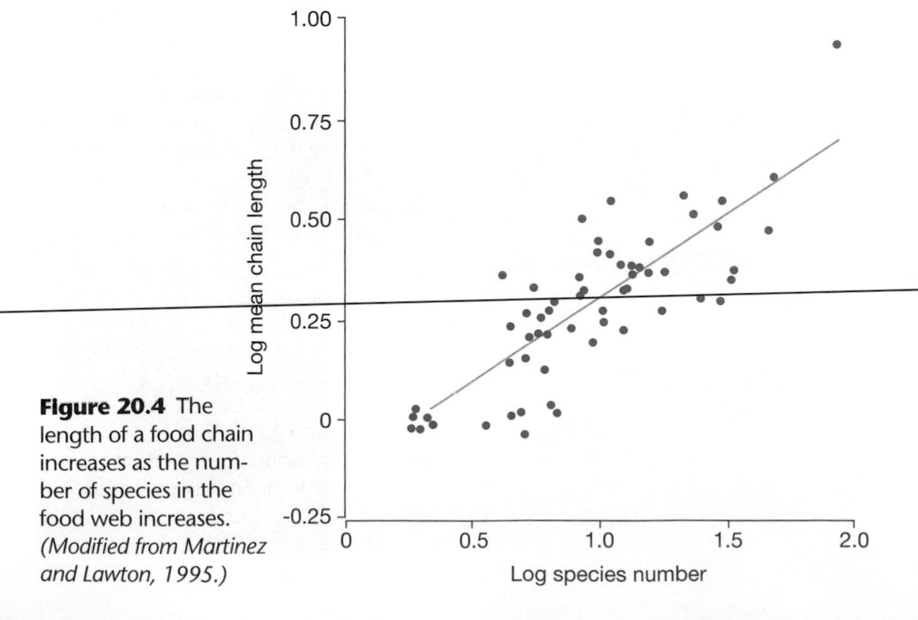

Figure 20.4 The length of a food chain increases as the number of species in the food web increases. *(Modified from Martinez and Lawton, 1995.)*

20.2 Food Web Patterns

Using data on connectance and linkage density, ecologists have put forth a number of generalizations about food webs. Most of these generalizations have been developed from an analysis of published works on food webs. Perhaps the most important assertion is that connectance remains constant as the number of species in the food web increases (Fig. 20.3). A constant level of connectance can be understood by imagining an insectivorous bird that feeds in two communities, A and B, with A having twice the number of insect species that B does. Neo Martinez, of San Francisco State University (1992), argued that it is then reasonable to suppose that the bird would eat twice as many species of insect in community A as in community B. Because this conclusion would likely apply to all species in the community, the connectance would remain unchanged as the species richness increased. Measures of connectance from 175 real food webs support this reasoning.

Another important pattern is that the mean chain length increases as the number of species in the web increases (Fig. 20.4). Although this relationship is intuitively obvious, it probably has an upper limit, so that the chain length reaches a maximum; we just don't know where this maximum is yet.

Finally, a long-known pattern is that top predators tend to be rather large and sparsely distributed, whereas herbivores are smaller and more common. This generalization is often termed the **pyramid of numbers** or, sometimes, *Elton's pyramid*, in deference to British ecologist Charles Elton, who first described it in 1927. Elton's example was that of a small pond, in which the numbers of protozoa may run into the millions and those of *Daphnia* and *Cyclops* (their predators) into the hundreds of thousands, whereas there will be possibly 10,000 beetle larvae and 100 fish (Fig. 20.5a). One can think of several exceptions to this pyramid. An oak tree, one producer, supports thousands of herbivorous beetles, caterpillars, and other primary consumers, which in turn support tens of thousands of predators and parasites (Fig. 20.5b). The best way to reconcile this apparent exception to Elton's pyramid is to weigh the organisms in each trophic level. The oak tree weighs 30,000 kg, all the herbivores on the tree total 5 kg, and the predators sum about 100 g. Looking at the mass, rather than numbers of organisms, at each trophic level shows that the oak tree is not a real exception (Fig. 20.5c). Nonetheless, inverted pyramids can still occur even when biomass is used as the

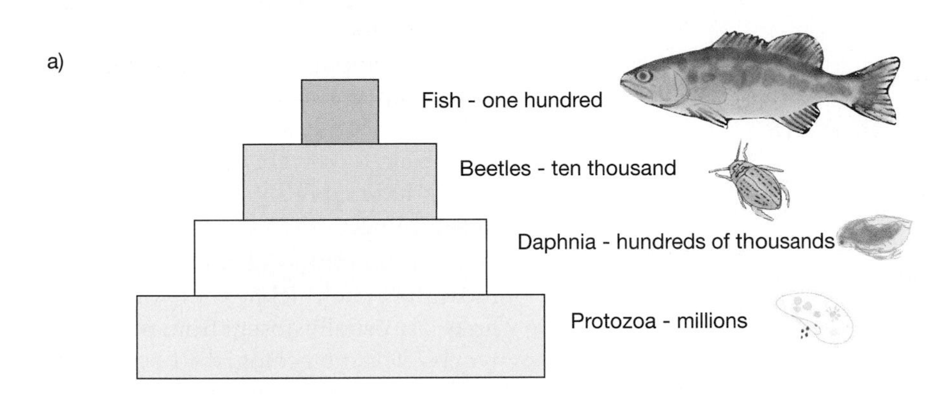

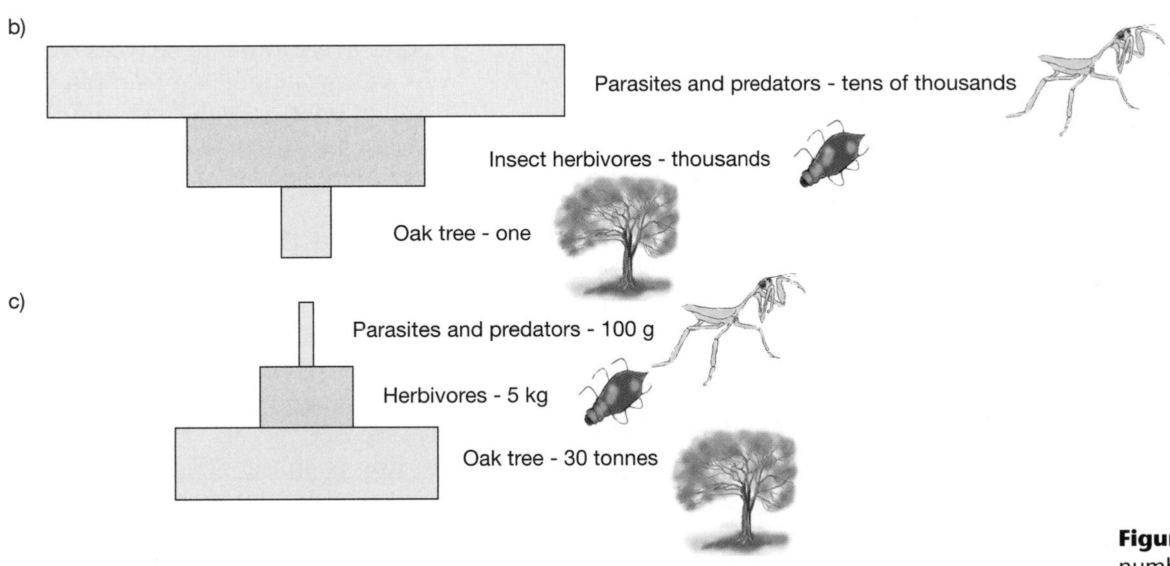

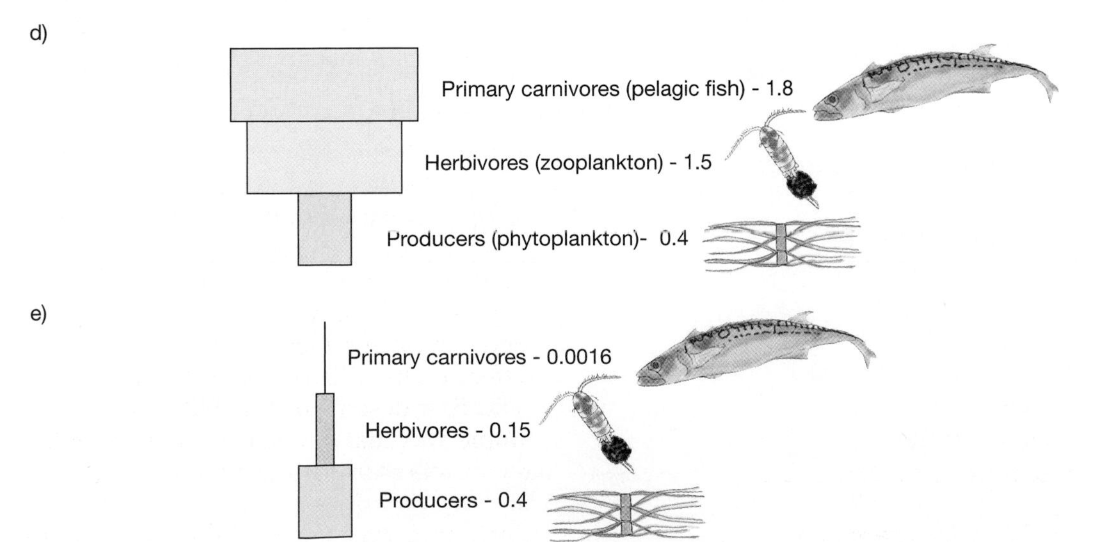

Figure 20.5 Pyramids of numbers. (a) Abundance of species in a typical British pond decreases with increasing trophic level. (b) Inverted pyramid of numbers, based on organisms living on an oak tree. This pyramid becomes right-side up (c) when biomass instead of abundance is used as a measure. (d) An inverted pyramid of biomass in the English Channel (g dry wt. m^{-2}) (e) Production rates of organisms in the English Channel (g dry wt. m^{-2} day^{-1}). Energy pyramids never become inverted. *(d and e from data in Harvey, 1950)*

measure: In the English Channel, the biomass of phytoplankton supports a higher biomass of zooplankton (Fig. 20.5d). This is possible because the production rate of phytoplankton is much higher than that of zooplankton, and the small phytoplankton standing crop (the biomass at any one point in time) processes large amounts of energy (Fig. 20.5e). The most realistic pyramid is thus the energy pyramid, which never becomes inverted.

20.3 Problems with Food Web Patterns

Problems with food web patterns focus on imperfectly known numbers and strengths of linkages.

While food web theory has undoubtedly led to an increase in the understanding of ecosystems, many authors have pointed out numerous problems with the data that may invalidate the observed patterns:

1. Predation on "minor" species is frequently omitted from the literature. Often, authors of published works on food webs are not sufficiently knowledgeable about organisms outside of their areas of specialty, so that some links in the web are never drawn. Gary Polis of Vanderbilt University in Tennessee (1991) argued that food webs in the real world are much more complex than is reported in the food web literature. In a relatively simple desert system in the Coachella Valley, California, Polis conducted an intensive 10-year study of the local ecosystem and found 174 species of plants, 138 vertebrate species, 55 species of spiders and scorpions, an estimated 2,000 to 3,000 species of insects, and an unknown number of microorganisms and nematodes, all in a single food web. Polis compared his food web with others reported in the literature, using statistics compiled by Joel Cohen and colleagues (1990; see Table 20.1). Chains were longer, omnivory was common, and linkage density was much greater in the intensively studied Coachella food web. Goldwasser and Roughgarden (1997) suggested that most food webs appearing in published works are too poorly sampled to be useful in developing theory. By using computer simulations, the authors were able to randomly sample a fully known Caribbean food web. Food web properties varied markedly with the intensity of sampling.
2. Data on quantities of food consumed, indicating the thickness of connecting links, are usually absent from published works. Marine ecologist Robert Paine (1992) provided the first example in which interaction strengths were calculated in a food web. Most of the connections were found to be weak, so the web was essentially very simple. This objection is virtually the opposite of the problem just discussed—there may be many more links than we realize, but most of them are unimportant. Benke and Wallace (1997) also showed how misleading a normal food web was for a riverine system in Alabama, because it implied the equivalence of all food resources. An energetic analysis that looked at ingestion rates of food actually showed great variation in the strengths of linkages (Fig. 20.6). The different ingestion rates argue for some

TABLE 20.1 Comparison of statistics from the Coachella Valley food web studied by Gary Polis with means of the statistics from food web summaries in Cohen, Briand, and Newman (1990). (*After Polis, 1991.*)

	Food Web Summaries	Coachella
Total number of links per web	31	289
Linkage density	2.0	9.6
Number of prey per predator	2.5	10.7
Number of predators per prey	3.2	9.6
Mean chain length	2.71	7.34
Basal species (%)	19	10
Intermediate species (%)	52.5	90
Top predators (%)	28.5	0
Omnivorous (%)	27	78

kind of weighting factor to be used in any analysis of connectivity. However, we cannot simply dismiss all weak interactions, because a superabundant species exhibiting numerous weak interactions could be important in food web dynamics.

3. In food web theory, species are often aggregated into "trophic species" (e.g., insect larvae or detritus). We see this in both food webs illustrated so far (Figs. 20.2 and 20.6). For example, various types of insects are often likely to be lumped together because of the difficulties of identifying them all. This grouping, however, disguises much important biology. Aggregation is rife in many published works on webs. A test of whether the practice affects web properties is provided by taking completely known webs, aggregating them, and comparing patterns. For instance, in Fig. 20.2, there are three pairs of biological species (10 and 11, 12 and 13, and 14 and 15) that have the same sets of predators and prey, and therefore the pairs could each be joined, resulting in three trophic species. The web could be further aggregated by joining species 7, 8, 12, and 13, which share similar species of predators and prey. The aggregation within food webs in this case affects the connectance and linkage density only slightly. In contrast, aggregation strongly affects connectance values in the marine intertidal zone (Paine, 1988).
4. Data on the importance of chemical nutrients are sparse. One apparently feeble link may be very important if it supplies a limiting chemical. (See Tilman, 1982; the topic is discussed further in Chapter 22.)
5. It is hard to define web boundaries. Some highly mobile species, such as predatory gulls, may be very important predators in food webs, but are almost always underrepresented in them.
6. Food web links may obscure positive effects of higher trophic levels on their food species. If pollinators are considered merely in the context of trophic interactions, then they are seen to consume energy that might otherwise have been used for other functions, negatively affecting the plants they pollinate. Obviously, that conclusion is misleading: Pollinators have a net positive effect on their hosts. The same argument could be made for seed dispersers or other mutualists. Therefore, links within the trophic network can portray a direct effect that does not accurately represent the real world.

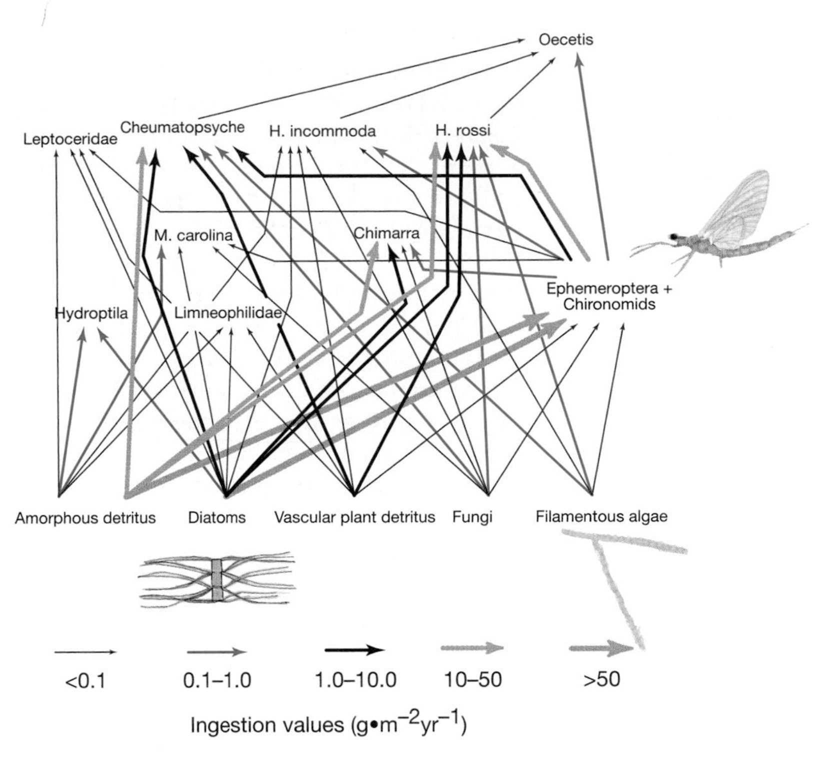

Figure 20.6 Quantitative food web in an Alabama stream ecosystem. Line thickness indicates the amount of energy transferred in that link. (*After Benke and Wallace, 1997.*)

There are obviously many disadvantages, as well as many advantages, to looking at communities in terms of their food webs. Perhaps we could gain more realism and more insights by looking at finer subsets of organisms within food webs—sets of organisms that often feed in the same way and that may be expected to interact closely. Such sets of species are known as *guilds*.

20.4 Guilds

A guild is a group of functionally similar organisms within a trophic level.

In the history of human society, craft guilds consisted of groups of craftsmen who helped each other. There were guilds of goldsmiths, wheelwrights, and other skilled artisans, each of whom made a living in the same way. In nature, the term **guild** seems to have departed subtly from this original meaning because, although guild members feed in the same way, a guild often connotes groups of potentially competing species, not cooperating ones. The term was originally coined by Cornell Professor Dick Root

(1967) in his studies of birds and was meant to describe a group of species that fed on the same resources and in the same way. For instance, in an insect community feeding on a plant, we may have the leaf-chewing guild, the sap-sucking guild, leaf miners, stem borers, stem gallers, root feeders, and flower feeders. There are five species that feed by boring into the stems of salt-marsh cordgrass, *Spartina alterniflora* (Fig. 20.7).

Why should we be interested in guilds? There are two main reasons. First, guilds might represent arenas of the most intense competition. Attention is focused on all competing species, regardless of their taxonomic relationship. For example, competition in a seed-eating guild may involve rodents, birds, beetles, and ants. Thus, for completeness' sake, investigations on competition should involve all guild members. This forces researchers to cross taxonomic lines. Second, and perhaps more important, guilds might represent the basic building blocks of communities. Eric Pianka of the University of Texas (1988) envisioned a periodic table of guilds, analogous to the periodic table of elements, with entries such as insectivores, granivores, and so on. Presumably, one might be able to predict interactions between guilds in much the same way as we can predict interactions between elements in different families of the periodic table, although ecologists haven't built up to this level of sophistication yet.

The main problem in guild theory is accurately defining the guild. How much overlap in diet does there have to be for species to belong to the same guild? For the *Spartina* stem-boring guild illustrated in Fig. 20.7, there is no problem: All species feed on the inside of the stem. However, in his original definition, Root (1967) omitted some species of birds from the foliage-gleaning guild because they took prey from

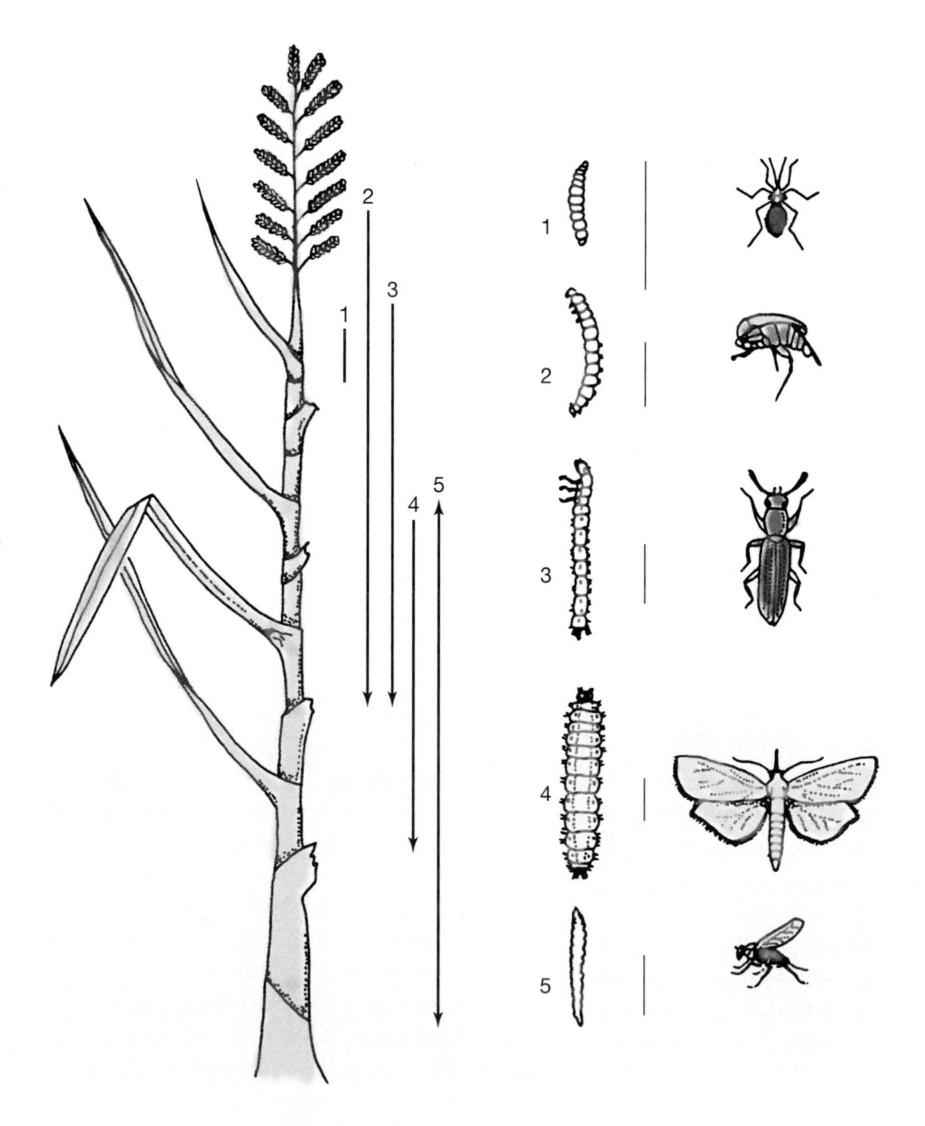

Figure 20.7 The stem-boring guild associated with salt-marsh cordgrass (*Spartina alterniflora*) on the Gulf Coast of North America. (1) *Calamomyia alterniflorae* (Diptera). (2) *Mordellistena splendens* (Coleoptera). (3) *Languria taedata* (Coleoptera). (4) *Chilo plejadellus* (Lepidoptera). (5) *Thrypticus violaceus* (Diptera). Arrows indicate where in the stem the larva of each species is found and the direction it bores. Adults, shown at extreme right, are free living. The length of each scale line between the larva and adult of each species represents 0.5 cm in the drawing.

foliage only occasionally. Similarly, a large raptor may overlap with a medium-sized raptor in the prey items it takes, and the medium-sized raptor may overlap with a small raptor in prey taken, but the large and small raptors may not overlap in their prey. Are all three raptors part of the same guild?

Jaksic and colleagues (Jaksic, 1981; Jaksic and Delibes, 1987) suggested a 50-percent overlap in diet as the minimum value for association in a guild. However, is the overlap 50 percent by types of prey or 50 percent by prey weight? For example, a red fox is likely to eat many species of insects, making up 90 percent of its types of prey, but the other 10 percent would be mammals—especially rabbits—which constitute 90 percent of the prey by weight. So is the fox a member of the insectivore guild or not? The answer depends on which resources one selects for analysis: The fox is an insectivore if you base diet overlap on species richness, but not if you base it on biomass of prey.

Patterns found in guild analysis.

Despite the difficulties associated with assigning guilds, some authors have suggested that guild structure may be more predictable and stable than either the abundance of individual species in a community or species composition. Density compensation within guilds could maintain overall guild abundance at or near carrying capacity, while the fortunes of different species within a guild vary individually in response to factors other than the availability of resources, such as weather or predators. For example, in the *Spartina* stem-boring guild, parasites could decrease the abundance of the beetle *Mordellistena*, but the occupancy of stems would stay the same if the frequency of the other species increased to compensate for this loss. What evidence is there that density compensation occurs in nature?

Recall from Chapter 15 the studies of John Lawton and colleagues (1984, 1993) of the herbivore community on bracken fern in England, Brazil, South Africa, New Mexico, the United States, Borneo, and Hawaii. Wherever insects were common, Lawton arranged them into four different guilds—chewers, suckers, miners, and gallers—and further divided them according to where they fed on the plant—the main stem, the leaves, or the leaf veins. Chewers were mainly caterpillars; suckers were species like aphids, which tapped into the phloem or xylem; miners were very small, flattened species that "mined" the tissue between the surfaces of the leaf; and gallers were mainly wasps and flies that induce tumor-like growths on plants, inside of which the wasp or fly larvae feed (Fig. 20.8). On bracken fern,

(a)

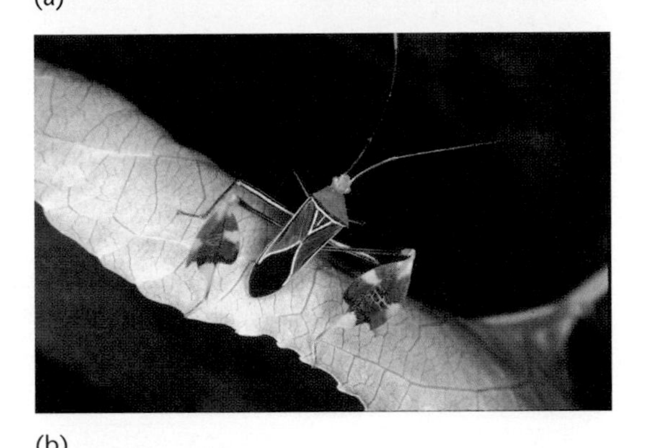

(b)

(c)

(d)

Figure 20.8 Different guilds of insects feed in different ways on plants. Shown are (a) leaf chewers, a frangipani hawk moth caterpillar on frangipani, (b) sap suckers, *Anisoscelis flavolineata*, a leaf-footed bug on passion fruit, (c) leaf miners, *Ophiomyia camarae*, an agromyzid fly larva on *Lantana*, and (d) stem gallers. *(Photos by Peter Stiling.)*

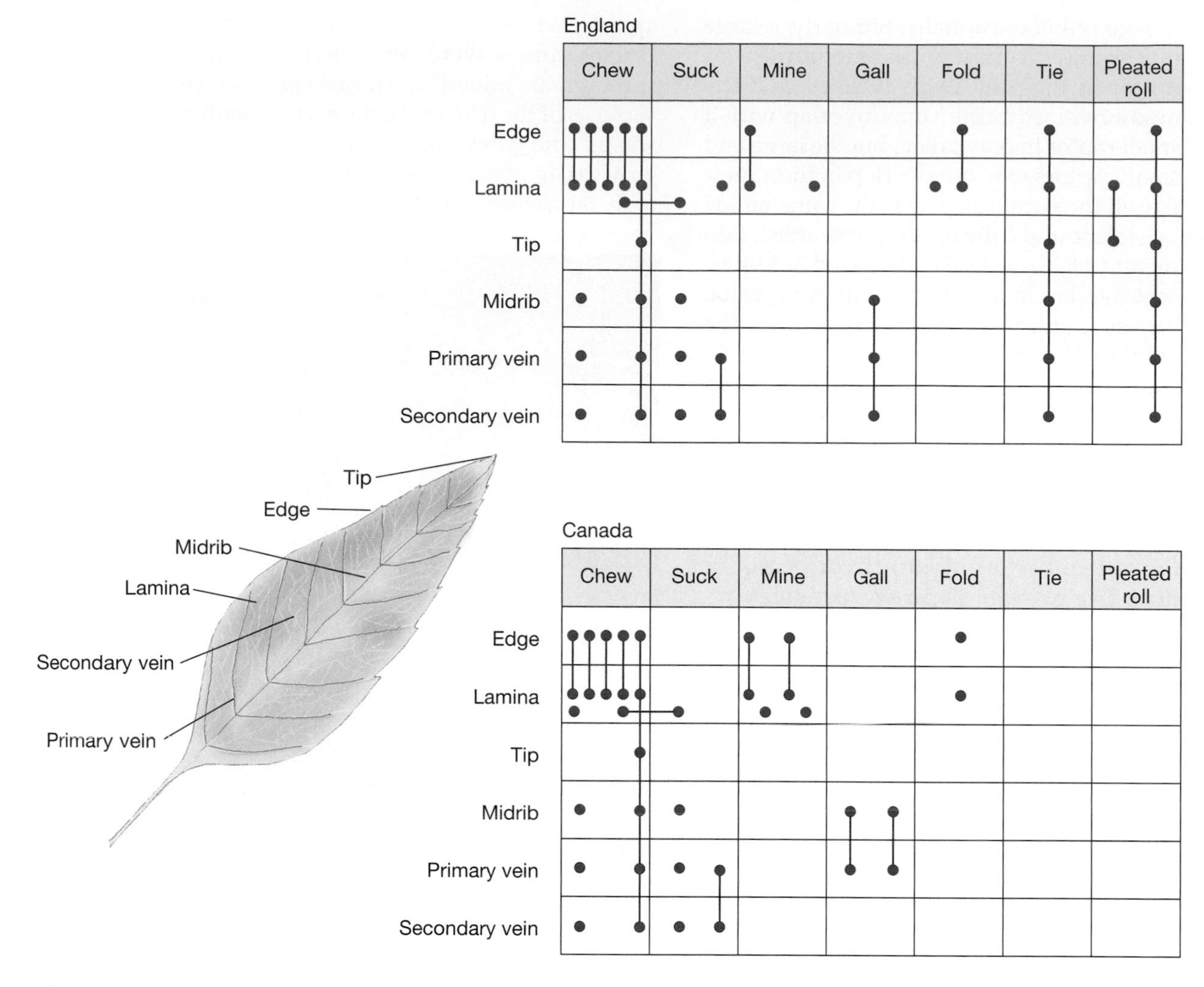

Figure 20.9 Guilds of insects feeding on red oak (*Quercus rubra*) in Canada and the United Kingdom. Dots represent species and areas of oak leaves on which they feed. Lines connecting dots represent species that feed in more than one area. There are many similarities between the two countries in the area of the oak leaves where each guild feeds and in the number of species in each guild. (*After Ashbourne and Putnam, 1987.*)

guilds did not exhibit any constancy between geographic areas. For example, there are no miners in England, but many in New Guinea (Fig. 15.10). The results of Lawton's study failed to support guild theory.

S. R. Ashbourne and R. J. Putnam of Southampton University in England (1987) did the same thing as Lawton, using the insect herbivores feeding on red oak (*Quercus rubra*) and aspen (*Populus tremuloides*) in Canada and the United Kingdom. They used the same four basic guilds as Lawton (chewers, suckers, miners, and gallers) and included three more: species that folded one leaf in half and fed inside, species that tied two leaves together and fed inside, and species that rolled leaves up and fed inside. Their result was the opposite of Lawton's, in that striking similarities were seen between the guild composition of the two regions (Fig. 20.9). The two results are typical of what has been found so far: Some studies find similarities in guild structure between ecosystems, and others do not. As yet, there is no consensus regarding which pattern is the more frequent.

Keystone species are members of a community that have an effect out of proportion to their abundance.

20.5 Keystone Species

Within ecosystems, food webs, and even guilds, some species may have an effect out of all proportion to their abundance or biomass. For example, the beaver, a relatively small species, can completely alter an ecosystem by building dams and flooding an entire river valley. All the trees in the valley die, and fish and wildfowl may become common in the resultant lake. This pattern contrasts with patterns in which the importance of species in an ecosystem varies randomly or in which each species is equally important (Fig. 20.10). The removal of species like beavers obviously has severe ramifications for a majority of community members, promoting fish die-offs, loss of waterfowl, and the death of trees adapted to waterlogged soil. The term for such species, **keystone species**, has enjoyed an enduring popularity in the ecological literature (much like the term *guild*) since its introduction by Robert Paine in 1969. It is important to distinguish a keystone species from a dominant species. A **dominant species** is a species that has a large effect in a community because it is very common. For example, *Spartina* cordgrass is a dominant species in a salt marsh in virtue of its large biomass and role in energy flow. Other common dominant species in different ecosystems include trees, prairie grass, corals, and giant kelp (Fig. 20.11).

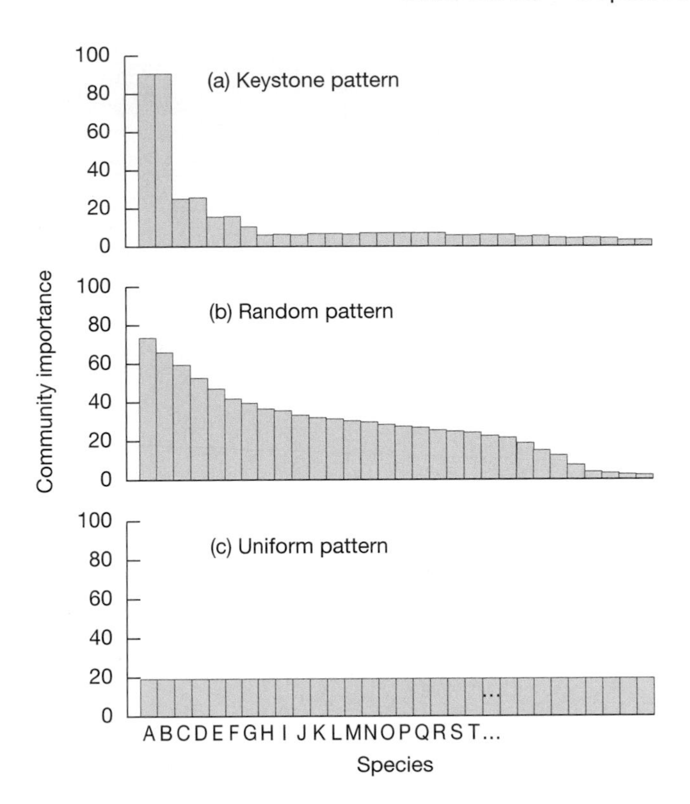

Figure 20.10 Community importance values (i.e., whether or not a species has a large impact on community diversity, nitrogen uptake, or some other function) for species in a hypothetical community, (a) based on the keystone-species model, (b) based on a random pattern, and (c) with all species equally important. (*After Mills et al., 1993.*)

Several different types of keystone categories have been recognized: keystone enemies, keystone prey, and keystone habitat modifiers. Among the most famous keystone predators are *Pisaster* starfish and predatory whelks (*Conchloepas*) in the rocky intertidal zone (Paine 1966, 1969). Removal of these carnivores leads to the nearly complete dominance of the substrate by one or two species of mussels that outcompete the other consumer species, resulting in greatly reduced species diversity, as we saw in Chapter 15. A rare rhinovirus that makes a wildebeest sneeze, V_R, would not qualify as a keystone species, because its impact is relatively low, but a rare distemper virus, V_D, that kills lions or wild dogs, would. Sea otters have been labeled as keystone predators because they limit the densities of sea urchins, which, in turn, eat kelp and can destroy this dominant species in the absence of the otters.

Tropical ecologist John Terborgh (1986) considered palm nuts, figs, and nectar to be keystone prey because they are critical to tropical forest fruit-eating guilds, including primates, rodents, and many birds. Together, these vertebrates account for as much as three-quarters of the forest bird and animal biomass. Without the fruit trees, wholescale extinction of frugivores would occur.

In the southeastern United States, gopher tortoises have been regarded as keystone species because their burrows provide homes for an array of mice, possums, frogs, snakes, and insects. Without tortoise burrows, many of these creatures would be unable to survive in the sandhill areas where they are found. The gopher tortoises act as keystone habitat modifiers (Fig. 20.12). In fact, both beavers and gopher tortoises have been called **ecosystem engineers** because they modify the habitat and cause ecological changes (Lawton and Jones, 1995), and this term seems to have caught on more than *keystone habitat modifier*. African elephants may also act as ecosystem engineers, through their browsing activity. Elephants destroy small trees and shrubs when browsing and can change woodland habitats into grasslands. Ungulates that graze the grasses are favored by the elephant's activities. More examples of ecosystem engineers and their effects on ecosystems are given in Table 20.2.

Are keystone species present in most communities? It is too early to tell. Indeed, few studies have analyzed the community importance, or interaction strength, of species. However, Paine (1988) suggests that humans, introduced pests, and diseases

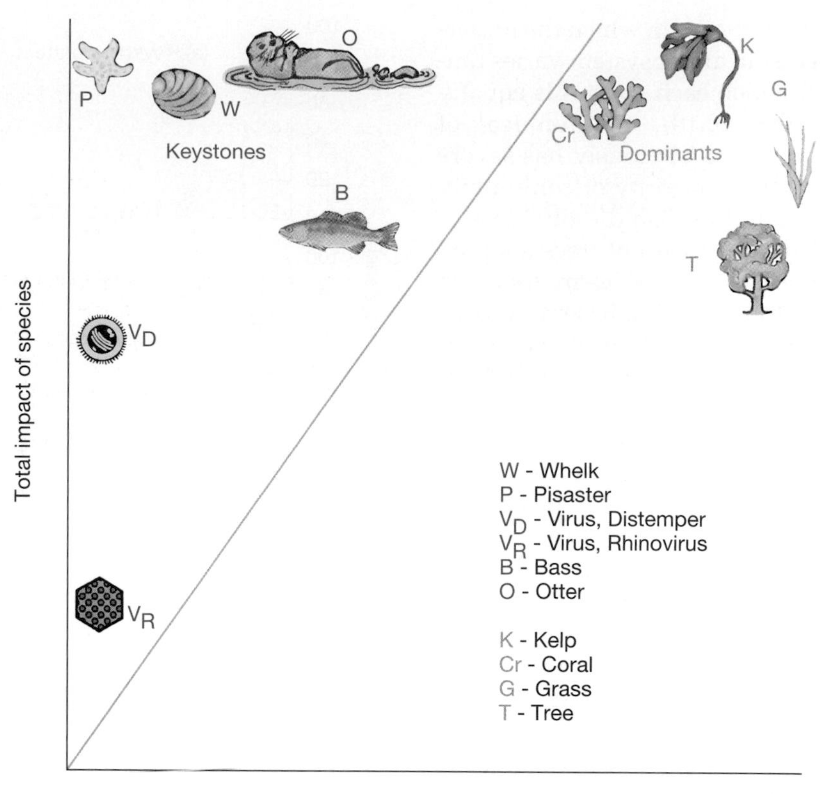

Figure 20.11 Keystone species are those species whose effects are out of all proportion to their biomass. Dominants are species that dominate community biomass and whose impacts on an ecosystem are large, but not disproportionate to their biomass. (*After Power and Mills, 1995.*)

Figure 20.12 Gopher tortoise burrows in U.S. southeast support a variety of other wildlife which uses them.

may commonly act as keystone species. The conservation community is keen to identify keystone species like the beaver and the gopher tortoise because, by conserving a keystone species, we are actually conserving the many species that live alongside it.

Food web theory has led to some valuable insights, such as that into the pattern of constant connectance in such webs. However, we should remain aware of the potential pitfalls of imperfectly known webs. To understand food webs better, researchers have focused on knowing more about certain parts of the webs, such as guilds. Keystone species offer great promise as focal points for conservation activities.

TABLE 20.2 Organisms as ecosystem engineers, some selected examples.

Engineer	Effect
Green plants	Oxygen production
Soil organisms	Increase and reduction of soil N and other nutrients
Beavers	Altered hydrology
Coral reef	Physical structure, alteration of flow rates
Trees	Physical structure
Gopher tortoises	Burrow
Woodpeckers	Nest cavity
Termite mounds	Homes for other arthropods
Vertebrate feces	Homes for dung beetles

Summary

Ecosystem ecology concerns the flow of energy and nutrients through communities. It offers the use of a common currency—energy or nutrients—by which to compare different ecosystems.

1. Many food webs can be compared by simple properties such as connectance (degree of connectivity) and linkage density (number of links per species).

2. The analysis of food webs has revealed some common patterns. Two are that, as the number of species in a food web increases, chain length tends to increase, but connectance remains constant. Another is that there is a pyramid of numbers, with fewer species on higher trophic levels than on lower trophic levels.

3. Because food webs are usually imperfectly known, the generalizations that have resulted from analyzing them may be incorrect. The inadvertent omission of minor or mobile species, lack of knowledge about the strength, importance, or beneficial effects of connecting links between species, and the taxonomic lumping of species into groups or guilds can all obscure true food web patterns.

4. It is possible to split up trophic levels into units called guilds. For example, among herbivores, we can recognize the leaf-chewing guild, the stem-boring guild, and the root-feeding guild. Guilds are a valuable analytical tool because they focus attention on groups of species most likely to be competing for resources.

5. Keystone species have a large influence on ecosystems out of all proportion to their abundance. A common keystone species is the beaver, whose activities can flood entire valleys and completely change the composition of the species that live there.

Food webs offer a potential shortcut in understanding how communities work. However, to yield reliable insights into the way a community functions, food webs must be thoroughly known.

Discussion Questions

1. How might food web properties change if we included all the decomposers in food webs? Much energy passes through decomposers.

2. Do you think food webs would be different in complexity between two-dimensional habitats, such as grasslands, versus three-dimensional habitats, like forests?

3. Would you expect differences in chain lengths between disturbed and undisturbed areas? How would disturbance affect connectance or linkage density?

4. Would fertilizing an ecosystem increase the length of a food chain, and if so, why?

5. Do you expect the position of a species in a food web to change as it ages and increases in size? If so, does this affect how we construct food webs?

C H A P T E R 21

Energy Flow

Road Map

1. The energy content of plants can be measured by calorimetry, harvesting, CO_2 production, O_2 production, or chlorophyll concentration.
2. The production of plants is limited by temperature, moisture, and nutrients such as nitrogen and phosphorus.
3. Primary production is highest in the tropics and decreases toward the poles.
4. The efficiency of the transfer of energy from plants to animals can be measured by three indices: production efficiency, consumption efficiency, and trophic-level transfer efficiency.
5. The production of animals is limited mainly by the production of plants.

Populations can be considered as energy transformers in a community. Calories can be used as a lowest common denominator, and populations can be reduced to caloric equivalents. In the search for generalizations in ecology, an ecosystems approach may provide insights into how biological systems function by closely examining the energy content of different trophic levels. Energy content is generally measured using dry biomass, or dry weight, because the bulk of living matter in most species is water and water content fluctuates widely, often according to wet or dry seasons. Dry weight is a more meaningful measure of calorific content. Of the dry weight, 95% is made up of carbon compounds, so measuring energy flow in ecosystems is in many ways equivalent to examining the carbon cycle.

Five ways of measuring primary production are by calorimetry, harvesting, carbon dioxide uptake, oxygen output, and chlorophyll concentration.

21.1 Measuring Production

The bulk of the Earth's living mantle, 99.9% by weight, is green plants; only a small fraction of life is animal. Therefore, when we measure the energy of an ecosystem, we are interested primarily in plants. Because plants represent the first, or primary, trophic level, we measure plant production as **gross primary production**, which is equivalent to the energy fixed in photosynthesis. **Net primary production** is gross primary production minus the energy the plant loses through respiration. Net primary production is the amount of biomass accrued by plants, measured as the weight of living plant material, including leaves, stems, and roots.

In order to examine ecosystems as energy transformers, we must have some way of determining the energy content of plants. This can be done in many ways, and we will examine five of the most common.

In *calorimetry*, a sample of dry material is burned in a small chamber, where it oxidizes completely to carbon dioxide and water. High-pressure oxygen is fed into the chamber to ensure complete combustion. A water jacket surrounds the chamber, and the rise in temperature of the water provides a measure of the heat generated by combustion of the sample. The greater the energy in a sample, the more heat will be generated.

In *harvesting*, samples of plant biomass are obtained at the beginning of a growing season, after which they are dried and weighed. At the end of the season, another sample is obtained and then dried and weighed. The difference of the two weights represents growth in plants in the area studied and provides a measure of production in a known unit of time.

In the *CO_2 uptake method,* plants are grown inside sealed chambers for short periods. Because we know that air contains 0.03% CO_2, any decrease in CO_2 inside the chambers is the result of CO_2 uptake by plants during photosynthesis. The higher the CO_2 uptake, the greater is the production of plant biomass. The exact plant biomass required to take up a given volume of CO_2 is then calculated.

We can estimate the production of many aquatic plants by their *O_2 output* instead of CO_2 intake. For example, to measure phytoplankton production, samples of phytoplankton are suspended in two bottles in a lake or an ocean. One of the bottles is dark in color, so photosynthesis is not possible and oxygen is used during respiration. The other bottle is light, and photosynthesis occurs, giving off oxygen beyond that used in respiration. The difference in oxygen concentration between the light and dark bottles yields the total oxygen produced by photosynthesis.

Finally, production can be estimated by measuring *chlorophyll concentration*. We know how much carbon is assimilated per gram of chlorophyll. By taking a sample of the plant under question and measuring its chlorophyll, we can calculate its carbon production. More modern methods take advantage of the reflectance of plants to measure the chlorophyll concentration. Thus, orbiting satellites can identify reflectance colors of different habitats and thus estimate their production.

Using these techniques, we begin to analyze the amount of energy contained in different ecosystem components, such as plants, herbivores, and carnivores. To analyze a complete ecosystem in this way, calculating the calorific content of all the component species, is a monumental task. However, some ecologists have concentrated on analyzing relatively simple systems such as salt marshes, in which there are few species. The patterns developed from such systems should be valid for other communities as well.

Among the pioneers of this type of work was John Teal, of the University of Georgia, who examined energy flow in a Georgia salt marsh (Fig. 21.1). In the marsh, most of the incident energy from the sun went to two types of plants: *Spartina* cord-

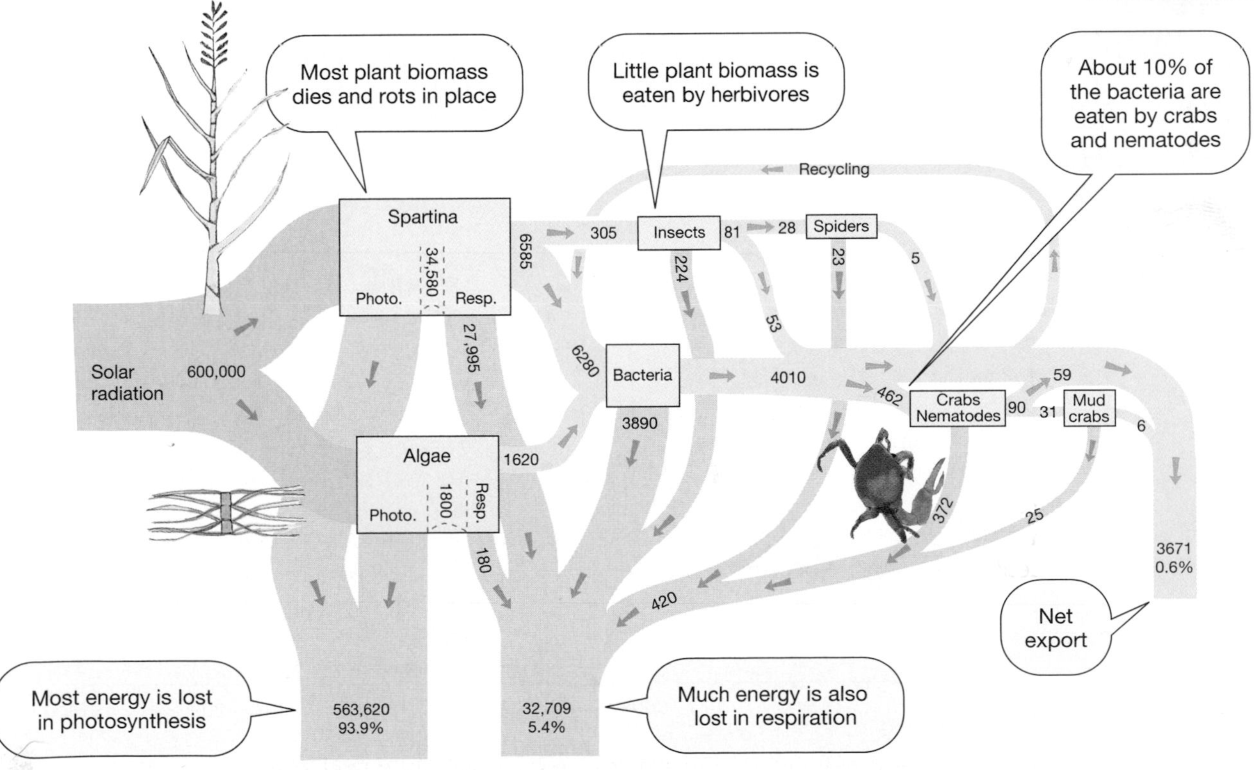

Figure 21.1 Energy-flow diagram for a Georgia salt marsh. Units are kilocalories per square meter per year. Thickness of flows are not exactly to scale. (*Redrawn from Teal, 1962.*) The photos shown above are a Georgia salt marsh and a fiddler crab.

grass and marine algae. The cordgrass plants were rooted in the ground, whereas the algae floated on the surface of the water or lived on the mud or on the *Spartina* leaves at low tide. The cordgrass and algae are the **producers** in the system. Most of the plant energy is used in the energetically costly process of photosynthesis. The remainder is lost mostly in **respiration**. The energy that is accumulated in plant biomass is small: 34,580 kcal m^{-2} yr^{-1} in *Spartina* and 1,800 kcal m^{-2} yr^{-1} in the algae, totaling 36,380 kcal m^{-2} yr^{-1}, or about 6% of the incident solar radiation. The herbivores take very little of this plant production, eating only 305 kcal m^{-2} yr^{-1} of the *Spartina* (0.8%) and none of the algae. The remaining 99.2% of the plant biomass dies in place and rots on the muddy ground, to be consumed by bacteria, the major **decomposers** in the system. In turn, the bacteria are eaten by crabs and nematodes, which sift through the mud. Very little of this dead material is removed from the system (exported) by the tide.

Primary production is limited mainly by light, water, nutrients, and temperature.

In the salt marsh, the producers are the most important consumers; in other words, most primary production (98.6%) is used in plant respiration and photosynthesis. The bacteria are next in importance: As decomposers, they degrade about 20% of the energy that plants use. As with many ecosystems, most plant biomass goes to the decomposers. Animal consumers are a poor third in importance, degrading only about 3.8% of the energy the bacteria use.

21.2 Limits to Primary Production

What limits primary production? In terrestrial systems, water is a major determinant, and production shows an almost linear increase with annual precipitation, at least in arid regions (Fig. 21.2). However, temperature is also important. Temperature affects production primarily by slowing or accelerating plant metabolic rates. For example, one study indicated a linear relationship between total net primary production and mean annual temperature, as measured on an elevational gradient in Hawaii: For each 1°C increase in mean annual temperature, total net primary production increased by 54 gm^{-2} yr^{-1} (Raich *et al.*, 1997). Michael Rosenzweig of the University of Arizona (1968) noted that the combined effects of temperature and moisture (i.e., the actual evapotranspiration rate) could predict the aboveground production with good accuracy in North America (Fig. 21.3). The evap-

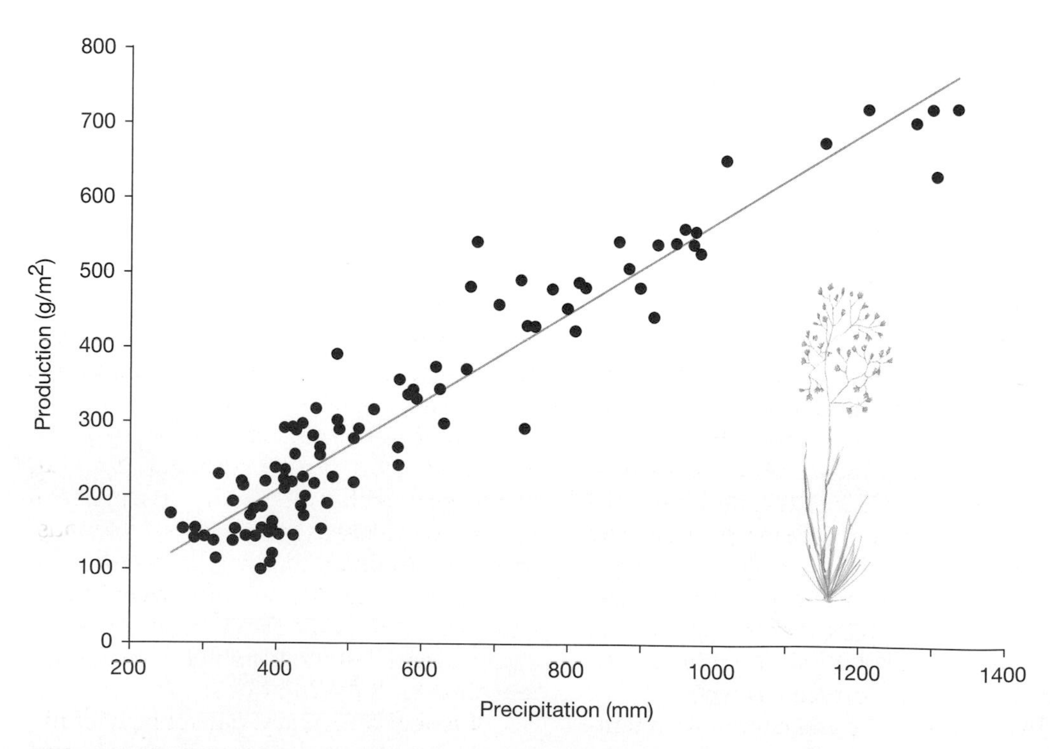

Figure 21.2 Relationship between mean annual precipitation and mean aboveground net primary production for 100 areas across the central grassland region of the Great Plains of the United States. (*Redrawn from Sala et al., 1988.*)

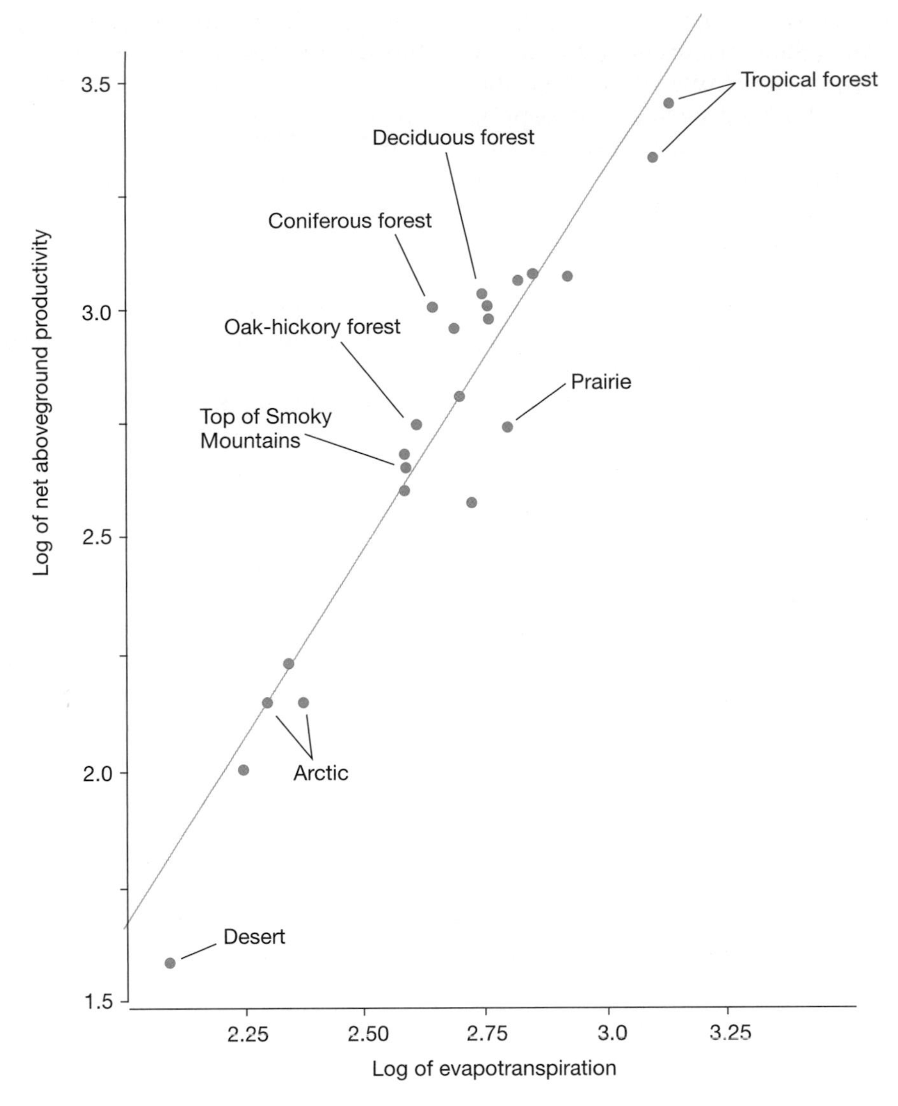

Figure 21.3 Plant production is positively correlated with evapotranspiration. Warm, humid environments are ideal for plant growth. Dots represent different ecosystems, some of which are indicated. (*Modified from Rosenzweig, 1969.*)

otranspiration rate is a measure of the amount of water entering the atmosphere from the ground and the vegetation. Water goes into the air as it evaporates from the ground in the heat of the sun, and also from the vegetation as it transpires. So, a desert will have low evapotranspiration because little water is available despite high temperatures. Only when both temperature and moisture are at high levels, as in hot tropical rain forests, are rates of evapotranspiration maximized. High evapotranspiration is ideal for plant growth, and primary production is then maximized as well.

In addition to its association with temperature and moisture, the net primary production of forests is well correlated with the length of the growing season. The more days available for forests to grow, the more wood they can put on. For this reason, southern Appalachian forests of Tennessee and North Carolina are more productive than their northern counterparts in Wisconsin and New England. The length of the growing season helps explain why tropical wet forests, with a very long growing season, are so productive and why deciduous trees, predominant in northern realms, are not. However, conifers at northern latitudes effectively extend the short growing season at high latitudes by retaining their leaves (needles) for long durations.

Nutrient deficiency, particularly of nitrogen and phosphorus, can limit primary production also, as agricultural practitioners know only too well. Fertilizers are commonly used to boost the production of annual crops. Stewart Cargill and Rob Jeffries (1984), of the University of Toronto, showed how both nitrogen and phosphorus were limiting to salt-marsh sedges and grasses in subarctic conditions in Hudson Bay, Canada (Fig. 21.4). Of the nutrients alone, nitrogen was the most limiting: The addition of phosphorus failed to increase production. However, once nitrogen was added and was no longer limiting to the system, phosphorus became the limiting factor, and the addition of phosphorus and nitrogen together increased production the most. This result supports Justus Leibig's (1842) law of the minimum, discussed in Chapter 7, which says that the abundance of species is controlled by the most limiting factor—in this case, nitrogen. The most limiting factor can change, as in the Hudson Bay marsh: When sufficient nitrogen is available, phosphorus becomes limiting. Once phosphorus becomes abundant, production is still likely limited by some other nutrient. Apparently, there is a sequence of limiting factors in ecosystems.

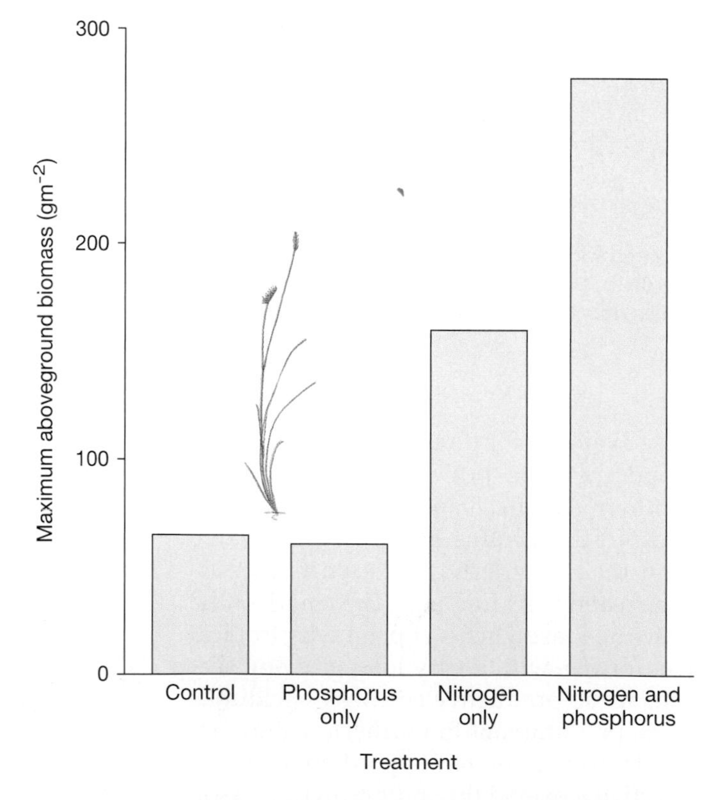

Figure 21.4 Net aboveground primary production of a salt-marsh sedge, *Carex subspathacea*, in Canadian salt marshes in response to the addition of nutrients. Nitrogen is more limiting than phosphorus, but once nitrogen becomes available, phosphorus then becomes limiting. *(From data in Cargill and Jeffries, 1984.)*

Aquatic systems.

Primary production in aquatic systems is limited mainly by the availability of light and nutrients.

Of the factors limiting primary production in aquatic ecosystems, among the most important are light and nutrients. Light is particularly likely to be in short supply, because water readily absorbs light. At one-meter depth, half the infrared energy has been absorbed. By 20 meters, only 5 to 10 percent of the radiation is left. Too much light can also inhibit the growth of green plants, by overheating them. Such a phenomenon can be found in tropical and subtropical surface waters throughout the year; maximum primary production occurs several meters beneath the surface.

The most important nutrients affecting primary production in aquatic systems are again nitrogen and phosphorus. Important only locally in terrestrial systems, both often limit production over large areas of the oceans, where they occur in very low concentrations. Few nutrients are tied up in the standing crop in aquatic systems, in contrast to terrestrial systems: Soil contains about 0.5% nitrogen, whereas seawater contains only 0.00005% nitrogen. Enrichment of the sea by the addition of nitrogen and phosphorus can result in substantial algal blooms. Such enrichment occurs naturally in areas of upwellings, where cold, nutrient-rich, deep water is brought to the surface by strong currents, resulting in highly productive ecosystems and plentiful fish. Such upwelling occurs in the Antarctic and along the coasts of Peru and California, and these are rich, productive waters.

Phosphorus is particularly important in limiting production in freshwater lakes. Some lakes in North America become polluted by runoff of rainwater enriched with phosphorus from fertilizers or from sewage. This input into the lakes causes huge blooms of algae, which clog the lake and increase its turbidity, a process termed **eutrophication**. One of the champions of the study of eutrophication is David Schindler (1974, 1977), who experimentally added phosphorus to lakes in northwestern Ontario. Not only does phytoplankton biomass increase with increased phosphorus (Fig. 21.5), but also, the composition of the algal community changes from green algae to blue-green algae. The blue-green algae can produce some secondary chemicals, so zooplankton

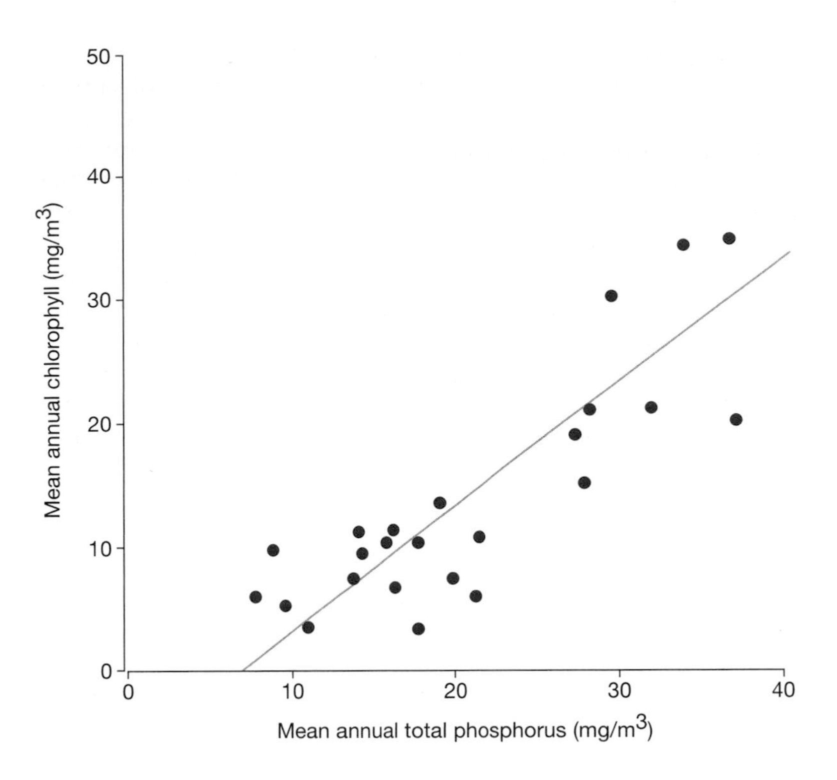

Figure 21.5 Increase in lake primary production, measured as chlorophyll concentrations, with phosphorus concentration. More algae grow as phosphorus levels increase. *(After Schindler, 1977.)*

prefer the green algae. As a result, the phosphorus-laden lakes become clogged with a scum of blue-green algae that is reviled by fish and humans alike.

The degree of eutrophication of large bodies of water like the Great Lakes shows a striking resemblance to maps of human population density along the shoreline. Industrial and urban wastewaters, especially those containing phosphorus, cause massive eutrophication. Even rural areas with high concentrations of livestock can cause eutrophication of water bodies as agricultural wastes seep into the water. The remedy, of course, is to divert wastewater away from lakes, or else to treat the wastewater, removing phosphorus and nitrogen before they discharge into lakes.

In marine systems, nitrogen and phosphorus also limit production, and iron is known to be limiting as well. Large-scale open-ocean fertilization experiments in the Sargasso Sea in the early 1960s showed that the addition of iron to surface water boosted production, as revealed by an increase in carbon uptake (Fig. 21.6).

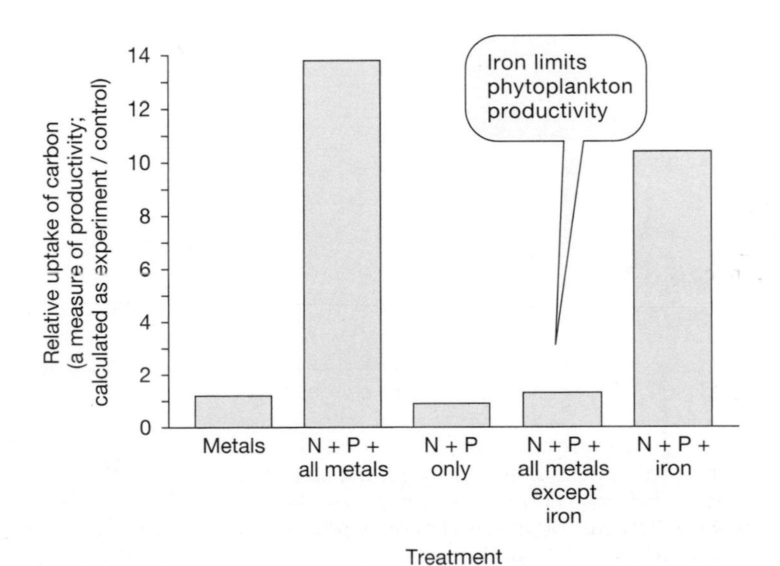

Figure 21.6 Results of experiments in which bottled seawater from the Sargasso Sea, south of Bermuda, was enriched with various nutrients. Production in experimental bottles was compared relative to that in controls to which no nutrients were added. Iron was the limiting factor. *(From data in Menzel and Ryther, 1961.)*

21.3 Patterns in Primary Production

Efficiency of primary production.

The efficiency of primary production can be calculated by comparing the energy produced by photosynthesis with the energy available in incident sunlight.

Armed with the knowledge that precipitation and temperature are limiting factors, we can make some predictions about the efficiency of primary production. For example, we could expect deserts to be not very efficient at converting huge amounts of incident energy from sunlight into plant biomass, because water is scarce. In fact, one can determine the efficiency of utilization of sunlight from the ratio

$$\text{efficiency of gross primary production} = \frac{\text{energy fixed by gross primary production}}{\text{energy in incident sunlight}}.$$

Phytoplankton communities have very low efficiencies—usually, less than 0.5%—because they are severely limited by the availability of nutrients (Fig. 21.7). The highest efficiencies occur in coniferous forests, because their numerous needles present a large surface area for photosynthesis. A higher efficiency generally leads to a high annual primary production. Thus, herbaceous and forest communities are efficient and highly productive, whereas desert and tundra ecosystems are not.

So far, we have looked mainly at patterns in gross primary production, but the patterns for net primary production are similar. Perhaps 50 to 70% of the energy fixed by photosynthesis is lost in respiration, so usually less than 1% of the sun's energy is actually converted into net primary production. In the temperate zone, this 1% is equivalent to 300 to 600 calories of primary production per square meter of land per year.

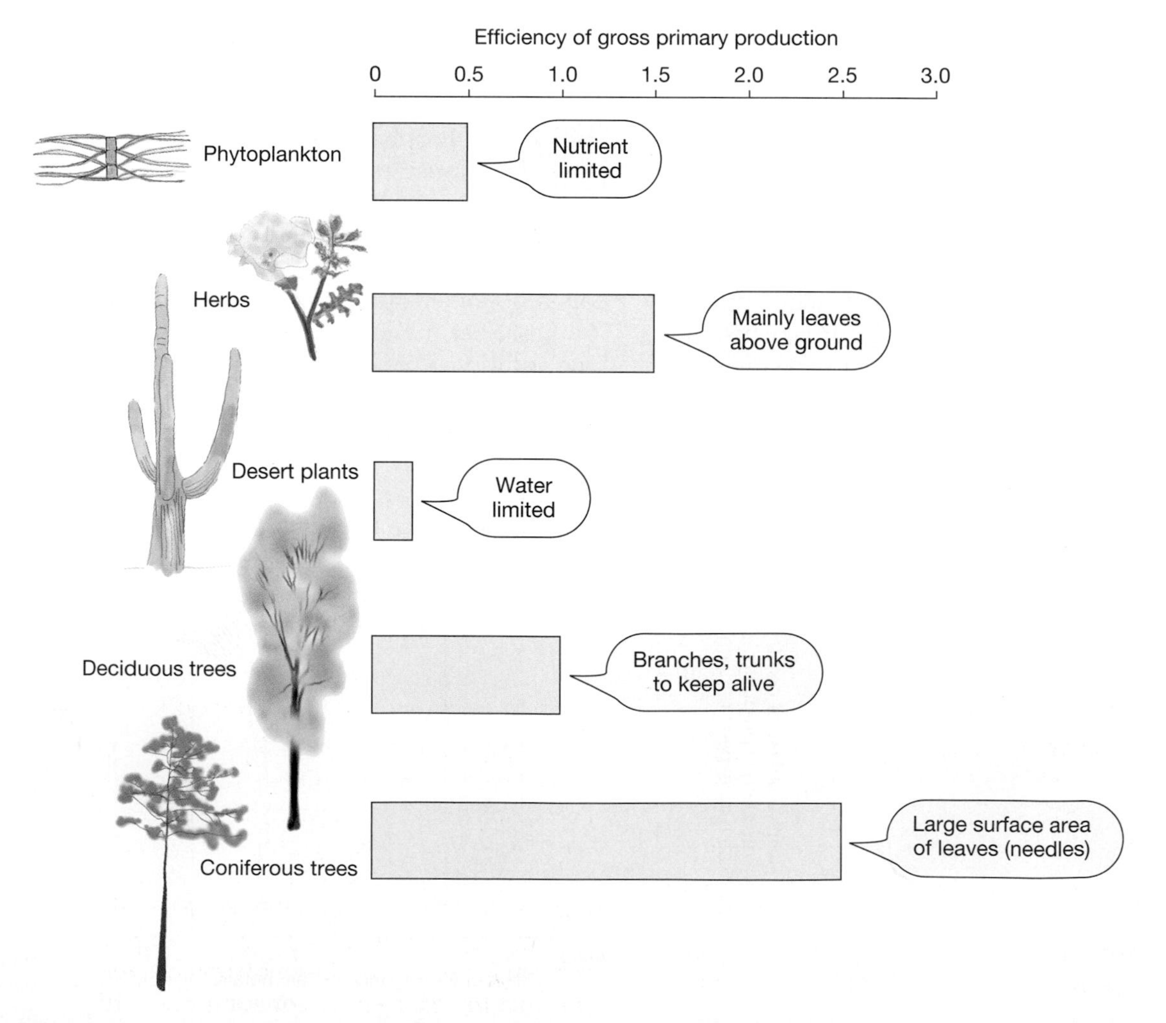

Figure 21.7 Relative efficiencies of primary production for various types of plants.

Global distribution of primary production.

On the basis of efficiencies of primary production, and knowing limiting factors, we can examine how mean net primary production varies across different types of vegetation on the Earth. In general, primary production is highest in the tropical rain forest and decreases progressively toward the poles (Table 21.1). This gradient occurs because temperatures decrease toward the poles and, as we saw, temperatures affect primary production greatly. The production of the open ocean is very low, somewhere between the production of deserts and that of the Arctic tundra. In effect, open oceans are the deserts of the marine world. Marine production is highest on coastal shelves, particularly in upwelling zones where the movement of water from the ocean bottom to the surface circulates nutrients. The global patterns of primary production can be helpful when one considers how to produce biomass fuels as an alternative to fossil fuels. (See "Applied Ecology: Biomass Fuels.")

Primary production is greater in areas with abundant warmth and moisture, such as tropical rain forests, swamps, and coral reefs.

TABLE 21.1 Primary production and plant biomass for the Earth. (*From Whittaker, 1975.*)

Type of Ecosystem	Mean Net Primary Production, $gm^{-2}\ yr^{-1}$
Tropical rain forest	2,200
Tropical seasonal forest	1,600
Temperate evergreen forest	1,300
Temperate deciduous forest	1,200
Boreal forest	800
Woodlands and shrub land	700
Savanna	900
Temperate grassland	600
Tundra and alpine	140
Desert and semidesert scrub	90
Extreme desert, rock, sand, and ice	3
Cultivated land	650
Swamp and marsh	2,000
Lake and stream	250
Total continental	**773**
Open ocean	125
Upwelling zones	500
Continental shelf	360
Algal beds and reefs	2,500
Estuaries	1,500
Total marine	**152**
Grand total	333

21.4 Secondary Production

Secondary production is the amount of new biomass produced by consumers.

The biomass of plants that accumulates in a community as a result of photosynthesis eventually goes in one of two directions: to herbivores or to detritus feeders. Most of the biomass dies in place and is available to the detritus feeders. This material is known as *dead organic matter*. Detritus feeders probably carry out 80–90% of the consumption of matter, often working it over on a number of occasions to extract the most energy from it. However, it is the herbivores that feed on the living plant biomass, and, as a result, they constitute the greatest selective pressure on plants. Let us therefore consider energy flow in secondary production and in herbivores in particular.

First, we need to be able to measure energy intake, which is not a difficult task. Generally, an herbivore is confined to a given area, and the mass of vegetation it eats per unit time can be recorded. For carnivores, too, the intake of prey items of known biomass and energy content can be measured. Measuring how much energy an animal assimilates from its prey is a little more tricky. Usually, we measure the nutrient content of the feces and urine the animal excretes to determine how much of the food was assimilated by the organisms. Of the energy assimilated, some goes to production (growth), but much goes to maintenance (respiration). To measure the amount going to production, we can measure the weight of an organism before and after a meal to get its increase in weight, but we have to wait a sufficient period—at least 24 hours—for the meal to have been completely absorbed into the body.

Efficiency of secondary production.

The efficiency of secondary production is estimated by comparing new tissue production with consumption or assimilation.

Following the gathering of data on food intake, energy uptake, and the growth of herbivores, we can examine the passage of energy through higher trophic levels. In viewing an herbivore or carnivore as an energy transformer, three main types of efficiency can be measured and used to compare species at different trophic levels: production efficiency, consumption efficiency, and trophic-level transfer efficiency.

Applied Ecology

Biomass Fuels

The energy content of plants can be recovered as heat in the combustion of biomass materials like wood, straw, and plant matter. Therefore, crops can be planted to produce biomass fuels as a renewable energy source, providing an alternative to fossil fuels. Crops that produce fuels can be grown almost anywhere on the Earth's surface, because photosynthesis continues at a slow and steady rate even when the intensity of sunlight is very low. There are, of course, some areas that would produce a far greater yield than others (Fig. 1). Also, fuel crops would be in severe competition for space with food crops in densely populated countries at high latitudes. Thus, the potential for producing biomass is greatest in humid, densely vegetated, less densely populated tropical countries. Rather than chopping down existing tropical forest, we would advocate choosing easily harvestable crops, such as sugarcane, to be used for fuel.

One of the main ways to generate biomass fuels is to use anaerobic glycolysis (fermentation) to break down the organic polymers into liquid alcohols. These concentrated fuels can be stored indefinitely, and they can be used to fuel vehicles. Since the late 1970s, a large part of the Brazilian sugar crop has been used to manufacture ethanol, which is added in a 1:3 ratio to gasoline for some cars and trucks. The changes in the fuel mixture require only small modifications to the engine design. The main practical disadvantages of ethanol as a fuel are that its calorific value is only a little over half that of petroleum and, at least in the 1990s, it was more expensive to produce. However, a big attraction of the alcohol fuels is that they burn cleanly compared with conventional fuels. Given their promise as clean-burning fuels, a good deal of research on alcohol fuels has been carried out in California, traditionally one of the worst smog areas of the world. Already one million cars that run purely on ethanol are used in Brazil.

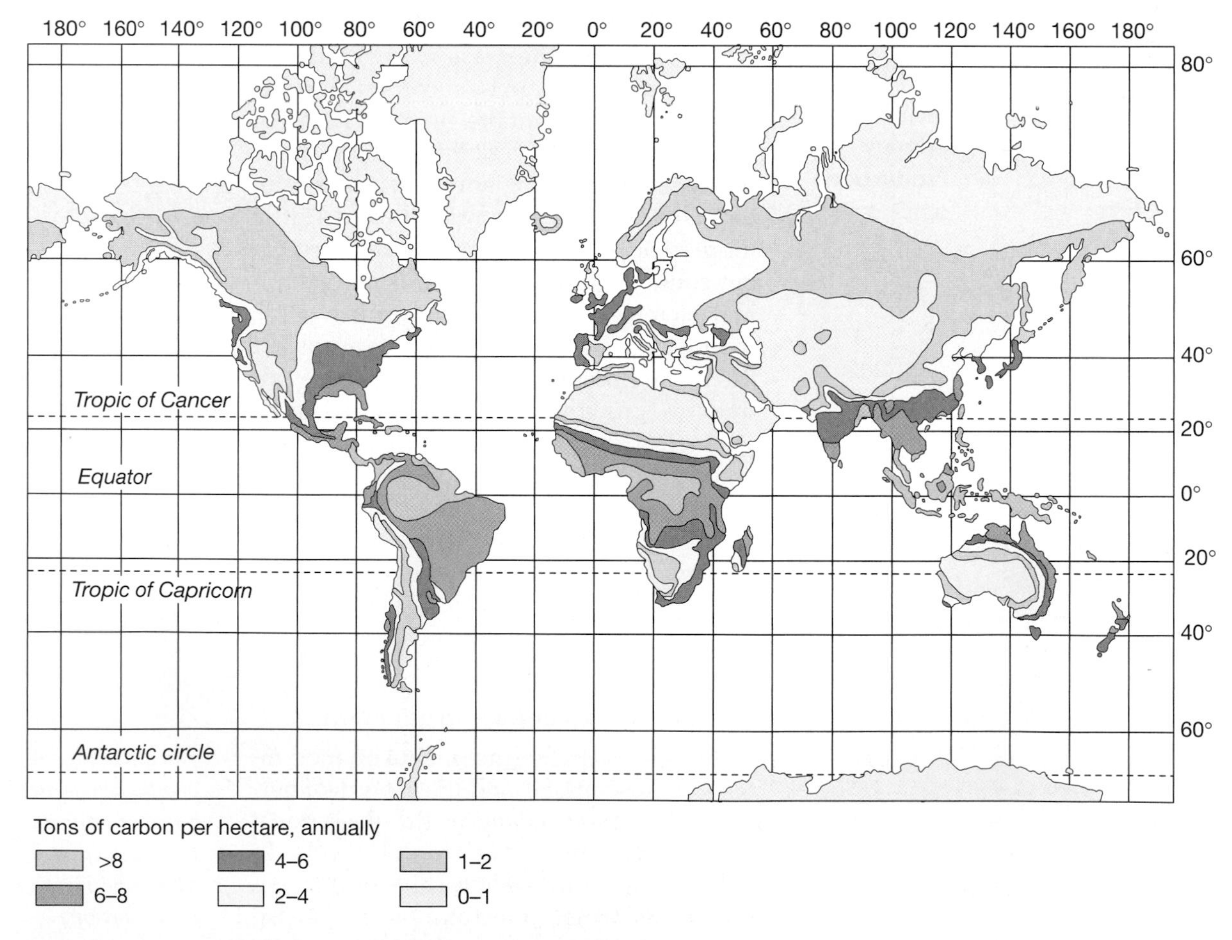

Figure 1 Geographical distribution of fuel production potential. Areas of greatest production are the tropical areas of South America, central Africa, and southeast Asia.

Production efficiency is defined by how well an organism uses the energy it assimilates from its food:

$$\text{product efficiency} = \frac{\text{net production}}{\text{assimilation}} \times 100.$$

Production efficiency in vertebrates is generally less than that in invertebrates because vertebrates use much more energy in respiration. Even within vertebrates, though, much variation occurs. In endotherms, over 98% of assimilation energy may be used in respiration, whereas in ectotherms, only about 90% of assimilation energy is respired. The difference reflects the energy cost of homeothermy. Thus, **production efficiencies** of around 10% are common for ectotherms and 1–2% for endotherms. Invertebrates generally have higher efficiencies—on the order of about 30 to 40%—losing much less energy as heat. Microorganisms also have very high production efficiencies. One consequence of these differences is that sparsely vegetated deserts can support healthy populations of snakes and lizards whereas mammals might easily starve. The large monitor lizard known as the Komodo dragon (Fig. 21.8) eats the equivalent of about a pig a month and its own weight every two months, whereas a cheetah consumes approximately four times its own weight in the same period of two months. It is interesting to note that production efficiencies also are higher in young animals (35%) than in old animals (3–5%)—hence the practice of raising young chickens and calves for meat. Further, smaller species often have higher growth rates than larger species: For a given amount of hay, rabbits produce the same quantity of meat as cattle do, but four times as quickly.

The second measure of efficiency between trophic levels is consumption efficiency, which measures the amount of one trophic level eaten by the trophic level above that is incorporated into the latter's biomass:

$$\text{Consumption efficiency} = \frac{\text{intake at trophic level } n}{\text{net production at trophic level } n - 1} \times 100.$$

Consumption efficiency values seem generally to fall in the range of 0–15% for terrestrial herbivores, meaning that 85–100% of the net terrestrial plant production goes into the decomposer chain. These low efficiency values may arise because most plants are chemically well defended or because herbivore densities are maintained at low levels by their natural enemies. In aquatic systems, zooplankton may be more efficient grazers, and consumption efficiency values of 30–40% percent have been reported. These higher consumption efficiencies may be due to the relatively low levels of chemical deterrents in phytoplankton and their higher digestibility because they lack complex structural cells. For carnivores, however, consumption efficiencies may reach 50–100%, showing that animals are much better equipped to digest meat than to detoxify and digest chemically protected plants.

Figure 21.8 The Komodo dragon is the world's largest lizard and lives on only a few small Indonesian islands. Its relatively high production efficiency, compared with that of an equivalent-sized mammal, means that the Komodo dragon can survive on relatively little food—one pig-sized animal per month.

The third measure of efficiency between trophic levels is trophic-level transfer efficiency, which is essentially a measure of the energy available at one trophic level that is acquired by the trophic level above:

$$\text{Trophic-level transfer efficiency} = \frac{\text{assimilation at trophic level } n}{\text{assimilation at trophic level } n - 1} \times 100.$$

Trophic-level transfer efficiency appears to average around 10 percent, although there is much variation (Fig. 21.9), and some data on marine food chains show that it can exceed 30 percent.

Trophic-level transfer efficiency is no higher than it is mainly for two reasons. First, many organisms cannot digest all parts of their prey. They take only the easily digestible muscles, guts, and skin, so that the hard, energy-rich bones are left behind. Only a very few organisms, such as hyenas, can successfully digest bones. Second, much of the energy assimilated by animals is used in maintenance, so most of their energy is lost as heat. The 10% transfer of energy from one trophic level to another necessitates short food chains, most with no more than four or five links; there simply is little energy to pass on to higher trophic levels.

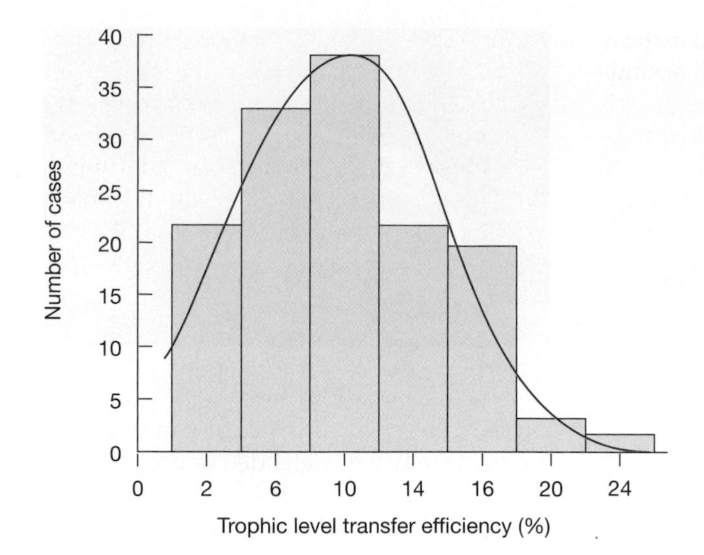

Figure 21.9 The frequency distribution of trophic-level transfer efficiencies calculated in 48 studies. The mean is 10.13%. (*After Pauly and Christensen, 1995.*)

21.5 The Limits to Secondary Production

Secondary production is limited by the availability and quality of primary production.

Exactly what controls secondary production is a complex issue, but it is generally thought to be limited largely by the available primary production. There is a strong relationship between primary production in a variety of ecosystems, on the one hand, and the biomass of, and consumption by, herbivores, on the other (Fig. 21.10). This means that more plant biomass leads to more herbivore biomass.

The existence of this relationship is not as obvious as one might think, because it means that secondary chemicals which make plants poisonous or distasteful to herbivores are much less important influences on consumption at the ecosystem level than they are at the population level. Therefore, when we examine the primary production of all plants in an ecosystem, we see that the few species with toxic chemicals that deter herbivores do not influence herbivore biomass much in comparison to all species combined.

Overall, if we view ecosystems as transformers of energy, then plants are by far the most important organisms on the planet. Limits to plant production caused by water shortages, temperature extremes, or the unavailability of nutrients control energy flow. After plant production, the next most im-

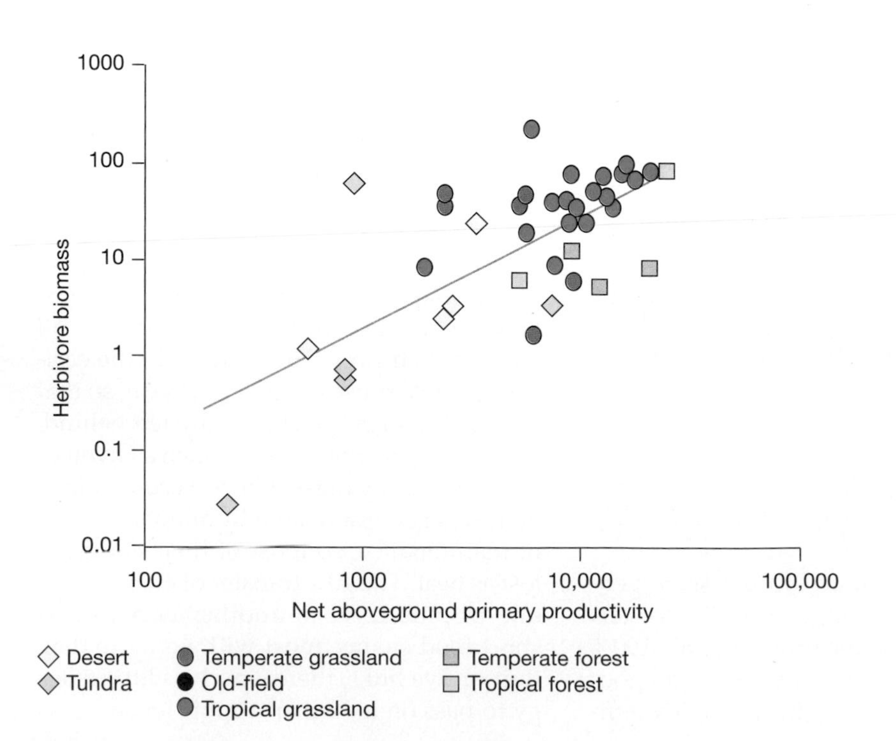

Figure 21.10 Herbivore biomass is positively correlated with net aboveground primary production. (*After Moen and Oksanen, 1991.*)

portant flow of energy is that utilized in the breakdown of plants by bacteria. At least 90% of plant biomass dies in place. In this general scheme, birds, mammals, and even insects contribute almost nothing to energy flow. And yet we cannot ignore these higher taxa, as they often act as keystone species and ecosystem engineers. Herbivores are the major selective forces acting on living plants. Despite their smaller roles in energy transfer specifically, herbivores and carnivores play key roles in ecosystems in general. Therefore, these higher taxa must also be studied. As we have seen over and over in this book, when we ignore the complexity of the ecological world, we risk misunderstanding it.

Summary

Ecosystems can be viewed as transformers of energy. In this view, many food web properties, such as a short chain length, can be explained in terms of the inefficiency of energy transfer between trophic levels.

1. In ecosystems, energy is lost with each transfer up the food chain. In contrast, chemicals are not dissipated; they remain in the ecosystem and often concentrate at higher trophic levels.

2. In most ecosystems, plant material goes not to herbivores, but to decomposers after the plant dies. Most primary production is used in plant respiration, so plants are the most important consumers. Bacteria are next in importance as decomposers, degrading about 20% of the energy that plants use. Herbivores degrade less than 10 percent of the energy the bacteria use.

3. In general, primary production is highest in tropical forests and decreases progressively toward the poles. There are exceptions to this trend, however: Temperate salt marshes are as productive as many tropical areas, and tropical deserts are not as productive as temperate grasslands. Thus, temperature and rainfall both limit primary production.

4. Nutrient deficiency, particularly of nitrogen and phosphorus, can limit primary production. In aquatic systems, the availability of light can also be important. In freshwater lakes, excess phosphorus can cause huge algal blooms and turbid water, a process known as eutrophication.

5. The limit to secondary production is available primary production. This means that, at a large scale, plant defenses do not effectively reduce consumption by herbivores.

The ecosystem view of ecology emphasizes the importance of plants as primary producers, because they play an overwhelming role in energy transformation. This view contrasts with the population and community viewpoints, which emphasize the importance of interactions between many trophic levels. However, it must be remembered that while herbivores process relatively small amounts of energy, they act on living plants and so are important in an evolutionary sense.

Discussion Questions

1. Eutrophication is a phenomenon most often observed in lakes, causing huge algal blooms. However, eutrophication can occur in marine environments, too, where it causes "red tide," a dinoflagellate bloom that is toxic to fish. What do you think might cause "red tide," and how could you test your ideas?

2. Based on your knowledge of what limits primary production, how could we improve crop yields? Are there any drawbacks to your suggestions?

3. What limits the efficiency of secondary production? Are native herbivore–rangeland systems such as bison–prairie likely to be more efficient than exotic systems like cattle–grassland?

CHAPTER 22

Nutrients

Road Map

1. Soils exhibit a characteristic series of layers or horizons: O, A, B, C, and E. The availability of nutrients is dependent on soil characteristics.
2. In most systems, nutrient availability is limited because the most limiting nutrients, such as nitrogen and phosphorus, are bound up in living plants, in the soil, or in lake sediments.
3. Light is an important limiting resource for plants, and different photosynthetic pathways—the so-called C_3 and C_4 pathways—have evolved to take advantage of low light levels and high light levels, respectively.
4. Through their feeding activities, herbivores can affect the degree of soil nutrients. There is a close relationship between nutrient availability and the abundance of organisms.

Many nutrients, such as nitrogen, phosphorus, water, and carbon dioxide, play an important role in the distribution of organisms. In addition to the basic building blocks of hydrogen, carbon, and oxygen, organisms require a wide variety of chemical elements (Table 22.1). The elements required in the greatest amounts—nitrogen, phosphorus, sulfur, potassium, calcium, magnesium, and iron—are known as macronutrients. In addition, a whole host of nutrients is needed in so-called trace amounts. Animals get most of their nutrients from their prey: plants or other animals. Plants get their nutrients from water, often in the soil. A knowledge of soils is therefore of critical importance to ecologists.

Nutrients are not readily available in certain soils, yet they are critical to the well-being of plants and animals.

22.1 Soils

A comparison of elements in the soil with those accumulated by vegetation (Table 22.2) suggests that certain elements are likely to be the most limiting. These limiting elements, like nitrogen, have relatively low availability in the soil, yet high uptake rates by plants, which leads to relatively low reserves in the soil. If microorganisms did not recycle nitrogen, it would be depleted from the soil in 40 years.

Soil is the complex loose terrestrial surface material in which plants grow. It is composed of (1) weathered fragments of parent material in organic matter in various stages of breakdown, (2) water, and (3) minerals and organic compounds dissolved in the water. Soil is also teeming with life, much of it microscopic. When an organism dies, the process of decay occurs in the soil through the activities of various other organisms, especially bacteria and fungi. Humus, finely ground organic matter, is produced, and eventually the minerals are absorbed by plant roots.

One of the best examples of the effects of soil on the distribution of plants is exhibited by the serpentine soils that occur in scattered areas all over the world, including California. Serpentine rock is basically a magnesium iron silicate. Plants that grow there must be tolerant of low nitrogen, low

TABLE 22.1 Macronutrients (nutrients needed in large amounts) and trace elements (nutrients needed in very small amounts) required by organisms, and a partial list of their functions. Carbon (C), Hydrogen (H), and Oxygen (O) are the basic constituents of all organic matter and are not listed.

Nutrient		Used in
Macronutrients		
Nitrogen	(N)	Nucleic acids, enzymes, and proteins
Phosphorus	(P)	Nucleic acids, phospholipids, bone, and energy transfer
Sulfur	(S)	Proteins
Potassium	(K)	Solute in animal cells, sugar formation in plants, protein synthesis in animals
Calcium	(Ca)	Bone, muscle contraction, blood clotting, plant cell walls, regulation of cell permeability
Magnesium	(Mg)	Chlorophyll and many enzyme activation systems
Iron	(Fe)	Hemoglobin and enzymes
Sodium	(Na)	Extracellular fluids of animals, osmotic balance, nerve transmission
Chlorine	(Cl)	Chlorophyll, osmotic balance
Trace elements		
Manganese	(Mn)	Chlorophyll, fatty acid synthesis
Zinc	(Zn)	Auxin production in plants, enzyme systems
Copper	(Cu)	Chloroplasts, enzyme activation
Boron	(B)	Vascular plants and algae
Molybdenum	(Mo)	Enzyme activation systems
Aluminum	(Al)	Nutrient for ferns
Silicon	(Si)	Nutrient for diatoms
Selenium	(Se)	Nutrient for planktonic algae
Cobalt	(Co)	Nutrient for mutualistic association of legumes and nitrogen-fixing bacteria in root nodules and in ruminants
Vanadium	(V)	Tunicates, echinoderms, and some algae
Iodine	(I)	Higher animals, thyroid metabolism
Fluorine	(F)	Bone and teeth formation

phosphorus, and high magnesium. Such conditions are lethal to many plants; hence, serpentine soils have little value for agriculture or forestry. However, many species have adapted to these conditions, forming stunted, endemic communities not found in normal soils (Fig. 22.1).

Because most soil-forming processes, like the fall of litter and the weathering of rocks, tend to act from the top down, soil

TABLE 22.2 Elemental uptake by plants and soil nutrient reserves for macronutrients. *(Modified from Bohn, McNeal, and O'Connor, 1979.)*

Element	Annual Plant Uptake kg ha^{-1} yr^{-1}	Soil Nutrient Reserves (yr) (Soil Content/Annual Plant Uptake)
Nitrogen (N)	30	40
Phosphorus (P)	7	150
Sulfur (S)	2	320
Potassium (K)	30	430
Calcium (Ca)	50	260
Magnesium (Mg)	4	2,000
Iron (Fe)	1	52,000
Sodium (Na)	2	4,600
Chlorine (Cl)	0.06	220
Manganese (Mn)	1	1,000

Figure 22.1 Serpentine soils are so deficient in nutrients that few trees will grow on them, as shown by the bare patch in Jeffrey Pine forest, Josephine County, Oregon, which looks like a clear-cut but is not. *(Photo by Susan Harrison.)*

develops a vertical structure referred to as the *soil profile* (Fig. 22.2). Soil scientists recognize five main layers, or horizons. Historically, these layers were designated in alphabetical order: A, B, C, etc., from the top surface down toward the bedrock. However, the original lettering system has undergone some modification. The O horizon is now the topmost layer, consisting of organic debris like leaves dropped from plants. It may be subdivided into O_i, the true litter layer, and O_a, the humus layer proper, where decomposition of the leaf occurs. The leaf litter layer may fluctuate seasonally, whereas the humus

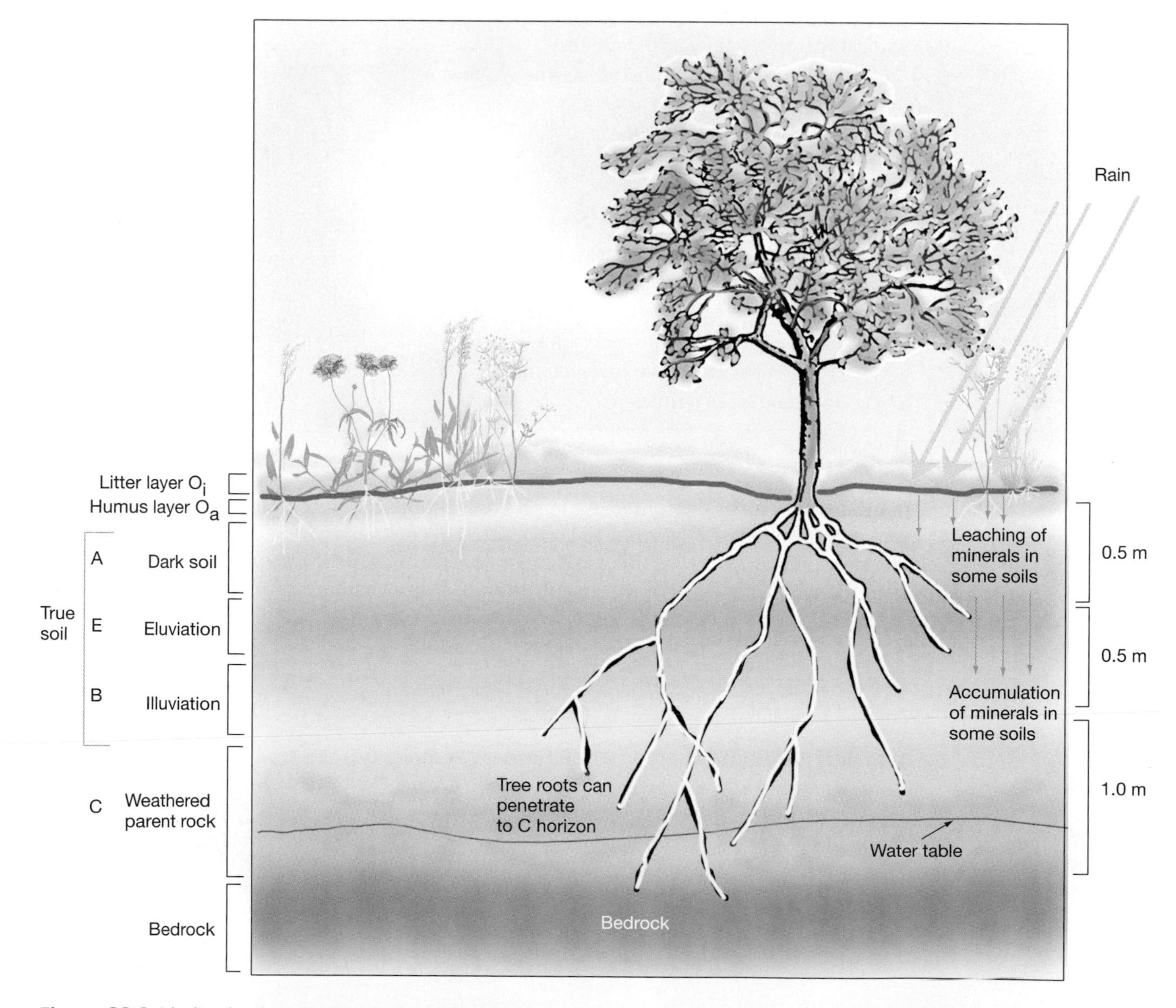

Figure 22.2 Idealized soil profile. O represents the organic layer, which may be divided into O_i, the loose litter, and O_a, the decomposing litter or humus. Beneath O is the dark colored A horizon, where many small plants have their roots and which has a high mineral and organic content. In the E layer, some minerals may be leaching out if the clay particles break down and there is sufficient rainfall resulting in soils called spodosols. These minerals may accumulate in the B layer, where iron, clay minerals, or organic matter may accumulate as well. Below the B layer is the weathered parent material (C layer) and below this the bedrock. Not all soils have this exact makeup, and the variations have different names (e.g., mollisols and oxisols).

layer does not. The next layer is the topsoil, or the A horizon, where most of the plant roots are located. Dead organic matter from the O horizon is added to this layer as it mixes with the topsoil in the A layer. The E layer is the zone of maximum leaching or eluviation—hence the designation E layer. Heavy rains can wash away or leach the nutrients from this layer of the soil and deposit them deeper down. Soils can have both A and E layers or just one layer—an A or an E layer. Organic material is largely absent from the next layer, the B horizon, and the mineral parent material is less thoroughly weathered. Materials leached from the A horizon may be deposited here (illuviation). Below the B horizon in some soils may be compact, slowly permeable layers like clay pans also leached out from layers above. Such layers sometimes cause severe waterlogging of the soil above the B layer. Below the B layer is the C horizon, generally consisting of weathered parent bedrock material.

Soil texture is based on the sizes of mineral particles making up the soil, classified as sand (<2 mm), silt (<0.05 mm), and clay (<0.002 mm). (See Fig. 22.3.) Soils made up mainly of small particles like clay are called *heavy soils*. Water soaks into heavy soils slowly, but is retained well there because of the myriad of small spaces between the tiny clay particles; such soils have the potential for being highly fertile. Light soils, such as sandy soils, are well aerated and allow free movement of roots and water, but are relatively infertile because water drains readily from them, leaching out their nutrients.

The difference in potential fertility between light and heavy soils results from the way minerals are retained in each of them. Minerals such as calcium and magnesium (specifically, most elements that form positive ions, or cations, when dissolved) are stored on the surfaces of particles. Anions, by contrast, are dissolved in soil water. Both are more prevalent in heavy soils. Plants' roots remove other cations and replace them with hydrogen ions. The potential fertility of a soil depends primarily on the soil's cation exchange capacity, a measure of the number of sites per unit of soil on which hydrogen can be exchanged for mineral cations. A soil with a high cation exchange capacity enables a plant to take up more nutrients and is therefore

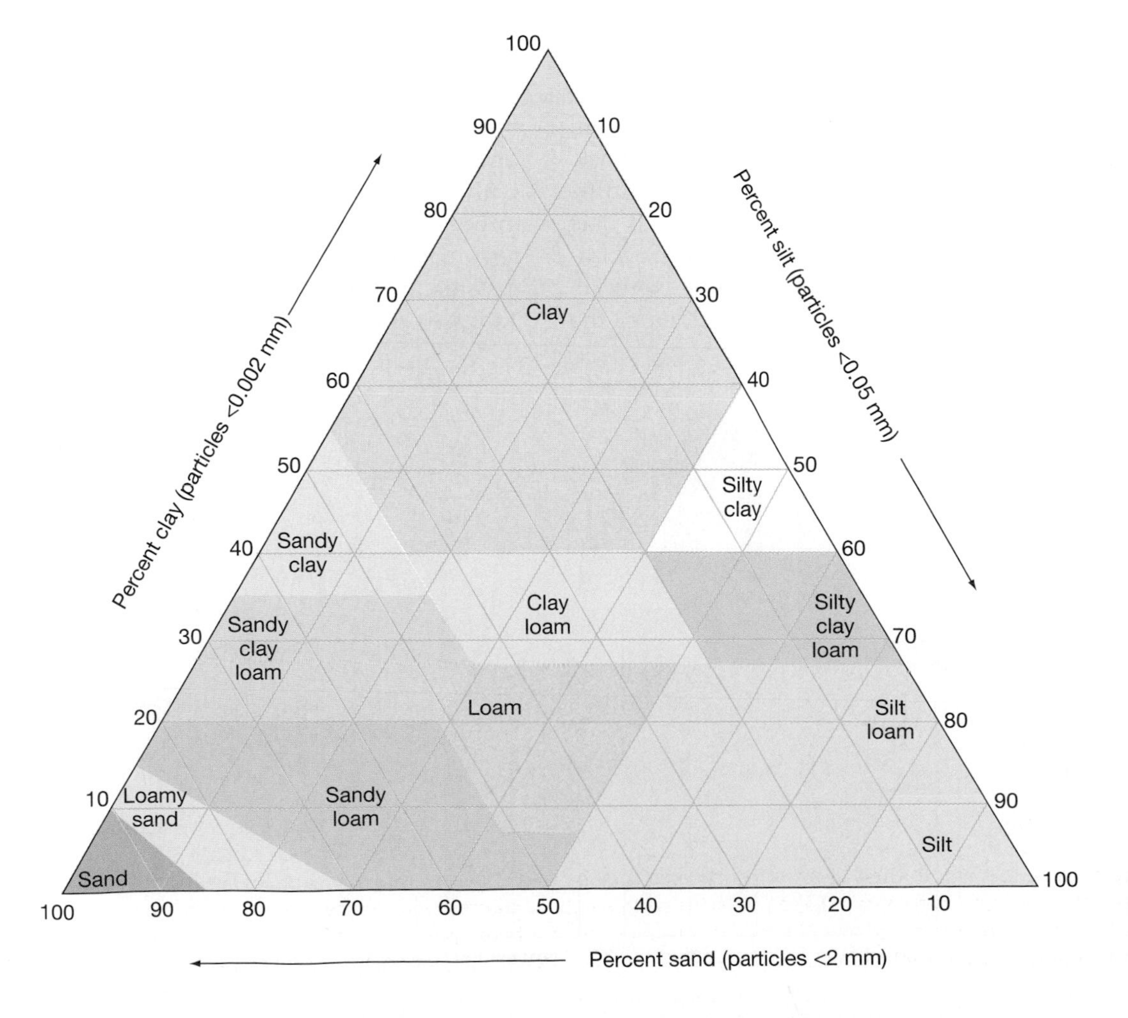

Figure 22.3 Soil texture classes according to the U.S. Department of Agriculture. The three sides represent baselines for sand, silt, and clay, with the apices representing 100% of each constituent. Percentages can be read off to give the textural name for any soil sample. For example, a soil with 10% clay, 50% sand, and 40% silt would be a loam.

(a)

(b)

(c)

Figure 22.4 Different types of soil horizons. (a) Mollisols, rich grassland soils with much organic matter. (b) Spodosols, where leaching of nutrients reduces fertility in the top or A horizon. (c) Oxisols, a characteristic red-colored soil of the tropics where abundant rainfall leaches material away from the top layers.

considered more fertile. Clay soil may have a cation exchange capacity from 2 to 20 or more times that of sand.

A soil with a high cation exchange capacity and, consequently, high potential fertility may actually be infertile if most of the sites on the particles are already filled with hydrogen ions. Such a soil will be acidic and will have little calcium to supply to the growing plant. This aspect of fertility is measured by the percentage base saturation—the percentage of the exchange capacity satisfied by calcium, magnesium, and similar elements. If the percentage base saturation is 60%, then 60% of the sites are filled by basic ions and 40% by hydrogen (acidic) ions.

Soil development is strongly influenced by the parent material, rainfall, evaporation, and temperature. However, vegetation can be important, too. In the U.S. northern Midwest, coniferous forest, deciduous forest, and grassland grow within a few miles of one another, and the differences in the soil are due entirely to the differing effects of the vegetation. Coniferous forest produces strongly acid soils with a heavy layer of undecomposed litter and hardpan of leached clay in the B horizon. Deciduous forest produces a thinner, less acidic litter that grades into the mineral soil below. In grassland the soil is less acid still, and each year the entire aboveground part of the plant dies back, allowing decomposers access to the nutrients at or near the surface of the soil. The grassland soil is dark because of the addition of this aboveground organic material, as well as the organic matter added to the soil at various depths by the death each year of many fine roots. These rich grassland soils are known as **mollisols** (Fig. 22.4). Thus, soils may have many variations in color and profile that are different from the idealized profile shown in Fig. 22.3, according to the type of vegetation and other features. There are at least 11 different types of soil, known as *orders*. For example **podsolization** happens in acid soils of coniferous forests wherein clay particles break down in the A horizon and the soluble ions are leached downward into lower horizons. This leaching reduces the fertility of the soil since the A horizon is where most plant roots grow. Such soils are known as **spodosols** and are found in New England, Michigan, and Wisconsin. In the tropics, the breakdown of clay and leaching of silica into lower layers because of the abundant rainfall causes **laterization**, a process in which only iron oxides and aluminum hydroxide are left in the A layer, giving soils a characteristic red coloration. These so-called **oxisols** are invariably of poor quality.

As we saw in Chapter 21, many studies have shown that nutrient limitations in the soil influence the abundance of plants, but relatively few studies have shown a similar influence on animals. A striking demonstration that nutrients limit animal distributions came from a study of African ungulates by Sam McNaughton of Syracuse University (1988), who showed that the mineral content of food is an important determinant of the spatial distribution of ungulates in the Serengeti National Park of Tanzania (Fig. 22.5). Areas of high concen-

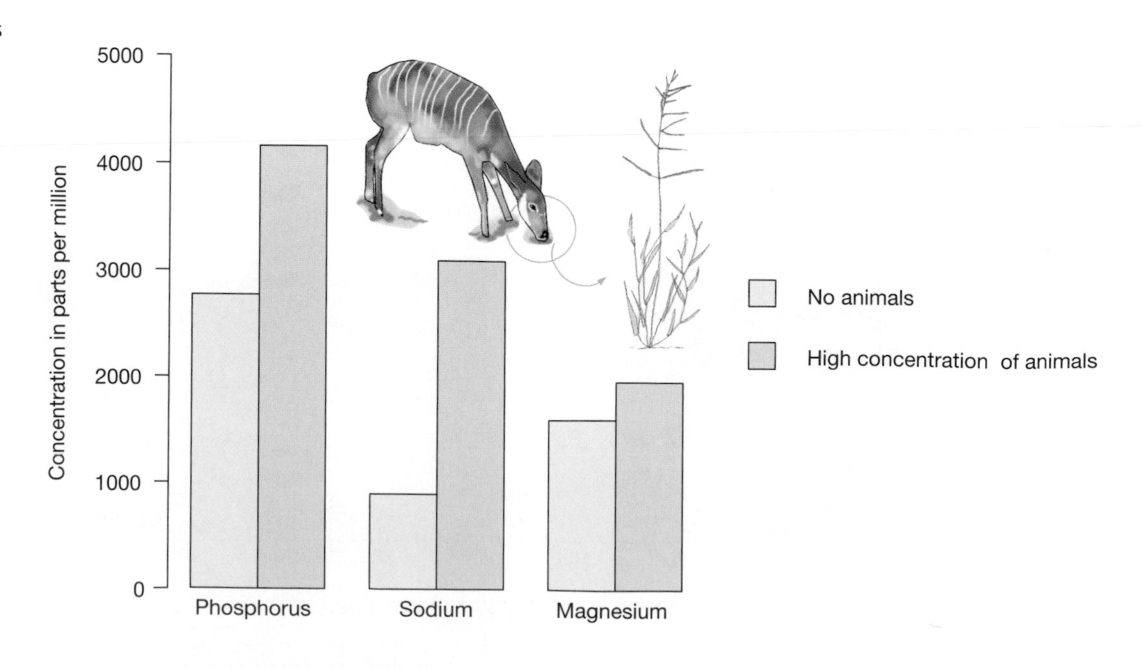

Figure 22.5 High concentrations of minerals in grasses of the Serengeti Plain lead to high concentrations of animals. *(From data in McNaughton, 1988.)*

tration of phosphorus, sodium, and magnesium supported higher densities of large herbivores than did areas of low concentrations of these minerals. Phosphorus is of particular importance to lactating animals.

22.2 Nutrient Availability

It is often argued that ecosystems can be well understood by following the paths of nutrients—the biogeochemical cycles. Unlike energy, chemicals are not dissipated, but remain in the ecosystem indefinitely. They therefore tend to accumulate in individuals or populations, which then act as "pools" of nutrients. Nutrients cycle between pools through meteorological, geological, or biological transport mechanisms. Meteorological inputs include dissolved matter in rain and snow, atmospheric gases, and dust blown by the wind. Meteorological sulfur alters the pH of lakes. The sulfur is emitted through smokestacks and becomes incorporated into rain, which then falls as "acid rain." Geological inputs include elements transported by surface and subsurface drainage, such as phosphorus being transported by subsurface flows and accumulating in lakes. Biological inputs result from the movements of animals or parts of animals between ecosystems. A more thorough knowledge of how nutrients accumulate in food webs would enable us to predict the indirect effects of the passage of novel substances in an ecosystem. For example, the toxic effects of DDT proved disastrous for higher organisms like birds and fish, even though these animals were never the intended targets. (See "Applied Ecology: Food Webs and the Passage of Pesticides.") A better understanding of nutrient cycles, food webs, and the passage of pesticides through ecosystems might have prevented such a catastrophe.

Nutrient cycles can be divided into two broad types:

1. *Local cycles*, such as the phosphorus cycle, which involve no long-distance transfer of elements.
2. *Global cycles*, which involve an interchange between the atmosphere and the ecosystem. Most elements, including nitrogen, carbon, oxygen, and water, are involved in global cycles. Global nutrient cycles unite all the world's living organisms into one giant ecosystem called the **biosphere**.

In most ecosystems, the availability of nutrients is limited because most nutrients are bound up in biological and geological pools.

We saw in Chapter 21 that many nutrients, such as nitrogen and phosphorus, limit the flow of energy in ecosystems. The **nutrient turnover time**—the time taken for a nutrient to pass through a complete cycle—is generally on the order of 10–50 years for most nutrients. In almost every case, the inputs and outputs of nutrients are small in comparison with the amounts held in biomass and recycled within the system, producing slow turnover rates. This relationship was illustrated by a study of the annual nutrient budgets in a forest called Hubbard Brook in the U.S. northeast. (Fig. 22.6). At Hubbard Brook, as in most ecosystems, nitrogen was added to the system only in precipitation

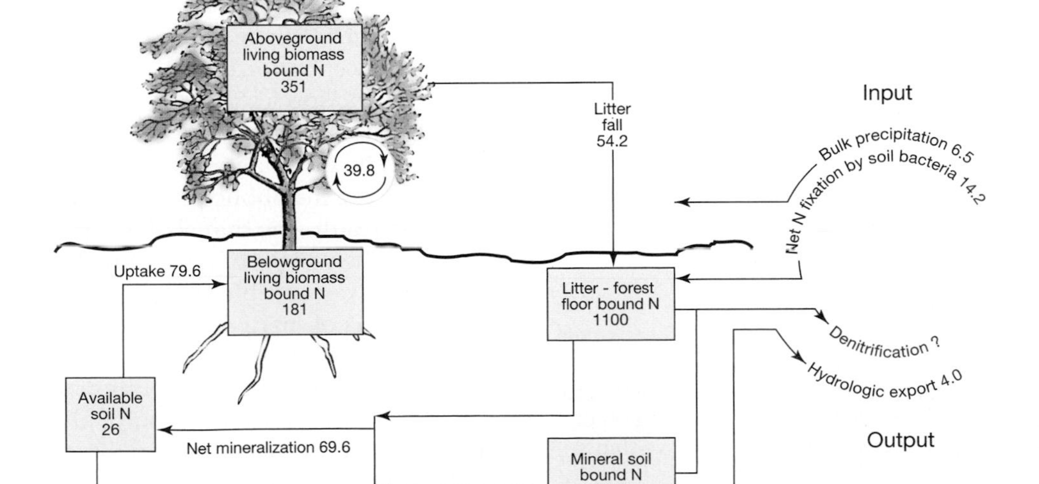

Figure 22.6 Simplified nitrogen budget for the undisturbed Hubbard Brook Experimental Forest. Values in boxes are the sizes of the various nitrogen pools, in kilograms of nitrogen per hectare. Values by arrows are transfer rates, expressed in kilograms of nitrogen per hectare per year. *(After Bormann et al., 1977.)*

Applied Ecology

Food Webs and the Passage of Pesticides

Dichloro-diphenyl-trichloroethane (DDT) was first synthesized in 1874. In 1939, its insecticidal properties were recognized by the Swiss scientist Mueller, who won a Nobel prize in 1948 for that discovery and subsequent research on the uses of the chemical. The first important application of DDT was in human health programs during and after World War II, and at that time its use in agriculture also began. The global production of DDT peaked in 1970, when 175 million kilograms (kg) were manufactured. The peak of production in the United States was 90 million kg in 1964. Most industrialized countries banned the use of DDT after the early 1970s, and only a few less developed countries still use it today.

DDT has several chemical and physical properties that profoundly influence the nature of its ecological impact. First, DDT is persistent in the environment: It is not easily degraded to other, less toxic chemicals by microorganisms or by physical agents such as the sun and heat. The typical persistence in soil of DDT is about 10 years, two to three times as long as the persistence of other organochlorine insecticides. The good news is that, following the outlawing of DDT in the United States in 1973, the amounts of the chemical in U.S. soils are by now negligible. Another important characteristic of DDT is its low solubility in water (less than 0.1 part per million [ppm]) and its high solubility in fats or lipids, a characteristic that it shares with other chlorinated hydrocarbons. In the environment, most lipids are present in living tissue; therefore, because of its high lipid solubility, DDT has a great affinity for organisms, and it tends to concentrate in biological tissues.

Figure 2 DDT can accumulate in fish which brown pelicans eat. In turn, this causes thin-shelled eggs which are easily crushed by the parent pelican.

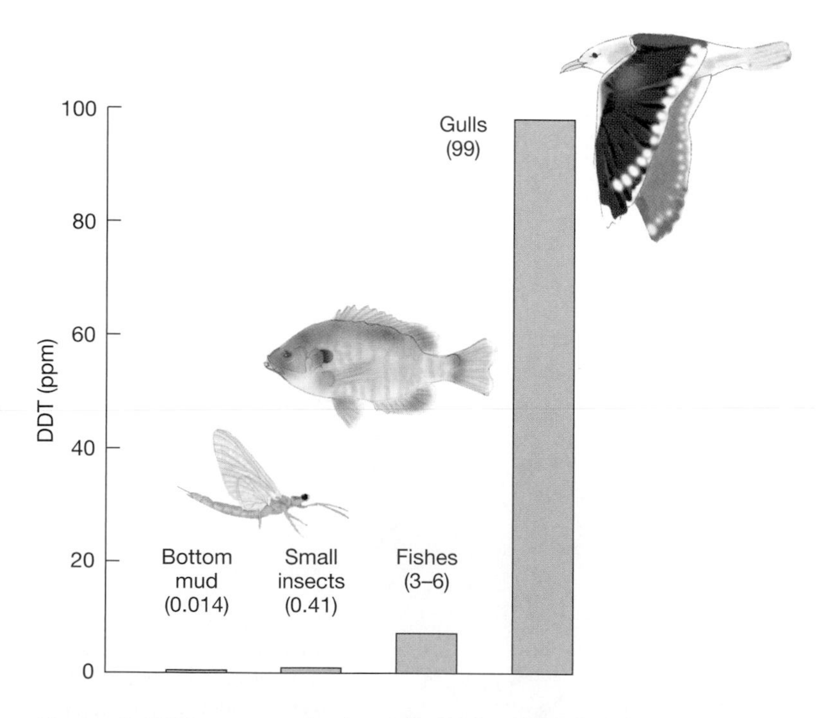

Figure 1 DDT concentration in a Lake Michigan food chain. The DDT load in gulls was about 240 times that in small insects. The bioconcentration of DDT in lipids causes its concentration to increase at each successive step in the food chain.

Furthermore, because organisms at the top of a food web are effective at accumulating DDT from their food, they tend to have an especially large concentration of DDT in their lipids. A typical pattern of food web accumulation of DDT is illustrated in Fig. 1, which summarizes the DDT residues found in a Lake Michigan food chain. The prime reason for the introduction of DDT there was to control mosquitoes. The largest concentration of DDT was found in gulls, an opportunistic species that often fed on small fish that, in turn, ate small insects. An unanticipated effect of DDT (and its metabolites) on birds was its interference with the metabolic process of eggshell formation. The result was thin-shelled eggs that often broke under the weight of incubating birds (Fig. 2), and DDT was responsible for a drastic decrease in the populations of many birds due to failed reproduction. Large residues also were present in some game fish, which became unfit for human consumption. Had scientists had a more thorough knowledge of the relevant food webs, some of these side effects might have been predicted.

(6.5×10^{-4} kg m^{-2} yr^{-1}) and atmospheric nitrogen fixation by microorganisms (14×10^{-4} kg m^{-2} yr^{-1}). The export, in streams, was only 4×10^{-4} kg m^{-2} yr^{-1}, although denitrification by other microorganisms, releasing nitrogen to the atmosphere, would also have been occurring, but was not measured. This low movement of nitrogen emphasizes how securely it is held and cycled within the ecosystem. Most nutrients are bound up in the soils or in living forest biomass. It is no wonder that nutrient availability is so limited!

These observational data were confirmed by a large-scale experiment in which all the trees were felled in one of the Hubbard Brook catchments (Fig. 22.7). The rate of export of dissolved nutrients from the experimental areas reached 13 times the rate from intact control areas (Fig. 22.8). There were two reasons for this huge increase in nutrients leaving the system. First, the enormous reduction in leaves that was a result of felling the trees led to 40 percent more water being discharged to streams, and this increased volume of water moving through the soil increased chemical leaching. Second, again because of the felling, there were no deciduous trees leafing out in the spring. The inorganic nutrients released by the activity of decomposers were therefore not taken up by the trees and instead leached out into the drainage water. Most other deciduous forests exhibit nutrient cycling patterns similar to those in Hubbard Brook, with inputs and outputs relatively low. Thus, nutrients are available in only limiting amounts in most ecosystems.

Figure 22.7 Deforested watershed 5 in the Hubbard Brook experimental area. Deforestation was used as an experimental means to see how much the removal of living biomass would speed up nutrient turnover in a terrestrial ecosystem.

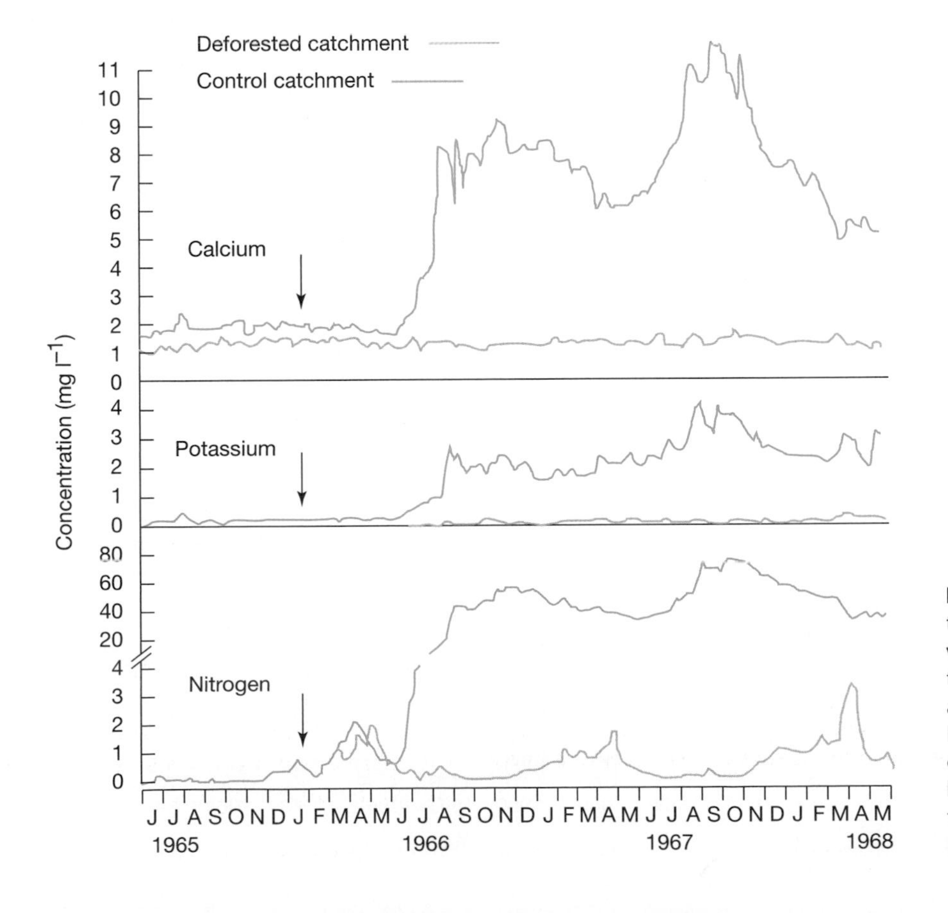

Figure 22.8 Concentrations of nutrients in stream water from the experimentally deforested catchment and a control catchment at Hubbard Brook. The timing of deforestation is indicated by arrows. Note that the nitrogen axis has a break in it. *(After Likens et al., 1970.)*

Figure 22.9 The inquisitive brown lemming, *Lemmus sibericus.*

The nutrient recovery hypothesis suggests that, following low abundance, herbivores recover to high densities only when nutrient availability to plants is high.

The efficiency with which organisms use nutrients can vary according to species and location. Evergreen plants generally use nutrients more efficiently than deciduous species do. The greater nutrient efficiency of evergreens is due partly to the greater longevity of their leaves and consequent lower loss of nutrients per unit time. Their greater retention of nutrients may explain why evergreen species are common in nutrient-poor environments. In fact, nutrients are used more efficiently in such environments. Also, plants in the Southern Hemisphere use nutrients more efficiently than do plants in the Northern Hemisphere. Most areas of the Northern Hemisphere have been glaciated, and the glaciers pulverized the rocks, creating a rich soil. However, in the Southern Hemisphere, the lands that arose from erstwhile Gondwanaland have very old, weathered, nutrient-poor soils. These lands include Australia, South America, and India. Many of the soils on them are deficient in nutrients and yet support healthy forests. However, deforestation of these systems takes most of the nutrients away, and little is left to promote crop production. In northern soils, many nutrients are still left in the soil and crops thrive.

In contrast to terrestrial systems, in freshwater streams and rivers only a small fraction of available nutrients are locked up in living biomass. Most nutrients flow into the lakes or the ocean. In lakes, however, there is often no outflow, and nutrients are exported via deposition to the sediment. No matter what the ecosystem, however, there is very rarely a surfeit of nutrients available to the plants and animals inhabiting it: Most nutrients are bound up in the soil or the living biomass in terrestrial systems or in the sediment in freshwater systems.

The nutrient recovery hypothesis.

A simple example from the Arctic shows how the release of limiting nutrients can affect the abundance of organisms. Phosphorus can be critical in limiting individual species and population cycles. One of the most striking examples of the effect of phosphorus was often thought to be that of the brown lemming, which lives in the tundra areas of North America and Eurasia (Fig. 22.9). The ecosystem of the tundra is simple compared with temperate or tropical ecosystems, and the lemming is often the major herbivore. Every three to four years, numbers of these small rodents build up, only to tumble again the next year (Fig. 22.10). This ecosystem has been studied in some detail on the Arctic coastal plain tundra near Point Barrow, Alaska. The traditional story is as follows: As lemming numbers increase, their feces and urine stimulate plant growth, which in turn increases lemming production. Eventually, the lemming numbers be-

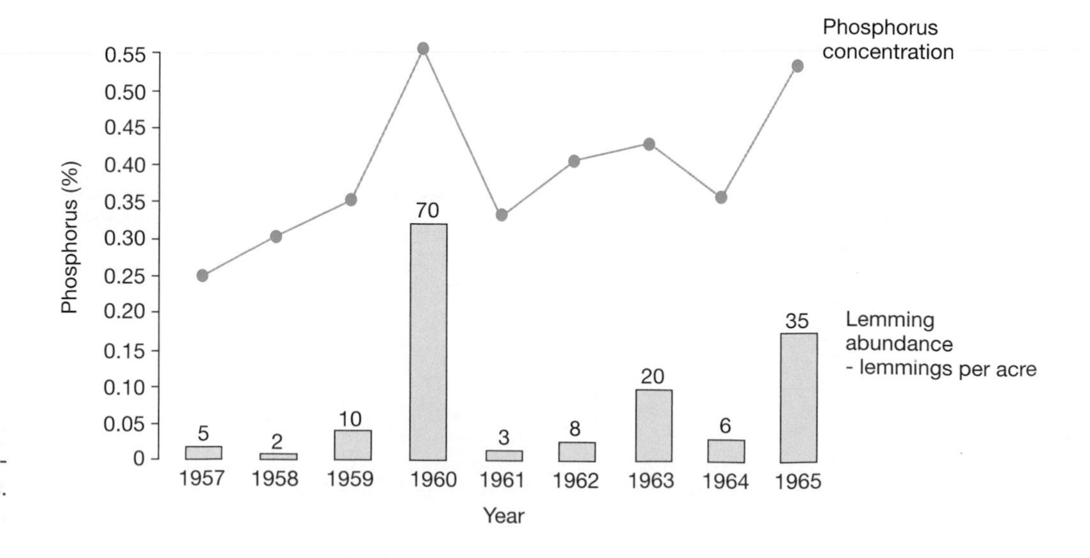

Figure 22.10 The influence of plant phosphorus content on lemming numbers. As phosphorus increases, so do lemming numbers. *(After Schulz, 1969.)*

come so great that their foraging thins out the vascular plant cover. More sunlight then reaches the ground, thawing out the soil, so nutrients leach down further into the soil and are not available to the plants. The intense grazing and lack of root nutrients soon lower the quality of forage, reducing lemming numbers drastically. Low nutrient quality also prevents breeding the following year. Gradually, during the next two to three years, the vegetation slowly recovers, nutrients increase, and the cycle begins again. This description of the process that controls population variation was termed the **nutrient recovery hypothesis** by Frank Pitelka (1964) and Arnold M. Schultz (1964) of Berkeley. It must be stressed that not all the elements of the cycle have been proven beyond question. Enemies—either predators or parasites—may play a role, and time lags in the system would also generate a four-year cycle, as we discussed in Chapter 6. However, the example at least shows how it is possible to relate an organism's abundance to the availability of nutrients.

22.3 Light

Light can be a limiting resource for photosynthesis. Some plants, such as oak and maple trees, reach maximal photosynthesis at one-quarter of full sunlight. Others, like sugarcane or the desert shrub *Larrea*, never reach a maximum, but continue to increase their photosynthetic rate as the intensity of light increases.

One reason the rate of photosynthesis varies among plants is that there are three different biochemical pathways by which the photosynthetic reaction can occur. Most plants use the C_3 pathway (or Calvin–Benson cycle), whereby CO_2 from the air is first converted to 3-phosphoglyceric acid, which is a three-carbon molecule—hence the name C_3. Until the mid-1960s, this pathway was believed to be the only important means of fixing carbon in the initial steps of photosynthesis. In 1965, sugarcane was discovered to fix CO_2 by first producing malic and aspartic acids (four-carbon molecules), and the C_4 pathway (Hatch–Slack) of photosynthesis was uncovered. C_4 plants have all the biochemical elements of the C_3 pathway as well, so they can use either method to fix CO_2. Typical C_4 plants do not reach saturation light levels even under the brightest sunlight, and they always produce more photosynthate per unit area of leaf than do C_3 plants. Therefore, C_4 plants are more efficient than C_3 plants. Most C_4 species are grasses and sedges, and because they do well in full sunlight, those species are more common in tropical areas than in temperate areas (Fig. 22.11).

Some desert succulents, such as cacti of the genus *Opuntia*, have evolved a third modification of photosynthesis: crassulacean acid metabolism (CAM). These plants are the opposite of typical plants in that they open their stomata to take up CO_2 at night, presumably as an adaptation to minimize water loss in the day. This CO_2 is stored as malic acid, which is then used to complete photosynthesis during the day. CAM plants have a very low rate of photosynthesis and can switch to the C_3 mode in daytime if the weather is cool. They are

Light is an important limiting resource for plants.

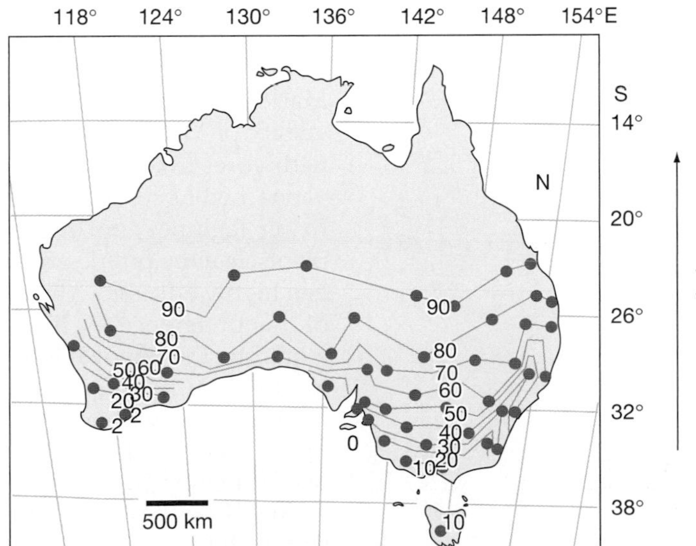

Figure 22.11 Approximate contour map of C_4 native grasses in Australia. Lines give percentages of C_4 species in total grass flora for different geographic regions. The percentage of C_4 species increases toward the tropics. *(From Hattersley, 1983.)*

adapted to live in very dry desert areas where little else can grow.

The C_3–C_4 dichotomy does not concern just light. C_4 plants can absorb CO_2 much more effectively than C_3 plants. Consequently, they do not have to open their leaf stomata as much, so they lose less water than C_3 plants. C_4 plants are therefore more frequent in arid conditions. Also, the concentration of Rubisco, the enzyme that converts CO_2 to either a C_3 or C_4 molecule, is generally lower in C_4 plants. Because Rubisco makes up 25% of the nitrogen in the leaf, the total leaf nitrogen content is generally lower in C_4 plants. As a consequence, C_4 plants are nutritionally less attractive to some herbivores. So why haven't C_4 plants come to dominate world vegetation? The answer is that they do not photsynthesize as efficiently in the shady, wet environments of northern latitudes. The few C_4 plants that have penetrated into temperate regions are often found in stressful environments (such as *Spartina* grasses growing in salt marshes) where osmotic conditions limit the availability of water.

22.4 Organismal Effects on Nutrient Availability

As well as responding to the availability of nutrients, both plants and animals can alter nutrient cycles on their own.

So far, much of our discussion has focused on how population densities of plants and animals can be affected by nutrient concentrations. Interestingly, some studies show how organisms themselves can affect the availability of these nutrients. This is certainly true for light. Forest trees differ in their tolerance to shade. Sun-loving species soon overtop the shade-loving species because of their faster rate of growth in full sun. However, they cast shade that is too dark for their own seedlings, and that inhibits their growth. Light levels below the crowns of sun-loving species are still sufficient for the growth of those species that do better in intermediate light, and the light below these intermediate species is in turn enough for species that are the most tolerant of shade. Thus, a stratified canopy may develop, with sun-loving species at the highest level, intermediate species below them, and shade-loving species lower still.

Returning to the Serengeti ecosystem, McNaughton and colleagues (1997) further investigated the relationship between ungulate densities and the availability of sodium in their forage. Just as they had showed earlier, ungulates—mainly Thompson's and Grant's gazelles, topi, and Kongoni antelopes—preferred areas rich in sodium. However, this time McNaughton and colleagues put a fence around certain areas to prevent grazing. They found higher soil sodium concentrations outside the fences than inside by an amount of 52.5 g of Na per square meter per year. To explain this finding, they argued that herbivores increase the transpiration of chewed grasses, and as the water evaporates through the cuts in plants, the salts are left behind. The process, analogous to the salinization of agricultural land, leads to increased sodium concentrations in the soil. In this way, McNaughton argued, some nutrient concentrations are raised by the activity of herbivores.

A different phenomenon was shown by Mark Ritchie and his colleagues at the University of Minnesota (1998), who excluded herbivores from an oak savanna. In the savanna, herbivores preferentially eat nitrogen-fixing legumes and nitrogen-rich woody plants, leaving prairie grasses that use nitrogen highly efficiently (Fig. 22.12). The effect of this preference for high-nitrogen plants would be the removal of nitrogen from the local system.

Finally, as we have stressed many times in this book, exotic species can affect ecosystem function by changing soil nutrient availabilities. At the northern end of the Great Plains in Canada, 20 miles north of

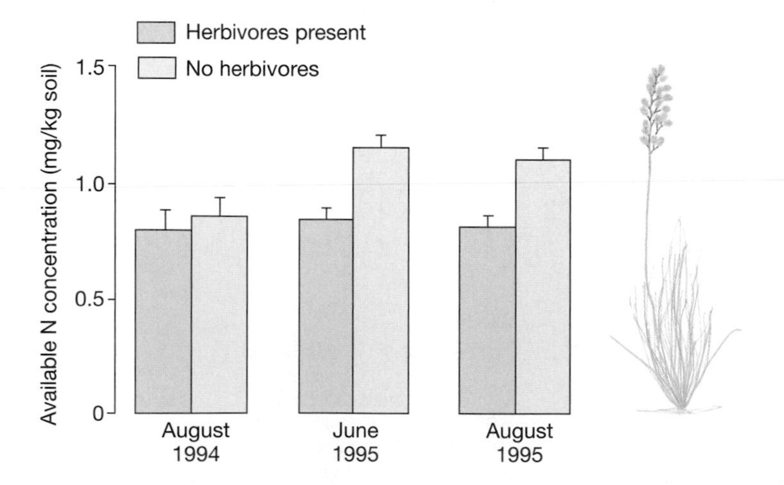

Figure 22.12 Increase in available soil nitrogen at three different times following the exclusion of herbivores from plots in Minnesota. Herbivores eat the nitrogen-rich plants, which absorb much of the soil nitrogen. *(After Ritchie et al., 1998.)*

the Montana state line, *Agropyron cristatum*, an introduced C_3 perennial grass, has been planted widely since the 1930s and dominates millions of acres of grassland. It has outcompeted native prairie grass over large areas. In these areas, soil carbon and available soil nitrogen are lower than in native prairie. Janice Christian and Scott Wilson of the University of Regina (1999) compared soil nitrogen levels of normal, undisturbed prairie, disturbed prairie that had been ploughed, but had returned to native prairie, and areas invaded by the exotic *Agropyron* (Fig. 22.13). Disturbance alone decreased soil nitrogen, but not significantly, whereas *Agropyron* depleted soil nitrogen greatly. The low soil nitrogen in turn reduced the diversity of plant species growing in the area.

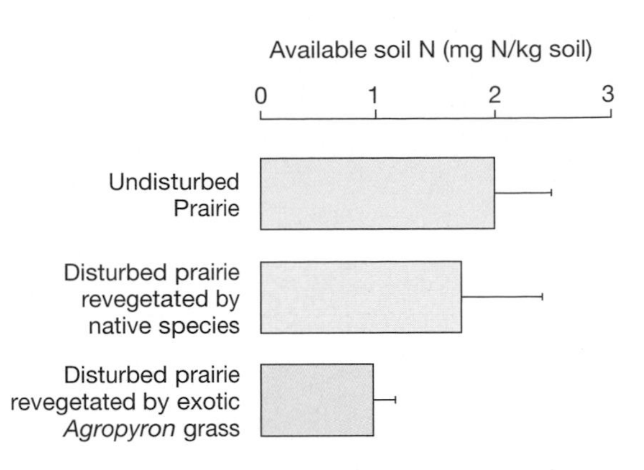

Figure 22.13 The decrease in soil nitrogen in ploughed soils in Canadian prairies that have been recolonized by native prairie plants is slight compared with that in areas which have been disturbed and planted with the exotic *Agropyron*. *(After Christian and Wilson, 1999.)*

In sum, nutrients, including light, water, sodium, and especially soil nitrogen, can influence the distribution of species. Conversely, higher trophic levels themselves can influence the availability of nutrients in soils. Further, the introduction of exotic species can upset the balance we see in native systems. As we have observed so often in this book, anthropogenic change, especially in the case of the introduction of exotics, is nearly always disadvantageous to the distribution and abundance of native species.

Summary

Nutrients may be important to species beyond their simple effect on boosting primary productivity.

1. The most critical limiting elements are nitrogen and phosphorus, which exist at relatively low levels in soils. Soil structure can greatly affect the availability of nutrients to organisms.

2. Nutrients cycle between biological, geological, and meteorological pools. Most nutrients are bound up in living things or in soils. The nutrient recovery hypothesis states that populations of some species can recover to high levels following the release of nutrients into the system.

3. Light can be limiting for plant photosynthesis. The C_3 and C_4 metabolic pathways in plants have evolved to take advantage of low-light and high-light environments, respectively.

4. Although the abundance of many organisms is affected by nutrient availability, organisms themselves can affect nutrient cycles. Herbivores can increase the availability of nutrients in the soil by feeding or decrease nutrient availability by eating nitrogen-rich plants.

There is a complex relationship between the abundance of organisms and the availability of nutrients. As we have seen many times, upsetting this relationship—for example, by the introduction of exotic species—can seriously affect the distribution and abundance of native species.

Discussion Questions

1. What sort of organisms live in soil? (You may need to do some research in the library to find out.) Where are their densities greatest—in the O, A, B, E, or C horizon? What characteristics do these species possess that enable them to live in the soil? What might govern their densities?

2. It is thought that human alterations have approximately doubled the rate of nitrogen input into the terrestrial nitrogen cycle. How has this happened? What effects will it have?

3. Slash-and-burn agriculture is common in tropical countries. Trees are cut down and burned in place, and crops are planted on the newly cleared fields. How does this practice affect nutrient cycles in general and soil nutrient levels in particular?

4. How would you determine whether herbivores increase or decrease soil nutrients?

Glossary

abiotic factors environmental influences produced other than by living organisms (e.g., temperature, pH, and other physical and chemical influences); contrast with **biotic factors**.

acclimation a collection of changes by an organism subjected to new environmental conditions that, together, enables the organism to withstand those conditions.

acid rain rainfall acidified by contact with sulfur dioxide (a by-product of the burning of fossil fuels) in the atmosphere.

adaptation a change in an organism's structure or habits that better enables the organism to adjust to the environment.

adiabatic cooling the cooling of air that results as the air is blown up over mountains and pressure is lessened.

aerobic having to do with organisms or processes that require oxygen.

age class collection of individuals of a particular age in a given population.

agroforestry the practice of mixing both forestry and agriculture, usually involving pastureland and tree farming.

allele one of two or more alternative forms of a gene located at a single point (locus) on a chromosome.

allelopathy the suppression of growth of one plant species by another through the release of toxic chemicals.

allogenic succession succession governed by abiotic disturbances.

allopatric speciation separation of a population into two or more evolutionary units as a result of reproductive isolation arising from geographic separation.

altruism enhancement of the fitness of a recipient individual by acts that reduce the evolutionary fitness of the donor individual.

amensalism competition in which one species has a large effect on another, but the other species has little effect on the first.

anaerobic pertaining to organisms or processes that occur in the absence of oxygen.

annual an organism, usually a plant, that completes its life cycle from birth through reproduction to death in a year.

aposematism coloration warning indicating that the bearer is toxic.

assimilation efficiency the percentage of energy ingested in food that is assimilated into the protoplasm of an organism.

association a group of species occurring in the same place.

autecology study of the individual in relation to environmental conditions; contrast with **synecology**.

autogenic succession succession governed by biotic factors, as in herbivores.

autotroph an organism that obtains energy from the sun and materials from inorganic sources; contrast with **heterotroph**.

balanced polymorphism the stable occurrence of more than one form in a population.

Batesian mimicry resemblance of an edible (mimic) species to an unpalatable (model) species to deceive predators.

benthic pertaining to aquatic bottom or sediment habitats.

benthos bottom-dwelling aquatic organisms (for example, burrowing worms, molluscs, and sponges).

Bergmann's (1847) rule among homeotherms, the tendency for organisms in colder climates to have larger body size (and thus smaller surface-to-volume ratio) than those in warm climates.

biochemical oxygen demand (BOD) the amount of oxygen that would be consumed if all the organic substances in a given volume of water were oxidized by bacteria and other organisms; reported in milligrams per liter.

biodegradable capable of being decomposed quickly by the action of microorganisms.

biogeochemical cycle the passage of a chemical element (such as nitrogen, carbon, or sulfur) from the environment into organic substances and back into the environment.

biogeography the branch of ecology that deals with the geographic distribution of plants and animals.

biological control the use of natural enemies (diseases, parasites, predators) to regulate populations of pest species.

biological magnification the concentration of a substance as it moves up the food chain from consumer to consumer.

biomass dry weight of living material in all or part of an organism, population, or community; commonly expressed as weight per unit area.

biome a major terrestrial climax community (e.g., coniferous forest or tundra).

biosphere the whole Earth ecosystem.

biota all the living organisms occurring within a certain area or region.

biotic factors environmental influences caused by living organisms; contrast with **abiotic factors**.

boreal occurring in the temperate and subtemperate zones of the Northern Hemisphere.

calcareous in soil terminology, rich in calcium carbonate and having a basic reaction.

canonical distribution Coined by Preston (1962), another term for a lognormal distribution of the numbers of individuals or species.

carcinogen a chemical or physical agent capable of causing cancer.

cardinal index an index that treats all species as equal in importance.

carnivore an animal (or plant) that eats other animals; contrast with **herbivore**.

carrying capacity the number of individuals that the resources of a habitat can support.

character displacement divergence in the characteristics of two otherwise similar species whose ranges overlap; caused by competition between the species in the area of overlap.

clear-cutting the practice of cutting all trees in an area, regardless of species, size, quality, or age.

climax community the community capable of indefinite self-perpetuation under given climatic and edaphic conditions.

clone a lineage of individuals reproduced asexually (e.g., vegetatively in plants).

coefficient of relatedness the proportion of the total genotype of one individual present in another as a result of shared ancestry.

coevolution development of genetically determined traits in two species to facilitate some usually mutually beneficial interaction between them.

coexistence occurrence of two or more species in the same habitat; usually applied to potentially competing species.

cohort those members of a population that are of the same age, usually in years or generations.

commensalism an association between two organisms in which one benefits and the other is not affected.

community a group of populations of plants and animals in a given place; used in a broad sense to refer to ecological units of various sizes and degrees of integration.

compensation level the depth within a lake at which the photosynthate production equals the energy used up by respiration.

competition the interaction that occurs when organisms of the same or different species use a common resource that is in short supply (exploitation competition) or when they harm one another in seeking a common resource (interference competition).

competitive exclusion principle the hypothesis that two or more species cannot coexist and use a single relatively scarce resource.

connectance in a food web, the actual number of links divided by the potential number of links.

conspecific belonging to the same species.

consumer an organism that obtains its energy from the organic materials of other organisms, living or dead; contrast with **producer**.

consumption efficiency the percentage of energy at one trophic level that is eaten by the next higher trophic level.

continental drift the movement of the continents, by tectonic processes, from their original positions as parts of a common landmass to their present locations.

continental island an island that is near to and geologically part of a continent (e.g., the British Isles or Trinidad).

continental shelf the shallow part of the seafloor immediately adjacent to a continent.

convergent evolution the development of similar adaptations by genetically unrelated species, usually under the influence of similar environmental conditions.

courtship any behavioral interaction between individuals of opposite sexes that facilitates mating.

crypsis coloration or appearance that tends to prevent detection of an organism by predators.

decomposers consumers—especially microbial—that reduce organic food into tiny particles and mineral nutrients.

deforestation removal of trees from an area without adequate replanting.

denitrification enzymatic reduction of nitrates to nitrogen gas by bacteria.

density the number of individuals per unit area.

density dependent having an influence on a population that varies with the number of individuals per unit area in the population.

density independent having an influence on a population that does not vary with the number of a individuals per unit area in the population.

desert a region receiving very small amounts of precipitation

deterministic model a mathematical model in which all relationships are fixed; contrast with **stochastic model**.

diapause a period of suspended growth or development during which the organism is resistant to unfavorable environmental conditions.

dimorphism the occurrence of two forms of individuals in a population.

diploid having two copies of each gene in a genotype.

direct competition exclusion of certain individuals from resources by other individuals through aggressive behavior.

dispersal movement of organisms away from the place of their birth or from centers of population density.

distribution the collection of areas in which a species lives and reproduces.

diversity a measure of the number of species and their abundance in a community or region; alpha diversity is the diversity of a particular habitat; beta diversity is the diversity of a region pooled across habitats.

dominance the influence or control exerted by one or more species in a community as a result of their greater number, coverage, or size.

ecology the branch of science dealing with the relationships of living things to one another and to their environment.

ecosystem a biotic community and its abiotic environment.

ecosystem engineer a keystone species (e.g., a beaver) that has a dramatic effect on an ecosystem by modifying the habitat.

ecotype a subspecies or race that is specially adapted to a particular set of environmental conditions.

edaphic pertaining to soil.

emergent property a feature of a community or system that is not deducible from the features of single species or lower order processes.

emigration the movement of organisms out of a population.

endangered species a species with so few living members that it will soon become extinct unless measures are taken to slow its loss.

endemic an organism that is native to a particular region.

endogenous produced from within; originating from or due to internal causes; contrast with **exogenous**.

environment the collection of all the biotic and abiotic factors that affect an individual organism at any one point in its life cycle.

epidemiology the study of disease in populations or groups.

epilimnion the upper layer of water in a lake, usually warm and containing high levels of dissolved oxygen.

epiphyte a plant that uses another plant for support, but that draws its own water and nutrients from natural runoff and the air.

equilibrium a condition of balance, such as that between immigration and emigration or birth and death rates in a population of fixed size.

euphotic zone that part of the water column which receives sufficient sunlight to support photosynthesis; usually limited to the upper 60 m.

eusociality relating to species which possess nonreproducing castes that assist the reproductive individuals.

eutrophic high in nutrients and organisms; high in productivity; contrast with **oligotrophic**.

eutrophication the normally slow aging process by which a lake fills with organic matter, evolves into a bog or marsh, and ultimately disappears.

evapotranspiration the sum of the water lost from the land by evaporation and plant transpiration. Potential evapotranspiration is the evapotranspiration that would occur if water were unlimited.

exogenous originating outside an organism; contrast with **endogenous**.

exponential growth the steepest phase in a growth curve—that in which the curve is described by an equation containing a mathematical exponent.

exponential rate of increase the rate at which a population is growing at a particular instant, expressed as a proportional increase per unit time.

extant a species currently represented by living individuals.

extinct a species no longer represented by living individuals.

extrafloral nectaries nectar-secreting glands found on leaves and other vegetative parts of plants.

extrinsic factors acting from outside a population, such as the weather, competition, or predation.

facilitation enhancement of a population of one species by another, often during succession; a type of one-way mutualism.

fecundity the potential of an organism to produce living offspring.

feral having reverted from domestication to the wild, but remaining distinct from other wild species.

fitness genetic contribution by an individual's offspring and their offspring to future generations.

fixation attainment by an allele of a frequency of 100 percent in a population, which in effect becomes monomorphic for that allele.

food chain Relationship exhibiting the feeding links of certain organisms on others, usually in a series beginning with plants and ending with the largest carnivores.

forest a region that, because it receives sufficient average annual precipitation (usually 75 cm [30 inches] or more), supports trees and small vegetation.

fossil fuels coal, oil, and natural gas, so-called because they are derived from the fossil remains of ancient plant and animal life.

fossorial living in burrows.

functional response a change in the rate of exploitation of prey by an individual predator resulting from a change in prey density. (See also **numerical response**.)

gene a unit of genetic information.

gene flow the exchange of genetic traits between populations by the movement of individuals, gametes, or spores.

generation time the time between the birth of a parent and the birth of its offspring.

genetic drift change in gene frequency caused solely by chance, usually unidirectional and more important in small populations.

genome the entire genetic complement of an individual.

genotype the genetic constitution of an organism, in contrast to its observable characteristics.

genus the taxonomic category above the species and below the family; a group of species believed to have descended from a common direct ancestor.

geometric growth growth in unlimited resources, which often yields a *J*-shaped curve.

geometric series a series in which each number is obtained by multiplying the previous one by the same factor (e.g., 1, 3, 9, 27, etc.)

global stability the ability to withstand perturbations of a large magnitude without being affected; contrast with **local stability**.

GNP the gross national product of a country.

grassland a region with sufficient average annual precipitation (25–75 cm [10–30 inches]) to support grass, but not trees.

greenhouse effect the heating effect of the atmosphere upon the Earth, particularly as the CO_2 concentration rises; caused by the atmosphere's ready admission of light waves, but its slower release of the heat they generate on striking the ground.

gross production production before respiration losses are subtracted; photosynthetic production for plants and metabolizable production for animals.

group selection elimination of a group of individuals with a detrimental genetic trait, caused by competition with other groups lacking the trait; contrast with **individual selection**.

guild a group of organisms that feed on a similar resource in the same manner (e.g., leaf chewers, sap suckers, or grain feeders).

habitat the sum of the environmental conditions under which an organism, population, or community lives; the place where an organism normally lives; the environment in which the life needs of an organism are supplied.

Hadley cell in climatology, the convection cell nearest to the equator, named in honor of British meteorologist George Hadley.

haplodiploidy the presence of haploid males and diploid females in the same species (e.g., in the Hymenoptera).

haploid having one set of genes in a genotype, as do sperm and eggs.

harem a group of females controlled by one male.

herbivore an organism that eats plants; contrast with **carnivore**.

heredity genetic transmission of traits from parents to offspring.

heterotroph an organism that obtains energy and materials from other organisms; contrast with **autotroph**.

home range the area in which an individual member of a population roams.

horizon in a soil, a major stratification or zone having particular structural and chemical characteristics.

Horse latitudes areas at about 30° latitude where cool air from the atmosphere sinks toward the Earth and where winds are relatively light.

host organism that furnishes food, shelter, or other benefits to an organism of another species.

hybridization breeding (crossing) of individuals from genetically different strains, populations, or, sometimes, species.

hypolimnion the layer of cold, dense water at the bottom of a lake.

hypothesis an explanation of a phenomenon.

immigration the movement of individuals into a population.

inbreeding a mating system in which adults mate with relatives more often than would be expected by chance.

inclusive fitness The total genetic contribution of an individual to future generations by way of its sons, daughters, and all other relatives, such as nieces, nephews, and cousins.

indirect competition exploitation of a resource by one individual that reduces the availability of that resource to others.

individual selection elimination of individuals with detrimental genetic traits; contrast with **group selection**.

inhibition type of succession whereby early colonists inhibit the establishment of later arriving species.

interference the physical interaction of one species with another, usually in competition over territory or resources.

interspecific between species; between individuals of different species.

intertidal zone the shallow zone where the land meets the sea, usually a rocky intertidal region but may be sandy shore.

intraspecific within species; between individuals of the same species.

intrinsic acting from within a population (e.g., crowding or intraspecific competition).

intrinsic rate of increase, *r* rate of increase of a population under controlled conditions.

isocline a line linking points with the same values.

iteroparity being able to breed continuously through a lifetime.

key factor a mortality factor that closely mirrors the overall population mortality.

keystone species a species having an effect on a community out of all proportion to its biomass.

kin selection a form of selection in which genetic individuals promote the survival of kin that carry the same alleles.

laterization the process of formation of a laterite, or red and yellow tropical soil, usually caused by intensive leaching in heavy rainfall in the tropics.

leaching the process by which soluble materials in the soil, such as nutrients, pesticides, or contaminants, are washed into a lower layer of soil or are dissolved and carried away by water.

lek a communal courtship area in which several males display to attract and mate with females.

lentic pertaining to standing freshwater habitats (ponds and lakes).

life table tabulation presenting complete data on the mortality schedule of a population.

ligase an enzyme that joins DNA together.

limiting resource the nutrient or substance that is in shortest supply in relation to organisms' demand for it.

linkage density in a food web, the average number of links per species.

local stability the tendency of a community to return to its original state when subjected to a small perturbation; contrast with **global stability**.

locus the site on a chromosome occupied by a specific gene.

logistic growth population growth described by a symmetrical *S*-shaped curve with an upper asymptote.

lognormal distribution a frequency distribution in which species abundance is expressed on a logarithmic scale on the *x*-axis.

lotic pertaining to running freshwater habitats (streams and rivers).

mollisols dark-colored rich soils with a dark A horizon.

monoculture cultivation of a single crop to the exclusion of all other species on a piece of land.

monogamy a mating system in which one male mates with only one female and one female mates with only one male.

monophagous feeding on one species or on two or three closely related species.

morph a specific form, shape, or structure.

Müllerian mimicry mutual resemblance of two or more conspicuously marked, distasteful species to reinforce avoidance by predators.

multivoltine having several generations during a single season; contrast with **univoltine**.

mutant an organism with a changed characteristic resulting from a genetic alteration.

mutation an alteration in the genetic makeup of an organism resulting from a chemical change in its DNA.

mutualism an interaction between two species in which both benefit.

natural selection the natural process by which the organisms best adapted to their environment survive and those less well adapted are eliminated.

neritic pertaining to the shallow, coastal marine zone.

net production production after respiration losses are subtracted.

net production efficiency percentage of assimilated energy that is incorporated into growth and reproduction.

net reproductive rate R for species with clearly defined discrete generations, the number of daughters a female can be expected to bear during her lifetime.

neutralism the occurrence together of two species with no interaction between them.

niche the place of an organism in an ecosystem; all the components of the environment with which an organism or population interacts.

nitrogen-fixing bacteria bacteria that can reduce atmospheric nitrogen to cell nitrogen.

nonrenewable resource a resource available in a fixed amount (such as minerals and oil) and that is not replaceable after use.

numerical response change in the population size of a predatory species as a result of a change in the density of its prey. (See also **functional response**.)

nutrient a chemical required for the growth, maintenance, or reproduction of an organism.

nutrient turnover time the time taken for a given nutrient (e.g., nitrogen) to pass through one complete cycle (the nitrogen cycle).

oceanic zone the open, deepwater marine biome.

oligotrophic low in nutrients and organisms; low in productivity; contrast with **eutrophic**.

omnivore an organism whose diet includes both plant and animal foods.

organic of biological origin; in chemistry, containing carbon.

oxisol old, highly weathered soils of the tropics, often with iron and aluminum oxides in the red-colored A layer.

parasite the organism that benefits in an interspecific interaction in which individuals of two species live symbiotically with one organism benefiting and the other being harmed. A parasite lives in intimate association with its host.

parasitoid a specialized insect parasite that is usually fatal to its host and therefore might be considered a predator rather than a classical parasite.

parthenogenesis reproduction without fertilization by male gametes, usually involving the formation of diploid eggs whose development is initiated spontaneously.

pathogen a microparasite that causes disease.

pelagic pertaining to the upper layers of the open ocean.

per-capita rate of population growth, ***r*** rate of population growth per individual; used for species with overlapping, nondiscrete generations.

permafrost a permanently frozen layer of soil underlying the Arctic tundra biome.

pH a measure of acidity or alkalinity.

phenology study of the periodic (seasonal) phenomena of animal and plant life (for example, flowering time in plants) and their relations to weather and climate.

phenotype the physical expression of a genotype in an organism; the outward appearance of an organism.

phoresy the transport of one organism by another of a different species.

photic zone the surface zone of a body of water that is penetrated by sunlight.

photosynthesis synthesis of carbohydrates from carbon dioxide and water with oxygen as a by-product.

phylum one of the primary divisions of the animal and plant kingdoms; a group of closely related classes of animals or plants.

phytoplankton the plant community, in marine water or fresh water, containing many species of algae and diatoms and that floats free in the water.

Pleistocene a geological epoch characterized by alternating glacial and interglacial stages and that ended about 10,000 years ago, lasted 1 to 2 million years, and is subdivided into four glacial and three interglacial stages.

podzolization the process of formation of a spodosol, often in cooler parts of the world under coniferous forests, in which acid humus breakdown is very slow.

pollutant any artificial substance that enters the ecosystem in such quantities that it does harm and makes a resource unfit for a specific purpose.

polygamy a mating system in which one individual of one sex mates with more than one individual of the opposite sex, such as when a male pairs with more than one female at a time (polygyny) or a female pairs with more than one male (polyandry).

polymorphism occurrence in a species of more than two different forms.

polyphagous feeding on many species; with a wide diet.

population a group of potentially interbreeding individuals of a single species.

predator an organism that benefits in an interspecific interaction in which it kills and feeds on prey. A predator lives in loose association with its prey.

primary production production by autotrophs, normally green plants; contrast with **secondary production**.

primary succession succession on completely sterile ground or water.

producer a green plant or chemosynthetic bacteria that converts light or chemical energy into organismal tissue; contrast with **consumer**.

production amount of energy (or material) formed by an individual, a population, or a community in a specific period.

production efficiency the percentage of assimilated energy that becomes incorporated into new biomass.

pyramid of numbers a spatial arrangement of numbers showing that population sizes decrease with increasing trophic level.

r **and** ***K*** **selection** alternative life history strategies that favor rapid population growth at low population density r or competititve ability at densities near the carrying capacity K.

recruitment addition, by reproduction, of new individuals to a population.

resilience the ability of a community to return to equilibrium following a disturbance, usually measured as elasticity—the speed of the return—or amplitude—the degree of disturbance the community can recover from.

resistance the size of a force needed to change a community structure.

resource a substance or an object required by an organism for normal maintenance, growth, or reproduction.

respiration the complex series of chemical reactions in all organisms by which stored energy is made available for use and that produces carbon dioxide and water as by-products.

riparian related to, living in, or located on the bank of a natural watercourse, usually a river, but sometimes a lake or tidewater.

runoff water entering rivers, lakes, reservoirs, or the ocean from land surfaces.

saprophyte a plant that obtains food from dead or decaying organic matter.

search image a behavioral selection mechanism that enables predators to increase their searching efficiency for prey that are abundant and worth capturing.

secondary production production by herbivores, carnivores, or detritus feeders; contrast with **primary production**.

secondary succession succession on partially cleared land.

selfish herd the concept of individuals banding together to use some members as protection against predators.

self-regulation population regulation by changes in behavior and physiology internal to the population, rather than by external forces such as predators.

semelparity having one reproductive episode per lifetime.

sere the series of successional communities leading from bare substrate to the climax community.

sessile of animals, attached to an object or fixed in place, as, for example, are barnacles.

siblings brothers or sisters

sigmoid curve an *S*-shaped curve (e.g., the logistic curve).

slash-and-burn cultivation a primarily tropical practice in which forest vegetation is cut, left to dry, and burned to add nutrients to the soil before the land is planted with crops and then abandoned after two to five years as a result of falling yields.

species one or more populations reproductively isolated from similar organisms and usually producing infertile offspring when crossed with them.

spodosol a soil with a strongly leached A horizon and having B horizons with an accumulation of iron or humus or both.

spring overturn the mixing of lake water as ice melts and storms churn up water from the bottom.

stability the absence of fluctuations in populations; the ability to withstand perturbations without large changes in the composition of community species.

stochastic model a mathematical model incorporating factors determined by chance and providing not a single prediction, but a range of predictions; contrast with **deterministic model**.

subsidence zone areas where cool air from the upper atmosphere falls toward the Earth, creating areas of high pressure.

succession replacement of one kind of community by another; the progressive changes in vegetation and animal life that tend toward climax.

symbiosis the living together of two or more organisms of different species.

sympatric occurring in the same place.

sympatric speciation formation of species without geographic isolation; reproductive isolation that arises between segments of a single population.

synecology the study of populations of interacting organisms; contrast with **autecology**.

synergism the situation in which the combined effect of two factors is greater than the sum of their separate effects.

taiga the northern boreal forest zone; a broad band of coniferous forest south of the Arctic tundra.

territory any area defined by one or more individuals and protected against intrusion by others of the same or different species.

thermocline the thin transitional zone in a lake that separates the epilimnion from the hypolimnion.

timberline the uppermost attitudinal limit of forest vegetation.

time lag delay in response to a change.

tolerance succession that is not affected by previous colonists.

topsoil the top few inches of soil, rich in organic matter and plant nutrients.

trophic level the functional classification of an organism in a community according to the organism's feeding relationships.

trophic-level transfer efficiency the percentage of energy of one trophic level that is incorporated into the bodies of individuals at the next trophic level.

tundra level or undulating treeless land having permanently frozen subsoil; characteristic of Arctic regions and high altitudes.

turnover rate of replacement of resident species by new, immigrant species.

univoltine having only one generation per year; contrast with **multivoltine**.

upwelling the process whereby, as a result of wind patterns, nutrient-rich bottom waters rise to the surface of the ocean.

vector an organism (often an insect) that transmits a pathogen (e.g., a virus, bacterium, protozoan, or fungus) acquired from one host to another.

watershed the land area that drains into a particular lake, river, or reservoir.

wilderness an undisturbed area, as it was before humans made changes to it.

zooplankton the animal community, predominantly single-celled animals, that floats free in marine and freshwater environments, moving passively with the currents.

Literature Cited

Abbott, I., and R. Black. 1980. Changes in species compositions of floras on islets near Perth, Western Australia. *Journal of Biogeography* 7: 399–410.

Adam, P. 1990. Saltmarsh ecology. Cambridge University Press, Cambridge, U.K.

Addicott, J. F. 1986. Variation in the costs and benefits of mutualism: The interaction between yuccas and yucca moths. *Oecologia* 70: 486–94.

Addicott, J. F., J. M. Aho, M. F. Antolin, D. K. Padilla, J. S. Richardson, and D. A. Soluk. 1987. Ecological neighborhoods: Scaling environmental patterns. *Oikos* 49: 340–46.

Alexander, R. D. 1974. The evolution of social behavior. *Annual Review of Ecology and Systematics* 5: 325–83.

Allendorf, F. W. 1994. Genetically effective sizes of grizzly bear populations. In G. K. Meffe and C. R. Carrol (eds.), *Principles of conservation biology* (pp. 155–56). Sinauer Associates, Sunderland, MA.

Anderson, R. M., and R. M. May. 1991. *Infectious disease of humans: Dynamics and control*. Oxford University Press, Oxford, U.K.

Andrewartha, H. G., and L. C. Birch. 1954. *The distribution and abundance of animals*. University of Chicago Press, Chicago.

Andrewartha, H. G., and L. C. Birch. 1960. Some recent contributions to the study of the distribution and abundance of insects. *Annual Review of Entomology* 5: 219–42.

Armitage, K. B. 1986. Marmot polygyny revisited: The determinants of male and female reproductive strategies. In D. I. Rubenstein and R. W. Wrangham (eds.), *Ecological aspects of social evolution: Birds and mammals* (pp. 303–331). Princeton University Press, Princeton, NJ.

Aron, W. I., and S. H. Smith. 1971. Ship canals and aquatic ecosystems. *Science* 174: 13–20.

Ashbourne, S. R. C., and Putnam. 1987. Competition, resource-partitioning, and species richness in the phytophagous insects of red oak and aspen in Canada and the U.K. *Acta Oecologia (Generalis)* 8: 43–56.

Ashmole, N. P. 1968. Body size, prey size and ecological segregation in five sympatric tropical terns. (Aves: Laridae). *Systematic Zoology* 17: 292–304.

Askew, R. R. 1971. *Parasitic insects*. Heineman, London.

Ayala, F. J. 1969. Experimental invalidation of the principle of competitive exclusion. *Nature* 224: 1076.

Ayala, F. J. 1972. Competition between species. *American Scientist* 60: 348–57.

Bailey, R. G. 1989. *Ecoregions of the continents*. U.S. Department of Agriculture, Forest Service, Washington, DC.

Baldwin, N. S. 1964. Sea lamprey in the Great Lakes. *Canadian Audubon Magazine*, November–December: 142–47.

Bates, H. W. 1862. Contributions to an insect fauna of the Amazon Valley. *Transactions of the Linnaean Society of London* 23: 495–566.

Bazely, D. R., M. Vicari, S. Emmerich, L. Filip, D. Lin, and A. Inman. 1998. Interactions between herbivores and endophyte-infected *Festuca rubra* from the Scottish Islands of St. Kilda, Berbecula and Rum. *Journal of Applied Ecology* 34: 847–60.

Bazzaz, F. A. 1979. The physiological ecology of plant succession. *Annual Review of Ecology and Systematics* 10: 351–71.

Beaver, R. A. 1985. Geographical variation in food web structure in *Nepenthes* pitcher plants. *Ecological Entomology* 10: 241–48.

Belt, T. 1974. *The naturalist in Nicaragua*. Dent and Sons, London.

Bengtsson, J., S. R. Baillie, and J. Lawton. 1997. Community variability increases with time. *Oikos* 78: 249–56.

Benke, A.C., and J. Bruce Wallace. 1997. Trophic basis of production among riverine caddisflies: Implications for food web analysis. *Ecology* 78: 1132–45.

Berenbaum, M. 1981. Patterns of furanocoumarin distribution and insect herbivory in the Umbelliferae: Plant chemistry and community structure. *Ecology* 62: 1254–66.

Berendse, F., M. Schmitz, and W. deVisser. 1994. Experimental manipulation of succession and heathland ecosystems. *Oecologia* 100: 38–44.

Berger, J. 1990. Persistence of different-sized populations: An empirical assessment of rapid extinctions in bighorn sheep. *Conservation Biology* 4: 91–98.

Berger, W. H., and F. L. Parker. 1970. Diversity of planktonic foraminifera in deep sea sediments. *Science* 168: 1345–47.

Bertness, M. D., and S. D. Hacker. 1994. Physical stress and positive associations among marsh plants. *American Naturalist* 144: 363–72.

Bertness, M. D., and G. H. Leonard. 1997. The role of positive interactions in communities: Lessons from intertidal habitats. *Ecology* 78: 1979–89.

Bertram, B. C. R. 1975. Social factors influencing reproduction in wild lions. *Journal of the Zoological Society of London* 177: 463–82.

Bertram, B. C. R. 1979. Serengeti predators and their social systems. In A. R. E. Sinclair and M. Morton-Griffiths (eds.), *Serengeti: Dynamics of an ecosystem* (pp. 221–48). University of Chicago Press, Chicago.

Bibby, C. J., M. J. Crosby, M. F. Heath, T. H. Johnson, A. J. Lang, A. J. Sattersfield, and S. J. Thirgood. 1992. Putting biodiversity on the map: Global priorities for conservation. ICBP, Cambridge, U.K.

Bigger, P. S., and M. A. Marvier. 1998. How different would a world without herbivory be? A search for generality in ecology. *Integrative Biology* 1: 60–67.

Birch, L. C. 1953. Experimental background to the study of the distribution and abundance of insects II. The relation between innate capacity for increase in numbers and the abundance of three grain beetles experimental populations. *Ecology* 34: 712–26.

Bohn, H., B. McNeal, and G. O'Connor. 1979. *Soil chemistry*. John Wiley, New York.

Bormann, F. H., G. E. Likens, and J. M. Melillo. 1977. Nitrogen budget for an aggrading northern hardwood forest ecosystem. *Science* 196: 981–83.

Boucher, D. H., S. James, and K. Kresler. 1984. The ecology of mutualism. *Annual Review of Ecology and Systematics* 13: 315–47.

Bright, C. 1998. *Life out of bounds: Bioinvasion in a borderless world*. W. W. Norton & Company. New York.

Bronstein, J. L. 1998. The contribution of ant–plant protection studies to our understanding of mutualism. *Biotropica* 30: 150–61.

Brower, L. P., W. M. Ryerson, L. L. Coppinger, and S. C. Glazier. 1968. Ecological chemistry and the palatability spectrum. *Science* 161: 1349–51.

Brown, A. A., and K. P. Davis. 1973. *Forest fire control and its use*, 2nd ed. McGraw-Hill, New York.

Brown, J. H. 1978. The theory of insular biogeography and the distribution of boreal birds and mammals. *Great Basin Naturalist Memoirs* 2: 209–27.

Brown, J. H. 1989. Patterns, modes, and extents of invasions by vertebrates. In J. A. Drake, H. A. Mooney, F. di Castri, R. H. Groves, F. J. Kruger, M. Rejmanek, and M. Williamson (eds.), *Biological invasions: a global perspective* (pp. 85–109). Wiley, Chichester, England.

Brown, J. H. 1997. An ecological perspective on the challenge of complexity. Ecoessay series number 1. *National Center for Ecological Analysis and Synthesis*, Santa Barbara, CA.

Brown, J. H., and A. Kodric-Brown. 1977. Turnover rates in insular biogeography: Effect of immigration and extinction. *Ecology* 58: 445–49.

Brown, J. H., P.A. Margret, and M. L. Taper. 1993. Evolution of body size: Consequences of an energetic definition of fitness. *American Naturalist* 142: 573–84.

Bryant, J. P., F. S. Chapin, III, and D. R. Klein. 1983. Carbon/nutrient balance of boreal plants in relation to vertebrate herbivory. *Oikos* 40: 357–68.

Cain, M. L., S. W. Pacala, J. A. Silander, and M. J. Fortin. 1995. Neighborhood models of clonal growth in the white clover, *Trifolium repens*. *American Naturalist* 145: 888–917.

Callaway, R., E. Delucia, D. Moore, R. Nowak, and W. H. Schlesinger. 1996. Competition and facilitation: Contrasting effects of *Artemisia tridentata* on desert versus montane pines. *Ecology* 77: 2130–41.

Callaway, R. M. 1995. Positive interactions among plants. *Botanical Review* 61: 306–49.

Callaway, R. M. 1997. Positive interactions in plant communities and the individualistic–continuum concept. *Oecologia* 112: 143–49.

Callaway, R. M. 1999. Are positive interactions species-specific? *Oikos* 82: 202–7.

Callaway, R. M., and E. T. Aschehoug. 2000. Invasive plants versus their new and old neighbors: A mechanism for exotic invasion. *Science* 290: 521–23.

Caraco, T. 1979. Time budgeting and group size: A test of theory. *Ecology* 60: 618–27.

Caraco, T., S. Martindale, and H. R. Pulliam. 1980. Avian flocking in the presence of a predator. *Nature* 285: 400–1.

Cargill, S. M., and R. L. Jeffries. 1984. Nutrient limitation of primary production in a sub-arctic salt marsh. *Journal of Applied Ecology* 21: 657–68.

Carpenter, S. R. 1996. Microcosm experiments have limited relevance for community and ecosystem ecology. *Ecology* 77: 677–80.

Caswell, H. 1978. Predator mediated coexistence: A nonequilibrium model. *American Naturalist* 112: 127–54.

Caughley, G. 1970. Eruption of ungulate populations, with emphasis on Himalayan Thar in New Zealand. *Ecology* 51: 53–72.

Chapin, F. S. III, L. R. Walker, C. L. Fastie, and L. C. Sharman. 1994. Mechanisms of primary succession following deglaciation at Glacier Bay, Alaska. *Ecological Monographs* 64: 149–75.

Charlesworth, B. 1984. The cost of phenotypic evolution. *Paleobiology* 10: 319–27.

Chitty, D. 1960. Population processes in the vole and their relevance to general theory. *Canadian Journal of Zoology* 38: 99–113.

Christian, J. M., and D. Wilson. 1999. Long-term ecosystem impacts of an introduced grass in the northern Great Plains. *Ecology* 80: 2397–2407.

Clarke, G. L. 1954. *Elements of ecology*. John Wiley and Sons, Inc.

Clay, K., and J. Holah. 1999. Fungal endophyte symbiosis and plant diversity in successional fields. *Science* 285: 1742–44.

Clements, F. E. 1905. *Research methods in ecology*. University Publishing Company, Lincoln, NE. Reprinted Arno Press, New York, 1977.

Clements, F. E. 1916. *Plant succession: Analysis of the development of vegetation*. Carnegie Institute of Washington Publication 242.

Clements, F. E. 1936. Nature and structure of the climax. *Journal of Ecology* 24: 252–84.

Cockburn, A. T. 1971. Infectious diseases in ancient populations. *Current Anthropology* 12: 45–62.

Cohen, J.E. 1995. *How many people can the earth support?* W.W. Norton, New York.

Cohen, J. E. 1997. Population, economics and culture: An introduction to human carrying capacity. *Journal of Applied Ecology* 34: 1325–33.

Cohen, J. E., F. Briand, and C. M. Newman. 1990. Community food webs: Data and theory. *Biomathematics*, vol. 20. Springer Verlag, Berlin.

Collins, S. L., A. K. Knapp, J. M. Biggs, J. M. Blair, and E. M. Steinauer. 1998. Modulation of diversity by grazing and mowing in native tall grass prairie. *Science* 280: 745–47.

Collins, S. L., and L. L. Wallace. 1990. Fire in North American tallgrass prairies. University of Oklahoma Press, Norman, OK.

Collinson, A. S. 1977. *An introduction to world vegetation*. George Allen and Unwin, London.

Colwell, R. K. 1973. Competition and coexistence in a simple tropical community. *American Naturalist* 107: 737–60.

Compton, S. G., J. H. Lawton, and V. K. Rashbrook. 1989. Regional diversity, local community structure, and vacant niches: The herbivorous insects of bracken in South Africa. *Ecological Entomology* 14: 365–73.

Connell, J. H. 1978. Diversity in tropical rain forests and coral reefs. *Science* 199: 1302–10.

Connell, J. H. 1980. Diversity and the coevolution of competitors, or the ghost of competition past. *Oikos* 35: 131–38.

Connell, J. H. 1983. On the prevalence and relative importance of interspecific competition: Evidence from field experiments. *American Naturalist* 122: 661–96.

Connell, J. H., and R. O. Slatyer. 1977. Mechanisms of succession in natural communities and their role in community stability and organization. *American Naturalist* 111: 1119–44.

Connell, J. H., and W. P. Sousa. 1983. On the evidence needed to judge ecological stability or persistence. *American Naturalist* 121: 789–824.

Constanza, R., R. d'Arge, R. de Groot, S. Farber, M. Grasso, B. Hannon, K. Limburg, S. Naeem, R. V. O'Neill, J. Paruelo, R. G. Raskin, P. Sutton, and M. van den Belt. 1997. The value of the world's ecosystem services and natural capital. *Nature* 387: 253–60.

Cook, R. E. 1969. Variation in species diversity of North American birds. *Systematic Zoology* 18: 63–84.

Cooper, W. S. 1923. The recent ecological history of Glacier Bay, Alaska: II. The present vegetation cycle. *Ecology* 4: 223–46.

Cote, I. M., and W. J. Sutherland. 1997. The effectiveness of removing predators to protect bird populations. *Conservation Biology* 11: 395–405.

Crawley, M. J. 1997. Plant–herbivore dynamics. In M. J. Crawley (ed.), *Plant Ecology*, 2d ed. (pp. 401–74). Blackwell Science, Oxford.

Cronin, E. W., and P. W. Sherman. 1977. A resource-based mating system: The orange rumped honey guide. *Living Bird* 15: 5–32.

Crow, J. F., and M. Kimura. 1970. *An introduction to population genetics theory*. Harper and Row, New York.

Currie, C. R., J. A. Scott, R. C. Summerbell, and D. V. Mulloch. 1999. Fungus growing ants use antibiotic-producing bacteria to control garden parasites. *Nature* 398: 701–4.

Currie, D. J., and V. Paquin. 1987. Large-scale biogeographical patterns of species richness of trees. *Nature* 329: 326–27.

Cyr, H., and M. L. Pace. 1993. Magnitude and patterns of herbivory in aquatic and terrestrial ecosystems. *Nature* 361: 148–50.

Darwin, C. 1859. *On the origin of species by means of natural selection*. John Murray, London.

Davidson, D. W. 1993. The effects of herbivory and granivory on terrestrial plant succession. *Oikos* 68: 23–35.

Davidson, J., and H. G. Andrewartha. 1948. The influence of rainfall evaporation, and atmospheric temperature on fluctuations in the size of a natural population of *Thrips imaginis*. *Journal of Animal Ecology* 17: 200–22.

Davis, M. G., and A. Zabinski. 1992. Changes in geographical range resulting from greenhouse warming: Effects on biodiversity in forests. In R. Peters and T. Lovejoy (eds.) *Global warming and biodiversity* (pp. 297–308). Yale University Press, New Haven, CT.

Dawkins, R., and J. R. Krebs. 1979. Arms races between and within species. *Proceedings of the Royal Society of London*, Series B 205: 489–511.

Dayan, T., and D. Simberloff. 1994. Character displacement, sexual dimorphism and morphological variation among British and Irish mustelids. *Ecology* 75: 1063–73.

Dean, T. A., and L. E. Hurd. 1980. Development in an estuarine fouling community: The influence of early colonists on later arrivals. *Oecologia* 46: 295–301.

Debach, P., and D. Rosen. 1990. *Biological control*, 2d ed. Cambridge University Press, Cambridge.

Debach, P. S., and R. A. Sundby. 1963. Competitive displacement between ecological homologues. *Hilgardia* 43: 105–66.

Deevey, E. S., Jr. 1947. Life tables for natural populations of animals. *Quarterly Review of Biology* 22: 283–314.

den Boer, P. J. 1968. Spreading of risk and stabilization of animal numbers. *Acta Biotheoretica* 18: 165–94.

den Boer, P. J. 1981. On the survival of populations in a heterogeneous and variable environment. *Oecologia* 50: 39–53.

Denno, R. F., M. S. McClure, and J. R. Ott. 1995. Interspecific interactions in phytophagous insects: Competition re-examined and resurrected. *Annual Review of Entomology* 40: 297–331.

Deshmukh, I. 1986. *Ecology and tropical biology*. Blackwell Scientific Publications, Oxford, U.K.

DeSteven, D. 1991. Experiments on mechanisms of tree establishment in old-field succession: Seedling survival and growth. *Ecology* 72: 1075–88.

De Vos, A., R. H. Manville, and G. Van Gelder. 1956. Introduced mammals and their influence on native biota. *Zoologica* 41: 163–94.

Diamond, J. M. 1969. Avifauna equilibria and species turnover on the Channel Islands of California. *Proceedings of the National Academy of Sciences of the United States of America* 74: 57–63.

Diamond, J. M. 1972. Biogeographic kinetics: Estimation of relaxation times for avifaunas of southwest Pacific islands. *Proceedings of the National Academy of Sciences of the United States of America* 69: 3199–3203.

Diamond, J. 1986. Overview: Laboratory experiments, field experiments, and natural experiments. In J. Diamond and T. J. Case (eds.), *Community ecology* (pp. 3–22). Harper and Row, New York.

Dixon, A. F. G., P., Kindlmann, J. Leps, and J. Holman. 1987. Why are there so few species of aphids, especially in the tropics? *American Naturalist* 129: 580–92.

Doak, D. F. 1992. Lifetime impacts of herbivory for a perennial plant. *Ecology* 73: 2086–99.

Doak, D. F., D. Bigger, E. K. Harding, M. A. Marvier, R. S. O'Malley, and D. Thomson. 1998. The statistical inevitability of stability–diversity relationships in community ecology. *American Naturalist* 151: 264–76.

Dobson, A. P., J. P. Rodriguez, W. M. Roberts, and D. S. Wilcove. 1997. Geographic distribution of endangered species in the United States. *Science* 275: 550–53.

Dobzhansky, T. 1950. Evolution in the tropics. *American Scientist* 38: 209–21.

Dodd, A. P. 1940. *The biological campaign against prickly pear*. Prickly Pear Board, Brisbane.

Downhower, J. F., and K. B. Armitage. 1971. The yellow-bellied marmot and the evolution of polygamy. *American Naturalist* 105: 355–70.

Drake, J. A. 1991. Community-assembly mechanics and the structure of an experimental species ensemble. *American Naturalist* 137: 1–26.

Drake, J. A., G. R. Huxel, and C. I. Hewitt. 1996. Microcosms as models for generating and testing community theory. *Ecology* 77: 670–77.

Durant, S. M. 1998. Competition refuges and coexistence: An example from Serengeti carnivores. *Journal of Animal Ecology* 67: 370–86.

Ebenhard, T. 1988. Introduced birds and mammals and their ecological effects. *Swedish Wildlife Research* 13: 1–107.

Edwards, C. A., and G. W. Heath. 1963. The role of soil animals in breakdown of leaf material. In D. Doiksen and J. van der Pritt (eds.), *Soil organisms*, (pp. 76–84). North-Holland, Amsterdam.

Edwards-Jones, G., and V. K. Brown. 1993. Successional trends in insect herbivore population densities: A field test of a hypothesis. *Oikos* 66: 463–71.

Egler, F. E. 1954. Vegetation science concepts: Initial floristic composition—a factor in old-field development. *Vegetatio* 4: 412–17.

Ehler, L. E., and R. W. Hall. 1982. Evidence for competitive exclusion of introduced natural enemies in biological control. *Environmental Entomology* 1: 1–4.

Ehrlich, P. R., and A. H. Ehrlich. 1981. *Extinction: The causes and consequences of the disappearance of species*. Random House, New York.

Ehrlich, P. R., and P. H. Raven. 1964. Butterflies and plants: A study in coevolution. *Evolution* 18: 586–608.

Ehrlich, P. R., and J. Roughgarden. 1987. *The science of ecology*. Macmillan, New York.

Ehrlich, P. R., and E. O. Wilson. 1991. Biodiversity studies: Science and policy. *Science* 253: 758–62.

Eisner, T., and D. J. Aneshansley. 1982. Spray aiming in bombardier beetles: Jet deflection by the Coanda effect. *Science* 215: 83–85.

Elton, C. 1927. *Animal ecology*. Sidgwick and Jackson, London.

Elton, C. 1958. *The ecology of invasions by animals and plants*. Methuen, London.

Elton, C., and M. Nicholson. 1942. The ten-year cycle in numbers of the lynx in Canada. *Journal of Animal Ecology* 11: 215–44.

Enright, J. T. 1976. Climate and population regulation. *Oecologia* 24: 295–310.

Errington, P. L. 1946. Predation and vertebrate populations. *Quarterly Review of Biology* 21: 144–77, 221–45.

Erwin, T. L. 1982. Tropical forests: Their richness in Coleoptera and other arthropod species. *Coleopterists Bulletin* 36: 74–75.

Erwin, T. L. 1983. Beetles and other insects of tropical forest canopies at Manaus, Brazil, sampled by insecticidal fogging. In S. L. Sutton, T. C. Whitmore, and A. C. Chadwick (eds.), *Tropical rain forest: Ecology and management* (pp. 59–75). Blackwell Scientific Publications, Oxford, U.K.

Facelli, J. M., and E. Facelli. 1993. Interactions after death: Plant litter controls priority effects in a successional plant community. *Oecologia* 95: 277–82.

Faeth, S. H., and K. E. Hammon. 1997a. Fungal endophytes in oak trees: Long-term patterns of abundance and associations with leafminers. *Ecology* 78: 810–19.

Faeth, S. H., and K. E. Hammon. 1997b. Fungal endophytes in oak trees: Experimental analyses of interactions with leafminers. *Ecology* 78: 820–27.

Farner, D. S. 1945. Age groups and longevity in the American robin. *Wilson Bulletin* 57: 56–74.

Fastie, C. L. 1995. Causes and ecosystem consequences of multiple pathways of primary succession at Glacier Bay, Alaska. *Ecology* 76: 1899–1912.

Feeny, P. 1970. Seasonal changes in the oak leaf tannins and nutrients as a cause of spring feeding by winter moth caterpillars. *Ecology* 51: 565–81.

Feeny, P. 1976. Plant apparency and chemical defense. *Recent Advances in Phytochemistry* 10: 1–40.

Fenner, F., and F. Ratcliffe. 1965. *Myxamatosis*. Cambridge University Press, Cambridge, U.K.

Ferson, S., P. Downey, P. Klerks, M. Weissburg, I. Kroot, S. Stewart, G. Jacquez, J. Ssemakula, R. Malenky, and K. Anderson. 1986. Competing reviews, or why do Connell and Schoener disagree? *American Naturalist* 127: 517–76.

Fischer, A. G. 1960. Latitudinal variation in organic diversity. *Evolution* 14: 64–81.

Fischer, J., N. Simon, and J. Vincent. 1969. *The red book—wildlife in danger*. Collins, London.

Fitzsimmons, A. K. 1996. Stop the parade. *BioScience* 46: 78–79.

Floyd, T. 1996. Top-down impact on creosotebush herbivores in a spatially and temporally complex environment. *Ecology* 77: 1544–55.

Fonseca, C. R., and G. Garade. 1996. Asymmetries, compartments and null interactions in an Amazonian ant–plant community. *Journal of Animal Ecology* 65: 339–47.

Foster, M. S. 1990. Organization of macroalgal assemblages in the Northeast Pacific: The assumption of homogeneity and the illusion of generality. *Hydrobiologia* 192: 21–23.

Fox, B. J. 1983. Mammal species diversity in Australian heathlands: The importance of pyric succession and habitat diversity. In F. J. Kruger, D. T. Mitchell, and J. U. M. Jervis (eds.), *Mediterranean-type ecosystems: The role of nutrients* (pp. 473–89). Springer-Verlag, Berlin.

Francis, W. J. 1970. The influence of weather on population fluctuations in California quail. *Journal of Wildlife Management* 34: 249–66.

Frank, D. A., and S. J. McNaughton. 1991. Stability increases with diversity in plant communities: Empirical evidence from the 1988 Yellowstone drought. *Oikos* 62: 360–62.

Fuller, C. A., and A. R. Blaustein. 1996. Effects of the parasite *Eimeria arizonensis* on survival of deer mice (*Peromyosus mariculatus*). *Ecology* 77: 2196–2202.

Gause, G. F. 1932. Experimental studies on the struggle for existence: I. Mixed population of two species of yeast. *Journal of Experimental Biology* 9: 389–402.

Gentry, A. H. 1988. Tree species of upper Amazonian forests. *Proceedings of the National Academy of Sciences of the United States of America* 85: 156.

Gilbert, F. S. 1980. The equilibrium theory of island biogeography: Fact or fiction. *Journal of Biogeography* 7: 209–35.

Gleason, H. A. 1926. The individualistic concept of the plant association. *Torrey Botanical Club Bulletin* 53: 7–26.

Goldwasser, L., and J. Roughgarden. 1997. Sampling effects and the estimation of food-web properties. *Ecology* 78: 41–54.

Goodman, D. 1975. The theory of diversity–stability relationships in ecology. *Quarterly Review of Biology* 50: 237–66.

Gotelli, N. 1995. *A primer of ecology*. Sinauer Associates, Sunderland, MA.

Grant, P. R. 1986. *Ecology and evolution of Darwin's finches*. Princeton University Press, Princeton, NJ.

Grant, P. R., and B. R. Grant. 1987. The extraordinary El Niño event of 1982–1983: Effects on Darwin's finches on Isla Genovesa, Galápagos. *Oikos* 49: 55–66.

Gray, J. S. 1981. *The ecology of marine sediments*. Cambridge Univ. Press, Cambridge, U.K.

Grime, J. P. 1977. Evidence for the existence of three primary strategies in plants and its relevance to ecological and evolutionary theory. *American Naturalist* 111: 1169–94.

Grime, J. P. 1979. *Plant strategies and vegetation process*. John Wiley, New York.

Grime, J. P. 1993. Ecology sans frontières, *Oikos* 68: 385–92.

Grinnell, J. 1918. The niche relationships of the California thrasher. *Auk* 34: 427–33.

Gronemeyer, P. A., B. J. Dilger, J. L. Bouzat, and K. N. Paige. 1997. The effects of herbivory on paternal fitness in scarlet gilia: Better moms also make better pops. *American Naturalist* 150: 592–602.

Grubb, P. J. 1992. A positive distrust to simplicity—lessons from plant defenses and from competition among plants and among animals. *Journal of Ecology* 80: 585–610.

Guiler, E. R. 1961. *Australian Journal of Science* 23: 207–10.

Gurrevitch, J., L. L. Morrow, A. Wallace, and J. S. Walsh. 1992. A meta-analysis of field experiments on competition. *American Naturalist* 140: 539–72.

Hacker, S. D., and M. D. Bertness. 1996. Trophic consequences of a positive plant interaction. *American Naturalist* 148: 559–75.

Haddad, N. M., J. Haarstad, and D. Tilman. 2000. The effects of long-term nitrogen loading on grassland insect communities. *Oecologia* 124: 73–84.

Hairston, J. G., Sr. 1989. *Ecological experiments: Purpose, design, and execution*. Cambridge University Press, Cambridge, U.K.

Hairston, N. G. 1969. On the relative abundance of species. *Ecology* 50: 1091–94.

Hairston, N. G., J. D. Allen, R. K. Colwell, D. J. Futuyma, J. Howell, M. D. Lubin, J. Mathias, and J. H. Vandermeer. 1968. The relationship between species diversity and stability: An experimental approach with protozoa and bacteria. *Ecology* 49: 1091–1101.

Hairston, N. G., F. E. Smith, and L. B. Slobodkin. 1960. Community structure, population control, and competition. *American Naturalist* 44: 421–25.

Hall, R. W., L. E. Ehler, and B. Bisabri-Ershadi. 1980. Rates of success in classical biological control of arthropods. *Bulletin of the Entomological Society of America* 26: 111–14.

Hamback, P. A., J. A. Agren, and L. Ericson. 2000. Associational resistance: Insect damage to purple loosestrife reduced in thickets of sweet gale. *Ecology* 81: 1784–94.

Hamilton, W. D. 1964. The genetical evolution of social behaviour: I, II. *Journal of Theoretical Biology* 7: 1–52.

Hamilton, W. D. 1967. Extraordinary sex ratios. *Science* 156: 477–88.

Hamilton, W. D. 1971. Geometry for the selfish herd. *Journal of Theoretical Biology* 31: 295–311.

Harborne, J. B. 1988. *Introduction to ecological biochemistry*. Academic Press, London.

Hardin, G. 1960. The competitive exclusion principle. *Science* 162: 1243–48.

Hardin, G. 1968. The tragedy of the commons. *Science* 162: 1243–48.

Harris, H. 1966. Enzyme polymorphisms in man. *Proceedings of the Royal Society of London*, Series B 164: 298–310.

Harrison, S. 1991. Local extinction in a metapopulation context: An empirical evaluation. *Biological Journal of the Linnean Society* 42: 73–88.

Harrison, S., D. D. Murphy, and P. R. Ehrlich. 1988. Distribution of the bay checkerspot butterfly, *Euphydryas editha bayensis*: Evidence for a metapopulation model. *American Naturalist* 132: 360–82.

Hart, D. D., and R. J. Horwitz. 1991. Habitat diversity and the species–area relationship: Alternative models and tests. In S. S. Bell, E. D. McCoy, and H. Mushinsky (eds.), *Habitat structure: The physical arrangement of objects in space* (pp. 47–68). Chapman and Hall, London.

Harvell, C. D., K. Kim, J. M. Burkholder, R. R. Colwell, P. R. Epstein, D. J. Grimes, E. E. Hofmann, E. K. Lip, A. D. M. E. Osterhaus, R. M. Overstreet, J. W. Porter, G. W. Smith, and G. R. Vasta. 1999. Emerging marine diseases—climate links and anthropogenic factors. *Science* 285: 1505–10.

Harvey, P. H., J. J. Bull, and R. J. Paxton. 1983. Looks pretty nasty. *New Scientist* 97: 26–27.

Hattersley, P. W. 1983. The distribution of C_3 and C_4 grasses in Australia in relation to climate. *Oecologia* 57: 113–28.

Hawkins, B. A. 1992. Parasitoid–host food webs and donor control. *Oikos* 65: 159–62.

Hawkins, B. A., N. J. Mills, M. A. Jervis, and P. W. Price. 1999. Is the biological control of insects a natural phenomenon? *Oikos* 86: 493–506.

Heatwole, H. 1965. Some aspects of the association of cattle egrets with cattle. *Animal Behavior* 13: 79–83.

Hector, A., and 33 other authors. 1999. Plant diversity and productivity experiments in European grasslands. *Science* 286: 1123–27.

Herbold, B., and P. B. Moyle. 1986. Introduced species and vacant niches. *American Naturalist* 128: 751–60.

Heywood, V. H. (ed.). 1995. *Global biodiversity assessment*. United Nations Environment Programme, Cambridge University Press, Cambridge, U.K.

Hiura, T. 1995. Gap formation and species diversity in Japanese beech forests: A test of the intermediate disturbance hypothesis on a geographic scale. *Oecologia* 104: 265–71.

Hixon, M. A., and W. N. Brostoff. 1996. Succession and herbivory: Effects of differential fish grazing on Hawaiian coral-reef algae. *Ecological Monographs* 66: 67–90.

Hjermann, D. O., and R. A. Ims. 1996. Landscape ecology of the wart-biter *Decticus verrucivorus* in a patchy landscape. *Journal of Animal Ecology* 65: 768–80.

Hokkanen, H., and D. Pimentel. 1984. New approach for selecting biological control agents. *Canadian Entomologist* 116: 1109–21.

Holdridge, L. R. 1967. *Life zone ecology*. Tropical Science Center, San Jose, CA.

Holt, R. D. 1997. Personal communication.

Holyoak, M., and S. P. Lawler, 1996. The role of dispersal in predator–prey metapopulation dynamics. *Journal of Animal Ecology* 65: 640–52.

Horn, H. S., and R. M. May. 1977. Limits to similarity among coexisting competitors. *Nature* 270: 660–61.

Howard, L. O., and W. F. Fiske. 1911. *The importation into the United States of the parasites of the gypsy-moth and the brown-tail moth.* U.S. Department of Agriculture, Bureau of Entomology, Bulletin 91.

Howard, W. E. 1949. Dispersal, amount of inbreeding, and longevity in a local population of prairie deer mice on the George Reserve, southern Michigan. *Contributions from the Laboratory of Vertebrate Biology of the University of Michigan* 43: 1–50.

Howarth, F. G. 1983. Classical biological control: Panacea or Pandora's box? *Proceedings of the Hawaii Entomological Society* 24: 239–44.

Hrdy, S. B. 1977. Infanticide as a primate reproductive strategy. *American Scientist* 65: 40–49.

Hubbell, S. P. 1979. Tree dispersion, abundance, and diversity in a tropical dry forest. *Science* 203: 1299–1309.

Huffaker, C. B. 1958. Experimental studies on predation: Dispersion factors and predator–prey oscillation. *Hilgardia* 27: 343–83.

Huffaker, C. B., and C. E. Kennett. 1969. Some aspects of assessing efficiency of natural enemies. *Canadian Entomologist* 101: 425–40.

Huffaker, C. B., K. P. Shea, and S. G. Herman. 1963. Experimental studies on predation: Complex dispersion and levels of food in an acarine predator–prey interaction. *Hilgardia* 34: 305–30.

Hughes, R. N., and Griffiths, C. L. 1988. Self-thinning in barnacles and mussels: The geometry of packing. *American Naturalist* 132: 484–91.

Hunter, M. D., and P. W. Price. 1992. Playing chutes and ladders: Heterogeneity and the relative roles of bottom-up and top-down forces in natural communites. *Ecology* 73: 724–32.

Hurtrez-Boussess, P. Perret, F. Renaud, and J. Blondel. 1997. High blowfly parasitic loads affect breeding success in a Mediterranean population of blue tits. *Oecologia* 112: 514–17.

Huston, M. A. 1997. Hidden treatments in ecological experiments: Re-evaluating the ecosystem function of biodiversity. *Oecologia* 110: 449–60.

Hutchinson, G. E. 1958. Concluding remarks. *Cold Spring Harbor Symposia on Quantitative Biology* 22: 415–27.

Hutchinson, G. E. 1959. Homage to Santa Rosalia, or why are there so many kinds of animals? *American Naturalist* 93: 145–59.

Iason, G. R., C. D. Duck, and T. H. Clutton-Brock. 1986. Grazing and reproductive success of red deer: The effect of local enrichment by gull colonies. *Journal of Animal Ecology* 55: 507–15.

Ichikawa, N. 1995. Male counterstrategy against infanticide of the female giant water bug *Lethocerus deyrollei* (Hemiptera: Bellostomatidae). *Journal of Insect Behavior* 8: 181–88.

Inouye, R. S., N. J. Huntly, D. Tilman, J. R. Tester, M. Stillwell, and K. C. Zinnel. 1987. Old-field succession on a Minnesota sand plain. *Ecology* 68: 12–26.

International Union for Conservation of Nature and Natural Resources. 1980. *World conservation strategy*. International Union for Conservation of Nature and Natural Resources, United National Environmental Program, World Wildlife Fund. Gland, Switzerland.

Jaccard, P. 1912. The distribution of the flora of the alpine zone. *New Phytologist* 11: 37–50.

Jaksic, F. M. 1981. Abuse and misuse of the term "guild" in ecological studies. *Oikos* 37: 397–400.

Jaksic, F. M., and M. Delibes. 1987. A comparative analysis of food–niche relationships and trophic guild structure on how assemblages of vertebrate predators differ in species richness: Causes, correlations, and consequences. *Oecologia* 71: 461–72.

Janzen, D. H. 1966. Coevolution of mutualism between ants and acacias in Central America. *Evolution* 20: 249–75.

Janzen, D. H. 1968. Host plants as islands in evolutionary and contemporary time. *American Naturalist* 102: 592–95.

Janzen, D. H. 1971. Escape of *Cassia grandis L.* beans from predators in time and space. *Ecology* 52: 964–79.

Janzen, D. H. 1973. Host plants as islands: II. Competition in evolutionary and contemporary time. *American Naturalist* 107: 786–90.

Janzen, D. H. 1979a. How to be a fig. *Annual Review of Ecology and Systematics* 10: 13–51.

Janzen, D. H. 1979b. Why fruit rots. *Natural History Magazine* 88(6): 60–64.

Jarvis, J. V. M., M. J. O'Riain, N. C. Bennett, and P. W. Sherman 1994. Mammalian eusociality: A family affair. *Trends in Ecology and Evolution* 9: 47–51.

Johnson, K. H., K. A. Vogt, H. J. Clark, O. J. Schmitz, and D. J. Vogt. 1996. Biodiversity and the productivity and stability of ecosystems. *Trends in Ecology and Evolution* 11: 372–77.

Johnston, D. W., and E. P. Odum. 1956. Breeding bird populations in relation to plant succession on the Piedmont of Georgia. *Ecology* 37: 50–62.

Karban, R. 1997. Evolution of prolonged development: A life table analysis for periodical cicadas. *American Naturalist* 150: 446–61.

Keever, C. 1950. Causes of succession on old fields of the Piedmont, North Carolina. *Ecological Monographs* 20: 230–50.

Keith, L. B. 1983. Role of food in hare population cycles. *Oikos* 40: 385–95.

Keller, M. A. 1984. Reassessing evidence for competitive exclusion of introduced natural enemies. *Environmental Entomology* 13: 192–95.

Kelly, D. 1994. The evolutionary ecology of mast seeding. *Trends in Ecology and Evolution* 9: 465–70.

Kenward, R. E. 1978. Hawks and doves: Factors affecting success and selection in goshawk attacks on wood-pigeons. *Journal of Animal Ecology* 47: 449–60.

Kerr, J. T., and D. J. Currie. 1995. Effects of human activity on global extinction risk. *Conservation Biology* 9: 1528–38.

Kerr, R. A. 1997. Greenhouse forecasting still cloudy. *Science* 276: 1040–42.

King, C. E. 1964. Relative abundance of species and MacArthur's model. *Ecology* 45: 716–27.

Knops, J., D. Tilman, N. M. Haddad, S. Naeem, C. Mitchell, J. Haarstad, M. E. Ritchie, K. M. Howe, P. B. Reich, E. Siemann, and J. Groth. 1999. Effects of plant species richness on invasion dynamics, disease outbreaks, insect abundancies and diversity. *Ecology Letters* 2: 286–93.

Knowlton, N. 1979. Reproductive synchrony, parental investment and the evolutionary dynamics of sexual selection. *Animal Behaviour* 27: 1022–83.

Korpimaki, E., and K. Norrdahl. 1998. Experimental reduction of predators reverses the crash phase of small rodent cycles. *Ecology* 79: 2448–55.

Krebs, J. R., and N. B. Davies. 1981. *An introduction to behavioural ecology*. Blackwell Scientific Publications, Oxford, U.K.

Kruuk, H. 1964. Predators and anti-predator behaviour of the black headed gull, *Larus ridibundus*. *Behaviour Supplement* 11: 1–129.

Lacey, R. C. 1987. Loss of genetic diversity from unmanaged populations: Interacting effects of drift, mutation, immigration, selection, and population subdivision. *Conservation Biology* 1: 143–58.

Lack, D. 1944. Symposium on "The ecology of closely allied species." *Journal of Animal Ecology* 13: 176–77.

Lande, R. 1976. The maintenance of genetic variability by mutation in a polygenic character with linked loci. *Genetic Research* 26: 221–35.

Landsburg, J., and C. Ohmart. 1989. Levels of defoliation in forests—patterns and concepts. *Trends in Ecology and Evolution* 4: 96–100.

Latham, R. E., and R. E. Ricklefs. 1993. Continental comparisons of temperate-zone tree species diversity. In R. E. Ricklefs and D. Schluter (eds.), *Species diversity in ecological communities* (pp. 294–314). University of Chicago Press, Chicago.

Laurie, H., P. J. Mustart, and R. M. Lowling. 1997. A shared niche? The case of the species pair *Protea obtusifolia–Leucadendron meridianum*. *Oikos* 79: 127–36.

Lawler, S. P. 1993. Species richness, species composition, and population dynamics of protists in experimental microcosms. *Journal of Animal Ecology* 64: 711–19.

Lawler, S. P., and P. J. Morin. 1993. Food web architecture and population dynamics in laboratory microcosms of protists. *American Naturalist* 141: 675–86.

Lawton, J. H. 1984. Non-competitive populations, non-convergent communities, and vacant niches: The herbivores of bracken. In D. R. Strong, D. Simberloff, L. G. Abele, and A. B. Thistle (eds.), *Ecological communities: Conceptual issues and the evidence* (pp. 67–100). Princeton University Press, Princeton, NJ.

Lawton, J. H. 1987. Are there assembly rules for successional communities? In A. J. Gray, M. J. Crawley, and P. J. Edwards (eds.), *Colonization, succession and stability* (pp. 225–44). Blackwell, Oxford, U.K.

Lawton, J. H. 1994. What do species do in ecosystems. *Oikos* 71: 367–74.

Lawton J. H., and M. P. Hassell. 1981. Asymetrical competition in insects. *Nature* 289: 793–95.

Lawton, J. H., and C. G. Jones. 1995. Linking species and ecosystems: Organisms as ecosystem engineers. In C. G. Jones and J. H. Lawton (eds.), *Linking species and ecosystems* (pp. 141–50). Chapman and Hill, New York.

Lawton, J. H., and S. McNeill. 1979. Between the devil and the deep blue sea: On the problem of being a herbivore. In R. M. Anderson, B. D. Taylor, and L. R. Taylor (eds.), *Population dynamics* (pp. 223–44). Blackwell Scientific Publications, Oxford, U.K.

Lawton, J. H., T. M. Lewinsohn, and S. G. Compton. 1993. Patterns of diversity for the insect herbivores on bracken. In R. E. Ricklefs and D. Schluter (eds.), *Species diversity in ecological communities* (pp. 178–84). University of Chicago Press, Chicago.

Le Boeuf, B. J., and S. Kaza (eds.). 1981. The Natural History of Año Nuevo. Boxwood Press, Pacific Grove, CA.

Leopold, A. S., S. A. Cain, C. M. Cottam, I. N. Gabrielson, and T. L. Kimball. 1963. Wildlife management in the national parks. *Transactions of the Northern American Wildlife and Natural Resources Conference* 28: 28–45.

Lerner, I. M. 1954. *Genetic homeostasis*. Oliver and Boyd, Edinburgh.

Leverich, W. J., and D. A. Levin. 1979. Age-specific survivorship and reproduction in *Phlox drummondi*. *American Naturalist* 113: 881–903.

Levin, D. A. 1981. Dispersal versus gene flow in plants. *Annals of the Missouri Botanical Garden* 68: 233–53.

Levins, R. 1969. Some demographic and genetic consequences of environmental heterogeneity for biological control. *Bulletin of the Entomological Society* 15: 237–40.

Liebig, J. 1842. *Animal chemistry or organic chemistry in its application to physiology and pathology*. 1964 reprint, Johnson Print Corporation, New York.

Lieth, H. 1975. Primary productivity in ecosystems: Comparative analysis of global patterns. In W. H. van Dobben and R. H. Lowe-McConnell (eds.), *Unifying concepts in ecology* (pp. 67–88). Dr. W. Junk, The Hague.

Likens, F. H., N. Bormann, N. M. Johnson, D. W. Fisher, and R. S. Pierce. 1970. The effects of forest cutting and herbivore treatment on nutrient budgets in the Hubbard Brook Watershed-Ecosystem. *Ecological Monograph* 40: 23–47.

Lill, A. 1974. Sexual behavior of the lek-forming white-bearded manakin (*Manacus manacus trinitatis* Hartett). *Zeitschrift für Tierpsychologie* 36: 1–36.

Limbaugh, C. 1961. Cleaning symbiosis. *Scientific American* 205: 42–49.

Lindeman, R. L. 1942. The trophic-dynamic aspect of ecology. *Ecology* 23: 399–418.

Loehle, C. 1987. Tree life history strategies: The role of defenses. *Canadian Journal of Forest Research* 18: 209–22.

Losey, J. E., A. R. Ives, J. Harman, F. Ballantyre, and C. Brown. 1997. A polymorphism maintained by opposite patterns of parasitism and predation. *Nature* 388: 269–72.

Losos, J. B., S. Naeem, and R. K. Colwell. 1989. Hutchinsonian ratios and statistical power. *Evolution* 43: 1820–26.

Lotka, A. J. 1925. *Elements of physical biology*. Reprinted, Dover Publications, New York, 1956.

Lovelock, J. 1991. *Healing Gaia: Practical medicine for the planet*. Harmony Books, New York.

Luckinbill, L. S. 1973. Coexistence in laboratory populations of *Paramecium aurelia* and its predator *Didinium nasutum*. *Ecology* 59: 1320–27.

Luckinbill, L. S. 1974. The effects of space and enrichment on a predator-prey system. *Ecology* 55: 1142–47.

Lynch, J. F., and N. K. Johnson. 1974. Turnover and equilibria in insular avifaunas, with special reference to the California Channel Islands. *Condor* 76: 370–84.

MacArthur, R. H. 1957. On the relative abundance of bird species. *Proceedings of the National Academy of Sciences of the United States of America* 43: 293–95.

MacArthur, R. H. 1958. Population ecology of some warblers of northeastern coniferous forests. *Ecology* 39: 599–619.

MacArthur, R. H. 1960. On the relative abundance of species. *American Naturalist* 94: 25–36.

MacArthur, R. H. 1972. Geographical ecology. Harper and Row, New York.

MacArthur, R. H., and J. W. MacArthur. 1961. On bird species diversity. *Ecology* 42: 594–98.

MacArthur, R. H., and E. O. Wilson. 1963. An equilibrium theory of insular biogeography. *Evolution* 17: 373–87.

MacArthur, R. H., and E. O. Wilson. 1967. *The theory of island biogeography. Princeton University Press, Princeton, NJ.*

Magurran, A. E. 1988. Ecological diversity and its management. Princeton University Press, Princeton, NJ.

Maiorana, V. C. 1978. An explanation of ecological and developmental constants. *Nature* 273: 375–77.

Margulis, L. 1976. Genetic and evolutionary consequences of symbiosis. *Experimental Parasitology* 39: 277–349.

Martin, T. E. 1993. Nest predation among vegetation layers and habitat types: Revising the dogmas. *American Naturalist* 141: 897–913.

Martinez, N. D. 1992. Constant connectance in community food webs. *American Naturalist* 139: 1208–18.

Martinez, N. D., and J. H. Lawton. 1995. Scale and food-web structure from local to global. *Oikos* 73: 148–54.

Massey, A. B. 1925. Antagonism of the walnuts (*Juglans nigra L.* and *J. cinerea L.*) in certain plant associations. *Phytopathology* 15: 773–84.

May, R. M. 1973. *Stability and complexity in model ecosystems.* Princeton University Press, Princeton, NJ.

May, R. M. 1976. Models for two interacting populations. In R. M. May (ed.), *Theoretical ecology: Principles and applications* (pp. 49–70). Sanders, Philadelphia.

May, R. M. 1988. How many species are there on earth? *Science* 241: 1441–49.

May, R. M. 1990. Taxonomy as destiny. *Nature* 347: 129–30.

May, R. M., and R. M. Anderson. 1979. Population biology of infectious diseases. *Nature* 280: 455–61.

McIntosh, R. P. 1985. The background of ecology: Concept and theory. Cambridge University Press, Cambridge, U.K.

McNab, B. K. 1973. Energetics and the distribution of vampires. *Journal of Mammalogy* 54: 131–44.

McNaughton, S. J. 1977. Diversity and stability of ecological communities: A comment on the role of empiricism in ecology. *American Naturalist* 111: 515–25.

McNaughton, S. J. 1986. On plants and herbivores. *American Naturalist* 128: 765–70.

McNaughton, S. J. 1988. Mineral nutrition and spatial concentrations of African ungulates. *Nature* 334: 343–45.

McNaughton, S. J., F. F. Banyikwa, and M. M. McNaughton. 1997. Promotion of the cycling of diet-enhancing nutrients by African grazers. *Science* 278: 1798–1800.

McNeely, J. A., K. R. Miller, W. V. Reid., R. A. Muttermeir, and T. B. Werner. 1990. *Conserving the world's biological diversity.* IUCN, Gland, Switzerland.

Mduma, S. A. R., A. R. E. Sinclair, and R. Hilborn. 1999. Food regulates the Serengeti wildebeest: A 40-year record. *Journal of Animal Ecology* 68: 1101–22.

Menge B. A., and T. M. Farrell. 1989. Community structure and interaction webs in shallow marine hard-bottom communities: Test of an environmental stress model. *Advances in Ecological Research* 19: 189–262.

Menge, B. A., and J. P. Sutherland. 1976. Species diversity gradients: Synthesis of the roles of predation, competition, and temporal heterogeneity. *American Naturalist* 110: 351–69.

Menzel, D. W., and J. H. Ryther. 1961. Nutrients limiting to the production of phytoplankton in the Sargasso sea, with special reference to iron. *Deep Sea Research* 7: 276–81.

Meretsky, V. J., N. F. R. Snyder, S. R. Beissinger, D. A. Clendenen, and J. W. Wiley. 2000. Demography of the California Condor: Implications for reestablishment. *Conservation Biology* 14: 957–67.

Miller, R. S. 1964. Larval competition in *Drosphilia melanogaster* and *D. simulans*. *Ecology* 45: 132–48.

Mills, K. H., S. M. Chalanchuk, L. C. Mohr, and I. J. Davies. 1987. Responses of fish populations in Lake 223 to 8 years of experimental eutrophication. *Canadian Journal of Fisheries and Aquatic Sciences*, Series 44 (Supplement 4): 114–25.

Mills, L. S., M. E. Soule, and D. F. Doak. 1993. The keystone-species concept in ecology and conservation. *Bioscience* 43: 219–24.

Minorsky, P. V. 1985. An heuristic hypothesis of chilling injury in plants: A role for calcium as the primary physiological transducer of injury. *Plant Cell and Environment* 18: 75–94.

Mittermeier, R. A. 1988. Primate diversity and the tropical forest: Case studies from Brazil and Madagascar and the importance of the megadiversity countries. In E. O. Wilson and F. M. Peter (eds.), *Biodiversity* (pp. 145–154). National Academic Press, Washington, DC.

Mittermeier, R. A., and T. B. Werner. 1990. Wealth of plants and animals unites "megadiversity" countries. *Tropicus* 4: 1, 4–5.

Mlot, C. 1993. Predators, prey, and natural disasters attract ecologists. *Science* 261: 1115.

Moen, J., and L. Oksanen. 1991. Ecosystem trends. *Nature* 355: 510.

Moore, J. 1995. The behavior of parasitized animals. *BioScience* 45: 89–96.

Moran, N. A., and T. G. Whitham. 1990. Interspecific competition between root-feeding and leaf-galling aphids mediated by host-plant resistance. *Ecology* 71: 1050–58.

Morin, P. J., and S. P. Lawler. 1995. Effects of food chain length and ominivory on population dynamics in experimental food webs. In G. A. Polis and K. O. Winemiller (eds.), *Food webs: Integration of patterns and dynamics* (pp. 218–230). Chapman and Hall, NY.

Morrell, V. 1993. Australian pest control by virus causes concern. *Science* 261: 683–84.

Morris, R. F. 1957. The interpretation of mortality data in studies on population dynamics. *Canadian Entomologist* 89: 49–69.

Morrison, C. W. 1997. The insular biogeography of small Bahamian cays. *Journal of Ecology* 85: 441–54.

Muller, C. H. 1966. The role of chemical inhibition (allelopathy) in vegetational composition. *Bulletin of the Torrey Botanical Club* 93: 332–51.

Muller, F. 1879. *Ituna* and *Thyridis*, a remarkable case of mimicry in butterflies, translated from the German by R. Meldola. *Proceedings of the Entomological Society of London* 27: 20–29.

Murdoch, W. W. 1975. Diversity, complexity, stability and past control. *Journal of Applied Ecology* 12: 795–807.

Murie, A. 1944. Wolves of Mount McKinley: Fauna of National Parks. U.S. Fauna Series Number 5, Washington, DC.

Myers, N., R. A. Mittermeir, C. G. Mittermeir, G. A. B. Fonseca, and J. Kent. 2000. Biodiversity hot spots for conservation priorities. *Nature* 403: 853–58.

Naeem, S., K. Håkansson, J. H. Lawton, M. J. Crawley, and L. J. Thompson. 1996. Biodiversity and plant productivity in a model assemblage of plant species. *Oikos* 76: 259–64.

Naeem, S., L. J. Thompson, S. P. Lawler, J. H. Lawton, and R. M. Woodfin. 1994. Declining biodiversity can alter the performance of ecosystems. *Nature* 368: 734–37.

Naeem, S., L. J. Thompson, S. P. Lawler, J. H. Lawton, and R. M. Woodfin. 1995. Biodiversity and ecosystem functioning: Empirical evidence from experimental microcosms. *Endeavour* 19: 58–63.

Nelson, T. C. 1955. Chestnut replacement in the southern Highlands. *Ecology* 36: 352–53.

Newsome, A. 1990. The control of vertebrate pests by vertebrate predators. *Trends in Ecology and Evolution* 5: 187–91.

Newsome, A. E., I. Parer, and P. C. Catling. 1989. Prolonged prey suppression by carnivores-predator-removal experiments. *Oecologia* 78: 458–67.

Neyman, J., T. Park, and E. L. Scott. 1959. Struggle for existence; the *Tribolium* model: biological and statistical aspects. *General Systems* 3: 152–79.

Nicholson, A. J. 1933. The balance of animal populations. *Journal of Animal Ecology* 2: 131–78.

Nieminen, M. 1996. Migration of moth species in a network of small islands. *Oecologia* 108: 643–51.

Nilsson, S. G., and U. Wästljung. 1987. Seed predation and cross-pollination in mast-seeding beech (*Fagus sylvatica*) patches. *Ecology* 68: 260–65.

O'Connor, R. J. 1991. Long-term bird population studies in the U.S. *Ibis* 133, supplement 1: 30–48.

O'Riain, M. J., J. U. M. Jarvis, and C. E. Faulkes. 1996. A dispersive morph in the naked mole-rat. *Nature* 380: 619–21.

Odum, E. P. 1969. The strategy of ecosystem development. *Science* 164: 262–70.

Oksanen, L., S. D. Fretwell, J. Arruda, and P. Niemela. 1981. Exploitation ecosystems in gradients of primary productivity. *American Naturalist* 118: 240–61.

Oosting, H. J. 1942. An ecological analysis of the plant communities of Piedmont, North Carolina. *American Midland Naturalist* 28: 1–126.

Opler, P. A. 1974. Oaks as evolutionary islands for leaf-mining insects. *American Scientist* 62: 67–73.

Oring, L. W. 1985. Avian polyandry. *Current Ornithology* 3: 309–57.

Owen, D. F., and J. Owen. 1974. Species diversity in temperate and tropical Ichneumonidae. *Nature* 249: 583–84.

Owen, D. F., and D. L. Whiteley. 1986. Reflexive selection: Moment's hypothesis resurrected. *Oikos* 47: 117–20.

Packer, C. P. 1997. Virus hunter. *Natural History* 106(9): 36–41.

Paine, R. T. 1966. Food web complexity and species diversity. *American Naturalist* 100: 65–75.

Paine, R. T. 1969. A note on trophic complexity and community stability. *American Naturalist* 103: 91–93.

Paine, R. T. 1988. Food webs: Road maps of interactions or grist for theoretical development? *Ecology* 69: 1648–54.

Paine, R. T. 1992. Food-web analysis through field measurement of per capita interaction strength. *Nature* 355: 73–75.

Paine, R. T., J. L. Reuesnik, A. Sun, E. L. Soulanille, M. J. Worham, C. D. G. Harley, D. R. Brumbaugh, and D. L. Secord. 1996. Trouble on oiled waters: Lessons from the *Exxon Valdez* oil spill. *Annual Review of Ecology and Systematics* 27: 197–235.

Park, T. 1948. Experimental studies of interspecies competition: I. Competition between populations of the flour beetles *Tribolium confusum* Duval and *Tribolium castaneum* Herbst. *Ecological Monographs* 18: 265–307.

Park, T. 1954. Experimental studies of interspecies competition: II. Temperature, humidity, and competition in two species of *Tribolium*. *Physiological Zoology* 27: 177–238.

Park, T., P. H. Leslie, and D. B. Metz. 1964. Genetic strains and competition in populations of *Tribolium*. *Physiological Zoology* 37: 97–162.

Parmessan, C., and 12 other authors. 1999. Poleward shifts in geographical ranges of butterfly species associated with regional warming. *Nature* 399: 579–83.

Passell, H. D. 2000. Recovery of bird species in minimally restored Indonesian tin strip mines. *Restoration Ecology* 8: 112–18.

Pauly, D., and V. Christensen. 1995. Primary production required to sustain global fisheries. *Nature* 374: 255–57.

Peakall, R., A. J. Beattie, and S. J. James. 1987. Pseudocopulation of an orchid by male ants: A test of two hypotheses accounting for the rarity of ant pollination. *Oecologia* 78: 522–24.

Pearl, R. 1927. The growth of populations. *Quarterly Review of Biology* 2: 532–48.

Pearl, R. 1928. *The rate of living*. Knopf, New York.

Pearl, R., and L. J. Reed. 1920. On the rate of growth of the population of the United States since 1790 and its mathematical representation. *Proceedings of the National Academy of Sciences of the United States of America* 6: 275–88.

Pellmyr, O., and C. J. Huth. 1994. Evolutionary stability of mutualism between yuccas and yucca moths. *Nature* 372: 257–60.

Pennings, S. C., and R. M. Callaway. 1996. Impact of a parasitic plant on the structure and dynamics of salt marsh vegetation. *Ecology* 77: 1410–19.

Peterson, R. O. 1999. Wolf–moose interaction on Isle Royale: The end of natural regulation? *Ecological Applications* 9: 10–16.

Pianka, E. R. 1986. *Ecology and natural history of desert lizards*. Princeton Press, Princeton, NJ.

Pianka, E. R. 1988. *Evolutionary ecology*, 4th ed. Harper and Row, New York.

Pielou, E. C. 1969. *An introduction to mathematical ecology*. Wiley Interscience, New York.

Pimentel, D. 1988. Herbivore population feeding pressure on plant hosts: Feedback evolution and host conservation. *Oikos* 53: 289–302.

Pimm, S. L. 1979. The structure of food webs. *Theoretical Population Ecology* 16: 144–58.

Pimm, S. L. 1980. Food web design and the effect of species deletion. *Oikos* 35: 139–47.

Pitelka, F. A. 1964. The nutrient-recovery hypothesis for Arctic microtine cycles: I. Introduction. In D. J. Crisp (ed.), *Grazing in terrestrial and marine environments*, pp. 55–56. Blackwell Scientific Publications, Oxford, U.K.

Pleszczynska, W. K. 1978. Microgeographic prediction of polygyny in the lark bunting. *Species* 201: 935–37.

Pockman, W. T., and J. S. Sperry. 1997. Freezing induced xylem cavitation and the northern limit of *Larrea tridentata*. *Oecologia* 109: 19–27.

Podoler, H., and D. Rogers. 1975. A new method for the identification of key factors from life-table data. *Journal of Animal Ecology* 44: 85–115.

Polis, G. A. 1991. Complex trophic interactions in deserts: An empirical critique of food-web theory. *American Naturalist* 138: 123–55.

Polis, G. A., and D. R. Strong. 1996. Food web complexity and community dynamics. *American Naturalist* 147: 813–46.

Power, M. E., and L. S. Mills. 1995. The Keystone cops meet in Hilo. *Trends in Ecology and Evolution* 10: 182–84.

Preston, F. W. 1948. The commonness and rarity of species. *Ecology* 29: 254–83.

Preston, F. W. 1962. The canonical distribution of commonness and rarity. *Ecology* 43: 185–215, 410–32.

Price, P. W. 1970. Characteristics permitting coexistence among parasitoids of a sawfly in Quebec. *Ecology* 51: 445–54.

Price, P. W. 1991. The plant vigour hypothesis and herbivore attack. *Oikos* 62: 244–51.

Prins, H. H. T., and F. J. Weyerhaeuser. 1987. Epidemics in populations of wild ruminants: Anthrax and impala, rinderpest and buffalo in Lake Manyara National Park, Tanzania. *Oikos* 49: 28–38.

Pugnaire, F., P. Haase, and J. Puigdefabregas. 1996. Facilitation between higher plant species in a semiarid environment. *Ecology* 77: 1420–26.

Quinn, J. F., and S. P. Harrison. 1988. Effects of habitat fragmentation and isolation on species richness: Evidence from biogeographic patterns. *Oecologia* 75: 132–40.

Rabinowitz, D. 1981. Seven forms of rarity. In H. Synge (ed.), *The biological aspects of rare plant conservation* (pp. 205–17). John Wiley, London.

Raich, J. W., A. E. Russell, and P. M. Vitusek. 1997. Primary productivity and ecosystem development along an elevational gradient on Mauna-Loa, Hawaii. *Ecology* 78: 707–21.

Ralls, K., and J. Ballou. 1983. Extinction: Lessons from zoos. In C. M. Schonewald-Cox, S. M. Chambers, B. MacBryde, and L. Thomas (eds.), *Genetics and conservation: A reference for managing wild animal and plant populations* (pp. 164–184). Benjamin/Cummings, Menlo Park, CA.

Reeve, H. K., D. F. Westneat, W. A. Noan, P. Wisherman, and C. F. Aquadro. 1990. DNA "fingerprinting" reveals high levels of inbreeding in colonies of the eusocial naked mole rat. *Proceedings of the National Academy of Sciences* 87: 2496–2500.

Reeve, J. D. 1988. Environmental variability, migration, and persistence in host–parasitoid systems. *American Naturalist* 132: 810–36.

Reice, S. R. 1994. Nonequilibrium determinants of biological community structure. *American Scientist* 82: 424–35.

Reid, W. V. 1998. Biodiversity hotspots. *Trends in Ecology and Evolution* 13: 275–80.

Rey, J. R. 1983. Insular ecology of salt marsh arthropods: Species-level patterns. *Journal of Biogeography* 12: 96–107.

Rey, J. R., E. D. McCoy, and D. R. Strong, Jr. 1981. Herbivore pests, habitat islands, and the species–area relation. *American Naturalist* 117: 611–22.

Rhoades, D. F., and R. G. Cates. 1976. Toward a general theory of plant antiherbivore chemistry. *Recent Advances in Phytochemistry* 10: 168–213.

Ricklefs, R. E., and I . J. Lovette. 1999. The roles of island area per se and habitat diversity in the species–area relationships of four Lesser Antillean faunal groups. *Journal of Animal Ecology* 68: 1142–60.

Ricklefs, R. E., and D. Schluter. 1993. Species diversity: Regional and historical influences. In R. E. Ricklefs and D. Schluter (eds.), *Species diversity in ecological communities* (pp. 350–64). University of Chicago Press, Chicago.

Rico-Gray, V. 1993. Use of plant-derived food resources by ants in the dry tropical lowlands of coastal Veracruz, Mexico. *Biotropica* 25: 301–15.

Ritchie, M. E., D. Tilman, and J. M. H. Knops. 1998. Herbivore effects on plant and nitrogen dynamics in oak savanna. *Ecology* 79: 165–77.

Ritland, D. B., and L. P. Brower. 1991. The viceroy butterfly is not a Batesian mimic. *Nature* 350: 497–98.

Rivard, D. H., J. Poitevon, D. Plasse, M. Carleton, and D. J. Currie. 2000. Changing species richness and composition in Canadian National Parks. *Conservation Biology* 14: 1099–1109.

Rohde, K. 1992. Latitudinal gradients in species diversity: The search for the primary cause. *Oikos* 65: 514–27.

Rohde, K. 1998. Latitudinal gradients in species diversity. Area matters, but how much? *Oikos* 82: 184–190.

Rood, J. P. 1990. Group size, survival, reproduction, and routes to breeding in dwarf mongooses. *Animal Behavior* 39: 566–72.

Root, R. 1967. The niche exploitation pattern of the blue-gray gnatcatcher. *Ecological Monographs* 37: 317–50.

Root, T. 1988. Energy constraints in avian distributions and abundances. *Ecology* 69: 330–39.

Rosenzweig, M. L. 1968. Net primary productivity of terrestrial communities: Prediction from climatological data. *American Naturalist* 102: 67–74.

Rosenzweig, M. L. 1971. The paradox of enrichment: Destabilization of exploitation ecosystems in ecological time. *Science* 171: 385–87.

Rosenzweig, M. L., and R. H. MacArthur. 1963. Graphical representation and stability conditions of predator–prey interactions. *American Naturalist* 97: 209–23.

Rowan, R. 1998. Diversity and ecology of zooxanthellae on coral reefs. *Journal of Phycology* 34: 407–17.

Royama, T. 1996. A fundamental problem in key factor analysis. *Ecology* 77: 87–93.

Saccheri, I., M. Kuussaari, M. Kankare, P. Vikman, W. Forteluis, and I. Hanski. 1998. Inbreeding and extinction in a butterfly metapopulation. *Nature* 392: 491–94.

Saffo, M. B. 1992. Coming to terms with a field: Words and concepts in symbiosis. *Symbiosis* 14: 17–31.

Sailer, R. I. 1983. History of insect introductions. In C. Graham and C. Wilson (eds.), *Exotic plant pests and North American agriculture* (pp. 15–38). Academic Press, New York.

Sala, O. S., W. J. Parton, L. A. Joyce, and W. K. Lauenroth. 1988. Primary production of the central grassland region of the United States. *Ecology* 69: 40–45.

Sanders, H. L. 1968. Marine benthic diversity: A comparative study. *American Naturalist* 102: 243–82.

Sang, J. H. 1950. Population growth in *Drosophila* cultures. *Biological Reviews* 25: 188–219.

Schaffer, M. L., and F. B. H. Samson. 1985. Population size and extinction: A note on determining critical population sizes. *American Naturalist* 125: 144–52.

Schall, J. J. 1992. Parasite-mediated competition in *Anolis* lizards. *Oecologia* 92: 58–64.

Scheffer, V. B. 1951. The rise and fall of a reindeer herd. *Scientific Monthly* 73: 356–62.

Schindler, D. W. 1974. Eutrophication and recovery in experimental lakes: Implications for lake management. *Science* 184: 397–99.

Schindler, D. W. 1977. Evolution of phosphorus limitation in lakes. *Science* 195: 260–62.

Schluter, D., and R. E. Ricklefs. 1993. Convergence and the regional component of species diversity. In R. Ricklefs and D. Schluter (eds.), *Species diversity in ecological communities* (pp. 230–40). University of Chicago Press, Chicago.

Schoener, T. W. 1974. Resource partitioning in ecological communities. *Science* 185: 27–39.

Schoener, T. W. 1976. Alternatives to Lotka–Volterra competition: Models of intermediate complexity. *Theoretical Population Biology* 10: 309–33.

Schoener, T. W. 1983. Field experiments on interspecific competition. *American Naturalist* 122: 240–85.

Schoener, T. W. 1985. Some comments on Connell's and my reviews of field experiments in interspecific competition. *American Naturalist* 125: 730–40.

Schultz, A. M. 1964. The nutrient-recovery hypothesis for Arctic microtine cycles: II. Ecosystem variables in relation to Arctic microtine cycles. In D. J. Crisp (ed.), *Grazing in terrestrial and marine environments* (pp. 57–68). Blackwell Scientific Publications, Oxford, U.K.

Schultz, A. M. 1969. *A study of an ecosystem: The arctic tundra in the ecosystem concept in natural resource management* pp. 77–93. Ed. G. van Dyne, New York, Academic Press.

Seghers, B. H. 1974. Schooling behaviour in the guppy *Poecilia reticulata*: An evolutionary response to predation. *Evolution* 28: 486–89.

Siemann, E., D. Tilman, J. Haarstad, and M. Ritchie. 1998. Experimental tests of the dependence of arthropod diversity on plant diversity. *American Naturalist* 152: 738–50.

Sih, A., P. Crawley, M. McPeek, J. Petranka, and K. Strohmeir. 1985. Predation, competition, and prey communities: A review of field experiments. *Annual Review of Ecology and Systematics* 16: 269–311.

Silvertown, J. W., M. Franco, and K. McConway. 1992. A demographic interpretation of Grime's Triangle. *Functional Ecology* 6: 130–36.

Silvertown, J. W., M. Franco, I. Pisanty, and A. Mendoza. 1993. Comparative planned demography: Relative importance of life-cycle components to the finite rate of increase in woody and herbaceous perennials. *Journal of Ecology* 81: 465–76.

Simberloff, D. S. 1976a. Experimental zoogeography of islands: Effects of island size. *Ecology* 57: 629–48.

Simberloff, D. S. 1976b. Species turnover and equilibrium island biogeography. *Science* 194: 572–78.

Simberloff, D. S. 1978. Colonization of islands by insects: Immigration, extinction and diversity. pp. 139–153, in L. A. Mound and M. Waloff (eds). Diversity of insect faunas. Blackwell Scientific Publications, Oxford, U.K.

Simberloff, D. 1988. The contribution of population and community biology to conservation science. *Annual Review of Ecology and Systematics* 19: 473–512.

Simberloff, D. S., and W. J. Boecklen. 1981. Santa Rosalia reconsidered: Size ratios and competition. *Evolution* 35: 1206–28.

Simberloff, D. S., B. J. Brown, and S. Lowrie. 1978. Isopod and insect root borers may benefit Florida mangroves. *Science* 201: 630–32.

Simberloff, D., and P. Stiling. 1996. How risky is biological control? *Ecology* 77: 1965–74.

Simberloff, D., and E. O. Wilson. 1969. Experimental zoogeography of islands: The colonization of empty islands. *Ecology* 50: 278–96.

Simberloff, D. S., and E. O. Wilson. 1970. Experimental zoogeography of islands: A two year record of recolonization. *Ecology* 51: 934–37.

Simpson, G. G. 1964. Species density of North American recent mammals. *Systematic Zoology* 13: 57–73.

Sinclair, A. R. E. 1977. *The African buffalo*. University of Chicago Press, Chicago.

Sinclair, A. R. E. 1989. Population regulation in animals. In J. M. Cherret (ed.), *Ecological concepts* (pp. 197–242). Blackwell Scientific Publications, Oxford, U.K.

Smith, B. R., and J. J. Tibbles. 1980. Sea lamprey (*Petromyzon marinus*) in Lakes Huron, Michigan, and Superior: History of Invasion and Control, 1936–1978. *Canadian Journal of Fisheries and Aquatic Science*. 37: 1780–1801.

Smith-Gill, S. J., and D. S. Gill. 1978. Curvilinearity in the competition equations: An experiment with ranid tadpoles. *American Naturalist*: 110: 849–60.

Soderstrom, T. R., and C. E. Calderon. 1971. Insect pollination in tropical rain forest grasses. *Biotropica* 3: 1–16.

Solomon, M. E. 1949. The natural control of animal populations. *Journal of Animal Ecology* 18: 1–32.

Sorensen, T. 1948. A method of establishing groups of equal amplitude in plant sociology based on similarity of species content. *Det. Kong. Danske Vidensk. Selsk. Biol. Skr.* (Copenhagen) 5(4): 1–34.

Sousa, W. P. 1979. Disturbance in marine intertidal boulder fields: The nonequilibrium maintenance of species diversity. *Ecology* 60: 1225–39.

Southwick, E. E. 1984. Photosynthate allocation to floral nectar: A neglected energy investment. *Ecology* 65: 1775–79.

Southwood, T. R. E. 1961. The number of species of insect associated with various trees. *Journal of Animal Ecology* 30: 1–8.

Southwood, T. R. E. 1978. *Ecological methods with particular reference to the study of insect populations*, 2d ed. Methuen, London.

Stearns, S. C. 1976. Life-history tactics: A review of the ideas. *Quarterly Review of Biology* 51: 3–47.

Stiling, P. D. 1980. Colour polymorphism in some nymphs of the genus *Eupteryx*. *Ecological Entomology* 5: 175–78.

Stiling, P. D. 1988. Density-dependent processes and key factors in insect populations. *Journal of Animal Ecology* 57: 581–93.

Stiling, P. D. 1990. Calculating the establishment rates of parasitoids in classical biological control. *Bulletin of the Entomological Society of America* 36: 225–30.

Stiling, P. 1993. Why do natural enemies fail in classical biological control programs? *American Entomologist* 39: 31–37.

Stiling, P. D., B. V. Brodbeck, and D. R. Strong. 1982. Foliar nitrogen and larval parasitism as determinants of leafminer distribution patterns on *Spartina alterniflora*. *Ecological Entomology* 7: 447–52.

Stiling, P., and A. M. Rossi. 1997. Experimental manipulations of top-down and bottom-up factors in an in-trophic system. *Ecology* 78: 1602–6.

Stiling, P. D., and D. Simberloff. 1989. Leaf abscission: Induced defense against pests or response to damage? *Oikos* 55: 43–49.

Stiling, P. D., and D. R. Strong. 1983. Weak competition among *Spartina* stem borers by means of murder. *Ecology* 64: 770–78.

Stork, N. E. 1988. Insect diversity: Facts, fiction and speculation. *Biological Journal of the Linnean Society* 35: 321–37.

Strauss, S. Y. 1988. Determining the effects of herbivory using damaged plants. *Ecology* 69: 1628–30.

Strickberger, M. W. 1986. *Genetics*, 3d ed. Macmillan, New York.

Strong, D. R. 1974. Nonasymptotic species richness models and the insects of British trees. *Proceedings of the National Academy of Sciences of the United States of America* 71: 2766–69.

Strong, D. R. 1988. Insect host range (special feature). *Ecology* 69: 885.

Strong, D. R., J. H. Lawton, and T. R. E. Southwood. 1984. *Insects on Plants: Community Patterns and Mechanisms*. Blackwell Scientific Publications, Oxford, U.K.

Strong, D. R., E. D. McCoy, and J. R. Rey. 1977. Time and the number of herbivore species: The pests of sugarcane. *Ecology* 58: 167–75.

Tansley, A. G. 1935. The use and abuse of vegetational concepts and terms. *Ecology* 16: 284–307.

Tapper, S. C., G. R. Potts, and M. H. Brockless. 1996. The effect of an experimental reduction in predator pressure on the breeding series and population density of grey partridges *Perdix perdix*. *Journal of Applied Ecology* 33: 965–78.

Terborgh, J. 1973. On the notion of favorableness in plant ecology. *American Naturalist* 107: 481–501.

Terborgh, J. 1986. Keystone plant resources in the tropical forest. In M. E. Soulé (ed). *Conservation biology: The science of scarcity and diversity* (pp. 330–44). Sinauer, Sunderland, MA.

Terborgh, J., and F. Faaborg. 1973. Turnover and ecological release in the avifauna of Mona Island, Puerto Rico. *Auk* 90: 759–79.

Thatcher, R. W., A. M. Berr, W. A. Lumbang, and V. Paul. 1998. Allelopathic interactions between sponges on a tropical reef. *Ecology* 79: 1740–50.

Thomas, C. D. 1990. Fewer species. *Nature* 347: 237.

Tilman, D. 1982. *Resource competition and community structure*. Princeton University Press, Princeton, NJ.

Tilman, D. 1996. Biodiversity: Population versus ecosystem stability. *Ecology* 77: 350–63.

Tilman, D. 1997. *Mechanisms of plant competition*. In M. J. Crawley (ed.), *Plant Ecology*, 2d ed. (pp. 239–61). Blackwell Scientific, Oxford, U.K.

Tilman, D., and J. A. Downing. 1994. Biodiversity and stability in grasslands. *Nature* 367: 363–65.

Tilman, D., C. L. Lehman, and C. E. Bristow. 1998. Diversity–stability relationships: Statistical inevitability or ecological consequences? *American Naturalist* 151: 277–82.

Tilman, D., D. Wedin, and J. Knops. 1996. Productivity and sustainability influenced by biodiversity in grassland ecosystems. *Nature* 379: 718–20.

Tokeshi, M. 1999. Species coexistence: Ecological and evolutionary perspectives. Blackwell Science, Oxford, U.K.

Trenberth, K. E. 1997. The use and abuse of climate models. *Nature* 386: 131–33.

Turner, J. R. G., C. M. Gratehouse, and C. A. Carey. 1987. Does solar energy control organic diversity? Butterflies, moths, and the British climate. *Oikos* 48: 195–205.

Turner, J. R. G., J. J. Lennon, and J. A. Lawrenson. 1988. British bird species distributions and the energy theory. *Nature* 335: 539–41.

Turner, T. 1983. Facilitation as a successional mechanism in a rocky intertidal community. *American Naturalist* 121: 729–38.

U.S. Congress Office of Technology Assessment. 1987. *Technologies to Maintain Biological Diversity*, OTA-F-330. U.S. Government Printing Office, Washington, DC.

Uvarov, B. P. 1931. Insects and climate. *Transactions of the Royal Entomological Society of London* 79: 1–247.

Valentine, J. F., K. L. Heck, J. Busby, and D. Webb. 1997. Experimental evidence that herbivory increases shoot density and productivity in a subtropical turtlegrass (*Thalassia testudinum*) meadow. *Oecologia* 112: 193–200.

Van der Heijden, M. G. A., J. N. Klironomos, M. Ursic, P. Moutoglis, R. Streitwolf-Engel, T. Boller, A. Wiemken, and I. R. Sanders. 1998. Mycorrhizal fungal biodiversity determines plant biodiversity, ecosystem variability and productivity. *Nature* 396: 69–72.

Van der Putten, A., C. Van Dijk, and B. A. M. Peters. 1993. Plant-specific soil-borne diseases contribute to succession in foredune vegetation. *Nature* 362: 53–56.

van Lenteren, J. C. 1980. Evaluation of control capabilities of natural enemies: Does art have to become science? *Netherlands Journal of Zoology* 30: 369–81.

Van Scoy, K., and K. Coale. 1994. Dumping iron in the Pacific. *New Scientist*, December 4, 1994, pp. 32–35.

Vane-Wright, R. I., C. J. Humphries, and P. H. Williams. 1991. What to protect? Systematics and the agony of choice. *Biological Conservation* 55: 235–54.

Varley, G. C. 1971. The effects of natural predators and parasites on winter moth populations in England. In *Proceedings, Tall Timbers Conference on Ecological Animal Control by Habitat Management*, No. 2 (pp. 103–16). Tall Timbers Research Station, Tallahassee, Florida.

Varley, G. C., and G. R. Gradwell. 1960. Key factors in population studies. *Journal of Animal Ecology* 29: 399–401.

Varley, G. C., G. R. Gradwell, and M. P. Hassell. 1973. *Insect Population Ecology: An Analytical Approach*. Blackwell Scientific Publications, Oxford, U.K.

Vince, S. W., I. Valiela, and J. M. Teal. 1981. An experimental study of the structure of herbivorous insect communities in a salt marsh. *Ecology* 62: 1662–78.

Vitousek, P. M., H. A. Mooney, J. Lubchenco, and J. M. Mellilo. 1997. Human domination of the earth's ecosystems. *Science* 277: 494–99.

Volterra, V. 1926. Fluctuations in the abundance of a species considered mathematically. *Nature* 118: 558–60.

Walker, B. H. 1992. Biodiversity and ecological redundancy. *Conservation Biology* 6: 18–23.

Walker, L. R. 1995. How unique is primary plant succession in Glacier Bay? In D. R. Engstrom, *Proceedings of the Third Glacier Bay Science Symposium, 1993* (pp. 137–146) National Park Service, Anchorage, AK.

Walker, L. R., and F. S. Chapin, III. 1987. Interactions among processes controlling successional change. *Oikos* 50: 131–35.

Wardle, D. A. 1999. Is "sampling effect" a problem for experiments investigating biodiversity–ecosystem function relationships? *Oikos* 87: 403–7.

Waring, G. L., and N. S. Cobb. 1992. The impact of plant stress on herbivore population dynamics. In E. Bernays (ed.), *Insect–plant interactions*, vol. 4 (pp. 168–226). CRC Press, Boca Raton, FL.

Weller, D. E. 1987. A re-evaluation of the –3/2 power rule of plant self-thinning. *Ecological Monographs* 57: 23–43.

Weller, D. E. 1991. The self-thinning rule: Dead or unsupported? A reply to Lonsdale. *Ecology* 72: 747–50.

West, C. 1985. Factors underlying the late seasonal appearance of the lepidopterous leaf-mining guild on oak. *Ecological Entomology* 10: 111–20.

West Eberhard, M. J. 1975. The evolution of social behaviour by kin selection. *Quarterly Review of Biology* 50: 1–33.

Westemeier, R. L., J. D. Brown, S. A. Simpson, T. L. Esker, R. W. Jansen, and J. W. Walk. 1998. Tracing the long-term decline and recovery of an isolated population. *Science* 282: 1695–98.

White, J. 1980. Demographic factors in populations of plants. In O. T. Solbrig (ed.), *Demography and evolution in plant populations* (pp. 21–48). Blackwell Scientific, Oxford, U.K.

White, J. A., and T. G. Whitham. 2000. Associational susceptibility of cottonwood to a box elder herbivore. *Ecology* 81: 1975–1983.

White, T. C. R. 1984. The abundance of invertebrate herbivores in relation to the availability of nitrogen in stressed food plants. *Oecologia* 63: 90–105.

Whitehead, D. R., and C. E. Jones. 1969. Small islands and the equilibrium theory of insular biogeography. *Evolution* 23: 171–79.

Whittaker, R. H. 1953. A consideration of climax theory: The climax as a population and pattern. *Ecological Monographs* 23: 41–78.

Whittaker, R. H. 1967. Gradient analysis of vegetation. *Biological Reviews* 42: 207–64.

Whittaker, R. H. 1970. *Communities and ecosystems*. Macmillan, London.

Whittaker, R. H. 1972. Evolution and measurement of species diversity. *Taxon* 21: 213–51.

Whittaker, R. H. 1975. *Communities and ecosystems*, 2d ed. Macmillan, New York.

Whittaker, R. J., M. B. Bush, T. Partomihardjo, N. M. Asquith, and K. Richards. 1992. Ecological aspects of plant colonization of the Krakatau islands. In I. Thorton (ed.), *Krakatau—a century of change* (pp. 201–11). *Geojournal* 28(2) 81–302.

Wiens, J. A. 1977. On competition and variable environments. *American Scientist* 65: 590–97.

Wiens, J. A., J. F. Addicott, T. J. Case, and J. Diamond. 1986. Overview: The importance of spatial and temporal scale in ecological investigations. In J. Diamond and T. J. Case (eds.), *Community ecology* (pp. 145–53). Harper and Row, New York.

Wilcove, D. S., D. Rothstein, J. Dubow, A. Phillips, and E. Losos. 1998. Quantifying threats to imperiled species in the United States. *BioScience* 48: 607–15.

William, G. S. 1966. *Adaptation and natural selection*. Princeton University Press. Princeton, NJ.

Williams, C. B. 1964. *Patterns in the balance of nature and related problems in quantitative ecology*. Academic Press, New York.

Williams, I. S. 1999. Slow-growth, high-mortality—a general hypothesis, or is it? *Ecological Entomology* 24: 490–95.

Williams, K. S., K. G. Smith, and F. M. Stephen. 1993. Emergence of 13-year periodical cicadas (Cicadidae: *Magicicada*): Phenology, mortality, and predator satiation. *Ecology* 74: 1143–82.

Williams, P., D. Gibbons, C. Margules, A. Rebelo, C. Humphries, and R. Presey. 1996. A comparison of richness hotspots, rarity hotspots, and complementary areas for conserving diversity of British birds. *Conservation Biology* 10: 155–74.

Williamson M. 1987. Are communities ever stable? In A. J. Gray, M. J. Crawley, and P. J. Edwards (eds.), *Colonization, succession and stability* (pp. 353–71). Blackwell Scientific Publications, Oxford, U.K.

Williamson, M. 1999. Invasions. *Ecology* 22: 5–12.

Willott, S. J., D. C. Lim, S. G. Compton, and S. L. Sutton. 2000. Effects of selective logging on the butterflies of a Bornean rainforest. *Conservation Biology* 14: 1055–65.

Wilson, J. B. 1993. Would we recognize a broken-stick community if we found one? *Oikos* 67: 181–83.

Witz, B. W. 1989. Antipredator mechanisms in arthropods: A twenty-year literature survey. *Florida Entomologist* 73: 71–99.

Woodell, S. R., J. H. A. Mooney, and A. J. Hill. 1969. The behavior of *Larrea divaricata* (creosote bush) in response to rainfall in California. *Journal of Ecology* 57: 37–44.

Woinarski, J. C. Z., O. Price, and D. P. Faith. 1996. Application of a taxon priority system for conservation planning by selecting areas which are most distinct from environments already reserved. *Biological Conservation* 76: 147–59.

Wolin, C. 1985. The population dynamics of mutualistic systems. In D. H. Boucher (ed.), *The biology of mutualism* (pp. 248–69). Oxford University Press, New York.

World Conservation Monitoring Centre. 1992. *Global biodiversity: Status of the earth's living resources*. Chapman and Hall, London.

Wright, D. A. 1983. Species-energy theory: An extension of species-area theory. *Oikos* 41: 496–506.

Wynne-Edwards, V. C. 1962. *Animal Dispersion in Relation to Social Behaviour*. Oliver and Boyd, Edinburgh.

Yoda, K., T. Kira., H. Ogawa, and K. Hozumi. 1963. Self-thinning in overcrowded pure stands under cultivated and natural conditions. Journal of the Institute of Polytechnics, Osaka City University, Series D, 14: 107–129.

Yu, D. W., and N. E. Pierce. 1997. A castration parasite of an ant–plant mutualism. *Proceedings of the Royal Society of London* 265: 375–82.

Zaret, T. M., and R. T. Paine. 1973. Species introduction in a tropical lake. *Science* 182: 449–55.

Photo Credits

Chapter 1 Figure 1-1, David S. Boyer/NGS Image Collection; Figure 1-4, Michael P. Gadomski/Photo Researchers, Inc.; Box 1, Figure 1, Terry Donnelly/Tom Stack & Associate; Figure Section 1, Gregory G. Dimijian/Photo Researchers, Inc.

Chapter 2 Figure 2-3, Simon Fraser/Photo Researchers, Inc.; Figure 2-8, Charles G. Summers Jr./Bruce Coleman Inc.; Figure 2-10, Roger Wilmshurst/Photo Researchers, Inc.; Box 1, Figure 1, Najlah Feanny/Stock; Figure Section 2, Tom & Pat Leeson/Photo Researchers, Inc.

Chapter 3 Figure 3-4a, Richard Ellis/Photo Researchers, Inc; Figure 3-4b, Michel Viard/Peter Arnold, Inc.; Figure 3-4c, Sullivan & Rogers/Bruce Coleman Inc.; Figure 3-4d, Stephen J. Krasemann/DRK Photo; Figure 3-6a, Jane Burton/Bruce Coleman Inc.; Figure 3-6b, Nick Gordon/Animals Animals/Earth Scenes; Figure 3-6c, Patti Murray/Animals Animals/Earth Scenes; Figure 3-6d, Edward Parker/Oxford Scientific Films Ltd.

Chapter 4 Figure 4-1, Peter D. Stiling; Figure 4-2, Dave Watts/Natural Selection Stock Photography, LLC.; Figure 4-3, K & K Ammann/Bruce Coleman Inc.; Figure 4-5, Gary Meszaros/MESZA/Bruce Coleman Inc.; Figure 4-7, Peter D. Stiling; Figure 4-10, Johann Schumacher Design/Peter Arnold, Inc.

Chapter 5 Figure 5-1a, Patti Murray/Animals Animals/Earth Scenes; Figure 5-1b, D. H. Thompson/Oxford Scientific Films Ltd.; Figure 5-3a, b, © 2001 D.E. Walter (C. Harvey, artist), The University of Queensland; Figure 5-4, John Cancalosi/Peter Arnold, Inc.; 5-6, Gregory G. Dimijian/Photo Researchers, Inc.; Figure 5-7, Patricio Robles Gil/Bruce Coleman Inc.

Chapter 6 Figure 6-1, Tom Bean/DRK Photo; Figure 6-6, Ken Brate/Photo Researchers, Inc.; Figure 6-8b, G.C. Kelley/Photo Researchers, Inc.; Figure Section 3, Tom Till/DRK Photo

Chapter 7 Figure 7-3b, Stuart Westmorland/Natural Selection Stock Photography, LLC.; Figure 7-4b, Dan Suzio/Photo Researchers, Inc.; Figure 7-5b, Zig Leszczynski/Animals Animals/Earth Scenes; Figure 7-6, Al Giddings/Al Giddings Images, Inc.; Figure 7-7a, Florida Park Service; Figure 7-7b, Peter Drowne/Color-Pic, Inc.; Figure 7-12, Peter D. Stiling; Figure 7-13, Virginia P. Weinland/Photo Researchers, Inc.; Figure 7-14, Runk/Schoenberger/Grant Heilman Photography, Inc.

Chapter 8 Figure 8-1, Peter D. Stiling; Figure 8-4, David Hall/Photo Researchers, Inc.; Figure 8-6a, E.R. Degginger/Photo Researchers, Inc.; Figure 8-13, Dwight R. Kuhn/DRK Photo; Figure 8-17, Runk/Schoenberg/Grant Heilman Photography, Inc.; Figure 8-18 Leonard Lessin/Peter Arnold, Inc.; Figure 8.UN.1, S.F./Animals Animals/Earth Scenes; Box 1 Figure 1, Runk/Schoenberger/Grant Heilman Photography, Inc.

Chapter 9 Figure 9-2, Kjell B. Sandved/Photo Researchers, Inc.; Figure 9-3, Rodger Jackman/Oxford Scientific Films Ltd.; Figure 9-4, Gregory G. Dimijian/Photo Researchers, Inc.; Figure 9-5, Adam Hart-Davis/Photo Researchers, Inc.; Figure 9-7, J.H. Robinson/Photo Researchers, Inc.; Figure 9-8, Gregory G. Dimijian/Photo Researchers, Inc.; Figure 9-9 Joel Sartore/Grant Heilman Photography, Inc.; Box 1 Figure 1, Bob Pool/Tom Stack & Associates; Figure 9.UN.1, Peter D. Stiling

Chapter 10 Figure 10-1a, Victoria McCormick/Animals Animals/Earth Scenes; Figure 10-3a, Ed Degginger/Color-Pic, Inc.; Figure 10-3b, Richard K. Laval/Animals Animals/Earth Scenes; Figure 10-3c, K.G. Vock/Okapia/Photo Researchers, Inc.; Figure 10-3d, E.R. Degginger; Figure 10-3e, Oxford Scientific Films Ltd.; Figure 10-3f, Klaus Uhlenhut/Animals Animals/Earth Scenes; Figure 10-13 Robert Maier/Animals Animals/Earth Scenes; Figure 10.UN.1, Peter D. Stiling; Figure 10.UN.2, Peter D. Stiling

Chapter 11 Figure 11-1a, Brenda Tharp/Photo Researchers, Inc.; Figure 11-1b, Janis Burger/BURGE/Bruce Coleman Inc.; Figure 11-1c, John Kaprielian/Photo Researchers, Inc.; Figure 11-1d, Adam Jones/Natural Selection Stock Photography, LLC.; Figure 11-3, Mike Amphlett/Holt Studios/Photo Researchers, Inc.; Figure 11-11, Peter D. Stiling; Box 1 Figure 1, Hans Reinhard/Bruce Coleman Inc.

Chapter 12 Figure 12-1, © Pieter Johnson, Roberts Enviromental Center, Claremont McKenna College; Figure 12-2, Bios (A. Compost)/Peter Arnold, Inc.; Box 1 Figures 1-5, John Bavosi/Photo Researchers, Inc.; Figure 12-7, Grant Heilman Photography, Inc.; Figure 12-8a, Jeff Lepore/Photo Researchers, Inc.; Figure 12-10, Peter D. Stiling; Figure 12.UN.1, Peter D. Stiling

Chapter 13 Figure 13-1, Peter Davey/Bruce Coleman Inc.; Figure 13-4b, Olivier Grunewald/Oxford Scientific Films Ltd.; Figure 13-4c , C. C. Lockwood/Animals Animals/Earth Scenes; Figure 13-9b, Leroy Simon/Visuals Unlimited; Figure 13.UN.1, 2, Peter D. Stiling

Chapter 14 Figure 14-10, Gregory G. Dimijian/Photo Researchers, Inc.; Figure 14-11, S.J. Krasemann/Peter Arnold, Inc.; Figure 14-12, Peter D. Stiling; Figure 14-14, Grant Heilman/Grant Heilman Photography, Inc.; Figure 14-15, Walter H. Hodge/Peter Arnold, Inc.; Figure 14-16, Charles Mauzy/Natural Selection Stock Photography, LLC.; Figure 14-17, Jim Zipp/Photo Researchers, Inc.; Figure 14-18, Jeff Rotman/Peter Arnold, Inc.; Figure 14-21a, David R. Frazier/Photo Researchers, Inc.; Figure 14-21b, William H. Mullins/Photo Researchers, Inc.; Figure Section 4, Ed Degginger/Color-Pic, Inc.

Chapter 15 Figure 15-1, Larry Allan/Bruce Coleman Inc.; Figure 15-5a, Dani/Jeske/Animals Animals/Earth Scenes; Figure 15-5b, Joe Mc Donald/Animals Animals/Earth Scenes; Figure 15-5c, Dani/Jeske/Animals Animals/Earth Scenes; Figure 15-7, Pat & Tom Leeson/Photo Researchers, Inc.; Figure 15-13, Valerie Giles/Photo Researchers, Inc.; Figure 15-15, Roger de la Harpe/Animals Animals/Earth Scenes; Figure 15-16, Shahid Naeem/Peter D. Stiling; Figure 15.UN.01, Peter D. Stiling;

Chapter 16 Figure 16-1, Peter Weimann/Animals Animals/Earth Scenes; Box 1 Figure 2, John Cancalosi/Peter Arnold, Inc.

Chaptet 17 Figure 17-1a, b, Roger Wilmshurst/Photo Researchers, Inc.; Figure 17-5a, b, Meckes/Ottawa/Photo Researchers, Inc.; Figure 17-8, Peter B. Kaplan/Photo Researchers, Inc.; Figure 17-12, Donald Brown/Animals Animals/Earth Scenes; Figure 17-13.UN.01, Sir Robert May; Box 1 Figure 1, Joel Bennett/Oxford Scientific Films Ltd.; Box 1 Figure 2, Vanessa Vick/Photo Researchers, Inc.

Chapter 18 Figure 18-1, Vanessa Vick/Photo Researchers, Inc.; Figure 18-5a, b, c, d, Tom Bean/DRK Photo; Figure 18-6a, Michael P. Gadomski/Animals Animals/Earth Scenes; Figure 18-6b /Grant Heilman Photography, Inc.

Chapter 19 Figure 19-6a, Jim Wark/Peter Arnold, Inc.; Figure 19-6b, Steven C. Kraufman/DRK Photo; 19-16a, Photo by Daniel Simberloff, Florida State University; Figure 19-16b, Photo by Daniel Simberloff, Florida State University

Chapter 20 Figure 20-20b, Fletcher & Baylis/Photo Researchers, Inc.; Figure 20-8a, b, c, d, Peter D. Stiling; Figure 20-12, Zig Leszczynski/Animals Animals/Earth Scenes; Figure Section 5,Nigel Cattlin/ Holt Studios International/Photo Researchers, Inc.

Chapter 21 Figure 21-1a, Wendell Metzen/Bruce Coleman Inc.; Figure 21-1b, Stephen J. Krasemann/Photo Researchers, Inc.; Figure 21-8 Gary Vestal/Natural Selection Stock Photography, LLC.

Chapter 22 Figure 22-1, Prof. Susan Harrison; Figure 22-4a, b, c, USDA Natural Resources Conservation Service, National Soil Survey Center; Figure 22-7, Provided courtesy of the USDA Forest Service, Northeastern Research Station; Figure 22-9, Tom McHugh/Photo Researchers, Inc.; Box 1 Figure 2, Betty Anne Schreiber/Animals Animals/Earth Scenes

Index

Q

R

S